I0885309

"Today military commanders across the world are experiencing widespread stress and mental health issues in their organisations. This book has outstanding psychological prescriptions for military commanders and psychologists alike. A compelling read."

*Lt. General Dushyant Singh, Commandant,*
*Army War College, India*

"This handbook, comprised of an international team of experts, is an extremely valuable resource of empirical, theoretical and practical knowledge for military psychology students and practitioners."

*Colonel (R) Reuven Gal, Ph.D., Former Chief Psychologist, Israeli Defense Forces*

"This outstanding work covers important topics often overlooked in the professional literature. Connecting theoretical concepts to practical applications, an impressive team of military psychology experts give readers a book they will want to keep close at hand."

*Colonel (R) Bruce E. Crow, Psy.D., Former U.S. Army Chief Psychologist*

"This volume is remarkable in its breadth, covering the ancient history of military psychology to the latest developments in assisting military service members preparing for combat and preparing to return to civilian life."

*Richard G. Tedeschi, Ph.D., Professor Emeritus, University of North Carolina, USA*

"A 'must-have' for every military psychologist, this book is a unique reference that transcends national borders, providing solid scientific research and thoughtful ethical discussions amid a perspective that captures the historical and prospective use of military psychology."

*Colonel (R) Sally C. Harvey, Ph.D., U.S. Army Psychologist*

# THE ROUTLEDGE INTERNATIONAL HANDBOOK OF MILITARY PSYCHOLOGY AND MENTAL HEALTH

Military psychology has become one of the world's fastest-growing disciplines with ever-emerging new applications of research and development. *The Routledge International Handbook of Military Psychology and Mental Health* is a compendium of chapters by internationally renowned scholars in the field, bringing forth the state of the art in the theory, practice and future prospects of military psychology.

This uniquely interdisciplinary volume deliberates upon the current issues and applications of military psychology not only within the military organization and the discipline of psychology, but also in the larger context of its role of building a better world. Split into three parts dedicated to specific themes, the first part of the book, "Military Psychology: The Roots and the Journey," provides an overview of the evolution of the discipline over the years, delving into concepts as varied as culture and cognition in the military, a perspective on the role of military psychology in future warfare and ethical issues. The second part, "Soldiering: Deployment and Beyond," considers the complexities involved in soldiering in view of the changing nature of warfare, generating a focal discourse on various aspects of military leadership, soldier resilience and post-traumatic growth in the face of extreme situations, bravery and character strengths and transitioning to civilian life. In the final section, "Making a Choice: Mental Health Issues and Prospects in the Military," the contributors focus on the challenges and practices involved in maintaining the mental health of the soldier, covering issues ranging from stress, mental health and well-being, through to suicide risk and its prevention, intervention and management strategies, moral injury and post-traumatic stress disorder.

Incorporating enlightening contributions of eminent scholars from around the world, the volume is a comprehensive repository of current perspectives and future directions in the domain of military psychology. It will prove a valuable resource for mental health practitioners, military leaders, policy-makers and academics and students across a range of disciplines.

**Updesh Kumar**, Ph.D., is Scientist 'G' and Head, Mental Health Division, Defence Institute of Psychological Research (DIPR), Defence Research and Development Organization (DRDO), Ministry of Defence, New Delhi, India.

## Routledge International Handbooks

**Routledge Handbook of the Belt and Road**
*Edited by Cai Fang and Peter Nolan*

**Routledge Handbook of Language Acquisition**
*Edited by Jessica S. Horst and Janne von Koss Torkildsen*

**Routledge Handbook of Identity Studies, 2e**
*Edited by Anthony Elliott*

**Routledge International Handbook of Poverty**
*Edited by Bent Greve*

**Routledge International Handbook of New Digital Practices in Galleries, Libraries, Archives, Museums and Heritage Sites**
*Edited by Hannah Lewi, Wally Smith, Dirk vom Lehn and Steven Cooke*

**Routledge International Handbook of Human Trafficking**
A Multi-Disciplinary and Applied Approach
*Edited by Rochelle L. Dalla and Donna Sabella*

**The Routledge Handbook of Comparative Rural Policy**
*Edited by Matteo Vittuari, John Devlin, Marco Pagani and Thomas Johnson*

**Routledge International Handbook of Masculinity Studies**
*Edited by Lucas Gottzén, Ulf Mellström and Tamara Shefer*

**Routledge Handbook of European Welfare Systems, 2e**
*Edited by Sonja Blum, Johanna Kuhlmann and Klaus Schubert*

**Routledge International Handbook of Heterosexualities Studies**
*Edited by James Joseph Dean and Nancy L. Fischer*

**Routledge Handbook of Contemporary European Social Movements**
Protest in Turbulent Times
*Edited by Cristina Flesher Fominaya and Ramón A. Feenstra*

**The Routledge International Handbook of Military Psychology and Mental Health**
*Edited by Updesh Kumar*

**Routledge International Handbook of Green Criminology**
*Edited by Avi Brisman and Nigel South*

**Routledge International Handbook of Contemporary Racisms**
*Edited by John Solomos*

**The Routledge Handbook of State-Owned Enterprises**
*Edited by Luc Bernier, Philippe Bance and Massimo Florio*

For more information about this series, please visit: www.routledge.com/Routledge-International-Handbooks/book-series/RIHAND

# THE ROUTLEDGE INTERNATIONAL HANDBOOK OF MILITARY PSYCHOLOGY AND MENTAL HEALTH

*Edited by*
*Updesh Kumar*

Routledge
Taylor & Francis Group
LONDON AND NEW YORK

First published 2020
by Routledge
2 Park Square, Milton Park, Abingdon, Oxon OX14 4RN

and by Routledge
52 Vanderbilt Avenue, New York, NY 10017

*Routledge is an imprint of the Taylor & Francis Group, an informa business*



*British Library Cataloguing-in-Publication Data*
A catalogue record for this book is available from the British Library

*Library of Congress Cataloging-in-Publication Data*
Names: Kumar, Updesh, editor.
Title: The Routledge international handbook of military psychology and mental health/edited by Updesh Kumar.
Description: Abingdon, Oxon; New York, NY : Routledge, 2020. | Includes bibliographical references and index.
Identifiers: LCCN 2019037539 (print) | LCCN 2019037540 (ebook) | ISBN 9780367237073 (paperback) | ISBN 9780367237066 (hardcover) | ISBN 9780429281266 (ebook)
Subjects: LCSH: Psychology, Military--Handbooks, manuals, etc. | War--Psychological aspects--Handbooks, manuals, etc. | Combat--Psychological aspects--Handbooks, manuals, etc. | Soldiers--Mental health--Handbooks, manuals, etc.
Classification: LCC U22.3 .R68 2020 (print) | LCC U22.3 (ebook) | DDC 355.001/9--dc23
LC record available at https://lccn.loc.gov/2019037539
LC ebook record available at https://lccn.loc.gov/2019037540

ISBN: 978-0-367-23706-6 (hbk)
ISBN: 978-0-429-28126-6 (ebk)

Typeset in Bembo
by Nova Techset Private Limited, Bengaluru & Chennai, India

*Dedicated to the Brave Soldiers of Indian Armed Forces*

# CONTENTS

# LIST OF FIGURES

# LIST OF TABLES

# EDITOR

**Updesh Kumar**, Ph.D., is Scientist 'G' and in the chair of the Head, Mental Health Division at Defense Institute of Psychological Research (DIPR), R & D Organization (DRDO), Ministry of Defense, Delhi. After obtaining his doctorate in the area of suicidal behaviour from Punjab University, Chandigarh, India, he has more than 28 years of experience as a scientist in an R&D organization. He specializes in the areas of military psychology, personality assessment, suicidal behaviour, health psychology, and test development for the selection of officers for the Indian Armed Forces. Currently, he is also the project director of a mega project titled "Comprehensive Soldier Fitness Program: Resilience Building" for the Indian Armed Forces. Dr. Kumar has been involved in the selection of officers and has been responsible for monitoring the selection system of the Indian Armed Forces for the last 28 years. Dr. Kumar has edited 11 volumes which include *Suicidal Behaviour: Assessment of People-at-Risk* (Sage, 2010), *Countering Terrorism: Psychosocial Strategies* (Sage, 2012), *Understanding Suicide Terrorism: Psychosocial Dynamics* (Sage, 2014), *Suicidal Behaviour: Underlying Dynamics* (Routledge, UK, 2015), *Positive Psychology: Applications in Work, Health and Well-Being* (Pearson Education, 2015), *The Wiley Handbook of Personality Assessment* (John Wiley & Sons, 2016), *The Routledge International Handbook of Psychosocial Resilience* (Routledge, 2016), *Handbook of Suicidal Behavior* (Springer Nature, 2017), and most recently *Emotion, Well-Being and Resilience: Theoretical Perspectives and Practical Applications* (Apple Academic Press, in press). *The highlighted books with Sage Publications; Routledge, UK; John Wiley & Sons; Springer Nature; and Apple Academic Press have been completed in collaboration with internationally acclaimed academicians/researchers from India, the United States, the United Kingdom, Australia, Singapore, Israel, France, Belarus, Germany, Italy, Hungary, Hong Kong, Norway, Ireland, Ukraine, Netherlands, Canada South Africa Croatia, Spain, and so on.* He is also the past editor of the *Journal of Indian Academy of Applied Psychology (JIAAP)*. Dr. Kumar has also authored manuals on *Suicide and Fratricide: Dynamics and Management* for defence personnel, *Managing Emotions in Daily Life & at Work Place* for the general population, *Overcoming Obsolescence & Becoming Creative in R&D Environment* for R&D organizations, and *Self-Help Techniques in Military Settings*. Besides completing many important research projects for the Indian Armed Forces, he has authored more than 60 other academic publications in the form of research papers, journal articles, and book chapters and represented his institute at the national and international level. Dr. Kumar has been a psychological assessor (psychologist) in various services selection boards for eight years for the selection of officers in the Indian Armed Forces. He is a certified psychologist by The British Psychological Society with level "A" and "B" Certificates of Competence in Occupational Testing. He was conferred with the DRDO's Best Popular Science Communication Award – 2009 by the Hon'ble Defense Minister of India. He was also the recipient of the DRDO Technology Group Award in 2001 and 2009, Professor Manju Thakur Memorial Award 2009 and 2012 by the Indian Academy of Applied Psychology (IAAP), and Professor N. N. Sen Best Paper Award for the year 2010 by the Indian Association of Clinical Psychologists (IACP). He has been conferred with Laboratory Scientist of the Year Award in 2012–2013 and the prestigious DRDO Scientist of the Year Award in 2013 by the government of India.

# CONTRIBUTORS

**Samridhi Ahuja** is a RCI Certified Counsellor trained in child and adolescent guidance and counselling from NIPCCD, New Delhi. She has pursued her Bachelors in psychology from Gargi College and Masters in psychology from North Campus, University of Delhi. Having previously practiced at NIPCCD, she is passionate about working with children with developmental disorders and groups with special needs to improve their quality of life. Currently, she is pursuing her Ph.D. in the area of clinical psychology. Her areas of interest also include Indian psychology, developmental psychology, military psychology, and counselling. As a Senior Research Fellow at the Defence Institute of Psychological Research, Defence R&D Organisation, Ministry of Defence, she is associated with the major research project "Comprehensive Soldier Fitness Program: Resilience Building" under the directorship of Dr. Updesh Kumar, Scientist 'G'. Apart from this, she has presented papers in various national and international conferences and has original publications in peer-reviewed journals.

**Archana**, Ph.D., is Scientist 'E' at Defence Institute of Psychological Research (DIPR), Ministry of Defence, India and is associated with R & D activities in the areas of organizational behaviour, mental health and well-being of soldiers in the peace as well as field locations. She has to her credit number of papers in national journals of repute. She has authored two manuals on 'stress management' and 'psychological well-being of soldiers at high altitude'. In addition, she has made valuable contributions in organizing many workshops and seminars for the soldiers in the area of military psychology. She has been the course coordinator in organizing training programs for service officers in the area of counselling. She has also been actively involved in delivering lectures on stress management and other related areas in active field environments.

**Markus Auschek** is a teaching psychologist at the Special Operations Training Center and has experience as an operational psychologist in a mountain brigade, including multiple deployments to Afghanistan and Mali. He is a trained sport psychologist.

**Nazanin H. Bahraini** is a Clinical Research Psychologist at the Rocky Mountain MIRECC for Suicide Prevention and an Associate Professor in the Departments of Psychiatry and Physical Medicine and Rehabilitation at the University of Colorado Anschutz Medical Campus.

**Col. (Retd.) Paul T. Bartone**, Ph.D., is a Visiting Senior Research Fellow at the Institute for National Security Policy, National Defense University. A Fulbright Scholar, Bartone has taught leadership and psychology at the Industrial College of the Armed Forces and at the U.S. Military Academy, West Point, where he also served as Director of the Leader Development Research Center. While on active duty, Bartone was the Senior Research Psychologist in the U.S. Army and served as Research Psychology Consultant to the Surgeon General and as Assistant Corps Chief for the Medical Allied Sciences. He is a former President of the American Psychological Association's Society for Military Psychology, a charter member of the Association for Psychological Science, and a life member of the American Psychological Association. He holds an M.A. and Ph.D. in psychology and human development from the University of Chicago.

**Col. (Dr.) Petrus C. Bester**, D.Phil., joined the then-South African Defence Force in 1987. He obtained a B. Mill. degree in 1991 and B. Mill. (Hons.) in industrial psychology in 1994 at the Military Academy in Saldanha. In 1998, he completed his master's degree in human resources management at the Rand Afrikaans University (RAU) and registered with the Health Professions Council as an industrial psychologist. He also completed the Doctoral Program Leadership in Performance and Change at the University of Johannesburg at the end of January 2007. In 2008 he completed the South African National Defence Force's Joint Senior Command and Staff Program at the South African National War College. Subsequently, in 2017, he underwent and completed the South African National Defence College's Security and Defence Studies program and postgraduate diploma in the management of security at the University of the Witwatersrand as well. His research interests include leadership, integrity, performance enhancement, test construction, and national security, and he is a Research Fellow at the Department of Industrial Psychology (Mil), Military Science, at the University of Stellenbosch.

**Bruce Bongar**, Ph.D., ABPP, FAPM, C.Psychol., C.Sci., is the Calvin Distinguished Professor of Psychology at Palo Alto University. Early in his career, Dr. Bongar developed strong interests in identifying the risk factors associated with suicidal behaviour and clinical emergencies and continues to be a prolific contributor to scholarly literature and research. He is a past president of the Section on Clinical Emergencies and Crises of Division 12 of the American Psychological Association. The American Association of Suicidology recognized Dr. Bongar for his work with the early career achievement Edwin S. Shneidman Award for outstanding contributions to research in suicidology and the Louis I. Dublin Award for significant lifetime career contributions to research in suicidology. In 2008, he was presented with the Florence Halpern award from the Division of Clinical Psychology of the APA for his distinguished contributions to the profession of clinical psychology. He also was awarded the Career Achievement Award from the Section on Clinical Emergencies and Crises of the Division of Clinical Psychology of APA for his work on suicide and life-threatening behaviours. In 2016, Professor Bongar's research and scholarly work on suicide was recognized by a lifetime achievement award in professional psychology by the California Psychological Association.

**Col. (Retd.) Stephen V. Bowles**, Ph.D., ABPP, is currently serving as a Visiting Research Fellow in the Institute for National Security Studies at the National Defense University in Washington, DC, and as an Adjunct Faculty at the Uniformed Services University of the Health Sciences in Bethesda, Maryland. He is the President Elect for the Society for Military Psychology and the American Psychology Association Council of Representatives for the District of Columbia Psychology. He served 27 years in the U.S. Army, is a veteran of Operation Iraqi Freedom, and is a graduate and Fellow of The Eisenhower School at the National Defense University. He is a Distinguished Practitioner and Fellow of the National Academy of Practice in Psychology, a Fellow in the American Psychological Association (Div 19), and board certified in clinical health psychology.

**Abigale Brady** is a second-year student in the trauma track of the Ph.D. program after earning her B.A. in psychology at Ashland University in Ohio. During her undergraduate career, she served at a transitional living centre and as a research assistant at Ashland University. She is interested in working with veterans and active duty military to treat combat-related PTSD and treatment effects. Her research interests include studying military populations and the development and treatment of phantom limb syndrome.

**Diana P. Brostow**, Ph.D., M.P.H., R.D.N., is a Research Nutritionist in the Rocky Mountain MIRECC and on the faculty of the Department of Physical Medicine and Rehabilitation at the University of Colorado Anschutz Medical Campus.

**Jacie Brown** is a second-year student in the clinical psychology Ph.D. program at Palo Alto University. Jacie is interested in pursuing the forensic area of emphasis with an interest in trauma. She received her Bachelor's of Arts in criminal justice and psychology from Ferris State University and completed an undergraduate internship at Berghuis Psychological Services working with adult outpatient sex offenders in a group setting. She also spent this time shadowing probation and parole officers. Her primary research interests include psychology of trauma within the forensic and veteran populations, police training, and adaptive strategies for veterans after returning home from combat.

**Shane S. Bush**, Ph.D., ABPP, is director of Long Island Neuropsychology, PC, a neuropsychologist with the VA New York Harbor Healthcare System, and an Adjunct Faculty member in the Department of Psychology at the University of Alabama. He is board certified in clinical neuropsychology, rehabilitation psychology, clinical psychology, and geropsychology. He has been awarded Fellow status in the American Psychological Association (Divisions 12, 18, 20, 22, 40, and 42). He is a Past President and Fellow of the National Academy of Neuropsychology. He has also held elected positions or served as a committee member for the American Academy of Clinical Neuropsychology, American Board of Professional Psychology, American Board of Professional Neuropsychology, American Board of Geropsychology, and multiple divisions of the American Psychological Association. He has served as an editorial board member for multiple journals and has presented on ethical and professional issues at national and international conferences. He has more than 130 publications, including journal articles, book chapters, position papers for professional organizations, and more than 20 books, some of which focus on psychological and neuropsychological practice with military personnel and veterans. He is a veteran of both the U.S. Marine Corps and Naval Reserve.

**Filipa Castanheira**, Ph.D., is Associate Professor at the Nova School of Business and Economics, Universidade Nova de Lisboa in Lisbon (Portugal). She has a Ph.D. in social psychology and teaches human resource management and organizational behaviour. She is member of the Nova Leadership for Impact Knowledge Center and has participated in several research projects. Her main research interests include stress and well-being at work, namely in a military context. She has published in *Military Psychology*.

**Maria José Chambel**, Ph.D., is Associate Professor at the Faculty of Psychology, University of Lisbon in Lisbon (Portugal). She has a Ph.D. in social psychology and teaches in work and organizational psychology. She is member of the CicPsi (Research Centre of Psychology Science of University of Lisbon), where she coordinates the group Adaptation Process in Context of applied psychology and has participated in several research projects. Her main research interests include subjects such as stress and well-being at work, namely in a military context. She has published in *Military Psychology* and *Armed Forces & Society*.

**Reoot Cohen-Koren** is currently Mental Health Officer, Air Force, IDF Mental Health Department, Medical Corps, Israel. She has an M.A. in clinical psychology from the Hebrew University in Jerusalem. Her M.A. thesis work was on the correlation between gender, post-traumatic growth, and ethnocentrism in adolescents that were exposed to terror. Her main research interest lies in the area of mental health among soldiers and combat military soldiers.

**Alanna Coady** is an experienced researcher of psychology and religion. She obtained her master's degree at Harvard Divinity School and is currently a research assistant in Dr. Brett Litz's lab at the Massachusetts Veterans Epidemiological Research and Information Center at the VA Boston Healthcare System. She is a project coordinator for an international mixed methods measurement development project, the Moral Injury Outcomes Study Consortium, and oversees numerous multisite randomized controlled trials of PTSD treatments for veterans and service members. Ms. Coady brings to these projects her expertise in examining cultural and religious factors in traumatic experiences, responses to trauma, and treatment approaches.

**Hermes de Andrade Jr**, Ph.D. Lieutenant Colonel of the Brazilian Army's active reserve has acted at all levels of leadership in several areas of the Ministry of Defense for 30 years. He graduated in international relations (UNESA, Brazil) and military sciences (AMAN, Brazil), with a master's degree in Sociology (UFRRJ, Brazil) and doctorate in Public Health (ENSP/FIOCRUZ, Brazil) with post-doctorate from Catholic University. He is senior researcher at the CEFH/FFCS/Catholic University, Portugal. He is also involved with Doctoral Program in Creativity, Social Innovation and Sustainability in University of Vigo, Spain. He has published many articles in various journals and has been a collaborator of INOVLIDER (Portugal).

**Marcos Aguiar de Souza** is a Ph.D. in psychology and currently is Associate Professor at the Federal University of Rio de Janeiro. He is interested in techniques of data collection and analysis, and construction and validation of measuring instruments in organizational psychology and studies of psychology in public safety and in military contexts.

**Yonel Ricardo de Souza**, Colonel of the Artillery (Retired) of the Brazilian Army, is a graduate of the Military Academy of Agulhas Negras (1990). He graduated in physical education from the University of Caxias do Sul (2010) and has a Master's degree in psychology from the Federal University of Rondônia/UNIR (2015). He was Professor of Psychology at the Military Academy of Agulhas Negras (2015). He is currently a student of the medicine course at the Federal University of Paraíba.

**Stephanie Ellickson-Larew** is a postdoctoral clinical research fellow working with Professor Brett T. Litz in the Massachusetts Veterans Epidemiological Research and Information Center division at the VA Boston Healthcare System. She is also currently a Ph.D. candidate working with David Watson, Ph.D., at the University of Notre Dame. Her research background is in psychometric and structural analyses of psychopathology broadly and of dissociation specifically, as well as their intersection with normal-range and pathological personality traits. She is currently interested in focusing on the impact of trauma, stress, and moral injury on developing and maintaining psychopathology and maladaptive personality traits, especially in the manifestations of dissociation, externalizing symptomatology, and suicidality.

**Gordana Nedic Erjavec**, Ph.D., is a research associate at Ruđer Boskovic Institute (RBI) Zagreb, Croatia. She works in the Laboratory for Molecular Neuropsychiatry at the Department of Molecular Medicine at RBI. She is a molecular biologist with a particular interest in biological psychiatry. As a collaborator on the offset project funded by Patria, she did her postdoctoral training on metabolomics of PTSD. Besides this project, Dr. Nedic Erjavec is a collaborator on a research project funded by Croatian Science Foundation, dealing with the role of specific genetic polymorphisms and glycome in the development of PTSD and its symptoms and comorbidities. She has coauthored 43 articles in journals and 15 chapters in scientific books and has been cited 832 times, with H-index = 14. In 2013 she was awarded the National Science Award of the Republic of Croatia – Annual Award for Junior Researchers for 2012 in the field of biomedicine.

**Marissa N. Eusebio** is a third-year student at Palo Alto University in the clinical psychology Ph.D. program. Prior to attending Palo Alto University, Marissa received a B.S. in psychology with an emphasis in neuroscience and behaviour from the University of California, San Diego. She is currently specializing in adult and child trauma, particularly focusing on military and military families. Marissa is currently a practicum student at the Martinez VA Health Care System where she works in the Neurocognitive Rehabilitation Clinic. She also works at the San Francisco VA Health Care System for the Stress and Health Research Program. Marissa has taken on a leadership role and has been involved in various research projects with the Clinical Crises and Emergencies Research Lab at Palo Alto University, predominantly focusing on suicide prevention strategies among marginalized populations.

**Anna Feinman** is a second-year student in the clinical psychology Ph.D. program at PAU, where she is looking forward to completing the forensic area of emphasis. She received her Bachelor of Arts in psychology and political science from University of California, Santa Barbara, and her masters in marriage and family therapy from University of Southern California. Anna completed her MFT Internship at Ventura County Behavioral Health working with SMI adolescents and their families. Anna is currently a member of Dr. Bongar's Clinical Crises and Emergencies Lab at PAU and is a student therapist at the Gronowski Clinic. She is also a research assistant at the UCSF Stress and Resilience Lab at the San Francisco VA. Her primary research interests include the psychology of terrorism and murder-suicide, firearm access and mental health, and police training.

**Fabio Biasotto Feitosa**, psychologist, graduated from Universidade Estadual Paulista/UNESP (2000) and is a Ph.D. in interpersonal relations at Federal University of São Carlos/UFSCar (2007). He has complementary training for the treatment of depression by the interpersonal approach in Leicestershire/NHS Trust, England. His postdoctoral degree is in treatment and psychological prevention at University College London/UCL (2015). He is Professor of the Department of Psychology/DEPSI of the Federal University of Rondônia/UNIR. He has been a member of the European Health Psychology Society/EHPS since 2009.

**Amber M. Fox** graduated from the University of Maryland in 2016 with a B.S. in psychology. During her time as an undergraduate, Amber Fox volunteered as a Research Assistant at the Culture Lab, directed by Dr. Michele Gelfand, as well as the Comprehensive Assessment and Intervention Program (CAIP) Lab, directed by Dr. Andres De Los Reyes. Currently, Amber Fox is working as a Research Assistant and Project Coordinator for the Suicide Care, Prevention, and Research (CPR) Initiative, directed by Dr. Marjan G. Holloway, at the Uniformed Services University of the Health Sciences. In this capacity, she engages in research on military suicide prevention, as well as curriculum development for a military program to enhance cognitive agility.

**Sheila Frankfurt**, Ph.D., is a counselling psychologist in the Department of Veterans Affairs, Veterans Integrated Service Network 17 Center of Excellence for Research on Returning War Veterans, in Waco, Texas. Dr. Frankfurt conducts research on trauma, post-traumatic stress disorder, and moral injury. She is currently working to develop high-quality assessments of moral injury events and outcomes. She is supported by a Career Development Award – 1 from the Department of Veterans Affairs Rehabilitation Research and Development Office (5IK1RX002427-02). Other current projects include a qualitative examination of veterans', service members', and care providers' needs and desires for moral injury treatments and an exploration of clinicians' decision-making processes in the treatment of PTSD. Dr. Frankfurt serves as Deputy Director of the Clinical Investigation Research Office at Carl R. Darnall Army Medical Center, Fort Hood, Texas. Dr. Frankfurt is currently on the editorial board of *The Counseling Psychologist.*

**Dror Garbi** is currently the Head of the Aeromedical Psychology Section, Air Force, IDF Mental Health Department, Israeli Medical Corps. Dror Garbi is an intern in clinical psychology; his main research interest is enhancement of cognitive performance using visual and cognitive training for aviators and developing sensitive monitoring tools for cognitive deterioration using the platform of flight simulators. Major Garbi also investigates individual differences in executive functions as part of his Ph.D. research at Ben-Gurion University.

**Shobit Garg** received his D.P.M. and M.D. from Central Institute of Psychiatry (CIP), Ranchi. He has served as an Assistant Professor in the Department of Psychiatry, Shri Guru Ram Rai Institute of Medical and Health Sciences (SGRRIM), since 2013. He was also a senior resident in CIP, Ranchi. He has various national and international publications and chapters to his name and has presented papers at national conferences.

**Shirley Gordon,** Ph.D., is currently the Head of the Psychology Branch in the Israeli Air Force, Mental Health Department, Israeli Medical Corps. Dr. Gordon is a clinical neuropsychologist (Ph.D.). Her main research interest is enhancement of cognitive performance using different training tools. Dr. Gordon has published several papers in journals of repute.

**Jacques J. Gouws**, MMM, C.Psych., Ph.D., is a former South African Air Force Officer and retired military psychologist. Dr. Gouws both researched and gained extensive first hand experience of the resourcefulness required from soldiers and their commanders when faced by the insurmountable obstacles posed by the stress of being deployed in conventional and non-conventional combat zones (complicated by the pressures from the international political arena, the reactions of the civilian population to the casualties of battle, and the strain placed on society in general) during long-term sustained military operations. As Chair of the Department of Industrial/Organizational Psychology, he established the first academic under- and postgraduate courses in military psychology at the Faculty of Military Science, University of Stellenbosch, South Africa, in 1990. He consults and lectures on the psychological effects and impact of war. He also provides psychological treatment to military, veterans, and first-line emergency responders, such as police and firefighters. He is a frequent speaker at international and local conventions and workshops. He is quoted in the media in particular on PTSD and how it manifests in soldiers. He is the author of several book chapters on these topics.

**Breanna Grunthal**, B.A., is a research technician for Dr. Brett Litz in the Massachusetts Veterans Epidemiological Research and Information Center division at the VA Boston Healthcare System. She obtained her Bachelor of Arts degree in psychology and health and wellness studies from State University of New York (SUNY), Binghamton University, in 2018 and plans to pursue a Ph.D. in clinical psychology. Although a new professional in the field, Breanna has already delved into multiple research projects and has embraced opportunities to work on her own independent projects. Her research interests include the impact of stressful life events and social pressure on internalizing symptoms and psychological resiliency.

**Catherine Hausman**, M.S., is a fifth-year student in the Trauma Emphasis of the clinical psychology Ph.D. program at Palo Alto University (PAU). She is currently on internship at the New Mexico VA where her primary rotations are in the Military Trauma Treatment Program and the Indian Health Service. She has completed prior clinical training with the Palo Alto VA and San Mateo County Behavioral Health and Recovery Services. For the Clinical Crises and Emergencies Research Group at PAU, advised by Bruce Bongar, Ph.D., Catherine works with Stanford School of Medicine developing a culturally based suicide screening protocol for primary care, with the army studying physical bravery in military service members and veterans, and studies training standards in behavioural emergencies for psychologists. Catherine's dissertation is examining physical bravery through original data collected from U.S. Army reservists. Also, Catherine has been involved in research on a clinical trial for veterans with PTSD and substance use as part of the National Center for PTSD's Substance and Anxiety Intervention Laboratory. Catherine is also involved with the Trauma Student Group and Student Veterans Organization at PAU. Her pre-graduate school research experience involved studying suicide prevention in veterans with PTSD with Dr. Herbert Hendin in NYC.

**Annett Heidler** is an operational psychologist in an armoured infantry brigade and has worked with Customs and Border Control.

**Claire A. Hoffmire** is an Epidemiologist in the Rocky Mountain MIRECC and Assistant Professor in the Department of Physical Medicine and Rehabilitation at the University of Colorado Anschutz Medical Campus.

**Ryan Holliday**, Ph.D., is an Advanced Psychology Postdoctoral Fellow in the Rocky Mountain MIRECC and Instructor in the Department of Psychiatry at the University of Colorado Anschutz Medical Campus.

**Marjan Ghahramanlou Holloway**, Ph.D., is an adjunct faculty member of the Beck Institute for Cognitive Behavior Therapy and an Associate Professor of Medical and Clinical Psychology and Psychiatry at Uniformed Services University of the Health Sciences. Dr. Holloway completed her postdoctoral training in 2005 at the University of Pennsylvania under the mentorship of Dr.

Aaron T. Beck. Dr. Holloway currently serves as the Director for the Suicide Care, Prevention, and Research Initiative. In this capacity, she conducts programmatic research on military suicide prevention, with primary emphasis on the application of cognitive behavioural principles to different settings and vulnerable patient populations.

**Tim Hoyt**, Ph.D., is Chief of the Connected Health Branch in the Defense Health Agency, Virginia. His primary expertise is in combat-related post-traumatic stress disorder, the integration of technology into health care, and treatment approaches for military sexual trauma.

**Faizan Imtiaz**, Ph.D., is an Assistant Professor of Organizational Psychology at Towson University. His research interests include resilience, organizational performance, and decision-making across the adult lifespan. Dr. Imtiaz has taught a wide range of courses in psychology, kinesiology, and business at Towson University, Queen's University, and the Royal Military College of Canada. His applied experience includes work as a performance consultant at the Queen's Smith School of Business and in the private sector.

**Col. (Retd.) Larry C. James**, Ph.D., ABPP, is a nationally recognized expert in national security, defence issues, clinical psychopharmacology, and clinical health psychology. He is currently the President and CEO of the Wright Behavioral Health Group, LLC, and a Professor at Wright State University. Previously he served as the Associate Vice President for Military Affairs at Wright State University in Dayton, Ohio. Prior to that assignment, he served as the Dean, School of Professional Psychology, Wright State University, from 2008 to 2013. He received his Ph.D. in counselling psychology at the University of Iowa and a Post-Doctoral Fellowship in behavioural medicine at Tripler Army Medical Center. He was the Chair, Department of Psychology, at Walter Reed Army Medical Center and the Chair, Department of Psychology, at Tripler Army Medical Center. He was awarded a Bronze Star and the Defense Superior Service Medal. In his nearly three-decade military career, Colonel James had many deployments. Most notably, he deployed to Abu Ghraib and Guantanamo Bay, Cuba. Colonel James distinguished himself during these deployments and is one of the nation's military officers who were credited with putting policies and procedures in place to prevent abuse. He has lectured internationally and has published eight books, with several others in press, and over 100 professional papers and conference presentations.

**E. Ann Jeschke**, Ph.D., is a bioethicist specializing in military medicine. Dr. Jeschke defended her dissertation on post-deployment reintegration in 2015 at St. Louis University's Center for Health Care Ethics. She is currently engaged in postdoctoral training under the mentorship of Dr. Marjan G. Holloway, Ph.D., in the Suicide Care, Prevention and Research Initiative. In this capacity, Dr. Jeschke is interested in exploring programmatic research that evaluates the effectiveness of behavioural health interventions that seek to increase cognitive agility and decrease stress related to major and minor role transition across the military life-cycle. Dr. Jeschke's current research portfolio also includes an ethical inquiry into the training needs of combat medics when attending death in combat casualty care.

**Li-Jun Ji**, Ph.D., is a Professor of psychology at Queen's University, Canada. Her research examines how culture shapes the way people think, reason, and make decisions, and the implications for behaviour, emotion, and motivation. Dr. Ji has published many articles in top psychology journals and books and has consulted for government institutions and businesses on issues related to culture, prediction, and decision-making.

**Ryaja Johnson**, M.B.A., Psy.M., is a doctoral student in clinical psychology at Wright State University's School of Professional Psychology in Dayton, Ohio. She is interested in Health Psychology and Rehabilitation Psychology. Ryaja's research and clinical training are currently in the areas of post-traumatic stress disorder and trauma-focused, evidence-based interventions. Her dissertation study is investigating the relationship between spirituality and cognitive processing therapy treatment outcomes to determine if spirituality is a non-targeted change mechanism. A former Captain in the United States Air Force, Ryaja served seven years on active duty as a Contracting Officer. She has negotiated and managed multimillion-dollar contracts stateside and while deployed. Ryaja continues to stay actively connected to veteran and military initiatives in her community. Her leadership and community engagement have been recognized and honoured at both national and local levels.

**K. C. Kalmbach**, Ph.D., is an Associate Professor of psychology and licensed clinical psychologist. Over the past two decades, she has held positions as a researcher, administrator, clinician, consultant, and educator. Her clinical work has revolved around high-risk/high-trauma populations. Before joining the faculty at Texas A&M University–San Antonio, she helped build the first juvenile offender mental health court in Houston, Texas. Prior to this, as faculty at California State University–Los Angeles, she conducted community needs assessments in high-risk neighbourhoods for the Los Angeles Mayor's Office. Before entering academia, Dr. Kalmbach served as a director of the Texas Regional Center for Policing Innovation, a federally funded training institute tasked with providing progressive training for community stakeholders and justice system personnel throughout the State of Texas. Currently, in addition to teaching, she has worked with hundreds of returning combat-deployed military members and families.

**Kalman J. Kaplan**, Ph.D., is Professor of psychiatry/Director of the Program for Religion, Spirituality and Mental Health at the University of Illinois at Chicago College of Medicine and Adjunct Professor at Spertus Institute of Jewish Studies. He has been Editor of the *Journal of Psychology and Judaism* and on the editorial board of *Omega*. Dr. Kaplan has published widely in the areas of interpersonal and international relations, the emerging field of biblical psychology, schizophrenia, and suicide/suicide prevention. Dr. Kaplan is a Fellow in the American Psychological Association, was co-recipient of the 1998 Alexander Gralnick Award for outstanding original research in suicide and schizophrenia, and was a 2006–2007 and 2011–2012 Fulbright Fellow at Tel Aviv University. Dr. Kaplan has published 15 books, many book chapters, and close to 100 published articles and has given over 150 presentations, both nationally and internationally. During 2007–2010, Dr. Kaplan was awarded a start-up grant from The John Templeton Foundation to develop an online program in religion, spirituality, and mental health at the University of Illinois College of Medicine. He is a member of the ongoing Faith Communities Task Force of the National Action Alliance for Suicide Prevention.

**Alexander G. Karayani**, Ph.D., is a principal researcher of the National Research Institute of the Ministry of Interior of the Russian Federation, Moscow. His research investigates how the perspective directions of development of military psychology depend on an image of future war. Dr. Karayani has published a set of scientific articles in the leading Russian magazines, monographs, and textbooks. He provides consultation to military management concerning psychological training of the military personnel, psychological support, and rehabilitation of combatants.

**Noah K. Kaufman**, Ph.D., ABN, ABPdN, earned a B.A. from the University of New Mexico and an M.S. and Ph.D. from the University of Oregon. He is board certified by both the American Board of Professional Neuropsychology and the American Board of Pediatric Neuropsychology. Dr. Kaufman is a Clinical Assistant Professor in the Department of Psychiatry at Texas Tech University Health Sciences Center in El Paso, Texas. He is the co-author of multiple peer-reviewed publications, has served as an invited reviewer for multiple journals, and was the recipient of the 2018 Distinguished Service Award from the American Board of Professional Neuropsychology. Dr. Kaufman currently focuses the majority of his time on running a clinical and forensic neuropsychology practice at The Center for Neuropsychological Studies in Las Cruces, New Mexico. Before pursuing his studies, Dr. Kaufman was a national champion bicycle racer in 1987 and became interested in psychology after being exposed to sport psychology while attending training camps at the Olympic Training Center in Colorado Springs, Colorado.

**Harprit Kaur,** Ph.D., is Associate Professor and Head of the Department at Punjabi University Patiala. She has her M.Phil. in clinical psychology from NIMHANS, Bangalore, and Ph.D. from Punjab University Chandigarh. She has worked in Government Medical College and Hospital (GMCH) and Post Graduate Institute of Medical Research (PGI) Chandigarh as clinical psychologist. For the past 15 years, she has worked as teaching faculty in the psychology department of Punjabi University, Patiala. She has published many research articles in national and international journals of repute. She has also conducted many workshops on stress management in various Army units, including Pathankot and Udampur in India.

**Mark Khei** is a Ph.D. candidate in social and cultural psychology at Queen's University, Canada. His research is devoted to better understanding the influence of culture and social environments in people's decision-making, as well as how suffering contributes towards meaning-making and resilience. Mark holds M.Sc. in cultural and social psychology from Queen's University and a Bachelor of Arts in social psychology from the Nanyang Technological University in Singapore.

**Marcela Konjevod**, M.Sc., is a molecular biologist and Ph.D. student in the Laboratory for Molecular Neuropsychiatry, Division of Molecular Medicine, at Ruđer Boskovic Institute. She is currently employed on the project funded by Croatian Science Foundation: "Genomic and Glycomic Biomarkers for PTSD," led by Nela Pivac. She started her education in the Department of Biology at Josip Juraj Strossmayer University in Osijek, and later at the Faculty of Science at the University in Zagreb, where she graduated and was promoted to M.Sc. of molecular biology in 2016. She is currently enrolled in Ph.D. studies (medicinal chemistry) at the University of Rijeka. Since April 2018, she has been in the pre-doctoral specialization at Centro de Metabolomica Bioanalisis, Universidad CEU–San Pablo, Madrid. She is conducting research on PTSD samples in order to find potential metabolomic biomarkers. She has published eight journal articles (IF = 3.45). She is a recipient of three Dean's awards, one Rector's award and two grants for FENS – Regional Meeting in Pécs, Hungary (2017) and for XVIII Scientific Meeting of the Spanish Society of Chromatography and Related Techniques (SECyTA), Granada, Spain (2018).

**Oliver Kozumplik**, M.D., Ph.D., Assistant Professor, is a psychiatrist working at the University Psychiatric Hospital Vrapče, Zagreb, and subspecialist in biological psychiatry. He organizes a postgraduate course with Suzana Uzun on "Side Effects of Psychopharmaceutic Drugs," School of Medicine, Zagreb University. He is a lecturer at the School of Medicine, University of Zagreb, and he participated (as a co-investigator) in 10 multicentric clinical psychopharmacological trials. He was a collaborator on the scientific project "Side Effects of Antipsychotics," led by Suzana Uzun, founded by the Ministry of Sciences, Education and Sports of the Republic of Croatia, and on the bilateral project with Slovenia, "The Role of 5-HT6 Receptors in Alzheimer's Disease," led by Suzana Uzun, and is collaborator on the project led by Nela Pivac, funded by Croatian Science Foundation: "Genomic and Glycomic Biomarkers for PTSD."

**Oliver Krueckel** is head of psychology in the German Army Training command and has 10 years of experience as an operational psychologist in the field, working in an armoured brigade, with the Special Forces, with the U.S. Defense Center of Excellence for Psychological Health and TBI, and in Training and Doctrine. He served as a reserve officer in an airborne brigade and has been deployed as a psychologist to Kosovo, Afghanistan, and Mali.

**Jessica M. LaCroix**, Ph.D., is a Social Psychologist and a Research Assistant Professor in the Department of Medical and Clinical Psychology at the Uniformed Services University of the Health Sciences in Bethesda, Maryland. She serves as Co-Principal Investigator at the Suicide Care, Prevention, and Research Initiative (Suicide CPR), where she works to design and implement a number of programmatic studies to advance the science and clinical practice of suicide prevention, intervention, and postvention for military populations. Her primary research interests include stigma, gender, attitudes toward suicide and mental health treatment, and the design, implementation, and evaluation of suicide prevention interventions for military personnel and their families.

**Brett T. Litz**. Ph.D., is a clinical psychologist and Professor in the Departments of Psychiatry and Psychology at Boston University. He is currently the Director of the Mental Health Core of the Massachusetts Veterans Epidemiological Research and Information Center at the VA Boston Healthcare System. Dr. Litz has been conducting clinical research on PTSD among war veterans at the Boston VA since 1987 and is internationally recognized as an expert on the assessment and treatment of war-related PTSD and moral injury. He is a principal investigator on numerous research projects funded by the Departments of Defense and Veterans Affairs. His recent work entails evaluating the social, psychological, and spiritual impact of the Iraq and Afghanistan wars among active-duty military personnel and veterans and conducting clinical trials of a psychotherapy called Adaptive Disclosure, which individualizes PTSD treatment by treating danger, loss, and moral injury traumas differently. Dr. Litz is a fellow of the Association of Behavioral and Cognitive Therapies, the American Psychopathological Association, and the Association for Psychological Science. He has over 300 publications and has authored three books.

**Sílvia Lopes**, Ph.D., is an Assistant Professor at the European University, Lisbon (Portugal), and is a member of the CicPsi (Research Centre of Psychology Science of University of Lisbon), where she has participated in several research projects. She has a Ph.D. in human resources, work, and organizational psychology and teaches human resource management and data analysis. Her main research interests include stress and well-being at work, namely in military context. She has published in *Military Psychology*.

**Col. Vasile Marineanu** is the Head of the Romanian National Military Center for Psychology and Behavioral Health, the President of Romanian Society for Military Psychology, member in the Board of the Romanian College of Psychologists, and Adjunct Professor at the University of Bucharest. He is a graduate of the Romanian Air Force Academy (1989), the University of

Bucharest (1996), and the European Centre for Strategic Security Studies "George C. Marshall," Germany (2001). Since 2014, he has been the national representative for mental health to NATO Center of Excellence for Military Medicine, and since 2015 a member of the Research Task Group on Sexual Violence within the Human Factors and Medicine Panel of NATO Science and Technology Agency. Dr. Marineanu is a licensed practitioner with over 20 years of experience in clinical and operational psychology, acquired during the operations in Bosnia-Herzegovina (1997), Kosovo (2000, 2002, 2004), and Afghanistan (2010). His scientific research includes publications and projects such as *The Psychology of Suicide Martyrdom in Afghanistan and Its Implications for Psychological Operations* (2017), Supervised Practice in Evidence-Based Clinical Psychology (2017), Crises Intervention and Psychological First Aid (2016), and Psychological Screening for Mental Health in Romanian Military (2015).

**Ipek Mete** is currently a doctoral candidate in the field of work and organizational psychology at the Department of Psychology in Middle East Technical University (METU). She also holds a Research Assistant position at the Department of Management in Ankara Yıldırım Beyazıt University. She received her MBA degree from METU in 2013, with a thesis concentrating on team composition and conflict. She spent a year of her doctoral studies as a visiting student at the Department of Psychology in University of South Florida, where she conducted research in team performance assessment. Her research interests include training and development in teams, team dynamics, team performance assessment, and selection and assessment in work and educational settings. Ipek Mete has presented her research on the topics of teams and educational assessment in international conferences. She has publications in national and international journals.

**Peter Mikoski**, B.A., is a graduate-level research intern at National Defence University under the supervision of Dr. Stephen Bowles. He is currently pursuing a M.P.S. degree in clinical psychological science at the University of Maryland. Peter is a member of both Psi Chi and Phi Beta Kappa. His research interests include anxiety disorders, affective disorders, well-being, and pseudoscience.

**Jyoti Mishra**, Ph.D., is working as a clinical psychologist at Government Medical College and Hospital (GMCH), Department of Psychiatry, Chandigarh. She received her M.Phil. in medical and social psychology (M.M&SP) from Central Institute of Psychiatry (CIP), Ranchi. She has also worked as an Assistant Professor in the Department of Clinical Psychology at Central Institute of Psychiatry (CIP). She has to her credit various publications and has presented papers at national conferences. She has two chapters in an international index book.

**Lindsey L. Monteith**, Ph.D., is a Clinical Research Psychologist in the Rocky Mountain Mental Illness Research, Education and Clinical Center (MIRECC) for Suicide Prevention and Assistant Professor in the Department of Psychiatry at the University of Colorado Anschutz Medical Campus.

**Bret A. Moore**, Ph.D., is a prescribing psychologist and board-certified clinical psychologist in San Antonio, Texas, and Vice Chair of Boulder Crest Institute. He is a former active duty Army psychologist and two-tour veteran of Iraq. He is the author and editor of 16 books, including the *Posttrauamtic Growth Workbook, Treating PTSD in Military Personnel: A Clinical Handbook, Wheels Down: Adjusting to Life after Deployment*, and *Taking Control of Anxiety: Small Steps for Overcoming Worry, Stress, and Fear*. Dr. Moore is a Fellow of the American Psychological Association and recipient of the Arthur W. Melton Award for Early Career Achievement in Military Psychology from Division 19 and the Early Career Achievement Award in Public Service Psychology from Division 18 of APA. His views on clinical and military psychology have been quoted in *USA Today*, *The New York Times*, and *The Boston Globe*, and on CNN and Fox News. He has appeared on NPR, the BBC, and CBC.

**Swati Mukherjee**, Ph.D., is Scientist 'E' at Defence Institute of Psychological Research (DIPR), Delhi. She is involved in many major research projects of the Institute, including research on suicide in the Armed Forces. She has to her credit a few publications in the form of journal articles and book chapters in books published by reputed publishers, including Taylor & Francis and Sage Publications. She has been the associate editor of a volume on "Recent Developments in Psychology" and has co-authored a manual on *Suicide and Fratricide: Dynamics and Management* for Armed Forces personnel and a manual on *Overcoming Obsolescence & Becoming Creative in R&D Environment* for R&D organizations. Her areas of interest are social psychology, positive mental health practices, and suicidal behaviour. She was a recipient of the Defence Research & Development Organization (DRDO) Best Performance Award in the year 2008.

**Thomas E. Myers**, Ph.D., ABPP-CN, is a board-certified clinical neuropsychologist who completed his Ph.D. at The Graduate Center/Queens College, City University of New York (CUNY) and an internship in rehabilitation neuropsychology at New York University's Rusk Institute for Rehabilitation. Dr. Myers completed a postdoctoral fellowship in clinical neuropsychology at Stony Brook University Hospital where he evaluated children and adults across the lifespan. Dr. Myers is an adjunct assistant professor at Queens College and Queens borough Community College, CUNY, where he has taught a variety of undergraduate psychology courses for over 12 years. He is an adjunct faculty member at Farmingdale State College and the State University of New York at Old Westbury. Currently, Dr. Myers has a part-time position as a paediatric neuropsychologist at St. Charles Hospital and has an independent practice consisting of clinical and forensic neuropsychological evaluations. Dr. Myers has conducted research in the areas of emotion processing and neuropsychological correlates of self-awareness and has published several peer-reviewed articles and book chapters. He is an ad hoc reviewer for the journals *Professional Psychology: Research and Practice* and *Translational Issues in Psychological Science*.

**Laura A. Novak** M.P.S. earned a Bachelor of Science in psychology from Towson University in 2010, and a Master's of Professional Studies in clinical psychological science from the University of Maryland, College Park, in 2016. Ms. Novak has over 7.5 years of experience working in psychological health research within the military health system. She previously worked as a contractor supporting the Defense Centers of Excellence for Psychological Health and Traumatic Brain Injury (DCoE) Deployment Health Clinical Center (DHCC). Ms. Novak currently works as a Scientific Program Coordinator for the Suicide Care, Prevention, and Research (CPR) Initiative, directed by Dr. Marjan G. Holloway, at the Uniformed Services University of the Health Sciences (USUHS). In this capacity, she engages in and supports programmatic research on military suicide prevention and intervention. Her research experience includes grant and portfolio management, clinical trial coordination, manuscript/presentation development, study recruitment, and data collection.

**Dan Nyaronga**, Ph.D., currently works at the State University of New York (SUNY)-Empire State College where he serves as an Associate Professor of Psychology. He is also a visiting faculty member at SUNY University at Buffalo, where he teaches Summer Study Abroad Program to

Africa "Community Development in Context: Mara Region, Tanzania," an interdisciplinary program with a focus on community development that offers innovative experiential learning for undergraduate and graduate students from SUNY Institutions in New York. His research focuses on students' experiential learning, risk and resilient factors, minority groups (e.g., military families, refugees, and immigrants), gender violence, and physical and mental health outcomes. Dr. Nyaronga completed both his doctorate and Masters of Science degree in human development and family studies (HDFS) from Iowa State University. His Bachelor's degree in community development is from Daystar University, Nairobi. Dr. Nyaronga previously taught at Des Moines Area Community College (DMACC) in Iowa. He also worked in California as a postdoctoral research fellow at UC Berkeley School of Public Health, focusing on the epidemiology of alcohol use and alcohol disparities. In addition, he worked as a research associate for the Military Family Research Institute (MFRI) at Purdue University (Indiana). His community affiliations include Flower Garden Childcare International (past president), HEAL International (past board Member), Buffalo Tanzania Education Project (Member), and Girls Education Collaborative (past board member). Dr. Nyaronga has authorship in a number of peer-reviewed journals and other publications.

**Vijay Parkash**, Ph.D., is Scientist 'D' at Defence Institute of Psychological Research, Defence R&D Organization, Delhi. His area of specialization includes health psychology, personality, and psychometrics, and he has 10 years of research experience. He has also served as a Psychologist at Air Force Selection Board, Dehradun, for around two years. He has been involved in many major research projects related to suicidal behaviour and test constructions for Personnel Selection in Indian Armed Forces and Paramilitary Forces. He has a co-editor of two edited volumes, *Recent Developments in Psychology* and *Counseling: A Practical Approach*, and has to his credit more than 20 other academic publications in the form of journal articles and book chapters. Defence Institute of Psychological Research conferred upon him the Scientist of the Year Award – 2011 and 2015 for his outstanding contributions.

**Matea Nikolac Perkovic**, Ph.D., is currently working as a research associate in the Laboratory for Molecular Neuropsychiatry at the Division of Molecular Medicine, Ruđer Bošković Institute, Zagreb (Croatia), where she is included in various national and international research projects. Most of her scientific work so far has been devoted to the study of molecular basis of various neurodegenerative, neuropsychiatric, and neurodevelopmental disorders. In 2013 she was awarded the most important Croatian national award in the field of science, the National Science Award of the Republic of Croatia–Annual Award for Junior Researchers in the field of Biomedicine. She was also acknowledged for her scientific achievements with the Annual Award of the Society of University Teachers, Scholars and Other Scientists – Zagreb for Young Scientists and Artists for 2013, and a few other national awards. She has published 45 papers in reputable journals and has co-authored 15 chapters in scientific books. Currently, she is at a postdoc training in Madrid, at The Centre of Metabolomics and Bioanalysis (CEMBIO) (the Pharmacy Faculty, San Pablo CEU), where she is expending her interests by obtaining up-to-date experience in a relatively new and very interesting field—metabolomics.

**Nela Pivac**, Ph.D., is a re-elected senior scientist and full professor at Ruđer Bošković Institute (RBI), Zagreb, Croatia. She is a head of the Laboratory for Molecular Neuropsychiatry at the Department of Molecular Medicine at RBI. Among other research topics in biological psychiatry, Dr. Pivac evaluates the neurobiological underpinning of PTSD, treatment response, and different biological/genetic markers in PTSD. She participated in pioneering studies showing that atypical antipsychotics are a good adjuvant therapy for treatment-resistant veterans with PTSD. She is a PI on research projects funded by the Croatian Ministry of Sciences, Education and Sport, Croatian Science Foundation, and a collaborator on the offset project funded by Patria regarding PTSD, and a PI of Croatian-USA or Croatian-Slovenian collaborative projects, evaluating biomarkers of schizophrenia and Alzheimer's disease. Her recent projects have evaluated the association between genetic variants and PTSD development, symptoms and comorbidities in PTSD, and the role of glycans, regulated by specific genes, or metabolomes, in combat-related PTSD. She has published 159 articles in journals and has authored/coauthored 43 chapters in scientific books and been cited 3,372 times, with H-index = 34, and was awarded the most prestigious Croatian National awards and MTA Distinguished Guest Scientists Fellowship Programme in Hungary.

**Laurel K. Cofell Rashti**, Ph.D., currently serves as the consulting psychologist for the U.S. Army Drill Sergeant Academy and Leader Training Brigade. In this capacity, she serves as the primary subject matter expert for Drill Sergeant Assessment and selection. She attained her doctorate in psychology from the Uniformed Services University and since that time has also served in clinical leadership positions. She is a member of Division 19, Society for Military Psychology, and Division 13, Society of Consulting Psychology.

**Carolina Rodrigues-Silveira** is a Captain in the Brazilian Army and a military psychologist. She holds a Bachelor's in psychology and a Latu Sensu Postgraduate Degree in military sciences. She was the Chief Organizational Psychologist in the 20th Contingent of the United Nations Peacekeeping Force in Haiti – MINUSTAH (2014) and carried out the psychological accompaniment of the 2nd National Force of the Portuguese Republic (Commands Battalion) during peacekeeping operations at Central African Republican – MINUSCA (2018). Presently, she is a Ph.D. student at the University of Lisbon and member of the CicPsi (Research Centre of Psychology Science of University of Lisbon).

**António Palma Rosinha**, Ph.D., is a Lieutenant-Colonel of the Portuguese Army and a military psychologist. He graduated in Military Sciences from Military Academy, with a Bachelor in psychology and a Ph.D. in human resources management and organizational psychology from Lisbon University. He is a Professor of Human Resources Management, Organizational Psychology and Team Leadership in Military Academy (AM) and also professor of Command and Leadership the Joint Staff Course at the Portuguese Military University Institute (IUM) and Deputy Director of the Higher Institute of Business Communication (ISCEM). Rosinha is a Researcher at the Research and Development Centre of the Military University Institute and at the Military Academy Research Center (CINAMIL). He integrates the Research Group of ANPEPP – Brazilian, Working Group Culture and Health in Organizations. He is also a founder of the Leadership Innovation Center (INOVLIDER). He has several research articles and is author of the book *Leadership Study Cases in Military Context: Practice Facing Theory* and co-author of the book *Performance Assessment*.

**Michael G. Rumsey**, Ph.D., worked at the U.S. Army Research Institute (ARI) for the Behavioral and Social Sciences from 1975, the year he received his doctorate in social psychology from Purdue University, to 2012. He became Chief of the Selection and Assignment Research Unit of the U. S. Army Research Institute (ARI) for the Behavioral and Social Sciences in 1989, and held that position until his retirement. Dr. Rumsey played a major role in two landmark projects to improve the Army's selection and classification system, Project A and Building the Career Force. A special interest of Dr. Rumsey's was meeting the Army's needs for quality soldiers in the 21st century. He launched two projects in this area, one on noncommissioned officers and one on junior enlisted soldiers. Dr. Rumsey has published extensively in the fields of selection and classification, performance measurement, and leadership, including editorship of the *2012 Oxford Handbook of Leadership* in 2012 and senior editorship of the 1994 Erlbaum book *Personnel Selection and Classification*. He is past president of the Society for Military Psychology and was a recipient of the division's John C. Flanagan Lifetime Achievement Award. He currently serves as Associate Editor of the journal *Military Psychology*.

**Marina Sagud**, M.D., Ph.D., is Assistant Professor in the School of Medicine, University of Zagreb, and also a Head of the Department for Schizophrenia and Psychotic Disorders in the Department of Psychiatry, University Hospital Centre Zagreb. She is currently leading the project: "The Influence of Acute Stress Reaction and Depressive Disorder on the Cardiovascular Pathology," sponsored by University of Zagreb. She has organized nine postgraduate courses in clinical use of antipsychotics and antidepressants at the School of Medicine, University of Zagreb. She has published 45 articles cited in the "Current contents" base, and has (as of January 31, 2019), 860 citations in the Web of Science, H-index = 17. She was awarded for scientific productivity by the University of Zagreb in 2011 and was a mentor of the student work which received a Dean's award in the School of Medicine, University of Zagreb, in 2018.

**Renata Sargon** is a fourth-year student in the clinical psychology Ph.D. program at Palo Alto University. She received her Bachelor of Science in psychology and Bachelor of Business Administration in marketing and Masters of Education in community counselling from Loyola University Chicago. Renata is a Licensed Professional Counsellor in the state of Illinois, where she worked in long-term care facilities for individuals diagnosed with severe mental illnesses as a Psychiatric Rehabilitation Services Director and consultant. Renata is currently a member of Dr. Bongar's Clinical Crises and Emergencies Research Lab at PAU and is a practicum student at Santa Clara Valley Medical Center. She is also an In-Custody Instructor for the New Opportunity Work program at Goodwill of Silicon Valley, where she completed her third-year practicum at their Wellness Center. Her primary research interests include crisis intervention training for police, the psychology of terrorism, and suicidality.

**Senior Master Sgt. Susan J. Scapperotti** is currently serving as the Air National Guard Recruiting and Retention Course Manager at Joint Base San Antonio, Lackland AFB, Texas. She is assigned to JBSA-Lackland on a Statutory Tour by the Air National Guard Readiness Center, Joint Base Andrews AFB, Maryland. She leads formal training for both ANG instructors and students within the Air Force Recruiting School located at the 344 Training Squadron. She entered into Active Duty Air Force in November of 1996 and began her recruiting career in January 2000. Her career advanced as she was selected for elevated operational positions. In 2006, she transferred to ANG recruiting duty and earned the Air Force Meritorious Medal for exemplary performance. Her notable lectures on generational understanding for recruiting and management provide relevant training throughout the Air Force.

**Bettina Schmid**, Ph.D., Clinical Psychologist, is an officer with the United States Army Reserve and has served on active duty in the Army and Air Force. During her time on active duty, she spent three years serving as a psychology consultant to leaders and soldiers in Army Recruiting. She also works at the Salt Lake City VA in telemental health. Her areas of expertise include PTSD, substance use disorders, and gerontology.

**Matthew B. Schwartz**, Ph.D., teaches ancient history and literature in the Departments of History and Near East Studies at Wayne State University. He is a Contributing Editor of *Menorah Review* and Associate Editor of the *Journal of Psychology and Judaism*. Dr. Schwartz has authored or co-authored six books and numerous articles that deal largely with Graeco-Roman and Jewish thought. The books include *Roman Letters: History from a Personal Point of View* and several works in partnership with Dr. Kaplan, including *A Psychology of Hope*; *Jewish Approaches to Suicide, Martyrdom, and Euthanasia*; and *The Family: Biblical and Psychological Foundations*.

**Leah Shelef**, Ph.D., is former Head of the Psychology Branch, Air Force IDF Mental Health Department, Israeli Medical Corps (reserve), Air Force, IDF Mental Health Department, Israeli Medical Corps. Dr. Shelef is a clinical social worker and a senior expert in the field of suicide in the IDF. Her doctorate field of research was "Distress, Personality Resources, Gender Subjective Experience and Suicide Facilitating Process among Soldiers Who Have Attempted Suicide," received from Bar-Ilan University in Israel. Dr. Shelef has published more than 30 papers; more than 20 are original papers, reviews, and book chapters—most of them are in the field of suicidal behaviour among Israeli soldiers.

**Yifat Erlich Shoham**, Ph.D., is currently the Surgeon General of the medical forces in the Israeli Air Force, IDF Israeli Medical Corps. Before her current commission, Dr. Erlich-Shoham was the head of the Aero-Medical Center, the head of the Medical Classification and Occupational Medicine Branch at the IDF Surgeon General Headquarters, and commander of the clinical wards at Tel Nof and Palmachim Air Force Bases. She is in active position at the IAF Medical Evacuation and Rescue Unit.

**Col. Tracy D. Smith** is the Chief of Recruiting and Retention for the Air National Guard, Joint Base Andrews, Maryland. She provides strategic direction, resources, training, policy, and guidance to support 90 Air National Guard Wings, 54 states, territories, and the District of Columbia's recruiting and retention programs. She gathers, analyzes, monitors, and reports metrics in support of Air National Guard recruiting and retention plans and programs to the Office of the Secretary of Defense, Headquarters Air Force, Chief National Guard Bureau, and Director of the Air National Guard. Colonel Smith is responsible for analysis of recruiting and retention trends, ensuring that national end strength meets operational mission requirements. She also oversees the development of national advertising and marketing initiatives and partnerships. Colonel Smith served on active duty as a traditional Guardsman and as an Active Guard Reservist in both staff and command positions at the wing, state, and national headquarters level. Prior to her current assignment, she was the Commander, 113th Mission Support Group, Washington, DC, Air National Guard.

**Matthias Soest** is currently acting head of the Psychological Service in the German Army and has more than 10 years of experience as a psychologist working for Multinational Joint Headquarters, for the Special Forces, the U.S. Marine Corps in Quantico, and the German Army Headquarters. He is a reserve officer (LRRP) and has had multiple deployments to Afghanistan.

**Jeanne Mager Stellman**, Ph.D., is Professor Emerita, Mailman School of Public Health, Columbia University. After retirement from Columbia, she became Professor and Chair, Environmental Health Sciences, and Associate Dean for Research at the School of Public Health at SUNY-Downstate Medical Center in Brooklyn, New York, for five years before returning to Columbia as a Special Lecturer. She is President of Columbia's Emeritus Professors in Columbia (EPIC) fellowship and continues to be actively engaged in research on Vietnam veteran health, generally in collaboration with Prof. Steven D. Stellman. They are currently carrying out a mortality study of a cohort of American Legionnaires they have followed for decades. With support of the National Library of Medicine, Stellman oversaw the creation of a popular website: Agent Orange Data Warehouse. Stellman has also worked in areas of environmental and occupational health, including research on World Trade Center recovery workers' health. She has written several seminal books on occupational health and many monographs and chapters and peer-reviewed publications. She created the Women's Occupational Health Resource Center at Columbia (1981–1991) and served as Editor-in-Chief of the *ILO Encyclopaedia of Occupational Health and Safety*, 4th edition (1998). Honours include a Guggenheim fellowship, Preventive Oncology Academic Awards, Principal Investigator of numerous grants, and being named one of Ms. Magazine's "80 Women to Watch in the 80's." Her papers are archived at Harvard's Schlesinger Library. Stellman holds a doctorate in physical chemistry from City University of New York (1972).

**Steven D. Stellman**, Ph.D., is Professor of Clinical Epidemiology at the Mailman School of Public Health, Columbia University. He is a cancer epidemiologist with over 40 years' experience designing and conducting cohort and case-control studies relating chronic disease outcomes such as cancer and heart disease to lifestyle and environmental risk factors. Prior to joining the Columbia University faculty, he worked in both government and non-profit organizations. As Chief of Epidemiology for the American Health Foundation, he directed studies of behavioural, metabolic, and genetic factors in causation of tobacco-related cancers and environmental risk factors for breast cancer. He was co-founder of Cancer Prevention Study II, an American Cancer Society study of over 1.25 million men and women. At the New York City Department of Health and Mental Hygiene, he served for three years as Assistant Commissioner for Epidemiology and Biostatistical Research and Director of the Bureau of Vital Statistics and for eight years as Research Director of the World Trade Center Health Registry, a cohort study of 71,000 survivors of the World Trade Center disaster. For the past 35 years, he has collaborated with his wife, Dr. Jeanne Stellman, in numerous studies of Vietnam veterans. Dr. Steven Stellman has a doctorate in physical chemistry.

**Dubravka Svob Strac**, Ph.D., works as Senior Research Associate at Laboratory for Molecular Neuropsychiatry, Division of Molecular Medicine, RudjerBoskovic Institute, Zagreb, Croatia, and holds a title of Assistant Professor at University of Osijek. She obtained her B.S., M.S., and Ph.D. degrees in molecular biology at Faculty of Science, University of Zagreb. Additionally, she graduated in project management at Accredited College of Business and Management, Zapresic. She also obtained various short-term training at home and abroad and received several awards and stipends. So far, she has published 59 (50 WoSCC) scientific papers (citations: 637; H-index: 16) and 15 book chapters. She coordinated 2 and actively participated in 10 national and 6 international projects. She is a lecturer at seven graduate and postgraduate courses at University of Zagreb, Osijek, and Rijeka, and mentor of 14 B.S. and 4 Ph.D. theses. Dubravka Svob Strac actively participates in the work of various scientific societies, organization of scientific meetings, and science popularization. She is member of the editorial board of several and reviewer for over 50 international scientific journals, several congresses/symposia, biology school textbooks, and Croatian Science Foundation projects. Her field of research is molecular basis of stress-related and neuropsychiatric disorders and molecular mechanisms of neuropsychoactive drugs.

**H. Canan Sümer** is a Professor at the Department of Psychology in Middle East Technical University (METU). She received her Ph.D. in the area of industrial and organizational psychology from Kansas State University in 1996. Her research includes development of personnel selection tests and techniques, personality assessment, performance management and culture, and leadership and gender-role stereotypes. Dr. Sümer is on the editorial board of a number of national and international journals in her field. She is a member of the Turkish Psychological Association and Society for Industrial and Organizational Psychology, and she has represented Turkey in EFPA's Assessment Board. She has led a number of large-scale projects on military personnel selection and performance management, and was an active member of NATO HFM Panel's task group on Military Recruitment and Retention. More recently she was involved in a project sponsored by the Scientific and Technological Council of Turkey (TÜBİTAK) in which a computerized selection system along with an assessment centre was developed to be used in the selection of blue-collar and white-collar employees. She is among the inventors of five national and two international patented inventions that came out of this selection system development project. Dr. Sümer has national and international publications on the topics of job analysis, personnel selection, performance management, and gender role stereotypes, and she is actively involved in training human resource management practitioners through the programs of METU Continuing Education Center.

**Swati**, Ph.D., is Scientist 'C' with Defence Research Development Organization (DRDO), Ministry of Defence, India. Previously, she has served as an Assistant Professor in SUS College of Research and Technology, Panjab. She is posted as a psychologist in Naval Selection Board, Visakhapatnam, for selection of Naval Officers. She has been awarded a commendation from Chief of Eastern Naval Command in December 2016 for her outstanding work.

**Lucija Tudor** finished her M.Sc. of molecular biology on the Faculty of Science, University of Zagreb, in 2016 and is currently a Ph.D. student of medicinal chemistry in the Department of Biotechnology, University of Rijeka, Croatia. She is currently working as an assistant/Ph.D. student in the Laboratory for Molecular Neuropsychiatry, Ruđer Boskovic Institute (RBI), on the project "Genomic and Glycomic Biomarkers for PTSD," led by Nela Pivac, financed by the Croatian Science Foundation. She has published seven journal articles and two chapters in scientific books. She was awarded a FENS-IBRO/PERC travel grant for the 11th FENS Forum of Neuroscience in Berlin in 2018. She actively presented preliminary results on glycomic and genetic biomarkers of PTSD, aging, and cognition on four international symposia and is included in popularization of science by active participation on the RBI manifestation "Open Days" for two years.

**Suzana Uzun**, M.D., Ph.D., Assistant Professor, is a psychiatrist working at the University Psychiatric Hospital Vrapče, Zagreb, and a subspecialist in biological psychiatry. She finished her residency in psychiatry in Psychiatric Hospital Karlovac and in Clinical Hospital Center Zagreb, Croatia. She was a leader of the scientific project "Side Effects of Antipsychotics," founded by the Ministry of Sciences, Education and Sports of the Republic of Croatia, a bilateral project with Slovenia "The Role of 5-HT6 Receptors in Alzheimer's Disease," and is collaborator on the project lead by Nela Pivac, funded by Croatian Science Foundation: "Genomic and Glycomic Biomarkers for PTSD." She has won several international awards provided by the European Workshop on Schizophrenia, Davos, Switzerland (in 1999, 2000, 2002, 2004, 2006, 2008), and Biennial Winter Workshop on Schizophrenia, Montroux (2006). She organizes with Oliver Kozumplik a postgraduate course, "Side Effects of Psychopharmaceutic Drugs," School of Medicine, Zagreb University.

**Kristen J. Vescera**, M.A., is a fourth-year graduate student at Palo Alto University. She received a Bachelor of Arts degree at Syracuse University in English and textual studies and geography.

After travelling to Latin American countries for a year assisting high school and college students in volunteer projects across Costa Rica, Honduras, and Nicaragua, she developed an interest in working with adolescents. She moved to Denver, Colorado, and received a Master of Arts degree from University of Colorado Denver in counselling psychology and counsellor education with an emphasis in school counselling. Following graduation, she enlisted in the United States Army Reserves as a Psychological Operations Specialist and immediately deployed to Afghanistan in support of a nine-month combat operation. Kristen is currently a reserve Staff Sergeant who teaches psychological operations doctrine to junior enlisted service members. She is interested in working with active duty and veteran populations within the larger context of policy and program development.

**Nicola von Luedinghausen** is a teaching and operational psychologist at the EOD Training Facility and also a trained sport psychologist with deployment experience in Afghanistan.

**Nirit Yavnai** is currently the Head of Medical Research Directorate, Academy and Research Branch, Israeli Medical Corps. Dr. Nirit Yavnai, D.M.D., M.P.H., joined the Academic Reserve Officers Program and received her D.M.D. degree from the Faculty of Dental Medicine, Hebrew University, in 2000. She was awarded a M.P.H. degree from the school of Public Health, Haifa University, in 2008. She finished her residency in public health dentistry and received her specialist license in 2012. Her military service started in 2000 as a dental surgeon in the Medical Corps and she served in command assignments in the military dental services and in the Israeli Navy. After finishing her residency in public health, she was assigned at the Academic & Research Branch as the head of the Research Promotion section. Her current assignment as the head of the research directorate in the IDF and IMOD includes a dual assignment with responsibility of prioritizing, funding, and promoting medical research in the IDF and MOD. Her publications and fields of interest include military medicine and public health education, epidemiology of infectious and chronic diseases, and health promotion.

# FOREWORD

Both the challenges and rewards of military psychology stem from the breadth of the discipline. If you introduce yourself as a clinical psychologist, it is pretty clear what you do for a living. A social psychologist presenting a paper at a social psychology conference will find him/herself surrounded by others who share a similar conceptual and methodological approach to the discipline. The same is true for just about all of the standard subdisciplines of psychology. But to tell someone that you are a military psychologist conveys little information beyond your focus on the population you study. In all likelihood, your doctoral degree is not in military psychology. Instead, it is in experimental, cognitive, social, clinical, counselling, or industrial psychology (just to name a few). Your degree may not even be from a psychology department. About half of engineering psychologists and human factors engineers obtain their education from graduate programs in engineering.

This interdisciplinary nature of military psychology is illustrated in Figure 0.1. For simplicity's sake, this Venn diagram has only three circles (there could be more) representing clinical, experimental, and applied psychology. Clinical psychology includes all flavours of that discipline, including counselling and neuropsychology. Experimental psychology includes cognitive, animal learning, social, cognitive neuroscience, and other areas that focus on the basic science of behaviour. Applied psychology includes industrial, organizational, and engineering psychology.

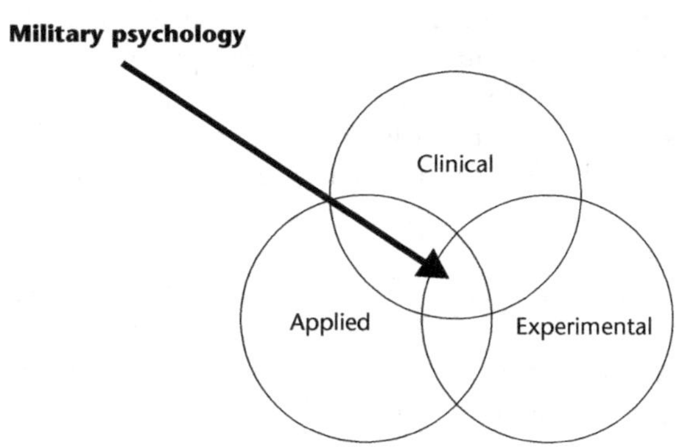


*Figure 0.1* Military psychology in context.

Military psychology resides in the shaded portion of the Venn diagram where the three circles of clinical, experimental, and applied psychology overlap.

Another feature of military psychology that distinguishes it from other disciplines is the international nature of the topics of interest within the field. In my 35-plus years as a military psychologist, I have observed time and again that issues of interest to U.S. military psychologists are of equal interest to military psychologists around the globe. Perhaps this is not surprising given that the economic, social, political, and technological factors that determine the nature of war are not isolated within a given country or region. Coupled with its interdisciplinary nature, this common bond of interest across different nations adds to the rich diversity of military psychology.

To paraphrase comments of the great early experimental psychologist Herman Ebbinghaus, military psychology has a long past but only a short history (Boring, 1950). Because of the intense nature of warfare, understanding the psychology of the soldier and factors that affect his or her performance were just as important to the armies of ancient Greece or China as they are to contemporary military leaders. Sun Tzu (1983) did not seek the counsel of psychologists, but a careful reading of his *Art of War* reveals deep insights into the psychology of war. Great generals have always been astute observers of human nature. Similarly, ancient literature such as Homer's *Iliad* and *Odyssey* reveal deep insights into the nature of war and of soldiers.

Military psychology as a distinct area of science and practice followed psychology's emergence as an independent discipline in the late 19th century. War drives developments in science in general, and this is true for psychology as well. The first great accomplishment of military psychologists occurred during World War I, when the U.S. Army called upon members of the American Psychological Association to devise aptitude tests to help it screen recruits and assign them to jobs fitting their abilities. World War II stimulated the origins of engineering psychology and provided a significant boost to clinical psychology. Vietnam led to the recognition of post-traumatic stress disorder (PTSD) as a diagnosable and treatable psychopathology. More recently, the Global War on Terror (GWOT) is giving rise to developments in cyber-psychology and mental health training (Matthews, 2014a).

A feature of contemporary military psychology is the further blurring of traditional disciplinary lines. Consider combat stress, for example. Once thought to be "psychological" in nature—that is, a learned, maladaptive response to the extreme stress of combat—it is now recognized that to fully understand PTSD, a range of factors must be considered. PTSD is more than a learned response. Vulnerability to the consequences of combat stress such as PTSD and depression may be influenced by neurochemistry (Chottekalapanda, Greengard, & Sagi, 2019). Genetic factors play a role in susceptibility (Martindale, 2019). Brain trauma, classified as mild traumatic brain injury (mTBI) or chronic traumatic encephalopathy (CTE), shares an overlap of symptoms with PTSD (Matthews, 2014b). Symptoms similar to PTSD may also occur in the absence of the experience of a traumatic event, per se, and are referred to as moral injury (Meagher & Pryer, 2018). Thus, the efforts of a diverse array of psychologists and other scientists are needed to fully understand and then to develop effective means of preventing and treating combat stress disorders.

Human nature has not changed, but the social and physical environment in which war occurs has. So-called Phase Zero operations involve shaping the social and political attitudes of people within a nation targeted by a hostile force.[1] Social media provides the perfect weapon to achieve this goal. Through the clever use of social media, hostile powers may directly manipulate the

1 For a discussion of Phase Zero operations, see https://www.cgdev.org/blog/phase-zero-pentagons-latest-big-idea

beliefs and attitudes of millions of people. These manipulations may create discord and confusion through the spread of misinformation or "fake news." It is imperative that military psychologists develop expertise in understanding this weaponisation of social media.

The physical environment may also offer new challenges. Megacities, defined as cities of 10 million or more inhabitants, are rapidly emerging. There are dozens such cities now, and many more are expected to develop in the coming decades. Many of these cities are seedbeds of discontent, with wide disparities of access to basic human needs such as healthcare or education. If only one percent of the inhabitants of a megacity are hostile forces, that means that in a city of 10 million people there would be 100,000 potential adversaries. And these adversaries would have an intimate knowledge of the terrain and culture of that city. The psychological impact of deploying an army to fight in such an environment is not currently well understood and represents an area of increasing concern for military strategists and military psychologists (Strategic Studies Group, 2015).

Rapid developments in neuroscience are also contributing to military psychology. A better understanding of the human brain, genetics, and neurobiology raises the very real possibility of creating an "engineered" soldier. That is, these sciences may be leveraged to produce a soldier that can fight longer and harder than normal human beings. Altering brain chemistry could inoculate soldiers against combat stress or allow them to function at peak effectiveness for extended periods of time. Brain implants may enhance learning and memory skills of soldiers. Epigenetic manipulations may allow enhancement of a wide spectrum of soldier capabilities (for a full treatment of human performance optimization science, see Matthews & Schnyer, 2019).

Historically, wars were planned and conducted by men. The growing acceptance of women in the military requires the continued attention of military psychologists. Nations differ in their specific policies concerning the utilization of women in the military, but they share the need for objective and honest information about this issue that military psychologists can provide. The same may be said for the inclusion of other groups of people including gay, lesbian, and transgender people in military service. As warfare evolves from an exercise in brute force (picture Roman centurions hacking away at opponents with swords) to an exercise in cognitive dominance (Matthews, 2014a), the rationale for war being a primarily male enterprise is eroding.

Another area of contemporary contributions of military psychology is in developing a better understanding of non-cognitive factors in soldier performance. Since World War I most of the efforts of military psychologists to understand soldier performance have centred on cognitive factors, such as intelligence and aptitude. In what I call the "25/75 Rule" (Matthews, Lerner, & Annen, 2019), cognitive factors account for about only a fourth of the variation in soldier performance. Non-cognitive factors such as motivation, grit, hardiness, and positive character strengths make up the remaining 75%. Unpacking this remaining 75% of variation in soldier performance represents an important challenge and a major contribution for today's military psychologists.

It is in this conceptual and historical space that *The Routledge International Handbook of Military Psychology and Mental Health* unfolds. In a single volume, these topics and more are addressed by the world's leading military psychologists. The contributors of this book represent nations from multiple continents with different histories and cultures. This diversity of perspective strengthens the content of this volume and adds to its credibility as an authoritative source of the broad and deep field of military psychology. A perusal of the Table of Contents illustrates the Venn diagram introduced above. This is not a book about clinical military psychology, experimental military psychology, or applied military psychology. Rather, it is the blending of all three conceptual and methodological approaches. The resulting *Gestalt* is, indeed, a case where the whole is greater than the sum of its parts.

## References

Boring, E. G. (1950). *A History of Experimental Psychology* (2nd edition). New York: Appleton Century Crofts.

Chottekalapanda, R. U., Greengard, P., & Sagi, Y. (2019). Molecular and cellular aspects of major depressive disorder. In M. D. Matthews & D. M. Schnyer (Eds.), *Human Performance Optimization: The Science and Ethics of Enhancing Human Capabilities* (pp. 62–89). New York: Oxford University Press.

Martindale, V. E. (2019). Optimizing cognitive performance: Genetic and epigenetic techniques. In M. D. Matthews & D. M. Schnyer (Eds.), *Human Performance Optimization: The Science and Ethics of Enhancing Human Capabilities* (pp. 287–301). New York: Oxford University Press.

Matthews, M. D. (2014a). Stress among UAV operators: Posttraumatic stress disorder, existential crisis, or moral injury? *Ethics and Armed Forces: Controversies in Military Ethics and Security Policy, 1*, 53–57.

Matthews, M. D. (2014b). Soldier's heart, shell shock, combat fatigue, posttraumatic stress disorder: Bio-behavioral disorder or social construction. *Paper presented at the Columbia University Conference on World War I*, 17 October 2014.

Matthews, M. D., Lerner, R. M., & Annen, H. (2019). Non-cognitive amplifiers of human performance: Unpacking the 25/75 rule. In M. D. Matthews & D. M. Schnyer (Eds.), *Human Performance Optimization: The Science and Ethics of Enhancing Human Capabilities* (pp. 356–382). New York: Oxford University Press.

Matthews, M. D., & Schnyer, D. M. (2019). *Human Performance Optimization: The Science and Ethics of Enhancing Human Capabilities*. New York: Oxford University Press.

Meagher, R. E. & Pryer, D. A. (2018). *War and Moral Injury: A reader.* Eugene, OR: Cascade.

Strategic Studies Group. (2015). *Chief of Staff of the Army, Strategic Studies Group-III, Final Report, Executive Summary*. Arlington, VA: CSA SSG.

Tzu, S. 1983. *The Art of War*. New York: Delacorte Press.

Michael D. Matthews, Ph.D.
*Professor of Engineering Psychology*
*U.S. Military Academy*

# PREFACE

Psychology, being the science of human behaviour, has always a significant role to play across all settings that involve humans. Accordingly, numerous branches of the discipline have flourished and keep flourishing as the multi-componential aspects of human involvement attract the attention of the scientific world. From child development, schools, societies, industries, and various organizations to clinics, everywhere the scientific role played by psychology is so vast and vivid that the setting itself justifies the existence of specific subdisciplines like child, developmental, organizational, or clinical psychology and so on and so forth. As nations exist, so do their militaries, which are considered the most crucial human assets. The critical significance of military personnel for nations and the complexities of this extreme profession deserve and have drawn a concrete focus of the behavioural scientific community in the form of military psychology. Though applications of psychological principles are inevitable from the inceptions of the first-ever thoughts of having militaries or similar security forces, military psychology as a full-grown scientific and research profession had grown not much earlier than the last three decades. With very scarce literary works in the past, the first major dedicated professional volume surfaced in 1992: *Handbook of Military Psychology* by Gal and Mangelsdorff (1992). Mingled with their personal military experiences and scientific acumen, the editors attempted to put together the applied domain of military psychology and its principles and practices. Around two decades later, Laurence and Matthews (2012) brought the *Oxford Handbook of Military Psychology*, presenting the discipline in a refreshed and more newly emerged perspective. An effort was further made by Bowles and Bartone (2017) to extend and add the specific direction of clinical and organizational practice to the discipline. With only these few major works and rarely any others, the scope of the ever-emerging field of military psychology remains vastly unaddressed. The present volume is an attempt to refresh the existing, extend to the required, and cover the uncovered domain of military psychology.

Taking the path initiated by the predecessors ahead, *The Routledge International Handbook of Military Psychology and Mental Health* is an endeavour to dedicate scientific focus to the service of those who dedicate their lives in the service of their nations. Systematically and concisely arranged in three broad section heads encompassing 35 chapters, the volume puts together the scholarly knowledge of internationally acclaimed experts in the field working across different corners of the world. The first section focuses on "The Roots and the Journey" of the discipline.

The section has eight chapters presenting deliberations on the beginning, continuation, and expansion of applications of psychological principles in militaries during war as well as peace times, evolutions of military-specific testing and some associated research areas, applications of culture and cognition in the domain, the significance of military psychology in future warfare, and its prospective directions as a scientific discipline, along with delving into various ethical issues and concerns at the global level.

The second section of the volume is loaded with professional discussions on "Soldiering: Deployment and Beyond." Though the section contains only 11 chapters, the scope of this section is much more elucidative and comprehensive. The concretely inherent roles and applications of military psychology in the form of sustainable leadership development through resilience building, hardiness training, and enhancement of virtues and character strengths among military personnel are at the centre of scholarly deliberations in this section, and the aspects of military recruitment, assessment, bravery, training, lone-wolf terrorism, military turnover, family reintegration, post-traumatic growth, and transition from military to civil life have been thoroughly enumerated and discussed in light of empirical evidence from across the globe. The section begins to provide insight into mental related issues in militaries and draws attention towards the applications of military psychology into the same, which is explicitly furthered in the third section of the volume.

The third section, "Making a Choice: Mental Health Issues and Prospects in Military," encompasses 16 chapters which bring together the efforts of military mental health professionals in explaining the mental health issues prevailing in military life. Ranging from work engagement, stress, coping, and burnout through other mental health and clinical concerns, like traumas, morbidities, post-traumatic experiences, bio-physiological abnormalities, moral injury, and suicidal behaviour related aspects in the military, the section delves into the requisite psychological interventions at the individual as well as family level in order to make resilient soldiers and veterans. The issues highlighted in this section, which are seemingly the most crucial for the attention of psychologists and other mental health professionals, have barely found such a synthesized collection at one place in any earlier work. The importance of this largest section of the volume is the reason which calls for the title *Handbook of Military Psychology and Mental Health*. The handbook intends to serve scientific professionals in the field to provide fruitful directions to their research efforts, military professionals to work towards the growth and flourishment of the personnel and the organization, mental health professionals to dedicate their services to the needed, policy-makers and strategists to devise various modalities in necessary directions, and lay readers to enlighten them about the complexities of military as a profession and military psychology as a scientific discipline.

Bringing this volume in such a literarily barren area has been quite a herculean task, but it has been a call of my duty as a *military psychologist* with experience of around three decades. What is delighting is the overwhelming response and forthcoming contribution of the globally renowned scholars, scientists, and stalwarts who have shared their in-depth knowledge in this endeavour of service to military personnel and the profession of military psychology. I am deeply indebted to all these contributing authors, although conveying a simple gratitude to them cannot be enough for such a giant cause. My gratitude goes to all those who have lent their support in any possible manner for this venture, and most importantly I offer my deep felt sense of indebtedness to Susannah Frearson, Heather Evans and Joanne Forshaw at Taylor and Francis Group for their unconditional support and relentless efficient strivings to publish this gigantic volume. I have a strong belief that this volume will be a ready reference to all those who are in any way connected with military and military psychology, and it will prove to be a milestone in the progress of the discipline of military psychology.

## References

Bowles, S. V., & Bartone, P. T. (2017). *Handbook of Military Psychology: Clinical and Organizational Practice.* Cham, Switzerland: Springer International Publishing.

Gal, R., & Mangelsdorff, A. D. (1992). *Handbook of Military Psychology.* New York: John Wiley and Sons.

Laurence, J. H., & Matthews, M. D. (2012). *The Oxford Handbook of Military Psychology.* New York: Oxford University Press.

Updesh Kumar, Ph.D.
*Scientist 'G' & Head Mental Health Division*
*Defence Institute of Psychological Research (DIPR)*
*Defence Research & Development Organisation (DRDO)*
*Ministry of Defence, India*

# PART I

# Military psychology: The roots and the journey

# 1

# MILITARY PSYCHOLOGY IN WAR AND PEACE

## An appraisal

*Swati Mukherjee and Updesh Kumar*

Psychology as a modern 'scientific' discipline has been a close ally of modern warfare, the two having established a mutually beneficial relationship, especially during the first half of the twentieth century and since. Military psychology has over the years established itself as a unique sub-discipline that determines its boundaries not through methodological concerns or subject content, but rather through its ability to optimally fulfil the requirements of the Armed Forces in specific contexts and under unique circumstances, making use of the advancements in the broader discipline of psychology. And in the endeavour, psychology has visibly made significant contributions to the military. Perhaps it is not an exaggeration to say, as Matthews says, 'psychology is more relevant and viable today for the military, than at any point in history.' (Matthews, 2014, P. 215). The growth of the sub-discipline of military psychology has been in tandem with the developments in the art and science of warfare. However, in the post–Cold War era with the international political equations rapidly changing, and in view of the unprecedented technological developments in the mechanics of warfare, the fundamental nature of war itself has undergone a metamorphosis. In this context, it becomes imperative for military psychologists to take stock of the discipline by asking the eternal question, as Bingham asked while striving to discern the pressing problems of military psychology, in the 'chaotic days between war and peace' after the Second World War – 'What are the continuing problems of military psychology?' (Bingham, 1947, P. 155). It is the same question that military psychology is faced with today even after the passage of half a century. The question seems to be gaining more relevance with increasing ambiguities and complexities of warfare, and the blurring boundaries between war and endeavours for peace. The present chapter begins by delineating the current standards of research and practice in military psychology in the context of the major historical trends in the discipline. Subsequently, it engages with the quintessential question of the dynamics between war and peace, and the precarious balance between the two, seeking to emphasise the need to enhance interaction between the domains of peace psychology and military psychology. The chapter concludes with an argument for structuring a nuanced ethical framework for research and practice of military psychology in order to prevent the discipline from becoming a mere cog in the machineries of war.

## Psychology during the World Wars

The primary role of psychology in the military is to support and assist the armed forces in achieving their goals by ensuring fitness of selection, capability building through training and sustained mental fitness for effective operation under varied circumstances. And in this vein, beginning from the days of the First World War, military psychologists have been consistently working in tandem with the militaries of the countries on both sides of the Atlantic, although some of the authors reviewing the development of the discipline are of the opinion that psychology had a very limited scope in the military in the early years of the twentieth century. Reviewing the role psychologists played during the Great War (1914–1918), Shephard (2015), for example, says that the military scarcely knew about psychology, especially in Britain. Providing a succinct account of contributions of psychology to the war effort during the First World War, Shephard (2015) says that in Britain the role of civilian psychologists was limited mostly to treating 'shell-shocked' soldiers in military hospitals, drawing on their medical expertise and experience. Dr. Charles Meyers, for example, was appointed the Consultant Psychologist to the first Royal Army Medical Corps, and established mental health hospitals in tents behind the enemy lines in France. It was only in the later years of the war that some of these pioneering psychologists (e.g. Myers, Rivers, Spearman) could find some more significant and relevant applications of psychology in various branches of the military. No attempt, however, was made to apply psychology to the testing of military recruits (Shephard, 2015, P. 944). At the same time, psychologists in Europe had a better say in military matters. Shephard (2015) describes how Wilhelm Wundt, among other prominent cultural and academic figures, supported the German army's stance against Belgium. Many of those involved in war were students of Wundt, due to which greater relevance was accorded to psychology in the war efforts on both shores of the Atlantic. Germany was a pioneer in establishing the discipline of psychology, and the German military benefitted from the industrial expertise gained by psychologists in the pre-war years. As a result, Germany started using aptitude tests for the selection of pilots, truck drivers, radio operators and other specialists as early as 1915. Psychologists performed many specified tasks for the German military, which did not have any long-term impact on the discipline – for example, a 'listening device' developed by Max Wertheimer for locating enemy artillery. On the other hand, certain explorations by psychologists while serving in uniform laid the foundations of later research in the fields of leadership, combat motivation, fatalism among soldiers and the like. Psychologists can be credited with making the German war effort 'more scientific, rational and modern' (Shephard, 2015, P. 945), and for demonstrating to the state the practical usefulness and applicability of the discipline, resulting in the creation of new positions of psychology in technology institutes and commercial academies (Geuter, 1992). Akin to Britain, France too had minimal involvement of psychologists in the war effort, wherein they devised some psycho-physical measurements of heart rate and respiration of machine gunners, but none of these countries employed large-scale military testing.

On the other side of the Atlantic, as the United States of America entered the war in the year 1917, psychologists with industry experience found useful employment in the military. The contributions of Walter Dill Scott and Robert M. Yerkes are remarkable in this era of early development of psychology in the military. By the end of the war, the Committee on the Classification of Personnel in the Army, which tested men for their aptitude in various fields of the military using Scott's techniques, had interviewed and classified thousands of men. Yerkes was a pioneer in initiating a more formal association of psychology with the military, and, along with Terman, Goddard and a few others, is credited with the development of Army General Classification Tests (Army Alpha and Army Beta), measures that allowed for testing of large samples simultaneously. Reviewing the status of military psychology, Melton (1957) remarks that during the First World War, psychologists also engaged in development of job-knowledge tests, training of naval gunners, analysis of aircraft pilot ability and perhaps many other areas, of which written records no longer exist. Among other consequences, the implementation of large-scale testing in the military gave impetus to psychology in the United States and eventually led to the establishment of the Division of Psychology in the office of the Surgeon General of the U.S. Army in the year 1917 (Mangelsdorff & Gal, 1991). On the whole, while some researchers go to the extent of claiming that 'social and clinical psychology received a significant boost from the First World War and a new branch of the profession emerged in the shape of military psychology' (Bourke, 2001), it is reasonable to conclude that the First World War at the minimum provided for conditions where psychology could envisage a new role for itself in the military.

The Second World War witnessed exponential expansion of the applications of psychology in the military, both in Britain and in the United States. Britain recruited a large number of psychologists in various branches of the armed forces, including the areas of personnel selection and health care systems (Hughes, McCauley, & Wilson, 2019). Significant contributions were made by psychology to the military efforts, and in recognition, the American Psychological Association (APA) included the Division of Military Psychology (Division 19) among its first group of formal subdivisions in the year 1945. A committee to evaluate the work of psychologists and psychiatrists in the armed forces during the war years was set up by the then-prime minister Winston Churchill. The committee reported upon the vital role played by the group in war, and as a result psychologists gained an official entry in the British civil services post war. Similarly, remarkable progress was being made in Germany in finding applications of psychology to military settings. Germany had made preliminary beginnings in this area during the First World War, wherein psychological technique was being applied for selection of drivers, pilots, wireless operators, sound detector operators and anti-aircraft personnel. After the war ended, the war ministry issued orders for development of psychology in the army, leading to rapid developments in the areas of personnel selection, resulting eventually in mandatory psychological testing using the 'whole personality approach' for selection of all armed forces officers (Fitts, 1946). Also, while much attention was given to selection in the armed forces by the German psychologists, not much contribution was made towards using psychology for structuring training. Similarly, clinical psychology found very limited application in an environment where decisions regarding psychogenic problems of soldiers were governed by military policy, rather than psychological advice. German psychologists differed significantly in their methods of analysis, as their focus was more on subjective analysis rather than objective testing.

Meanwhile, in the United States of America, with the beginning of the Second World War, a large number of psychologists were recruited into the Armed Forces for research studies and technical applications. The single largest programme implemented by the Army Air Force focused primarily on psychological selection of personnel, and in the later years included problems of training and proficiency measurement (Melton, 1957). Research programmes in other domains of the military were also focused primarily upon personnel selection. Utilisation of the expertise and services of psychologists was on a piecemeal basis, and was yet to find a systematic application in military operations. Over the years, psychology gradually made a valued place for itself within the military sciences, finding vast and varied applications across domains. Personnel selection by ensuring person-job fit continues to be a prime focus in general in most of the nations, while clinical intervention, man-machine interface and human factors engineering have expanded to study adaptations to complex machine environments. With ever-newer applications of information technology, especially in the domain of artificial intelligence, cognitive-behavioural applications of psychology have expanded immensely. Along with personnel selection, training, leadership and maintaining the morale of the troops continue to be core areas of research in conventional units as well as in non-conventional military formations and deployments.

## Post-World War military psychology

The world plunged into a prolonged era of uneasy peace post the Second World War. International relations remained tense and nations continually strived to achieve a balance. As war was replaced by diplomacy, the nations strived to find alternative means of attaining and defending national interests. As the distinction between wartime and peacetime targets got blurred, so did the distinction between civil and military targets. As the boundaries became fuzzy, the services of experts, including the psychologists utilised by the military, were deemed indispensable (Herman, 1995). With innovative research paradigms emerging in the broader subject domain of psychology, and with the emergence of a dynamic interface between social and behavioural sciences, military psychology rapidly expanded its horizons and created an important space for itself in military policy in the context of changed geo-political global arrangements. Postwar years produced newer kinds of warfare, e.g. guerrilla or jungle warfare and countless insurgencies across many Third-World nations. Psychology emerged as a mainstay for formulation of policies and strategies in this context. As Herman (1995) puts it, 'the combination of unchecked weapons of technology and underdeveloped social technology was poisonous. Psychological expertise was among the only antidotes' (P. 125). Psychological research flourished during these years, especially in the United States, where the military wholeheartedly supported it. A major domain of psychological research effort in the military was dedicated to the designing of an ideological war against perceived enemies, and psychology increasingly became an essential component in military strategy and foreign policy. Lessons and insights learnt from the experiences of the U.S. military during the Korean war in 1950 had reiterated the usefulness and reliability of the military-psychology combination owing to research in domains such as brainwashing, mass communication, persuasive communication and the like. The U.S. military reemphasised the definition of war as a psychological struggle, where victory depended not only on military superiority, but also on psychological strategising to weaken the enemy, and a laboratory that provided for ever-newer opportunities for testing psychological theories and propositions. (Herman, 1995). Military research support to psychology in the Cold War years and the resultant unflinching support of psychology to the American military, though immensely beneficial and boosting for the development of the discipline, also drew strong criticism on moral and ethical grounds, especially after the Department of Defense's Project Camelot came to light in the shape of an international scandal in the year 1965. Project Camelot was a major plan with a strong psychology component to involve behavioural experts in predicting and controlling Third-World revolutions and development in order to gain an upper hand internationally (Herman, 1995).

The recent years have seen major changes not only in the nature and techniques of warfare, but also in the very paradigm of war. With the rise of terrorism and the 'war on terror' in the post-9/11 years, the concept of war has undergone a metamorphosis. Conventional warfare that had a definitive enemy force as the opponent, most often controlled by another nation-state, is being replaced by an ambiguous enemy and non-state actors controlling these forces across international borders. Military forces of most developed nations and many of the developing nations are involved in or are contributing towards engaging these non-state actors in controlling most acts of terrorist violence. This involvement is either direct (e.g. U.S. military forces in the Middle-East or Asia) or as a part of the United Nations Peace Keeping contingent (e.g. Indian Armed Forces deployed in South Sudan). In both instances, the stated purpose of military action is to establish or maintain peace by defeating or subduing the violence perpetuated by non-state actors. The world has been witnessing perpetual and multiple armed conflicts in recent years. At the time of the present lines being written, there are at least 25 armed conflicts taking place across the world, as reported by the Council of Foreign Relations, an American non-partisan think tank. Changing rules of engagement and increasing ambiguities have resulted in innumerable controversies and contentions regarding the role of the military in dealing with terror groups and organisations, insurgents and non-state actors. Under the circumstances, two major themes need to be highlighted and deliberated upon in order to structure the future directions the discipline of military psychology would take.

The first theme pertains to the need to evolve a perspective stance for military psychology, and the nature of its contributions towards military endeavours. Although it is rational to assert that 'Supporting the military's understanding of human behaviour is not tantamount to advocating war' (Laurence & Matthews, 2012, P. 3), the increasingly complex definition of warfare and the ambiguity of sociopolitical environments inherent in most contemporary conflicts make it imperative for the discipline of military psychology to be well informed about the ethical pitfalls of advocating value-free scientific knowledge. The second equally pertinent theme that demands attention of contemporary military psychologists across nations is regarding the scope and nature of the subject matter of military psychology. Though defined and usually delimited by the operational concerns of the militaries of the home countries, the scope and applications of military psychology must be expanded beyond understanding war. As Matthews (2014) says, both war and peace are essentially human phenomena, and scientific understanding of human behaviour can help explain and structure both the conditions. As psychology has made a move from understanding disease and pathology to understanding resilience and positive behaviour, in the same vein military psychology can no longer ignore making a move from understanding war to understanding peace.

## Military as a peace-builder: Role of psychology

About a century ago, Clausewitz (1918) famously defined war as continuation of politics by other means, and viewed it as an act of violence intended to compel our opponent to fulfil our view. In the contemporary world scenario, with ever-increasing ambiguities in making a distinction between friend and foe, psychology has found a renewed relevance owing to its potential to influence policy and politics. As it is true for the present epoch that it is a period of perpetual conflicts, so it is also true that never before have so many nations and groups come together in the pursuit of world peace. Peace is not mere absence of conflict, but is marked by combined effort of people and nations in working towards common good, social cooperation, empathy and pro-social behaviour (Matthews, 2014). Matthews (2014), in his aptly title book *Head Strong*, emphasises the potential of psychology in bringing about this shift from war to peace. He describes in some detail the increasing interface between the APA division of Military Psychology (Division 19) and the APA division for Peace (Division 48 – Society for the Study of Peace, Conflict and Violence), and the shared vision of a peaceful world by the members of both divisions. Emphasising the need to recognise the significance of resolving disputes and conflicts through peaceful dialogue and negotiations, he also questions the disparities in the funding of military research and research on peace psychology. Extending his argument and calling for action in the same direction, he suggests certain concrete domains where military psychology can contribute in order to facilitate the move from war to peace. Quoting from Matthews (2014, Pp. 217–221), some such domains are: changing mindsets of populations to replace 'fixed' mindsets with 'growth' mindsets that have been shown to promote favourable intergroup relations and amenability to negotiation; facilitating intergroup cooperation using social psychological techniques of common goals, as demonstrated by the classic Robber's Cave experiments done by Sherif and colleagues (1988); using the operant conditioning paradigm of rewards and punishments to enhance the likelihood of desirable political behaviour and furthering peaceful behaviour by using tools developed by social psychology over the years.

In the years after the Second World War, the United Nations (UN) organisation has played a significant role in delimiting conflicts across the globe. Along with using military force drawn from its member nations for peacekeeping, a few years ago, the UN called for a greater willingness on the part of the international community to address the basic economic, social, political and ethnic causes of conflict (Boutros-Ghali, 1992). This implied recognising the structural forms of violence that lie at the roots of most violent conflicts. An ideological understanding of the structural roots of violence can be facilitated through psychological research in the domains of relative deprivation, intergroup dynamics, prejudice and stereotypes and identity dynamics.

One of the most significant domains where psychology has grown in the post–World War years, especially during the Cold War era, is the use of cognitive and social psychological research for designing and disseminating publicity material and propaganda for influencing public opinions in favourable directions, initially by the American military, and subsequently by the Central Intelligence Agency (CIA). Though being utilised since the days of the First World War, these psychological operations (psy-ops) reached their peak propensity during the Cold War years, when a bipartisan agreement was reached among U.S. political leaders about the essentiality of beating the Soviet Union and countering its communist propaganda using any means possible (Wolfe, 2018). These vast array of intellectual and cultural instruments being utilised for spreading 'American values' in countries with potential communist influence during the Cold War years were largely discontinued by the United States after the fall of communism in the post-1991 world (Leonard, 2002). However, in view of the 'war on terror' in the post-9/11 world, need for 'public diplomacy', i.e. persuasive communication (or propaganda) with the populations overseas for the purpose of convincing them of the legitimacy of the U.S. military actions against terror outfits, has gained ever more importance. Leonard (2002) provides many examples across nations (e.g. governments in Kosovo, Greece and Italy faced uncertainty in the face of volatile public opinions) in recent decades, where popular public perceptions rather than governments set the pace for international diplomacy. In this context, international public diplomacy has attained a new perspective and significance, though in the current era of easily accessible information, a much greater finesse would be required to make an impact on public perceptions and opinions. Military psychology thus has a potent area of renewed research and application open to its efforts in making a move from assisting war to building peace by finding the psychological process of 'building relationships, starting from understanding other countries' needs, cultures, and peoples, and then looking for areas to make common cause' (Leonard, 2002, P. 50).

A related area of concern while military psychology looks forward to expanding its scope to influence peace processes is evolving a critical and nuanced awareness of desirable ethical standards of research and application.

## Psychology for the military: Focus on ethics of the profession

Ethics have been defined as 'the agreed upon standards of aspirational and mandatory behaviours and practices' (Leong, Altmaier, & Johnson, 2008, P. 182) by the members of a specific group. Professional ethics set the aspirational standard of behaviour for the member of a profession in a manner that distinguishes the *malum in se* (bad in itself) from *malum prohibitum* (wrong only because law prohibits it). In simple words, this means that the members of the group are expected to internalise the prescribed standards of professional practice as inherent values of the group and adhere to these under all circumstances, and not only in order to avoid punishment. In this sense ethics are placed on a higher level than law, as ethical practice implies unflinching adherence to the 'good' and refraining from the 'bad' voluntarily, without any external influence or compulsion. The crucial question of defining the 'good' and the 'bad' in professional contexts lies upon the professional body and the ideological stance taken by it.

While many researchers recognise and elaborate upon the ethical issues and dilemmas in the field of research and applications of military psychology (e.g. Frey, 2017; Johnson & Landsinger, 2017; Staal & King, 2000), most discussions pertain to the ground-level problems of practicing psychology in clinical or organisational settings within complex operational environments. Looking for a broader ideological stance for the discipline, that would define and shape the research endeavours and delimit the contours of scientific application of knowledge thus generated, one comes across a void. There are hardly any discussions on the ontology of military psychology, though the apparent epistemological stance it takes is largely positivist. Devoid of philosophical debates about the nature of knowledge and lacking any critical examination of assumptions underlying theories and constructs, military psychology in its current form exemplifies stark positivist science, that is unambiguously based on observation of 'facts' or 'data', linked by rigorous logic to hypotheses or theories (Manicas & Secord, 1983). Military psychology as a part of the traditional field of military studies has always stood as an apolitical, atheoretical and largely quantitative approach to scientific understanding (Basham, Belkin, & Gifkins, 2015).

Military psychology has grown exponentially in the last century as a discipline majorly rooted in the American ethos. Despite a declared adherence to the highest American values of freedom and democracy and bound by the professional ethic of protecting and promoting the mental health and well-being of those they serve, military psychologists (as represented by the APA) have faced severe criticism in recent years for certain stark actions and decisions taken in apparent disregard of any critical self-appraisal. Murray, Lyubansky, Miller and Ortega (2014) have critically examined the APA's refusal to condemn the use of torture or of acknowledging the involvement of its members in designing and implementing abusive interrogation and detention techniques used by the military and CIA against terror suspects. Though they attribute this blind adherence of the APA with the military to the historical closeness of American psychology with the military, it is also reflective of the ideological shortsightedness of the discipline. While looking for friends and allies in the 'global war against terror' to achieve the elusive target of a peaceful world, such ideological unawareness is proving detrimental.

In order to break out of the current mould and move towards peace building, the discipline of military psychology needs to evolve a reflexive and critical awareness of its ideological positioning and value preferences. The first and foremost requirement in this regard is to develop an acceptance of the inherently interpretive nature of the peace-building ventures, which must rely upon understanding cultural diversities and co-constructing peace as an ongoing process. A value-based perspective rooted in critical self awareness is the first step towards developing a psychology for the military that truly serves the purpose of making the world a more peaceful place. The Herculean but essential task of evolving a psychology for the military that goes beyond finding utilitarian solutions for the forces and helps in creating a value-based critical awareness of actions and decisions in an environment of perpetual ambiguities might be the stepping stone of the path that leads towards a peaceful world. A move from militarism to peace building would inevitably rely upon an ideologically aware and critically reflexive avatar of military psychology that accepts the simultaneous existence of multiple perspectives, does not divide the world into dichotomies and breaks down hierarchies to minimise power asymmetries and perceived moral superiorities of certain groups and nations.

## Conclusion

The present chapter focuses upon the need for reconceptualizing the role and subject matter of military psychology in the context of the contemporary socio-political realities fraught with perpetual violence and multilayered conflicts. It situates the discipline in historical context by tracing its major developments since the First World War up until the present era. It highlights the intertwined nature of military and the discipline of psychology especially in the USA. As the discipline of military psychology has come to be dominated by the American perspectives, these historical influences are now shaping the structure and ideological stance of the discipline in general. The chapter concludes by highlighting the anomalies in two concomitant domains where the authors see possibilities for change in the positive direction. First domain is about redrawing disciplinary boundaries and redesignating disciplinary priorities in order to make a move from facilitating war to establishing peace as the ultimate aim. The second domain which arises from this shift, both as a prerequisite and as an outcome of the same, is the grave need for the discipline to achieve clarity about its ideological positioning.

The idea of replacing war with an alternative is not new. William James (1910/1995, p. 17), questioning the necessity of war says, "The war against war is going to be no holiday excursion or camping party. The military feelings are too deeply grounded to abdicate their place among our ideals until better substitutes are offered than shame and glory that come to nations as well as to individuals from the ups and downs of politics and the vicissitudes of trade." It is heartening to note that even in the early days of psychology he saw a potent role for the discipline for making a transition from war to peace. Realizising that militarism is too deeply rooted in the psyche of the individual and the society to even imagine a military-less world, he emphasises the need to find an alternative that would create opportunities for the human virtues to emerge and shine, as militarism does. This goal of finding a substitute for militarism that creates opportunities for a goal-oriented, hardy, selfless and resolute human character to emerge can very well be fulfilled by psychologists by creating alternative pathways that channel human energies in a positive and constructive manner.

The core point propagated by the present chapter, thus, is that the discipline of military psychology inherently holds the potential for transforming the very nature of war to the extent of eliminating it altogether at some point in the future, replacing it with peaceful cooperative striving. Such transitions have become imperative, and psychology with its immense propensity to bring about change can become the flag bearer in transforming the world altogether.

## References


Basham, V. M., Belkin, A., & Gifkins, J. (2015). What is critical military studies? *Critical Military Studies, 1*(1), 1–2.

Bingham, W.V. (1947). Military psychology in war and peace. *Science, 106*, 155–160.

Bourke, J. (2001). Psychology at war, 1914–1945. In G. C. Bunn, A. D. Lovie, & G. D. Richards (Eds.). *Psychology in Britain: Historical Essays and Personal Reflections* (pp. 133–149). Leicester, UK: BPS Books.

Boutros-Ghali, B. (1992). An agenda for peace: Preventive diplomacy, peacemaking and peace keeping. *International Relations, 11*(3), 201–218.

Clausewitz, C.V. (1918). *On War, Vol. 1*. London: Kegan Paul, Trnch, Trubner & C.

Fitts, P. M. (1946). German applied psychology during World War II. *American Psychologist, 1*(5), 151–161.

Frey, R. A. M. (2017). Ethical challenges for military psychologists: When worlds collide. *Ethics & Behavior, 27*(4), 283–296.

Geuter, U. (1992). *The Professionalization of Psychology in Nazi Germany*. Cambridge: Cambridge University Press.

Herman, E. (1995). *The Romance of American Psychology: Political Culture in the Age of Experts*. Berkley: University of California Press.

Hughes, H., McCauley, M., & Wilson, L. (2019). History of military psychology. *Journal of the Royal Army Medical Corps, 165*(2), 68–70.

James, W. (1910/1995). The moral equivalent of war. *Peace and Conflict: Journal of Peace Psychology, 1*(1), 17–26.

Johnson, W. B., & Landsinger, K. L. (2017). Ethical issues in military psychology: Clinical and organizational practice. In S.V. Bowles & P. T. Bartone (Eds.). *Handbook of Military Psychology* (pp. 105–224). Cham, Switzerland: Springer International Publishing.

Laurence, J. C. & Matthews, M. D. (2012). The handbook of military psychology: An introduction. In J. H. Laurence & M. D. Matthews (Eds.). *The Oxford Handbook of Military Psychology* (pp. 1–3). New York: Oxford University Press.

Leonard, M. (2002). Diplomacy by other means. *Foreign Policy*, *132*, 48–56.

Leong, F. T. L., Altmaier, E. M., & Johnson, B. D. (2008). *Encyclopaedia of Counseling, Vol. 1.* Thousand Oaks, CA: Sage.

Mangelsdorff, D. A. & Gal, R. (1991). Overview of military psychology. In R. Gal & D. A. Mangelsdorff (Eds.). *Handbook of Military Psychology* (pp. xxv–xxxvi). New York: John Wiley & Sons Ltd.

Manicas, P. T. & Secord, P. E. (1983). Implications for psychology of the new philosophy of science. *American Psychologist*, *38*, 399–413.

Matthews, M. D. (2014). *Head Strong: How Psychology Is Revolutionizing War.* New York: Oxford University Press.

Melton, A. W. (1957). Military psychology in the United States of America. *American Psychologist*, *12*(12), 740–746.

Murray, H., Lyubansky, M., Miller, K., & Ortega, L. (2014). Toward a psychology of non-violence. In E. Mustakova-Possardt, M. Lyubansky, M. Basseches, & J. Oxenberg, (Eds.). *Toward a Socially Responsible Psychology for a Global Era* (pp. 151–182). New York: Springer.

Shephard B. (2015). Psychology and the great war, 1914–1918. *The Psychologist*, *28*, 944–946.

Sherif, M., Harvey, O. J., Hood, W. R., Sheriff, C. W., & White, J. (1988). *The Robbers Cave Experiment: Intergroup Conflict and Cooperation.* Middletown, CT: Wesleyan University Press.

Staal, M. A. & King, R. E. (2000). Managing a multiple relationship environment: The ethics of military psychology. *Professional psychology: Research and practice*, *31*(6), 698–705.

Wolfe, A. (2018). *Freedom's Laboratory: The Cold War Struggle for the Soul of Science.* Baltimore, Maryland: Johns Hopkins University Press.

# 2

# WAR, PEACE AND THE MILITARY IN BIBLICAL AND ANCIENT GREEK SOCIETIES

*Matthew B. Schwartz and Kalman J. Kaplan*

This chapter offers a comparison and contrast as to how ancient Greek and Biblical civilizations viewed peace and war and soldiering. Wars always involve violence, suffering and death, often of innocent people. Yet Biblical and ancient Greek views on warfare differ notably. Wars in the Bible can be very violent and destructive. Good fighters were honored, as were people who were successful in other endeavors. "Saul has slain his thousands and David his ten thousand," sang the women, and II Samuel 22 tells of the exploits of some of David's best warriors. Yet war is not glorified for its own sake by the Bible. Wars are fought at best under certain guidelines and for aims that the Bible describes, largely to achieve God's purposes for human history. On some occasions, the Bible will mandate the destruction of a culture that has deviated too far from morally acceptable behavior. Yet this seems fundamentally unlike the picture of fighting in ancient epics like the Iliad in which violence seems to be admired as an end in itself and which delight in describing in gory detail the slaughter on the battlefield, totally disconnected to any purposes of God's.

The contrast between Biblical and ancient Greek attitudes towards heroism and violence in battle can be seen in the combats between Achilles and Hector and between David and Goliath. These two narratives seem paradigmatic of different attitudes underlying the nature of conflict in these two cultures. Goliath and the Philistines of his day, ca. 1030 BCE, appear to have been one group among the mass of so-called Sea Peoples, who moved from various points west to the Eastern Mediterranean ca. 1200 BCE. One group came close to toppling the Egyptian Empire in about 1170 BCE. Another group destroyed the Hittite Empire. Others conquered parts of the Eastern Mediterranean seaboard, including Ugarit and the Gaza area. It is almost certain that some of the Sea Peoples came from Greece and spoke Greek. By the mid-eleventh century, they were adjusted to the languages and cultures of their new homelands, but it is very possible that Goliath and his contemporaries still maintained certain ties to their Greek past, its lifestyle and traditions. Perhaps Goliath knew the story of the siege of Troy and of Achilles (approximately 1300 BCE), and perhaps aspired to be like him. If, as some ancient writers suggest, Homer was possibly Phoenician or Egyptian, it is not impossible that Homer, who composed the Iliad in the eighth century BCE, knew something of the story of David and Goliath. In any case, there are certainly phrases in I Samuel 17, especially the description of Goliath's armor, that seemed Iliadic. Let us compare the two stories.

## David and Goliath

Goliath is described as six cubits and a span tall, perhaps over nine feet. His armor alone weighed 5000 shekels. Descriptions of armor are not the norm in Scripture, and one is reminded of Homer's detailed description of Achilles's armor in the Iliad. David, in contrast, was a young shepherd with no armor, carrying a staff, a shepherd's pouch and a sling with five stones (I Samuel 17: 38–40). The description of their combat further confirms the difference between the two protagonists (I Samuel 17: 42–54). Goliath is a warrior from his youth who flaunts the Israelites with a personal challenge, which he knows no one will accept. It is all a ploy to demoralize the Israelite army, and it is highly successful. For 40 days, every morning and evening, the giant comes forth and offers the same challenge, mocking the Israelites and their God. King Saul has offered rewards of wealth plus his daughter's hand in marriage to the man who can meet the giant, but there are no takers. David now arrives at the camp. He is a ruddy, handsome youngster, but he sees immediately the growing demoralization of the Israelite soldiers. He goes to King Saul and tells him that he will fight Goliath. David has no promise of victory from God, but he feels that someone has to stand up and fight for God and for His people. The outcome will be up to God, but someone has to try. Though young, David is a person of deep faith that God will do what He wants. David goes out quickly to meet Goliath, carrying his sling with five stones and an ability to use them expertly. Goliath seems surprised and shaken. He goes through his usual pre-combat speech, threatening to give David's body to the birds of prey and the cattle, but David detects the uncertainty in Goliath's hesitating approach and also notes that Goliath threatened to give his body to the cattle (*behema*). Cattle, of course, are not normally carnivorous, and David senses that Goliath is confused. He responds that the combat will not be decided by human weapons but by God. David seizes the right moment and fires a stone into Goliath's forehead, killing him. He then completes his astonishing victory by severing Goliath's head, at which the Philistines flee in panic, and the Israelites pursue them vigorously (see Appendix).

A number of points should be noted in this story. First, violence is not glorified, nor is brute strength. Goliath is experienced, well-armed and huge. David moves swiftly. More important, Goliath has no purpose higher than just fighting, while David feels he is fighting for a higher purpose. David prevails, and cuts off the head of Goliath, not as a show of valor but to intimidate the Philistines.

## Achilles and Hector

Let us compare this to the Iliad's narrative of the combat of the Greek Achilles and the Trojan Hector (Homer, 1860). Hector is described as family man, very human and aware of his obligations to his family and to society; he has serious doubts about the behavior of his brother Paris in abducting Helen from her husband King Menelaus of Sparta, thus instigating the war, but Hector does nothing about it. Unlike David, Hector is terrified of his much stronger adversary.

Achilles is described almost in Goliath-like terms. While not as huge, he is super-human and invulnerable except for his heel. He is moody and ferocious as a warrior and seems to express himself through violence. When the mortally wounded Hector proposes that the victor treat the vanquished's body with respect, Achilles angrily refuses, saying that he only wishes he could bring himself to eat Hector's body. The Achaean soldiers stab and mock Hector's corpse. Then

Achilles ties the body to his chariot and drags it around the walls of Troy, where the Trojans can see it (see Appendix).

The Greeks and the Philistines seem addicted to violence, while the Israelites are not.

## War and peace in the Hebrew Bible

The Hebrew Bible considers peace one of the greatest virtues, whether it be between friends, husband and wife or nations. Zevin (1946) cites several important ideas *(Le'or HaHalacha ch 1)*. God tells a man as great as King David that he should not build the Jerusalem Temple because he has fought many wars and has shed blood. Nor can the Israelites use iron to build the altar to the Jerusalem Temple because the altar is created to lengthen life, while iron shortens life. Do not bring the sacred Temple and war together. Only the wars commanded by God to Moses and later to Joshua, particularly those against Amalek and the Canaanites, do not require the consent of the High Court. All other wars do. The king must also be consulted, and during the First Temple, the *Urim Vetumim,* the prophetic breastplate of the High Priest would also be consulted. Wars of self-defense also do not require consultation.

As Ecclesiastes says, there is a time for peace and a time for war. Despite the desire for peace, there have always been wars and likely there always will be, some justified and others unjust. Biblical Israel was in a "tough neighborhood," constantly at odds with aggressive Philistines, Amorites, Aramaeans, Canaanites, Amalekites, Ammonites, Moabites, Edomites and others. Egypt was on the southern border and the Assyrians to the east, and war was a part of life.

This is the basic view of the Hebrew Scriptures to war. However, each of the many wars has its own story and its own lessons to teach. We shall use several wars as case studies, and shall open with a look at Deuteronomy 20, a general view of how a monotheistic Israel prepared for battle.

Deuteronomy 20 presents a number of rules for Israelite armies to follow in conducting a war. Do not fear the enemy's advantage in numbers or weapons, for God is with you, that same God Who brought you up miraculously out of Egypt. It is God Who will determine the outcome of the battle, "He does not desire a horse's strength, nor does He takes delight in a man's musculature" (Ps 147), and the Israelites must approach the battle as God's soldiers. This is not to imply that they are fanatical "holy warriors" or like Viking berserkers, but that war, like peace, must be understood and treated as part of God's plan for history. Soldiers should be aware that when they fight, it is not for pillage or glory, but to sanctify God's name (Roter, 1998–1996, 679c).

On their approaching the fighting, a priest specially designated for this duty will address the men. Ancient history recounts many speeches given by generals to their men as they faced battle. Robert Burns immortalized the speech given by Robert the Bruce of Scotland to his men before the Battle of Bannockburn in 1314. "Scots wha hae wi Wallace bled, Scots wham Bruce has often led, Welcome to your gory bed or to victory…"

The words of the Israelite priest focus on preparing the soldiers morally and spiritually. "Hear O Israel…" he begins, in words deliberately reminiscent of the great prayer expressing Israel's faith in one God, and he goes on to recall the miraculous saving of the Israelites at the Red Sea (Deut 20.3). They must not fear before their enemy, for God will fight for them.

Officers will then address the men, telling them that anyone who had recently built a new house and not yet dedicated it, let him go home lest he be overly concerned that he will die and someone else will live in it. Similarly, one who has planted a new vineyard but not consecrated it, or become engaged to a woman but has not married her, let him return home. Commentaries offer several reasons. The soldier may be so preoccupied with these matters that he will not be able to fight at his best. Or he may be afraid of what he will lose if he never returns home. This is not only in a physical sense but more in not having fulfilled needs that are of great importance and meaning in human life. In addition, a soldier who hesitates in battle may rouse fear in his fellows and cause them to fight less well. If he has sinned in some way and has lost confidence, the officer's announcement gives him a cover to return home without undue embarrassment. And indeed, the officer ends by saying, "Any man who is afraid and weak-hearted, let him go and return home and not weaken his brothers' hearts as his own" (Deut 20.5–8). His fellows will not know if he is leaving because of fear or because of his sins.

Biblical Scripture goes on. Before attacking or besieging a city, the Israelites must offer peace. Only if the city does not choose peace, then they may make war against it. They shall kill the men and spare women and children. The people of Canaan must be killed if they choose to fight, women and children as well, as these people are inveterate enemies of the Israelites, vulgar idolaters, practicing temple prostitution and child sacrifice and the like. If they live, they will constantly be a thorn in Israel's side and will seek to attract them to their evil practices –"So that they will not teach you to do like all their abominations…" (Deut 20.18). When the Israelites prepared to invade Canaan, Joshua offered the above choices to the inhabitants. The Girgashites left the country, the Jebusites made peace, the others all fought and were conquered over a period of time.

When besieging a city, the army should not wantonly destroy. It may eat of the local fruit trees but must not destroy them. (This may remind us that when the Spartans were besieging Athens early in the Peloponnesian War, they tried to destroy Athens's olive trees, although they in fact found that olive trees are difficult to ruin permanently.)

There are further lessons. It is dangerous to be too merciful with a bitter enemy. I Kings tells of Ahab, a powerful king of Israel, who had it within his grasp to destroy the power of a constant and bitter enemy, King Ben Hadad of Aram. Ahab foolishly decided to disregard the prophet's instruction and spared Ben Hadad. The war was renewed, and Ahab was killed in battle. (There is an opposite concern – lest Israel's young men should begin to become like killers. For this reason, Scripture next raises the matter of how to treat a dead body found near a city in time of peace.)

Sometimes it can become necessary to destroy a plant, root and branch, rather than to let it grow and damage other plants around it. So God instructed Moses, who tried to reason with the rebellious Korah and his allies Dathan and Abiram (Num 16–17), who would not even talk to Moses about their supposed grievances other than in insulting, personal attacks. This was not a matter of simply bringing peace or attending to someone who had a legitimate need. These were people bent on the destruction of Moses and of what God had planned and commanded for his chosen people. God responded by splitting open the ground and having them fall through the crevices all the way down to Sheol, they along with their wives and some of their children. God could see that these children would carry on being perverse, argumentative and rebellious as their fathers were. Indeed, history has shown over and over the futility of trying to appease evil people and the need often to destroy them before they can do terrible harm. (Think Neville Chamberlain.)

## *A first Biblical war*

The first war recounted in the Hebrew Bible (Gen 14) tells of four powerful Mesopotamian kings, including Amrafel (probably the famous Hammurabi), who conquered much of Canaan and ruled over it for a dozen years. At that point, five local kings rebelled and refused to pay the regular tribute, and the four kings returned in force to teach the rebels a lesson. The four overran Canaan, winning a series of victories over local peoples – the Refaim in Ashterot Karnaim, the Zuzim in Ham, the Emim in Shaveh Kiryataim, the Hurrians in Mt. Seir to Eil Paran, the entire field of the Amalekites and the Amorites in Hatsatson Tamar. Then the King of Sodom and his four allies were soundly beaten in the Valley of Sidim. The Kings of Sodom and Gomorra fled and hid in slime pits. The four winners thoroughly looted their cities and took many captives, including Abraham's nephew, Lot, who had settled in Sodom after parting ways with Abraham.

A refugee brought the news to Abraham that Lot had been carried away. Abraham was living quietly in the Plains of Mamre, but now he did not hesitate. He did not dwell on the fact that Lot's parting from him had not been entirely harmonious. He did not plead for time to gather further details and to check out the information. He quickly armed 318 men and called on his three friends to gather their retainers and join him in pursuit of the powerful enemy force to rescue Lot. Moving very quickly (in a manner used so successfully millennia later by Nathan Bedford Forrest), Abraham divided his force and attacked the Mesopotamians by night near Damascus, a long way from his home. The surprise attack was highly successful, and Abraham's victorious force pursued the enemy to some distance. The captives, Lot among them, were rescued, and the booty retaken.

Abraham did not slowly contemplate the dangers involved or his chances of victory vs. defeat. It does not seem that he consulted any authority as to the righteousness or wisdom of his decisions. He knew what was right, and he had to move quickly to rescue Lot.

As Abraham was returning, he was greeted by Melchizedek, the righteous King of Salem, who properly welcomed the army with food and wine. Melchizedek blessed Abraham in the name of the "Most High God" and thanked God for giving Abraham's enemy into his hand. Abraham recognized Melchizedek as a true priest and gave him a tithe of all he had.

Now that Abraham had driven off the Four Kings and had restored order and safety, the King of Sodom emerged from his hiding place to gain what he could and also to in some way minimize Abraham. Instead of thanking and praising Abraham and certainly not thanking G-d, and instead of bringing food for Abraham's tired warriors and allies, the king demanded the return of his townsmen who had been carried off. He suggested that Abraham could keep the spoils he had taken, which by the usual customs of war were Abraham's anyway.

Abraham saw through the wicked king, and his response was blunt. He would not take for himself even as much as a *khut,* possibly a small hair ornament or a shoe strap. Abraham would take nothing from the boorish king, so that the king could never claim that it was not G-d but "I that made Abraham rich." Abraham would not speak for his allies who were entitled legitimately to claim their rightful shares of the spoils and certainly their expenses. Even after Abraham had saved him, the king needed in some way to stain and diminish the sanctity that he could sense in Abraham. Besides what if Abraham's spirituality and ethics rubbed off on the people of Sodom? Where would that leave the king? Abraham's refusal to take the loot showed his faith in a divinely ordered world history. It is G-d who determines victory or defeat. It was G-d who had promised to give Abraham wealth. Let no one think the King of Sodom could do this. In Abraham's view, this war, like all other events, had a meaning for history, perhaps a purpose far higher in the long term than merely the actual fighting itself. Abraham would not soil his mission by accepting anything from the evil king.

## A second Biblical war

In a second event (Numbers 25. 16f.), God commanded the Israelites to attack Midian in return for Midian's attempt to damage the Israelites by leading them into low behavior at Baal Peor as they were approaching the Promised Land after almost 40 years in the Sinai. The Midianites had not attacked the Israelites in a battle but struck instead at the core Biblical values by which God commanded the Israelites to live. A nation can lose a battle and still win a war. To come back from a moral fall is much more difficult. Such enemies cannot merely be fought, and there can be neither appeasement nor compromise. In the Bible's view, this was God's decision. The animosity of such enemies must be destroyed.

The Book of Numbers 22–25 and 31 tells the story of perhaps the greatest danger that the Israelites faced during the 40 years they lived in the Sinai Desert on their journey from Egypt to the Promised Land. Near the end of the 40 years, they approached the border of Moab. Balak, King of Moab, feared the approach of the large nation with so many fighting men. Perhaps not knowing or not believing that God had commanded the Israelites not to harm Moab, a very frightened Balak appealed to the Midianites for support, although the Midianites apparently did not live that close by and could have nothing to fear from Israel. The two nations invited Balaam ben Beor, a soothsayer from Ptor in Mesopotamia, to curse the Israelites, promising him riches and honors. Perhaps a curse would be effective where military force seemed inadequate.

A cagey and greedy man, Balaam came to Moab. However, God would not let him curse the Israelites but made him bless them instead. Balaam was made to acknowledge among other things that Israelite family life was strong, "How goodly are they tents, O Jacob" (Num 24. 5) and that God "saw no sin in Jacob and no fault in Israel" (23.21). The great soothsayer could not harm God's people by cursing but tried instead to earn his rewards by suggesting a different strategy – if the Israelites could be led into bad behavior, they would become vulnerable. The Midianites sent their women to seduce Israelite men into the disgusting orgiastic worship of their deity Baal Peor.

The plan succeeded up to a point, creating a deep trauma for the Israelites, the effect of which was felt even years later. The peak was reached when Zimri ben Salu, a prince of the tribe of Simeon, brought Cozbi, a Midianite princess, to his tent in full view of Moses and the Israelite elders. It is noteworthy that the name Cozbi is associated both with an Akkadian cognate term for beautiful and with a Hebrew term meaning deceitful. Moses seemed stunned and unsure how to react. But Pinhas, grandson of Aaron the High Priest, reacted quickly and decisively. Taking a spear in hand, he followed the couple into Zimri's tent and speared them, deliberately catching them *in flagrante delicto*. The danger was ended, and God's anger quieted, but not before 24,000 Israelites died in a Heaven-sent plague (Num 25.6–9).

With calm restored, God now commanded Moses to "harass the Midianites because they are your enemies deceiving you with entrapments," (Numbers 25. 17–18) and the Midianites also wanted revenge for the killing of their princess. The Midianites still hated the Israelites and would still try to destroy them if they had the chance. Israel had to remove that danger. Moab, however, seems to have left off their opposition to Israel and to have dropped out of the picture. Perhaps they came to realize that the Israelites would indeed not attack them. The Bible's account does not fault the Moabites for wanting to protect themselves. It does fault the Midianites for becoming involved in the way they did in matters not relevant to them. Although not threatened, they nevertheless used even their daughters, including the princess Cozbi, to lure the Israelites into orgiastic idolatry (Viz. Abravanel, on Num 25).

Soon after, God spoke to Moses, commanding the Israelites to revenge themselves on the Midianites. This would be Moses's last mission before his death. Nevertheless, he carried it out with full diligence (Num 31). Moses called on the people to carry out God's vengeance and to supply 1,000 men from each of the 12 Israelite tribes: 12,000 men, a rather small force designed perhaps as a raid rather than a total war. The tribe of Simeon was deliberately included, despite their large role in the encounter with Baal Peor.

The army would be led by Pinhas – the man who had brought the first success would carry the work to the finish. Some commentators suggest that another 12,000 or even 24,000 managed supplies for the fighters but did not themselves join in the battle.

The Midianites were motivated by cynical hatred, not by legitimate fear (Roter, 1988–1996 991a). They saw only a small force approaching and expected an easy victory. They were wrong. Pinhas and his 12,000 men completely defeated the Midianites, killing all their men and capturing the rest (31. 7f.). Their five kings were among the slain, and so was Balaam the sorcerer, who had stopped in Midian to collect his reward for the 24,000 Israelites who had died in the plague (31.12).

The army returned to the Israelite camp, and Moses and the High Priest Elazar came out to meet them, whether to instruct them in certain procedures required of returning soldiers or to honor their success (Roter, 1988–1996, 994a). The original instructions may have been incomplete or unclear, and many prisoners had been brought back. Moses was not cruel or bloodthirsty, but God had apprised him of the dangerous hatred of the Midianites toward the Israelites, and the prisoners would have to be killed, particularly the women, as it was they who had lured the Israelite men into sin. Also, the victorious soldiers had taken no personal booty and brought everything they had taken to the High Priest to decide what to do with it (31). Again, a war of destruction was carried out at God's direct command. There is no mention of glory, and the soldiers did not take booty for themselves.

## *A third Biblical war*

As the Israelites under Joshua prepared to enter and conquer Canaan, they offered its nations three choices. (1) They could depart the land in peace; (2) they could make peace and continue to dwell in Canaan, accepting the Noahide laws and paying tribute to the Israelites; or (3) they could fight to the death. The Talmud says that the Girgashites left Canaan, and the Jebusites (Joshua 6) made peace (Talmud 1:14). The other Canaanite peoples chose to fight.

Joshua and the Israelites, after crossing the River Jordan, won a spectacular success in the conquest of Jericho, a city ancient even in those days, protecting the entry to Canaan. The walls miraculously collapsed or, in another interpretation, sank into the earth. The Israelites wiped out the population, except for Rahav the innkeeper and her family who had helped the two Israelite spies whom Joshua had sent days before. Joshua treated Jericho as a divine gift, forbidding his people to take any booty and placing a curse on anyone who would try to rebuild it. The site of Jericho was to remain an everlasting monument to God's great miracle.

Now the Israelites looked toward Ai, a city of only 12,000 inhabitants but as well fortified as Jericho and strategically important. Joshua sent spies to scout the area. They reported back that the Israelites would conquer Ai easily. It was not even necessary to send the full army. A few thousand men would take the city. Accepting this report, Joshua sent just 3,000 men. Their frontal attack failed, and 36 men were killed, a defeat unexpected and dangerously disconcerting. The Israelites could lose confidence and the momentum of their earlier successes. The Canaanites would be encouraged to fight. What had gone wrong?

Joshua learned that one Israelite man, Achan, had transgressed by taking booty from Jericho, a fine Babylonian cloak and gold and silver. By doing so, he had undermined the spiritual dedication of the Israelites, and he was punished with death. A second and perhaps related difficulty was that the victory at Jericho had perhaps given rise to some over-confidence. Joshua had not scouted Ai himself, and his spies had thought it could be conquered by a small force; there would be no need to send the full army.

Joshua now took the campaign against Ai more in hand, and he formed an elaborate plan. First, a small contingent would fake another frontal attack on the city and then pretend to flee, drawing Ai's defenders out of the city in pursuit. A much larger force posted in hiding on the other side of Ai would move in quickly as soon as the gate was undefended, and they would set the city on fire. Ai's army would now be caught between two Israelites forces of overwhelming numbers with the city already in flames.

A third force was ready to block intervention if any neighboring cities, like Bethel, tried to send troops to reinforce Ai. The Israelites took Ai and killed all its populace as well as its king.

The victory was complete. Scripture does not mention Ai again, and it seems to have ceased to exist.

Because the Israelites had been shamed in their first attack on Ai, God wanted them to take it by fighting, not simply by visible divine intervention. The Israelites needed a large army for the ambush and also to demonstrate the actual power of the army, as well as to discourage attacks from neighboring cities. Joshua had shown his strong leadership first by executing Achan then by sleeping in the army camp with his soldiers to calm any fears they might have had the night before the successful attack.

## *A fourth Biblical battle*

Forty years passed after the great victory of Deborah and Barak over the Canaanites, but Israel fell into the worship of Baal and Asherah, and the people were severely harassed by Midianites, Amalekites and others, who came like great swarms of locusts and stole or ravaged their crops and their herds. The Israelites would hide in caves or fortified towns and many were starving. Finally, the people cried out to God, Who sent a prophet and then an angel to rebuke and teach them. Gideon ben Joash of the Israelite tribe of Manasseh was preparing wheat to flee from the next raid. The angel addressed him, "God be with you, Man of Valor." Gideon answered, if God is with us, why has He given us into the hand of Midian?

God was informing Gideon that he was God's chosen messenger to save Israel from Midian. Gideon protested that he was unworthy, the least of his tribe and of his own family, but God said that He would be with him and Gideon would smite Midian. The battle was in God's hands. Gideon felt still unsure. He prevailed on the angel to stay until he could make a proper offering to God. When Gideon put meat and cakes on a rock, fire came forth from the rock and devoured the offering, and the angel disappeared. Following God's order, Gideon then destroyed the altar of Baal and the sacred tree of Asherah and built a new altar to God on which Gideon would sacrifice to God a bullock, using the broken wood of the Asherah for fuel.

The townsmen awoke the next morning, found their idols destroyed and learned that Gideon was responsible. They were ready to execute him, but Joash, Gideon's father, intervened, saying that Baal should fight his own battles; you do not need to fight for him. From this, Gideon came to be called Jerubaal, meaning one who fights with Baal (Judges 6.22).

The Midianites and their allies now gathered in great force in the Plain of Jezreel. A spirit of divine strength cloaked Gideon, and he sent to gather men from the Israelite tribes of Manasseh, Asher, Zevulun and Naftali. Gideon asked for a sign from God that He would indeed support the Israelites. He apparently believed in God's ability to save Israel but was unsure of his own worthiness. Gideon left a small woolen cloth out at night. In the morning the cloth was drenched with dew, while around it all was dry. This symbolized that Gideon himself was worthy of receiving divine help even if his enemies were not (so Abravanel). Malbim interpreted the sign as meaning that Gideon was indeed fit to receive God's help even if Israel in general was less fit. The second night showed a cloth dry and the ground around it wet, indicating that God would help Israel even with just a small Israelite force. Let people not think that wars are won by armies. Wars are won by God.

The battle was now imminent: 32,000 men had come to Gideon, but God told him that there were too many. It should be clear to all that the victory would be of God, not of people. So God instructed Gideon to send home the men who were afraid either because of the battle or because they had worshipped idols. Better they should leave now quietly than flee from the battle. Twenty-two thousand went home. God again came to Gideon to tell him that the 10,000 remaining were too many. The divinely ordained victory did not depend even on 10,000.

God told Gideon to let the men drink at the nearby stream. Most bowed or lay down to drink. Only 300 brought the water up to their mouths. Those who bowed to drink were those who had bowed before Baal or who drank with too great desire. Bringing the water to one's mouth was the practice of a fighter, not lazy or self-indulgent (Malbim). Three hundred men passed the test, and the other 9,700 went home. With these 300, God would bring victory.

Gideon was yet unsure if his own merits were sufficient. God gave him another sign. Gideon and one other soldier sneaked into the enemy camp at night and heard one man tell another of his dream in which he saw a bread roll into the camp and turn it upside down. When his listener interpreted the dream as a sign that God would defeat them, Gideon saw this as an indication of a failure of confidence among the Midianites and as a sign therefore of victory. Gideon planned an attack that would suit his force of only 300, far fewer than the hordes of Midianites, Amalekites and "Easterners." First, he attacked at night when his small numbers would be less recognized. He divided his men into three units, which would approach the enemy camp from three sides. Each man would carry a shofar (horn) and a flame concealed in a pitcher. At a signal, all 300 would blow their shofars, break their pitchers revealing their flames and shout, "The sword of the Lord and of Gideon." This would be done at the changing of the first watch of the night when sleepy guards were going off duty and the second watch was not fully alert and ready.

Gideon's plan worked beautifully. The camp was thrown into utter confusion. In the dark, even those Midianites willing to fight were stabbing at each other. They fled in panic, and Gideon and his men pursued them. Gideon then sent ahead to the men of Ephraim to block the fugitives at the fords of the Jordan, which they did effectively, killing the two Midianite generals, Orev and Zeev. Trouble then arose from within the ranks, but Gideon handled it wisely. The Ephraimites took Gideon's call so late in the battle as a deep insult and quarreled bitterly with Gideon. Gideon defused the tension, avoiding a civil war. What have I done, he answered them, compared to what you have done. I began the victory, but it was you who completed it. "Are not the last grapes of Ephraim better than the early grapes of Aviezer (Gideon's home)?" The Ephraimites were pacified (Judges 8.1–3, with Metzudot Tsion gloss).

Gideon and his 300 crossed the Jordan in pursuit of the fleeing Midianites. Tired and hungry, Gideon approached the Israelite inhabitants of the town of Succot and asked for food, but they refused. He then approached the city officials, who answered that they could not safely help him while the Midianites were still a threat (8. 4–7. Malbim). Gideon threatened severe punishment when he would return victorious. Almost the same happened at Penuel.

The two Midianite kings Zevah and Zalmuna remained with a force of 15,000. Still operating with only his 300, Gideon attacked by night again and destroyed the enemy force, capturing the two kings. He then punished the people of Succot for their refusal to supply food. He did not punish the officials, as he felt their fears were understandable (8, 14–16; see Malbim). Then he went on to destroy Penuel.

With victory complete, Gideon turned to the Midianite kings. "What manner of men were those you slew at Tabor?" He asked them. Apparently, while in flight from Gideon, the kings had stopped at Mt Tabor where they had killed Gideon's brothers. They explained to Gideon that they regarded them as rebels and worthy of death and added that they would have killed Gideon himself as a rebel if they could have. Gideon responded that if the kings had spared the lives of his brothers, he would have spared their lives, and he executed them.

The grateful Israelites asked Gideon to rule over them, him and his children and descendants forever, but he refused, saying that only God should rule over Israel. He also did not take the large share of the booty to which a winning general was entitled. Instead he asked only for gold nose rings, which the Midianites wore, and made from them a sort of jacket as a remembrance of God's deliverance of His people.

Yet violence is not glorified in Hebrew thinking and should be avoided if at all possible. Consider the story in this regard of the "wise woman" of Abel-Beth-Maacha and Sheba ben

Bichri who sought to undermine the rule of King David (II Samuel 20). The very dangerous revolt of Absalom had just been quelled, and David was returning to Jerusalem with much work to do in order to restore the stability of the nation and government. The Israelite tribes were quarreling with David's own tribe, Judah, as to who should lead the welcome of the king to Jerusalem, the capital. At that critical moment, Sheba ben Bichri stepped up. Scripture describes him as a "son of Belial," i.e. a rabble rouser and troublemaker. He blew the ram's horn and proclaimed that "we have no portion with David, nor any inheritance with the son of Jesse. Each man to his tent, O Israel." His point was that the tribes should not argue over who should welcome David, for no one owed him allegiance, and he was the son of Jesse, not of a king.

Many followed Sheba, and David could count only on the tribe of Judah. He recognized that Sheba's threat was serious. If Sheba could gain a hold on several fortified cities, a long, disastrous war could follow (20. 9). The new disturbance had to be put down quickly. David's force led by Joab marched to Abel Beth Maacha, to which Sheba had fled, and they began to build up siege mounds and to batter the city's defenses.

A bloody all-out attack was imminent. The most notable role in the story is played by the "wise woman," as she is termed. She called to Joab from the city wall, wisely and bluntly. "Are you Joab?" seeming to imply, "Are you Joab who thinks he is so smart?" Surely, people will say that they should have offered peace to Abel, and thus they would have had peace. She was reminding him of the Biblical injunction to offer peace to a city before attacking it. She went on to point out that she and her fellow citizens were peaceful and loyal and to ask how Joab could seek to destroy a city that was a "city and a mother in Israel": "Why will you swallow the inheritance given by God?" (20, 16–19). The wise woman was pleading for the city, yet rebuking Joab's quickness to fight in such a way that Joab could only be impressed with her honesty and intelligence.

Joab answered, "Far be it from me that I should swallow or destroy." However, Sheba ben Bichri has raised his hand against King David. Give him over, and I will depart. The woman replied that Sheba's head would be thrown to Joab over the wall. She knew both by law and by good sense what had to be done. The text does not tell exactly how she persuaded the city leaders to hand Sheba over. However, Abel was a Benjaminite city, and Sheba was of that tribe, perhaps a kinsman. A midrashic source suggests that she approached the city leaders and told them Joab was demanding that they turn over 50 people to be executed. She then pretended to have bargained with him, gradually reducing the number of people to be executed to only Sheba ben Bichri himself. At that point, the city fathers were grateful to get away so lightly, and they readily agreed. The Talmud also uses the story to discuss the question of what people should do in a situation where enemies demand that a group give one of their number to be murdered or raped. In Sheba's case, the Talmudic law would agree that he could be given over to save the rest because (1) the enemies demanded Sheba specifically, and (2) he was guilty of rebellion against a legitimate king and therefore in any case subject to a death penalty.

The city fathers were won over by the woman's wisdom, and they threw Sheba's head over the wall to Joab. They had ended the war before it began and had also prevented the besieging army from entering the city, which is always risky even if there is peace.

## War and peace among the ancient Greeks

Herodotus, the fifth-century historian often called the father of history, reports on many early wars of the Greeks and their neighbors. Herodotus introduces his history, saying that he wants to record the achievements of both Greeks and non-Greeks and also to show how the conflict between the two races led to the Persian Wars (Herodotus, 1961).

A most entertaining writer, Herodotus's perspective on history is very different than the Bible's. In Herodotus's world, the gods are capricious as well as of limited power. This appears especially in his account of the life of King Croesus of Lydia. Probably the world's wealthiest man, Croesus determined at one point to try to conquer the growing Persian power of Cyrus the Great. He sought the advice of the best of oracles and after careful testing determined that the Oracle of Delphi would best answer his question – Shall I go to war against the Persians? The oracle's answer was an ambiguous riddle as usual: Go to war and a mighty kingdom will fall. Croesus accepted the oracle's word and attacked. A mighty kingdom did fall – his own. As Croesus was being led away by guards to be burned to death, they heard him muttering the word "Solon" over and over. Cyrus's curiosity was aroused and he stopped the execution. Croesus told him the story of how the Athenian wise man Solon had explained to Croesus that no man could consider himself happy until he could see all of his life, i.e. until his death. Croesus had ridiculed the notion that anyone could be happier than himself, given his immense wealth. Now he realized that Solon was right.

The story carries with it deep implications of the pessimistic tone often paramount in Greek historical thinking. Where were the gods? Herodotus seems to have believed in them, and often includes them in his narratives. What is missing here that so differentiates his view from the Bible's is a sense that any god cares about the world or about humankind and has any plan for its history. And, of course, the gods did not create the world, as the God of Genesis did. Even so brilliant a man as Solon did not see the world in this way.

A story Herodotus relates from the Second Persian War can be seen as pointing to another difference between Biblical and Greek thinking. In 480 BCE, the Persian leader Xerxes invaded Greece with a massive army and fleet. A small Greek force including 300 Spartans held a key pass in the mountains at Thermopylae.

Xerxes asked Demaratus, an exiled Spartan, for his opinion of his city. He replied, interestingly.

> For though they (the Spartans) are free, they are not free in all respects, for they are "actually ruled by a lord and master: law is that master, and it is the law that they inwardly fear—much more so than our men fear you. They do whatever it commands... which is to win or fight to the death." (Herodotus, VII. 104)

Xerxes laughed at this, wrote Herodotus, but was stunned when the Greeks held off his army for days. Three hundred Spartans remained and fought to the death.

What does this tell us? Certainly, the Spartans were known as excellent and loyal soldiers. Indeed, Socrates seemed to prefer Sparta to Athens. Yet an element seems to be absent in the description of the creed of the Spartan soldier as compared to that of a soldier in the army of ancient Israel. The Spartan soldier is a highly disciplined, well-trained fighting machine. He is brave and loyal and feels a strong sense of duty and will not leave his post. He is the best of Greek soldiers. Yet, for what is the resolute Spartan solider really fighting? Does he have the same transcendent sense of historical purpose as an Israelite soldier?

Let us note Socrates's guard-post allegory regarding suicide. A soldier must stay at his post until he is relieved (*Phaedo*, 62b-c). Though life may hold no meaning or even pleasure for him, he must not leave it out of a sense of duty. For the Bible, God gives life and takes it away. This is not a human decision. What is a human decision is to live fully and follow God's commandments. "See I have put before you life and death, blessing and curse, and you shall choose life so that you and your seed shall live" (Deuteronomy 30:19). The Biblical soldier believes in the transcendent value of what he is doing, and also life itself. Unlike the Spartan soldier, no matter how brave, the Biblical soldier fights not simply out of duty but because he believes in the cause for which he is fighting.

Thucydides (1978), the other great Greek historian of the fifth century, wrote the history of the Peloponnesian War. Tellingly, he never mentions the gods in his work and may not have believed in gods at all. This is in contrast to Socrates, a deeply religious man who often expressed his belief in gods and oracles, though he was repelled by the vicious narratives of earlier Greek mythology and seemed to have had little interest in historical study.

Biblical history starts with God creating the world and proceeds toward a messianic denouement. Greek writers tended to see the world more in terms of cycles. Yet there were some exceptions, notably Pericles, who saw history in a larger context. Pericles's funeral oration, as reported by Thucydides in *The Peloponnesian War*, perhaps signals the peak moment of Athenian history, a moment when its greatest statesman in its Golden Age explained what Athens meant to him. The Peloponnesian War, 431–404 BCE, was into its third year. There had been many difficulties, but Pericles's strategies seemed to be working. The Athenians held a public funeral to honor their dead and invited Pericles to speak.

Pericles opened by noting that speeches on such occasions must of necessity be inadequate, but that it is Athenian tradition to hold such ceremonies along with an oration. "I shall begin by speaking about our ancestors…" that by their courage have handed down to us a free and strong nation, and he goes on to praise the Athenian spirit and way of life that make Athens so superior. Indeed, we need not copy other cities; they need to copy us. Athens is a democracy where effort and hard work will be rewarded. Poverty need not hold one back, and the only shame is not to try to rise from it. Athens offers opportunity to its citizens as well as beauty and recreation. We are an open society, not secretive or fearful of others.

In war, Athens is superior to its enemies because it relies not on secret weapons but on courage and loyalty. Whereas the Spartans stress constant military preparations from early boyhood, the Athenians live without all the preparations and are just as ready to face battle. Athenians love beauty and culture, but do not become soft. Others are brave out of ignorance; Athenians know what is good in life and still enter battle with courage. Altogether then, Athens is an education to all of Greece. It was for such a city that our men fought and some so gallantly died. "You should fix your eyes on the greatness of Athens as she really is and should fall in love with her." He concludes: Great men have the whole earth as their memorial.

Pericles presents a positive, optimistic, far-sighted view of his city and its way of life. This is not lugubrious and tragic like Sophocles's view of the world or carping and argumentative like Socrates's. While it is not the Biblical view, it does offer a sense of purpose and a love of something (in this case Athens) that for Pericles is genuine. Pericles's successors in leadership were of much lower character and Athens went on to be badly defeated in the long and destructive war.

We wish to comment on a final curiosity emerging in Homer's *Odyssey* (1950). Why did it take so long (10 years) for Odysseus to return to his hone to Ithaca after the fall of Troy? Biblical warriors who survived typically returned home to their daily lives; Homer's *Odyssey* records his many adventures:

a. Leaving Troy, he attacks the Cicones, but is beaten off.
b. He is caught by a storm and visits the country of the Lotophagi (Lotus Eaters). They are not hostile people, but eating of the lotus plant causes loss of memory and sense of purpose.
c. Odysseus is able to pull his men away and resume his journey and wanders to the land of the Cyclopes. One of them, Polyphemus, traps Odysseus's party in his cave and devours two of them at each meal. To escape, Odysseus blinds the one-eyed monster, incurring the wrath of Poseidon.

d. Escaping from Polyphemus, Odysseus next visits the home of Aeolus, god of the winds, who gives him a closed sack containing all the winds. But Odysseus falls asleep as he nears Ithaca, and his crew unties the sack, releasing all the other winds, which blow the ship back to Aeolia where Aeolus refuses to help them further, speculating they must be cursed by the gods.
e. After six days sailing, Odysseus reaches the home of the cannibal Laestrygonians, who sinks all the Greek ships in the harbor, save that of Odysseus, and eats all of the crews, leaving Odysseus's own ship as the sole survivor.
f. Odysseus and his remaining crew reach Aeaea, the home of the enchantress Circe, who turns Odysseus's men into pigs. With Hermes's advice, Odysseus become Circe's lover. Circe then releases Odysseus's men from the spell. Odysseus and his men leave and go to Hades to confront the ghost of the blind Greek prophet Tiresias, who tells Odysseus he must sacrifice to the god Poseidon to appease the god's anger toward him because he blinded Polyphemus.
g. Resuming his further voyage; Odysseus passes the Sirens. He then must choose between confronting Scylla, a six-headed monster who eats six of Odysseus's remaining men, and Charybdis, a monster whirlpool that swallows everything near it three times a day.
h. Odysseus and his remaining crew come to an island of the sun god Helios where his cattle pasture. Warned not to eat the sun god's cattle, the famished men are short of provisions and disobey this warning during Odysseus's brief absence. Zeus is furious and destroys their ship as the Greeks depart, leaving Odysseus as the lone survivor. Odysseus is washed ashore on the Island of Calypso, where he becomes her lover, until released by her on Zeus's order seven years later on a boat of Odysseus's own making.
i. Poseidon, still angry, wrecks his boat and Odysseus swims ashore at Scheria, where the Phaecians follow their tradition of returning shipwrecked strangers to their homelands, and send him back to Ithaca.
j. Odysseus and his son Telemachus kill the "suitors" of his wife Penelope.

The question we are left with is why Odysseus takes so long to return to Ithaca. It is as if the mundane life is of little value to the Greeks, who only find meaning in adventure, even if such adventure might destroy them. (See the poem "Ulysses" by Alfred, Lord Tennyson.)

## Job versus Zeno: Living purposiively versus searching for meaning

This exaggerated search for meaning is exemplified in Diogenes Laertius's story of the death of Zeno the Greek Stoic, who attempts to find meaning in the wrenching of his toe on the way back from the Stoa where he has been teaching. He cannot interpret this simply as a minor mishap but instead must give cosmic meaning to it.

> The manner of his (Zeno's) death was as follows. As he was leaving the school he tripped and fell, breaking a toe. Striking the ground with his fist, he quoted the line from the *Niobe*. "I come, I come, why dost thou call for me?" And died on the spot through holding his breath. (Diogenes Laertius, 1853, *Lives of Eminent Philosophers*, VII, 28).

The medically impossible depiction of his suicide begs further consideration. Zeno holding his breath until death negates the involuntary aspect of breathing and the Biblical conception that God breathes life into man and takes it away (Gen. 2:7). Zeno, like other Greek and Roman stoics, needs to control the conditions around his death, equating this control with a tragically tinged sense of freedom (Seneca, De Ira, 3.15.34). Why should Zeno kill himself after so seemingly minor an annoyance as wrenching his toe, while Job is able to withstand much greater stressors? Zeno's over-interpretation may represent his attempt to find meaning and purpose in an otherwise hopeless and meaningless world.

The Biblical figure of Job, in contrast, does not commit suicide despite far greater misfortunes – the loss of his wealth, his family and his health. His friends tell him to admit he deserves his punishment, but he refuses because he knows it is not true (Job 4–32). He certainly complains bitterly but does not break his relationship with God. On the opposite extreme, his wife tells him to "curse God and die" (Job 2: 9). But he refuses this response as well. Job is deeply grieved and indeed wrestles with suicide, stressing the same method of death, *strangling*, as did Zeno. "So that my soul chooseth strangling, and death, rather than these bones" (Job 7: 15). Strikingly, however, he differs from Zeno in seeing life as a gift from God, who decides when life will begin and end. Job reaffirms his relationship with his Creator: "Though He slays me, yet will I trust in Him" (Job 13: 15).

Job is anchored in a sense of a personal Creator who is with him from the moment of his birth and will be with him to death and beyond. Thus, he can withstand far greater misfortune than can Zeno without the need to attribute cosmic meaning to it. This does not make Job less rational, but simply anchors his interpretive structure in his desire to live. He overcomes major losses in his life by deepening his faith in his Creator, who provides him with an inherent meaning for his life. The book of Job stresses that it is God himself that has given Job the breath of life. "The spirit of God hath made me, and the breath of the Almighty giveth me life" (Job 33: 4). When God restores the breath, man rises again and renews the face of the earth. Job is not obsessed with death, nor does he need to control it or to worry that it is timely. Job does not interpret each event as a signal to exit, but as a challenge to live the life that has been given to him in dignity. The implications of this difference for differential view of wars cannot be overestimated.

Combat itself seems to provide Greeks with a sense of meaning that is absent for them in everyday life. Life itself is seen as competition. In his book "The Glory of Hera," sociologist Philip Slater makes the following observations. "Nothing defines the quality of Greek culture more than the way in which competition was extended from physical prowess to the realms of the intellect, to feats of poetry and dramatic composition." Nothing seemed to have any meaning to the Greeks unless it involved the defeat of another (Slater, 1968, p. 36). The Biblical view is very different. Competition is not valued in itself. It is far better to focus on fulfilling God's purpose on earth. Consider the famous axiom of Hillel: "If I am not for myself, who will be for me? If I am only for myself, what am I?" (Mishna Avot 1:14). There is nothing in this directive that implies one's own worth is dependent on defeating another. Better that people live together cooperatively if they all follow God's sense of morality.

## Conclusions

What can we conclude from this brief comparison of war, peace and violence in these two bedrock cultures of Western civilization? Let us examine a number of categories to make a comparison.

1. Precipitation of Conflict. While the Biblical world seems to want to avoid conflict if possible, the ancient Greek world seems to relish in it. Everything seems to be a competition in Greek thinking. A wins only if B loses.

2. Purpose of the Conflict. Biblical stories emphasize that war must be conducted in terms of God's plan for history. Greek wars are typically not fought in terms of any divine plan for history.
3. Actions towards the Environment. Greek armies typically destroyed the environment, while Biblical armies were forbidden by Biblical law to destroy fruit trees.
4. Glorification of violence. While the Biblical world is certainly not passive and will wage serious war, there seems to be little love of violence in itself. The ancient Greek mindset seems to be inherently competitive and enjoy war for war's sake. The Biblical mindset is willing to wage war in terms of defense of the Biblical way of life.
5. Flexibility of the Military: While the Greek approach to battle often seems to involve a fairly stationary test of brute strength and numbers, the Biblical approach typically seems to depend on flexibility and a sense of transcendent purpose. Military battles are fought for a reason, not just as a test of strength or a sense of duty.
6. Willingness to physically destroy the enemy. While both the Biblical and ancient Greek mind-sets seem willing to completely annihilate the enemy, the reasons are different. The Greek mind set seems to enjoy the battle and inflicting severe pain upon one's adversary. The Biblical mind-set, in contrast, does not focus on violence, but is willing to use violence to destroy nations who are attacking the Israelite covenant with God.
7. Preference for Peace over War. The Biblical mind-set, while willing to engage in war, seems to prefer a more peaceful cooperative solution if possible. The Greek mind-set seems to actually enjoy physical conflict as a form of competition.
8. Depiction of Military Heroes. While the ancient Greeks often seem to idealize military heroes, the Biblical mindset is more mixed. While honoring military achievement, the Biblical attitude points more to peace. The Greek mind seems to love conflict.
9. Perception of the Human Being. The Greeks continuously search for life meaning through grandiosity, often involving military glory. The Israelites, in contrast, believed that man is inherently an exalted being created in the image of God himself, not to seek glory but to live according to God's instructions.

In summary, then, this chapter has attempted to contrast Biblical and ancient Greek attitudes towards the military with regard to peace, war and violence, and indeed the meaning of life itself.

## Appendix

1. The Hebrew Scriptures (I Samuel 17) presents the battle of David and Goliath

    1. And the Philistines assembled their camps to war; and they assembled at Sochoh which belonged to Judah, and they encamped between Sochoh and Azekah in Ephes-dammim.
    2. And Saul and the men of Israel assembled, and they encamped in the Valley of the Terebinth, and they set the battle in array against the Philistines.
    3. And the Philistines were standing on the mountain from here, and Israel was standing on the mountain from here, and the valley was between them.
    4. And the champion emerged from the Philistines' camp, named Goliath, from Gath; his height was six cubits and a span.
    5. And a helmet of copper was on his head, and he was wearing a coat of mail, and the weight of the coat was five thousand shekels of copper.
    6. And greaves of copper were on his legs, and a copper javelin was between his shoulders.

7. And the shaft of his spear was like a weaver's beam, and the spear's head was six hundred shekels of iron, and the shield-bearer went before him.
8. And he stood and called to the array of Israel, and he said to them, "Why should you come out to wage war? Am I not the Philistine, and you the servants of Saul? Choose for yourselves a man, and let him come down to me.
9. If he is able to fight with me and kill me, we shall be slaves to you, and if I overcome him, and kill him, you shall be slaves to us and serve us."
10. And the Philistine said, "I taunt the ranks of Israel this day; give me a man, and let us fight together."
11. And Saul and all Israel heard these words of the Philistine, and they were panic stricken and very much afraid.
12. And David was the son of this Ephrathite man from Bethlehem of Judah, whose name was Jesse; and he had eight sons; and the man was old in Saul's time, coming among men.
13. And Jesse's three eldest sons who had followed Saul, went to war; and the names of his three sons who went to war, were: Eliab, the firstborn, and the second to him, Abinadab, and the third, Shammah.
14. And David was the youngest, and the three eldest followed Saul.
15. Now David went to and fro from Saul, to tend his father's sheep in Bethlehem.
16. And the Philistine drew near morning and evening, and presented himself forty days.
17. And Jesse said to David his son, "Take now to your brothers, an ephah of this parched corn and ten loaves of this bread, and rush them to the camp, to your brothers.
18. And you shall bring these ten cheeses to the captain of the thousand, and you shall see how your brothers are faring, and you shall take the tidings of their welfare."
19. And Saul and they, and all the men of Israel in the Valley of the Terebinth, were at war with the Philistines.
20. And David arose early in the morning, and he left the sheep with a keeper; and he carried off and went, as Jesse had commanded him; and he came to the barricade, and the army which was going out to the battle array, shouted in the battle.
21. And the Israelites and the Philistines put the battle in array, army against army.
22. And David left the baggage in the hand of the keeper of the baggage, and he ran to the battle array, and he came and greeted his brothers.
23. And while he was speaking with them, behold, the champion, named Goliath the Philistine, from Gath, was coming up from the ranks of the Philistines, and he spoke the same words (as before), and David heard.
24. And all the men of Israel, upon seeing the man, fled from before him, and were exceedingly frightened.
25. And the men of Israel said, "Have you seen this man who is coming up, for he is coming up to taunt Israel? And it will be, that the man who will kill him, the King will enrich him with great riches, and he will give him his daughter, and he will make his father's house free in Israel."
26. And David said to the men who were standing before him, saying, "What shall be done to the man who slays that Philistine, and takes away reproach from Israel, for who is this uncircumcised Philistine, that he should taunt the ranks of the living God?"
27. And the people said to him, "So shall be done to the man who slays him."
28. And Eliab, his eldest brother, heard when he spoke to the men, and Eliab's wrath was kindled against David, and he said, "Why have you come down? With whom have you left those few sheep in the desert? I know your impetuousness, and the evil of your heart, for you have come down in order to see the war."

29. And David said, "What have I done now? Is it not but a word?"
30. And he turned around from him, toward another. And he spoke according to the same manner, the people answered according to the former manner.
31. And the words which David had spoken, were heard. And they related (them) before Saul, and he took him.
32. And David said to Saul, "Let no man's heart fail because of him. Your bondsman will go, and battle with the Philistine."
33. And Saul said to David, "You are unable to go to this Philistine to battle with him, for you are a lad, and he is a warrior since his youth."
34. And David said to Saul, "Your bondsman was a shepherd of sheep for his father, and there came a lion and also a bear, and carried off a lamb from the flock.
35. And I went out after him and smote him, and saved it from his mouth. And he arose upon me, and I took hold of his jaw, and I smote him and slew him.
36. Both the lion and the bear has your bondsman slain, and the uncircumcised Philistine shall be like one of them, for he taunted the armies of the living God."
37. And David said, "The Lord Who saved me from the paw of the lion and from the paw of the bear, He will save me from the hand of this Philistine." And Saul said to David, "Go, and may the Lord be with you."
38. And Saul dressed David with his garments, and he placed a copper helmet on his head, and he dressed him with a coat of mail.
39. And David girded his sword over his garments, and he did not want to go (in them), for he was not accustomed, and David said to Saul, "I cannot go with these, for I am not accustomed." And David took them off.
40. And he took his staff in his hand, and he chose for himself five smooth pebbles from the brook, and he placed them in the shepherds' bag which he had, and in the sack, and his slingshot was in his hand, and he approached the Philistine.
41. And the Philistine was continuously drawing nearer to David, and the man who bore the shield was before him.
42. And the Philistine looked and saw David, and he despised him, for he was a youth, and ruddy, with an attractive appearance.
43. And the Philistine said to David, "Am I a dog, that you come to me with sticks?" And the Philistine cursed David by his god.
44. And the Philistine said to David, "Come to me, and I shall give your flesh to the fowl of the air, and to the cattle of the field."
45. And David said to the Philistine, "You come to me with sword, spear and javelin, and I come to you with the Name of the Lord of Hosts, the God of the armies of Israel which you have taunted.
46. This day, the Lord will deliver you into my hand, and I shall slay you, and take off your head, and I shall give the carcasses of the camp of the Philistines this day, to the fowl of the air and to the beasts of the earth, and all the earth shall know that Israel has a God.
47. And all this assembly shall know that not with sword and javelin does the Lord save, for the battle is the Lord's, and He will deliver you into our hand."
48. And it was, when the Philistine arose and drew closer to David, that David hastened and ran to the battle array, toward the Philistine.
49. And David stretched his hand into the bag, and took a stone therefrom, and slung it, and he hit the Philistine in his forehead, and the stone sank into his forehead, and he fell on his face to the ground.

50. And David overpowered the Philistine with the slingshot and with the stone, and he smote the Philistine and slew him: and no sword was in David's hand.
51. And David ran and stood up to the Philistine, and he took his sword and drew it from its sheath, and slew him, and with it, severed his head. And the Philistines saw that their hero had died, and they fled.
52. And the men of Israel and Judah arose and shouted, and pursued the Philistines until you come to the valley, and until the Gates of Ekron; and the slain of the Philistines fell on the road of Shaaraim, and up to Gath and up to Ekron.
53. And the children of Israel returned from pursuing the Philistines, and plundered their camps.
54. And David took the head of the Philistine, and brought it to Jerusalem, and his weapons, he put in his tent.
55. And when Saul saw David going out toward the Philistine, he said to Abner, the general of his army, "Whose son is this youth, Abner?" And Abner said, "By your life, O King, I know not."
56. And the king said, "You ask whose son this youth is."
57. And as David returned from slaying the Philistine, Abner called him, and brought him before Saul, and the Philistine's head was in his hand.
58. And Saul said to him, "Whose son are you, young man?" And David said, "The son of your bondsman, Jesse the Bethlehemite."

2. Book 22 of The Iliad presents the epic battle of Hector and Achilles.

They soon, approaching each the other, stood
Opposite, and huge Hector thus began.

Pelides! I will fly thee now no more.
Thrice I have compass'd Priam's spacious walls
A fugitive, and have not dared abide
Thy onset, but my heart now bids me stand
Dauntless, and I will slay, or will be slain.
But come. We will attest the Gods; for they
Are fittest both to witness and to guard
Our covenant. If Jove to me vouchsafe
The hard-earn'd victory, and to take thy life,
I will not with dishonor foul insult
Thy body, but, thine armor stripp'd, will give
Thee to thy friends, as thou shalt me to mine.
To whom Achilles, lowering dark, replied.
Hector! my bitterest foe! speak not to me
Of covenants! as concord can be none
Lions and men between, nor wolves and lambs
Can be unanimous, but hate perforce
Each other by a law not to be changed,
So cannot amity subsist between
Thee and myself; nor league make I with thee
Or compact, till thy blood in battle shed
Or mine, shall gratify the fiery Mars.

Rouse all thy virtue; thou hast utmost need
Of valor now, and of address in arms.
Escape me more thou canst not; Pallas' hand
By mine subdues thee; now will I avenge
At once the agonies of every Greek
In thy unsparing fury slain by thee.
He said, and, brandishing the Pelian ash,
Dismiss'd it; but illustrious Hector warn'd,
Crouched low, and, overflying him, it pierced
The soil beyond, whence Pallas plucking it
Unseen, restored it to Achilles' hand,
And Hector to his godlike foe replied.
Godlike Achilles! thou hast err'd, nor know'st
At all my doom from Jove, as thou pretend'st,
But seek'st, by subtlety and wind of words,
All empty sounds, to rob me of my might.
Yet stand I firm. Think not to pierce my back.
Behold my bosom! if the Gods permit,
Meet me advancing, and transpierce me there.
Meantime avoid my glittering spear, but oh
May'st thou receive it all! since lighter far
To Ilium should the toils of battle prove,
Wert thou once slain, the fiercest of her foes.
He said, and hurling his long spear with aim
Unerring, smote the centre of the shield
Of Peleus' son, but his spear glanced away.
He, angry to have sent it forth in vain,
(For he had other none) with eyes downcast
Stood motionless awhile, then with loud voice
Sought from Deiphobus, white-shielded Chief,
A second; but Deiphobus was gone.
Then Hector understood his doom, and said.
Ah, it is plain; this is mine hour to die.
I thought Deiphobus at hand, but me
Pallas beguiled, and he is still in Troy.
A bitter death threatens me, it is nigh,
And there is no escape; Jove, and Jove's son
Apollo, from the first, although awhile
My prompt deliverers, chose this lot for me,
And now it finds me. But I will not fall
Inglorious; I will act some great exploit
That shall be celebrated ages hence.
So saying, his keen falchion from his side
He drew, well-temper'd, ponderous, and rush'd
At once to combat. As the eagle darts
Right downward through a sullen cloud to seize
Weak lamb or timorous hare, so brandishing
His splendid falchion, Hector rush'd to fight.

Achilles, opposite, with fellest ire
Full-fraught came on; his shield with various art
Celestial form'd, o'erspread his ample chest,
And on his radiant casque terrific waved
The bushy gold of his resplendent crest,
By Vulcan spun, and pour'd profuse around.
Bright as, among the stars, the star of all
Most radiant, Hesperus, at midnight moves,
So, in the right hand of Achilles beam'd
His brandish'd spear, while, meditating wo
To Hector, he explored his noble form,
546 Seeking where he was vulnerable most.
But every part, his dazzling armor torn
From brave Patroclus' body, well secured,
Save where the circling key-bone from the neck
Disjoins the shoulder; there his throat appear'd,
Whence injured life with swiftest flight escapes;
Achilles, plunging in that part his spear,
Impell'd it through the yielding flesh beyond.
The ashen beam his power of utterance left
Still unimpair'd, but in the dust he fell,
And the exulting conqueror exclaim'd.
But Hector! thou hadst once far other hopes,
And, stripping slain Patroclus, thought'st thee safe,
Nor caredst for absent me. Fond dream and vain!
I was not distant far; in yonder fleet
He left one able to avenge his death,
And he hath slain thee. Thee the dogs shall rend
Dishonorably, and the fowls of air,
But all Achaia's host shall him entomb.
To whom the Trojan Chief languid replied.
By thy own life, by theirs who gave thee birth,
And by thy knees, oh let not Grecian dogs
Rend and devour me, but in gold accept
And brass a ransom at my father's hands,
And at my mother's an illustrious price;
Send home my body, grant me burial rites
Among the daughters and the sons of Troy.
To whom with aspect stern Achilles thus.
Dog! neither knees nor parents name to me.
I would my fierceness of revenge were such,
That I could carve and eat thee, to whose arms
Such griefs I owe; so true it is and sure,
That none shall save thy carcase from the dogs.
No, trust me, would thy parents bring me weigh'd
Ten—twenty ransoms, and engage on oath
To add still more; would thy Dardanian Sire
547 Priam, redeem thee with thy weight in gold,

Not even at that price would I consent
That she who bare should place thee on thy bier
With lamentation; dogs and ravening fowls
Shall rend thy body while a scrap remains.
Then, dying, warlike Hector thus replied.
Full well I knew before, how suit of mine
Should speed preferr'd to thee. Thy heart is steel.
But oh, while yet thou livest, think, lest the Gods
Requite thee on that day, when pierced thyself
By Paris and Apollo, thou shalt fall,
Brave as thou art, before the Scæan gate.
He ceased, and death involved him dark around.
His spirit, from his limbs dismiss'd, the house
Of Ades sought, mourning in her descent
Youth's prime and vigor lost, disastrous doom!
But him though dead, Achilles thus bespake.
Die thou. My death shall find me at what hour
Jove gives commandment, and the Gods above.

He spake, and from the dead drawing away
His brazen spear, placed it apart, then stripp'd
His arms gore-stain'd. Meantime the other sons
Of the Achaians, gathering fast around,
The bulk admired, and the proportion just
Of Hector; neither stood a Grecian there
Who pierced him not, and thus the soldier spake.
Ye Gods! how far more patient of the touch
Is Hector now, than when he fired the fleet!

Thus would they speak, then give him each a stab.
And now, the body stripp'd, their noble Chief
The swift Achilles standing in the midst,
The Grecians in wing'd accents thus address'd.
Friends, Chiefs and Senators of Argos' host!
Since, by the will of heaven, this man is slain
Who harm'd us more than all our foes beside,
Essay we next the city, so to learn
The Trojan purpose, whether (Hector slain)
They will forsake the citadel, or still
548 Defend it, even though of him deprived.
But wherefore speak I thus? still undeplored,
Unburied in my fleet Patroclus lies;
Him never, while alive myself, I mix
With living men and move, will I forget.
In Ades, haply, they forget the dead,
Yet will not I Patroclus, even there.
Now chanting pæans, ye Achaian youths!
Return we to the fleet with this our prize;
We have achieved great glory, we have slain

Illustrious Hector, him whom Ilium praised
In all her gates, and as a God revered.
He said; then purposing dishonor foul
To noble Hector, both his feet he bored
From heel to ancle, and, inserting thongs,
Them tied behind his chariot, but his head
Left unsustain'd to trail along the ground.
Ascending next, the armor at his side
He placed, then lash'd the steeds; they willing flew
Thick dust around the body dragg'd arose,
His sable locks all swept the plain, and all
His head, so graceful once, now track'd the dust,
For Jove had given it into hostile hands
That they might shame it in his native soil.
Thus, whelm'd in dust, it went. The mother Queen
Her son beholding, pluck'd her hair away,
Cast far aside her lucid veil, and fill'd
With shrieks the air. His father wept aloud,
And, all around, long, long complaints were heard
And lamentations in the streets of Troy,
Not fewer or less piercing, than if flames
549 Had wrapt all Ilium to her topmost towers.
His people scarce detain'd the ancient King
Grief-stung, and resolute to issue forth
Through the Dardanian gates; to all he kneel'd
In turn, then roll'd himself in dust, and each
By name solicited to give him way.
Stand off, my fellow mourners! I would pass
The gates, would seek, alone, the Grecian fleet.
I go to supplicate the bloody man,
Yon ravager; he may respect, perchance,
My years, may feel some pity of my age;
For, such as I am, his own father is,
Peleus, who rear'd him for a curse to Troy,
But chiefly rear'd him to myself a curse,
So numerous have my sons in prime of youth
Fall'n by his hand, all whom I less deplore
(Though mourning all) than one; my agonies
For Hector soon shall send me to the shades.
Oh had he but within these arms expired,
The hapless Queen who bore him, and myself
Had wept him, then, till sorrow could no more!
So spake he weeping, and the citizens
All sigh'd around; next, Hecuba began
Amid the women, thus, her sad complaint.

Ah wherefore, oh my son! wretch that I am,
Breathe I forlorn of thee? Thou, night and day,
My glory wast in Ilium, thee her sons

And daughters, both, hail'd as their guardian God,
Conscious of benefits from thee received,
Whose life prolong'd should have advanced them all
To high renown. Vain boast! thou art no more.
So mourn'd the Queen. But fair Andromache
Nought yet had heard, nor knew by sure report
Hector's delay without the city gates.
She in a closet of her palace sat,
A twofold web weaving magnificent,
With sprinkled flowers inwrought of various hues,
And to her maidens had commandment given
Through all her house, that compassing with fire
An ample tripod, they should warm a bath
For noble Hector from the fight return'd.
Tenderness ill-inform'd! she little knew
That in the field, from such refreshments far,
Pallas had slain him by Achilles' hand.
She heard a cry of sorrow from the tower;
Her limbs shook under her, her shuttle fell,
And to her bright-hair'd train, alarm'd, she cried.
Attend me two of you, that I may learn
What hath befallen. I have heard the voice
Of the Queen-mother; my rebounding heart
Chokes me, and I seem fetter'd by a frost.
Some mischief sure o'er Priam's sons impends.
Far be such tidings from me! but I fear
Horribly, lest Achilles, cutting off
My dauntless Hector from the gates alone,
Enforce him to the field, and quell perhaps
The might, this moment, of that dreadful arm
His hinderance long; for Hector ne'er was wont
To seek his safety in the ranks, but flew
First into battle, yielding place to none.
So saying, she rush'd with palpitating heart
And frantic air abroad, by her two maids
Attended; soon arriving at the tower,
And at the throng of men, awhile she stood
Down-looking wistful from the city-wall,
And, seeing him in front of Ilium, dragg'd
So cruelly toward the fleet of Greece,
O'erwhelm'd with sudden darkness at the view
Fell backward, with a sigh heard all around.
Far distant flew dispersed her head-attire,
Twist, frontlet, diadem, and even the veil
By golden Venus given her on the day
When Hector led her from Eëtion's house
Enrich'd with nuptial presents to his home.
Around her throng'd her sisters of the house

Of Priam, numerous, who within their arms
551 Fast held her, loathing life; but she, her breath
At length and sense recovering, her complaint
Broken with sighs amid them thus began.
Hector! I am undone; we both were born
To misery, thou in Priam's house in Troy,
And I in Hypoplacian Thebes wood-crown'd
Beneath Eëtion's roof. He, doom'd himself
To sorrow, me more sorrowfully doom'd,
Sustain'd in helpless infancy, whom oh
That he had ne'er begotten! thou descend'st
To Pluto's subterraneous dwelling drear,
Leaving myself destitute, and thy boy,
Fruit of our hapless loves, an infant yet,
Never to be hereafter thy delight,
Nor love of thine to share or kindness more.
For should he safe survive this cruel war,
With the Achaians penury and toil
Must be his lot, since strangers will remove
At will his landmarks, and possess his fields.
Thee lost, he loses all, of father, both,
And equal playmate in one day deprived,
To sad looks doom'd, and never-ceasing-tears.
He seeks, necessitous his father's friends,
One by his mantle pulls, one by his vest,
Whose utmost pity yields to his parch'd lips
A thirst-provoking drop, and grudges more;
Some happier child, as yet untaught to mourn
A parent's loss, shoves rudely from the board
My son, and, smiting him, reproachful cries—
Away—thy father is no guest of ours—
Then, weeping, to his widow'd mother comes
Astyanax, who on his father's lap
Ate marrow only, once, and fat of lambs,
552 And when sleep took him, and his crying fit
Had ceased, slept ever on the softest bed,
Warm in his nurse's arms, fed to his fill
With delicacies, and his heart at rest.
But now, Astyanax (so named in Troy
For thy sake, guardian of her gates and towers)
His father lost, must many a pang endure.
And as for thee, cast naked forth among
Yon galleys, where no parent's eye of thine
Shall find thee, when the dogs have torn thee once
Till they are sated, worms shall eat thee next.
Meantime, thy graceful raiment rich, prepared
By our own maidens, in thy palace lies;
But I will burn it, burn it all, because

Useless to thee, who never, so adorn'd,
Shalt slumber more; yet every eye in Troy
Shall see, how glorious once was thy attire.
So, weeping, she; to whom the multitude
Of Trojan dames responsive sigh'd around.

## References


Babylonian Talmud, *The*, Pirkei Avot 1: 14.

Hebrew Bible, The. (1996). Translated by A. J. Rosenberg. *The Book of Samuel One*. New York: Judaica Press.

Herodotus. (1961). *The Histories*. Aubrey DeSelincourt (Translator). Baltimore, MD: Penguin Books.

Homer. (1860). *The Iliad*. William Cowper (Translator). New-York: D. Appleton & Co.

Homer. (1950). *The Odyssey*. A. H. Butcher and A. Lang (Translators). New-York: Random House: The Modern Library.

Roter, A. (1988–1996). *Shaarei Aharon* (sixteen volumes). Bnei Brak, Israel: Seforim Center.

Slater, P. E. (1968). *The Glory of Hera: Greek Mythology and the Greek Family*. Boston, MA: Beacon Press.

Thucydides. (1978). *History of the Peloponnesian War*. Translated by Rex Warner. NY: Penguin.

Zevin, S.Y. (1946). *Le'or Hahalacha: Be'ayot u Verurim*. Jerusalem, ISRAEL: Mossad Harav Kook.

# 3
# FOUR STAGES IN THE EVOLUTION OF MILITARY ENLISTMENT TESTING

*Michael G. Rumsey*

Military personnel decision-making has become more scientific, expansive, and effective as the environment in which the military operates has become more complex and demanding. This progression is particularly evident in the history of psychological testing for enlisted selection and classification in the United States military. This chapter focuses on how this history has gone through three progressive stages: 1) screening for basic cognitive ability in a single service, 2) matching individual cognitive capabilities to job requirements separately for each service, and 3) developing a single cognitive battery to enable person-job matching for all services, and how it is currently embarking on a fourth stage: developing the capacity to select and classify applicants using a combination of cognitive and noncognitive measures. This last phase, while continuing the evolution, is revolutionary in terms of both character and pace (Rumsey, in press). These advancements have been facilitated by a number of technical advances, including the development and administration of group ability tests, the development of computer adaptive tests, the development of a model of job performance, and the development of fake-resistant personality tests.

While focusing on these four stages, it is important to acknowledge related developments that have unavoidably been left out in this chapter, including those with respect to officer selection, selection and classification outside the United States, and personnel allocation systems to augment person-job classification. While these developments are outside the scope of the current paper, they are nevertheless important. While one may search in vain for an inclusive, integrative treatment of the vast number of military selection programs throughout the world, proceedings from a North Atlantic Treaty Organization meeting (Research and Technology Organization, 2000) did provide a snapshot of various officer selection programs at the time.

## Basic ability tests: Army Alpha and Beta

While both the Revolutionary War and the Civil War required the mobilization of large armies, the process of enlisting troops during these periods might be characterized more as desperation than any type of systematic selection and classification. When the United States entered World War I in 1917, the situation was similarly desperate, with the need to rapidly mobilize large numbers of Army troops to avoid a disastrous outcome at the European front, but this time the reaction was more thoughtful. Robert Yerkes, president of the American Psychological

Association, initiated a project to develop a test to ensure that new enlistees had sufficient ability to perform basic tasks. Such luminaries as Henry Goddard, Walter Bingham, Lewis Terman, and Edward Thorndike participated in this effort. The concept for this test was influenced by the Binet Simon scales developed to measure intelligence. Ultimately, two test batteries were developed. These contained a variety of tasks, but the common element of all was a requirement that the applicant demonstrate general cognitive ability. One of the new batteries, the Army Alpha, was composed of verbal tasks for those who could read and write. Representative tests in this battery included general information, arithmetic, and analogies. The other, the Army Beta, was composed of non-verbal items, such as mazes, picture completion, and number checking (Ramsberger, 2012).

Ideally, selection occurs before one is actually placed in a job. However, perhaps partly because an infrastructure for pre-enlistment testing had not yet been created, this is not the way it worked in World War I. Soldiers were tested after enlistment by Commanders in the field. Based on the results of these tests, a limited number were discharged or transferred.

World War I was a time of transition, during which the Army gave tentative approval to limited use of testing for decision-making, but with clear ambivalence over whether such testing actually provided better information than a Commander's judgment. After August 14, 1918, Commanders were given the option of not using the results at all. Nevertheless, the Army did allow 1,750,000 men to be tested on either the Army Alpha or the Army Beta during this unprecedented World War (Ramsberger, 2012).

Even in these early days of standardized testing, the Army recognized the need to determine how well test scores related to other measures. To the extent that could be measured, the results were quite impressive. For example, it was reported that rankings by officers who had known their subordinates 6 to 12 months correlated an average of .536 with ranked scores on the Army Alpha for an enlisted sample of 965 men (Yoakum & Yerkes, 1920). This extraordinarily high correlation would appear to be an anomaly, as more recent correlations between performance on cognitive tests and performance ratings have tended to range between –.02 and.20 (Borman, White, Pulakos, & Oppler, 1991; Vineberg & Taylor, 1972). However, the data reported by Yoakum and Yerkes was not, strictly speaking, based on ratings of performance. Two of the three rating categories specifically mentioned "ability," while only the third referenced "efficiency" (Yoakum & Yerkes, 1920, p. 31).

Army Alpha scores were also found to correlate with another cognitive measure, the Stanford-Binet, in the range of .80 to .90, and the Army Alpha was found to correlate with the Army Beta at about .80. The Army Alpha scores were also related to success in training, rank, and later success in field operations (Yoakum & Yerkes, 1920).

## Classification batteries

Although the Army Alpha and Beta test batteries were not completely irrelevant to person-job matching, as those scoring in the upper ranges might be considered best qualified to perform particularly cognitively challenging jobs, this was not their primary function. Later tests, such as the Army General Classification Test (AGCT; Staff, Personnel Research Section, 1945), were only marginally more suited to matching individuals to jobs, a function known as personnel classification. The next major development was to create test batteries designed specifically to support such matching. Since each service had somewhat different needs, separate test classification batteries were developed for each one (Rumsey, 2012).

The problem of matching individuals to jobs is a complicated one when the numbers of jobs and individuals are as great as they are in each military service. First, consider that there might be a number of jobs that are very similar in terms of the attributes needed to perform successfully. For example, there might be a number of electronics jobs that are distinguished primarily by the type of equipment used in each. Does there need to be a separate combination of test scores to determine an individual's qualifications for each? Is it even possible to identify how each job might be different in terms of individual attributes needed? Is it feasible to develop an instrument that would reliably make the distinctions necessary to make differential assignments across such jobs?

Second, consider how each person-job assignment affects every other potential person-job assignment. If Person A is assigned to Job X, because that is the best fit for Person A, that has implications for other people who also might have been assigned to Job X, as well as for other jobs to which Person A might have been assigned. Person B might not be as well qualified as Person A for Job X, but because Person A is so much better qualified for Job Y than Person B, the Army might be better served by having Person B in Job X and Person A in Job Y. These are just some examples of complications associated with large-scale person-job classification.

Faced with such conundrums, and lacking the high-speed computational power available today, the services opted for a partial solution. Rather than developing a test composite for every single job, the services developed composites, or combinations of tests, for groups of jobs. A 1972 version of the Army Classification Battery (ACB), for example, included composites for such job categories, or Aptitude Areas, as Combat, Electronics Repair, General Maintenance, and Clerical. Of the 16 scores yielded by the ACB, a subset was selected for each of nine Aptitude Areas. The composite for Electronics Repair, for example, consisted of the sum of scores on Arithmetic Reasoning, Electronics Information, Trade Information, Mechanical Comprehension, and a self-reported Electronics inventory measure, each equally weighted (Maier & Fuchs, 1972).

Maier (1993) noted the great care taken by the services in developing their classification tests at each stage in the process, including defining content, constructing test items, collecting data to refine the items, and testing the correspondence between test scores and course grades. Validities for predicting course grades were consistently high, such as .63 for the Air Force's 1960 version of the Airman Qualification Examination (Weeks, Mullins, & Vitola, 1975) and .57 for the Navy's Basic Test Battery (Thomas, 1970).

## *Joint-service battery*

The story of the development of a joint-service selection and classification instrument, the Armed Services Vocational Aptitude Battery (ASVAB), is in some respects a cautionary one. Ultimately, the outcome was a battery that has persevered in various forms for roughly 50 years. Along the way, however, the developmental process encountered severe problems that proved disruptive and embarrassing for each of the military services.

The development of the very first iterations of the ASVAB proceeded without notable controversy. In 1966, the Assistant Secretary of Defense for Manpower and Reserve Affairs directed that a common aptitude battery be developed that all services could use for high school testing. Content areas of the existing Army, Navy, and Air Force classification tests that were determined to be interchangeable (e.g., word knowledge) across the services were identified, and items were drawn from each of the service tests in that content area to form a new subtest. New content was added as needed. Appropriate analyses were conducted to ensure the new battery met scientific and practical standards (Bayroff & Fuchs, 1970). The first version of the ASVAB was administered from 1968 to 1973 to high school students for vocational counseling and recruiting purposes. Two parallel forms of a second version were administered from 1973 to 1975 (Seeley, Fischl, & Hicks, 1978). One form was used for high school testing and the other for Air Force and Marine Corps operational screening (ASVAB Working Group, 1980).

## *Armed Services Vocational Aptitude Battery misnorming*

The problems arose with the third iteration of the ASVAB. In 1974, the Assistant Secretary of Defense for Manpower and Reserve Affairs directed that the ASVAB be used as a common instrument for all the services—not just for high school students, as before, but now for operational enlisted selection and classification as well (ASVAB Working Group, 1980). Along with this decision came three immediate implications for the next iteration of the ASVAB: 1) as a determinant of selection and classification decisions for all the services, it would be more consequential than previous versions; 2) there was a perceived need to implement the new forms quickly, in part because the single form then used for operational screening was viewed as inadequate by the Marine Corps and in part by a desire by the Department of Defense to make the transition to joint testing expeditiously (ASVAB Working Group, 1980); and 3) for the first time, with the draft having ended in 1972 (Rostker, 2006), the test would be developed in an all-volunteer environment.

An ambitious schedule for fielding the new ASVAB collided with numerous delays. In the end, this meant little time was available for determining how the raw scores on the ASVAB could be interpreted in making selection and classification decisions (ASVAB Working Group, 1980). This involved an estimation of how individuals in general would score on this test, not just those individuals tested at this particular time and place. To make this estimation, scores on the ASVAB were linked to another test, a reference test, already calibrated against a representative population. However, later analyses suggested that the linkage process was flawed in multiple respects (Maier & Truss, 1983). The test used for linkage in the Army was the one then being used for operational selection and classification decisions. Thus, examinees were motivated to perform higher on the reference test and recruiters were more likely to coach examinees how to perform higher on this test than would have been the case for the earlier representative population. Also, the reference test used for the Navy and Air Force was found to be incorrectly scored (Maier & Truss, 1983).

The new ASVAB was fielded in all services in 1976. Almost immediately, there were suspicions that the standards being applied with respect to the new forms were incorrect, but pinpointing the problem proved to be difficult (ASVAB Working Group, 1980). Finally, research was conducted that generated standards that the services could agree on, and tests based on these were implemented in 1980 (Maier & Grafton, 1981). By then, it had been determined that the scores used for selection and classification had been seriously inflated, leading to the enlistment of many more with marginal scores on the ASVAB than had been intended. To Army General Maxwell Thurman, who was "arguably the single most important person in the history of the All-Volunteer Force" (Rostker, 2006, p 387), the result was a "calamity" (Rostker, 2006, p. 398). The entire episode, known as the ASVAB misnorming, was described as a "tragedy of errors. A travesty of sound psychometric practices and common sense" (Laurence & Ramsberger, 1991, p. 72, quoted in Rostker, 2006, p. 481).

However, the ultimate impact of the misnorming was that sound psychometric practices and common sense came back to the fore as never before. A Defense Advisory Committee (DAC) on Military Personnel Testing was established in 1981 composed of outside experts to oversee the process of updating military test practices. The DAC and technical and policy military groups met multiple times each year to provide oversight to the enlistment testing program (Maier, 1993).

## *Computerized Armed Services Vocational Aptitude Battery*

As enlisted selection and classification was becoming a joint-service enterprise, a research program was initiated to generate a computerized version of the ASVAB (Wiskoff, 1997).

Computerization offered some benefits in terms of efficiency, but more importantly it could permit tailoring test items to an applicant's ability (Sellman & Arabian, 1997). An item answered correctly could trigger the administration of a more difficult item; an item answered incorrectly could trigger the administration of an easier item. Each item would provide a more precise estimate of the applicant's true ability level. Time would not be wasted on the administration of items far above or below that level (Sands & Waters, 1997).

Interest in a computer adaptive test heightened as a result of the misnorming. The computer adaptive technology was viewed as a partial antidote to the coaching and scoring issues that emerged during this crisis (Wiskoff, 1997). It is difficult to coach a candidate on the correct answers to a test that has no set sequence of items and presents different items to different applicants. Also, computer scoring would reduce the likelihood of human error.

One lesson that apparently had not been learned from the ASVAB misnorming was that pressure to meet an ambitious schedule may have unforeseen consequences. After preliminary research had been completed in 1982, the Navy was identified as the lead service, and completion of the project was scheduled for 1985 (Wiskoff, 1997). However, it soon became apparent that this schedule could not be met. Available computer hardware and technical capabilities were not adequate for the requirements of full-scale enlisted testing in multiple sites across the nation (McBride, 1997). Funding and management issues (Martin & Hoshaw, 1997) also emerged. Finally, in 1993, implementation of CAT-ASVAB was approved (Martin & Hoshaw, 1997). While the development time was much longer than anticipated, the services, under the Navy's lead, persevered to overcome enormous technical and management problems.

## *Job Performance Measurement Project*

As the misnorming revealed that the services were unintentionally not meeting the standards they had set for themselves, the question arose of what importance these standards actually had. What did it mean for someone to receive a low score on the ASVAB? This was a question Congress was very much interested in, and it laid the groundwork for what was to become known as the Job Performance Measurement Project (JPM). Each of the services was directed to conduct research to determine what the link between ASVAB enlistment standards and job performance was (Knapp, 2006).

The Department of Defense determined that the gold standard for performance assessment was hands-on performance. Each service was to include a measure of hands-on performance in its validation strategy. However, beyond this, the services diverged in terms of other criteria they employed (Knapp, 2006). The most ambitious strategy was the one used by the Army. It was driven by the philosophy that no single type of measure truly captured the complexity of performance. Hands-on performance reflected the maximum capability of an individual on certain well-defined tasks. Job knowledge was another measure of capability. However, ratings reflected performance over a longer period of time, and were more indicative of an individual's motivation to perform, even when not being tested. Similarly, administrative measures, such as awards or disciplinary history, provided a view of performance that could not be completely represented by a task-oriented proficiency measure (each of these types of measures was described in Knapp, Campbell, Borman, Pulakos, & Hanson, 2001).

The Army's research effort, known as Project A, was conducted over a 13-year period, from 1982 to 1995, and completely revamped conceptions of job performance and individual differences related to such performance. These individual differences were measured not just by ASVAB, a cognitive measure, but also by measures of personality, vocational interests, spatial ability, and psychomotor ability (Russell & Peterson, 2001). For the first time since the early

1970s, there was active service interest in using a broader perspective than cognitive ability alone to guide selection and classification decisions.

This multidimensional approach to both performance and the predictors of performance proved highly illuminating. It generated results that revealed that one could not simply ask whether a measure predicted performance, but rather one needed to ask the question in terms of what type of performance was being predicted. The combination of hands-on and job knowledge yielded one type of performance, sometimes referred to as can-do or task performance. Ratings that consisted of dimensions that reflected behavior that essentially transcended specific task performance, combined with administrative measures of behavior, yielded another category, which could be labeled will-do performance (Rumsey, Peterson, Oppler, & Campbell, 1996).

To those vested in the credibility of ASVAB, the results were a relief. ASVAB was a strong predictor of can-do performance. To those who believed that a broader combination of predictors was needed for selection, there was also support. ASVAB did not predict will-do performance as well as it predicted can-do performance, but personality and interests showed promise in being able to add to the ASVAB in predicting this type of behavior (McHenry, Hough, Toquam, Hanson, & Ashworth, 1990). In combination with the results from the other services participating in the JPM (Sellman, Russell, & Strickland, 2017), the prevailing conclusion was that ASVAB was doing the job intended for it, predicting technical performance, but that the selection system could likely be improved by the addition of noncognitive individual difference measures.

### *Armed Services Vocational Aptitude Battery today*

Since 1976, the ASVAB has been a dual-use instrument. Its core is a combination of verbal and quantitative tests used for selection. These tests and more specialized technical tests are combined in various ways by the different services for job placement decisions. The ASVAB's utility for selection has been well demonstrated. Recently, its utility for classification has become somewhat controversial. Because it consists of cognitive tests, and such tests have been found to be highly correlated with one another, some have concluded (e.g., Murphy & Davidshofer, 1991) that the different ASVAB tests measure essentially the same thing: general cognitive ability. However, other scientists have conducted simulations showing that, even with high correlations between different ASVAB components, substantial classification utility is still possible (e.g., Zeidner & Johnson, 1994).

## The rise of the noncognitives

### *Personality measures*

Project A stimulated interest in measurement of individual differences that could predict will-do performance, particularly personality variables. Despite this interest, progress toward including noncognitive variables in selection moved slowly at first. Early efforts to introduce a joint-service personality measure were unsuccessful due to concerns about faking (Rumsey, 2012). Faking was indeed a serious issue. If an individual responded to a personality item in terms of what he or she thought would impress the examiner, then the test was not measuring what it was intended to measure.

For a number of years, researchers labored somewhat inconspicuously to overcome this obstacle. Two researchers in particular, Leonard White and Mark Young, at the Army Research Institute for Behavioral and Social Sciences, explored building an instrument using a forced choice methodology. The idea was to disguise each item so it would not be clear what response would put the examinee in the most favorable light. The result of their endeavors was a measure known as the Assessment of Individual Motivation (AIM) (White & Young, 1998). Their work came to sudden prominence when the Army needed a new approach to the screening of those without high school diplomas, whose proclivity towards attrition made them poor risks for the Army. Research indicated that the AIM could be useful in identifying those non–high school diploma graduates who did not present the same level of risk (White, Rumsey, Mullins, Nye, & LaPort, 2014). The AIM was implemented, but early results were discouraging (White, Young, Hunter, & Rumsey, 2008). The researchers persevered, identified and replaced problematic items, and a revised version of the AIM proved to be a successful component of a new screening system directed specifically at nongraduates (White et al., 2014).

Following this success, further developments in personality testing proceeded with new momentum. Another group of researchers working in the Drasgow Consulting Group (DCG) developed a new personality test technology that could be applied to an entire applicant pool. A paper-and-pencil instrument with limited items could, with repeated and frequent use, be particularly vulnerable to compromise. The DCG technology streamlined the item development process to facilitate production of a massive number of items. Further, it employed a computer adaptive process (Stark et al., 2014). The ASVAB had already been converted to a computer adaptive format, but applying this approach to a personality measure was something new. The basic idea is the same, however: approach an individual's true position on a scale by iteratively selecting the items administered to that individual based on his or her responses to previous items until the individual's position is identified with sufficient accuracy.

The combination of an enormous item pool and the lack of a predetermined ordered list of items made the new adaptive tests highly fake resistant (Stark et al., 2014). Also, as noted above, with a computer adaptive approach, items could be limited to those that provided the most information about the individual on the attribute in question. Thus, a computer adaptive test could be administered in much less time than a traditional paper-and-pencil test. For these reasons, this approach was highly desirable for screening large numbers of applicants in a limited period of time.

Both the Navy and the Army have had success implementing computer adaptive personality measures. The Navy's version, the Navy Computer Adaptive Personality Scales (NCAPS), was introduced as a tool used in selection for Special Operations training (Stark et al., 2014). The Army version, the Tailored Adaptive Personality Assessment System (TAPAS), was provisionally implemented as a supplement to the ASVAB in enlisted selection. It replaced the AIM as a tool for identifying non–high school diploma graduates who were acceptable attrition risks (White et al., 2014) and was also used as a screen for high school diploma graduates whose ASVAB scores put them in jeopardy of missing the cut (Knapp, Heffner, & White, 2011).

TAPAS has been found to predict performance in training (White et al., 2014) and on the job (Kirkendall, Bynum, Kennedy, & Hughes, in press), as well as attrition (Hughes, O'Brien, Reeder, & Purl, in press; Kirkendall, Nye et al., in press). Both the Army (Wolters, Heffner, & Sams, 2015) and Air Force (Trent, Barron, Rose, & Carreta, in press) have found it useful for predicting counterproductive work behaviors. The Air Force has conducted research demonstrating its resistance to faking and its stability over time. Navy recruits have also participated in TAPAS research (Stark et al., 2014).

## *Vocational interest measures*

The new noncognitive advances in military testing have also extended to vocational interest measurement. An early version of the ASVAB included interest items carried over from the Army Classification Battery, but the interest measures were subsequently dropped "due to low predictive validity" (Greenston, 2012). Each of the services has been conducting sporadic interest measurement research since 1949 (Ingerick & Rumsey, 2014) but with little to show for it in terms of operational success up to the turn of the century. Since then, however, such research has gained momentum. The Navy has, having identified key job characteristics as a basis for determining applicant interest in particular Navy jobs (Watson, in press), developed an interest measure that of this writing is scheduled for implementation for all Navy enlisted accessions in October 2018. The Air Force has developed a tool based on the Navy's work called the Air Force Work Interest Navigator (AF-WIN; Johnson, Romay, & Barron, in press). The Army has built upon past work to identify a new instrument covering 20 basic interest categories (Kirkendall, Nye et al., in press).

The controversy over the ASVAB's classification potential suggests the need for further examination of how much can be gained through optimal combinations of cognitive measures for job placement. Since, as shown in Project A (Peterson et al., 1992), cognitive measures and noncognitive measures are not highly correlated, it also suggests that incorporating noncognitive measures into the classification process may be particularly helpful. Preliminary research provides some support for the potential value of TAPAS as a classification tool (Nye et al., in press).

## Conclusion

Military enlistment testing has gone through a number of progressive changes. First, with the Army Alpha and Beta, there were service-specific tests of general cognitive ability. Such tests were primarily useful for selection but not well designed for person-job matches. Then each service developed its own classification test battery, facilitating classification decisions.

The development of a joint-service battery was another step in the maturation of military testing. Initial stumbles of disastrous proportions threatened the continuing existence of joint-service testing, but the services pulled together and took a number of steps that made their testing program stronger than ever. From the misnorming disaster came increased oversight, improved testing technology, and a demonstration of the efficacy of the ASVAB for predicting job performance, as well as a better understanding of such performance and the relationship between different types of individual differences and different types of performance.

New technological advances and accumulation of evidence of the validity of noncognitive tests have led to renewed attention to personality and vocational interest measures, not only as enlisted selection tools, but also as tools for Reserve Officers' Training Corps scholarship screening (Legree, Kilcullen, Putka, & Wasko, 2014) and in-service selection (Muhammad, Wolters, & Jane, in press; Nye et al., in press) tools. They may also have utility in matching individuals to jobs. Despite this, with their limited operational history, their acceptance remains tenuous.

The history of change in the military with respect to selection and classification is consistent with Gersick's (1991, p. 12) paradigm of organizational evolution: "relatively long periods of stability (equilibrium), punctuated by compact periods of qualitative, metamorphic change (revolution)." Poor organizational performance or major environmental changes "often precipitate discontinuous periods of change" (Colarelli, 1998, p. 1049).

Major environmental changes have been associated with each of the four evolutionary stages of military enlistment testing. The first scientifically developed group selection tests, the Army Alpha and Beta, came as a direct response to the personnel needs triggered by World War I. The first true service-specific classification batteries, the Navy Basic Test Battery (completed in 1943; Odell, 1947), the Air Force's Airman Classification Battery (completed in 1948; Weeks, Mullins, & Vitola, 1975), and the Army Classification Battery (completed in 1949; Zeidner, Harper, & Karcher, 1956), were all developed during or after World War II, again in response to critical personnel needs.

The move toward an all-volunteer force was a significant incentive for the services to develop a joint-service selection and classification battery. The prospect of discouraging recruits by requiring separate enlistment tests for each service was not an enticing one. The shift from the Cold War to a world of multiple, diverse challenges could be seen as a factor in the rise of noncognitive testing. The new environment required adaptable service members who could handle uncertainty and ambiguity. Accordingly, personnel managers' perspective shifted from a focus on cognitive ability alone to a broader, "whole person" orientation.

The progressive evolution of military testing has never been, and will never be, inevitable. Not all organizations adapt to change. "Once a social technology becomes routine, a tacit consensus emerges that this is the way things are done" (Colarelli, 1998, p. 1048), and thus resistance to change, especially discontinuous change, is natural. Continued exploration of alternatives and a willingness to scientifically evaluate these alternatives and to commit to the alternative determined to be most functional is necessary to ensure the military will continue to adapt.

Military enlistment testing has thus far adapted well. The ASVAB misnorming was a great setback but one it overcame. Questions about the value of ASVAB for classification and about the value of noncognitive testing for selection or classification will limit the acceptability of these tools until these questions can be definitely answered. Yet even these questions are catalysts for positive change. They should stimulate further research to sharpen our classification tools and improve our noncognitive instruments.

A key element of the success of the military testing program has been its scientific rigor. Colarelli (1998, p. 1048) noted the "low use of scientifically valid technologies in organizations." Not only does the military take validation seriously, but it has a history of developing highly valid tests. The selection portion of the ASVAB was found to correlate with hands-on job performance at a level of .40 across all the services in the JPM project (Sellman, Russell, & Strickland, 2017).

The military has benefited from research in civilian testing technology, but it has also led the way in terms of its own research advancements, whether in terms of large-scale cognitive testing, computerized testing, performance measurement, personality measurement, or vocational interest measurement. It has developed a state-of-the-art selection and classification system but will constantly need to refine and advance its procedures to meet the challenges of a complex and evolving threat environment and to maintain acceptability from both its own leadership and the general public.

## References


ASVAB Working Group. (1980). *History of the Armed Services Vocational Aptitude Battery, 1974–1980* (AD-E750743). Washington, DC: ASVAB Working Group.

Bayroff, A. G., & Fuchs, E. F. (1970). *The Armed Services Vocational Aptitude Battery* (Tech. Research Rep. No. 1161). Arlington, VA: U. S. Army Behavior and Systems Research Laboratory.

Borman, W. C., White, L. A., Pulakos, E. D., & Oppler, S. H. (1991). Models of supervisory job performance ratings. *Journal of Applied Psychology*, *76*(6), 863–872.

Colarelli, S. M. (1998). Psychological interventions in organizations: An evolutionary perspective. *American Psychologist*, *53*(9), 1044–1056.

Gersick, C. J. G. (1991). Revolutionary change theories: A multilevel exploration of the punctuated equilibrium paradigm. *Academy of Management Review*, *16*(1), 10–36.

Greenston, P. (2012). Classification research. In P. F. Ramsberger, N. R. Wooten, & M. G. Rumsey (Eds.), *A History of the Research into Methods for Selecting and Classifying U. S. Army Personnel* (pp. 275–338). New York: Mellen Press.

Hughes, M., O'Brien, E., Reeder, M., C., & Purl, J. (in press). Attrition and reenlistment in the Army: Using the Tailored Adaptive Personality Assessment System to improve retention. *Military Psychology*.

Ingerick, M., & Rumsey, M.G. (2014). Taking the measure of work interests: Past, present, and future. *Military Psychology, 26*(3), 165–181.

Johnson, J., Romay, S., & Barron, L. G. (in press). Air Force Work Interest Navigator (AF-WIN) to improve person-job match: Development, validation, and initial implementation. *Military Psychology*.

Kirkendall, C., Bynum, B., Nesbitt, C., & Hughes, M. (in press). Validation of the TAPAS for predicting in-unit soldier outcomes. *Military Psychology*.

Kirkendall, C., Nye, C., Rounds, J., Drasgow, F., Chernyshenko, O. S., & Stark, S. (in press). Adaptive vocational interest diagnostic: Informing and improving the job selection process. *Military Psychology*.

Knapp, D. (2006). The U.S. Joint-Service Job Performance Measurement Project. In W. Bennett, C. E. Lance, & D. J. Woehr (Eds.), *Performance Measurement: Current Perspectives and Future Challenges* (pp. 113–140). Mahwah, NJ: Erlbaum.

Knapp, D. J., Campbell, C.H., Borman, W. C., Pulakos, E. D., & Hanson, M. A. (2001). Performance assessment for a population of jobs. In J. P. Campbell & D. J. Knapp (Eds.), *Exploring the Limits in Personnel Selection and Classification* (pp.181–235). Mahwah, NJ: Erlbaum.

Knapp, D. J., Heffner, T. S., & White, L. (2011). *Tier One Performance Screen Initial Operational Test and Evaluation: Early Results* (Tech. Rep. No. 1285). Arlington, VA: U. S. Army Research Institute for the Behavioral and Social Sciences.

Legree, P. J., Kilcullen, R. N., Putka, D. J., & Wasko, L. E. (2014). Identifying the leaders of tomorrow: Validating predictors of leader performance. *Military Psychology, 26*(4), 292–309.

Maier, M. H. (1993). *Military Aptitude Testing: The Past 50 Years* (Technical Report No. 93–007). Defense Manpower Data Center.

Maier, M. H., & Fuchs, E. F. (1972). *An Improved Differential Army Classification System* (Tech. Rep. No. 1177). Arlington, VA: U S. Army Behavior and Systems Research Laboratory.

Maier, M. H., & Grafton, F. C. (1981). *Scaling Armed Services Vocational Aptitude Battery (ASVAB) Form 8AX* (Research Rep. No. 1301). Alexandria, VA: U. S. Army Research Institute for the Behavioral and Social Sciences.

Maier, M. H., & Truss, A. R. (1983). *Original Scaling of ASVAB Forms 5/6/7: What Went Wrong* (CRC 457). Alexandria, VA: Center for Naval Analyses (AD-A129499).

Martin, C. J., & Hoshaw, C. R. (1997). Policy and program management perspectives. In W. A. Sands, B. K. Waters, & J. R. McBride (Eds.), *Computerized Adaptive Testing: From Inquiry to Operation* (pp. 11–20). Washington, DC: American Psychological Association.

McBride, J. R. (1997). Technical perspective. In W. A. Sands, B. K. Waters, & J. R. McBride (Eds.), *Computerized Adaptive Testing: From Inquiry to Operation* (pp. 11–20). Washington, DC: American Psychological Association.

McHenry, J. J., Hough, L. M., Toquam, J. L., Hanson, M. A., & Ashworth, S. (1990). Project A validity results: The relationship between predictor and criterion domains. *Personnel Psychology, 43*(2), 335–354.

Muhammed, R., Wolters, H., & Jayne, B. S. (in press). Using personality testing to enhance in-service selection. *Military Psychology*.

Murphy, K. R., & Davidshofer, C. O. (1991). *Psychological Testing, Principles and Applications*. Englewood Cliffs, NJ: Prentice-Hall.

Nye, C., White, L., Drasgow, F., Prasad, J., Chernyshenko, O., & Stark, S. (in press). Examining personality for the selection and classification of soldiers: Validity and differential validity across jobs. *Military Psychology*.

Nye, C., White, L., Horgen, K., Drasgow, F., Stark, S., & Chernyshenko, O. (in press). Predictors of attitudes and performance in U. S. Army recruiters: Does personality matter? *Military Psychology*.

Odell, C. E. (1947). Selection and classification of enlisted personnel. In D. B. Stuit (Ed.), *Personnel Research and Test Development in the Bureau of Naval Personnel* (pp. 21–30). Princeton: Princeton University Press.

Peterson, N., Russell, T., Hallam, G., Hough, L., Owens-Kurtz, C., Gialluca, K., & Kerwin, K. (1992). Analysis of the experimental battery: LV sample. In J. P. Campbell & L. M. Zook (Eds.), *Building and Retaining the Career Force: New Procedures for Accessing and Assigning Army Enlisted Personnel: Annual Report, 1990 Fiscal Year* (Tech. Rep. No. 952). Alexandria, VA: U. S. Army Research Institute for the Behavioral and Social Sciences.

Ramsberger, P. F. (2012). Selection and classification in World War I: Personnel psychologists lend a hand. In P. F. Ramsberger, N. R. Wooten, & M. G. Rumsey (Eds.), *A History of the Research into Methods for Selecting and Classifying U. S. Army Personnel* (pp. 70–94). Lewiston NY: Edwin Mellen Press.

Research and Technology Organization. (2000). *Officer Selection* (Meeting Proceedings No. 55). Neuilly-Sur-Seine, Cedex, France: Research and Technology Organization, North Atlantic Treaty Organization.

Rostker, B. (2006). *I Want You: The Evolution of the All-Volunteer Force*. Santa Monica, CA: The Rand Corporation.

Rumsey, M. G. (2012). Military selection and classification in the United States. In J. H. Laurence & M. D. Matthews (Eds). *The Oxford Handbook of Military Psychology* (pp. 129–147). New York: Oxford University Press.

Rumsey, M. G. (in press). Introduction to the special issue on noncognitive testing. *Military Psychology*.

Rumsey, M. G., Peterson, N. G., Oppler, S. H., & Campbell, J. P. (1996). What's happened since Project A: The future career force. *Journal of the Washington Academy of Sciences, 84*, 94–110.

Russell, T. L., & Peterson, N. G. (2001). The experimental battery: Basic attribute scores for predicting performance in a population of jobs. In J. P. Campbell & D. J. Knapp (Eds.), *Exploring the Limits in Personnel Selection and Classification* (pp. 269–306). Mahwah, NJ: Erlbaum.

Sands, W. A., & Waters, B. K. (1997). Introduction to ASVAB and CAT (pp. 3–9). In W. A. Sands, B. K. Waters, & J. R. McBride (Eds.), *Computerized Adaptive Testing: From Inquiry to Operation* (pp. 11–20). Washington, DC: American Psychological Association.

Seeley, L. C., Fischl, M. A., & Hicks, J. M. (1978). *Development of the Armed Services Vocational Aptitude Battery (ASVAB) Forms 2 and 3* (Tech. Paper No. 289). Alexandria, VA: U. S. Army Research Institute for the Behavioral and Social Sciences.

Sellman, W. S., & Arabian, J. M. (1997). Foreword (pp. xv–xvii). In W. A. Sands, B. K. Waters, & J. R. McBride (Eds.), *Computerized Adaptive Testing: From Inquiry to Operation* (pp. 11–20). Washington, DC: American Psychological Association.

Sellman, W. S., Russell, T. L., & Strickland, W. J. (2017). Selection and classification in the U. S. military. In J. L. Farr & N. T. Tippins (Eds), *Handbook of Employee Selection* (pp. 697–721). NY: Taylor and Francis.

Staff, Personnel Research Section, Classification and Replacement Branch, the Adjutant General's Office. (1945). The Army General Classification Test. *Psychological Bulletin, 42*, 760–768.

Stark, S., Chernyshenko, O. S., Drasgow, F., Nye, C. D., White, L. A., Heffner, T., & Farmer, W. L. (2014). From ABLE to TAPAS: A new generation of personality tests to support military selection and classification decisions. *Military Psychology, 26*(3), 153–164.

Thomas, P. J. (1970). *A comparison between the Armed Services Vocational Aptitude Battery and the Navy Basic Test Battery in predicting Navy School performance* (Tech. Bulletin No. STB 70-4). San Diego: Navy Personnel and Training Research Laboratory.

Trent, J. D., Barron, L. G., Rose, M. R., & Carreta, T. R. (in press). Tailored Adaptive Personality Assessment System (TAPAS) as an indicator for counterproductive work behavior: Comparing validity in applicant, honest, and directed faking conditions. *Military Psychology*.

Vineberg, R., & Taylor, E. N. (1972). *Performance in Four Army Jobs by Men at Different Aptitude (AFQT) Levels: 3. The Relationship of AFQT and Job Experience to Job Performance*. Alexandria, VA: Human Resources Research Organization.

Watson, S. E. (in press). Job Opportunities in the Navy (JOIN). *Military Psychology*.

Weeks, J. L., Mullins, C. J., & Vitola, B. M. (1975). *Airman Classification Batteries from 1948 to 1975: A Review and Evaluation*. (Tech. Paper No. 75–78). Lackland Air Force Base, TX: Air Force Human Resources Laboratory.

White, L. A., Rumsey, M. G., Mullins, H. M., Nye, C. D., & LaPort, K. A. (2014). Toward a new attrition screening paradigm: Latest Army advances. *Military Psychology, 26*(3), 138–152.

White, L. A., & Young, M. C. (1998). *Development and validation of the Assessment of Individual Motivation*. Paper presented at the meeting of the American Psychological Association, San Francisco.

White, L. A., Young, M. C., Hunter, A. E., & Rumsey, M. G. (2008). Lessons learned in transitioning personality measures from research to operational settings. *Industrial and Organizational Psychology*, 1, 291–295.

Wiskoff, M. F. (1997). R&D laboratory management perspective. In W. A. Sands, B. K. Waters, & J. R. McBride (Eds.), *Computerized Adaptive Testing: From Inquiry to Operation* (pp. 11–20). Washington, DC: American Psychological Association.

Wolters, H., Heffner, T., & Sams, M. (2015, October). *Overview and Introduction of ARI's Non-Cognitive Selection and Assignment Research: Enlisted Personnel*. Paper presented at the meeting of the Manpower Accession Policy Working Group, Seaside, CA.

Yoakum, C. S., & Yerkes, M. (1920). *Army Mental Tests*. New York: Holt.

Zeidner, J., Harper, B. P., & Karcher, E. K. (1956). *Reconstitution of the Aptitude Areas* (PRB Tech. Research Rep. No. 1095). Washington, DC: Adjutant General's Office (Army).

Zeidner, J., & Johnson, C. D. (1994). Is personnel classification a concept whose time has passed? In M. G. Rumsey, C. B. Walker, & J. H. Harris, (Eds.), *Selection and Classification Research: New Directions* (pp. 377–410). Hillsdale, NJ: Erlbaum.

# 4
# POLEMOLOGY
## Orphan of military psychology

*Jacques J. Gouws*

Historians may one day look back upon the twentieth century with equal importance as the Italian Renaissance because of the massive paradigm shift in how humankind perceives the world and its place therein. The Renaissance, a reawakening to the science, art, and culture of the time, laid the foundation for the Industrial Revolution of the 1700s and 1800s and related economic paradigms. Technology enhanced military armaments, the application of which brought about another paradigm shift. The doctrine of warfare changed rapidly as the wars of the late 1800s and early 1900s benefited from the application of these new technologies to create ever more destructive weapons of war. These new weapon systems resulted in the slaughter campaigns of World War I – the war that gave birth to modern military psychology. Finally, these new technologies also brought about the ultimate weapon of self-destruction, the nuclear bomb. Artificial intelligence is yet another new technology that will bring about another paradigm shift in warfare.

In its infancy, military psychology's first role was the selection and placement of soldiers based on cognitive tests. As the discipline matured, this role expanded with the aim to create soldiers whose skills sets surpassed those of their predecessors in dealing with the ever-increasing complex demands of the modern battlefield. The Cold War was an arms race that ravaged whole economies, yet brought lucrative contracts to arms dealers. The arms race also negatively impacted small nations involved in regional conflicts where new weapon systems were field tested by opposing forces. Throughout these conflicts, military psychology and its sub-disciplines worked hand in hand with national military forces to enhance the capabilities of the modern fighting soldier psychologically, physically, and technologically.

Zinchenko, Veraksa, and Leonov (2011, p. 53) quoted the definition of military psychology in Mangelsdorff and Gall (1991): *Military psychology can be defined as an area of the study and application of psychological principles and methods to the military environment*, and then expanded this by referencing Cronin (2003): *Cronin compares military psychology with a microcosm which embraces psychological disciplines and which affects almost all aspects of military setting.* These authors also compared the classical to the post-classical type of rationality in the development of military psychology, with the former being physiological research in the military sphere. This was most significant with regard to the latter:

> Non-classical psychological science introduced new criteria of cognition: necessity to consider the method used to obtain data, the adoption of relativism in the reference of existing scientific paradigms. It stresses the dependence of the resulting knowledge

> from the social context and number of other factors that regulate the organization of complex self-regulating systems. According to V.S. Stepin "if within the mechanical picture of the world the things (bodies) served as something primary, e.g. a substrate, processes would be interpreted as the effect of one thing to another by transfer of power that in a new system of representations of everything appeared as procedural system, self-reproducing as a result of its interaction with the environment and through self-regulation." (Zinchenko et al., 2011, pp. 54–55)

This is a vital perspective in light of the radical world changes since the Berlin Wall fell in 1989, ending the Cold War and creating the vacuum subsequently filled by a Jihad War and a War on Terror. The twenty-first century Renaissance plays out in a world barely conscious of the consequences of its all-new social media structures utilising technology to communicate and interact at levels beyond government control and even general comprehension. What was thought impossible a mere decade ago is now fingertip-accessible in real time. One example: the rise of additive manufacturing (using 3D printers) allows anyone to create any solid object in the confines of a living room or bedroom from a digital model. It is not unlikely that future wars could be fought against governments and countries by local civilians and/or revolutionaries using such technologies to arm themselves against statutory forces, aided by the bottomless fount of real-time social media intelligence.

Even if this scenario never fully materialises, conventional combat deployments with soldiers on the ground will become obsolete and be replaced by robots, either controlled by humans or acting independently through artificial intelligence. The latter option is a much more likely scenario given, for example, the documented psychological impact on personnel who work as "drone pilots" operating unmanned aerial vehicles (UAVs) or remotely piloted aircraft (RPAs) in the United States Air Force. Current thinking still supports human control in these weapon systems designed to neutralise enemy forces and reduce casualties among land forces. However, long-distance remote warfare will not be sufficient to attack enemy targets that are primarily civilian-based, non-statutory militia.

The wave of research papers and books on strategies to counter the "asymmetrical" or "hybrid" warfare phenomenon is testimony to the failure of conventional approaches to war that had served the military well until the end of the Cold War. Remnants of this remain the action of choice despite the face of warfare changing so dramatically that military might is no longer a guarantee of success. At the time of writing of this chapter, the United States and its allies have been in Afghanistan and Iraq for 17 and 15 years, respectively, with no end in sight because these mighty modern armies are held at bay by small numbers of militia fighters with improvised devices and even outdated weapons.

Military psychologists grapple with the impact of long, multiple deployments on soldiers, looking for ways to enhance resilience, foster hardiness, and reduce stress reactions in these theatres of operation where the "Arab Spring" brought about radical change in decades-old dictatorships, replacing them with little-known leaders and political office incumbents. Turmoil had spread throughout the Middle East and North Africa, and a self-proclaimed Islamic Caliphate[1] had quickly conquered vast territory. This latter conquest only ended (but is far from over) when the Syrian regime's resistance and brutal counterattacks against opposition fighters (albeit with Russian and Iranian backing) turned the tables on their civil war.

These examples illustrate the necessity for military psychology to urgently direct its focus to the psychological factors at play in the world's current conflict areas. Regardless of the back-and-forth

1 See, for example, Adam Chandler's article: "What is an Islamic Caliphate and why did ISIS make one?" In: *The Atlantic*, June 30, 2014.

propaganda from all sides involved in these conflicts, there are psychological motives that play a major role in perpetuating these conflicts which require thorough analysis so as to prevent even greater a catastrophe arising from ongoing ill-informed international military intervention decisions. Polemology, when added to the traditional military psychology research paradigm, could be a primary tool to assist military leadership and government decision-makers in countering the psychological forces (with their potential to destroy the world) that are feeding current conflicts.

## What is polemology?

In the experience of this chapter's author, there is a paucity of even rudimentary knowledge of polemology in the traditional Western military alliances (USA, NATO). This disinterest in the traditional West to develop a sound understanding of the science of war remains puzzling to this author, who had taught polemology based on its European roots as an integral part of academic military psychology training in early 1990s South Africa.

Although polemology is generally ignored in the West, it is of interest in the traditional East. For example, Dodonov, Kovalskyi, Dodonova, and Kolinko (2017) described in the abstract to their monograph that:

> Polemology deals with the wars and armed conflicts of the past, present and future. Novel hybrid wars take a respective place among them. They involve using all available warfare, regular and irregular, cyber and those allowing for the use of weapons of mass destruction, and also information, psychological and propaganda war using the latest information and media technologies...
>
> For the purpose of studying wars and armed conflicts from the polemological perspective it means the need to focus on social changes in all the areas of human life, on considering various elements of the political, economic or even technological context, which influence the war as a social phenomenon. (p. 98)

Ironically, across the centuries an integral part of the "Art of War" would today be considered "psychological operations." As a science, however, psychology only formally became useful to the military when it devised scientific tests to assist in the placement of troops in modern war settings (Johnson 1991, p. xxi). With its focus on particular subfields, military psychology remained a fragmented service simply because it neglected the big picture and so lost valuable opportunities that could have guided political decision-makers and military planners (Gouws, 2014). Tragically, lessons not learned cannot be applied to prevent future mistakes. As Napoleon Bonaparte (2018) reportedly stated:

> There are but two powers in the world: the sword and the mind. In the long run the sword is always beaten by the mind.[2]

2 Napoleon originally said this somewhat differently in 1808. It is most quoted in the form indicated in the text in military training settings and online web pages. According to Herold (1955, p. 76), Napoleon had said to his education minister: "do you know Fontanes, what astonishes me most in this world? The inability of force to create anything. In the long run, the sword is always beaten by the spirit." Borowitz (1985, p. 63) quoted it in the more common form: There are only two powers in the world, the sword and the mind.... In the long-run the sword is always defeated by the mind. The original quote in French is reportedly: Il n'y a que deux puissances dans le monde, le sabre et l'esprit. J'entends par l'esprit les institutions civiles et religieuses.... A la longue, le sabre est toujours battu par l'esprit. (*Revue des deux mondes, Vol. 4* (1838). p. 645): See: https://en.wikiquote.org/wiki/Talk:Napoleon_I_of_France.

Napoleon's wise words still apply, perhaps more than ever before. Gouws (1990), in his work integrating polemology and military psychology in the 1990s, reviewed various sources on the causes of war and its prevention, including Watson (1978, p. 1), who had opined that psychology in war is seen first and foremost to mean "propaganda." Gouws (1990) emphasised that although military psychology progressed to include selection and training, design of military hardware, and a myriad of other activities, these took a back seat in comparison to other scientists. Gouws (2014) still asserts today that, while the contribution of psychology to the military in its war efforts is considerable, because of its fragmented nature it lacks an overall integrated scientific structure. Even more significant is the absence of a structured and systemised study of the war phenomenon itself which should be applied towards the reduction, if not resolution, of potential conflict even before military action is embarked upon.

A foundation for the structured and systemised study of the war phenomenon was created in 1945. Most military psychologists in the Western Alliances are unaware of its existence. Polemology, a term coined by Gaston Bouthoul after the end of World War II, is the study of war and its precursors and consequences; given the ability that humankind had developed to wipe out itself and the earth with nuclear weapons. Bouthoul, a sociologist, initiated the systematised study of the phenomena of aggression and wars in the context of strategy, tactics, economics, politics, ideology, and cultural contexts. Bouthoul postulated that by a better understanding of conflicts and wars, one can find positive ways to prevent them. These lofty and worthy aspirations died with Bouthoul in 1980. Fortunately, a revival in the interest in this work came about in the 2012 landmark publication, *The Ashgate Research Companion to War*, edited by Hall Gardner and Oleg Kobtzeff (both at The American University of Paris, France). Ironically, this work has not piqued the interest of Western Alliance military psychologists, perhaps for the same reasons the work of Bouthoul was ignored:

> By the time postmodernism became very influential in France as well as in the world, around the 1980s and 1990s, marking the decline of the Annales School (which had itself reached the peak of its influence in the French intelligentsia and public in the 1970s), war was still not a subject that the New Historians had been able to interpret. The original prejudice against war studies in particular and political science in general still prevailed after three generations.
>
> This probably explains the isolation of French sociologist Gaston Bouthoul (1896–1980) who devoted his career to fusing the knowledge and methodologies of all social sciences into a new interdisciplinary field of research and reflection for which he invented a new word: polemology… The word "polemology" remains either almost unheard of or is mistakenly associated with the kind of practical academic programs delivered by military institutions. The tens of millions of casualties made it difficult to accept that war can be quietly studied as an object of science, an "ordinary" social phenomenon… Also, when Bouthoul and his team of researchers founded, in 1945, a modest research center named the French Institute of Polemology (Institut Français de Polémologie) they were met with suspicion which was typical of the late forties to mid-to-late 1960s. On one hand, the now dominating Annales school suspected too much political comment over war; on the other, an increasingly radical leftwing intelligentsia rejected any activity in the arts or social sciences that could not justify itself as politically and socially useful for the Revolution (peace studies, like hippies, risked demobilizing the working masses). The fact that Bouthoul's institute cooperated with the French military academic system made him even more suspicious. Nevertheless, Bouthoul, a socialist who loathed war, strongly argued that his scientific activities

> were pacifist in nature. Although he conceived polemology as a discipline based on scientific methods, academic objectivity was not meant to abolish the scientist' ability to function as a thinking and responsible citizen striving for peace. Fortunately, other scholars initiated comparable efforts in other parts of the world. At a time when nuclear holocaust appeared to be a real possibility, it became evident that conflict prevention or resolution was hampered by the dramatic lack of research structures that could offer the necessary broad array of conflict analysis tools. The urgency of the need for more research on war and peace became evident after such dramatic moments as the Cuban missile crisis, or under the growing pressure from peace movements in the 1960s. The emerging disciplines or sub-disciplines of sociology or political science became better known as "Peace Studies" with the creation of (such) groups and institutions... (Gardner & Kobtzeff, 2012, pp. 2–3)

Unfortunately it seems as if, as indicated by the above quotes, the general reluctance to scientifically study war and to use knowledge gained from such research to effect change in a conflict-ravaged world remains unchanged. It is unfortunate that military psychology had remained disinterested in this important endeavour for so long.

## Bouthoul on polemology

The work of Bouthoul is monumental in its scope and application, and it is impossible to summarise it in brief. However, even a cursory review of his work provides impetus for different thinking and interpretation of the phenomenon of war and the role military psychology can play in bringing about a reduction in war conflict through analysis of the underlying drives and motives that fuel the bellicose spirit of nations around the world.

Probably the most significant application of the work of Bouthoul was offered by Franco Fornari (1921–1985), an Italian psychiatrist and psychoanalyst. He became interested in polemology in the 1960s. He subsequently founded the first "anti-H group" and established the Italian Institute of Polemology (ISTIP). Fornari also published an excellent book in1966 titled *Psicoanalisidella Guerra,* followed by an English translation in 1974: *The Psychoanalysis of War.*

Although psychoanalytic in nature, this early work in polemology provides important insights that may be applied to the processes by which decision-makers make determinations on whether to go to war. Fornari, relying on the work of Bouthoul and also Durkheim, explains the origins of much of the military action and subsequent rituals still taking place today. One can even illustrate the applicability of Fornari's thesis to the United States' military response to the attack on the Twin Towers on September 11, 2001:

> The splitting of the world into friend and enemy represents an extreme simplification whereby the good and the bad are no longer integrated in the same instinctual situation and in the same object relationship, but the same situation assumes different characters according as it is consummated upon the self, or upon the other, in the paranoid aphorism of 'morstua vita mea.[3]' The enormous weight of human ambivalence is suddenly lifted when love and hate find two separate objects of attention. (Fornari, 1974, p. 23)

Hence, war is a method to bring about a radical transformation of values, to make people think, feel, and act differently from before, to make them distinguish between "good" and "bad."

3 Latin, literally translated: "your death is my life" or more commonly expressed in English: "your loss is my gain."

However, to bring about this transformation, the initiator's powerbase must first be coaxed into action and then become unified in its aggression against the opposition. This is achieved through mass behaviour taking place in a specific context, analogous to a feast:

> In Bouthoul's opinion, war has all the characteristics of a feast, whose principal function, according to Durkheim, is to unify the group. The most typical psychological aspects of a feast, in the sociological sense of the word, are the following: (1) It brings about a meeting of the members of a group; (2) it is a rite of expenditure and dissipation; (3) it is accompanied by a modification of certain moral laws; (4) it is a rite of collective exaltation; (5) it brings about a state of physical insensibility; (6) it is accompanied by sacrificial rites. As every one of these characteristics is found in war, war might be regarded as the **supreme feast**. (Fornari, 1974, p. 23)

There is thus a commonality between the characteristics of a feast, as described above, and many public ritualistic activities that still take place in war. Fornari made the connection between primitive peoples' war preparations, including dances, that appear preserved in military parades in advanced societies (Fornari, 1974, pp. 23–24). Furthermore, the manner in which the rules of war transcend civilian law parallels the modification of morals at a feast in that the tunnel vision and lack of feeling and sensation often found in battle is likewise a state of physical insensibility. Finally, in the aftermath of war, it is the ritual ceremonies that recall the sacrifice to memory:

> Among primitive peoples, in fact, war is intimately joined to the idea of a human sacrifice that is pleasing to the gods, i.e., to the cult of death… Bouthoul notes that in western civilization, concurrently with the decline of religions, the cult of the fallen in war has been revived in the form of monuments and parks. He interprets this situation as the return of the archaic customs in our civilization. The cult of the fallen in war therefore tends to replace the cult of the saints. (Fornari, 1974, p. 24)

## Towards an integration of military psychology and polemology

War and its destructive aims are in stark contrast to humankind's achievements of conquering land, sea and air, and, after landing on the moon, targeting the stars and planets. These are in addition to the many efforts aimed at conquering disease, poverty, slavery, subjugation, and even class and social differences. Underlying these achievements is a belief that conquering these inequities will create healthy, affluent societies, where people would be free to express their previews and premises absolved from prejudice and intimidation. The notion for some is that once this state of affairs is achieved, there would be lasting peace and security for all.

Yet, almost two decades into the twenty-first century, the world again finds itself as close to another world war as a century ago and with as clear a threat of nuclear war as during the Cold War.[4] It seems as if humankind, nations, politicians, and military leaders are completely unable to break free from the violence and destruction imposed by the powerful on the weak, the rich on the poor, the aggressive on the passive, and the devious on the ignorant. They seem intent

4 The Doomsday Clock (the imaginary worldwide nuclear war apocalypse) was set at two minutes to midnight on January 25, 2018, this for only the second time since it was created in 1947. The first time was in 1953 based on the then-U.S. and USSR thermonuclear arsenals, the use of which in that tense period could completely destroy the planet. There was no change in this setting on January 24, 2019.

on bringing down their war temple upon themselves, and just like Samson[5] of old, they would surely die, as would the whole of humankind with them.

In the past, psychology's attempts to explain war and its role in human endeavour focussed on reasons for the continuous eruption of wars whether on grand scale or small, often to little avail. This is why it is imperative that military psychology must include polemology as a core research topic and application in the psychological study of war and its causes in order to promote conflict resolution and prevention of unnecessary war operations. However, this is not the same as pacifism. Pacifist literature places the blame for war squarely on the shoulders of the soldiers themselves with arguments that the commanders of forces, i.e. the generals, are the warmongers, the ones who really give an essence, a right to existence, for armies, and thus war. They blame "old men who sent the sons of others to their untimely ends" and calls for the dissolving of all armed forces. The pacifist argument is that if the military dissolves, then war and violence would disappear, because it is soldiers who foster and enhance the spirit of conflict. Likewise, if societies were to disarm, violence would disappear. If only it was this easy to end to violence and war!

Polemology realised that the reasons for war lie within deeper human drives and needs, which can explain the war phenomenon. There is a simple reality about the complexity of war, pointed out by Boatner (1956, p. 3): "soldiering is not only the third oldest profession, but since the dawn of mankind the soldier himself is the only element that has varied least through the ages." He demonstrated this from a quote of Ardent du Picq:

> Centuries have not changed human nature. Passions, instincts - particularly the most powerful instinct of self-preservation - these things are expressed differently in accordance with the times, the circumstances, the character and temperament of races.... But, beneath it all, the same man is to be found. (Boatner, 1956, p. 3)

This must be an important focus in both psychology and polemology: as the above quotation indicates, humans have something within them, call it the "psyche" for convenience's sake, which will always drive them towards some form or another of conflict or war in their ceaseless quest for peace. This contradiction between a desire for both war and peace is important to both understand and resolve to prevent humans from destroying the world with their modern weapons of mass destruction. While war is the "legalised" method by which conflict may be resolved, there are better choices. What prevents these better choices from being implemented is that humans, despite striving towards the achievement of lasting peace, are prohibited by their nature of perceiving others as threats to their continued existence of implementing this ideal. Even as they attempt to secure peace through mutually agreed-upon treaties, individual human groups continue to prepare for war to deter the perceived enemy. To deter also means to pre-empt and to engage in war, as the defensive measures taken by one will always be interpreted by the other as a build-up of the threat to the other. Thus, a subjective threat

5 This refers to the ancient story of Samson, a renowned Israelite warrior, who lost his immense strength when Delilah, a Philistine woman with whom he had a relationship, cut his hair (the source of his strength) while he was asleep. He was taken prisoner and enslaved. The Philistines, celebrating their victory over Samson, brought him to the Temple of Dagon, placing him between the temple's supporting pillars so that he could amuse them. While in captivity his hair had grown back and thus, in a final act of revenge, he used his immense strength to pull down these pillars, causing the temple's collapse and killing all inside, including Samson. For the full story of Samson, see: Judges 13–16, New King James Version Bible (NKJV).

perception eventually culminates in the rationale for war with all the violence and destruction associated with it.

Over the years, psychologists have studied various phenomena in depth that resulted in violent outcomes. Some such controversial research studies tested the limits beyond which ordinary individuals would not follow instructions, with the prime example of the Milgram experiments[6] on obedience to authority in the early 1960s (Milgram, 1974). Zimbardo's 1971 Stanford prison experiment[7] less than a decade later actually materialised in the real world in the Abu Ghraib jail scandal[8] three decades later. While politicians and the public expressed their shock and dismay, there should have been no lack of understanding of the role of psychological factors in bringing about these very "uncharacteristic" behaviours by otherwise "normal" soldiers. Yet it seems that examples like these serve to deter the psychological study of the all-too-human failures to live up to modern civilised codes, this especially in the military. There is clearly a contrast between that which is the ideal that humans strive for and their actual behaviours when they go to war, regardless of how they may act in civilian settings. Yet, the role of the psyche is paramount in determining the course of conflict resolution:

> It is now quite universally recognised that the superficial and naive statement that war maintains the balance of nature in humans in the same way that predators do in the animal world is a myth, since the present studies show that the expression of naked aggression, manifested through the destruction of other human beings, is largely led and initiated by psychologically abnormal individuals.
>
> One can say with certainty that the time-honoured human habit, unfortunately not rare, of placing nations under the control of psychologically abnormal, aggressively perverted persons, together with the modern factor of the ever-increasing new invention of massive weapons of destruction, is certain to be suicidal for the human race unless interrupted. This tendency of societies to select psychiatrically and pathologically charismatic abnormal individuals, of the nature of aggressive perverts, to control nations and to lead them into war indicates that intra-specific aggression in the human race has gotten out of hand and has become a serious threat to the survival of the species. This can only mean that natural selection in the human is being interfered with, and that unconscious forces, basically responsible for the existence of aggressive perversions as well as for the uncanny ability of such pathologically charismatic individuals to seduce normal human beings into selecting them as leaders and following them to death, can effectively upset the balance of human nature and can brutalize and destroy human beings en masse with enormous damage to the painfully attained process of civilisation. (Walsh, 1971, p. 6)

Walsh's statement in the above quotation has important implications for understanding the reasons for war as being based on both the individual and the collective psyches of humans when making war decisions. An understanding of the human drives, fears, and anxieties that influence especially leaders' perceptions of given situations will also explain much about their decisions on war engagement and disengagement.

6 Described by Stanley Milgram in his book in 1974.

7 Described by Zimbardo et al. in the September 1973, Naval Research Reviews.

8 See: http://www.udel.edu/PR/UDaily/2006/dec/zimbardo120705.html.

If anything, it is the engagement in war, rather than war itself, that feeds the fanatic zeal of the proponents of war. As long as war can be engaged in, a reason will be found to wage such war. However, when eventually these fanatic war urges have been released and to some extent expended, it becomes possible to engage in peace or truce negotiations. That also requires, once war engagement comes to an end, that a new state of mind be created to make the incomprehensible war actions comprehensible and to justify the sacrifices made during the war. The "victorious fallen" are remembered for their "heroic struggle," while the "vanquished" are brought to "justice and repentance for their wrongs." This process takes place as a very real and visible phenomenon:

> Describing the psychological transformation in the aftermath of war, Bouthoul states that the most important fact observable during the period immediately following the establishment of peace is a rapid decline of the bellicose spirit. Having insisted that their rivalry or disagreements were intolerable, nations suddenly find that they can come to terms with each other. The aggressiveness subsides, as if war were a sort of orgasm followed by relaxation. Mental states and actions, clearly understood by everyone only a short time before, suddenly become incomprehensible. Bouthoul compares the euphoria of peace to the silent, secret joy that prevails, when someone dies, among his prospective heirs, and to the relaxation of tension experienced by students after examinations. The vanquished people often find a sort of satisfaction in the thought that war has freed them from their errors and show a distinct tendency to attend the school of the victor. The vanquished people's institutions are despised and their defeat rationalized as punishment for their crimes. In this manner defeat is a source of repudiations. The vanquished people deserved to be defeated on account of their errors: the vanquished are guilty. While sacrifice participates in war as a propitiatory procedure for future misdeeds, the expiatory attitudes following war are reparation for past misdeeds. The sacrificing of scapegoats seems to be a part of the rites of atonement. (Fornari, 1974, pp. 28–29)

It is indeed a strange trait of humankind that people find it impossible on both the individual and collective levels to accept responsibility for their deeds and actions. Following individual aggression, the perpetrator blames someone else or circumstances, but refuses to accept responsibility for his or her own flaws. Although such individual transgressions may have consequences brought to bear by an offended society, the same does not apply at the collective level. This is because society, through its leaders, will rationalise the evil perpetuated or the threat posed by the opposition, placing the blame for the resultant conflict on the inherent unjustness of a system which had to be opposed, ignoring the fact that:

> War is something that people do, not something that happens: activity with a military dimension is activity, not mere blind process and event. (Paskins & Dockrill 1979, p. 210)

The rationalisation that the aggression of the foe elicited the intervention by own forces softens the sacrifice of lives and material, wastefulness which otherwise cannot be admitted to. The ritual of entering into war and then remembering the fallen afterwards justifies the "counter-aggression" which "saved the nation from destruction by the evil forces of the foe." It is in this context that a distinction should be made between individual and collective outbursts of aggression, as these two activities are markedly different:

> One of the essential differences between individual aggressiveness and the warlike impulse, according to Bouthoul, is that while individual aggressiveness is momentary, transient, felt specifically as such and usually limited to one individual, the warlike impulse is a generalised, profound emotional state of acceptance and approval of future violence rather than a manifestation of violence itself. The warlike state corresponds, that is, to a sense of the need for a period of violence and destruction rather than to genuine aggressive excitation. Before it is an action, it is a conviction; sometimes it is merely a resignation to a calamity considered inevitable. The apparent absence of aggressive excitation in war as a festival destructive of humanity is certainly a peculiar problem… (Fornari, 1974, p. 30)

When war is seen as being more of an emotional state of acceptance and approval of future violence, rather than a manifestation of violence itself, it is also brings acceptance that "some might die so that others can live." This makes the ritual of deciding on going to war vitally important on two levels. The first level arises from the emotional reaction and how that is rationalised to motivate others to engage in war. The level provides the justification for entering into war that entails the sacrifice of one's own who died or were the wounded and maimed victims of war, but who also were all loved ones, the love objects that had formed part of one's own clan. It is an incomprehensible paradox that, in order to save and preserve something, that same something, or at least a part thereof, must be sacrificed and destroyed:

> But an even more surprising fact is that war as sacrificial duty, though performing essentially destructive functions, has for men the significance of a destruction put into the service of the preservation of what they love. Fanatic behavior, the idealization of the leader, the need to sacrifice oneself in the name of an ideal, the giving and receiving of martyrdom, and the soldier's being both the sacrificer and the victim in the sacrificial rite… – all these psychological aspects pose a series of very complex problems. (Fornari, 1974, p. 32)
>
> The psychological aspects of war, namely the crises of the instinct of self-preservation and the idealisation of the need for sacrifice, as well as the idealisation of the leader, all seem to be phenomena that occur on the basis of the fact that individuals form a group on the basis of identification with a common love object. Since the group ideal (as an object of love and identification) is fantasised as that which gives life to the individuals within the group, the preservation of the common love object is felt as a primary function as compared with the preservation of the individual. A situation of this sort is specifically human and justifies in a manner the need for sacrifice. Nevertheless, I believe that I have succeeded in showing that the group ideal, translating into an illusory dimension the concrete and vital function of the primary love object – in a way usurping it – causes the evolution of the original need for sacrifice (expressed by the primitive need for guilt) into an illusory and unauthentic dimension.
>
> In war, in fact, sacrifice is not aimed at mastering the bad parts of the self, as a function of preservation of the original love object threatened by the self, as happens in the ethics of the individual. In war, sacrifice becomes sadistic through the projection onto the enemy of the destructive tendencies originally directed against one's love object. (Fornari, 1974, p. 36)

War, then, is not waged about right and wrong, good or evil, nor is it about just and unjust; it is about the externalising of the inner aggression of self felt towards one's love objects onto those with whom one ironically should have no quarrel. War, in essence, is therefore the displacement of the expression of vengeance and hate, found deep inside the self, projected onto others who do not form part of the own love objects, making third parties the scapegoats and therefore "legitimate targets" of the unacknowledged, inner aggression towards the love objects. In other words: war becomes the ritual by which humans atone for the real guilt they bear for hating their own and even themselves. A better description is difficult to find than stated by Albert Camus (1963):

> We used to wonder where war lived, what it was that made it so vile. And now we realize that we know where it lives, that it is inside ourselves. (September 7, 1939)

Thus war and the psychological impact thereof must be seen for what it is: a fierce, unrelenting paranoid struggle of humans within themselves. There are many remnants of primitive, unconscious beliefs and accompanying rituals that even today play a vital part in forming people's attitudes towards war and the ceremonies that accompany these actions. Present-day weapons of war scare the hell out humankind, bringing into play even more powerful, illusory fears. These have become a vicious cycle that encroaches more and more upon humans to the point where reality and illusion are fused into one indiscernible entity of Absolute Terror. Humans created it; humans must destroy it or be destroyed by their own creation. There is no question that military psychology must be cognisant of the role it plays in the perpetuation of conflict and war through its involvement in creating and maintaining the bellicose spirit.

It is true that there are very real threats posed by nations to each other. To allow these threats, whether real or imagined, to escalate into war when other solutions could be found is unacceptable in modern times. The answer to the prevention of this destruction is found in knowledge: knowing about humankind's deepest, unconscious fears and drives. Humankind cannot afford to allow the superweapons of today to destroy all humankind and the world because of the uncontrollable urges of leaders, both military and political, who, for some undefined fanatical cause, will lead their people and the world into an all-destructive war. The soldier is the only one that knows the real horror of war; the soldier is also the only one that can remedy this horror. The remedy is found in knowledge and understanding of the real, often unconscious issues that direct the actions of many countries and their leaders. The world is ready to address the issue of international and national conflicts by the political diplomacy taking place. However, as military psychology maintains its focus on the psychological factors related to war and operations it, as easily as everyone else, succumbs to its own individual and collective fears and biases. However, by military psychology adding a focus on the real causes of war, it could provide the data and the critical thinking required for the decision-making leadership to objectively analyse the military threats and propose measures to counter that threat appropriately, without necessarily engaging in war. Given its unique ability to comprehend humans' deeper motives and fears, military psychology has a major role to play in the eventual taming of the bellicose spirit by applying some of the principles developed in polemology:

> The polemological approach argues that there is rarely a mono-causal factor for conflict and war, but interrelated causes, of which some may be more significant or relevant than others depending upon the confluence of forces in specific situations and historical periods. (Gardner & Kobtzeff, 2012, p. 3)

## Conclusion

In a rapidly changing world, it is imperative to recognise that the changes of the past three decades brought a major shift in how people see the world. The most important tools in a government's hands are its military. However, the military and its members are as much subject to human frailties as any other societal structure.

In the new world with its new demands, military psychology in the twenty-first century should integrate polemology into its active study and application of lessons learned from the historical behaviours and actions of humans, governments, organisations, institutions, and corporations that ultimately resulted in armed conflict in all its forms, ranging from low-level skirmishes to full-scale war, and the active applications of these insights in order to prevent, limit, and end hostilities between opposing parties, in favour of bringing about negotiated resolution and tolerant co-existence of prevailing political, economic, religious, ideological, and psycho-social differences.

Furthermore, the role of military psychology should expand to support the principle of deterrence as a strategy to prevent armed conflict and facilitate resolution of differences. This should include not only psychological factors, training of soldiers, and support of operations but the education of politicians and military leaders on other options that should first be pursued to prevent unnecessary and irrational engagement in war.

War as the last choice of action can be achieved by military psychology researching solutions for the management of challenging political and military situations. All governments posture within the new Global Village on the old "us–them" paradigm. However, it is the failure of governments to adequately distinguish reality from their own propaganda that often results in the tragic and unnecessary waste of human life and destruction of property. Sometimes elements in the military may even be complicit in the war mongering that takes place. From a soldier's perspective, it cannot be said any better than this:

> What all of us know is that we are soldiers still. Some of us revisit the battlefield in nightmares. Some of us wear scars, visible and invisible, that mark us as changed men who walk unseen among our neighbours, who have never known what it is like to hold a dying boy in their arms and watch the life fade from his questioning eyes. The world may now know something of the events that changed us, but thankfully most are spared the experiences that are ours and the burden that is the province of men who have killed other men at the bidding of political leaders more concerned with personal pride and national honor than with peace.
>
> Yes, we were soldiers once, when we were young. Now that we are old we are soldiers still. We are soldiers who mourn for young men and women dying on other battlefields in other parts of our world four decades and more after our war ended so badly. A generation of political leaders who studiously avoided service in our generation's war seemingly learned nothing from that history and thus consign a new generation of soldiers to "preemptive" wars of choice, condemning them to carry their own memories of death and dying through their lives.
>
> May God bless and keep all soldiers, young and old, and may that same God open the eyes of all political leaders to the truth that most wars are a confession of failure – the failure of diplomacy and negotiation and common sense and, in most cases, leadership.
>
> We who still dream of war in our troubled nights hope against hope for peace and its blessings for all. (Moore & Galloway, 2008, pp. xix, xx)

Military psychology must work within the massive paradigm shifts that had followed the end of the Cold War. The world's nations again face military threats that could make Einstein's prophetic statement a reality:

> I know not with what weapons World War III will be fought, but World War IV will be fought with sticks and stones. (Calaprice, 2005, p. 173)

The conflict motives illustrated in this chapter require serious and urgent military psychological research to manage the challenging political and military situations in which the world finds itself. Starting with the psychological component of polemology as a research focus, military psychology has much to offer that may bring about strategies to not only manage and resolve conflict, but to limit war risks by providing the necessary perspective on which decision-makers have to base their decisions. If not, the world will destroy itself while military psychology continues to contribute to the conflicts, but not their resolution.

## Chapter end note

The author expresses his deep appreciation and gratitude to the many individuals, whether they were colleagues, peers, superiors, professors, teachers, students, or friends, who across a lifetime of service in military and civilian life have expressed their opinions, brought their criticisms to bear, and often through their vocal interaction in long discussions encouraged the author to express on paper the multitude of insights gained through stringent scientific research combined with his personal experience in various parts of the world regarding war, peace, and conflict resolution. Without these inputs, spanning close to five decades, this work would not have been possible. Any factual misunderstandings or misinterpretations, where such may be identified, are solely those of the author, who would gladly review them for correction in a future publication. The same would apply to this author's interpretations of quotations from the work of those "giants on whose shoulders we stand"[9] from a variety of disciplines who studied and wrote about the elusive solutions to prevent war and maintain peace in our world.

## References

Boatner III, M. M. (1956). *Military Customs and Traditions*. New York: David McKay Company.

Bonaparte, N. (2018). *Napoléon Bonaparte quotes*. Retrieved from http://www.goodreads.com/author/quotes/210910.Napoleon_Bonaparte

Borowitz H. O. (1985). *The Impact of Art on French Literature: From de Scudéry to Proust*. Newark: University of Delaware Press.

Calaprice, A. (2005). *The New Quotable Einstein*. Princeton, NJ: Princeton University Press.

Camus, A. (1963). *Carnets, 1935–1942*. London: H. Hamilton. (Translated from the French and with an introduction and notes by Philip Thody).

Cronin, C. C. (2003). *Military Psychology: An Introduction* (2nd ed.). Pearson Learning Solutions.

Dodonov, R., Kovalskyi, H., Dodonova, V., & Kolinko, M. (2017). Polemological Paradigm of Hybrid War Research. *Philosophy and Cosmology*. *19*, 98–109.

Doomsday, C. (2018). *It is now two minutes to midnight*. Retrieved from https://thebulletin.org/2018-doomsday-clock-statement/

Doomsday, C. (2019). *A new abnormal: It is still 2 minutes to midnight*. Retrieved from https://thebulletin.org/2019-doomsday-clock-statement/

9 Edge inscription on the 1997–2015 British £2.00 coin, attributed to Sir Isaac Newton, although some sources trace this back to the twelfth century to Bernard of Chartres.

Fornari, F. (1974). *The Psychoanalysis of War*. Canada: Fitzhenry & Whiteside. (Translation by Doubleday).
Gardner, H. & Kobtzeff, O. (2012). Introduction. In H. Gardner & O. Kobtzeff (Eds.), *The Ashgate Research Companion to War* (pp. 1–34). New York, NY: Ashgate Publishing.
Gouws, J. J. (1990). *Military Psychology Lecture Handout Notes*. University of Stellenbosch, Faculty of Military Science, Department of Industrial Psychology (Mil). 1990–1995.
Gouws, J. J. (2014, June). A model for the defragmentation of military psychology. *Paper presented at the International Applied Military Psychology Symposium*, Tallinn, Estonia.
Herold, C. J. (1955). *The Mind of Napoleon: A Selection from His Written and Spoken Words*. New York: Columbia University Press.
Johnson E. M. (1991). Foreword. In R. Gal & A. David Mangelsdorff (Eds.), *Handbook of Military Psychology*. London: Wiley.
Milgram, S. (1974). *Obedience to Authority: An Experimental View*. New York: Harper and Row Publishers, Inc.
Moore, H. G., & Galloway, J. L. (2008). *We are Soldiers Still. A Journey to the Battlefields of Vietnam*. New York: Harper Collins.
Paskins, B. & Dockrill, M. (1979). *The Ethics of War* (p. 210). London: Duckworth.
*Revue des deux mondes*, Vol. 4 (1838). As quoted on: https://en.wikiquote.org/wiki/Talk:Napoleon_I_of_France.
Walsh, M. N. (1971). *War and the Human Race* (p. 6). New York: Elsevier Publishing Company.
Watson, P. (1978). *War on the Mind: The Military Uses and Abuses of Psychology*. London: Hutchinson.
Zinchenko, Y. P., Veraksa, A. N., & Leonov, S. V. (2011). Methodological foundations of military psychology and psychological security. *Psychology in Russia: State of the Art*, *4*, 53–61.

# 5

# THE APPLICATION OF CULTURE AND COGNITION WITHIN A MILITARY CONTEXT

*Faizan Imtiaz, Mark Khei, and Li-Jun Ji*

Imagine a scene where a few soldiers are walking through a forest. When visualizing this image, do you focus your attention on the soldiers only, or do you also focus on the surrounding forest? If one soldier is walking ahead of the others, do you perceive him to be the leader, or is he one of the subordinates? In large part, the answers to these questions depend on one's cultural background.

Though culture has been defined in various ways (Chiu & Hong, 2006), most definitions emphasize culture as shared norms, customs, and meaning systems among people who inhabit a common language and life space at any given time. According to Hong and Khei (2014), culture is a special knowledge system that is (1) shared amongst interconnected individuals who are delineated by race, ethnicity, or nationality; (2) expressed in various symbols, artifacts, and social norms; (3) used to facilitate successful communication between in-group members; (4) transferred from one generation to the next; and (5) constantly being adapted to fit modern social orders.

This chapter examines how culture influences various psychosocial factors such as attention and memory, information processing and prediction, temporal focus, and goal regulation. Moreover, it aims to extend this knowledge base by examining its potential application in a military context.

## Cultural influences on attention and memory

Culture influences how people attend to their direct environment, as well as how they encode and memorize this information. One of the earliest psychological studies demonstrating this effect was conducted by Segall, Campbell, and Herskovits (1966), who examined how visual illusions such as the Muller-Lyer could be perceived differently across 15 unique countries. The Muller-Lyer illusion refers to the phenomenon in which people perceive line A to be longer than line B, even though the two lines are indeed identical in length. Such an illusion is caused by the different directions of the arrow heads on the ends of the vertical lines.

Segall and colleagues found that European individuals were more likely than individuals from the Zulu tribe in South Africa to be susceptible to the illusion. Such cultural differences, according to the researchers, may be due to the more modernized world that Europeans inhabit compared to the Zulu people. Specifically, contemporary urban landscapes regularly expose

Europeans to sharp corners in their eco-cultural environments, whereas the circular huts that the Zulu tribe live in are less likely to expose tribe members to lines and angles such as the ones in the Muller-Lyer illusion. This experiment serves as one of the first empirical findings highlighting the relationship between one's culture and attentional focus.

Culture also influences how much attention individuals pay to the context. Ji, Peng, and Nisbett (2000) examined cultural differences between East Asians (including Chinese, Japanese, and Koreans) and Euro-Americans in sensitivity to context. They tested participants with the rod-and-frame test (Witkin et al., 1954), in which a rod and a frame can be rotated independently.

In this study, participants' task was to make perceptual judgments about the position of the rod, irrespective of the position of the frame. Ji and her colleagues reported that East Asians, as compared to Euro-Americans, made more errors while judging the position of the rod, indicating that they were more distracted by the surrounding frame. These findings indicated that East Asians were spontaneously more attentive to their surroundings (i.e., the frame) compared to Euro-Americans and were therefore more field dependent in their perceptual judgments.

In 2003, Kitayama, Duffy, Kawamura, and Larsen replicated and extended these findings using a frame-line task. In this task, participants were presented with a square frame with a vertical line inside of it. Next, they were given another square of a different size and were asked to draw a vertical line inside of it. In the absolute task condition, they were told that the length of the line should be identical to the original vertical line (thus ignoring visual context or field independency). In the relative task condition, they were told that the line should have the same proportional relationship with the new square as the original line had with the original square, thus taking visual context into account or field dependency. The results showed that Americans performed better on the absolute task, whereas Japanese performed better on the relative task. This suggests that the Japanese were better at incorporating contextual information, whereas Americans were better at focusing on the target, detached from its context.

Culture influences not only what people pay attention to, but also what people remember. To this end, Masuda and Nisbett (2001) presented Japanese and American participants with animated underwater scenes, followed by a question period regarding what they could recollect about the scene. In each picture, a few focal fish (large with salient colors) were moving in front of a background. Results indicated that Japanese participants reported the background and the relationship between objects in the background significantly more than American participants. In a succeeding recognition task, participants were presented with either objects that had been previously included in the underwater scenes or novel objects that were absent from the scenes, each with the original or a new background. They were then asked to indicate whether they had seen any of these stimuli before. The researchers found that Japanese participants recognized the previously seen objects more accurately when the objects were shown with the original background than when the objects were presented with novel backgrounds. This suggests that Japanese participants attended to the objects *together* with their contextual environment as a whole, treating them as interdependent. Americans' recognition, on the other hand, was less affected by the paired background, presumably because they decontextualized the objects from their background (Yap, Ji, & Hong, 2018).

## Military application

A direct military application of the attentional differences across cultures involves what military personnel may pay attention to on the battlefield. Related to this, research in social psychology has demonstrated that although we often erroneously believe that we are paying attention to everything in our direct environment, human beings actually pay attention to a relatively small percentage of their visual field (Simons, 2010). This phenomenon, called inattentional blindness, usually stems from the simple fact that we are typically more motivated to pay attention to certain things, and less motivated to pay attention to others. This combined with the fact that we have finite cognitive resources that are easily depleted leads to us prioritizing certain stimuli at the cost of ignoring others. Indeed, people seldom experience the world through a truly neutral attentional focus, and there is usually some motive that is directing our attention, regardless of whether we are consciously aware of it.

A major implication of inattentional blindness is that we may end up missing a significant amount of information from our visual field. Linking this to culture, the differences in attentional focus covered in the previous section may not only lead to individuals from varying cultures focusing on distinct sets of stimuli, but perhaps more importantly missing unique information from the broader environment. These ubiquities may lead to individuals from varying cultures evaluating the exact same scene completely differently. In the military, these small differences may produce massive outcomes as high-stakes decisions regarding what one is witnessing (e.g., enemy or compatriot) and how to appropriately respond (shoot or refrain from shooting) are often being made with major time constraints and under extreme pressure.

The rationale above may also be extended to international conflicts if two units are focusing on completely different aspects of the same situation. For instance, individuals from cultures emphasizing context and background may focus on the incident at hand as well as the contextual circumstances that led to it. Meanwhile, people from cultures that emphasize the focal point of interest may want to focus solely on the issue at hand.

A real-life example of the scenario above occurred in 2001, when an American spy plane collided with a Chinese fighter jet, leading to the death of the Chinese pilot. After the event, the Chinese, who tend to focus on the context and background (Masuda & Nisbett, 2001), demanded an apology from the Americans not only because of the crash itself, but also due to other historical indiscretions (e.g., American-led NATO bombing of the Chinese Embassy in Belgrade in 1999) where the Americans had been perceived to be aggressors (Aberman, 2001). The Americans, who tend to focus more on central point of interest, discredited the importance of the previous events as they viewed them to be independent from the current incident, and refused to apologize in lieu of their assertion that their crew was not at fault in this specific case. This practical example highlights how small differences in attentional focus can have major ramifications in real-world settings. Future research should aim to tease apart these intricacies in applied contexts such as the military in order to advance our knowledge about how this variability influences our daily interactions, decisions, and relationships.

## Cultural impact on perception and prediction

Culture not only influences what information we attend to, but also how we process information (Ji, 2005). For example, East Asians tend to be more dialectical in their thinking than European North Americans (Nisbett, Peng, Choi, & Norenzayan, 2001; Peng & Nisbett, 1999). Dialectical thinkers are more inclined to view everything in the world as comprising contradictory elements (principle of contradiction), and more likely to believe that every element in the universe is in a constant flux (principle of change; Peng & Nisbett, 1999).

In a seminal study examining these issues, Peng and Nisbett (1999) reported that Chinese preferred dialectical proverbs containing contradictory elements (e.g., "Beware of your friends not your enemies") to a higher degree compared to Americans. Moreover, Chinese favored a middle-path approach when solving social conflicts by considering the issues from both sides and trying to resolve the problem by using a compromising strategy. In contrast, Americans opted to solve the same issues by identifying which party was more at fault and demanding change from the guilty party. Finally, Peng and Nisbett found that Chinese inclined more towards dialectical arguments and were more tolerant of opposing viewpoints when two contradicting positions were presented, whereas Americans favored logical arguments and responded to contradictory statements by gravitating towards the one that they believed to be more correct and polarizing their views. These findings illustrate how Eastern individuals tend to be more holistic in the thinking styles, while Western individuals gravitate more towards an analytical approach to problem solving.

Another way in which culture shapes dialectical thinking is by influencing people's predictions about the future. Ji, Nisbett, and Su (2001) presented American and Chinese participants with hypothetical scenarios, and asked them to make an estimation at the likelihood of an opposite future event taking place. They found that Chinese participants, compared to American participants, predicted a greater likelihood that the future would be different. For example, Chinese participants predicted greater likelihoods for two kids fighting in kindergarten to become future lovers, and for a chess champion to lose in the next game in the midst of winning streak. Along the same lines, Americans and Chinese also made different judgments about the stock market (Ji, Guo, Zhang, & Messervey, 2009). That is, when a stock was going up, Americans believed that it would continue to rise, and when a stock was going down, they believed that it would continue to decline. In contrast, Chinese participants expected that the opposite trend would occur (i.e., an upward-moving stock would be more likely to experience a decline in the future). These findings indicate that culture has a major bearing on expectations and the belief that a situation can change and transform over time.

## *Military application*

One intriguing military application for dialectical thinking across cultures is in the realm of stress management, where researchers have long argued that one's subjective perception of a situation is often just as important, if not more important, than the objective reality of the circumstance (James, 1984). Indeed, contemporary research has also shown that the body's physiological response to stress is influenced significantly by how one subjectively perceives the stressor. Specifically, in a large-scale ($n = 28{,}753$) prospective study examining the longitudinal impact of stress, Keller et al. (2012) reported that individuals who experienced a significant amount of stress, and believed that the stress would have a major impact on their health, experienced a 43% increased risk of premature death, even after controlling for socioeconomic status, health behaviors, and access to healthcare.

Relating these findings to the military, there is no doubt that challenge, adversity, and suffering play an inherent role in the military experience (Imtiaz, Khei, & Ji, 2017). As for coping with this stress, researchers and clinicians alike have argued that meaning-making plays a crucial role in how one acclimatizes to highly stressful situations (Davis, Wortman, Lehman, & Silver, 2000; Gillies & Neimeyer, 2006; Park & George, 2013). That is, voluntarily engaging the stressful experience (as opposed to ignoring or suppressing it) in an attempt to understand the experience, resolve the crisis, make sense out of what happened, and find some benefits in it can be both meaningful and curative (Calhoun, Cann, Tedeschi, & McMillan, 2000).

Interestingly, this meaning-making process may be mediated by culture, as dialectical and non-dialectical belief systems may lead to individuals perceiving their stressful experiences in unique ways. For instance, an individual who approaches a stressor with a dialectical mindset may apply the principles of contradiction and change in order to extract optimism from the situation. Indeed, if one believes that there is a strong possibility that the negative experience will transform into a positive one in the future, this may allow a person to stay the course during times of difficulty. Likewise, a dialectical belief that things exist in a perpetual cycle of change and renewal may lead to one realizing that similar negative experiences are likely to reoccur in the future, and thus whatever can be gained from the current experience may be useful in navigating future stressors.

Though there may be significant stress management benefits associated with adopting a dialectical mindset, it must be noted that there may also be an alarming dark side to applying this mindset in times of war. Specifically, in wars of attrition, a dialectical mindset may increase one's willingness to remain engaged in a war, even in the face of bleak prospects. Thus, the same principles of contradiction and change that lead to enhanced optimism in times of difficulty could also lead to armies remaining in battles for too long, with potentially catastrophic results.

A closer examination of Japan, a dialectical culture, and the events that unfolded at the end of the Second World War provides credence to this notion. After the absolutely devastating effects of the first atomic bomb on the city of Hiroshima on August 6, 1945, the Japanese did not immediately surrender in spite of the fact that it was clear that they had virtually no path to victory (Stimson & Truman, 1947). This prolongment led to the United States deciding to drop a second atomic bomb on the city of Nagasaki three days later, a decision that raised the number of already high casualties to well over 100,000 people (Stimson & Truman, 1947). Though world wars are incredibly complicated and we are by no means arguing that the Japanese's dialectical beliefs were the sole reason for them not surrendering earlier, we are highlighting that this important element of culture undoubtedly influences our thinking and decision-making in everyday life, and there is no reason to believe that it would not have a similar impact in military settings. As such, it is our hope that these ideas can stimulate future empirical research devoted to addressing these incredibly important questions.

In contrast to the dialectical approach, a non-dialectical mindset may lead to military personnel becoming more confident in times of prosperity if they believe that things will continue to progress in a linear manner. However, this same approach could also backfire and spin individuals into a perpetual cycle of despair when things are not going well. Unfortunately, empirical research devoted to answering these important questions is sparse. In one of the few studies related to these issues, Grossman and Kross (2010) investigated the role of culture in how people reflect on their negative affect and experiences. The researchers reported that individuals from an interdependent culture (Russia) tended to adopt a more external perspective when reflecting on their negative experiences, which led to less-detrimental outcomes and more adaptive coping. In contrast, individuals from an independent culture (America) showed a tendency to get more absorbed in their emotions, which led to less-adaptive outcomes. These findings suggest that culture may shape coping and resilience in meaningful ways, and more research in this field is required, especially in how factors such as dialectical and non-dialectical thinking may influence the way people deal with their negative affect and experiences.

## Impact of culture on Temporal Focus

Cultural differences in attention may also lead to different orientations towards the past and the future. Compared to European North Americans, East Asians' greater attention to the context and background in perceptual tasks (e.g., Ji et al., 2000; Masuda & Nisbett, 2001) may be generalized to the temporal dimension, where the past and future serve as the context or background for the present. Indeed, researchers have found that, compared to European North Americans, East Asians attend more to the past and future, perceive the past and future to be closer to the present, and have greater access to the past and future (Ji, Hong, Guo, Zhang, Su, & Li, in press).

For instance, Ji and colleagues (2009) examined how much consideration Canadian and Chinese individuals gave to past and present information when solving a hypothetical criminal case about theft. Results indicated that Chinese participants perceived information that was related to the past (e.g., "three years ago, one student spent all her money on lottery tickets") to be more pertinent to the present decision compared to Canadians. In a follow-up study, the researchers also found that Chinese participants reported significantly more details about events that occurred in their past compared to Canadians. Lastly, Chinese participants also perceived the past to be more connected to the present compared to Canadian participants, even though the actual temporal distance did not vary. That is, the same objective time period (e.g., 1 year ago) felt subjectively closer for the Chinese participants. Together, this research highlights how the past may be perceived as more relevant, easier to recall and closer in subjective time to the present for Chinese participants.

Cultures differ not only in how much they attend to the past and the future, but in also in how much they value these temporal periods. To this end, research has shown that Western individualsvalue the future more than the past. Specifically, Guo, Ji, Spina, and Zhang (2012) reported that European-Canadians were more appreciative (as indexed by the amount of money they spent on a thank-you gift) of a favor that an acquaintance was scheduled to do for them in the future compared to an identical favor that was done in the past.In similar research, Caruso, Gilbert, and Wilson (2008) demonstrated that American individuals placed a higher value ($125) on work that they imagined doing in the future compared to the exact same work done in the past ($62). Moreover, these findings emerged in spite of the fact that participants did not perceive the past and future work to be different in terms of difficulty and demand. These findings highlight that the future is perceived as intrinsically more valuable for individuals in the west.

In trying to pinpoint why Western individuals appear to be more future oriented, researchers have pointed to the fact that the future is often referenced as relatively superior to the past in Western culture (Kluckhohn & Strodtbeck, 1961). As to why the future is framed more positively than the past in the West, Van Boven and Ashworth (2007) have postulated that anticipating future positive emotions may be the driving factor behind this effect. Across a series of experiments, the researchers found that North American individuals reported more intense positive emotions while anticipating a positive event in the future (e.g., upcoming thanksgiving dinner) than recollecting the same positive event in the past. Similarly, Caruso et al. (2008) reported that reflection over future events produced significantly greater affect compared to reflection over past events in American subjects, indicating that Western individuals tend to display stronger emotional reactions when looking forward than when looking backwards.

In contrast to their Western counterparts, Eastern individuals tend to place a higher value on the past compared to the future. For instance, Guo et al. (2012) reported that Chinese individuals were more grateful for a hypothetical favor that an acquaintance had done for them in the past compared to a similar favor done in the future, indicating that they perceived the past favor to be relatively more valuable. The researchers also reported that Chinese individuals placed a higher monetary value on work that was done in the past compared to the same work scheduled to be done in the future. Along the same lines, Levinson and Peng (2007) asked East-Asian and North American participants to place a monetary value on antique furniture that had been valued at $350 USD in 1985. Compared to the North American participants, whose evaluations were 2.83 times greater than that of the original 1985 value, East-Asians estimated that the chair would now be worth 12.02 times more than the 1985 value. Similarly, research in consumer behavior has shown that Chinese participants tend to display significantly more brand loyalty than Americans towards companies that they have done business with in the past (Robinson, 1996), a proclivity that some researchers have attributed to their robust orientation towards the past (Yau, 1988).

In trying to understand why Eastern cultures may have this positive affinity towards the past, one can begin by examining the broader culture, which has always valued its rich history and held the past in high regard (Brislin & Kim, 2003). This is reflected in the respect that is often given to ancestors and communal traditions (Kluckhohn & Strodtbeck, 1961). Another potential explanation involves differences in temporal focus, which is the amount of attention individuals devote to the past and the future. Specifically, Guo et al. (2012) have postulated that Eastern cultures value the past more because they allocate more attention towards the past, while Western cultures value the future more due to the greater attention they devote to the future. Evidence for this argument comes from the fact that when induced to focus on the future, Chinese begin to value the future more than the past. Likewise, when induced to focus on the past, Euro-Canadians begin to value the past more than the future. The fact that these natural inclinations can be reversed suggests that temporal focus may indeed be an important determinant of how cultures value different time periods.

### *Military application*

Military commanders may use cultural idiosyncrasies in temporal focus to motivate their troops to "fight for their future" or "fight for their tradition" Based on the literature outlined in the previous section, we might expect Western soldiers to respond more favorably to calls advocating for the future, while Eastern soldiers may respond better to similar calls framed with a focus on past history and tradition. Research by Van Boven and Ashworth (2007), which has demonstrated that Western participants show greater positive emotions when anticipating the future versus recollecting about the past, certainly gives credence to this argument.

Related to this, Heine and Lehman (1995) reported that, compared to Japanese participants, Canadian samples were significantly more likely to be more optimistic about the future, sometimes even to unrealistic degrees. Interestingly, the researchers also found that Canadians were actually more optimistic (i.e., they felt relatively invulnerable) when there was an element of threat associated with a future event. Lastly, this unbridled optimism further led to the Canadians perceiving that they had more control over the future event, especially as the perceived severity of the threat increased. As many military operations often contain an inherent element of threat and danger, these findings may prove to be particularly important in this setting, and more research on this question is certainly warranted. If Western militaries can tap into these narratives strategically and motivate their personnel to think about the future success that would come about through a successful operation, this may drastically improve morale and performance on the battlefield. East Asians, on other hand, may respond better when prompted to think about the lives they have left behind and that are waiting for them upon successful completion of their missions.

## Culture and regulatory focus

An interesting final extension of the differences in temporal focus across cultures involves the pursuit of various social goals. To this end, regulatory focus theory argues that individuals vary in their pursuit of promotion and prevention goals in order to accommodate their unique self-regulatory strategies and needs (Higgins, 1997). In other words, while some tend to focus more on promotion or approach goals (i.e., chasing your aspirations and trying to maximize gains), others find greater motivation in prevention or avoidance goals (i.e., fulfilling your obligations and trying to minimize potential loses).

Relating this to culture, Lee and colleagues reported that North Americans preferred promotion goals, while their East Asian counterparts preferred prevention goals (Lee, Aaker, &

Gardner, 2000). As it relates to the military, these differences may have important implications on how leaders go about motivating and getting the best out of their units, an application that we outline with more detail in the next section.

### *Military application*

Cultural differences in regulatory focus may influence the overarching goals that military units adopt. For instance, should leaders aim to go for great achievements and accomplish unprecedented feats, or simply carry out their operations in a manner such to limit potential harm and not disappoint others? The literature on regulatory fit theory would suggest that the answer may depend on the cultural dispositions that individuals bring to the operation. To this end, Kitayama, Markus, Matsumoto, and Norasakkunkit (1997) asked American and Japanese participants whether experiences that were classified as successes or failures were more important to their self-esteem. The researchers found that American participants believed that success experiences (e.g., "getting a good grade on a test when I study hard") were more pertinent to their self-esteem than failure experiences (e.g., "when your employer tells you that you are not performing well on the job"). In contrast, Japanese participants actually rated the failure experiences (e.g., "when I was jilted by someone I was thinking of marrying") as more relevant to their self esteem than success experiences (e.g., "when I remember a difficult job in the past that I managed to carry through"). Furthermore, findings also revealed that Americans believed that their self-esteem would increase more after successful experiences than it would decrease following failure, whereas the Japanese showed the opposite pattern by reporting that their failures would be more impactful to their self-esteem than their successes.

These results are in line with other research illustrating that many East Asian cultures do not share the same desire for admiration and positive self-regard that is commonplace in the West, and a self-critical focus centered around not making egregious errors and maintaining overall harmony is more desired in these individuals (Heine, Lehman, Markus, & Kitayama, 1999). Together, these findings suggest that aiming for grandeur is by no means a universal motive, and many individuals may perform better in every tasks (e.g., school, work) when simply trying to avoid failure. This assertion has been supported in other domains such as healthcare, where Uskul, Sherman, and Fitzgibbon (2009) found that Western individuals, who had a stronger promotion regulatory focus, responded more favorably to gain-framed health promotion messages, while Eastern participants, who had a stronger prevention regulatory focus, responded more positively to loss-framed messages. Relating this back to our discussion, it seems that Western militaries would be well served to motivate their teams by encouraging them to strive for major victories and success, whereas Eastern militaries may be able to achieve a similar positive outcome by motivating their squads to simply get the job done in order to avoid falling short. Again, researchers should aim to empirically test these hypotheses across diverse military units.

## Conclusion

This chapter aimed to bridge the gap between existing research on cultural differences in cognition and its potential applications in a military context. Generally, culture can be defined as the shared norms and values of a certain group of individuals that sets them apart from other groups. Much research has been conducted to compare cross-national differences in cognition, but little research has examined the implications for military, which is arguably a culture by itself. Effective knowledge translation of this work will ensure that it has a meaningful impact on international military operations around the world, the importance of which cannot be overstated in a world rife with global conflict. Successful consideration and implementation of this knowledge may prove to be a critical step in enhancing international negotiations and eradicating some of the most important issues in the world today.

## References

Aberman, S. (2001). *Anatomy of a crisis*. Retrieved from http://www.pbs.org/newshour/extra/features/jan-june01/china_plane.html

Brislin, R. W., & Kim, E. S. (2003). Cultural diversity in people's understanding and uses of time. *Applied Psychology: An International Review*, *52*(3), 363–382.

Calhoun, L. G., Cann, A., Tedeschi, R. G., & McMillan, J. (2000). A correlational test of the relationship between posttraumatic growth, religion, and cognitive processing. *Journal of Traumatic Stress: Official Publication of the International Society for Traumatic Stress Studies*, *13*(3), 521–527.

Caruso, E. M., Gilbert, D. T., & Wilson, T. D. (2008). A wrinkle in time: Asymmetric valuation of past and future events. *Psychological Science*, *19*(8), 796–801.

Chiu, C. Y., & Hong, Y. Y. (2006). *Social Psychology of Culture*. Michigan, Ann Arbor: Psychology Press.

Davis, C. G., Wortman, C. B., Lehman, D. R., & Silver, R. C. (2000). Searching for meaning in loss: Are clinical assumptions correct? *Death Studies*, *24*(6), 497–540.

Gillies, J., & Neimeyer, R. A. (2006). Loss, grief, and the search for significance: Toward a model of meaning reconstruction in bereavement. *Journal of Constructivist Psychology*, *19*(1), 31–65.

Grossmann, I., & Kross, E. (2010). The impact of culture on adaptive versus maladaptive self-reflection. *Psychological Science*, *21*(8), 1150–1157.

Guo, T., Ji, L. J., Spina, R., & Zhang, Z. (2012). Culture, temporal focus, and values of the past and the future. *Personality and Social Psychology Bulletin*, *38*(8), 1030–1040.

Heine, S. J., & Lehman, D. R. (1995). Cultural variation in unrealistic optimism: Does the West feel more vulnerable than the East? *Journal of Personality and Social Psychology*, *68*(4), 595–607.

Heine, S. J., Lehman, D. R., Markus, H. R., & Kitayama, S. (1999). Is there a universal need for positive self-regard? *Psychological Review*, *106*(4), 766–794.

Higgins, E. T. (1997). Beyond pleasure and pain. *American Psychologist*, *52*(12), 1280–1300.

Hong, Y., & Khei, M. (2014). Dynamic multiculturalism: The interplay of socio-cognitive, neural, and genetic mechanisms. In Benet-Martínez, V., Hong, Y. (Eds.), *The Oxford Handbook of Multicultural Identity* (pp. 11–34). New York, NY: Oxford University Press.

Imtiaz, F., Khei, M., & Ji, L. J. (2017). Resilience through suffering: A potential military application. In MacIntyre, A., Lagacé-Roy, D., & Lindsay, D. R. (Eds.), *Global Views on Military Stress and Resilience* (pp. 37–54). Kingston, ON: CDA Press.

James, W. (1984). *Psychology, Briefer Course* (*Vol. 14*). Harvard University Press.

Ji, L. J. (2005). Culture and lay theories of change. In Sorrentino, R. M., Cohen, D., Olson, J., & Zanna, M. (Eds.), *Culture and Social Behavior: The Tenth Ontario Symposium* (pp. 117–135). Hillsdale, NJ: Erlbaum.

Ji, L. J., Guo, T., Zhang, Z., & Messervey, D. (2009). Looking into the past: Cultural differences in perception and representation of past information. *Journal of Personality and Social Psychology*, *96*(4), 761–769.

Ji, L. J, Hong, E., Guo, T., Zhang, Z., Su, Y., & Li, Y. in press. Culture, psychological proximity to the past and future, and self-continuity. *European Journal of Social Psychology*.

Ji, L. J., Nisbett, R. E., & Su, Y. (2001). Culture, change, and prediction. *Psychological Science*, *12*, 450–456.

Ji, L. J., Peng, K., & Nisbett, R. E. (2000). Culture, control, and perception of relationships in the environment. *Journal of Personality and Social Psychology*, *78*(5), 943–955.

Keller, A., Litzelman, K., Wisk, L. E., Maddox, T., Cheng, E. R., Creswell, P. D., & Witt, W. P. (2012). Does the perception that stress affects health matter? The association with health and mortality. *Health Psychology*, *31*(5), 677–684.

Kitayama, S., Duffy, S., Kawamura, T., & Larsen, J. T. (2003). Perceiving an object and its context in different cultures: A cultural look at New Look. *Psychological Science*, *14*(3), 201–206.

Kitayama, S., Markus, H. R., Matsumoto, H., & Norasakkunkit, V. (1997). Individual and collective processes in the construction of the self: Self-enhancement in the United States and self-criticism in Japan. *Journal of Personality and Social Psychology*, *72*(6), 1245–1267.

Kluckhohn, F. R., & Strodtbeck, F. L. (1961). *Variations in Value Orientations*. Evanston, IL: Row, Peterson.

Lee, A. Y., Aaker, J. L., & Gardner, W. L. (2000). The pleasures and pains of distinct self-construals: The role of interdependence in regulatory focus. *Journal of Personality and Social Psychology*, *78*(6), 1122–1134.

Levinson, J., & Peng, K. (2007). Valuing cultural differences in behavioral economics. *ICFAI Journal of Behavioral Finance, 4*, 32–47.

Masuda, T., & Nisbett, R. E. (2001). Attending holistically versus analytically: Comparing the context sensitivity of Japanese and Americans. *Journal of Personality and Social Psychology, 81*(5), 922–934.

Nisbett, R. E., Peng, K., Choi, I., & Norenzayan, A. (2001). Culture and systems of thought: Holistic vs. analytic cognition. *Psychological Review, 108*(2), 291–310.

Park, C. L., & George, L. S. (2013). Assessing meaning and meaning making in the context of stressful life events: Measurement tools and approaches. *The Journal of Positive Psychology, 8*(6), 483–504.

Peng, K., & Nisbett, R. E. (1999). Culture, dialectics, and reasoning about contradiction. *American Psychologist, 54*(9), 741–754.

Robinson, C. (1996). Asian culture-the marketing consequences. *Market Research Society. Journal, 38*(1), 1–8.

Segall, M. H., Campbell, D. T., & Herskovits, M. J. (1966). *Influence of Culture on Visual Perception*. Indianapolis, IN: Bobbs-Merrill.

Simons, D. J. (2010). Monkeying around with the gorillas in our midst: Familiarity with an inattentional-blindness task does not improve the detection of unexpected events. *Iperception, 1*(1), 3–6.

Stimson, H. L., & Truman, H. S. (1947). The decision to use the atomic bomb. *Bulletin of the Atomic Scientists, 3*(2), 37–67.

Uskul, A. K., Sherman, D. K., & Fitzgibbon, J. (2009). The cultural congruency effect: Culture, regulatory focus, and the effectiveness of gain- vs. loss-framed health messages. *Journal of Experimental Social Psychology, 45*(3), 535–541.

Van Boven, L., & Ashworth, L. (2007). Looking forward, looking back: Anticipation is more evocative than retrospection. *Journal of Experimental Psychology: General, 136*(2), 289–300.

Witkin, H. A., Lewis, H. B., Hertzman, M., Machover, K., Meissner, P. B., & Wapner, S. (1954). *Personality through Perception*. New York: Harper.

Yap, S., Ji, L. J., & Hong, E. (2018). Culture and cognition. In Wixted, J. T., & Tompson-Schill, S. L. (Eds.), *The Steven's Handbook of Experimental Psychology and Cognitive Neuroscience Language and Though volume* (4th Ed.) (pp. 491–520). New Jersey: John Wiley & Sons, Inc.

Yau, O. H. M. (1988). Chinese cultural values: Their dimensions and marketing implications. *European Journal of Marketing, 22*(5), 44–57.

# 6

# MILITARY PSYCHOLOGY AND THE FOURTH INDUSTRIAL REVOLUTION

## Implications for the South African National Defence Force's Directorate psychology

*Petrus C. Bester*

"The world we now inhabit is changing at a pace and in a manner that is unprecedented in human history. Revolutionary advances in technology are reshaping the way people work and live. They are transforming the way people relate to each other, the way societies function and the way they are governed" (Ramaphosa, 2019). These are the words of the South African president, Mr Cyril Ramaphosa, emphasising the importance of the Fourth Industrial Revolution (4IR) when announcing the appointment of a Presidential Commission on the 4IR to serve as a national overarching advisory mechanism on digital transformation. He warned that if it does not adapt to new technologies, South Africa faces the grim prospect of being rendered irrelevant and facing stagnation as a country (University of Johannesburg, 2019a).

Technology is correspondingly vital in war because Symonds, as cited in The Economist (2018), emphasises that although war is still a contest of wills, its character is changing because of geopolitical competition and technology. Military organisations profoundly rely on technology, and as one of the elements of state power and more particularly "hard power" (Nye, 2009), the South African National Defence Force (SANDF) needs to be aware of the importance of technology. The defence industries are usually among the early adopters of new technology, specifically in the aerospace, automotive and manufacturing industries (Lele, 2019; von Scheel, 2016). The way humans use this technology changes war, not the technology in itself (Tuck, 2019).

It can thus be postulated that both the SANDF and its opponents can benefit from new technologies emanating from the 4IR. Bester's (2016) definition of military psychology describes it as the application of research technology, principles and methods of psychology in the military environment to address challenges, which leads to improved behavioural capability of own forces and countering the potentially diminishing effects the opposing forces' activities might have on the human capabilities of own forces. This indicates that military psychology as a discipline would be able to provide support in dealing with the challenges and benefits emanating from the changed 4IR environment to ensure optimised human performance.

Therefore, this chapter aims to address a number of questions, of which the first is what exactly the 4IR is, how it will influence warfare in general and more specifically the SANDF, how the 4IR will affect the SANDF's Directorate Psychology and, lastly, how it should respond. This discussion is based on the assumption that the SANDF will reach a point in future where it cannot be successful without responding to the 4IR by embracing and participating in it.

## Contextualising military psychology in the South African National Defence Force

As mentioned in the introduction, the SANDF is a manifestation of South Africa's hard power. It consists of various services and divisions, of which four are noteworthy, namely the South African Army (SA Army), South African Navy (SAN), South African Air Force (SAAF) and the South African Military Health Service (SAMHS). The SAMHS delivers military healthcare to all services and divisions within the SANDF, including a military psychological service.

The embodiment of this military psychological service can be found in the SAMHS's Directorate Psychology, which ensures the application of military psychological principles and practices in the culturally diverse military community to ensure the operational effectiveness of the SANDF's various user systems in terms of physical and mental combat readiness. In a discussion document compiled by the SANDF's Deputy Surgeon General, it is postulated that in future the main purpose of Military Health Psychological Services will be: "*... mental and cognitive superiority over our opponents ...*" (Ndhlovu, 2019, p. 19).

The Director Psychology's draft statement of intent for 2019–2024 (Burgess, 2018) emphasises that the Director Psychology (and his directorate) is responsible for providing the strategic direction and policy required for the governing of a comprehensive military psychological service, for the management of the pool of military psychologists and for exercising functional control over all processes and services regarding psychology in the Department of Defence (DOD). The vision of the military psychological service is "*A psychologically healthy and effective military organisation*" and the mission is "*To support the SANDF by providing a quality psychological service for the full range of military deployments and to sustain and promote the health and mental functioning of the organisation, members of the armed forces, their families and others eligible for care by the SAMHS*" (Burgess, 2018, p. 8). Hence, it implies ensuring mental and cognitive superiority over the SANDF's opponents in the present state and/or a future state characterised by the 4IR.

The services centralised under the SAMHS' Directorate Psychology include *inter alia* a clinical service and specialised operational military psychological service for deployed soldiers, as well as psychological assessment related to human resources (HR) for selection and placements, organisational development, development and learning leadership assessment and finally general psychological research. These services manifest themselves in several programmes focusing on force health protection and force health support (Burgess, 2018). These programmes are: primary psychological care; tertiary psychological care (including neuropsychological assessment; inpatient and outpatient support to hospital departments and patients and specialised clinics); aviation psychology (for the SAAF); maritime psychology (for the SAN); special forces and parachute specialised contingency services; selection and separation assessments; development and learning assessment and training; organisational development; behavioural research and development; operational psychology/human factor combat readiness services; Defence Intelligence psychology; hostage and escape evasion and survival psychology and a number of accredited internship programmes for industrial, clinical, research and counselling psychology. With the positioning of military psychology and the services it presents within the SANDF as background, the next step is to gain understanding of what the 4IR is.

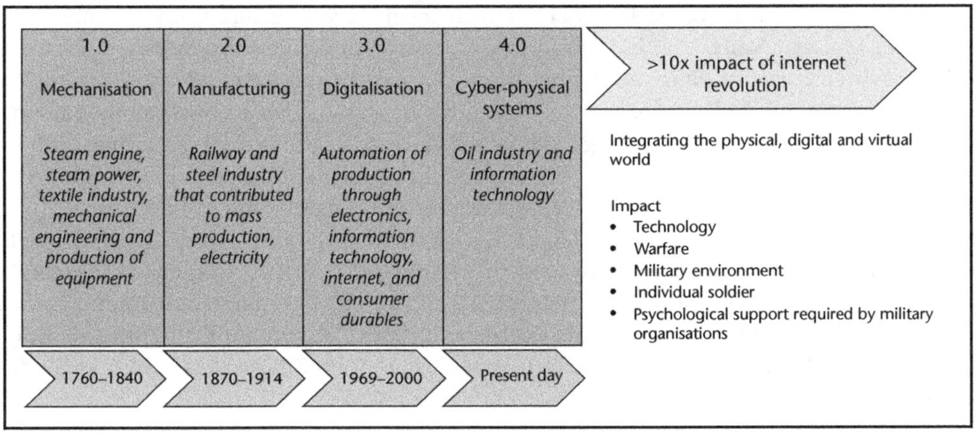


*Figure 6.1* The four industrial revolutions.

*Source:* Adapted from Society for Industrial and Organisational Psychology [SIOPSA], 2019.

## Fourth industrial revolution defined

The term "Industry 4.0" was first coined in 2011 at the Hanover Fair, Germany (Lele, 2019), signifying that an industrial revolution[1] is about the beginning of a new industrial age. The Society for Industrial and Organisational Psychology in South Africa (SIOPSA, 2019), Lele (2019) and von Scheel (2016) broadly demarcate the different revolutions as the *first*, driven by mechanisation made possible by the steam engine, steam power, the textile industry, mechanical engineering and production of equipment from about 1760 to 1840; the *second*, made possible by electricity and the railway and steel industries, with production lines that enabled mass production (manufacturing) from 1870–1914; the *third*, which made use of electronics and information technology to automate production (digitalisation) from 1969–2000, and lastly the 4IR, enabled by the oil industry and information technology that developed machines that are more human-like (cyber-physical systems) from about 2000. Campbell (2017), Lele (2019) and African Daily Voice (2019) refer to the last mentioned as a digital revolution, but Mostert (2019) observes that it is rather seen as a variation of the 3IR. For the purpose of this discussion, the four revolutions are categorised as mechanisation, manufacturing, digitalisation and a cyber-physical system. As illustrated in Figure 6.1, there is thus general acceptance in the literature that the 1IR was characterised by steam and water, the 2IR by electricity being used for mass production, the 3IR by internet and communication technologies and adoption of major digitalisation processes and the 4IR as the integration of man and machine, hence the cyber-physical system. According to the SIOPSA (2019) the 4IR is likely to have ten times the impact of the internet revolution (digitalisation) and will consequently have an impact on technology, warfare, the military environment, the individual soldier and the psychological support required by military organisations.

The 4IR, referred to as cyber-physical systems, is essentially a very interesting phase in integrating the digital, physical and virtual worlds (Lele, 2019; von Scheel, 2016). An overview of 4IR literature (Campbell, 2017; Lele, 2019; von Scheel, 2016) identifies the following as the

1 Revolution implies change of great magnitude.

core elements or domains of the 4IR: robotics (man-to-machine, machine-to-man, machine-to-machine), mobile connectivity, bioinformatics, nanotechnology, mass customisation, advanced connectivity, internet of things, autonomous vehicles, cybersecurity, neurotechnology, genetic engineering, wearable technology, blockchain software, cloud services, quantum computing, 3D printing (additive manufacturing), energy storage, advanced material, sensoring, artificial intelligence (AI) and big data 2.0. In many instances it is not so much about developing new technologies, but about continually finding new ways of using existing technologies.

One of the areas in which fundamental change is occurring is that of speed, with interconnectedness dominating (University of Johannesburg, 2019b). It implies, for example, that AI will replace the notion that computers will only react to human instructions; they will start to make their own decisions (Mostert, 2019). Consequently, the differentiation between the real and technological or virtual worlds becomes diffuse and blurred, as alluded to above.

In the military environment, it implies that the distinction between humans and machines will become more diffuse and soldiers and "things" (such as weapons and combat systems) will connect at any time or in any place for any need, whether written or verbal communication, data, information, intelligence, satellite navigation systems or neurobiological monitoring systems. Therefore, the next section will address the 4IR's impact on warfare, with specific reference to the SANDF. Thomas (2019) emphasises that by forecasting the shape of future wars, one can determine the capabilities a military, like the SANDF, requires to impede its opponents and what issues to include in budget requests.

## 4IR and future warfare

The conduct of war is a continuously evolving phenomenon (Andresky & Henderson, 2018; Del Monte, 2018; Hughes, 2018; Kamieński, 2017; Lele, 2019; Miller, 2019; O'Hanlon, 2018; The Economist, 2018; Thomas, 2019) and is often assumed to be a particularly technological human enterprise (Tuck, 2019). Also, *The Economist* (2018) observes that combat speed will be so high that humans will not be able to keep up. It is therefore no surprise that Christian Brose, the former staff director of the United States Senate Armed Forces Committee, refers to the coming changes in warfare as a "revolution in military affairs" (Miller, 2019). Consequently, new technologies such as AI, machine learning and big data will be some of the driving forces influencing future military operations. Military forces will operate in complex environments against peer competitors across all domains (cognitive, land, air, outer space, maritime and cyberspace), the electromagnetic spectrum, and the information environment (Andresky & Henderson, 2018; Thomas, 2019) where own forces and/or their opponents are likely to be roboticised units (Del Monte, 2018) or a mixture of humans and robots. Human factors such as will, fear, decision-making and even human brilliance may become less evident (The Economist, 2018).

One can consequently, contrary to what one is used to, expect fewer humans and more (swarms of) machines as part of a larger-scale heterogeneous force delivering a greater volume and higher velocity of firepower than ever before on the battlefield. There is also going to be biotechnological integration of man and machine (Kamieński, 2017), where bio-enhancement and eksoskeletons (Gutsche, 2018) will enable soldiers to perform super-human tasks, such as lifting very heavy objects or to see at night and/or over long distances.

The view of fewer humans is supported by Mostert (2019), who observes that not only will fewer humans be required, but they will also be excluded from various interactions between machines. For example, a situation may occur where self-driving or automated tanks will challenge each other on the battlefield without any soldiers physically opposing one another.

Even the maintenance of this equipment will be done by robots with sophisticated diagnostic tools, which can exchange information on solving mechanical problems through robotics and autonomous systems. Militaries are thus likely to increase the use of digital technologies such as big data, cloud computing, mobile devices and social networking on an organisational level.

This automation through AI has its advantages, as AFRIBLOG (2019) postulates that certain life-threatening jobs, such as dealing with bombs and terrorist threats, will be performed by robots. This will minimise the loss of life or human casualties should the check-point be attacked by, for example, a suicide bomber. Guards can be replaced by efficient remote monitoring and access control systems. The same type of technology can also be used by military health personnel who have to deal with patients infected with deadly diseases, such as Ebola haemorrhagic fever. These systems will by themselves, or by merging humans and systems, ensure interoperability, decentralisation, analysis in real time, flexible services, high transmission speeds and rapid reactions. According to Thomas (2019), the main goal will be speed, synchronisation, concurrency and overwhelming all-over superiority, while relying in real time on computers, telecommunications and satellite communications. Therefore, Gareyev, cited in Thomas (2019), postulates a high level of dynamism and manoeuvrability along non-coherent fronts.

In support of the above, Del Monte (2018) emphasises the use of AI, automated weapon systems and what he refers to as genius weapons, where the decision to act and kill will be made by a machine without the interference of man. These machines will depend on data, information and intelligence that will form part of a "global brain" from which a military user system can get information, for example, on weather conditions on the battlefield. Such autonomous systems will be able to independently compose and select among different courses of action to accomplish objectives based on their knowledge of the situation, themselves and world (The Economist, 2018). Campbell (2017) and von Scheel (2016) conclude that 4IR technology will be a game-changer, bringing about significant improvement in operational efficiency, productivity, analysis and system integration.

From the above, one can thus conclude that military user systems such as anti-aircraft, unmanned aerial vehicles (UAVs), artillery and anti-tank weapons will be a network of people, data and machines. Moreover, these will continuously collect data, display it to the operator (soldier) either before or after processing it through advanced analytics into real-time intelligence to provide actionable information that allows predictive analysis for the operator to make more accurate decisions, accompanied by improved maintenance and operational processes. In some instances, there will be no operator and the machine will do everything.

These higher levels of automation, greater speed and accuracy that characterise the 4IR will, singularly or jointly, offer various options to improve a force's operational readiness (Lele, 2019). Consequently, this kind of warfare will be propagated because it is economical in that decisions can be made at a higher speed, it reduces cost and personnel, dependence on communication, human error and fatigue, and it minimises human losses. Del Monte (2018) even claims, for example, that killer robots can respect international humanitarian laws better than humans.

As implied above, the 4IR will have a significant impact on the cyber domain and consequently on information warfare. *The Economist* (2018) states that cyber warfare will become a contest between algorithms. Disinformation, for example, can affect societies in general and more specifically those where reliable information is in short supply. An aggressor can rely on loosely defined networks of pro-aggressor actors using traditional and social media to influence opinions both at home and abroad (Robinson, 2018). Advanced technologies, such as bots and cyborgs enhanced with deep fake technology, result in a proliferation of sources spreading false information that seems authentic and is difficult to verify. Thomas (2019, p. 90) identified the concept of "behavioural wars," which refers to: "*manipulating behavior* [sic] *algorithms, habits, activity stereotypes etc, that have been installed in us by our social group, and also by our biographies and cultural environment.*" It will consequently have a noteworthy impact on cybersecurity and the authentication of information used in psychological operations against own forces, which in turn can be countered by technology emanating from the 4IR. Hence the contest between algorithms.

Future warfare will thus be characterised by a high operational tempo where war is conducted at any time by night or day, at breathtaking speed, non-stop, mostly by machines, although man-machine integration will become the norm, and usually beyond the biological limitations of the mind and body of humans.

South Africa's current economic situation requires the SANDF to function with a greatly reduced budget (Kings, 2019), and it is very unlikely that it will right away reap the benefits that the 4IR can provide to the military. Conversely, the contextual application of the SANDF is the African battle space (ABS) that is described as the "*environment in which the South African soldier would increasingly be placed in asymmetric situations against opponents who are not easily identifiable, are probably better armed and equipped and have access to better communications and technology*" (Bester & O'Neil, In press). This compels the SANDF to acquire more advanced technology and equipment to maintain a competitive edge over its adversaries, despite financial constraints. It is thus expected that this will happen, although not at the pace seen in more advanced economies such as the United States and China. This supports the view of Lele (2019), who states that the manufacturing of defence equipment is usually a technologically advanced and innovation-driven system and takes time for a state to implement.

Mostert (2019), however, refers to a maxim among futurists that the future is already here, but it is not yet equally distributed. One can thus expect that although the SANDF is already in the 4IR, it will not change overnight into an "SANDF 4.0",[2] although nothing prevents a state from importing technologies once they exist while still developing its own capabilities to become self-reliant. The technology will thus have to be "equally distributed". The 4IR is a reality with which the SANDF will have to deal, sooner rather than later. Thus, when adding the military applications of 4IR technology to the ABS, one can postulate that it will have a significant influence on the South African soldier and his or her workplace.

## The impact of the 4IR on the soldier and his or her workplace

The 4IR will be a "game-changer" in the conduct of war and it can thus be hypothesised that the SANDF will be affected in three domains, namely military hardware (weapons), its organisational structures (architecture of the organisation) and what Uzzell, cited in Campbell (2017), refers to as the "human footprint".

Although initially the 4IR will not immediately influence South African soldiers, as it takes time for new technology to be integrated into the military, one can expect that they might first come across such technologies and especially military hardware on the battlefield, whether in conventional warfare or during asymmetric conflict.

When referring to the impact on organisational structures, one can foresee that existing "hard" hierarchical structures are likely to be replaced by more "soft" and less hierarchical or hybrid organisational structures, the requirement of new competencies, an integration of different workplaces across international borders, virtual workers, complex matrix structures, open work hours and agility in the workplace (Mostert, 2019; Visser, 2019). Therefore, one must start to anticipate the challenges and opportunities that arise from using such technologies in new structures, with specific reference to the impact on the soldier.

2 "SANDF 4.0" suggests a fully functional military force based on the technologies of the 4IR.

Kamieński (2017, p. 205) emphasises the impact on the human footprint when he refers to the growing pressure on soldiers who will be the "weakest element" on the battlefield, where war will be fought at a murderous operational tempo. This suggests not only focusing on kinetic matters, but also on the behavioural sciences. This requires changes in the workforce, whether through interventions such as training, development, socialisation in a 4IR society or through recruitment and selection of a new kind of soldier for future wars. These soldiers will be required to function on the edge of human capability with greater stamina, improved alertness and cognitive skills and the proficiency to function in an environment where their own abilities are enhanced through fused man-machine interfaces (Kamieński, 2017).

Military psychology consequently has a role to play in all these three domains, whether considered individually or in combination. It is thus clear that the pace and scale of technological change in the 4IR has a significant impact on humankind in general and more specifically military personnel. Burgess (2018) emphasises the importance of human beings by indicating that despite tremendous growth in military technology and doctrine, individual human beings remain the main resource of the SANDF. He emphasises the importance of military psychology when he states that new combat and support systems still have a human interface and the right (healthy and competent) man or woman will be required for the job.

## *4IR and military psychology*

Annen, Nakkas and Gehring (2017) and Bester (2016) identify several tasks or activities in which military psychologists are usually involved when dealing with the "human footprint" in military organisations such as the SANDF. The next part contextualises some of those military psychology-related activities in the 4IR.

The 4IR will affect *recruitment and selection* criteria for SANDF soldiers to successfully participate in future wars. The SANDF HR division for which the military psychologists work must review current recruitment and selection criteria to obtain the best talent for its future needs. Information on competencies identified during job analysis on future posts need to be fed back to this division, which is responsible for the recruitment and appointment of future soldiers. In this way, the new competencies can be used to search for the right talent.

Subject matter knowledge is not enough to thrive, and Kriel (2018) emphasises that those hiring should look beyond education received from a traditional university. Practitioners should search for candidates who are prepared for the future world of work (4IR-related) and who can apply knowledge practically in a creative way. This also emphasises the important role of the Faculty of Military Science at the South African Military Academy of the University of Stellenbosch in bridging the gap between theory and practice when educating the SANDF's future leaders. Military psychologists will thus play a vital role in *competency design* and the training of SANDF personnel to enable them to do targeted recruitment, to manage and establish selection boards for efficiently identifying the best candidates with veracious competencies for the future.

Moreover, the Directorate Psychology should explore the use of 4IR technology in its processes, such as selection, training and diagnosis of *mental health-related matters*, especially when testing for mental health fitness. Within the next 20 years, technology such as AI will no longer be an isolated industry, but will be integrated into all facets of life (Zhou, in AFRIBLOG, 2019). AI has machine learning as a subsystem, in which machines teach themselves to solve problems (AFRIBLOG, 2019). In this way, algorithms can be compiled that make more accurate diagnoses of psychopathology than a psychologist could do. Phillips (2019) refers to recent research at the New York University School of Medicine that created an algorithm which, through analysing patient interviews, sorts through tens of thousands of variables in their speech that are imperceptible to the human ear to identify audio markers of post-traumatic stress disorder (PTSD) to make an accurate diagnosis. This type of technology is costly to develop, but will cut costs through early identification of the disorder, excluding practitioner bias in shaping diagnosis, thus yielding more accurate diagnosis, as disorders such as insomnia and anxiety share many symptoms with PTSD, and making it possible to reject false claims for compensation based on PTSD. This can be done by an application on a device such as a smart phone. Furthermore, Kott (2018) emphasises that future resilience training should include dealing with adverse situations due to technological surprise in warfare when confronted with a weapon system that was not expected.

Rosenbaum (2019) states that AI employed by IBM is 95% accurate in predicting which employees will leave a job by using its "predictive attrition program". Information obtained can be used to predict better who is likely to be successful in a particular job and who is not. This kind of information can be used to retain employees. Military psychologists are also likely to deal with people's fear that AI may cause them to lose their jobs (AFRIBLOG, 2019). INVESTEC (2019), however, emphasises that technology replacing jobs is nothing new, but based on history, technology has the power to create as much work as it displaces, though Mostert (2019), postulates that there will be job losses. Military psychologists are already facing *new kinds of psychopathology*, such as internet addiction and remotely acquired post-traumatic disorder related to the use of technologies making it possible to kill people in a foreign country over a long distance by an operator manning remotely controlled weapon systems in his or her home country.

SANDF soldiers use mobile devices to communicate, access the internet for shopping, read the news, participate in various forums, complete surveys and fill in tax returns. Consequently, a wide range of varied activities have excellent potential for various types of data to be collected. This creates an opportunity for military research psychologists to do *psychological research* on various matters, in huge volumes that was impossible or very difficult to research in the past. Consequently, owing to the vast scale and scope, more complete, in-depth and updated studies can be done of various phenomena. In addition, Pinter, Toninelli and de Pedraza (2015), point out that in human daily activity, data are constantly flowing through satellites, cameras, sensors, radio frequencies, cars, private appliances, tablets, mobile phones and the like. Smart watches that can monitor various psychophysiological activities of humans, such as sweat and heart rate, can be included in this list. Furthermore, more advanced software enables data collection through these devices that can keep a record of human behaviour in its various facets. This will have an impact on the way in which informed consent is obtained, as research has the potential to be very intrusive in the private lives of individual soldiers.

Aside from the *ethical issues* arising in the psychological research domain and opinions on whether machines must exclude humans from decision-making, the human will still be affected. Emotional strain will be placed on soldiers, especially in cases with humans in the decision-making loop of weapons with much more destructive and remote killing capabilities – even across different continents (The Economist, 2018). Consequently, military psychologists need to anticipate and identify future stressors that require new ethical considerations.

Klaus Schwabb (in World Economic Forum, 2018) emphasises that individual workers will have to engage in lifelong learning to remain employable. Companies should thus focus on reskilling and upskilling strategies for selected talent in new appointments and job transitions. Therefore, most, if not all, soldiers will have to be retrained (Rosenbaum, 2019), which suggests another important role for the South African Military Academy. This training may include more training on simulators, coaching for cognitive superiority where commanders make better decisions, and hardiness or mental toughness training that makes soldiers more resistant to stress and trauma (Bar-Gil, 2018). Furthermore, where training or retraining is required, new technologies can also be used, thus having a significant influence on *development and learning*. The SIOPSA (2019) postulates that the 4IR could introduce much development and creativity in the areas of learning and development. Therefore, military commanders must link learning and development with the military strategy of the SANDF, which must also be aligned with the 4IR. Development and learning will assist soldiers to adapt their existing behaviour, and these new behaviours will assist the SANDF to achieve excellence during deployments.

The SIOPSA (2019) reports that a 2018 survey by the international company, Deloitte, found that the so-called "soft skills" will become a critical future priority and 80% of the respondents indicated that the so-called Generation Z[3] is increasingly "tech savvy"[4] but may lack the soft skills required in the digital world. Although the study was conducted in the private sector, one can expect that the same can be said for the military, as the military is usually a representation of the broader community. Furthermore, it was found that learner development remained a key retention tool for employees, but it needs to be more learner-centric than content-centric. One of the matters raised about learning is what is referred to as "learning on demand" (SIOPSA, 2019), which implies that it is difficult to make time for training and it will thus be important to take the learning to where the soldiers are. Gutche (quoted by AFRIBLOG, 2019) emphasises the use of augmented reality in training, where various scenarios can be role-played as if they are actually being experienced. Prospective military leaders can be placed in scenarios where real-life-like decision-making would be required, which can be assessed in real time, thus enhancing training to prepare more realistically for war.

From a behavioural sciences perspective, the military psychologists of the SANDF's Directorate Psychology will be best equipped to facilitate the *changed approach to learning*. They will also have to identify opportunities for learning, education and development that will make the transition to the digital environment easier. The Chinese entrepreneur Jack Ma made the following statement at the World Economic Forum: "*If we do not change the way we teach, thirty years later we will be in trouble*" (2018). He furthermore stated that people must move away from knowledge-based teaching and teach their children something unique to ensure that machines can never catch up with them.

Consequently, the SANDF will have to plan, implement, embed and evaluate change interventions, which is addressed under the discipline of *change management* (Busby, 2017). In many instances one can expect that machines will take over the functions that humans used to do. Military psychologists will facilitate the emotional impact on humans where the physical, mental and cyber domains start to overlap and interact. As mentioned above, there will always be the fear that machines will take over their functions and duties, as machines can be more creative. Although machines lack social interaction skills, they are likely to fare better in activities that require physical dexterity and mobility (Mostert, 2019; SIOPSA, 2019). Considering Ma's (2018) warning above, humans will consequently have to develop what Mostert (2019) refers to as agile competencies in areas such as continuous learning, and to focus on not acting like computers but rather nurture and develop specific human skills, such as imaginativeness, creativity, the ability to make connections that cannot otherwise be made (systems thinking) and the general ability to think about the future (anticipation) that makes them competitive to machines.

The SANDF 4.0 will require a new approach to the *career management* of its HR. Mostert (2019) emphasises that workers have significantly different expectations in the workplace because of the lack of stability and predictability due to the dynamic nature and the speed at which change occurs

3 People born from the mid-1990s to the early 2000s.

4 Well-informed about the skilful use of modern technology.

in the 4IR. The psychological contract between a soldier and the SANDF will also take on another form and in some cases it will need to be renegotiated. Military psychologists will thus have to anticipate and plan for dealing with the new challenges and processes mentioned above and will in most cases drive change management in the SANDF or at least provide advice on the matter.

Change management also deals with *changes in organisational culture* and especially with changes in structure towards a matrix or hybrid organisational structure with remote or virtual workers, workplaces across international borders, flexible working hours and a constantly changing work environment. With the introduction of virtual workers with flexible work hours, roll call may be done online, where employees log on to a computer and their work is electronically or digitally monitored. An HR practitioner may, for example, work from home, doing leave administration that is submitted electronically with customised software. There might also be new ways of establishing work teams, although some areas in the military will be more flexible than others. More agility and flexibility will be required from workers, involving less bureaucratic red tape, but more digitally monitored outcome-based performance assessments. This will have a psychosocial impact on virtual or remote workers who must deal with loneliness when they are unable to interact personally with their co-workers.

The changed work environment is likely to cause changes in the *physical work space* of soldiers that will require new thinking and advice on design thinking for work flow and work processes. In this regard, knowledge of ergonomics will be essential. Military psychologists will thus not focus on the human only, but also on the design of the environment in which the soldier functions.

The 4IR and its changing environment, sophisticated technology, information accessibility and cultural challenges will complicate the challenges facing future leaders (Cooper, 2009). Thus, when considering that Seiler and Pfister (quoted by Annen, Seiler & Jonas, 2010) explained leadership behaviour in terms of complex interaction between individual factors, group processes, organisational factors, context variables and situational factors, it is clear that there is a requirement for a fresh look at leadership at the tactical, operational and strategic level of the SANDF. This consequently requires renewed focus on *leadership development and assessment,* such as adaptability, agility and self-awareness, as well as critical and creative thinking, to name a few (Cooper, 2009; Mostert, 2019). Cooper (2009) suggests creating early opportunities to gain experience, professional education at an in-residence service college (such as the South African Military Academy and the various Service Schools), and an approach of life-long learning. This includes the competencies of imaginativeness, creativity, systems thinking and anticipation that make people different from machines, as stated above. Military psychology will thus play a vital role in equipping and preparing future military leaders to meet future challenges.

It is also likely that military psychologists will be required to assist with the management of *new threats in the information warfare domain.* Recent research by the British Broadcasting Cooperation refers to "deep fake technology" that uses machine learning and AI to create, for example, convincing videos representing events that never happened (Robinson, 2018). In this regard military psychologists can play a role in assisting computer engineers from the Council for Scientific and Industrial Research with the development of machine learning to help recognise and counter manipulated information, especially that coming through the media.

New 4IR-related technologies can also assist with *monitoring* and *evaluation.* The Directorate Psychology can seize the opportunity to monitor and evaluate in real time some of the services provided. These can include both health- and non-health-related interventions. It can be used to optimise service delivery and to ensure that services are delivered in the most efficient and cost-effective manner possible.

All of the above will lead to *requirements for a new policy and regulatory framework* within the DOD in general and more specifically the Directorate Psychology. All the above changes enforced by the 4IR will create several policy challenges. In this regard military psychology is likely to become a significant role-player in providing policy advice, developing policy and regulatory frameworks for new user systems that will be implemented.

## Conclusion and recommendations

The 4IR with its integrated cyber-physical systems is a reality, and future warfare will be radically different in terms of speed, duration, intensity and human involvement. The SANDF must come to terms with the accompanying internal dynamics and physical changes and optimise its forces within the allocated budget. Both the SANDF and the Directorate Psychology need to ride the coming wave of cutting-edge technology with its vast potential and vast risks. Technology and physical advances are, however, not the only means to improve the force, as it requires highly trained and effective soldiers with different skills and competencies to take advantage of technology. Military psychology as a discipline is ideally equipped to work both with and against technology to develop optimally functioning soldiers for an efficient and capable future force.

In some instances, military psychology will retain its "traditional" roles, but it must look at these roles in new ways and adopt new roles after reorienting itself in terms of the development and internal dynamics of the SANDF, looking at the future and focusing on force design, force development, force support and force deployment (operations) as field practitioners. In addition to the recommendations cited above, the recommendations below are suggested for the Directorate Psychology. Implementation will ensure delivery of the best quality and a highly responsive military psychological service to the 4IR military community. In the process, military psychologists will work side by side with engineers and physical scientists to advance soldier and system performance.

First, military psychologists need to understand what future wars require from humans and the impact these will have on them to gain insight into future human behaviour in the military. Moreover, knowledge of the new generations entering the military from society (generations X and Y, millennials and others) is essential to prepare the organisation for them. In this way those responsible will know what to recruit and select for. Consequently, new ways of recruitment and selection should be explored, with the possibility of migrating from a more hierarchical structure to a more matrix-like organisation when considering the notion of virtual workers, flexible working hours and organisational culture change.

Second, since postulating future configurations of technology and humans, military psychologists should not only react to technology, but should also utilise it. These new technologies can be integrated into their current and future service delivery processes, such as training on demand, online assessments and diagnosis supported by technology. Considering the skills required, as well as the complexity and cost of developing new technologies, the Directorate Psychology should consider public-private partnerships for developing technology that has both military and commercial benefit.

Third, new research possibilities can be explored and the collection of formerly unknown structures of data will give new insight into human behaviour. It is essential for the current way in which informed consent for research is obtained to change, as there is potential for more intrusion into the privacy of soldiers. Research areas should include competencies for future posts and ergonomic design to integrate humans with new technologies. Military psychologists involved in leadership development should also determine future leadership requirements.

Fourth, in addition to private-public partnerships, the Directorate Psychology should also involve stakeholders from academia as an integrated and comprehensive way to respond to the challenges and opportunities emanating from the 4IR. These role-players can contribute significantly to policy formulation. Fifth, the Directorate Psychology needs to develop more responsive governance by identifying and recommending policies, strategies and plans that will position it as an irreplaceable role player in supporting the broader SANDF to position itself as a competitive military force in fulfilling its obligations at national, regional and continental level. These must include state-of-the-art psychological support for all aspects of missions.

Sixth, besides these governance issues, the Directorate Psychology also has to provide strategic direction and guidance to subordinate psychologists in providing a comprehensive military psychological service in the context of the 4IR. Moreover, informing the various role-players on the SAMHS's Futures Forum[5] on the 4IR will stimulate them to seek creative solutions. Lastly, it is worth noting that the military psychologists of today need to become the experts of tomorrow if they want to make a difference in the future SANDF.

**Note**: This is an academic document and contains facts and opinions that the author alone considered appropriate and correct for the subject. It does not necessarily reflect the official policy or the opinion of the South African Government or Department of Defence and Military Veterans.

## References

AFRIBLOG. (2019, March 24). *How AI (artificial intelligence) will change our lives.* Retrieved from AFRIBLOG: http://blog.afrihost.com/tech/how-ai-artificial-intelligence-will-change-our-lives/

African Daily Voice. (2019). *South Africa appoints commission on fourth industrial revolution.* Retrieved from African Daily Voice: https://africandailyvoice.com/en/2019/04/09/south-africa-appoints-commission-fourth-industrial-revolution/

Andresky, N. L., & Henderson, J. (2018). *U.S. Army future warfare division White Paper: Operationalizing robotic and autonomous systems in support of multi-domain operations.* Retrieved from Public Intelligence: https://info.publicintelligence.net/USArmy-RoboticAutonomousMultiDomainOps.pdf

Annen, H., Nakkas, C., & Gehring, T. M. (2017). "What if?" The Swiss Armed Forces' approach to military psychology. In S.V. Bowles, & P.T. Bartone (Eds.), *Handbook of Military Psychology: Clinical and Organizational Practice* (pp. 539–548). Washington: Springer.

Annen, H., Seiler, S., & Jonas, K. (2010). Editorial: Military psychology in Switzerland: A short story with a long history. *Swiss Journal of Psychology, 69*(2), 75–82.

Bar-Gil, O. (2018). Head strong and the future of military psychology. *The Military Psychologist, 33*(1), 13–14.

Bester, P. (2016). Military psychology for conventional operations in Africa. In G.A. van Dyk (Ed.), *Military Psychology for Africa* (pp. 1–42). Stellenbosch: Sun Press.

Bester, P. C., & O'Neil, J. W. (in press). The military leadership training landscape of the South African National Defence Force as related to extreme situations. In M. Holeweger (Ed.), *Leadership in Extreme Situations.* Zurich: Switzerland.

Burgess, W. B. (2018, July 16). *Draft D Psych Statement of Intent 2019–2024.* Pretoria: Directorate Psychology.

Busby, N. (2017). *The Shape of Change: A Guide to Planning, Implementing and Embedding Organizational Change.* New York: Routledge.

Campbell, K. (2017, October 27). *The fourth industrial revolution is upon us and South African industry must adapt.* Retrieved from Creamer Media's Engineering News: http://www.engineeringnews.co.za/article/the-fourth-industrial-revolution-is-upon-us-and-south-african-industry-must-adapt-2017-10-27/rep_id:4136

Cooper, W. (2009, May 04). What does the military need to ensure that our operational leaders are prepared to meet the challenges of the 21st century? *Unpublished paper submitted to the Faculty of the Naval War College in partial satisfaction of the requirements of the Department of Joint Operations.* Newport: Naval War College.

Del Monte, L. A. (2018). *Genius Weapons: Artificial Intelligence, Autonomous Weaponry, and the Future of Warfare.* New York: Prometheus Books.

Gutsche, J. (2018, November 26). *AI and the super future: AI keynote speaker Jeremey Gutsche at the Fun Festival.* Retrieved from https://youtu.be/OzBJigTfGeE

5 This is an internal think tank discussing the future of the SAMHS.

Hughes, G. (2018, February 02). *Predicting future trends in warfare.* Retrieved from Defence in depth: htps://defenceindepth.co/2018/02/21/predicting-future-trends-in-warfare

INVESTEC. (2019, March 11). *Five myths about the future work.* Retrieved from Business Day: https://www.businesslive.co.za/bd/business-and-economy/2019-03-11-investec-dispels-five-myths-about-the-future-of-work/

Kamieński, Ł. (2017). Toward robotic warfare. In S. N. Romaniuk, & F. Grice (Eds.), *The Future of US Warfare* (pp. 193–207). New York: Routledge.

Kings, S. (2019, April 05). *Inside the SANDF's budget crisis.* Retrieved from Mail & Guardian: https://mg.co.za/article/2019-04-05-00-inside-the-sandfs-budget-crisis

Kott, A. (2018). Technological surprise and resilience in military systems. In B. D. Trump, M.-V. Florin, & I. Linkov (Eds.), *IRGC Resource Guide on Resilience: Domains of Resilience for Complex Interconnected Systems (Vol. 2)*. Lausanne, CH: *EPFL International Risk Governance Centre.* Retrieved from https://irgc.epfl.ch

Kriel, P. (2018). *Why it is time to review recruitment criteria to land the right candidates.* Retrieved from HR future: https://hrfuture.net/articles/talent-management/hiring-process/why-is-it-time-to-review-recruitment-criteria-to-land-the-right-candidates

Lele, A. (2019). *Disruptive Technologies for the Militaries and Security.* Singapore: Springer. doi:10.1007/978-981-13-3384-2

Ma, J. (2018). *Panel discussion at the World Economic Forum.* Retrieved from https://www.youtube.com/watch?v=pQCF3PtAaSg

Miller, A. (2019). *The military will soon be facing a defeat far worse than the French suffered at Agincourt.* Retrieved from The Trumpet: https://www.thetrumpet.com/20599-former-defense-official-military-revolution-to-unseat-us-superpower?utm_source=fb&utm_medium=article%20boost%20ad&utm_campaign=military%20funnel%20us%20uk%20former%20defense%20official:%20military%20revolution%20to%20unsea

Mostert, M. (2019). *Vierde industriële rewolusie [Fourth industrial revolution]. (I. du Plessis, Interviewer) South African Broadcasting Corporation.* Retrieved from http://www.rsg.co.za/potgooi-soek.asp

Ndhlovu, N. (2019). Discussion paper on enabling the South African Military Health Service to deliver an affordable and best-quality military healthcare service. Pretoria.

Nye, J. (2009). Get smart: Combining hard and soft power. *Foreign Affairs, 88*(4), 160–163.

O'Hanlon, M. E. (2018). *The role of AI in future warfare.* Retrieved from Brookings: https://www.brookings.edu/research/ai-and-future-warfare/

Phillips, D. (2019). *The military wants better tests for PTSD. Speech analysis could be the answer.* Retrieved from The New York Times Magazine: https://www.nytimes.com/2019/04/22/magazine/veterans-ptsd-speech-analysis.html?em_pos=small&emc=edit_war_20190425&nl=tbd&nl_art=0&nlid=91109412mc%3Dedit_war_20190425&ref=headline&te=1

Pinter, R., Toninelli, D., & de Pedraza, P. (2015). Mobile research methods: Possibilities and issues of a new promising way of conducting research. In D. Toninelli, R. Pinter, & P. de Pedraza (Eds.), *Mobile Research Methods: Opportunities and Challenges of Mobile Research Methodologies* (pp. 1–10). London: Ubiquity Press. doi:dx.doi.org/10.5334/bar.

Ramaphosa, M. C. (2019). *President Cyril Ramaphosa: 2019 State of the Nation Address.* Retrieved from South African Government: https://www.gov.za/speeches/president-cyril-ramaphosa-2019-state-nation-address-7-feb-2019-0000

Robinson, O. (2018). *Malicious Use of Social Media: Case Studies from BBC Monitoring.* Riga: NATO. Retrieved from https://www.stratcomcoe.org/malicious-use-social-media-case-studies-bbc-monitoring

Rosenbaum, E. (2019). *IBM artificial intelligence can predict with 95% accuracy which workers area about to quit their jobs.* Retrieved from CNBC: https://www.cnbc.com/amp/2019/04/03/ibm-ai-can-predict-with-95-percent-accuracy-which-employees-will-quit.html

Society for Industrial and Organisational Psychology [SIOPSA]. (2019). *Industrial revolution 4.0 and disruptive learning-Learning {re}imagined.* Retrieved from Society for Industrial and Organisational Psychology: https://www.siopsa.org.za/1983-2-4/

The Economist. (2018, January 27). Special report: The future of war. *The Economist*, pp. 1–16.

Thomas, T. L. (2019). *Russian forecasts of future war.* Retrieved from Army University Press: https://www.armyupress.army.mil/Portals/7/military-review/Archives/English/MJ-19/Thomas-Russian-Forecast.pdf

Tuck, C. (2019). *Technology, uncertainty, and future war.* Retrieved from War in depth: https://defenceindepth.co/2019/03/11/technology-uncertainty-and-future-war/

University of Johannesburg. (2019a). *Ignore revolutionary technological advances at your own peril: President Ramaphosa.* Retrieved from University of Johannesburg: https://www.uj.ac.za/newandevents/Pages/Ignore-new-revolutionary-technological-advances-at-your-own-peril--President-Ramaphosa-.aspx

University of Johannesburg. (2019b). *Join the debate: Is the academic thesis obsolete in the Fourth Industrial Revolution?* Retrieved from Business Day: https://www.businesslive.co.za/bd/national/2019-04-04-is-the-academic-thesis-becoming-obsolete-in-the-fourth-industrial-revolution/

Visser, A. (2019). *Die werksplek en die werksmag van die toekoms [The workplace and the workforce of the future].* Retrieved from Netwerk 24: https://www.netwerk24.com/Finweek/die-werksplek-en-werksmag-van-die-toekoms-20190214

von Scheel, H. (2016). *Keynote address delivered at the World Trade Symposium on mastering the fourth industrial revolution, in London on 8 June 2016.* Retrieved from Researchgate: https://www.researchgate.net/publication/309591344_Mastering_the_Fourth_Industrial_Revolution

World Economic Forum. (2018). *Towards a Reskilling Revolution: A Future of Jobs for All.* Geneva: World Economic Forum.

# 7

# WAR OF THE FUTURE AND PROSPECTIVE DIRECTIONS OF MILITARY PSYCHOLOGY

*Alexander G. Karayani*

The ability to look forward and to imagine horizons of own possibilities is a cardinal condition of existence and development of any science, including military psychology. A view of the prospect, like a distant light of car headlights, allows military psychology to see the right direction of movement, to highlight the most important landmarks and possible obstacles and be prepared for meeting the challenges of the future.

Forecasting is not just a look forward. It is a scientifically grounded attempt to imagine possible conditions, directions of development and prospective tasks of military psychology. It is necessary to analyze the main determinants of military psychology development in order to avoid reducing the prediction to meaningless divination.

Analysis of scientific literature shows that predicting the future is not a favorite theme of military psychologists. The bibliography on this problem is extremely scarce and includes only a few modern scientific works containing a general analysis of the prospects of military psychology (Bowles & Bartone, 2017; Cronin, 2003; Karayani, 2016; Kennedy & Zillmer, 2006; Laurence & Matthews, 2012; Mukherjee, Kumar & Mandal, 2009) and only one work directly dedicated to forecasting in the field of military psychology and practice (Matthews, 2014).[1] Studying of the mentioned works allows one to state that scientific prediction of military psychology prospects is connected by authors with their representations about future wars directly or indirectly.

Thus, Cronin (2003) allocates four factors as the most important variables of military psychology's prospective development: (1) achievements in technologies, (2) budget reduction, (3) political changes and (4) demographic changes. These factors reflect the state's perceptions of war threats and wars of the future. Laurence and Matthews (2012) and Matthews (2014) consider the type/types of future wars, modern technologies aspiring to the future and the science which has a decisive influence on the results of combat action as the main determinants of prospective trends of military psychology. Kennedy and Zillmer (2006) and Bowles and Bartone (2017) believe that the future of military psychology will depend on real characteristics of fighting,

1 Of course, there are works with limited access, which contain the forecast of military psychology's directions of development. However, for known reasons these works are not analyzed in this chapter. In addition, a number of works contain predictions about the future of individual military-psychological industries: psychological training (McNab, 2001), military neuropsychology (Kennedy & Moore, 2010), etc.

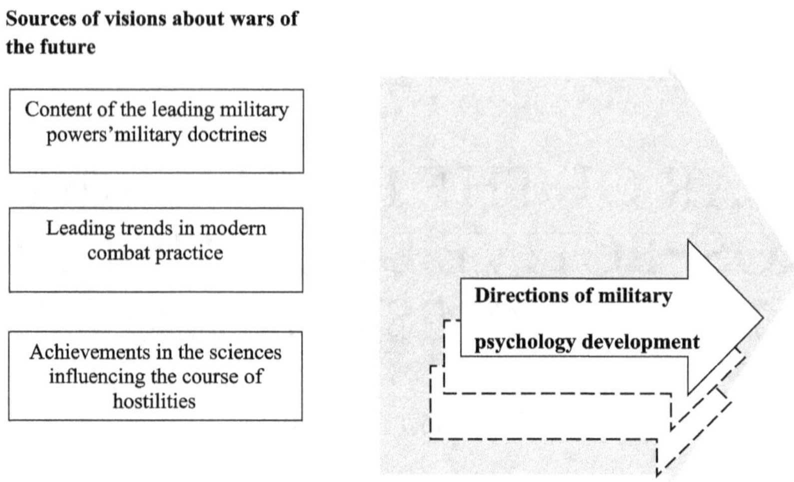


*Figure 7.1* The influence of perceptions of future wars on the vision of prospective directions of military psychology development.

determining the level of psychological pressure on the military. Karayani (2016) points to the importance of containing states' military doctrines in the determination of sustainable trends of military psychology.

Generalization of the above approaches allows allocating those sources where the future of military psychology is already born today (Figure 7.1).

The link between official visions of future war and the prospective directions of military psychology is not accidental. A strong and mutually useful connection between military psychology, on the one hand, and military science and practice, on the other hand, was established. In early stages of its development, military science and martial arts grew out of the psychological insights, ideas and generalizations of talented generals. Their martial art was inextricably linked with the military's cunning and ability to preempt, frighten and deceive the enemy. With military psychology's acquisition of the status of a science, it began to develop, focusing on current and future needs of military science and combat practice. Precisely, military science and combat practice of troops are customers of specific research and practical activities of military psychologists. Military science predicts the kinds, means and ways of future wars, to which it is necessary to prepare servicemen, combat and the psychological qualities of warriors necessary for victory in the future battles (Karayani, 2016).

Let us consider the highlighted determinants of military psychology development (Figure 7.1).

## Military doctrines of states about future wars

Scientifically grounded views on future international relations, determining the most significant threats to the country's security, outlining possible wars, formulating basic theses of military policies, and objectives for the protection of national interests are expressed in military doctrines of states. They take into account political and demographic trends, as well as the budget capacity of states.

Studying military doctrines of leading military powers shows that the world is entering an era of high-risk, large-scale wars again. A large-scale war is a war between coalitions of states or between the largest states of the world community (Military Doctrine of the Russian

Federation, 2014). In the military-doctrinal document of China, "White Book," a new world war is considered inevitable (Cheng, 2013; Shishkina, 2015); in the military doctrines of the United States (National Security Strategy of the United States of America, 2017) and Russia (Military Doctrine of the Russian Federation, 2014), the danger of new world war, as it was earlier, is not excluded. In all recent editions of military doctrinal documents of these countries, the main source of danger that is identified is not international terrorism, but the world's leading powers.

As is known, in large-scale wars the sides pursue radical goals and mobilize all their material and spiritual resources. Such wars, as the history of two world wars shows, can be perceived by people as a catastrophe of planetary scale, which in a moment destroys the basic beliefs and values of millions of people. War veterans can be subjected to extreme psychological stress. The immense space-time scale of the war, monstrous destructive power of weapons and gigantic loss of human lives in acting armies can lead to unpredictable psychological consequences (Karayani & Karayani, 2016).

However, large-scale wars of future in military, technological and psychological terms will differ significantly from wars of the past. According to the views of U.S. and Chinese military experts, understanding of modern and future wars is based on the imperishable methodological ideas of Sun Tzu (2003) and Carl von Clausewitz (2002) that the wars are becoming more and more psychological. According to their logic, war is understood not as the physical extermination of the enemy, but as a process of purposeful change of perception, consciousness, worldview, emotions, will and behavior of the enemy. The criterion of victory is considered the scrapping of the opponent's will, forcing it to the desired actions and behavior. Such an understanding of war opens up a large arsenal of means of influencing the enemy. In essence, the war of today and the war of the future are hybrid wars.

Hybrid warfare combines features of all known types of wars: traditional wars of different generations and different scales (local, regional, large-scale), network-centric, behavioral, diplomatic, political and economic wars. At the beginning of the 21st century, the role of psychological methods of warfare (psychological operations, provoking civil disobedience, "color revolutions", separatism, terrorist and guerrilla activities) noticeably increased in the structure of war.

War loses spatial and state borders, lines of distinction between war and peace, front and rear, law and lawlessness. The battlefield turns into a "combat space" (Messneras, as cited in Marchenkov, 2005, pp. 76–81), and a war itself can be remote, "silent", invisible to inhabitants. Hybrid warfare can crawl "underground" for a long time like peat fire, systematically breaking out in the form of local and civil wars.

Political and religious organizations, financial and industrial corporations, private military companies, self-proclaimed quasi-states and civilians are becoming increasingly involved in achievement of military objectives. According to the views of NATO military strategists, the ideal of hybrid warfare is getting the enemy masses out of emotional balance and inducing them to struggle with their own government (U.S. Department of the Army, 2011). This way of achieving military goals is called the "Trojan Horse strategy". Similar ideas are contained in the Chinese doctrine of political war "Three Warfares", representing the triunity of war for public opinion, psychological warfare and legal warfare (Cheng, 2013).

The existence of civil war features in hybrid war can fill it with special intolerance, hatred and cruelty. Inflammatory ideas and propaganda slogans can penetrate into military units, disturb soldiers and destroy their values and moral guidelines.

The split in the enemy camp allows it to conduct war by remote methods without endangering its own warriors, causing panic and despair in the enemy ranks.

However, hybrid warfare is not limited to using "indirect actions". In recent years, a peculiar "resurrection" of nuclear weapons has been observed. The possibility of using nuclear weapons is provided in the military-doctrinal documents of China, Russia and other countries. In the United States, a special doctrine that defines a strategy of using nuclear weapons was developed—the nuclear posture review (U.S. Department of Defense, 2018), which defines a course for tactical nuclear weapon development. Considering that terrorist organizations have more and more abilities to produce "dirty nuclear bombs", the extent of nuclear weapon use in the wars of the future can be enormous.

In the literature (Cronin, 2003; Karayani, 2016; Kennedy & Zillmer, 2006; Ritchie, 2011), it is shown that the psychological consequences of nuclear weapon use are due to the specificity of its main damaging factor action, as well as the necessity to act in the means of protection. The range of psychological reactions of military personnel to the use of nuclear weapons is very extensive: from disorientation and panic caused by loss of vision due to eye damage by light radiation, impaired perception and thinking associated with brain injury caused by shock wave, to radio phobia generated by penetrating radiation and radioactive contamination of the terrain.

If, in future wars, nuclear weapons will become a tactical means of confrontation and their use will be almost daily, the scale of mental trauma of the military and the number of persons who need psychological assistance is difficult to imagine today.

Along with weapons of mass destruction, in future wars, high-precision, hypersonic weapons and weapons based on new physical principles, electronic warfare, information and control systems, unmanned aerial and autonomous marine apparatus and controlled robotic models of armament and military equipment will be widely used (Military Doctrine of the Russian Federation, 2014). The means of defeat will become even faster, high-precision, possessing intellectual properties of search, identification, choice of attack method and guaranteed defeat.

All of this will make fighting extremely dangerous, tense, complex and stressful.

## Leading trends in modern combat practice

Martial art changes faster than theses of military doctrines. To some extent, military art is a process and result of the creativity of many combat actions' participants, both in regular armies and irregular paramilitary formations. The traits of future wars can be reflected in the live combat experience of commanders and soldiers as well as in developing methods of tactical action.

An analysis of stable tendencies in strategy and tactics of troop combat actions in modern military conflicts shows that the strategy of multi-domain confrontation (U.S. Department of the Army, 2018) is realized more and more actively. The essence of this strategy is that the enemy is affected in all areas of its presence and depth of its location.

Analysis of military activity spheres of various states' armies allowed us to allocate 10 spaces of modern and future wars (Figure 7.2).

In Figure 7.2, it is seen that wars can be unfolded in 10 spaces of military confrontation: ground, underground, surface, underwater, air, cosmic, psychological, social, informational and cyberspace. The boundaries between these spaces are very conditional. Moreover, these spaces penetrate each other and unite in different constellations, making each war unique. At the same time, the conduct of hostilities in each of these spaces is characterized by pronounced psychological specificity.

Thus, in the *ground space*, warriors can face both high-tech opponent and irregular formations, which use tactics of sabotage, terrorist acts, ambushes, guerrilla actions, wide using of traps and improvised explosive devices (IEDs). Additional negative psychological effects can be caused by the presence of women and children in enemy ranks. According to statistics, about 300,000

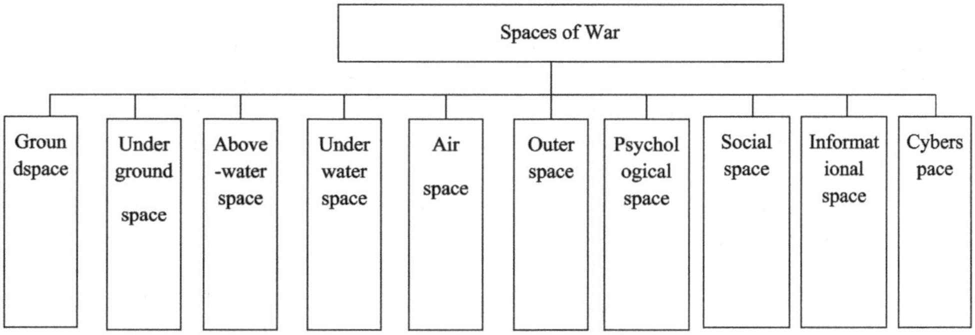


*Figure 7.2* Spaces of future wars.

children participated in 30 military conflicts in the last 10 years (Inaeva, 2017). Children do not have a paralyzing horror of death. Often they perceive taking part in war like a game. Children more quickly acquire new activities, they are more easy, mobile, not burdened by property and obligations. Therefore, their number in irregular military formations can increase in the future.

The ground space is heterogeneous. In the ground space specific zones where human sensations, perceptions and feelings function in a special way are distinguished qualitatively, and maximum involvement of neuro-psychic reserves is required. Such zones include mountains, deserts, jungle, arctic regions and so on.

The analysis of wars of the late 20th century–early 21st century shows that in cases where warring parties are determined to fight to the end and none of them succeed in realizing the strategy of remote war, the outcome of confrontation will depend on a soldier-infantryman. In these circumstances, the role of small, highly mobile, professionally trained, technologically equipped combat units will increase significantly. Such units will travel long distances in a short time to different natural-geographical and weather-climatic zones. This will require the ability of military to survive in a complex environment and maintain psychological stability and group cohesion in isolation from main forces.

The *water surface space* creates a special combat reality, which generates the "effect of a showcase". There is nowhere to hide, so almost every battle on the water surface will be decisive.

The *underground space* of war is extremely difficult psychologically. Battle tunnels and underground communications were widely used in the wars in Korea (1950–1953), Vietnam (1965–1974), Afghanistan (1979–1989) and a number of Arab-Israeli conflicts. Today dozens of kilometers of combat tunnels are used in Syria not only for movement of militants, but also for some specimens of military equipment. In hybrid wars, irregular formations will use the action tactics of underground war increasingly actively. Such tactics allow parties to save forces under distant blows of the technologically superior enemy opponent, to seem invisible, omnipresent, owning the initiative. It can cause anxiety, distress and psychological tension in the military. The soldier thus is required to have the ability to act alone, in confined space, in conditions of poor visibility, constantly waiting for ambushes and traps for confrontation in underground communications.

The importance of *underwater space* in confrontation will increase due to active exploration of natural resources of the Arctic, laying of the Northern Sea Route, constructing of ocean oil platforms, holding of strategic gas pipelines and trunk cables of communication on the bottom of the seas. In underwater combat, soldiers must be ready to act alone, as part of small militant groups, in conditions of diminishing perceptual capabilities and lack of communication and in situations of uncertainty.

In hostilities in the *airspace*, increasingly widespread use of unmanned reconnaissance and shock aircraft (UAVs) creates new psychological effects. Sometimes more than 50 UAVs a day cruised in the sky of Syria. Swarm technologies used by drone groups are at the stage of testing. In such conditions, vigilance, observation and discipline of disguise become especially important qualities. The psychological features of reconnaissance activities and percussion drone operators require special studies.

War between technologically equipped adversaries can spread into outer space. We have yet to comprehend the psychological effects and consequences of hostilities in space.

Already today many experts assess the consequences of war in *cyberspace*, including psychological, as the most dramatic of possible consequences of military confrontation. According to experts' assessments, disruption of intelligence system functioning, analysis, combat management and communications can cause a sense of catastrophe, disorganization of management, abandonment, betrayal and panic among soldiers.

Informational space is closely connected with cyberspace. Through information, people create false realities that have a significant impact not only on the outlook, will and behavior of military, but also on world public opinion. Thus, one staged video about the use of poisonous chemicals against civilians can produce a more powerful social, political and even military effect than actually using chemical weapons.

According to experts' opinion, achievement of victory in wars of the information age is determined by the ability to provide informational dominance (Cheng, 2013).

In the middle of the 20th century, military theorist and futurologist Messner (as cited in Marchenkov, 2005, pp. 78) proved that war had gained a special *psychological space,* where issues of victories and defeats will be solved in the future. War is considered an intellectual, motivational, emotional and volitional confrontation. Such psychological phenomena as vision and understanding of fight, desire to fight, will to victory, self-confidence, confidence in justice of own actions, confidence in fighting comrades, belief in the commander, discipline, cohesion, pride for one's own unit and psychological stability of opposing sides' warriors are faced here.

Analysis of modern military events allows us to suggest that the *social space* will be the priority space of confrontation in future wars. Leaders, social groups and movements, electorate and masses will fight for this space. Thus, in Syria, negotiations with political, military and religious leaders and humanitarian actions for the population achieved more than bombing and ground-based hostilities.

Today it is obvious that in future wars hostilities will be waged simultaneously in several spaces. Attacks of warriors can be expected from space, from the air, from the ground, from underground, from near and distant waters and from underwater. Not only enemy soldiers, but also civilian populations, including women and children, will be a threat. Danger will come not only from the enemy's combat assets, but also from the mass media, computers, smartphones, televisions and radio receivers. Only soldiers who have a belief in the fair mission of their country, their army, their commander, combat comrades and their own forces and who have high psychological stability and the will to victory can act in this situation effectively.

## Modern technologies are conductors towards the future

Modern combat technologies make a battlefield more and more similar to the circus arena, with numerous observers, powerful spotlights and dangerous predators.

Spacecraft groupings, heavy and light drones and remote surveillance cameras view each meter of a modern battlefield; numerous motion and noise sensors "listen" to every rustle; small-sized, noiseless and barely noticeable combat robots have already begun "palping" everything that causes suspicion. Uninhabited underwater stations, underwater motion sensors and unmanned underwater combat vehicles will gradually transform the underwater space into peculiar huge "aquariums". GoPro action cameras placed on helmets of warriors will make a large number of spectators witnesses to each warrior's movement. The emerging "effect of a showcase" is a serious new challenge for participants in fighting (Karayani, 2016). It adds a sense of extraordinary responsibility, inevitable external control and lack of personal space to the feeling of danger, uncertainty and ambiguity.

At the same time, military and civil technologies develop in the direction of maximal expansion of a person's natural abilities: eyesight, hearing, force, speed of movement and so on. Such technologies take an important place in the fighting equipment of soldiers, which is developed within the programs "Soldier of the Future" in the United States, Russia and some other armies. Judging by their content, the warriors of the near future will have superpowers, such as high speed of thinking, foresight, power, endurance, high security and survival.

The famous science popularizer Kaku (2011) notes that a chip in today's musical greeting card surpasses all computing power available in the world in 1945. In today's cell phone, there's computing power that surpasses all the computing resources that NASA had in 1969 when two astronauts first stepped on the Moon.

Today, computer stations that possess considerable power and the ability to calculate many options for the development of combat situations and to offer effective ways of action are at disposal of warriors. The opportunity to be constantly aware of the situation, to make informed combat decisions and to carry out staging of combat tasks and target designations to subordinates in real time is given to commanders.

Built into the warrior's combat equipment, the CCTV system allows them to detect the enemy for miles, in the day and night, in conditions of poor visibility. Nano-drones and planning mines with video cameras, emerging in a number of armies, provide warriors with the opportunity to peek behind obstacles, into various shelters, and view in advance paths of intended movement. So, the military equipment of the Russian warrior "Ratnik-3" will not only protect the warrior with reliable armor, but also will significantly increase its strength and endurance and allow him to make long marches on heavily rough terrain with a large combat load due to the exoskeleton. A masking system will automatically adjust the color of a warrior's outfit to the background of surrounding area. The system of remote monitoring of psychophysiological condition, fatigue degree and sleep quality will allow commanders and military-medical workers to detect unfavorable physiological conditions and wounds of a warrior in time. All of this will increase the confidence of a warrior in his abilities, such as the fact that in case of injury, he will be provided with targeted and quick help (Karayani, 2016).

Technologies that save lives of soldiers and return them to active life after serious injuries and traumas will positively affect the mental state of warriors and their confidence in them. This will be facilitated by achievements of telemedicine and effective analgesics, narcotizing servicemen for the time of delivery to medical institutions, and perfect bioprosthetic limbs or exoskeletons, allowing restoration of their lost mobility (Matthews, 2014).

The above circumstances will allow tomorrow's soldiers to make combat decisions using artificial intelligence, to detect enemies on the distant outskirts in the day and at night, to detect attempts of their hidden displacement and to feel great self-confidence and confidence in their colleagues and commanders. Professionally trained, tight-knit small units and individual warriors with high fighting spirit will be able to solve combat problems of historical scale. However, the presence of similar funds in the enemy will turn many military episodes into fierce intellectual and emotional confrontations.

## Prospective vectors of psychology and related science development

The main features of the future war can be predicted by studying tendencies in the sciences, which develop prospective military technologies.

American military futurologist general Scales asserts that a scientific branch was a part of each significant war of the 20th century, radically influencing rapid development in fighting and making decisive contributions to victory. So, in World War I, chemistry allowed the development of combat poisoning formulations and explosives. In World War II, the leading science was physics which was the basis of radar and atomic bomb development, which had wide impact on the results of fighting. During the Cold War, which Scales calls World War III, information technologies which underlie powerful computers and advanced command and control systems development dominated. In the era of the Global War on Terrorism (World War IV), it has become impossible to win wars between political, religious and social ideologies solely by kinetic means. The way to victory, according to Scales, lies through use of firepower in close unity with the achievements of modern social sciences: psychology, sociology, anthropology, culturology (Matthews, 2014).

Today the military is increasingly interested in achievements in the field of psychology in its interaction with cognitive, social and neuro-sciences, first of all in cognitive neuroscience, neuropsychology, psychophysiology and social neuroscience. Achievements in integrated scientific fields can significantly change our views of particular perceptions by military of the combat situation and information, mechanisms of influence of combat stressors on mental states and combat activity of warriors, social and cultural variables of behavior and human interaction.

Recommendations, which are expected from neuropsychology, can make significant changes in the practice of psychological training, support and rehabilitation of fighting participants and recovery of wounded soldiers (Kennedy & Moore, 2010).

In-depth understanding of social and psychological regularities of interpersonal and inter-group communication allows specialists to transform a large number of social networks users of the opposite side into active distributors and even into producers of informational and psychological materials. Tactics of "crowdsourcing" allow the spread of ideology, worldview, values and culture among the population of the enemy through its population due to "soft power". Thanks to this, social networks (Twitter, Facebook, YouTube, etc.) and virtual networks of messengers (WhatsApp, Viber, Telegram, etc.) can become the arena of civil informational and psychological warfare (Karayani, Karayani, & Tsygankov, 2018).

Thus, analysis of the content of military doctrines of the leading military powers of the world, stable tendencies in the practice of past and modern wars, trends in the development of military and potential military technologies and sciences aspiring to the future allows the suggestion that wars of the future can have a variety of faces and scenarios. If none of the opposing military forces will achieve absolute superiority in military force or its speed of application and retaliation, wars are likely to have a hybrid nature. Along with the most advanced military means, "asymmetrical", "soft" and "indirect" strategies for coercing the enemy to the desired behavior will be used. Wars of the future can consistently or simultaneously leak in different spaces on tactical, operational and strategic forefronts.

Hybrid wars, gaining more and more psychological character, will give the soldiers a sense of uncertainty, obscurity, variability, contradiction, chaos, danger, poor self esteem, doubts about the fairness of their own actions and extraordinary personal responsibility. The force of the psychological pressure of the fight upon the military can surpass the limits of their opportunities.

In these conditions, psychology as a science about uncertainty, ambiguity, variability, danger, thinking, will and behavior, in community with other sciences should become a force that helps soldiers to remain fighters and carry out the combat tasks, maintaining their mental and physical health.

## The image of future war as a predictor of military psychology development

Shumkov (1905), at the beginning of the 20th century, stressed that military psychology will become a necessary science for a commander only when it will show him ways of increasing the psychological strength of his troops and reducing the psychological capabilities of the enemy in practice.

Will military psychology become a vital science to commanders in the future? The answer to this question depends on how military psychology will take into account prospective tendencies in military science, military art and military technologies, what it sees as its mission and tasks in future battles, how critically it will assess its real and potential opportunities and prospects and how it will build up relations with frontier social and neurosciences.

Analysis of the available military-psychological literature allows one to state that military psychologists of different countries (Karayani, 2016; Kennedy & Zillmer, 2006; Laurence & Matthews, 2012; Miao et al. 2017; Mukherjee, Kumar & Mandal, 2009), reflecting on the directions of military psychology in different terms, generally agree with their main list, placing different emphases only on their content. Military psychologists daily solve the most important tasks related to the study and development of soldiers' and commanders' personalities, consolidation of military teams, prevention of suicide, drug addiction, crimes in the army, development of new military equipment and so on. But their main mission is to train effective fighters and make them psychologically stronger, able to solve complex combat problems effectively and survive in modern combat.

This mission is implemented in the following areas: psychological selection of military personnel, psychological training of servicemen, psychological prevention of destructive combat stress and post-traumatic stress disorders (PTSDs), psychological assistance and rehabilitation of combat operation participants and veterans.

The concepts of psychological components of future wars described above will inevitably require improvement of the system of *psychological selection of military personnel*. Currently, the main focus is on assessing the conformity of military personnel's psychological qualities to one or the other military specialty. Therefore, cognitive methods, allowing measurement of the quality of perception, attention, memory, imagination and thinking dominate in the structure of psychodiagnostic methods.

However, according to experts (Matthews, 2014), effectiveness of a soldier depends only 25% on the degree of his or her cognitive quality development. In their "pure form", these qualities are necessary mainly only to those military experts whose activity is connected with tense cognitive activity—scouts; snipers; professionals managing complex hardware complexes and systems; operators of unmanned aerial vehicles, submarines and combat robots and so on. The absolute majority of combat action participants do not need "pure" cognitive qualities, but their concrete derivatives: not perception, but observation, not attention, but attentiveness, not memory, but mindfulness, not thinking, but resourcefulness.

Apparently, active equipping of a soldier with powerful computational means will slightly reduce requirements for some of his cognitive qualities. Indeed, devices of augmented reality, tactical communicators and other regular gadgets will take on complex cognitive functions. A soldier on a battlefield will be a "neuron" of a powerful "neural" network of combat control, articulated with groups of combat satellites, UAVs and detection and control sensors. Devices mounted to the set of his combat equipment will allow him to see and hear better, to think more effectively and to make more informed decisions. Soon they will assume functions of storing and "carrying" the knowledge which is needed in battle. Interface of these devices is maximally simple and convenient that will allow one to receive and transfer necessary information almost instantly. However, such devices are very vulnerable. Therefore, a soldier must see, hear and understand a fight and act without gadgets' help.

Predictive characteristics of future wars show that warriors of the future must have an unwavering faith in the justice of their country's affairs, in their commander and their own forces (Copeland, 1951), a developed sense of debt, will, endurance and high resilience, creativity, initiative and leadership qualities to defeat the enemy.

Psychological selection should be directed toward the assessment of such abilities in warriors (Laurence & Matthews, 2012; Matthews, 2014).

To reveal the degree of development of intuition, emotional, volitional and motivational sphere of warriors by blank methods and diagnostic software-hardware complexes is not easy. It will be necessary to introduce to psychological selection methods of evaluation by means of simulating situations similar in their psychological structure to real combat situations. Such assessment can be carried out on a certain tactical background, during the process of working out tactical and fire tasks, military-sports games, working at combat simulators and so on. Virtual reality complexes and specially developed computer games can be actively used to simulate diagnostic situations in the future (Karayani, 2016).

The necessity of *psychological preparation* as a priority task in preparation for hostilities was understood for the first time at the beginning of the 20th century. Analyzing the psychological consequences of the Russo-Japanese War (1904–1905), Shumkov (1905) concluded that in addition to combat and technical training for a fight, there should also be psychological preparation. Even back then, the task of psychological preparation was seen not only to expand the psychological capabilities of own warriors (psychological training), but also to shake and undermine the psychic forces of the enemy.

Currently, many armies are aware of the need to abandon the priority of the "disease-treatment" strategy in favor of the strategy of prevention of psychological problems in participants in hostilities (Laurence & Matthews, 2012; Matthews, 2014) through their psychological preparation to change the focus in military psychology from retroactive management to proactive preparedness (Mukherjee, Kumar & Mandal, 2009).

Psychological training is considered the unity of general, special and target psychological training. Such architecture of psychological preparation allows the realization of universal, special and situational tasks, to make a soldier a universal fighter and a narrow specialist. Thus, *general psychological training* is conducted with all soldiers throughout their military service and is aimed to form their psychological preparedness for participation in hostilities and resistance to general combat stresses (Karayani & Korchemny, 2016).

The most ambitious, scientifically and methodically substantiated program of general psychological training of servicemen—the program "Comprehensive Soldier and Family Fitness"—is realized today in the United States (Laurence & Matthews, 2012; Matthews, 2014; Sinclair & Britt, 2013). With spiritual, emotional, social, family and physical dimensions in its structure, this program is designed to maximally psychologically prepare military personnel for performing tasks in the conditions of hybrid wars. The main goal in the implementation of the program is to make soldiers more resilient to stress factors of modern and future wars by forming their common beliefs, values, principles, mutual trust, friendly relations, optimism, self-awareness, strong character, self-regulation skills and ability to build reliable social connections.

*Special psychological training* is carried out with servicemen, who solve special tasks or act in specific combat spaces in specific natural and social conditions. *Targeted psychological training* is designed to configure people for performing a specific combat task, taking into account characteristics of the existing combat situation.

Special and targeted psychological training should be carried out on the basis of pre-developed psychological models of specific combat tasks. Servicemen should prepare for autonomous fighting against an extraordinary thinking enemy, in difficult weather-climatic and ethno-social conditions, in the day and at night, on the ground, saturated with means of observation and control, explosive devices and traps (Karayani & Korchemny, 2016).

Psychological training is designed to develop the military's rapport, skills of effective combat interaction and cohesion. Methods of learning, which cause trainee interest and competitiveness, providing instant feedback, applied on a specific tactical background, simulating confrontation with the enemy, interaction with the aggressively minded population and activities in zones of radioactive and chemical contamination, are well demonstrated in the development of such qualities. Modern complexes of virtual reality, simulators, team sports and tactical games such as "Paintball", "Airsoft", "Hardball" and related computer games can be used more widely for this purpose.

In the psychological training of unit commanders, it will be necessary to develop their ability to simplify a situation, to bring logic into illogical events, to act in multitasking mode and to make decisions in a matter of seconds. Commanders should be able not only to manage the fighting of their subordinates effectively, but also to take measures for misleading the enemy and to prevent its informational and psychological actions, to assess the possible behavior of the population in the area of a combat task performance, to interact with troops of allied armies and to be ready to act in conditions of communication and control means failure.

If we look into the future, we can assume that deployed, deeply layered systems of psychological assistance (Karayani, 2018; Ritchie, 2011) to military personnel will be claimed mainly in large-scale military conflicts with massive use of troops.

In a dynamic fighting confrontation with the wide use of small autonomous units, using psychological assets will be an effective form of providing psychological support to the military. The experience of psychologically trained military personnel providing psychological assistance to colleagues in overcoming distress in combat situations has been accumulated in the British armed forces. The Trauma Risk Management System (TRiM) is being used here. It is considered that the timely response to traumatic situations with the support of colleagues helps to prevent subsequent development of PTSD among combat action participants (Hughes & Neville, 2014). In our opinion, the TRiM practice contains great heuristic potential for development of the system of psychological assistance to participants of combat operations.

The approach, developed by Hoge (2010), in which PTSD is considered not a disease but a kind of inertia in manifestation of combat reflexes among war veterans, seems very promising in prevention of PTSD. This approach allows one to overcome the military servicemen's fear of the stigma and resistance to seeking psychological help. It promotes transformation of PTSD from mental disorder to inertia of a warrior's heroic past.

"Virtual" therapists, tablets and smartphones with psychological self-help programs can help to remove the negative impact of stigma on war veterans. In our opinion, an interactive complex of virtual reality, BRAVEMIND, developed by Rizzo, contains great potential in working with PTSD (BRAVEMIND, 2019). It allows the realization of many forms of exposition psychotherapy of war veterans.

Future wars will require military psychologists to carefully assess the effectiveness of various forms of fight participants' socio-psychological re-adaptation. The reduction of the duration of the military presence in the combat zone and the rotational nature of many specialists' participation in solving of combat tasks require additional studies of warrior re-adaptation practice, called in some NATO countries "Third Location Decompression" (De Soir, 2011; Karayani, 2018).

Today military psychology is in the prime of its strength. Military psychologists work in countries with different state apparatuses, with different ideologies and religions. They gain more and more authority among soldiers and commanders. In some armies, combat planes do not take off into the air, submarines are not set off to military campaigns and warriors do not go to perform military tasks without the approval of psychologists.

Military psychologists cannot prevent war, but they perform a high moral mission. They help compatriots who defend the national interests of their country with weapons in their hands to survive, to maintain mental and physical health on the battlefields and to return to peaceful life as full-fledged citizens.

Even a brief analysis of military psychology prospective problems, arising from ideas about future wars, shows that in the near future, military psychology will face grandiose challenges. Matthews (2014), evaluating the contribution of military psychology in various branches of social practice, showed that this contribution is weighty where military psychology actively interacts with other sciences—social, neuro-sciences, informatics. Apparently, it is in the integration with other sciences that military psychology will be able to comprehend the psychological features of wars of the future, find the key to understanding the psychological phenomena of combat, give warriors recommendations about turning fear into fury, indecision into anger, weakness into endurance and heroism and the ways of defeating yourself to become invincible (Karayani, 2016).

As military psychology becomes more international, international conferences are held, international textbooks are being developed and interpersonal contacts of different countries' military psychologists are carried out. In such international cooperation, first of all concerning the area of psychological assistance, rehabilitation and re-adaptation of fight participants, there is a powerful potential for the development of theory and practice of military psychology in the future.

## Conclusion

Forecasting the future of military psychology is an important condition of its successful development. The right forecast allows definition of the most important directions of military and psychological research, lines of interaction with related sciences, the content of training of military psychologists and resources necessary for development. The correct vision of the prospect allows military psychology to be the science vital to commanders and soldiers in the face of new challenges.

Objective anticipation of the prospects of military psychology is possible only on the basis of the scientific analysis of future wars. The main character of any war is the person with concrete intellectual, emotional, motivational and strong-willed qualities and a certain level of psychological stability. Research into these qualities is a prerogative of military psychology. Therefore, war acts as the most important customer of military and psychological researches and technologies.

The image of future wars as in a mirror is reflected in military doctrines of the states, in issues of military scientific research, in steady trends of modern fighting events and in long-term goals of the sciences developing perspective military technologies. The face of future war which is depicted today contains such major lineaments as novelty, uncertainty, unpredictability and variability. These characteristics of the war make a request for soldiers, differing watchfulness, ingenuity, resourcefulness, positive thinking, ability to act in a team, dedicated to their profession and loving their country. These inquiries are a specific compass that defines for military psychology the correct course into the future.

## References


Bowles, S.V., & Bartone, P.T. (2017). *Handbook of Military Psychology. Clinical and Organizational Practice.* New York, USA: Springer.

BRAVEMIND. (2019). *Medical Virtual Reality.* Retrieved from http://medvr.ict.usc.edu/

Cheng, D. (2013). *Winning Without Fighting: The Chinese Psychological Warfare Challenge.* Retrieved from https://www.heritage.org/global-politics/report/winning-without-fighting-the-chinese-psychological-warfare-challenge

Clausewitz, C. (2002). *About the War: in 2 V.* Moscow, Russia: Ast, V. 1.

Copeland, N. (1951). *Psychology and the Soldier.* Harrisburg, USA: Military Service Publishing Co.

Cronin, Ch. (2003). *Military Psychology: An Introduction.* Boston, USA: Pearson.

De Soir Erik, L. J. L. (2011). *The Belgian End of Mission Transition Period: Lessons Learned from Third Location Decompression after Operational Deployment.* Retrieved from https://www.researchgate.net/publication/273831888_The_Belgian_End_of_Mission_Transition_Period_Lessons_Learned_from_Third_Location_Decompression_after_Operational_Deployment

Hoge, Ch.W. (2010). *Once a Warrior Always Warrior: Navigating the Transition from Combat to Home Including Combat Stress, PTSD, and mTBI.* Guilford, UK; Connecticut, USA: Lyons Press.

Hughes, J. H., & Neville, M. (2014). *Battle Against Stigma.* 2-volume tome. Barcelona, Spain: Grafos.

Inaeva, D. D. (2017). Children in armed conflict. *Young Scientist, 4,* 594–596.

Kaku, M. (2011). *Physics of the Future. How Science Will Shape Human Destiny and our Daily Lives by the Year 2100.* New York, London, Toronto, Sydney, Auckland: Doubleday.

Karayani, A. G. (2016). *Military Psychology.* 2-volume tome. Moscow, Russia: Youright.

Karayani, A. G. (2018). *Psychological Rehabilitation of the Participants of the Fighting (Psychology of Combat Stress and Stress Management).* Moscow, Russia: Military University.

Karayani, A. G., & Karayani, Y. M. (2016). Military psychology during the First World War. *Psychological Journal, 37*(2), 122–135.

Karayani, A. G., Karayani, Y., & Tsygankov, S.V. (2018). Crowd sourcing technologies as part of information-psychological warfare via social networks. *Military Academic Journal, 17*(1), 75–82.

Karayani, A. G., & Korchemny, P. A. (2016). *Psychological Preparation of Combat Actions.* Moscow, Russia: Military University.

Kennedy, C. H., & Moore, J. L. (2010). *Military Neuropsychology.* New York, USA: Springer.

Kennedy, C. H., & Zillmer, E. A. (2006). *Military Psychology: Clinical and Operational Applications.* New York, USA; London, UK: Guilford press.

Laurence, J. H., & Matthews, M. D. (2012). *The Oxford Handbook of Military Psychology.* New York, USA: Oxford University Press.

Marchenkov, V. I. (2005). *You Want Peace, Conquer the war of Rebellion! E. E. Messner's Creative Heritage.* Moscow, Russia: Military University.

Matthews, M. D. (2014). *Head Strong: How Psychology Is Revolutionizing War.* New York, USA: Oxford University Press.

McNab, C. (2001). *Endurance Techniques.* London, UK: Brown Books.

Miao D., Wang H., Liu X., Zhu X., Xiao W., & Wu Sh. (2017). Development of Military Psychology in China. In S.V. Bowles, & P.T. Bartone (Eds.), *Handbook of Military Psychology. Clinical and Organizational Practice* (pp. 481–488). New York, USA: Springer.

Military Doctrine of the Russian Federation. (2014, December 30). *Russian newspaper,* pp. 1–3.

Mukherjee S., Kumar U., & Mandal M. K. (2009). Status of Military Psychology in India: A Review. *Journal of the Indian Academy of Applied Psychology, 2,* 181–194.

National Security Strategy of the United States of America. (2017). Retrieved from https://www.whitehouse.gov/wp-content/uploads/2017/12/NSS-Final-12-18-2017-0905.pdf

Ritchie, E. C. (2011). *Combat and Operational Behavioral Health.* Fort Detrick, USA: Borden Institute.

Sinclair, R. R., & Britt T. W. (2013). *Building Psychological Resilience in Military Personal.* Washington, DC, USA: APA.

Shishkina, N. I. (2015). *Military Strategy of China: The Ninth "White Book".* Retrieved from http://rusrand.ru/events/voennaja-strategija-kitaja-devjataja-belaja-kniga

Shumkov, G. E. (1905). *Psyche Fighters during Battles*. St. Petersburg, Russia: Russian Disabled Person.
Sun, T. (2003). *The Art of War. Strategy and Tactics of the Winner*. Moscow, Russia: Eksmo.
U.S. Department of Defense. (2018). *Nuclear Posture Review*. Retrieved from https://media.defense.gov/2018/Feb/02/2001872886/-1/-1/1/2018
U.S. Department of the Army. (2018). *The U.S. Army in Multi-Domain Operations 2028*. Retrieved from https://publicintelligence.net/usarmy-multidomain-ops-2028/
U.S. Department of the Army. (2011). *U.S. Army Field Manual: 3-13.2. Military Information Support Operations*. Retrieved from https://www.globalsecurity.org/military/library/policy/dod/joint/jp3_13_2_2011.pdf

# 8

# ETHICAL ISSUES IN MILITARY PSYCHOLOGY

## Promoting international ethical readiness

*Thomas E. Myers and Shane S. Bush*

The field of psychology has a long relationship with the military, and the military was critical in the development of applied/clinical psychology. Psychologists have played key roles during times of war. They developed tests to select recruits and identify their duties, and the work of psychologists in the military ultimately resulted in psychologists becoming involved in the treatment of mental health problems in military personnel (Seligman & Fowler, 2011). Military psychology has been defined as the subdiscipline within psychology that is concerned with the conduct of research and application of psychological principles within a military environment. The roles of a psychologist within the military vary, but Page (1996) stated that the military psychologists' overall function is "to promote the fighting power and combat readiness of individual personnel and the military collective" (p. 383). In the United States, military psychologists, broadly defined, may be (a) military personnel on active duty or in the reserves, (b) contracted civilians who work with the military, (c) civilians who work with veterans, or (d) civilians who work within the Department of Defense. Specific services provided by psychologists within the military may include providing direct clinical care in the form of assessment and treatment, advising military commands, consulting, teaching/training (e.g., psychology interns and residents), and conducting research, among others.

Military psychologists may also have officer responsibilities of a more general nature, such as performing administrative duties for the unit. Given the diversity of roles and activities that psychologists may engage in, there are a wide range of ethical issues that may arise while performing the various activities and in different roles. The goals of the present chapter are to describe the relevant ethical considerations in military psychology and provide clinicians with a structured approach to ethical decision-making. Because the manner in which mental health professionals practice with military personnel differs both within and among countries, an attempt is made to cover issues that are relevant broadly, with the expectation that practitioners will be able to apply general principles to their specific contexts.

## Ethical considerations in military psychology

Psychologists benefit the military and its service members in many ways, ranging from identification of service members who are at risk for, or are experiencing, psychological problems, to enhancing the resilience, strength, and performance of military service members. Psychological and neuropsychological evaluation and treatment of military personnel promote the well-being of service members and enhance military readiness. To do so most effectively, a variety of ethical issues must be considered, and ethical practice must begin with an awareness of the relevant issues.

To understand the specific ethical challenges experienced by military psychologists, it is helpful to understand the broader ethical issues experienced by psychologists more generally. A random sample of members of the American Psychological Association surveyed by Pope and Vetter (1992) revealed the most common ethical dilemmas faced by psychologists within the United States included confidentiality (reported by 18%); blurred, dual, or conflictual relationships (reported by 17%); and payment sources, plans, settings, and methods (reported by 14%). Assessment and competence were reported by 4% and 3% of respondents, respectively.

Regarding the practice of military psychology, Johnson and Landsinger (2017) identified the following ethical issues (not in order of importance) as occurring most frequently or resulting in the most acute ethical conflicts: (a) boundaries of competence, (b) confidentiality, (c) sudden role shifts, (d) multiple relationships, (e) presenting accurate research findings and clinical diagnoses, (f) consultation to detainees or intelligence-gathering operations, and (g) the psychologist's own psychological fitness. Within the subdiscipline of military neuropsychology, Bush and Cuesta (2010) identified the following ethical concerns as common: (a) professional competence, (b) third-party request for services/informed consent, (c) privacy and confidentiality, (d) roles/relationships (dual/multiple), (e) avoiding harm, (f) assessment, (g) conflicts between civilian and military ethical and legal requirements, (h) record keeping, (i) test security/release of raw test data, (j) objectivity, (k) third-party observer requests for services, and (l) psychological services delivered to or through organizations. Bush, Russo, and Cuesta (2012) further identified ethical issues of particular relevance to neuropsychologists working in VA settings, which overlap substantially with the issues previously listed.

## Professional competence

Professional competence serves as the foundation of the general bioethical principles of beneficence and nonmaleficence. Psychologists cannot "do good" (beneficence) or "avoid doing harm" (nonmaleficence) if they are not practicing competently. Practicing competently requires certain knowledge, skills, and experiences (professional and supervised) based on education and training that permit psychologists to practice in a manner consistent with standards of practice. Without possessing adequate competence, it is likely that psychologists' activities will result in harm to others due to inappropriate behaviors, judgments, or recommendations.

In addition to professional competence to perform clinical activities, it is critical that military psychologists possess cultural competence of the context in which services are provided. Military personnel share a system of beliefs and values, as well as experiences, unique language, dress, manners, and behavioral norms (Bush & Cuesta, 2010). It is important for psychologists, particularly civilians working with military personnel, to become familiar with those factors that may affect professional activities in working with the military. Familiarity with the military language, including acronyms, rank, and behavioral norms, is an ethical responsibility, as it is required for the competent provision of psychological services (Bush & Cuesta, 2010).

Further complicating the ethical issue of competence, psychologists often function as the only mental health professional and may be asked to provide services outside their area of competence (Johnson, 2016). The Ethics Code of the American Psychological Association (APA, 2017) recognizes that psychologists may be in the best position (i.e., when no one else could provide needed services) to provide services even when they may lack the proper training and experience in order to ensure that such services not be denied (Ethical Standard 2.02; Providing Services in Emergencies). Such services should be discontinued when the emergency ends or the appropriate services or clinicians become available. While it is not possible for a psychologist to be knowledgeable about every clinical service or condition, it is important for psychologists to recognize those conditions that are known to occur more frequently among military populations (Orme, 2012).

## Roles, relationships, and obligations

The specific roles that psychologists are asked to perform in the military vary, and sudden role shifts may occur (Johnson & Landsinger, 2017). As a result, psychologists may be asked to perform duties (e.g., treating clinician and fitness for duty examiner) that may result in the psychologist taking on multiple roles, relationships, and obligations simultaneously. The APA Ethics Code advises against entering into multiple relationships when such relationships could be expected to cause impairment or harm to an individual. The Ethics Code (3.04 Multiple Relationships) specifically states, "A psychologist refrains from entering into a multiple relationship if the multiple relationship could reasonably be expected to impair the psychologist's objectivity, competence, or effectiveness in performing his or her functions as a psychologist, or otherwise risks exploitation or harm to the person with whom the professional relationship exists." In the case of taking on the role of both treating clinician and objective examiner for the same service member, it is very likely that a psychologist's objectivity would be impaired.

Perhaps the greatest opportunity for ethical challenges to arise stems from the simultaneous roles held by psychologists in the military (i.e., clinician and commissioned military officer), which results in a unique potential for ethical conflicts. Specifically, military officers have a responsibility to "promote the combat readiness and fighting power of the military and to support the immediate military mission" (Johnson & Landsinger, 2017, p. 105), which may result in conflicts between the interests of the patient (e.g., their personal well-being) and the interests of the referral source (e.g., the military mission). This has been referred to as dual or mixed agency (Carr, 2005; Howe, 2003), whereby conflicts can arise between obligations to different parties. Mixed agency can create problems identifying the client, balancing the interests of relevant parties, and avoiding harm resulting from multiple relationships (Zur & Gonzalez, 2002).

## Privacy and confidentiality

Privacy and confidentiality are often considered essential components of the therapeutic relationship between psychologists and their patients. The assurance of confidentiality, with certain limitations explained, is what allows patients to openly express their inner thoughts, feelings, and behaviors that they may otherwise be unwilling to disclose in the absence of a confidential relationship. Because of this fundamental expectation, the APA Ethics Code states that, "Psychologists have a primary obligation and take reasonable precautions to protect confidential information" (4.01, Maintaining Confidentiality), and that "Psychologists include in written and oral reports and consultations, only information germane to the purpose for which the communication is made" (4.04, Minimizing Intrusion on Privacy).

Confidentiality was described by Johnson and Landsinger (2017) as, "one of the most persistent and pressing ethical quandaries for military psychologists" (p. 108). The confidential nature of most clinical interactions in a civilian context may not be protected in a military context (Kennedy & Johnson, 2009). In fact, Johnson and Landsinger (2017) noted that Department of Defense (DoD) directives indicate that "a legitimate military authority… may have access to client records on a 'need to know basis.' Leave a thin space" Additionally, the DoD's "privacy directives include exceptions that allow dissemination of private information without the service member's authorization when necessary for the completion of the military mission (e.g., fitness for duty, illegal activities)" (Bush & Cuesta, 2010, p. 37). It is important to note as well that military psychologists do not maintain control over their patients' records and so have limited control over confidentiality for this reason. Because of the limits of confidentiality in military settings, it is critical that the limits of confidentiality be discussed with patients as early as possible. Having knowledge of this information before disclosure is part of what makes informed consent truly "informed" and is the best way to prevent ethical conflicts from arising as a result of the limitations of confidentiality.

## Selection and use of methods and measures

When considering the selection and use of assessment methods and measures in a military context, psychologists consider the purpose of the evaluation, the evidence base supporting test use with the specific population, and any contextual factors (patient or environment) that could affect the psychologist's ability to obtain valid data regarding constructs of interest. APA Ethical Standard 2.04 (Bases for Scientific and Professional Judgments) states, "Psychologists' work is based upon established scientific and professional knowledge of the discipline," and Standard 9.01 (Bases for Assessments) states that "Psychologists base the opinions contained in their recommendations, reports, and diagnostic or evaluative statements, including forensic testimony, on information and techniques sufficient to substantiate their findings."

Given the sometimes unpredictable and uncontrollable environmental factors that can be present during attempts to administer psychological tests, ethical concerns may arise with regard to psychologists' ability to draw accurate conclusions when standardized administration procedures are not able to be followed (e.g., administering a neuropsychological test to assess the impact of a traumatic brain injury [TBI] when a private room is unavailable and interruptions are present). When such influences are present, psychologists "indicate any significant limitations of their interpretations" (APA Ethical Standard 9.06 Interpreting Assessment Results).

## Diversity

Awareness of the importance of diversity has received increased attention in recent years as the diversity of the U.S. population increases. The movement toward increasing the diversity of the military in the United States is exemplified by the Department of Defense (DoD) Diversity and Inclusion Strategic Plan, which outlined the President's Executive Order 13583 (EO 13583) to promote diversity in the federal workforce (including the military). This document states, "An all-volunteer force must represent the country it defends. We will strengthen our commitment to the values of diversity and inclusivity, and continue to treat each other with dignity and respect. We benefit immensely from the different perspectives, and linguistic and cultural skills of all Americans" (DoD, 2012, p. 4). Although the diversity of both the general U.S. population and the military has been increasing, concerns have been raised with regard to the cultural competence of psychologists to provide services to such diverse populations (Sue, 1998). This focus on diversity within military psychology mirrors the increased emphasis on the importance of diversity within the field of psychology in general (Atuel & Castro, 2018). The profession increasingly understands that the ability to effectively understand a diverse military or patient population requires an ongoing commitment to improving one's knowledge of the experiences of diverse persons.

Several sections of the APA Ethics Code provide guidance to psychologists on the ethical issues related to diversity. General Principle E (Respect for People's Rights and Dignity), states the following:

> "Psychologists are aware of and respect cultural, individual, and role differences, including those based on age, gender, gender identity, race, ethnicity, culture, national origin, religion, sexual orientation, disability, language, and socioeconomic status, and consider these factors when working with members of such groups. Psychologists try to eliminate the effect on their work of biases based on those factors, and they do not knowingly participate in or condone activities of others based upon such prejudices."

The APA has also published a set of guidelines for psychologists who work with ethnically, culturally, and linguistically diverse populations (APA, 2017). Several articles, book chapters, and books have been published that address diversity issues related to psychological and neuropsychological assessment (Brickman, Cabo, & Manly, 2006; Iverson, & Slick, 2003; Mindt Byrd, Saez, & Manly, 2010; Yamada, Atuel, & Weiss, 2013), including lesbian, gay, bisexual, and transgender military personnel (Glotfelter, Georgemiller, & Banderman, 2017) and women warriors (Saitzyk et al., 2017).

## Detention and interrogation support

The role of military psychologists in interrogation is controversial. Some military psychologists believe that there is an important role for psychologists in the interrogation process (James & Pulley, 2013). Research has shown that rapport-based techniques during interrogation can provide "valuable and actionable intelligence from detainees without causing harm" (Johnson & Landsinger, 2017, p. 110). Additionally, preventing detainees from accessing appropriate psychological care is an ethical concern. Nevertheless, in October 2015, the APA released a policy statement banning psychologists from national security interrogations (APA, 2015). This policy banned psychologists from working at "detention settings operating in violation of the U.S. Constitution or international law (as deemed by specific U.N. authorities), including the Guantanamo Bay detention facility, unless providing treatment to military personnel or working directly for the detainees or for an independent third party to protect human rights." Johnson and Landsinger's (2017) review of APA policy on the role of psychologists in detention facilities and interrogation support "has unequivocally: (a) prohibited torture, cruel, inhuman, and degrading treatment; (b) prohibited harsh ("enhanced") interrogation techniques; (c) prohibited psychologists from mixing roles with detainees or disclosing clinical healthcare records to intelligence entities" (p. 110). The appropriate roles of psychologists in detention centers and interrogation procedures remain a topic of discussion.

## Maintaining one's own mental fitness

As is the case for all military personnel, military psychologists may engage in military operations that are physically and psychologically demanding. Military psychologists are susceptible to many of the same stressors as other military personnel, which include separation from loved ones, the threat of injury and loss of life, exposure to harsh living conditions and stressful and traumatic experiences, and actual or perceived mistreatment by seniors or other military personnel. These challenges can lead to a variety of psychological symptoms, deficiencies in performance, or undesirable behaviors. For military psychologists to be able to serve others effectively, they must first maintain their own mental fitness.

Unfortunately, there is evidence that psychologists may have difficulty in performing self-assessments (Davis, et al., 2006; Karpen, 2018). For this reason, it is important that psychologists consult with colleagues regularly and be aware that they may have difficulty accurately assessing themselves. A failure to ensure their own mental fitness may have adverse effects on psychologists' own performance and thus result in harm to others. Consistent with the general bioethical principles of beneficence and nonmaleficence, military psychologists should try to maintain awareness of their own mental and physical well-being and the potential impact of any personal problems on their professional responsibilities.

## Positive ethics and the 4 As of ethical practice

Ethical practice requires an active commitment from clinicians to ensure they possess the knowledge and skills to perform their professional responsibilities. Similar to the way in which clinical knowledge must be constantly assessed and updated, ethical competence must be continuously assessed and updated as well (Bush, 2009). Most ethical standards and legal requirements represent the minimum that psychologists are required to comply with in order to avoid ethical misbehavior. This approach to ethics represents a "floor-based" approach, and is also described as remedial ethics, which is the basis for which psychologists may be disciplined for having violated such standards. In contrast to a reactive approach, in which psychologists who violate standards may have actions taken against them, positive ethics is proactive, and involves pursuit of ethical ideals and the highest standards of one's professional conduct (Knapp, VandeCreek, & Fingerhut, 2017). In positive ethics, clinicians are encouraged to strive toward their highest professional potential.

Bush (2009) outlined four ways in which psychologists can facilitate high standards of ethical practice and prepare for sound ethical decision-making, which were referred to as the 4 A's of ethical practice (Anticipate, Avoid, Address, and Aspire). By *anticipating* the ethical issues that may arise in a given situation, psychologists are typically well positioned to develop sound ethical practices. This in turn puts psychologists in the best position to *avoid* ethical misconduct. That is, by knowing in advance what the issues are, a psychologist will be better able to avoid engaging in unethical behavior. A lack of knowledge of ethical standards and practice is a recipe for engaging in unethical behavior despite a lack of intention to behave unethically. Not all ethical challenges are completely avoidable. However, anticipating what ethical challenges are most likely to arise puts psychologists in the best position to *address* the challenges. It is ideal when ethical issues can be addressed prior to becoming dilemmas, but if they cannot, they should be addressed immediately when they are encountered. Finally, consistent with the concept of positive ethics, clinicians are encouraged to *aspire* toward the highest standards of ethical practice.

## Evidence-based ethical practice: Identifying and using resources

Psychology is a scientific discipline. Clinicians rely on empirical evidence in their practice. The APA Ethics Code, first published in 1953, was initially developed using an empirical process that involved identifying the ethical challenges that were experienced by its members (Pope & Vetter, 1992). That is, rather than a committee of experts determining the ethical issues to be addressed, it was the broad membership of the APA that led to the development of the Code. As such, the Code was empirically derived, and the modifications of the code were also based, at least in part, on consideration of the ethical challenges experienced by its members.

Evidence-based practice (EBP) in psychology has been increasingly emphasized in recent years (APA Presidential Task Force on Evidence-Based Practice, 2006; Chelune, 2010). While most psychologists recognize the importance of EBP in guiding assessment and treatment, fewer psychologists seem to approach the ethical decision-making process using such an approach. Bush (2018 emphasized the importance of making ethical decisions using an evidence-based approach, as opposed to an "I think" approach. There are many resources available to military psychologists when considering the evidence base underlying ethical practices and decisions.

Beauchamp and Childress (2013) described the following four core principles of bioethics that underline moral reasoning in health care: respect for autonomy, nonmaleficence, beneficence, and justice. Respect for autonomy refers to the rights of the individual to make their own decisions without undue influence that would reduce the voluntariness of a decision; this principle forms the basis for informed consent. Nonmaleficence, meaning "do no harm," refers to clinicians' responsibility to avoid harming their patients or others who receive their services. Beneficence, meaning "do good," refers to clinicians' responsibility to promote and defend the rights and welfare of their patients. The principle of justice refers to the rights of patients to receive fair and appropriate treatment based on what is deserved; the concept of "fairness" forms the core of this principle. Justice can be subdivided into distributive justice, which refers to the equitable distribution of healthcare resources, and formal justice, which refers to "equal treatment for those who are equals and unequal treatment for those who are not equals" (Bush, 2009, p. 13). In the context of mental health, the additional principles of fidelity (i.e., the responsibility to be truthful and faithful, keep promises, and maintain loyalty) and general beneficence (i.e., a responsibility to defend and advance the welfare of society at large) have been proposed (Knapp et al., 2017). These bioethical principles underlie the more specific ethical standards and can often be used to arrive at sound ethical decisions when other ethical resources conflict with each other or are silent on a specific matter.

Within the United States, the most widely used resource for consultation related to ethical matters is the American Psychological Association's Ethical Principles of Psychologists and Code of Conduct (APA Ethics Code; APA, 2017). Psychologists must also abide by the Code of Conduct of the Association of State and Provincial Psychology Boards, which publishes their own Rules of Conduct (ASPPB, 2018). Because one of the most common activities of psychologists involves assessment, psychologists should be familiar with the Standards for Educational and Psychological Testing (American Educational Research Association, American Psychological Association, & National Council on Measurement in Education, 2014). Because military psychologists frequently work within VA and/or Department of Defense settings, they must consult with VA/DoD regulations with regard to handling ethical issues. Most states have their own organizations, which provide a number of resources for consultation when handling ethical matters. Liability insurance carriers have experts within their company that are available for consultation and guidance. In addition to the APA, there are professional organizations (e.g., American Academy of Clinical Neuropsychology, National Academy of Neuropsychology) that have published guidelines for psychologists to ensure that clinicians remain abreast of the current best practices in the field.

Outside of the United States, professional organizations in many countries also have their own ethics code or adopt the codes of other countries. Leach and Harbin (1997) examined the similarities between the APA Ethics Code (the 1992 version) and the ethical codes used by 23 other countries (representing 19 codes because some countries adopt the codes of other countries). They found that four countries (Canada, Israel, Australia, and South Africa) shared all of the principles in the APA Ethics Code; one country (China) was found to have no overlap in principles. This finding is not surprising given that ethics codes reflect the shared values of professionals, which reflect the larger society. They found that 10 of the APA Ethics Code individual standards "approach a universal standard," which was defined as greater than 75% overlap among codes. The principles

Disclosures and Maintaining Confidentiality were found to be most consistent across countries, with 100% and 95% consistency for these principles across the ethics codes in different countries, respectively. Importantly, it was also noted that each country analyzed in their study had ethical requirements that were not found to match any of the U.S. standards. Thus, while there is clearly a great deal of overlap in ethical principles among countries, there is also diversity among countries. Thus, consultation with the specific codes and resources within a given country is needed. Military psychologists deployed to other countries benefit from being familiar with both their own ethics code and the code enforced in the country to which they are deployed.

In an effort to come up with universal guidelines that could be utilized by the psychology community across the world, the Universal Declaration of Ethics Principles for Psychologists (Universal Declaration) was developed by the International Union of Psychological Science (IUPsyS) and International Association of Applied Psychology (IAAP). Their goal was "to provide a moral framework and generic set of ethical principles for psychology organizations worldwide" (Ad Hoc Joint Committee, 2008). The Universal Declaration was adopted in 2008 and was designed to describe ethical principles that are based on shared human values (Gauthier, 2008).

The Universal Declaration contains four principles, in addition to the preamble. The preamble states that, "Psychologists are committed to placing the welfare of society and its members above the self-interest of the discipline and its members." The first principle (Principle I: Respect for the Dignity of Persons and Peoples) is "the most fundamental and universally found ethical principle across geographical and cultural boundaries, and across professional disciplines." This principle embraces the inherent worth of all human beings, and the value that all humans deserve equal moral consideration. The second principle, "Competent Caring for the Well-Being of Persons and Peoples," is described as working for the benefit of all people, and "above all, doing no harm." Thus, this principle is similar to Beneficence and Nonmaleficence. The third principle, "Integrity," refers to honesty and truthful, open, and accurate communications. The fourth principle, "Professional and Scientific Responsibilities to Society," includes contributing to the knowledge of human behavior and to persons' understanding of themselves and others to improve the condition of individuals, families, groups, communities, and society.

The specific resources drawn upon to make evidence-based decisions regarding ethical issues vary from one context to another. For example, references that may be considered by psychologists working within the United States Veterans Administration (VA) psychologists include (a) jurisdictional laws, (b) VA/Department of Defense Regulations, (c) the APA Ethics Code, (c) the Association of State and Provincial Psychology Boards (ASPPB; 2018) *Code of Conduct*, (d) ethics committees and state licensing boards, (e) professional liability insurance carriers or attorneys, (f) scholarly publications and position papers of professional organizations, (g) general bioethical principles, and (h) colleagues or supervisors (Bush, 2018). Regardless of the specific context, several of these resources would be applicable, and those that are specific to the VA context will likely have a counterpart in other contexts. Psychologists benefit from reviewing the available resources based on their specific situation and the nature of the ethical issue or dilemma.

## A structured approach to ethical decision-making

A variety of models have been proposed to assist psychologists in the ethical decision-making process. Of the existent models, those proposed by Stephenson and Staal (2007) and Bush and Cuesta (2010) were adapted within a military context, and Bush, Russo, and Cuesta (2012) adapted the model for neuropsychological practice with veterans. Although not an ethical decision-making model per se, Johnson and Landsinger (2017) provided several "brief recommendations for military psychologists designed to reduce or mitigate ethical tensions so that they do not escalate into ethical conflicts" (p. 112). Their recommendations include the following: (a) establish strong consultative relationships with senior members of the military organizations that solicit your services, (b) remember that your commissioned military status does not override your obligations to the Ethics Code, (c) provide rigorous and ongoing informed consent to all clients, (d) know ethical standards and federal statutes and abide by an ethical decision-making process, (e) appreciate the distinction between mixed-agency tensions and conflicts, (f) assume that every member of the military organization is a potential client, (g) increase your own tolerance for routine boundary crossings, (h) establish and maintain ongoing external consulting relationships, (i) remain attuned and responsive to your own levels of distress and competence, and (j) always ask yourself: In this situation, what is in my client's best interest? The following model describes the structured approach proposed by Bush and colleagues (Bush & Cuesta, 2010; Bush, Russo, & Cuesta, 2012).

## Identify and clarify the problem(s)

Some ethical problems are obvious, and the most appropriate course of action is apparent, while other ethical issues may be much less clear. Identification and clarification of the problem permit psychologists to distinguish between aspects of a problem that represent an ethical dilemma versus those that involve legal requirements or professional preferences. For example, clinicians who choose one assessment measure over another may or may not present an ethical dilemma, but rather a difference of opinion or preference. There are typically multiple assessment measures that are well validated to answer a particular clinical question. In contrast, such differences in assessment choices may represent ethical problems. For example, psychologists who use tests that are obsolete for the purpose at hand are violating ethical standards (Ethical Standard 9.08, Obsolete Tests and Outdated Test Results), and this would not represent a simple difference of opinion or preference.

## Consider the significance of the context and purpose of the service

Military psychologists provide services in a variety of contexts, including military theaters, as members of deployed units, on military bases, and within VA hospital settings, among others. The degree of uncontrollable environmental influences will vary greatly across these settings, and thus influence the validity and integrity of the assessment and treatment services provided by psychologists.

The importance of considering the context is also evident in the differences in assessment findings and treatment recommendations across settings. For example, the symptoms of PTSD (e.g., hypervigilance) within an active combat zone may be adaptive, and not necessarily symptomatic of psychopathology (Bryan, 2013), whereas the reporting of such symptoms in a military veteran who returned from combat six months earlier may reflect clinically significant symptoms of PTSD. In this case, a failure to take into account the proper context could result in improper conclusions and treatment recommendations.

## Consider the obligations owed

The mixed-agency dilemma is one of the most unique and salient ethical issues experienced by military psychologists, who serve in the role of both psychologist and military officer (Jeffrey, Rankin, & Jeffrey, 1992). These mixed-agency dilemmas result in obligations owed to two or more entities and thus may involve ethical conflicts related to the best interests of the different parties. Within the military, the "client" may be the Department of Defense, the branch of the military in which the psychologist serves, or an entity to which a psychologist is assigned within the military. As military officers, the psychologist's ultimate duty is to support the military mission and ensure the combat readiness of service members.

Johnson (2013) described the difficulty that may be involved in identifying the primary "client" within military contexts. Psychologists in the military provide services to service members who are typically referred by officers who are senior in rank to a psychologist, and there may be specific referral questions related to the mission at hand. Psychologists often feel obligations owed to the individual being assessed/treated, the referring command, and also those affected by the psychologist's assessment (e.g., other services members involved in the same mission). When it is clear to the psychologist that obligations are owed to multiple parties, psychologists must attempt to clarify this to all parties as early as possible and focus on minimizing harm and maximizing benefits to all parties involved. Because military psychologists' roles with clients may shift with little or no notice, anticipation of assuming multiple roles and obligations is needed (Johnson, 2013).

## Identify and utilize ethical and legal resources

The specific ethical and legal resources will vary from one context to another. Some resources provide a foundation for understanding ethical principles and thus provide guidance across a variety of settings (e.g., Beauchamp and Childress's four core principles of bioethics; national and international ethics codes). Other resources that are more context specific should be utilized to provide further guidance for psychologists working in very specific contexts (e.g., military ethical and legal resources, state mental health law, VA regulations, and mental health chain of command, professional practice guidelines and other scholarly publications; books and book chapters). When there are conflicts between ethical and legal resources, or even between ethical resources themselves, consultation with knowledgeable colleagues can be invaluable.

## Consider personal beliefs and values

There is evidence for both universal and more culture-specific values and beliefs related to ethical issues. Individual psychologists also have their own personal beliefs and values that influence how they view ethical issues and the ethical decision-making process. When psychologists conduct assessments, they typically attempt to do so in the most objective manner possible and utilize psychological tests that are specifically designed to assess psychological constructs in an objective way that is not possible without the use of such tests (e.g., IQ tests, memory tests). Nevertheless, there is always the potential for psychologists' personal beliefs and values to influence both their conclusions from less standardized assessments (e.g., an unstructured clinical interview) as well as their interpretations of objective assessment measures (e.g., when a score falls in a range that suggests "impairment"). The only way that psychologists can take into account the influence of these personal beliefs and values is to be aware of their existence and potential to influence their clinical activities. Bush (2018) noted that practitioners have a responsibility to assess the extent to which their personal morality is consistent with that of larger society, the profession, and the institutions with which they are involved.

## Develop possible solutions to the problem

Multiple possible solutions exist for most ethical challenges. To develop sound ethical solutions, psychologists need to consult with the available resources. The process of developing possible solutions forces psychologists to think through the ethical situation in a manner that is consistent with evidence-based ethical practice.

## Consider the potential consequences of various solutions

Based on the solutions generated, clinicians are encouraged to assess the possible positive and negative consequences of each solution. Consequences should be weighted based on the potential impact. For example, there are often positive and negative consequences when facing ethical dilemmas; choosing a course of action in which the positives outweigh the negatives is preferable. The anticipated consequences for all parties involved should be considered.

## Choose and implement a course of action

After thoughtful and thorough consideration of the consequences of the possible solutions identified, psychologists may choose and implement the most appropriate course of action.

## Assess the outcome, implement changes as needed, and document the process

It is important for psychologists to recognize that the ethical decision-making process does not end once a course of action has been chosen and implemented. Consistent with a positive ethics approach, psychologists should be proactive in assessing the outcomes of their choices and actions. Depending on the outcome, additional action may be warranted. When this is the case, going through the steps of this model will likely assist the psychologist in obtaining a more favorable and ethically desirable outcome. Further, assessment of the outcome permits psychologists to learn from the experience, resulting in additional information to draw upon in the future when faced with ethical challenges. In addition, such knowledge can be shared with students and/or colleagues to assist in others' ethical decision-making.

The final step of this ethical decision-making process is to document the efforts made and the steps taken by the psychologist. Documentation of one's ethical decision-making efforts is important in case a psychologist needs to provide a record of the steps they took to ensure their actions were consistent with ethical requirements. A common phrase known to many healthcare professionals is, "If it wasn't documented, it didn't happen." Thus, documentation is important when the actions of a psychologist must be defended, which can occur when ethical dilemmas result in less than favorable outcomes for some parties involved. Bush (2018) recommended that such documentation follow the ethical decision-making model and includes the details of the specific situation that would likely meet the requirements of most interested parties. Of particular importance is documentation of the resources consulted and the reasoning behind the psychologist's choice and implantation of the action(s) taken (Bush, 2018).

## Conclusions

Military personnel make many sacrifices in the service to their country and deserve the highest level of care commensurate with the sacrifices they make. Psychologists have a long history of providing valuable services within the military, and both psychology and the military have benefited from this relationship. Psychological and neuropsychological evaluation and treatment of military personnel has tremendous potential to promote the well-being of service members and enhance military readiness. However, the unique contexts in which military psychologists work present a number of ethical challenges that must be considered for psychological services to be effective. Military psychologists are encouraged to follow an evidence-based, positive ethics approach, and to utilize a systematic and structured ethical decision-making process. Although many ethical issues and principles are relevant broadly, military psychologists need to consider the issues and apply the principles as they relate to their specific contexts. Regardless of the specific context, by having adequate knowledge of the ethical issues most frequently encountered by military psychologists, utilizing the proper resources, and following an evidence-based, positive ethics approach, psychologists are likely to produce good ethical solutions and outcomes, and thus continue to be of benefit to the military and all those who serve.

## References


Ad Hoc Joint Committee. (2008). *Universal declaration of ethical principles for psychologists*. Retrieved from http://www.iupsys.net/about/governance/universal-declaration-of-ethical-principles-for-psychologists.html

American Educational Research Association, American Psychological Association, & National Council on Measurement in Education. (2014). *Standards for educational and psychological testing*. Washington, DC: Author.

American Psychological Association. (2015). *APA alerts federal officials to new policy banning psychologists from national security interrogations*. Retrieved from https://www.apa.org/news/press/releases/2015/10/banning-psychologists-interrogations.

American Psychological Association. (2017). *Ethical principles of psychologists and code of conduct*. Retrieved from www.apa.org/ethics/code/index.aspx.

American Psychological Association. (2017). *Multicultural guidelines: An ecological approach to context, identity, and intersectionality*. Retrieved from apa.org/about/policy/multicultural-guidelines.pdf.

American Psychological Association Presidential Task Force on Evidence-Based Practice. (2006). Evidence-based practice in psychology. *American Psychologist, 61*, 271.

American Psychological Association Presidential Task Force on Psychological Ethics and National Security. (2005). Report of the American Psychological Association Presidential Task Force on Psychological Ethics and National Security.

Association of State and Provincial Psychology Boards. (2018). *Code of conduct*. Retrieved from https://cdn.ymaws.com/asppb.net/resource/resmgr/guidelines/asppb_Code_of_Conduct_Januar.pdf.

Atuel, H. R., & Castro, C. A. (2018). Military cultural competence. *Clinical Social Work Journal, 46*(2), 74–82.

Beauchamp, T. L., & Childress, A. F. (2013). *Principles of Biomedical Ethics* (7th ed.). New York: Oxford University Press.

Brickman, A., Cabo, R., & Manly, J. (2006). Ethical issues in cross-cultural neuropsychology. *Applied Neuropsychology, 13*(2), 91–100.

Bryan, C. J. (2013). Psychologists on the frontlines. In B. Moore & J. E. Barnett (Eds.), *Military Psychologists' Desk Reference* (pp. 127–132). New York, NY: Oxford University Press.

Bush, S.S. (2009). *Geriatric Mental Health Ethics: A Casebook*. New York: Springer Publishing Company.

Bush, S. S. (2018). *Ethical Decision Making in Clinical Neuropsychology* (2nd ed.). New York: Oxford University Press.

Bush, S. S., & Cuesta, G. M. (2010). Ethical issues in military neuropsychology. In C. H. Kennedy & J. Moore (Eds.), *Military Neuropsychology* (pp. 29–56). New York: Springer Publishing Co.

Bush, S., Russo, A., & Cuesta, G. (2012). Ethical considerations in the neuropsychological evaluation and treatment of veterans. In S. S. Bush (Ed.), *Neuropsychological Practice with Veterans* (pp. 333–356). New York, NY: Springer.

Carone, D., & Bush, S. (2013). *Mild Trauma Brain Injury: Symptom Validity Assessment and Malingering*. New York, NY: Springer Publishing Company.

Carr, V. F. (2005). Dual agency and fiduciary responsibilities in modern medicine. *Physician Executive, 31*(6), 56–58.

Chelune, G. J. (2010). Evidence-based research and practice in clinical neuropsychology. *The Clinical Neuropsychologist, 24*(3), 454–467.

Davis, D. A., Mazmanian, P. E., Fordis, M., Harrison, R. V., Thorpe, K. E., & Perrier, L. (2006). Accuracy of physician self-assessment compared with observed measures of competence: A systematic review. *JAMA, 296*(9), 1094–1102.

Department of Defense. (2012). *Department of Defense diversity and inclusion strategic plan 2012–2017*. Retrieved from http://www.defence.gov.au/Diversity/strategy/.

Gauthier, J. (2008). The universal declaration of ethical principles for psychologists presented at the United Nations DPI/NGO Conference in Paris. *Psychology International, 19*(4). Retrieved from http://www.apa.org/international/pi/2008/10/gauthier.aspx.

Glotfelter, M. A., Georgemiller, R. J., & Banderman, K. M. (2017). Lesbian, gay, bisexual, and transgender service members: Clinical practice considerations. In S. Bowles & P. T. Bartone (Eds.), *Handbook of Military Psychology: Clinical and Organizational Practice* (pp. 333–356). Cham, Switzerland: Springer International Publishing.

Howe, E. G. (2003). Mixed agency in military medicine: Ethical roles in conflict. In E. Pellegrino (Ed.), *Textbook of Military Medicine: Military Medical Ethics* (pp. 331–365). Washington, DC: Office of the Surgeon General, U.S. Department of the Army.

Iverson, G. L., & Slick, D. J. (2003). Ethical issues associated with psychological and neuropsychological assessment of persons from different cultural and linguistic backgrounds. In I. Z. Schultz & D. O. Brady (Eds.), *Psychological Injuries at Trial* (pp. 2066–2087). Chicago, IL: American Bar Association.

James, L. C., & Pulley, L. (2013). Military psychologists' roles in interrogation. In C. H. Kennedy & J. Moore (Eds.), *Military Neuropsychology* (pp. 158–161). New York, NY: Springer Publishing Co.

Jeffrey, T. B., Rankin, R. J., & Jeffrey, L. K. (1992). In service of two masters: The ethical-legal dilemma faced by military psychologists. *Professional Psychology: Research and Practice*, *23*(2), 91–95.

Johnson, W. B. (2013). Mixed-agency dilemmas in military psychology. In B. Moore & J. E. Barnett (Eds.), *Military Psychologists' Desk Reference* (pp. 103–107). New York, NY: Oxford University Press.

Johnson, W. B. (2016). Military settings. In J. Norcross, G. R. VandenBos, D. K. Freedheim & M. M. Domenech Rodríguez (Eds.), *APA Handbook of Clinical Psychology: Volume 1. Roots and Branches* (pp. 495–507). Washington, DC: American Psychological Association.

Johnson, W. B., & Landsinger, K. L. (2017). Ethical issues in military psychology. In S. Bowles & P. T. Bartone (Eds.), *Handbook of Military Psychology: Clinical and Organizational Practice* (pp. 105–114). Cham, Switzerland: Springer International Publishing.

Karpen, S. C. (2018). The social psychology of biased self-assessment. *American Journal of Pharmaceutical Education*, *82*(5), 6299.

Kennedy, C. H., & Johnson, W. B. (2009). Mixed agency in military psychology: Applying the American Psychological Association ethics code. *Professional Psychology: Research and Practice*, *6*(1), 22–31.

Knapp, S. J., VandeCreek, L.D., & Fingerhut, R. (2017). *Practical Ethics for Psychologists: A Positive Approach* (3rd ed.). Washington, D.C.: American Psychological Association.

Knapp, S., Younggren, J. N., VandeCreek, L., Harris, E., & Martin, J. N. (2013). *Assessing and Managing Risk in Psychological Practice: An Individualized Approach* (2nd ed.). Rockville, MD: The Trust.

Leach, M. M., & Harbin, J. J. (1997). Psychological ethics codes: A comparison of twenty-four countries. *International Journal of Psychology*, *32*(3), 181–192.

Mindt, M. R., Byrd, D., Saez, P., & Manly, J. (2010). Increasing culturally competent neuropsychological services for ethnic minorities: A call to action. *The Clinical Neuropsychologist*, *24*(3), 429–453.

Orme, D. R. (2012). Neuropsychological conditions experienced solely or more commonly by military populations. In S. S. Bush (Ed.), *Neuropsychological Practice with Veterans* (pp. 257–278). New York: Springer Publishing Company.

Page, G. D. (1996). Clinical psychology in the military: Developments and issues. *Clinical Psychology Review*, *16*(5), 383–396.

Pope, K. S., & Vetter, V. A. (1992). Ethical dilemmas encountered by members of the American Psychological Association: A national survey. *American Psychologist*, *47*(3), 397–411.

Saitzyk, A.R., Harvey, S., Landes, A., Long, C., & Porter, R. (2017). Becoming and beginning: The journey of the woman warrior. In S. Bowles & P. T. Bartone (Eds.), *Handbook of Military Psychology: Clinical and Organizational Practice* (pp. 399–415). Cham, Switzerland: Springer International Publishing.

Seligman, M. E., & Fowler, R. D. (2011). Comprehensive soldier fitness and the future of psychology. *American Psychologist*, *66*(1), 82–86.

Stephenson, J. A., & Staal, M. A. (2007). An ethical decision-making model for operational psychology. *Ethics & Behavior*, *17*(1), 61–82.

Sue, S. (1998). In search of cultural competence in psychotherapy and counseling. *American Psychologist*, *53*(4), 440–448.

Yamada, A., Atuel, H. R., & Weiss, E. L. (2013). Military culture and multicultural diversity among military service members: Implications for mental health providers. In F. A. Paniagua & A. Yamada (Eds.), *Handbook of Multicultural Mental Health: Assessment and Treatment of Diverse Populations* (2nd ed.) (pp. 389–410). San Diego, CA, US: Elsevier/Academic Press.

Zur, O., & Gonzalez, S. (2002). Multiple relationships in military psychology. In A. A. Lazarus & O. Zur (Eds.), *Dual Relationships and Psychotherapy* (pp. 315–328). New York, NY, US: Springer Publishing Co.

# PART II

# Soldiering: Deployment and beyond

# 9

# SHAPING MILITARY LEADERS

## Role of character strengths and virtues

*Archana, Samridhi Ahuja, and Updesh Kumar*

Leadership has been defined as a process through which an individual influences others to achieve a goal by equipping them with a purpose and course to help in realizing a common mission (Northouse, 2007). It is illustrated by varied demands that are placed on the leaders in multiple settings to bring forth their attributes, capabilities and experiences to exercise control over the unfolding of the vision and mission of an organization. Leadership then is considered one of the key components that predict and facilitate organizational success. The leadership effectiveness criterion depends on multiple factors that include the performance and growth of team members, their capability to deal with setbacks, accomplishment of set goals, contentment and commitment of group members to goals, psychological well-being of the team and the number of tasks performed and attained successfully by both the leaders as well as their followers (Yukl, 1989).

Leaders don't function in a void. They require a set of adherents who support them in leading towards a desired goal. They boost the confidence level of their followers, make them more creative and enhance their risk-taking ability. Such leaders are likely to bring changeover processes in the organization by providing vision, motivation and intellectual stimulation to the team members. The effectiveness of any leader depends upon the leadership style being followed by the leaders. Leadership is about managing people who bring change and compel transformations in organizations. According to Kotter (2001), both leadership and management are 'complementary systems of action' wherein management focuses on handling adversities and leadership relates to dealing with change. It is difficult to have a precise definition of leadership as the concept has been defined by varied researchers in their own ways. Kouzes and Posner (2002) stated that leadership is a process wherein an ordinary person generates the finest for themselves as well as others. They further suggested that leaders share certain characteristics like being reliable, enabling others to act and also appreciating the hard work put in by the followers.

Any organization, in order to flourish, requires a leader with the capability of inspiring and influencing people. However, when approaching the concept of 'military leadership', it becomes imperative to acknowledge that military leadership strategies qualitatively differ from leadership strategies in any other organization. Military organization operates in two exclusively different environmental conditions, i.e., peace and war. In peace conditions, the goals of the organization are different as the environment is stable and reliable. In the peace environment, the leadership is focused on providing training to the troops. The leadership then follows a standardized norm of rules and regulations and a systematic approach for resolving any organizational issues

(Kieser & Ebers, 2006). On the other hand, a war environment witnesses a different leadership model at play. Depending on the intensity of the conflict, the organization may become chaotic, which results in decreased compliance with the pre-defined bureaucratic organizational norms and standards. It is during this chaos that the personal style of the leaders emerges with spontaneity (Weber, 2009). Leadership remains the most important parameter that can turn the path of a conflicting situation and change losses into victory, be it in a conventional war or an operation. Leaders intend to channel the anxieties of the troops into passion to succeed and utilize the group energy to achieve the objectives of the mission (Rothkopf, 2012).

Military leadership demands certain attributes, which involves having an ability to judge situations, mental flexibility, creativity for doing something innovative, maintaining healthy interpersonal relations with others and possessing military-specific knowledge and skills for attaining organizational goals. The entire process of influencing subordinates builds morale and unit esprit de corps, improving facilities, training as a unit and enhancing resources that require a good amount of time on the part of leader. The time, effort and interest invested by the leaders towards their subordinates further add to developing hope, trust and confidence in relations.

Just as the concept of leadership is not restricted to a particular organization, so are the leadership theories. There are varied leadership theories that are in place at a given time in the organizations. The next section aims to explore some of these leadership theories.

## Leadership theories

Leadership behaviours or styles that relate to the basic psychological needs of followers tend to produce positive behaviour outcomes, whereas a lack of fulfilment of psychological needs is likely to produce negative outcomes (Baard, Deci & Ryan, 2004). A leadership style is considered healthy if it is able to promote the needs of autonomy, relatedness and competence within employees. Autonomy is when individuals choose to engage in a particular behaviour which is in accordance with their values and reflects an individual's interest. Relatedness is defined as a sense of connection with others, a feeling of belongingness to a group or an organization. Competence relates to a sense of effectiveness and proficiency in one's work (Ryan & Deci, 2002). After these basic psychological needs are satisfied, leaders are able to create an environment that enables employees to perform effectively and experience higher job satisfaction (Baard, 2002). Actions undertaken by leaders to promote satisfaction of basic employee needs make the organization a more conducive place for growth of employees and their well-being (Ryan & Deci, 2002). Leadership theories focus on ways to improve employee motivation and satisfaction. Several theories have emerged in the past. This section will look at various leadership theories that contribute to employee well-being and growth.

## Charismatic leadership theory

According to the charismatic leadership theory, a charismatic leader portrays unorthodox behaviours relating to the status quo, expresses a clear vision for the organization and facilitates the employees to share and participate in achieving the vision (Conger & Kanungo, 1987). Over time, two theories of such leadership have been proposed; the attribution theory (Conger & Kanungo, 1987) and the self-concept theory (House, 1977). The attribution theory of charismatic leadership defines charisma as an attribution that employees or followers make based on the various behaviours the leader exhibits (Conger & Kanungo, 1987). The self-concept model focuses on the observed leadership behaviours that the leader exhibits rather than any attributions made (House, 1977). A charismatic leader motivates his employees through various strategies that involve the leader's capability to boost the level of satisfaction of the employees, improve their levels of judgment and motivate the employees to let go of personal interests in favour of the interests of the organization (Gilbert & Kelloway, 2014). This type of leadership tends to understate the value of extrinsic rewards while applauding the idea of work being a reward in itself, thereby augmenting the intrinsic value of work. In order to enhance the intrinsic value of work, the leader sets forth goals that are consistent with the value of his employees and propels them to work towards realizing those goals. Charismatic leaders promote autonomy by setting goals that are in line with the value of their employees and by giving them a sense of purpose and meaning. When the employees perceive the goals of the organization as compatible with their own values, they are more inclined to self-regulate their behaviours, which in turn contributes to improving their self-concept (Shamir, House, & Arthur, 1993) and promoting a sense of connectedness with the group. Charismatic leaders also come across as sensitive to the needs of employees, leading to a positive relationship between the leader and the employee (Barling, Christie, & Hoption, 2010). The self-concept theory of charismatic leadership posits that charismatic leaders tend to have higher expectations for their employees and reflect a confidence in their employees to fulfil the goals that they set out to achieve, leading to an increase in employee self-efficacy and confidence.

## Authentic leadership theory

Authentic leadership has been defined as 'a pattern of leader behaviour that draws upon and promotes both positive psychological capacities and a positive ethical climate. Authentic leadership fosters greater self-awareness and positive self-development' (Walumbwa, Avolio, Gardner, Wernsing, & Peterson, 2008). Authentic leadership is composed of four components: self-awareness, relational transparency, balanced processing of information and positive moral perspective. Self-awareness is defined as the leader's knowledge of their own weaknesses as well as one's views about oneself. Relational transparency refers to the ability of the leader to share their true self with the employees without any façade or frills. Balanced processing encompasses processing of the relevant information before reaching a conclusion to make a decision and examining information that is not consistent with one's perceptions. Internalized moral perspective talks about regulation of behaviour guided by the moral values of the individual (Walumbwa et al., 2008).

Authentic leaders act as role models for the employees, which serves as motivation for them to exhibit authentic behaviours themselves (Ilies, Morgeson, & Nahrgang, 2005). Authentic leadership promotes organizational identification, as the employees tend to personally identify with the leaders. This personal identification tends to extend into organizational identification, as the leaders reflect the value of the organization. Identification with both the leader and the organizational values leads to establishing a sense of meaning for the employees. This type of leadership promotes relatedness as it strives to build open, trusting and cooperative relations among the employees and leaders. The trusting and cooperative relationship that is built between the employees and the leaders also helps in enhancing the competence of the employees by endorsing the unrestricted flow of knowledge and information that contributes to boosting the employee performance, thereby providing the employees with further opportunities to learn and enhance their skill sets (Walumbwa, Luthans, Avey, & Oke, 2011).

## Leader-member exchange theory

This theory proposed by Dansereau, Graen, and Haga (1975) stresses the role of the interactions between the leaders and the employees, labelled exchanges. The relationship that each employee shares with the leader is characterized by a 'dyad' and therefore each employee shares a dyadic relationship with the leader. Within this dyadic relationship, each member possesses the power to influence the behaviour of the other member.

Relationships that are shared amongst the members of the dyad determine the work outcome and environment. When the relationship is high quality, there exists mutual respect, trust and responsibility towards each other and the organization. A low-quality relationship is characterized by distrust, lack of respect and poor sense of responsibility (Graen & Uhl-Bien, 1995). This theory acknowledges the autonomy needs of the employees by sharing a high-quality relationship with the employees, where they are trusted to manage and be responsible for their assignments and duties. High-quality leader-member exchange (LMX) relationships also tend to promote relatedness, as the relationship in itself is characterized by respect and trust towards each other, which encourages the partners to do well and go beyond their personal interests to meet the needs of the organization (Howell & Hall-Merenda, 1999). The supportive relationship that employees and leaders share with each other tends to facilitate the process of accomplishing the goals of the organization, thereby promoting competence among employees.

## Transformational leadership theory

In the recent past, leadership paradigms have been broadened with the emergence of transactional and transformational leadership, which are considered the most effective styles of leadership. Bass (1985) categorized these two styles of leadership, proposing that transactional leadership depends upon a chain of exchanges and negotiations between both leaders and their followers. Although this kind of leadership focuses on individual as well as organizational objectives, it fails to enhance the capabilities of the followers. Transactional leadership is composed of two important dimensions: contingent reward and management by exception. In the case of contingent reward, the relationship between leader and follower is based on a kind of bargain between the two, where the leaders promote and support good work by their followers in the form of encouragement and promotions and rebuke their undesirable behaviour. Management by exception relates to the kind of relationship wherein leaders intrude by providing negative feedback to their followers only in situations when followers divert from their goals (Bass, 1985). On the other hand, transformational leadership is practiced more frequently in organizations, as it results in better, flourishing output at the workplace. It moves a step ahead of the basic principles of bargaining. Transformational leadership refers to a process which brings about key changes in attitudes of employees and builds a sense of commitment for the organization's goals and mission (Yukl, 2001). These leaders hold an optimistic outlook and motivate their followers to attain the goals of the organization. They foster positive changes at the workplace by moving beyond their self-interest and widening the scope of their followers' growth and development. Some of the traits unique to transformational leaders include: (i) idealized influence that reflects elevated levels of willpower, self sacrifice and a clear future vision, stating that the benefits of the group are much more essential than the benefits of an individual; (ii) inspirational motivation that offers challenges to the employees for working together as a team; (iii) intellectual stimulation, supporting resourcefulness and creativity in looking at things from different perspectives; and (iv) individualized consideration which relates to coaching, mentoring and guiding followers by understanding their feelings to prosper and grow in the organization (Pearce & Sims, 2002). All these traits contribute to a more vigorous, dependable and trustworthy leadership. Such leaders gain respect from their followers by fulfilling their need to feel significant and worthy towards the progress and advancement of the organization.

Research has shown that all these traits of transformational leadership are found to be associated with the objective as well as the subjective determinants of performance (Lowe, Kroeck, & Sivasubramaniam, 1996). Scholarly views have revealed that although transformational leadership is not a substitute for transactional leadership, it adds to it by motivating employees to give their best by working hard and executing more towards attaining their goal (Howell & Avolio, 1993). It takes into account the aspects related to role modelling, maintaining interpersonal relationships with the employees, work-directed approaches and critical thinking (Singh & Krishnan, 2007). It also focuses on enhancing the moral values of the followers. One of the prerequisites of transformational leadership is to provide a focused vision that enhances performance at the workplace and helps in sustaining motivation in the long run.

Transformational leadership is found to be positively related to job satisfaction and organizational commitment. Within the literature, it has been observed that there are certain variables that tend to moderate the relationship between transformational leadership and followers' organizational commitment. These include culture, structural distance between transformational leaders and followers and psychological empowerment (Avolio, Zhu, Koh, & Bhatia, 2004). Leaders serve the base for organizational success, express a compelling vision for the followers, operate as a captivating role model, promote goal-setting behaviour and motivate performance by offering individualized support and intellectual stimulation to the employees (Podsakoff, MacKenzie, Moorman & Fetter, 1990). These leaders make their employees understand the importance of the desired goals along with the need to shift their focus from self interest to collective interest. They also provide intellectual inspiration to their followers by boosting their thinking processes and encouraging them to solve problems in an innovative manner.

Transformational leaders tend to depend on intellectual stimulation which encourages autonomy among the employees as they learn to take decisions for themselves and strategize according to the needs of the situation (Barling et al., 2010). It also fulfils the need for relatedness of the employees by promoting leader-employee relationships marked by mutual respect, support and care. Communication of a shared vision also plays an important role in creating a sense of connectedness. Transformational leaders tend to present their employees with challenges and goals that are achievable so as to satisfy their need of competence (Gilbert & Kelloway, 2014). Transformational leaders, then, play an important role in influencing employee behaviour so as to facilitate the achievement of organizational goals and values. However, leadership behaviour is not limited to employee motivation; it extends to the ethics and values of the leader which contribute to creating an authentic workplace and flourishing organizations.

## Ethical leadership at workplace

Every organization is composed of both external and internal good (MacIntyre, 1985). External goods are observed in the form of wealth, possessions, assets or status, and internal goods take into account things like seeking pleasure at work, feeling of pride in achievements and job satisfaction. A leader giving immense importance to external goods builds immoral organizational character (Moore, 2005). In contrast, valuing internal goods leads to a virtuous organizational character that contributes to flourishing organizations (Wright & Goodstein, 2007).

Leaders engaging in ethical leadership behaviour serve as virtuous agents in enhancing an ethical climate (Flynn, 2008). Ethical leadership relates to 'the demonstration of normatively appropriate conduct through personal actions and interpersonal relationships, and the promotion of such conduct to followers through two-way communication, reinforcement, and decision-making' (Brown, Trevino & Harrison, 2005, p. 120). Ethical leadership promotes an ethical climate, influences the attitude of members of the organization towards work, creates job satisfaction and builds attachments and commitment towards the workplace.

Leaders following an ethical code of conduct in organizations serve as role models and attachment figures for encouraging character among subordinates. According to social learning theorists, individuals' behaviour is strongly swayed by observing role models, as they learn suitable behaviour by discerning what is being rewarded and what actions are discarded (Bandura, 1986). Power and status are two factors that promote credibility and attractiveness. Leaders high on these factors strongly influence members within organizations. The attachment theory of leadership proposes that leaders promote relational attachments between themselves and organizational members. Relational attachments relates to the emotional connection formed between the members of an organization wherein the members interact in close association with each other, thereby satisfying each other's needs (Davidovitz, Mikulincer, Shaver, Izsak & Popper, 2007). Leaders also stimulate relational attachments through thoughtful and honest aspects of ethical leadership behaviour.

Ethical leaders operate keeping in mind the best interests of their subordinates which involves displaying faithful behaviour, taking care of employees' needs, making reasonable decisions and avoiding violations of ethical standards (McCloskey, 2008). The ethical climate at the workplace is mainly shaped by organizational leaders (Schminke, Ambrose & Neubaum, 2005). Organizations also support leaders in ascertaining and upholding an ethical climate, putting into practice ethical codes, audits and standardized procedures (Weaver, Trevino & Cochran, 1999). Ethical codes and principles are mostly revealed in actions rather than words.

Leaders play a vital role in shaping ethical climate, and this is illustrated in their behaviour and interactions (Dickson, Smith, Grojean & Ehrhart, 2001). Due to the power and status in the organization, leaders contribute to enhancing ethical behaviour by communicating ethical expectations, highlighting and adhering to policies and practices that strengthen ethical behaviour and holding subordinates responsible for ethical actions (Brown & Trevino, 2006). Leaders also promote interpersonal relations in terms of forming strong associations and bonds with members (Davidovitz et al., 2007). Leaders displaying ethical leadership behaviour tend to model suitable behaviour in a consistent, reasonable and honest manner. Such behaviour diminishes the level of anxiety associated with uncertainty in ethical situations by being thoughtful, understanding, truthful and reliable and by holding high ethical values and principles (Trevino, Brown & Hartman, 2003). The culture and climate of any organization influence the ethical decision-making, behaviour, perception and attachment of the employees towards their jobs. Research findings support the influence of ethical climate on organizational commitment and job satisfaction (Martin & Cullen, 2006). Organizations characterized by ethical conduct, trust, care for each other's needs and social bonds tend to keep employees happy and committed to their jobs.

## Understanding military as an organization

Military life is physically, mentally, psychologically and socially challenging. Military personnel are often exposed to numerous operational stressors that involve a wide range of traumatic incidents, extended deployments, fellow soldiers' injuries and death, unpredictable working conditions, high levels of uncertainty and harsh environmental conditions. These stressors can lead to long-term psychological problems like post-traumatic stress disorder, depression or anxiety, which are likely to influence the well-being and productivity of soldiers. Bliese and Castro (2003) have proposed the Soldier Adaptation Model (SAM) that presents an array of potential effects of stressors on military personnel. This model takes into account three elements: stressors, strains and outcomes. Stressors are those factors in the environment that produce demand on the soldiers. These stressors include factors like operating in extreme conditions, separation from family, physical threat and extended working hours. Strain refers to the collection of negative responses to stressors and includes aspects related to memory dysfunctions, frustration, emotional fatigue, sleep disturbance, alcohol dependence and withdrawal. Outcomes relate to the end result of strain resulting in illness, diminished performance and lack of motivation.

The military is a high-risk organization and the soldiers serving in this organization have to face unpredictable, difficult and stressful situations in routine life. That means these soldiers are required to deal with situations occurring unexpectedly, with unknown content, where the outcomes of actions are unforeseen (Torgersen, Steiro & Sæverot, 2013). Quality of life for military personnel largely resides in their ability to adjust to multiple demands in the organization. Military life demands soldiers respond to stressors effectively, and this promotes their level of adjustment and operational effectiveness. The basic training imparted to the soldiers emphasizes making them physically, mentally and psychologically resilient in order to withstand constant pressures. Military operations also require their leaders to be well equipped with skills and abilities to cope with dangerous and difficult situations. Such leaders play a vital role in developing soldiers' morale and character. The foundation of military leadership relies on basic pillars of values, beliefs, character and ethics. Developing these pillars requires robust training which brings success and achievement in the organization. Military leadership is a process of applying these core beliefs, values and knowledge to attain military-specific operations.

## Positive psychology of character strengths and virtues

Positive psychology emphasises making our lives worth living (Seligman & Csikszentmihalyi, 2000). According to Peterson (2006), positive psychology takes into account: (i) positive subjective experiences, including happiness and satisfaction; (ii) positive individual traits covering character strengths and virtues; and (iii) positive institutions involving families and organizations. Positive institutions facilitate positive traits, like character strengths and virtues, which in turn cultivate positive experiences (Peterson 2006). Character can be thought of as, 'those impenetrable and habitual qualities within individuals, and applicable to organizations that both constrain and lead them to desire and pursue personal and societal good' (Wright & Lauer, 2013, p. 27). Researchers have argued that behaviours stimulating high character assist in attaining greater products and profits (Peterson & Park, 2006). Peterson and Seligman's (2004) concept of character strengths offer a strong theoretical platform for shaping characters and virtues in military populations. The authors suggest that all these virtues must be present at above threshold levels for a person to be deemed to have a 'good character'. Good character does not denote having a single attribute; rather, it takes into account a collection of positive strengths. Peterson and Seligman (2004, p. 74) propose that character strengths 'embody values when the behaviour they organize and direct becomes habitual'. Values are the 'conceptions of the desirable that influence the way people select action and evaluate events' (Schwartz & Bilsky, 1987, p. 550).

The work environment is one of the most important places for fostering application of character strengths (Park & Peterson 2007). Character strengths represent varying components of good character. Peterson and Seligman (2004) have established a classification of 24 such character strengths divided into six virtues: (i) wisdom and knowledge, including creativity, curiosity, love of learning, open-mindedness and perspective; (ii) courage, covering bravery, integrity, persistence and vitality; (iii) humanity, taking account of love, kindness and social intelligence; (iv) justice, involving citizenship, fairness and leadership; (v) temperance, composed of self-control, prudence, forgiveness and humility and (vi) transcendence, incorporating spirituality, hope, appreciation of beauty and excellence, gratitude and humour. Each of these character strengths has strong linkages with emotional, psychological, physical, moral, spiritual and organizational outcomes. Research findings by Gentry, Cullen, Sosik, Leoplod and Tonidandel (2013) assert that some character strengths like integrity, bravery and social intelligence reflect organisational effectiveness and prosperity. A few other researchers have established an association between self-control and dynamic interpersonal relations. According to Peterson and Seligman (2004), a majority of people have between three and seven core or 'signature' strengths. Signature strengths are possessed by individuals and are found to be related to overall positive experiences like life satisfaction, well-being and meaning in life (Wood, Linly, Maltby, Kashdan & Hurling, 2011). Such positive experience at the work front is enhanced when individual signature strengths are applied in the organization.

Military leadership demands robustness in order to cope effectively with complex and tough situations. That means these leaders have to be resilient to perform in military combat operations. Since it is possible to develop resilience, the same is applicable for character strengths as well. The principles of positive psychology are very much pertinent to the military organization as they focus upon the culture that promotes personal growth. Personal growth can be enhanced through developing character strengths. To this end, a military leader plays a significant role in influencing a soldier's morale and character.

## Role of character strengths and virtues in shaping military leaders

There has been very limited research that has examined the role of character strengths in military organizations. In one of the studies conducted by Boe, Bang and Nilson (2015), nine character strengths, namely leadership, integrity, persistence, bravery, citizenship, open-mindedness, social intelligence, self-regulation and creativity, have been observed to be the most important character strengths for military officers. According to Gayton and Kehoe (2015), people high in character strengths reflect integrity, a strong work ethic, an innovative and intelligent bent of mind and the ability to work in a group.

Strength refers to a 'natural capacity for behaving, thinking and feeling in a way that allows optimal functioning and performance in the pursuit of valued outcomes' (Linley & Harrington, 2006, p. 39). In other words, psychological strengths relate to the 'ways of behaving, thinking or feeling that an individual has a natural capacity for, enjoys doing and which allows the individual to achieve optimal functioning while they pursue valued outcomes' (Govindji & Linley, 2007, p. 143). According to Zenger and Folkman (2002), successful leaders are defined not by the absence of their weaknesses, but rather by the presence of their clear strengths.

Today military organizations have been very focused and attentive in shaping the character of the leaders, since they serve as a foundation element for combat operations (Michelson, 2013). Several military doctrines in different countries, including the United States, New Zealand, Australia and Canada, have emphasised the role of character strengths in leadership. Military organizations all over the world share certain basic core military values. For instance, the U.S. Army focuses on core values of honour, respect, selfless service, courage, loyalty, duty and integrity in its leaders (Exum & Coll, 2008). Among British forces, the six core values of loyalty, integrity, courage, discipline, respect for others and selfless commitment are considered vital for effective leadership. The Norwegian Army focuses on the core values of respect, responsibility and courage (Dyrhaug Moe, 2015), whereas Argentinean military doctrine emphasises specific traits like consistency, audacity, composure, energy, initiative and honesty in its leaders. Peterson and Seligman's (2004) character strengths reveal that most of the military core values mentioned above are shared among the 24 character strengths as put forth by them. The Australian Defence Force highlights that leadership is based on character and competence. It states that 'character, built on an individual's values and attitudes, includes things like social capacity, interpersonal skills, personal integrity, conscientiousness, self-assurance and trustworthiness' (Australian Defence Doctrine Publication, 2012, p. 14). The U.S. Army's leadership doctrine places emphasis on three qualities of leaders: character, presence and intellect (Army Doctrine Reference Publication, 2012). According to this doctrine, 'character' is described as 'a person's moral and ethical qualities, [which] helps a leader determine what is right and gives a leader motivation to do what is appropriate, regardless of the circumstances or consequences' (p. 15).

In modern-day society, the media is flooded with news that shows lapses in ethical standards, alterations of truth and diminishing humanity and value systems, causing moral, emotional and psychological damage to people's well-being. Under such circumstances, it becomes important for the military to take the initiative in empowering subordinates to build character and core values. Placing character strengths into practice in military organization reflects successful accomplishment of a mission in a positive direction. Military leaders exhibiting character instil strengths, passion, motivation, zest and ethical principles in their subordinates. In recent years, researchers have shifted their interest to character-based leadership processes that portray the role of leaders in lifting the level of moral, emotional, social, psychological and spiritual development of employees, thereby strengthening the ethical climates in organizations. Recognising one's character strengths, promoting them and living in harmony with them to attain a higher purpose leads to attaining meaning in life. Character strengths at work result in job satisfaction, pleasure, engagement and meaning at work (Littman-Ovadia & Steger, 2010). At times the prevailing norms at the workplace may not promote behaviour that matches with individuals' traits. Therefore, under such circumstances, the applicability of character strengths of an individual depends upon the situational factors.

Excellent leadership is all about influencing the members of the organization in a positive direction. The way leaders present themselves leaves a huge impact on members. Leaders transmit character through their actions, behaviours and ethical code of conduct. Among all styles of leadership, transformational leadership has been found to be most effective in influencing the members of any organization, be it an industrial organization, military establishment, educational setting or any religious organization all over the world (Sosik & Jung, 2018). Transformational leadership involves four kinds of behaviours: inspirational motivation, idealized influence, intellectual stimulation and individualized consideration. These behaviours strengthen faith and confidence among subordinates, promote them to work efficiently on missions, increase their work output and provide happiness, contentment and satisfaction with the leader (Bass, 1985).

Research scholars have proposed that the character strengths of bravery, social intelligence, integrity and self-control can be attained by exhibiting behaviours associated with transformational leadership (Gentry et al., 2013). Bravery relates to not giving up during threat, challenge or any complexities; having the courage to raise one's voice for what is right even when others show resistance and operating on one's belief system even if it does not match with others. This definition looks beyond physical courage and accounts for courage at the moral level, laying emphasis on what is right and acceptable in one's professional and personal life (Park, Peterson & Seligman, 2004). Bravery underlies the significance of being fearless and bold for successful attainment of mission-specific goals. Bravery is exhibited through idealized influence and intellectual stimulation behaviours. Bravery facilitates leaders to operate constantly with their values and a strong belief system and motivate others to attain organizational expectations. It inspires subordinates to contribute willingly and enthusiastically in producing high-quality performance in the organization.

Social intelligence relates to understanding one's emotions, taking care of others' feelings and intentions, holding respect for self and others and valuing and treating others with humility and dignity. Social intelligence is transmitted through the display of idealized influence, inspirational motivation and individualized consideration behaviours. Leaders high in social intelligence are highly equipped to understand the emotions, sentiments and feelings of their subordinates. These leaders also have the knack of better identifying the potential and aptitude of their subordinates. Such personal involvement of leaders inspires subordinates to put in more efforts towards the attainment of goals.

Integrity refers to the real projection of oneself without any deception and assuming responsibility for one's emotions and behaviour (Park, Peterson & Seligman, 2004). Integrity means having coherence in one's words and actions and being honest, clear and genuine in behaviour and trustworthy in establishing human bonds. Integrity is transmitted through the display of idealized influence behaviour. Integrity relates to sincerity, truthfulness and authenticity. Leaders high in integrity have strong moral principles and are trustworthy, honest and kind. They take responsibility for their actions. Idealized influence reflects high core values like being true to oneself and others, and leaders' integrity can be manifested to subordinates through the display of idealized influence.

Self-control refers to regulating one's emotions, cognition and behaviour in a socially acceptable manner, thus resulting in adaptive coping. It also involves modulating the emotional reaction that depends upon the way the emotions are experienced and expressed. Self-control serves a regulatory function with respect to our traits emphasizing who we are, emotions pertaining to how we feel, cognition focussing on what we think and behaviours resulting in what we do (Baumeister & Exline, 2000). Self-control as one of the character strengths is also displayed through the idealized influence and intellectual stimulation behaviours.

The reason character strengths are important for military leaders is that they are based on values. According to Gayton & Kehoe (2015), values have been found to play an important role in leadership, adaptability and achievement. Military organizations demand their leaders possess good character to bravely encounter their enemies. Character is determined by the core values and strong belief system. A leader's belief is a driving force that strengthens the willpower of their subordinates in attaining any mission or duty. Character as one of the most important assets of an individual is built with dedication and hard work. Leaders high on character live by strong value systems. Military leadership demands robustness for thinking clearly and dealing effectively with one's feelings while encountering difficult and challenging situations. In the present-day scenario, leaders in high-risk military organization are being selected based on their measurements of personality (Picano & Roland, 2012). According to Biswas-Diener, Kashdan & Minas (2011), personality is about differences between individuals in terms of how one reacts to situations, whereas character is about the values that govern actions and behaviour. Military leadership focuses on individuality and intellect of the leader, which are considered vital to lead. Military leaders are required to be equipped with skills and abilities to face and cope with demanding situations. A sound character, a firm presence and a dedicated intellect help the leader perform with greater impact in the military organization.

## Concluding remarks

Military leadership begins with a leader having values and attributes that shape character. These attributes are internal, as they make up a leader's identity. Having a good character translates into effective military leadership. The quality of character is established by a strong belief system and the values one holds towards others. These beliefs are the facts that allow a person to use their strengths in the best possible way. A leader's belief helps in strengthening the willpower to accomplish military-specific missions. Values, on the other hand, place importance on people and things around them. Military requirements are very specific as they demand sacrifices during military operations. It is essential for leaders to understand that qualifications are just a tool for establishing a good profession. They fail to develop an individual as a whole. A complete person is symbolised by character, discipline, a positive attitude and a strong value system. Character is built with dedication and hard work. It is a combination of numerous attributes that include trustworthiness, high morals, self discipline, honesty and integrity. Military operations require their leaders to be well equipped with skills and abilities to face adversities. A strong leadership develops soldiers' morale and ethics. Building upon character strengths and virtues in military leaders will go a long way towards enhancing the growth, prosperity and effectiveness of the organization as a whole. Although there is a vast body of literature that indicates a strong association between personality traits and job performance, the need of the hour demands development of tools that focus on measuring the character strengths of personnel in order to facilitate the process of developing better military leaders. Inclusion of these character strengths in selection will be an asset for identifying the most resilient leaders for military operations. Applying the principles of positive psychology for strengthening and educating military leaders will contribute to shaping leaders who do not just govern with effective leadership styles but also have a strong hold on various character strengths.

## References


Army Doctrine Reference Publication, (2012). *Army Leadership*. Washington, D.C. Retrieved from https://fas.org/irp/doddir/army/adp6_22.pdf.

Australian Defence Doctrine Publication (2012). *Foundations of Australian Military Doctrine*. Canberra, Australia. Retrieved from http://www.navedu.navy.mi.th/stg/databasestory/data/laukniyom/workjob/bigcountry-workjob/Australia/ADDP-D-FOUNDATIONS-OF-AUSTRALIAN MILITARY DOCTRINE.pdf.

Avolio, B. J., Zhu, W., Koh, W., & Bhatia, P. (2004). Transformational leadership and organizational commitment: Mediating role of psychological empowerment and moderating role of structural distance. *Journal of Organizational Behaviour, 25*(8), 951–968.

Baard, P. P. (2002). Intrinsic need satisfaction in organizations: A motivational basis of success in for-profit and not-for-profit settings. In E. L. Deci & R. M. Ryan (Eds.), *Handbook of Self-Determination Research* (pp. 255–275). Rochester, NY: University of Rochester Press.

Baard, P. P., Deci, E. L., & Ryan, R. M. (2004). Intrinsic need satisfaction: A motivational basis of performance and well-being in two work settings. *Journal of Applied Social Psychology, 34*(10), 2045–2068.

Bandura, A. (1986). *Social Foundations of Thought and Action*. Englewood Cliffs, NJ: Prentice-Hall.

Barling, J., Christie, A., & Hoption, C. (2010). Leadership. In S. Zedeck (Ed.), *Handbook of Industrial and Organizational Psychology* (pp. 183–240). Washington, DC: American Psychological Association.

Bass, B. M. (1985). *Leadership and Performance beyond Expectations*. New York: The Free Press.

Baumeister, R. F., & Exline, J. J. (2000). Self-control, morality, and human strength. *Journal of Social and Clinical Psychology, 19*(1), 29–42.

Biswas-Diener, R., Kashdan, T. B., & Minhas, G. (2011). A dynamic approach to psychological strength development and intervention. *Journal of Positive Psychology, 6*(2), 106–118.

Bliese, P. D., & Castro, C. A. (2003). The Soldier Adaptation Model (SAM): Applications to behavioral science peacekeeping research. In T. W. Britt, & A. B. Adler (Eds.), *The Psychology of the Peacekeeper: A Multinational Perspective* (pp. 185–203). Westport, CT: Praeger.

Boe, O., Bang, H., & Nilsen, F. A. (2015). Selecting the most relevant character strengths for Norwegian Army officers: An educational tool. *Procedia-Social and Behavioral Sciences, 197*, 801–809.

Brown, M. E., & Trevino, L. K. (2006). Ethical leadership: A review and future directions. *The Leadership Quarterly, 17*(6), 595–616.

Brown, M. E., Trevino, L. K., & Harrison, D. A. (2005). Ethical leadership: A social learning perspective for construct development and testing. *Organizational Behaviour and Human Decision Processes, 97*(2), 117–134.

Conger, J. A., & Kanungo, R. N. (1987). Towards a behavioural theory of charismatic leadership in organizational settings. *Academy of Management Review, 12*(4), 637–647.

Dansereau, F., Graen, G., & Haga, W. J. (1975). A vertical dyad linkage approach to leadership with formal organizations: A longitudinal investigation of the role making process. *Organizational Behaviour and Human Performance, 13*(1), 46–78.

Davidovitz, R., Mikulincer, M., Shaver, P. R., Izsak, R., & Popper, M. (2007). Leaders as attachment figures: Leaders' attachment orientations predict leadership related mental representations and followers' performance and mental health. *Journal of Personality and Social Psychology*, *93*(3), 632–650.

Dickson, M. W., Smith, D. B., Grojean, M. W., & Ehrhart, M. (2001). An organizational climate regarding ethics: The outcome of leader values and the practices that reflect them. *The Leadership Quarterly*, *12*(2), 197–218.

Dyrhaug Moe, H. (2015). *Uniforms, medals and values*. Retrieved from https://forsvaret.no/en/facts/uniforms-and-medals/values.

Exum, H., & Coll, J. E. (2008). *A Civilian Counselor's Primer for Counseling Veterans*. New York: Linus Publications, Inc.

Flynn, G. (2008). The virtuous manager: A vision for leadership in business. *Journal of Business Ethics*, *78*(3), 359–372.

Gayton, S. D., & Kehoe, E. J. (2015). A prospective study of character strengths as predictors of selection into the Australian Army special force. *Military Medicine*, *180*(2), 151–157.

Gentry, W. A., Cullen, K. L., Sosik, J. J., Leopold, C. R., & Tonidandel, S. (2013). Integrity's place in the character strengths of middle-level managers and top-level executives. *The Leadership Quarterly*, *24*(3), 395–404.

Gilbert, S. L., & Kelloway, E. K. (2014). Leadership. In M. Gagné (Ed.), *The Oxford Handbook of Work Engagement, Motivation, and Self-Determination Theory* (pp. 181–198). NY: Oxford University Press.

Govindji, R., & Linley, P. A. (2007). Strengths use, self-concordance and well-being: Implications for strengths coaching. *International Coaching Psychology Review*, *2*(2), 143–153.

Graen, G. B., & Uhl-Bien, M. (1995). Relationship-based approach to leadership: Development of a leader-member exchange (LMX) theory of leadership over 25 years: Applying a multi-level multi-domain perspective. *Leadership Quarterly*, *6*(2), 219–247.

House, R. J. (1977). A 1976 theory of charismatic leadership. In J. G. Hunt, & L. L. Larsen (Eds.), *Leadership: The Cutting Edge* (pp. 189–207). Carbondale, IL: Southern Illinois University Press.

Howell, J. M., & Avolio, B. J. (1993). Transformational leadership, transactional leadership, locus of control and support for innovation: Key predictors of consolidated business-unit performance. *Journal of Applied Psychology*, *78*(6), 891–902.

Howell, J. M., & Hall-Merenda, K. E. (1999). The ties that bind: The impact of leader-member exchange, transformational, and transactional leadership, and distance on predicting follower performance. *Journal of Applied Psychology*, *84*(5), 680–694.

Ilies, R., Morgeson, F. P., & Nahrgang, J. D. (2005). Authentic leadership and eudaemonic well-being: Understanding leader-follower outcomes. *The Leadership Quarterly*, *16*(3), 373–394.

Kieser, A., & Ebers, M. (Eds.). (2006). *Organisationstheorien*. Kohlhammer Stuttgart.

Kotter, J. P. (2001). What leaders really do? *Harvard Business Review*, *79*(11), 85–96.

Kouzes, J., & Posner, B. (2002). *The Leadership Challenge*. San Francisco, CA: Jossey-Bass.

Linley, P. A., & Harrington, S. (2006). Strengths coaching: A potential-guided approach to coaching psychology. *International Coaching Psychology Review*, *1*(1), 37–46.

Littman-Ovadia, H., & Steger, M. (2010). Character strengths and well-being among volunteers and employees: Toward an integrative model. *The Journal of Positive Psychology*, *5*(6), 419–430.

Lowe, K., Kroeck, K., & Sivasubramaniam, N. (1996). Effectiveness correlates of transformational and transactional leadership: A meta-analytic review of the MLQ literature. *The Leadership Quarterly*, 7(3), 385–425.

MacIntyre, A. (1985). *After Virtue: A Study in Moral Theory*. London: Duckworth.

Martin, K. & Cullen, J. (2006). Continuities and extensions of ethical climate theory: A meta-analytic review. *Journal of Business Ethics*, *69*(2), 175–194.

McCloskey, D. (2008). Adam Smith, the last of the former virtue ethicists. *History of Political Economy*, *40*(1), 43–71.

Michelson, B. M. (2013). Character development of U.S. Army leaders: The laissez faire approach. *Military Review*, *93*(5), 30–39.

Moore, G. (2005). Corporate character: Modern virtue ethics and the virtuous corporation. *Business Ethics Quarterly*, *15*(4), 659–685.

Northouse, P. G. (2007). Transformational leadership. *Leadership: Theory and practice*, *4*, 175–206.

Park, N., & Peterson, C. (2007). Methodological issues in positive psychology and the assessment of character strengths. In A. D. Ong & M. H. M. van Dulmen (Eds.), *Handbook of Methods in Positive Psychology* (pp. 292–305). New York, NY: Oxford University Press.

Park, N., Peterson, C., & Seligman, M. E. P. (2004). Strengths of character and well-being. *Journal of Social and Clinical Psychology, 23*(5), 603–619.

Pearce, C. L., & Sims, H. P. (2002). Vertical versus shared leadership as predictors of the effectiveness of change management teams: An examination of aversive, directive, transactional, transformational, and empowering leader behaviors. *Group Dynamics: Theory, Research, and Practice, 6*(2), 172–197.

Peterson, C. (2006). *A Primer in Positive Psychology*. Oxford: Oxford University Press.

Peterson, C., & Park, N. (2006). Character strengths in organizations. *Journal of Organizational Behavior, 27*(8), 1149–1154.

Peterson, C., & Seligman, M. E. P. (2004). *Character Strengths and Virtues: A Handbook and Classification*. Washington, DC: American Psychological Association.

Picano, J. J., & Roland, R. R. (2012). Assessing psychological suitability for high-risk military jobs. In J. H. Laurence & M. D. Matthews (Eds.), *The Oxford Handbook of Military Psychology* (pp. 148–157). New York: Oxford University Press.

Podsakoff, P. M., MacKenzie, S. B., Moorman, R. H., & Fetter, R. (1990). Transformational leader behaviours and their effects on followers' trust in leader, satisfaction, and organizational citizenship behaviours. *Leadership Quarterly, 1*(2), 107–142.

Rothkopf, D. (2012). A new challenge for our military: Honest introspection. *Foreign Policy, 192*, 45–48.

Ryan, R. M., & Deci, E. L. (2002). Overview of self-determination theory: An organismic dialectical perspective. In E. L. Deci, & R. M. Ryan (Eds.), *Handbook of Self-Determination Research* (pp. 3–33). Rochester, NY: University of Rochester Press.

Schminke, M., Ambrose, M. L., & Neubaum, D. O. (2005). The effect of leader moral development on ethical climate and employee attitudes. *Organizational Behavior and Human Decision Processes, 97*(2), 135–151.

Schwartz, S. H., & Bilsky, W. (1987). Towards a universal psychological structure of human values. *Journal of Personality and Social Psychology, 53*(3), 550–562.

Seligman, M. E. P., & Csikszentmihalyi, M. (2000). Positive psychology: An introduction. *American Psychologist, 55*(1), 5–14.

Shamir, B., House, R. J., & Arthur, M. B. (1993). The motivational effects of charismatic leadership: A self-concept based theory. *Organization Science, 4*(4), 577–594.

Singh, N., & Krishnan, V. R. (2007). Transformational leadership in India: Developing and validating a new scale using grounded theory approach. *International Journal of Cross - Cultural Management*, 7(2), 219–236.

Sosik, J., & Jung, D. (2018). *Full Range Leadership Development: Pathways for People, Profit, and Planet*. New York: Routledge.

Torgersen, G. E., Steiro, T. J., & Sæverot, H. (2013). Strategic education management: Outlines for a didactic planning model for exercises and training of the unexpected in high risk organizations. *Proceedings of the 22nd Society for Risk Analysis Europe (SRA E) Conference*.

Trevino, L. K., Brown, M., & Hartman, L. P. (2003). A qualitative investigation of perceived executive ethical leadership: Perceptions from inside and outside the executive suite. *Human Relations, 55*, 5–37.

Walumbwa, F. O., Avolio, B. J., Gardner, W. L., Wernsing, T. S., & Peterson, S. J. (2008). Authentic leadership: Development and validation of a theory based measure. *Journal of Management, 34*(1), 89–126.

Walumbwa, F. O., Luthans, F., Avey, J. B., & Oke, A. (2011). Authentically leading groups: The mediating role of collective psychological capital and trust. *Journal of Organizational Behaviour, 32*(1), 4–24.

Weaver, G. R., Trevino, L. K., & Cochran, P. L. (1999). Corporate ethics programs as control systems: Influences of executive commitment and environmental factors. *Academy of Management Journal, 42*(1), 41–57.

Weber, M. (2009). *Wirtschaft und Gesellschaft: Entstehungsgeschichte und Dokumente (24)*. Mohr Siebeck.

Wood, A. M., Linley, A. P., Maltby, J., Kashdan, T. B., & Hurling, R. (2011). Using personal and psychological strengths leads to increases in well-being over time: A longitudinal study and the development of the strengths use questionnaire. *Personality and Individual Differences, 50*(1), 15–19.

Wright, T. A., & Goodstein, J. (2007). Character is not "dead" in management research: A review of individual character and organizational-level virtue. *Journal of Management, 33*(6), 928–958.

Wright, T. A., & Lauer, T. L. (2013). What is character and why it really does matter. *Organizational Dynamics, 42*(1), 25–34.

Yukl, G. (1989). Managerial leadership: A review of theory and research. *Journal of Management, 15*(2), 251–289.

Yukl, G. (2001). *Leadership in Organizations*. Upper Saddle River, NJ: Prentice-Hall.

Zenger, J. H., & Folkman, J. (2002). *The Extraordinary Leader: Turning Good Managers into Great Leaders*. New York: McGraw-Hill.

# 10

# POSTTRAUMATIC GROWTH IN MILITARY POPULATIONS

## Theory, research, and application

*K. C. Kalmbach and Bret A. Moore*

As far back as human recorded history, the theme of the 'hero's journey' has been reflected in numerous religious, philosophical, and historical texts. Many such stories detail the hero's struggle through extraordinary trauma and tribulation to emerge with profound and transformational wisdom and personal growth. Indeed, this simple idea has long been reflected in the seminal works of many diverse and influential thought leaders, namely that adversity provides an opportunity for personal growth (Frankl, 1963; Maslow, 1954; Yalom, 1980). A related theme is reflected in the musings of 19th-century nihilist philosopher Friedrich Nietzsche: 'From life's school of war: that which does not kill us, makes us stronger' (Nietzsche & Large, 1998, p. 5). But is this commonly traded platitude in fact true? Can the experience of life-altering trauma result in positive psychological outcomes and transformational personal growth?

Although the human experience of it is not new, it was not until the mid-90s that the term *posttraumatic growth* (PTG) was coined (Tedeschi & Calhoun, 1996). Other terms, related but not necessarily interchangeable, have been used to describe the construct (e.g., *adversarial growth, benefit finding, stress-related growth*). However, over the past decade, many scholars have begun to unify around the term PTG in order to share a common research language. PTG is defined as positive changes in emotion, cognition, and behavior occurring as a result of a complex and dynamic personal struggle with traumatic events or circumstances (Tedeschi & Calhoun, 2004). In fact, it is the struggle which defines the construct. Within the PTG framework, growth derives *directly* from the process of grappling with the aversive experience and is influenced by a range of environmental (e.g., social support, cultural values) and individual factors (e.g., level of distress, disposition/personality, coping style, comfort level with self-disclosure). Over the past two decades, accumulating research suggests that the PTG phenomenon is common (see Linley & Joseph, 2004 for review) and is reflected worldwide in personal narratives of diverse people and regions, albeit with cultural variation (Kashyap & Hussain, 2018; Splevins, Cohen, Bowley, & Joseph, 2010).

Some scholars embrace the idea of PTG as both a distinct process and a unique outcome (e.g., Cann et al., 2011; Calhoun & Tedeschi, 2004, 2006; Morris & Shakespeare-Finch, 2011;

O'Leary, Alday, & Ickovics, 1998; Tedeschi & McNally, 2011). Others are less persuaded and question what role common resilience plays or whether posttraumatic changes are simply self-enhancing 'illusory' distortions—necessary to reduce the cognitive dissonance which occurs following a cataclysmic rupture in assumptive worldviews (Affleck & Tennen, 1996; Bonanno, 2004; Coyne & Tennen, 2010; Frazier et al., 2009; Hobfoll et al., 2007; Taylor, 1983). Yet it is possible that the PTG model may allow for both positions to be true. Early phases of automatic coping with trauma may be inherently self-serving, as a desperate attempt to restore one's sense of safety and world order. Over time, with deliberative cognitive processing, a more nuanced and 'authentic' growth may take place (Maercker & Zoellner, 2004; Tedeschi & McNally, 2011; Zoellner & Maercker, 2006).

However, the experience of PTG is by no means universal; not all, nor even a majority, of trauma victims will necessarily manifest PTG. Clinicians emphasize that no traumatic event, nor any aspect of it, is a good or desirable experience. To the contrary—there are decades' worth of research documenting the damage done by early, repeated, or severe trauma (see, for example, Carr, Martins, Stingel, Lemgruber, & Juruena, 2013; Sperry, 2016). Therefore, the absence of a PTG-like experience in trauma survivors does *not* imply poorer coping or recovery status and should not be considered a clinical failure or a less-valued outcome (Tedeschi, Shakespeare-Finch, Taku, & Calhoun, 2018). In fact, research indicates the most resilient individuals may manifest PTG the *least*, perhaps due to more robust coping skills (leaving less room for the impact of trauma, the destruction, and subsequent reconstruction required in PTG) (Levine, Laufer, Stein, Hamama-Raz, & Solomon, 2009). This seeming paradox might be reflected in an observation by Holocaust survivor Viktor Frankl (1959), who noted that 'some prisoners of a less hardy make-up often seemed to survive camp life better than did those of a robust nature [perhaps] because these sensitive people … were able to retreat from their terrible surroundings to a life of inner riches and spiritual freedom' (p. 35). Nonetheless, for those who wish to understand PTG, an appreciation for the theory upon which it is premised is necessary.

## The role of the assumptive worldview

At the heart of the PTG model is an organizing principle captured in the *shattered assumptions theory* (Janoff-Bulman, 1989, 1992). Humans, it is argued, have a very complex, deeply rooted, and well integrated schema about the world and themselves. Specifically, we hold three basic, generally positive assumptions that are rarely consciously examined or challenged: (a) the world is benevolent, predictable, and controllable; (b) events occur for a reason; (c) the self (and most others) are generally good. These beliefs work together to create the foundation of a complex system that governs much of who we are as individuals, all the while operating largely outside conscious awareness or examination. These beliefs serve an important function in maintaining a psychological sense of well-being, and are self- and security-enhancing in nature. In short, our *assumptive world* allows us to believe that we (and others) are good and worthy people, who live in a (relatively) just and benevolent world, where (most) things that happen make sense. These assumptions allow us to move through the world with a generally confident, coherent, and stable outlook. Extreme trauma—of any sort—shakes the very foundation of this core belief system, revealing our beliefs to be delusory, and forcing us to confront the terrifyingly random, arbitrary, cruel, or capricious nature of life. In fact, when traumatic events are experienced, they cause great cognitive dissonance and distress that may be apparent even years or decades later (Beder, 2005; Janoff-Bulman, 1989, 1992, 2004). How trauma survivors attempt to resolve this cognitive dissonance forms the starting point for PTG theory and research.

## Defining posttraumatic growth

Three aspects of PTG bear explication as terms used vary somewhat from common usage and definitions for trauma and stress-related disorders in the *Diagnostic and Statistical Manual of Mental Disorders—Fifth Edition* ([DSM-5], American Psychiatric Association, 2013). What is meant by the term *trauma,* how is the word *growth* conceptualized, and when does it occur?

### *Trauma*

Much debate exists over the definition of the construct of trauma, especially following recent DSM-5 revisions to posttraumatic stress disorder (PTSD) criteria (May & Wisco, 2016). Traditionally, scholars have defined trauma as a highly stressful event or situation that overwhelms the individual's capacity to cope, and "pose[s] an extreme psychological or physical threat or harm to an individual" (Sperry, 2016, p. 162). For scholars in PTG, however trauma is defined not so much by the event itself as its ability to shatter existing schemas and assumptive worldviews (Calhoun & Tedeschi, 2004, 2006; Linley & Joseph, 2004). The trauma event is often conceptualized as a 'turning point,' a 'break in the life narrative,' or a 'watershed moment' (Tedeschi & McNally, 2011). Thus, it is the *survivor's interpretation* of the event as catastrophic that will define an event as traumatic. Central to the construct is loss; the actual loss of someone or something, but even more so, the loss of one's identity, role, or assumptions about self, others, and the world more broadly. In this context, trauma is defined more broadly than in the DSM-5 (i.e., threat to life/physical integrity); however, the event must be considered 'extremely undesirable' (O'Leary et al., 1998; Tedeschi, Park, & Calhoun, 1998) and represent a 'seismic' disruption in core beliefs or worldview (Calhoun & Tedeschi, 1999). Less emphasis is placed on a fear-based or physiological response to the threat than to the cognitive and emotional sequelae of the trauma.

Unlike DSM-5 criteria for PTSD, there is no expectation that the trauma occurred in a single or easily definable temporal space. Thus, a traumatic *event* in the PTG model may be a single, easily definable occurrence (e.g., a violent assault) but may also be an experience that happens over time and is embedded in a complex number of other successive events (e.g., natural disaster and its after-effects). Many traumatic experiences unfold over time (e.g., a chronic, progressively debilitating illness) and others simply cannot be neatly extricated from their larger context (e.g., a particularly violent firefight within an ongoing combat deployment).

### *Growth*

In simple terms, growth often suggests a progressive improvement, refinement, or maturing of existing conditions. Indeed, many people report positive growth and development over time simply as a result of maturing and acquiring wisdom. For the PTG model, however, the growth must be *transformative*. PTG is not age-related, and growth is earned only in the arduous process of meaning-making following deeply life-altering experiences of trauma. It is the growth that manifests *as a result of the struggle* which can be defined as PTG. Normative developmental processes may be occurring alongside of PTG, but are distinguished from PTG by the absence of any type of hard-won personal struggle with traumatic content. PTG is the result of deliberative cognitive processing or rumination focusing on issues related to the 'shattered worldview' in an attempt to modify, adapt, or revise cognitive schemas which have been violated by the trauma. Simplistic, binary, or dichotomous thinking must give way to more nuanced and flexible analysis. In other words, it is not a typical anxiety-fueling ruminative style, but rather a consciously cognitive approach to understanding what happened, re-examining core beliefs, and perhaps modifying or creating new belief systems that can incorporate the trauma into a new, expanded worldview that makes sense to the survivor (Taku, Cann, Tedeschi, & Calhoun, 2015; Tedeschi & Calhoun, 2004; Triplett, Tedeschi, Cann, Calhoun, & Reeve, 2012).

Finally, it is important to note that PTG does *not* preclude the presence of concurrent distress, or even active trauma-related symptoms. A number of studies suggest both PTG and PTSD symptoms can co-occur in complex ways (Blix, Birkeland, Hansen, & Heir, 2016; Engelhard, Lommen, & Sijbrandij, 2015; First, First, Stevens, Mieseler, & Houston, 2018; Frazier et al., 2009). Especially in military populations, it would be unsurprising to witness PTSD (Steenkamp, Litz, Hoge, & Marmar, 2015), moral injury (Litz et al., 2009; Nash, 2007), and/or complicated grief reactions (Lichtenthal, Cruess, & Prigerson, 2004) co-occurring alongside the process of PTG. This fact underscores the uniqueness of the PTG construct and may simultaneously defeat the argument that the growth reported is simply 'illusory' or self-protecting denial of distress or suffering (Tedeschi & McNally, 2011).

### *Post-trauma growth period*

As important as the definition of the trauma is an understanding of the critical phase in PTG. Important for the PTG model is the assumption that significant growth or changes will occur in people *after*, and sometimes long after, the traumatic event. Much of immediate responding to a traumatic event is unthinking or automatic and involves typical coping strategies. (This may be the period that some scholars refer to as 'illusory' where survivor gains are more of a self-serving nature simply to stabilize their world and be able to function within it.) PTG, on the other hand, is thought to be the result of more deliberate cognitive processing and is often an unplanned or unexpected process of *meaning-making* that seeks to create a new narrative consistent with an altered worldview. Thus, the focus of PTG is not so much about the experience or cognitions during or immediately after the trauma as on the long-term consequences and outcome (which are, of course, different for each individual). This timeline suggests that capturing PTG may require a much more flexible window than is often available in research paradigms. On one hand, a 30-year retrospective may lose too much to the fog of time (Dekel, Mandl, & Solomon, 2011); on the other hand, assessing for PTG two months posttrauma may be too soon to detect significant change (Frazier et al., 2009). The idiosyncrasy of each individual's growth pattern is just one of the many factors which make this complex and nuanced topic challenging to research.

## Posttraumatic growth in context

It bears noting that the vast majority of people who report experiencing PTG have accomplished this transformational change without the assistance of a mental health professional. Often, social support circles surrounding and supporting the trauma survivor are sufficient to provide the necessary scaffolding to create conditions conducive to personal growth. In fact, research shows that 50%–70% of the U.S. population will experience at least one violent or traumatic event during their lifetime and yet only 5%–10% will develop PTSD (Ozer, Best, Lipsey, & Weiss, 2003). This fact suggests most trauma survivors will resist, recover, and/or thrive posttrauma—and will do so without the assistance of a therapist or doctor. Nonetheless, clinical research has identified important factors which appear to facilitate PTG, and mental health professionals are encouraged to utilize this developing knowledge base to support clients in their meaning-making journey posttrauma. However, it is equally important to understand what PTG is *not*, as well as what it *is*.

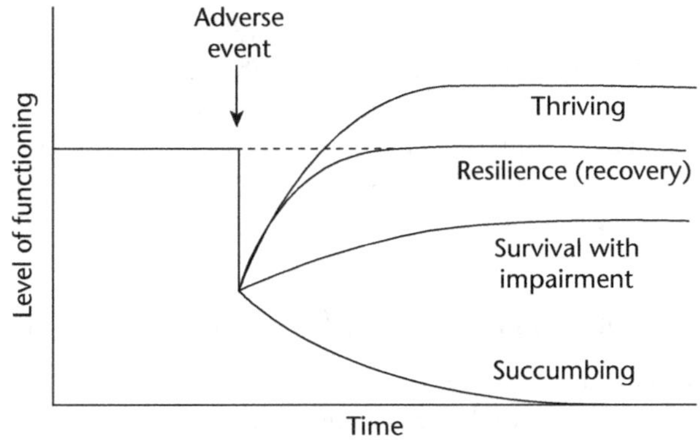


*Figure 10.1* Responses to adversity.

*Source:* Adapted by O'Leary & Ickovics, 1994, 1995 and used with permission.

## *Posttraumatic growth is not resilience*

Tedeschi and Calhoun (1995, 2004) have described PTG as transformational positive psychological changes experienced as a result of the struggle with traumatic or highly challenging life-altering circumstances. *Resilience*, on the other hand, is often defined as the ability to adaptively return to baseline functioning (O'Leary & Ickovics, 1995) or to resist being compromised by trauma in the first place (Bonanno, 2004). While some insist on clearly distinguishing the three related terms (Bonanno, 2004), others have tried to integrate multiple aspects into a global definition (i.e., *resistance, recovery*, and *reconfiguration*) (Lepore & Revenson, 2006). To be clear, PTG is *not* simply a return to baseline; it is to surpass a previous baseline on the way to something even better: more nuanced, more heartfelt, and more hard-won than 'normal.' The end state of such a transformational struggle is sometimes conceptualized as *thriving* (as opposed to surviving, resisting, or recovering). A model by O'Leary and Ickovics (1994, 1995) depicts the distinction between the common understanding of resilience (recovery or return to normal function/ baseline) and thriving (see Figure 10.1). Although somewhat oversimplified, it could be said that PTG most closely resembles the idea of thriving in this model.

## The posttraumatic growth model

The current corpus of research, taken as a whole, suggests the emerging picture of PTG is much more complicated than first imagined; the field is still very much evolving. For example, some studies have found resilience to be negatively correlated with PTG and PTSD symptoms (Levine et al., 2009), whereas others have found resilience to be positively associated with PTG, but negatively with PTSD symptoms (Bensimon, 2012). Nonetheless, the factor structure of PTG (Tedeschi & Calhoun, 1996) has been extensively researched and enjoys wide support. Additionally, there are early indicators of modest agreement on a number of factors which seem to facilitate or be associated with PTG, although other variables are related in complex and sometimes contradictory ways.

## *Measuring posttraumatic growth*

There are nearly a dozen measures designed to tap into domains similar or related to PTG. The most prominent measures include the Stress Related Growth Scale (Park, Cohen, & Murch, 1996),

Perceived Benefit Scale (McMillen & Fisher, 1998), Thriving Scale (Abraído-Lanza, Guier, & Colón, 1998), Changes in Outlook Questionnaire (Joseph, Williams, & Yule, 1993), and Posttraumatic Growth Inventory ([PTGI]; Tedeschi & Calhoun, 1996, 2004). A discussion of all measures is beyond the scope of this chapter; however, factor analysis of the PTGI, long considered the gold standard in the field, has yielded five empirically derived domains: *greater appreciation of life and changed priorities; warmer, more intimate relations with others; greater sense of personal strength; recognition of new possibilities;* and *spiritual development.*

### *Greater appreciation of life and changed priorities*

As a result of the trauma and its aftermath, survivors often report valuing 'the little things,' which they may have previously taken for granted. Now fully aware that they have little control over some events, survivors may report feeling 'lucky' or 'blessed' and frequently report that their life priorities have shifted towards more meaningful and relational-based goals and values. Acknowledging the pain of loss has allowed them to appreciate the value of many other things previously overlooked (e.g., relationships, time, health, resources).

### *Warmer, more intimate relations with others*

Especially in the case of bereavement, survivors often experience a shift in values towards building, maintaining, and cherishing closer and more intimate social relationships. By way of increased self-disclosure, survivors may experience the support and emotional engagement which might have been lacking in their lives previously. Empathy and compassion may be more manifest in close relationships, and reports of being better able to understand others' pain are common. Social connections can shift: new relationships may be formed, and existing relationships that were not supportive or productive may be ended.

### *Greater sense of personal strength*

Paradoxically, the PTG experience often leaves survivors with an enhanced sense of self efficacy (i.e., *If I got through that, I can get through anything.*) Although they may be fully aware of and humbled by their own vulnerability, there is often a more nuanced understanding that strength lies in the ability to be flexible and adaptive to changing life circumstances. Having overcome previously unimaginable trauma or challenges, survivors often have a much deeper and more accurate understanding of what they are able to personally cope with and overcome.

### *Recognition of new possibilities*

Often the experience of trauma opens doors to avenues never previously considered. For example, survivors might consider a new or different occupation (e.g., losing a child to a drunken driver prompts a mother to create an anti-drinking campaign). For others, the PTG process helps them to see that an entirely new life path or goal set is attainable, and, given the uncertainty of life, it may be reasonable to 'take the leap' now (whereas previously it seemed too risky or irresponsible). Constructing a new philosophy about life can lead to the recognition of new possibilities in a wide range of areas (e.g., relationships, professional endeavors, health and lifestyle).

### *Spiritual development*

Expressed in various ways, the common theme in this domain is an enhanced sense of closeness with, or understanding of, larger forces and their role in the survivor's life. For example, if the survivor endorses a belief system with a 'personal God' (more common in the West), they will often report that the deity was responsible for them 'getting through it all.' For individuals who are not necessarily religious, or who are adherents of other belief systems, this feeling is often expressed as a larger existential awareness or an increased orientation towards humanism or connectedness to others. For obvious reasons, research findings depend largely on the survivor's culture as well as prior orientation towards religion and/or spirituality. A recent revision of the PTGI scale (PTGI-X) has expanded this section and now allows for much more flexibility and inclusion of those who do not endorse traditional religious perspectives (Tedeschi, Cann, Taku, Senol-Durak, & Calhoun, 2017).

## Universal and ubiquitous: The experience of posttraumatic growth

There has long been evidence that many survivors of traumatic or highly challenging events report transformational and positive psychological outcomes. In fact, between 40%–70% of trauma survivors—including, for example, natural disaster victims, sexual assault survivors, chronically ill and cancer patients, vehicular crash and accident victims, prisoners of war and combat veterans—report some positive psychological benefit as a result of trauma experienced (Abraído-Lanza, et al., 1998; Calhoun & Tedeschi, 1999; Cryder, Kilmer, Tedeschi, & Calhoun, 2006; Dohrenwend et al., 2004; Feder et al., 2008; First et al., 2018; Frankl, 1963; Frazier, Tashino, Berman, Steger, & Long, 2004; Joseph et al., 2005; Maguen, Vogt, King, King, & Litz, 2006; Ramos & Leal, 2013; Shakespeare-Finch & Armstrong, 2010; Shakespeare-Finch & de Dassel, 2009; Sledge, Boydstun, & Rabe, 1980; Snape, 1997; Solomon & Dekel, 2007; Stanton, Bower, & Low, 2006; Tedeschi & Calhoun, 2004; Tsai, El-Gabalawy, Sledge, Southwick, & Pietrzak, 2015). Although there are hundreds of studies investigating PTG in various types of trauma survivor populations, many of the researchers in this area are military affiliated with specialized expertise as a result of working with a population at high risk for experiencing trauma. Indeed, veterans from eras going back to World War II have reported experiencing PTG as a result of their wartime experiences.

### *Posttraumatic growth in military and related settings*

Remarkably, a significant proportion of combat veterans and POWs consistently report psychological benefits resulting from their period of captivity and maltreatment. Israeli researchers (Solomon & Dekel, 2007) working with war veterans and POWs have reported on the most extensive longitudinal study to date. Three key findings emerged from the 30-year follow up: (a) POWs reported significantly higher PTG scores and higher levels of PTSD symptoms when compared to controls (combat experienced, but non-POWs); (b) PTG and PTSD symptoms were positively correlated, meaning they were co-existent; and (c) PTG manifested a curvilinear relationship (inverted U) with severity of PTSD symptoms, indicating that an intermediate level of severity was associated with the highest levels of PTG. Furthermore, PTG was significantly and positively associated with *combat exposure, loss of control, suffering while in captivity*, and *belief in a just world* (Dekel, Mandl, & Solomon, 2011). In a study of former East German political prisoners, researchers found 72% of survivors spontaneously reported at least one aspect of perceived growth gained from the experience of being imprisoned. The growth area most frequently pertained to a *changed philosophy of life* and increased *understanding of human nature* (Maercker &

Zoellner, 2004). Research with Dutch peacekeeper ISAF personnel deployed to Afghanistan indicated PTG was positively associated with *perceived threat* while in-country and inversely correlated with *rank* (Boermans, Kamphuis, Delahaij, van den Berg, & Euwema, 2014). However, a recent study of survivors of the Oslo terrorist bombing in Norway found an initially strong correlation between PTG and PTSD, which gradually became nonsignificant after 22 months. Perhaps more importantly, this cross-lagged design revealed that PTG was both a consequence *and* antecedent of PTSD, meaning that PTG predicted later PTSD and vice versa (Blix, et al., 2016). In another recent cross-lagged design, researchers in the Netherlands were concerned to see that higher levels of PTG five months post-deployment were positively correlated with PTSD symptoms, and were also predictive of even higher levels of stress symptoms 15 months later, suggesting that PTG may be a harbinger of worse outcomes (Engelhard, et al., 2015).

In the United States, a great deal of research has been done with Vietnam-era veterans. Based on a national database (the National Vietnam Veterans Readjustment Study; Kulka et al., 1990), Dohrenwend and colleagues (2004) found that 71% of veterans reported the effect of their wartime service in Vietnam was mainly positive. Similarly, aviators who had been taken prisoner by the North Vietnamese reported high rates of PTG; in one study, nearly two-thirds (63%) reported moderate or greater levels of growth, and 37% of that sample indicated a high or very high degree of PTG as a result of their suffering. Interestingly, *dispositional optimism* and *duration of captivity* were significant predictors of PTG in this POW population (Feder et al., 2008). Other POWs who survived captivity and torture reported that the experience had increased their *self-confidence* and reorganized *life priorities*. Specifically, 61% reported that they had benefited psychologically from their traumatic experience; the more severe the maltreatment suffered, the more likely the survivor was to report experiencing PTG (Sledge et al., 1980). In a study of Gulf War veterans, *perceived threat* while in the war zone was the strongest predictor of *appreciation for life* (one aspect of PTG); however, overall, PTG was best predicted by *social connectedness* post-deployment (Maguen, Vogt, King, King, & Litz, 2006). More recently, researchers drawing on a large national database of post-9/11 veterans confirmed the curvilinear relationship between PTG and PTSD severity and also reported that half of the total sample, and 72% of those who screened positive for PTSD, reported at least moderate levels of PTG as a result of their worst traumatic event. They found PTG to be positively correlated with *social connectedness*, *intrinsic religiosity*, and having a *life purpose* (Tsai et al., 2015). In another U.S. Army sample, PTG was significantly and positively correlated with *number of combat experiences*; it was negatively correlated with *recent suicidal ideation* (Gallway, Millikan, & Bell, 2011).

The preceding discussion of research in military and related settings reveals the fundamental paradox present in PTG, namely that "out of loss there is gain" (Tedeschi & Calhoun, 2004, p. 6). The experience of PTG can help survivors construct meaning, personal significance, and even hope—in the aftermath of suffering. However, given that not all trauma survivors experience PTG, what do those who report PTG have in common? Research has revealed a handful of factors that are associated with PTG. However, many of the studies await replication, and variables under investigation sometimes manifest complex and seemingly counterintuitive relationships with each other. Therefore, the following discussion of findings should be considered tentative at this time.

## Factors associated with posttraumatic growth

A wide range of factors or variables have been connected to PTG over the years (Linley & Joseph, 2004). The factors can be categorized generally at the environmental or individual level, which, together, create a unique profile for each survivor of trauma.

## Environmental factors

Environmental factors are those outside forces or conditions which act upon or influence the experience of the survivor, but are not internally located nor necessarily under their direct control.

### Social support and connectedness

Research shows that a lack of social support is implicated in a wide range of adverse outcomes for both physical and mental health; conversely, the presence of a strong and supportive social network functions as a powerful protective factor against psychopathology (Price, Pallito, & Legrand, 2018). In fact, social support was the strongest predictor in a meta-analysis of studies examining risk and protective factors for PTSD (Ozer et al., 2003). It seems clear that those who have experienced a traumatic loss often fare better if they perceive themselves within a strong social network of support (Park et al., 1996; Tsai et al., 2015). Complicating the issue, however, is the fact that personality and ruminative style appear to interact with perceived social support and PTG such that ruminators are more likely to seek support and benefit from it, but will actually report receiving *less* social support than non-ruminators (Nolen-Hoeksema, & Davis, 1999). Generally, it is reasonable to assume that PTG would be facilitated by strong social support; however, it is presently unclear whether the relationship between social support and PTG is bidirectional and direct or whether one precedes the other or manifests via indirect pathways (Schaefer & Moos, 1998).

### Religiosity/religious participation

Perhaps more than other domains, research in this area struggles with operationalizing terms and avoiding cultural bias (Kashyap & Hussain, 2018). Spirituality is a challenging construct to assess and even more difficult to interpret. Nonetheless, some scholars have found a positive association between PTG and spirituality/religiosity (i.e., religious beliefs, religious participation) (Calhoun, Tedeschi, Cann, & McMillan, 2000; Koenig, Pargament, & Nielsen, 1998; Park et al., 1996; Tsai et al., 2015). For adherents of any major faith tradition, the idea of facing a challenge and triumphing over adversity is a well-known and cherished theme reflected in dozens of parables in the guiding documents of every religious tradition. It may be then, that the idea of growth following a personal tragedy or trauma is something for which followers already possess a schema, a language, and an expectation of future success. Combined with the strong social support system that many religions foster, this inculcation of values and expectations may serve as a powerful protective or health-promoting factor when faced with a personal trauma. However, others have noted an inverse relationship, with the trauma event precipitating the death of previously held spiritual or religious beliefs, presumably because the rupture which destabilized the survivor's worldview was too great to overcome or reconcile with the religious belief system (Pargament, Desai, & McConnell, 2006).

### Nature of the traumatic event

There is no clear evidence that a specific *type* of traumatic event is more likely to foster PTG; however, the perceived *controllability* and *severity* of the trauma do appear to play a role. Specifically, there appears to be a curvilinear (inverted U) relationship between PTG and trauma severity such that intermediate levels of trauma are most strongly associated with growth (Dekel et al., 2011; Fontana & Rosenheck, 1998; Park et al., 1996; Schnurr, Rosenberg, & Friedman, 1993; Solomon & Dekel, 2007; Tsai et al., 2015). It may be that lesser degrees of trauma are not sufficiently disruptive to cause the survivor to examine or reorganize belief systems or worldview. On the other end of the spectrum, extreme or repeated trauma may simply overwhelm the survivor's ability to cognitively process the event or even recover baseline following the trauma.

## Individual factors

Individual factors are those attributes which reside within the survivor, including attitudes, knowledge, skills, genetics, and dispositional/personality or other personal characteristics.

### Demographics

It appears there is a tendency for younger people to report PTG more frequently than older individuals. Women are more likely than men to report PTG, as are those individuals with higher levels of education (Linley & Joseph, 2004; Tedeschi & Calhoun, 2004). Although these findings are interesting, a number of alternative explanations should be considered. It is possible that older people have already acquired significant wisdom and life experience or that their perspective has shifted, having less time to live. Older individuals may be more resilient simply due to life experience and thus manifest less PTG as has been found in some studies with resilient survivors. Women tend generally to be more comfortable with self-disclosure as well as therapeutic engagement; it may be that women are more likely to report PTG simply because they are more likely to possess many of the factors associated with PTG (e.g., social support networks; willingness to self-disclose) and are more comfortable or familiar with the 'language' of therapy or growth. Finally, a higher level of education may be correlated with a range of variables which better position a survivor to experience PTG (e.g., cognitive capacity, more resources, greater freedom or opportunity).

### Distress or trauma symptoms

At first glance it would seem intuitive that PTG would be associated with lower levels of distress. However, the research contains a host of contradictory findings. In some cases, PTG is negatively correlated with distress such as depression, anxiety, and/or trauma symptoms (Cadell, Regehr, & Hemsworth, 2003; Linley, Joseph, & Goodfellow, 2008; Tomich & Helgeson, 2004); in others, PTG and distress/trauma symptoms co-exist (Calhoun & Tedeschi, 2006; Solomon & Dekel, 2007). In yet other studies, researchers have found PTG to be both an antecedent and consequence of PTSD symptoms (Blix, et al., 2016). The research seems to suggest a complicated relationship between the presence of distress/trauma symptoms and PTG. Nonetheless, it is clear that the two are not mutually exclusive and that the presence of one may tell us little or nothing about the other.

### Personality/dispositional traits

Although much work remains to be done in this area, studies do generally find that four of the five 'Big Five' traits (*extraversion, openness to experience, agreeableness,* and *conscientiousness*) are positively associated with PTG, while *neuroticism* (i.e., negative emotionality, such as anxiety, fear, anger, guilt) is often negatively correlated with PTG (Linley & Joseph, 2004; Tedeschi & Calhoun, 2004), but not always (Engelhard, et al., 2015). Perhaps not surprisingly, *optimism* (Zoellner & Maercker, 2006) and *extraversion* are associated with PTG (Engelhard, et al., 2015). However, once again, the relationship is complicated by the fact that optimistic people generally have stronger/wider social support networks, are more likely to engage in self-disclosure, express their feelings more effectively, and employ adaptive coping—all of which bode well for the PTG process. Whether optimism is a predictor or simply a marker variable for PTG remains to be determined (Bostock, Sheikh, & Barton, 2009). Other characteristics positively associated with PTG include *self-efficacy* and *self-esteem* (Linley & Joseph, 2004). Finally, a core characteristic associated with growth is *emotional disclosure.* Individuals who engage in this behavior not only report PTG, also manifest improved physical function, reduced distress, and better immune system functioning (Lepore, Fernandez-Berrocal, Ragan, & Ramos, 2004). Again however, the relationship of emotional disclosure to PTG may not be a direct one but rather conflated with a cluster of other pro-health behaviors, traits, or characteristics.

### *Rumination mode*

In traditional clinical parlance, a 'ruminative style' is associated with a wide range of poor mental health outcomes (e.g., increased risk for depression, anxiety) (Johnson et al., 2016). Rumination is typically defined as an automatic tendency to obsessively and automatically focus on negative content, repeatedly recalling and re-experiencing the material without processing in a healthy or helpful manner. However, researchers in the PTG field have made a distinction between that style of rumination which occurs shortly after a trauma or highly challenging event (i.e., automatic, obsessive, intrusive)—and another more deliberate and focused mode which generally manifests much later in the process (Lindstrom, Cann, Calhoun, & Tedeschi, 2013; Nolen-Hoeksema & Davis, 1999). They argue that both ruminative modes may be present in the PTG process, but that it is the latter one, a deliberate processing of the traumatic event with the purpose of understanding and accommodating new realities, which will produce the transformational changes in the trauma survivor (Calhoun & Tedeschi, 2004). In this context, the term *rumination* takes on an expanded meaning, to include a later-stage, more deliberate and productive cognitive processing of traumatic material.

### *Coping style*

Coping style refers to the manner in which individuals respond to stressors or traumatic events. There are generally considered to be three coping *techniques*: active, passive, or avoidance. At the individual level, scholars have proposed dozens of coping *strategies*. The more prominent approaches reference a handful of strategies such as *appraisal-focused* (adaptive cognitive), *problem-focused* (adaptive behavioral), *emotion-focused,* and *avoidance coping* (Billings & Moos, 1981). Some research suggests that problem- and emotion-focused coping are positively associated with PTG, while avoidance coping is negatively correlated (Linley & Joseph, 2004). It is unclear why a cognitive coping style would fail to be associated with PTG given the central importance of cognitive processing in the PTG model. However, it may be that the more important elements of PTG involve an emotional rather than intellectual understanding.

## Therapeutic approaches

In current practice, treatment of traumatic psychological injuries is considered to be successful if clinical symptoms are eliminated or ameliorated. As is common in Western medicine, this model assumes the healthy natural state to be the absence of pain or suffering. However, this conceptualization stands in contrast to the rich and complex process of meaning-making in PTG, which may occur alongside symptom reduction, but often coexists with symptoms of distress (see Figure 10.2). In fact, the PTG construct may be better aligned with Eastern philosophies that reject such dualistic thinking and instead embrace the necessity of complementary light and dark forces (Zheng & Gray, 2015). Nonetheless, it is clear that in the process of operationalizing the construct it has developed a distinct Western flavor (Kashyap & Hussain, 2018; Splevins et al., 2010). The Eastern model embraces the malleability of the self and acknowledges both suffering and growth can be present in the same moment. These core beliefs are reflected in modern schools of thought such as *contemplative psychotherapy* or *transpersonal psychology* more generally (Darnall, 2007; Friedman, 2018). Tedeschi and Calhoun (1995, 1996, 2004) have been refining a model of the PTG process over the past two decades (Figure 10.2). Over time, it has become apparent that there are many routes to achieving PTG, and much depends upon environmental and individual factors, which vary for each individual.

## Adaptive disclosure therapy

As both a constructivist process and a clinical outcome, PTG is argued to be distinct from, yet potentially overlapping, the typical recovery process from traumatic stress exposure as is commonly seen in traditional therapeutic interventions for PTSD (e.g., cognitive behavioral

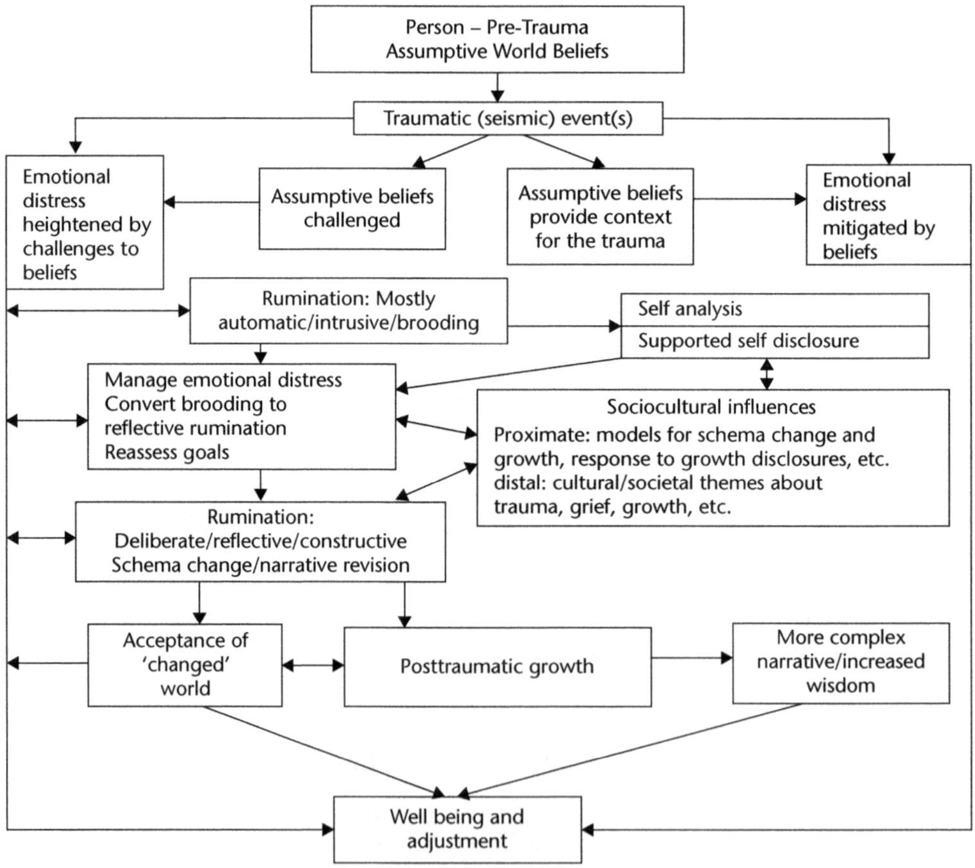


*Figure 10.2* Posttraumatic growth pathways.

*Source:* Tedeschi, 2012, used with permission.

therapy; prolonged exposure). However, a novel intervention specifically designed for active-duty military members was recently developed; preliminary findings suggest it may actually facilitate PTG (Gray et al., 2012). *Adaptive disclosure* is an exposure-based intervention that posits combat stress-related disorders are multifaceted injuries that may also include traumatic loss and *moral injury* (Litz et al., 2009). Moral injury is a construct that encompasses a complicated syndrome of demoralization, shame, anger, and grief that results from a violation of deeply held ethical and moral values or beliefs. A common experience underlying both moral injury and PTG may be the violent rupture of the survivor's assumptive worldview. Findings from a study of 44 U.S. Marines showed that adaptive disclosure therapy appeared to have fostered significant decreases in ratings of depression, PTSD symptoms, and negative posttraumatic appraisals. Interestingly, the intervention also appeared to promote PTG (Gray et al., 2012).

## *Integration into existing models*

Although adaptive disclosure therapy has shown great promise with military members, it may not be appropriate or available to many therapists working with trauma survivors. Therefore, clinicians are encouraged to work flexibly within the appropriate modality, matching client needs to therapist skills—while integrating the core features of PTG promotion or facilitation. Distilling best practices down to simple, pithy points, Tedeschi and McNally (2011) have offered a number of recommendations. Rather than assuming a traditionally paternalistic or expert role, they suggest adopting the persona of an 'expert companion,' someone who is willing to learn from the client and not feel compelled to foster, direct, or dictate the terms of growth. The therapist should nonetheless be aware of and actively listening for PTG themes emerging, without ever attempting to shape or modify the therapeutic process. The manifestation of distress or active symptoms is entirely consistent with the PTG process and therapists should neither expect nor deny distress when it arises. It is important to allow the client to assign their own interpretation and understanding of events; regardless of whether such interpretations are judged 'illusory' or not, the client's experience should be validated (assuming they are not delusional). Finally, not all trauma survivors will experience PTG; this should not be interpreted as treatment failure, as research has suggested that the most resilient may not manifest PTG, and that the nature of the traumatic event may play a role in the ability of the individual to experience PTG.

## Conclusion

Research into PTG as a clinical construct has steadily grown over the past two decades since its introduction and operationalization in the literature; progress has been made. Two basic tenets have been consistently supported in the research. First, research has revealed that PTG and PTSD/distress symptoms frequently coexist. This relationship is neither unexpected nor unexplainable—assuming familiarity with the PTG model. Adopting a non-binary and holistic viewpoint allows one to see that it is precisely the struggle that is essential in order to produce the growth. Second, the relationship between PTG and trauma/threat severity is likely curvilinear. Consistent with what we know of stress/arousal more generally, research supports the idea that low levels of stress/trauma are unlikely to be sufficiently motivating or to cause significant ruptures of existing schemas or worldviews necessary for PTG. At the same time, extremely high levels of stress/trauma may completely overwhelm coping mechanisms and cognitive processing capacity. In such a case, it would be difficult to construct a new and positive meaning for events. (Nonetheless, it's important to highlight the subjective nature of the lived experience of trauma, as research with POWs has revealed.) Finally, a number of factors that appear to be associated with or predictive of PTG have been identified, but efforts continue to clarify and replicate findings which are complex and often paradoxical.

Although many questions remain to be answered, much progress has been made since the PTG term was first introduced. The next decade will witness a further refining of terms, methodology, and a wider cultural adaptation. Our understanding of PTG, a seemingly ageless and universal construct, will continue to evolve. Perhaps in the final analysis, the best aphorism to describe the PTG phenomenon is not one from a 19th-century nihilist, but rather one offered by a 20th-century healer and Holocaust survivor:

> We must never forget that we may also find meaning in life even when confronted with a hopeless situation, when facing a fate that cannot be changed. For what then matters is to bear witness to the uniquely human potential at its best, which is to transform a personal tragedy into a triumph, to turn one's predicament into a human achievement....When we are no longer able to change a situation ...we are challenged to change ourselves. (Frankl, 1963, p. 112)

## References


Abraído-Lanza, A. F., Guier, C., & Colón, R. M. (1998). Psychological thriving among Latinas with chronic illness. *Journal of Social Issues*, *54*, 405–424.

Affleck, G., & Tennen, H. (1996). Construing benefits from adversity: Adaptational significance and dispositional underpinnings. *Journal of Personality*, *64*, 899–922.

American Psychiatric Association. (2013). *Diagnostic and Statistical Manual of Mental Disorders*. (5th Ed.). Washington, DC: Author.

Beder, J. (2005). Loss of the assumptive world—How do we deal with death and loss? *Omega*, *50*(4), 255–265.

Bensimon, M. (2012). Elaboration on the association between trauma, PTSD and posttraumatic growth: The role of trait resilience. *Personality and Individual Differences*, *52*(7), 7782–7787.

Billings, A. G., & Moos, R. H. (1981). The role of coping responses and social resources in attenuating the stress of life events. *Journal of Behavioral Medicine*, *4*(2), 139–157.

Blix, I., Skogbrott Birkeland, M., Bang Hansen, M., & Heir, T. (2016). Posttraumatic growth—An antecedent and outcome of posttraumatic stress: Cross-lagged associations among individuals exposed to terrorism. *Clinical Psychological Science*, *4*(4), 620–628.

Boermans, S., Kamphuis, W., Delahaij, R., van den Berg, C., & Euwema, M. (2014). Team spirit makes the difference: The interactive effects of team work engagement and organizational constraints during a military operation on psychological outcomes afterwards. *Stress and Health*, *30*, 386–396.

Bonanno, G.A. (2004). Loss, trauma, and human resilience: Have we underestimated the human capacity to thrive after extremely aversive events? *American Psychologist*, *59*, 20–28.

Bostock, L., Sheikh, A. I., & Barton, S. (2009). Posttraumatic growth and optimism in health-related trauma: A systematic review. *Journal of Clinical Psychology in Medical Settings*, *16*, 281–296.

Cadell, S., Regehr, C., & Hemsworth, D. (2003). Factors contributing to posttraumatic growth: A proposed structural equation model. *The American Journal of Orthopsychiatry*, *73*(3), 3279–3287.

Calhoun, L. G., & Tedeschi, R. G. (2004). The foundations of posttraumatic growth: New considerations. *Psychological Inquiry*, *15*, 93–102.

Calhoun, L. G., & Tedeschi, R. G. (1999). *Facilitating Posttraumatic Growth: A Clinician's Guide*. Mahwah, NJ: Lawrence Erlbaum Associates.

Calhoun, L. G., & Tedeschi, R. G. (2006). The foundations of posttraumatic growth: An expanded framework. In L. G. Calhoun, & R. G. Tedeschi (Eds.), *Handbook of Posttraumatic Growth: Research and Practice* (pp. 03–23). Mahwah, NJ: Lawrence Erlbaum Associates.

Calhoun, L. G., Tedeschi, R. G., Cann, A., & McMillan, J. (2000). A correlational test of the relationship between posttraumatic growth, religion, and cognitive processing. *Journal of Traumatic Stress*, *13*(3), 521–527.

Cann, A., Calhoun, L. G., Tedeschi, R. G., Triplett, K. N., Vishnevsky, T., Lindstrom, C. M. (2011). Assessing posttraumatic cognitive processes: The event related rumination inventory. *Anxiety, Stress, & Coping*, *24*, 137–156.

Carr, C. P., Martins, C. M. S., Stingel, A. M., Lemgruber, V. B., & Juruena, M. F. (2013). The role of early life stress in adult psychiatric disorders: A systematic review according to childhood trauma subtypes. *Journal of Nervous and Mental Disease, 201*(12), 1007–1020.
Coyne, J. C., & Tennen, H. (2010). Positive psychology in cancer care: Bad science, exaggerated claims, and unproven medicine. *Annals of Behavioral Medicine, 39*, 16–26.
Cryder, C. H., Kilmer, R. P., Tedeschi, R. G., & Calhoun, L. G. (2006). An exploratory study of posttraumatic growth in children following a natural disaster. *The American Journal of Orthopsychiatry, 76*, 65–69.
Darnall, K. T. (2007). Contemplative psychotherapy: Integrating Western psychology and Eastern philosophy. *The Behavior Therapist, 30*(7), 156–160.
Dekel, S., Mandl, C., & Solomon, Z. (2011). Shared and unique predictors of post-traumatic growth and distress. *Journal of Clinical Psychology, 67*(3), 241–252.
Dohrenwend, B. P., Neria, Y., Turner J. B., Turse, N., Marshall, R., Lewis-Fernandez, R., & Koenen, K. C. (2004). Positive tertiary appraisals and posttraumatic stress disorder in U.S. male veterans of the war in Vietnam: The roles of positive affirmation, positive reformulation, and defensive denial. *Journal of Consulting and Clinical Psychology, 72*, 417–433.
Engelhard, I. M., Lommen, M. J., & Sijbrandij, M. (2015). Changing for better or worse? Posttraumatic growth reported by soldiers deployed to Iraq. *Clinical Psychological Science, 3*(5), 789–796.
Feder, A., Southwick, S. M., Goetz, R. R., Wang, Y., Alonso, A., Smith, B. W., … Vythilingam, M. (2008). Posttraumatic growth in former Vietnam prisoners of war. *Psychiatry, 71*, 359–370.
First, J., First, N., Stevens, J., Mieseler, V., & Houston, J. B. (2018). Post-traumatic growth 2.5 years after the 2011 Joplin, Missouri tornado. *Journal of Family Social Work, 21*(1), 5–21.
Fontana, A., & Rosenheck, R. (1998). Psychological benefits and liabilities of traumatic exposure in the war zone. *Journal of Traumatic Stress, 11*(3), 485–503.
Frankl, V. (1959). *From Death-Camp to Existentialism: A Psychiatrist's Path to a New Therapy*. Boston, MA: Beacon Press.
Frankl, V. (1963). *Man's Search for Meaning: An Introduction to Logotherapy*. Oxford, England: Washington Square Press.
Frazier, P., Tashino, T., Berman, M., Steger, M., & Long, J. (2004). Correlates of levels and patterns of positive life changes following sexual assault. *Journal of Consulting and Clinical Psychology, 72*(1), 119–130.
Frazier, P., Tennen, H., Gavian, M., Park, C., Tomich, P., & Tashiro, T. (2009). Does self-reported posttraumatic growth reflect genuine positive change? *Journal of Psychological Science, 20*(7), 912–919.
Friedman, H. L. (2018). Transpersonal psychology as a heterodox approach to psychological science: Focus on the construct of self-expansiveness and its measure. *Archives of Scientific Psychology, 6*(1), 230–242.
Gallaway, M. S., Millikan, A. M., & Bell, M. R. (2011). The association between deployment-related posttraumatic growth among US Army soldiers and negative behavioral health conditions. *Journal of Clinical Psychology, 67*(12), 1151–1160.
Gray, M. J., Schorr, Y., Nash, W., Lebowitz, L., Amidon, A., Lansing, A., … Litz, B. T. (2012). Adaptive disclosure: An open trial of a novel exposure-based intervention for service members with combat-related psychological stress injuries. *Behavior Therapy, 43*(2), 407–415.
Hobfoll, S. E., Hall, B. J., Canetti-Nisim, D., Galea, S., Johnson, R. J., & Palmieri, P. A. (2007). Refining our understanding of traumatic growth in the face of terrorism: Moving from meaning cognitions to doing what is meaningful. *Applied Psychology: An International Review, 56*(3), 345–366.
Janoff-Bulman, R. (1989). Assumptive worlds and the stress of traumatic events: Applications of the schema construct. *Social Cognition, 7*(2), 113–136.
Janoff-Bulman, R. (1992). *Shattered Assumptions: Towards a new Psychology of Trauma*. New York, NY, US: Free Press.
Janoff-Bulman, R. (2004). Posttraumatic growth: Three explanatory models. *Psychological Inquiry, 15*, 30–34.
Johnson, D. P., Rhee, S. H., Friedman, N. P., Corley, R. P., Munn-Chernoff, M. A., Hewitt, J. K., & Whisman, M. A. (2016). A twin study examining rumination as a transdiagnostic correlate of psychopathology. *Clinical Psychological Science, 4*(6), 971–987.
Joseph, S., Linley, A., Andrews, L., Harris, G., Howle, B., Woodward, C., & Sheylin, M. (2005). Assessing positive and negative changes in the aftermath of adversity: Psychometric evaluation of the changes in outlook questionnaire. *Psychological Assessment, 17*, 70–80.
Joseph, S., Williams, R., & Yule, W. (1993). Changes in outlook following disaster: The preliminary development of a measure to assess positive and negative responses. *Journal of Traumatic Stress, 6*, 271–279.
Koenig, H. G., Pargament, K. I., & Nielsen, J. (1998). Religious coping and health status in medically ill hospitalized older adults. *Journal of Nervous and Mental Disease, 186*, 513–521.

Kashyap, S., & Hussain, D. (2018). Cross-cultural challenges to the construct 'posttraumatic growth.' *Journal of Loss and Trauma, 23*(1), 51–69.

Kulka, R., Schlenger, W., Fairbank, J., Hough, R., Jordan, B., Marmar, C., … Grady, D. (1990). *Trauma and the Vietnam War Generation: Report of Findings from the National Vietnam Veterans Readjustment Study.* New York: Brunner/Mazel.

Lepore, S. J., Fernandez-Berrocal, P., Ragan, J., & Ramos, N. (2004). It's not that bad: Social challenges to emotional disclosure enhance adjustment to stress. *Anxiety, Stress, and Coping, 17*, 341–361.

Lepore, S., & Revenson, T. (2006). Relationships between posttraumatic growth and resilience: Recovery, resistance and reconfiguration. In L. G. Calhoun & R. G. Tedeschi (Eds.), *The Handbook of Posttraumatic Growth: Research and Practice* (pp. 24–46). Mahwah, NJ: Lawrence Erlbaum.

Levine, S. Z., Laufer, A., Stein, E., Hamama-Raz, Y., & Solomon, Z. (2009). Examining the relationship between resilience and posttraumatic growth. *Journal of Traumatic Stress, 22*, 282–286.

Lichtenthal, W., Cruess, D., & Prigerson, H. (2004). A case for establishing complicated grief as a distinct mental disorder in DSM-IV. *Clinical Psychology Review, 24*, 637–662.

Lindstrom, C. M., Cann, A., Calhoun, L. G., & Tedeschi, R. G. (2013). The relationship of core belief challenge, rumination, disclosure, and sociocultural elements to posttraumatic growth. *Psychological Trauma, 5*(1), 150–155.

Linley, P. A., & Joseph, S. (2004). Positive change following trauma and adversity: A review. *Journal of Traumatic Stress, 17*, 11–21.

Linley, P. A., Joseph, S., & Goodfellow, B. (2008). Positive changes in outlook following trauma and their relationship to subsequent posttraumatic stress, depression, and anxiety. *Journal of Social and Clinical Psychology, 27*(8), 877–891.

Litz, B. T., Stein, N., Delaney, E., Lebowitz, L., Nash, W. P., Silva, C., & Maguen, S. (2009). Moral injury and moral repair in war veterans: A preliminary model and intervention strategy. *Clinical Psychology Review, 29*(8), 695–706.

Maercker, A., & Zoellner, T. (2004). The Janus face of self-perceived growth: Toward a two-component model of posttraumatic growth. *Psychological Inquiry, 15*, 41–48.

Maguen, S., Vogt, D. S., King, L. A., King, D. W., & Litz, B. T. (2006). Posttraumatic growth among Gulf War I veterans: The predictive role of deployment-related experiences and background characteristics. *Journal of Loss and Trauma, 11*, 373–388.

Maslow, A. (1954). *Motivation and Personality*. Oxford, England: Harpers.

May, C. L., & Wisco, B. E. (2016). Defining trauma: How level of exposure and proximity affect risk for posttraumatic stress disorder. *Psychological Trauma: Theory, Research, Practice, and Policy, 8*(2), 233–240.

McMillen, J. C., & Fisher, R. H. (1998). The Perceived Benefit Scales: Measuring perceived positive life changes after negative events. *Social Work Research, 22*, 173–187.

Morris, B. A., & Shakespeare-Finch, J. (2011). Rumination, post-traumatic growth, and distress: Structural equation modelling with cancer survivors. *Psycho-Oncology, 20*, 1176–1183.

Nash, W. P. (2007). Combat/operational stress adaptations and injuries. In C. R. Figley & W. P. Nash (Eds.), *Combat Stress Injury: Theory, Research, and Management* (pp. 70–102). New York: Routledge.

Nietzsche, F. W., & Large, D. (1998). *Twilight of the Idols, or, How to Philosophize with a Hammer*. Oxford: Oxford University Press.

Nolen-Hoeksema, S., & Davis, C. G. (1999). 'Thanks for sharing that': Ruminators and their social support networks. *Journal of Personality and Social Psychology, 77*(4), 801–814.

O'Leary, V. E., Alday, C. S., & Ickovics, J. R. (1998). Models of life change and posttraumatic growth. In R. G. Tedeschi, C. L. Park, & L. G. Calhoun (Eds.), *The LEA Series in Personality and Clinical Psychology. Posttraumatic Growth: Positive Changes in the Aftermath of Crisis* (pp. 127–151). Mahwah, NJ, US: Lawrence Erlbaum Associates Publishers.

O'Leary, V. E., & Ickovics, J. R. (1994). *Women's resilience: A heuristic model.* Paper presented at the American Psychological Association Conference on Psychosocial and Behavioral Factors in Women's Health. Washington, DC.

O'Leary, V. E., & Ickovics, J. R. (1995). Resilience and thriving in response to challenge: An opportunity for a paradigm shift in women's health. *Women's Health, 1*(2), 121–142.

Ozer, E. J., Best, S. R., Lipsey, T. L., & Weiss, D. S. (2003). Predictors of posttraumatic stress disorder and symptoms in adults: A meta-analysis. *Psychological Bulletin, 129*, 52–71.

Pargament, K. I., Desai, K. M., & McConnell, K. M. (2006). Spirituality: A pathway to posttraumatic growth or decline? In L. G. Calhoun & R. G. Tedeschi (Eds.), *The Handbook of Posttraumatic Growth: Research and Practice* (pp. 121–137). Mahwah, NJ: Lawrence Erlbaum.

Park, C. L., Cohen, L. H., & Murch, R. L. (1996). Assessment and prediction of stress-related growth. *Journal of Personality, 64*, 71–105.

Price, M., Pallito, S., & Legrand, A. C. (2018). Heterogeneity in the strength of the relation between social support and post-trauma psychopathology. *Journal of Psychopathology and Behavioral Assessment, 40*(2), 334–343.

Ramos, C., & Leal, I. (2013). Posttraumatic growth in the aftermath of trauma: A literature review about related factors and application contexts. *Psychology, Community & Health, 2*(1).

Schaefer, J. A., & Moos, R. H. (1998). The context for posttraumatic growth: Life crises, individual and social resources, and coping. In R. G. Tedeschi, C. L. Park, & L. G. Calhoun (Eds.), *Posttraumatic Growth: Positive Changes in the Aftermath of Crisis* (pp. 99–125). Mahwah, NJ: Lawrence Erlbaum.

Schnurr, P. P., Rosenberg, S. D., & Friedman, M. J. (1993). Change in MMPI scores from college to adulthood as a function of military service. *Journal of Abnormal Psychology, 102*, 288–296.

Shakespeare-Finch, J., & Armstrong, D. (2010). Trauma type and post-trauma outcomes: Differences between survivors of motor vehicle accidents, sexual assault, and bereavement. *Journal of Loss and Trauma, 15*, 69–82.

Shakespeare-Finch, J., & de Dassel, T. (2009). The impact of child sexual abuse on victims/survivors. *Journal of Child Sexual Abuse, 18*, 623–640.

Sledge, W. H., Boydstun, J. A., & Rabe, A. J. (1980). Self-concept changes related to war captivity. *Archives of General Psychiatry, 37*, 430–443.

Snape, M. C. (1997). Reactions to a traumatic event: The good, the bad and the ugly? *Psychology Health and Medicine, 2*, 237–242.

Solomon, Z., & Dekel, R. (2007). Posttraumatic stress disorder and posttraumatic growth among Israeli ex-POWs. *Journal of Traumatic Stress, 20*, 303–312.

Sperry, L. (2016). Trauma, neurobiology, and personality dynamics: A primer. *The Journal of Individual Psychology, 72*(3), 161–167.

Splevins, K., Cohen, K., Bowley, J., & Joseph, S. (2010). Theories of posttraumatic growth: Cross-cultural perspectives. *Journal of Loss and Trauma, 15*, 259–277.

Stanton, A. L., Bower, J. E., & Low, C. A. (2006). Posttraumatic growth after cancer. In L. G. Calhoun & R. G. Tedeschi (Eds.), *The Handbook of Posttraumatic Growth: Research and Practice* (pp. 138–175). Mahwah, NJ: Lawrence Erlbaum.

Steenkamp, M. M., Litz, B. T., Hoge, C. W., & Marmar, C. R. (2015). Psychotherapy for military-related PTSD: A review of randomized clinical trials. *Journal of the American Medical Association, 31*(4), 489–500.

Taku, K., Cann, A., Tedeschi, R. G., & Calhoun, L. G. (2015). Core beliefs shaken by an earthquake correlate with posttraumatic growth. *Psychological Trauma: Theory, Research, Practice, and Policy*, 7, 563–569.

Taylor, S. E. (1983). Adjustment to threatening events: A theory of cognitive adaptation. *American Psychologist, 38*, 1161–1173.

Tedeschi, R. G. (2012). *Posttraumatic growth: Psychological reconstruction in the aftermath of disaster.* Keynote delivered to 9th Annual Conference of the Institute for Disaster Mental Health (IDMH), New Paltz, New York.

Tedeschi, R. G., & Calhoun, L. G. (1995). *Trauma and Transformation: Growing in the Aftermath of Suffering.* Thousand Oaks, CA: Sage.

Tedeschi, R. G., & Calhoun, L. G. (1996). The Posttraumatic Growth Inventory: Measuring the positive legacy of trauma. *Journal of Traumatic Stress, 9*, 455–471.

Tedeschi, R. G., & Calhoun, L. G. (2004). Posttraumatic growth: Conceptual foundations and empirical evidence. *Psychological Inquiry, 15*, 1–18.

Tedeschi, R.G., Cann, A., Taku, K., Senol-Durak, E., & Calhoun, L.G. (2017). The Posttraumatic Growth Inventory: A revision integrating existential and spiritual change. *Journal of Traumatic Stress, 30*(1), 11–18.

Tedeschi, R. G., & McNally, R. J. (2011). Can we facilitate posttraumatic growth in combat veterans? *American Psychologist, 66*, 19–24.

Tedeschi, R. G., Park, C. L., & Calhoun, L. G. (1998). Posttraumatic growth: Conceptual issues. In R. G. Tedeschi, C. L. Park, & L. G. Calhoun, (Eds.), *Posttraumatic Growth: Positive Changes in the Aftermath of Crisis* (pp. 1–22). Lawrence Erlbaum Associates.

Tedeschi, R. G., Shakespeare-Finch, J., Taku, K., & Calhoun, L.G. (2018). *Posttraumatic Growth: Theory, Research and Applications.* New York: Routledge.

Tomich, P. L., & Helgeson, V. S. (2004). Is finding something good in the bad always good? Benefit finding among women with breast cancer. *Health Psychology, 23*(1), 116–123.

Triplett, K. N., Tedeschi, R. G., Cann, A., Calhoun, L. G., & Reeve, C. L. (2012). Posttraumatic growth, meaning in life, and life satisfaction in response to trauma. *Psychological Trauma: Theory, Research, Practice, and Policy*, *4*, 400–410.

Tsai, J., El-Gabalawy, R., Sledge, W., Southwick, S., & Pietrzak, R. (2015). Post-traumatic growth among veterans in the USA: Results from the National Health and Resilience in Veterans Study. *Psychological Medicine*, *45*(1), 165–179.

Yalom, I. (1980). *Existential Therapy*. New York, NY: Basic Books.

Zheng, P., & Gray, M. J. (2015). Posttraumatic coping and distress: An evaluation of Western conceptualization of trauma and its applicability to Chinese culture. *Journal of Cross-Cultural Psychology*, *46*, 723–736.

Zoellner, T., & Maercker, A. (2006). Posttraumatic growth in clinical psychology: A critical review and introduction of a two component model. *Clinical Psychology Review*, *26*, 626–653.

# 11

# BUILDING RESILIENCE AND HARDINESS IN MILITARY LEADERS – ROBUSTNESS TRAINING PROGRAMS OF THE GERMAN ARMY

*Oliver Krueckel, Annett Heidler, Nicola von Luedinghausen, Markus Auschek, and Matthias Soest*

## Hardiness, resilience and military leadership training

When soldiers and military leaders speak of "toughness" and "robustness" of their soldiers, they often envision them as mentally and physically invincible: men of steel that are able to cope with almost any dangerous situation, overcome any threat, or defeat any enemy. The most common questions the military has are how to select and most importantly how to train soldiers to become that tough super-soldier they imagine.

The psychological concepts that come closest to what the military envisions are those of hardiness and resilience, the latter having become a buzzword in the academic and civilian world.

First described by Werner in 1971 in her famous longitudinal study on the resilience of Hawaiian children, resilience is probably best described as the capability of an individual – or a society – to adapt to stressful and adverse life events and recover from them.

In contrast to the common metaphor of being "as tough as nails," so that nothing is able to harm the outer and inner structure of the individual, resilience refers more to the ability to deform and bounce back when stress occurs. Hence, resilience is considered a process of the interaction of an individual with the environment; therefore from a psychological point of view, resilience is considered a "state" rather than a "trait" (Bartone & Hystad, 2010). Coping, stress management, "salutogenesis" and post-traumatic growth are psychological theories related to this state of resilience.

Hardiness (or mental toughness) on the other hand, as first described by Maddi (1967) and later by Kobasa (1979), is considered the key element within an individual for understanding resilience. Key traits of hardiness include the inter-related tendencies of commitment (versus alienation), control (versus powerlessness) and challenge (versus need for security) (Bartone, 2008).

Although there is still an ongoing academic discussion as to what extent resilience and hardiness can be taught, the fact that both are trainable seems undisputed. Many programs in the military, such as the U.S. Comprehensive Soldier Fitness Program, focus on resiliency, mitigating the effects of stressful and traumatic events (Castro & Adler, 2011, Escolas, Pitts, Safer, & Bartone,

2013) and being seen as the "future of psychology" and as part of a positive psychology (Seligman & Fowler, 2011).

The German Army has come up with a comprehensive integrated program which faces today's challenge of recruiting the right people for the right job in the military. It seeks to develop physically and psychologically fit young soldiers who are able to keep up with the challenges of the battlefield.

The program focuses on the fitness of soldiers throughout their military lifecycle, relying on courses and training already in existence in the Army for years or even decades, such as the ranger training course. This is in line with quite a few other armies in the world who are trying to adapt their training to a new generation of recruits, putting the focus on motivation and physical training. The Swiss Army, for example, has set up a program for the development and strengthening of the leadership personality and has integrated selected elements of resilience training into leadership training. Positive effects manifest in stress perception, stress management and emotion regulation, thereby reducing drop-out rates and attrition in army basic training (Annen, Niederhauser, & Zueger, 2018; Wyss & Annen, 2012).

Interestingly, some studies show that it is not overall physical fitness in soldiers that has been declining over time but specific factors such as running performance. This is possibly due to increased body weight, reduced experience with running or lower motivation (Knapik, Sharp, Darakjy, Jones, Hauret, & Jones, 2006; Knapik, Sharp, & Steelman, 2017). At the same time, operational physical challenges like carrying heavy gear have increased over time (Knapik, Reynolds, & Harman, 2004), as well as the need to operate high-tech equipment.

The basic principle in German Army training is the same for officers and non-commissioned officers (NCOs) alike. Classroom orientation to teach basic regulations and principles is followed by rock-drill training combined with simulation-based training to improve leadership skills. The most indispensable part and highlight of education is the real combat training and live firing exercise in one of the training centers of the Army. Subsequently, team training provides the units from platoon to company and even brigade/division level with the required training level.

"Robustness training," as the German Army calls it, is included in all levels of military leadership training. Some training like the ranger training course has been in existence since the founding of the Bundeswehr. Structured robustness training sets the framework for physical and psychological training over a soldier's lifecycle from basic training to retirement, meeting state-of-the-science standards in sports, nutrition and mental training. The program focuses on persevering in combat, since this is a key attribute of the individual soldier and the basis for mission success of the Army.

## German army training structure

When a young NCO or officer candidate enters the Army, he or she will start with basic training and will slowly get accustomed to the challenges of military life. A special program has been designed to bring the young recruits up to speed, both mentally and physically, for meeting the standards of the German Army. From a psychological point of view, it is important that the first lessons include stress identification, stress management and psychological first aid as mandatory topics.

Later in their leadership training, officers and NCOs alike will have to pass their leadership training courses (officer and platoon leader course or senior NCO and section commander courses, respectively), as well as their military occupational specialty training (infantry, armor, logistics, etc.). Psychological training is included in some of the military occupational specialty training, for example, communication and negotiation for Human Intelligence troops (HUMINT), special training for Explosive Ordnance Disposal personnel (EOD) and Special

Forces training. Robustness training is a highlight of the ranger training course, which is mandatory for leaders in the infantry, as well as the increased infantry competencies course for non-infantry officers and NCOs, which is currently being reexamined and revised.

Mental training, which is common in sports psychology, can also be used in marksmanship training and during sniper courses, completed by lessons and discussions about the psychological effects of killing, getting killed or wounded and losing comrades in combat. For military leaders, theoretical lessons on how to convey a message of death are mandatory and supported by practical training exercises and role play. The military psychologists work hand in hand with social workers and military chaplains during those training courses, working closely together in a so called "psychosocial network of support." Ethical issues are also consequently discussed, as mandatory in the German Army concept of mission type tactics ("Auftragstaktik") and the German Army leadership development and civic education concept ("Innere Fuehrung").

The highest level of leadership training regarding resiliency training is the combined arms operations training on the Task Force (battalion) level which occurs at the German Army Combat Training Center. Military leaders on all levels, but especially the officers on platoon, company and battalion levels, are trained and given feedback on their performance, supported by technical engagement simulators and video and voice-recorded assessment and evaluation. The focus of the feedback from the psychologist is the practical training that is mentally challenging such as leading in combat under stress, including mass casualty situations and direct engagement with civilian population.

## Academic leadership training

Future trends and developments will look deeper into leadership development training as a continuous process in a holistic approach, integrating theoretical knowledge about leadership theory taught at the military academies into practical exercises and training. The focus will be on the ability to self-reflect on one's own leadership skills and behavior in peacetime and under stress. Coaching and mentoring can be some of the methods of choice in a future continuous leadership development training program and will fall in line with the efforts of life-long learning programs in the German Army.

## Psychological and physical fitness in basic training

German Army personnel must be physically and psychologically hardy and resilient to fulfill their mission. This is why, in future, the fitness of recruits is to be monitored and improved through a holistic program from day one of their service. To this end, basic training has been redesigned as part of a pilot project whose key element is the physical and psychological fitness of recruits. The aim is to improve the initial fitness of recruits through tailored training methods in separate performance groups, so that, after the first six weeks of training, all recruits will be able to successfully pass the German Army's basic fitness test in order to fulfill the requirements for participating in combat training.

The pilot training course was conducted in an armored infantry battalion. In order to implement professional physical training that meets modern standards, the trainers were supported by additional sports instructors as well as trainers for military and physical fitness. Apart from the sports facilities available at the barracks (fully equipped gym including a workout room, a sports field with tartan track, several smaller sports fields and training grounds), a TacWork container and a variety of additional sports equipment (such as TRX sling trainers, Flexi Bars, medicine balls, gym mats, marking cones, kettlebells) were procured to ensure a wide variety of training methods.

Before the pilot course started, the instructors attended a "train-the-trainer week," where they were instructed on the new objectives of basic training and received information on issues like heat prevention, healthy diet, sports and motivation by different Bundeswehr agencies.

In their very first week of training, the new recruits completed the German Army's basic fitness test for the first time to assess their initial fitness level. Based on the results, the recruits were sorted into three performance groups.

The focus during the first six weeks was on physical training. In the morning, a platoon-level sports session was integrated into the duty roster, and in the afternoon a longer sports session was conducted in the performance groups. The training comprised elements of activation, mobilization, strength and endurance enhancement as well as active regeneration.

In the sixth week of training, the recruits repeated the basic fitness test. At the same time, they completed the soldier's basic fitness tool for the first time. The results showed a significant improvement in performance for all recruits across all performance groups. Furthermore, all participants now met the standards of the basic fitness test and, by implication, fulfilled the requirements for participating in combat training. As the recruits entered the physically demanding combat training, the afternoon training sessions in the performance groups were reduced. The morning sports routine remained unchanged during the whole time of basic training. In total, the pilot course comprised 70 net hours of physical training. In the 12th week of training, the recruits again repeated the basic fitness test and the soldier's basic fitness tool. The results once again showed an increase in performance for all recruits, although the improvement was not as pronounced as the one achieved after the first six weeks.

During the pilot run, the military psychology focus for the improvement of psychological fitness was on the issue of motivation. The aim was to raise awareness of this topic among recruits and instructors, to initiate a process of reflection and to build competence in selected motivation methods. During classes, the influence of motivation on human performance was explained to the recruits, and the basic motivational processes were described. Furthermore, reflection was encouraged among the recruits, for example, on their personal motives with regard to basic training. Scientific evidence has shown that setting specific goals has a positive influence on the motivation and performance of human beings (Locke & Latham, 1990). The recruits were therefore encouraged to establish individual goals for basic training.

For the recruits, the instructor is a key figure in their training. For this reason, the focus of military psychology during the training was on "training the trainers." In the instructors' seminar, the motivational process was explained in more detail, compiling internal and external factors that affect motivation and reflecting on the possibilities and limitations of influencing the motivation of recruits. The following aspects of motivation were specifically dealt with: incentives, objectives and positive feedback. Methodologically, both classes used a combination of theoretical instruction, practical elements and reflection.

At the end of basic training, a final workshop was held with the instructors, where they reflected retrospectively on the motivational factors that had influenced recruits and instructors during the pilot run. It was found that while the physical strain was relatively high, the motivation of the recruits remained at a continuously high level. The low dropout rates and the comparably low sickness rates were cited as indicators for a high level of motivation. Nevertheless, it is difficult to make a definitive statement on the cause of this positive effect since various interacting factors come into play. The intense and professional sport sessions in the performance groups and the resulting improvement of individual fitness is certainly one key influencing factor. The excellent instructor-recruit ratio was identified as another factor. The small size of the learning groups made it possible to convey the learning content and the physical training elements in an individual and goal-oriented manner. From the instructors' point of view, the daily sports sessions in the morning led to improved activation and receptiveness of the recruits. This was especially demonstrated by the fact that training levels were reached more quickly than in the traditional training runs. Less repetition was required. Another factor was the strong group cohesion among the recruits, which was noticeable to the instructors from the second week of training. With the measures taken, it was thus possible to teach and exemplify comradeship as an important element for capable armed forces right from the start.

To sum up, the newly structured basic training with its intensive sports sessions had a positive influence on physical and psychological fitness (motivation, receptiveness and group cohesion). For future projects on the issue of enhancing performance capability, however, our recommendation and objective is to conduct a scientific evaluation of the motivation factor in order to come to reliable and measurable conclusions.

When it comes to increasing psychological fitness during basic training, we see the future role of military psychology in the teaching of recruits and, most importantly, in the training of instructors. In their key function, the instructors play a central role in influencing the individual and group dynamic psychological processes among recruits. Swiss colleagues (Wyss & Annen, 2012) have shown that a transformational style of leadership (characterized by goodwill, convincing communication and setting a good example) has a positive influence on the recruits' performance motivation. The aim is therefore to further increase awareness of psychological topics among instructors, to strengthen their capability for targeted observation and critical reflection and to improve their competence in applying psychological methods. During the pilot course the following topics were identified: motivation, communication, group dynamic processes, stress resilience and mental techniques. Another effect of such regular training courses is that they offer instructors an opportunity to practice mental hygiene, which in turn plays an important role in maintaining their motivation and psychological resilience in the long run. In addition, selective psychological supervision during the practical training is considered useful in order to assess and support the implementation of the psychological elements of training.

In conclusion, the new concept for basic training contributed to the improvement of physical and psychological performance capability. We therefore recommend continuing with the new concept – despite the fact that its implementation throughout the Bundeswehr will certainly give rise to challenges in the areas of personnel, equipment and organizational structures.

## Psychological training at the German Explosive Ordnance Disposal training unit

Kabul, Afghanistan/ISAF, 6 March 2002: Five NATO soldiers, including two German Explosive Ordnance Disposal technicians, died in an accident while disarming a surface-to-air missile. Additionally, German and Danish soldiers were wounded severely.

After an investigation and appraisal of the circumstances, significant deficiencies in leadership, procedures, education and training of the EOD specialists were identified and discussed (EOD Training Unit, 2018a,b). Consequently, the German Explosive Ordnance Disposal Training Center was established in 2005.

The goal of the newly established unit was to create a central training facility. Previously, EOD technicians had been employed piecemeal throughout the German Armed Forces. Bringing all the personnel together meant collecting and concentrating the specific knowledge and experience of both young and experienced specialists to the benefit of all. In this unit, noncommissioned officers and officers were to be taught and trained to equal standards at a high level before they returned to their various units. Thereby, further accidents and injuries were to be prevented.

Today the EOD Training Unit conducts the training for all basic, advanced and qualified explosive ordnance defense capabilities as well as the unique asset of K9 teams.

## Explosive Ordnance Defense

Explosive Ordnance Defense is a function of all armed forces: it covers all measures to counter any threat posed by explosive ordnance in order to protect own units and installations and promote mobility and freedom of action operations.

We can define the knowledge and skills of Explosive Ordnance Defense at different levels. In basic training, every soldier in the German army is introduced to the threat posed by unexploded ordnance and mines (UXO/mine awareness), as well as improvised explosives devices (IEDs). In subsequent advanced training, every soldier is taught to conduct subtasks of Explosive Ordnance Reconnaissance (EOR), including searching, localizing, describing, marking the site and reporting the find.

There is also a requirement for more qualified personnel in every unit who not only can conduct EOR, but who can also advise the commanding officer and recommend and execute appropriate safety measures. If necessary, EOD assistance will be requested by them. They also train the soldiers of their unit in basic Explosive Ordnance Defense.

The highest level of training is Explosive Ordnance Disposal, which covers EOR, the identification of explosive ordnance, tactical site exploitation and the disposal of all sorts of explosive ordnance – including IED (EOD information briefing, 2018a,b).

Only a few specialists are able to delicately defuse manually without any detonation. This is especially important if area evacuation is impossible or (collateral) damage due to an explosion is unacceptable.

Nonetheless robots are used wherever possible and at a safe distance. But the task remains high risk and dangerous.

## Psychological training at the Explosive Ordnance Disposal training unit

It is hardly surprising that there are not many such trained personnel.

Firstly, it is not everybody's dream job to disarm explosives. Secondly, the courses and the final exam are extremely difficult as well as physically and mentally demanding.

Candidates must not only be good team players and flexible, but also be persistent and accept responsibility not only for their own lives but also for those of other people. While grit, nerves of steel and a need for adrenaline are vital, "the right mindset" is the most crucial quality to perform these kinds of tasks successfully.

From the beginning, it was clear that the physical qualities alone are not enough. Like for winning a gold medal in a race, there is also a strong mental component in explosive ordnance disposal. Mind and body complement each other, so one has to train both for a best outcome.

That is why the psychologist is involved in the training of the specialists: he/she trains them from the beginning in foundation classes up to the elite training in manual neutralization. Only the best candidates are selected for their ability to perform with willpower, awareness and intelligence.

The psychologist trains his or her students not only on the basis of well-established psychological research and theory but also knows recent literature. He/she is always up to date with the newest trends in the field, psychological methods and publications. He/she constantly upgrades his/her interdisciplinary knowledge to provide the best, most suitable and most effective education and training. In that way, he/she reaches the highest professional standards and supports the students as much as possible in developing their full mental potential.

In the foundation class, the psychologist educates the future EOD officers and EOD NCOs in what stress is, what the various forms and indications of stress are and how to minimize and reduce the negative effects of stress. He/she also educates them about the consequences of long-term stress (for stress management techniques, see Kaluza, 2011).

He/she advises them about different coping techniques and teaches them stress-management methods like special breathing techniques, mini-meditations, imagination, power poses or progressive muscle relaxation according to the Jacobson technique. These techniques can be practiced regularly or used spontaneously in situations of extreme pressure. Using them daily could even shift one's mindset completely to a more relaxed state.

Mindfulness-based techniques are taught as well. This includes noticing and observing the environment and surroundings (situational awareness), the body, the feelings and the mind in an accepting and enquiring manner without judging (for mindfulness techniques, see Kabatt-Zinn, 2011).

The student also learns ways to stop brooding and how to rephrase negative thinking into positive and motivating soliloquies and slogans. As a saying attributed to Henry Ford goes, "Whether you think you can do a thing or you think you can't do a thing – you're right." (University of Michigan, 1962). So what one says to oneself will influence one's attitude, one's view of the world and one's self-confidence and self-efficacy.

The goal of the stress-management training is that the soldiers learn to understand and identify their triggers and acquire a deeper understanding of how to control and better handle their physical and emotional reactions during and after stressful situations. In that way, they learn to improve their physical and emotional well-being and are better prepared for the demanding situations they will get into as soldiers.

The psychologist also introduces the students to mental training methods and performance enhancement strategies that are already used successfully in sports (see Eberspaecher, 2011).

For EOD personnel, visualization training combined with real exercise is very helpful. Only dummy explosives are used during training so the only pressure for the students is failing the exam. Using visualization techniques, they can play out different kinds of tasks and try out different solutions. Thereby they experience different outcomes and familiarize themselves with pressure and learn to control emotions and excitement. For these specialists it is very important to find their own optimal level of heightened mental state and physical readiness in order to perform optimally. Experienced technicians often reach a flow-like state where they know exactly what to do with very clear, highly concentrated, calm thinking and being perfectly in the moment. Too little as well as too much heightening of the mental state will lead to suboptimal manual coordination, trigger nervousness and distress; cause a lack of concentration and finally lead to poor performance (Yerkes-Dodson-Law, 1908). To get in the right state of mind, building a routine before starting a task seems beneficial for many of them.

Also related to training are occasional visits to the department of forensic medicine. There the students are directly confronted with dead bodies and the phenomenon of death. They can reflect on the transitory nature of life and reflect on their own death as well as that of a possible squad partner. This kind of information and preparation is crucial to performing their tasks and dealing with life-and-death situations.

The psychologist also supports the specialist staff of the EOD Operations Company who are trained in manual neutralization with the aim of being able to best deploy their knowledge in mentally critical tasks such as dealing with suicide bombers or talking to hostages to calm them down and stabilize them.

Finally, he/she oversees all business at the Training Unit as the go-to person for psychological matters including counseling on demand.

### *Chaos-Driven Situations Management Retrieval System*

Also used in the psychological training and mental preparation of the EOD technicians is the so called CHARLY (Chaos-Driven Situations Management Retrieval System) stress-management computer program.

CHARLY combines multimedia simulations and videos, biofeedback via skin response and heart rate variability measurement, computer gaming and a virtual coach who guides the student though 12 modules. The latest version also includes a smartphone application.

The soldiers are first educated about stress and the fight-or-flight response and learn how to calm themselves using stress-management techniques. They have to play different stressful computer games and then apply these techniques. Their heart rate is measured before and during the games as well as after using the relaxation techniques. CHARLY provides a clear visualization of the results and shows that stress-management techniques really work by changing attitude towards and knowledge about stress (Krueckel, 2018; Wesemann et al., 2016).

In 2002, a disastrous accident occurred as EOD specialists were trying to dispose of an air-to-ground missile. There were deaths and injuries. Learning from that, the German EOD Training Center was established in 2005. That unit consists of selected robust specialists and experts who are educated, taught and trained in one central training facility. Because it is perfectly obvious that explosive ordnance disposal is a mental activity, these experts are also intensively trained by a psychologist.

Stress-management training helps the soldiers to be better prepared when they are suddenly confronted with incidents. Thus, they are well prepared to work in all kinds of operational theatres. Mental training, which is also used in sport and exercise psychology, is of particular benefit to these elite soldiers in realizing their full potential and better preparing them to successfully perform their critically important tasks.

## Survival, Evasion, Resistance and Extraction (SERE) training as a tool to develop resilience

### *SERE as a tool to develop resilience*

"We don't know what we are capable of until we have to survive. Each of us is stronger than we know – the key is to tap the power at will" (O'Grady, n.d., Motivational Speeches).

The power O'Grady refers to is resilience. He was an F16 pilot and was downed over Bosnia. Owing to his faith, his positive attitude and his training he was able to survive behind enemy lines for six days. A Survival, Evasion, Resistance and Extraction (SERE) course had been part of his training. In SERE courses, soldiers learn how to survive until they are rescued after getting isolated behind enemy lines and being left to their own devices, how to evade enemy forces and how to resist exploitation by interrogation in the event they get captured. In SERE training, soldiers are confronted and must cope with the extreme physical and mental challenges associated with such situations in order to train and enhance their resilience (Southwick & Charney, 2012). Starting from the quote from Scott O'Grady, this part intends to describe the psychological components in SERE training in the context of resilience. For a better understanding, a brief summary of the contents of SERE training will be provided in the first part of this section. This is followed by a description of the development of the psychological components of SERE training and of the methods used for conveying knowledge on the latter.

### *Survival, Evasion, Resistance and Extraction training in the German armed forces*

In the German Armed Forces, soldiers who are at a high risk of getting isolated and/or being subject to enemy exploitation – namely Special Forces (SF), Special Operations Forces (SOF) and Special-Purpose Forces personnel in particular – undergo Level C SERE training. In these courses, knowledge is taught in a classroom environment and applied in realistic exercise scenarios (Salas, Priest, Wilson, & Burke, 2006). Over a period of 4–5 weeks, students acquire knowledge on survival, namely immediate measures in situations of emergency, priority survival principles that can be broken down into Protection, Location, Water and Food (PLWF). Furthermore, they learn why it is necessary to take care of these priorities both for physiological and for psychological reasons, and they are taught how to best use their individual equipment. Dealing with the component of evasion, students are taught how to prepare and apply evasion plans, avoid detection by enemy forces and navigate safely and accurately. As to the component of resistance, one aspect covered is the legal framework of captivity as a prisoner of war and hostage. On the other hand, students gain knowledge on the physical and mental challenges posed by captivity, isolation and interrogation and acquire a set of coping techniques. In the last field covered, namely extraction, students learn all procedures necessary for supporting a safe extraction and recovery by own forces. Owing to the increase in practical training elements, intensity gradually increases over time and culminates in a final exercise going on for several days. During this final exercise, students are required to retrieve and apply the skills they have learned in a real operational environment. The exercise comprises all phases relevant to survival in a military environment, namely survival, evasion, resistance and extraction. The aim is to impart knowledge to students and to have them successfully apply it under conditions simulating an operational environment in order to enhance the students' competence and resilience in such a way that they can return from a real SERE event without having lost their dignity.

### *Psychological components of Survival, Evasion, Resistance and Extraction training*

Psychological aspects are incorporated in SERE training at two different levels. As to the first level, psychological knowledge and techniques are taught in order to enable the students to cope with the above general challenges associated with each of the SERE phases. The students are expected to draw upon their general abilities and military skills under most adverse conditions. As to the second level, the students acquire specific knowledge and techniques helpful with regard to Resistance to Interrogation (RTI). The knowledge is about the challenges associated with isolation, captivity as a hostage or as a POW, interrogation and torture and about the techniques which can be applied for coping with these challenges. This section describes the techniques that are used and also explains why they are used.

If you look for suitable techniques with regard to providing the best possible training to soldiers for future SERE events, you have to look for techniques enabling individuals to consistently retrieve abilities and skills in situations requiring maximum performance. "Be good when it matters" (Eberspaecher, 2011) is a credo of sports psychology. Top athletes are expected to deliver a specific performance in a competition, that is, at a specific time and under specific conditions. Conditions prior to, during and after a competition, such as safe and comfortable accommodation, regular meals, a safe venue for the competition, the availability of material and social resources meet a high standard.

During the past decades, top athletes have increasingly used sports psychology techniques as a response to expectations requiring them to repeatedly deliver their top performance and to develop maximum performance capabilities and expertise. These techniques have an influence on mental factors such as emotions (Tenenbaum & Sacks, 2007), perception (Farrow & Abernethy, 2007), concentration (Moran, Toner, & Campbell, 2018), decision-making (Tenenbaum 2003; Yates & Tschirhart, 2006) and motivation (Beckmann & Elbe, 2007).

Considering the operational reality for Special Forces, Special Operations Forces/Special-Purpose Forces and the challenges arising from this operational reality, the wording of Eberspaecher's aforementioned credo must be expanded as follows: "Be good when it matters, even under most adverse conditions" (Eberspaecher, 2011). A top level of proficiency is required from SF/SOF/Special-Purpose Forces personnel not only for a range of fundamental military skills but also for specific skills associated with their respective specialization. Conditions to be coped with may range from lack of food, water and sleep, unclear situations, hostile climatic conditions, flora and fauna to being under an extreme threat from enemy forces. In a SERE scenario, all this is complemented by isolation and an even greater degree of uncertainty, as well as by physical and psychological violence.

With regard to the mental capabilities and skills SF/SOF/Special-Purpose Forces personnel need in SERE situations, and to the aim to increase their resilience by means of special training (Greene & Staal, 2017), it is only logical to make use of sports psychology techniques adapted to a military context. A conceptual similarity (Wagstaff & Leach, 2015) suggests a use and adaptation of sports psychology techniques in the military. The replies obtained in a non-representative survey from soldiers who successfully completed the aptitude test for future assignments to the German Special Forces KSK (Kommando Spezialkraefte) and long-range reconnaissance forces showed that such personnel intuitively made use of the entire portfolio of sports psychology. This includes breathing techniques, stress-relief techniques, routines, self-talk, goal setting, visualization and so on. Prior to the beginning of the course, the students successfully underwent selection procedures pushing them to the limits of their physical fitness and mental resilience. Requirements to be met resemble the challenges arising in a military survival situation. In order to get through such a situation, many soldiers use mental techniques without being aware of it. Lessons taught mainly focus on making students aware of these techniques used intuitively and to categorize them before the students are made to train and apply them in a structured and target-oriented manner. The advantage is the students' high commitment to the techniques taught.

As mentioned above, resistance is particularly important. The focus lies on making the students live through a realistic situation of being held hostage, including interrogations and Conduct after Capture (CAC). Appropriate psychological training is the first thing necessary for ensuring that this challenging training achieves its aim, namely to train soldiers in such a way that they can survive with dignity (Eid, Brun, Laberg, & Johnsen, 1998; Laberg, Johnsen, Eriksen, & Zachariassen, 2000). The first part of the psychological theoretical education comprises interrogation methods, exploitation, possible sexual violence and so on and their psychological effects on individuals. In the second part, students tackle the motivational basis of survival, starting from a quote from Victor Frankl (2006), namely "Those who have a 'why' to live, can bear with almost any 'how'." In the third part of classroom training, the sports psychology techniques covered in the general SERE training are dealt with in greater depth and adapted to the needs of SF/SOF/Special-Purpose Forces personnel. The final part deals with procedures such as reintegration (Personnel Recovery) which are applied in order to help soldiers recover from a kidnapping ordeal and return to normal life as quickly as possible. Knowing that you will be given good support, care and counseling in a real emergency will also strengthen your morale and "will to go on." Furthermore, the training situation and the attitude of the instructors/Conduct after Capture Instructors (CACIs) must be conducive to enabling the students to apply what they have learned and thus experience positive things, small victories for increasing their self-efficacy. CACIs undergo specific psychological training with regard to this task.

Future topics relevant to the further development of SERE training include research on the issue of efficiency in resilience training. "Do the techniques taught and the design of the practical phases serve their purpose?" and "In what way do the instructors' personalities influence the success of training?" are only two among many other possible working questions. The use of biofeedback procedures is another topic. The purpose is to clarify the benefit of the respective procedures in SERE training. In conclusion, one can say that the SERE C course for German SF/SOF and Special-Purpose Forces provides demanding and realistic training to be considered a part of the resilience training the aforementioned assets are required to undergo.

## Summary and outlook

Even though there is still a lot of scientific research and dispute on the trainability of hardiness and resilience going on, the trend of armies finding or developing the particularly resilient soldier is presumably as old as warfare itself.

This chapter gave an insight on how applied military psychology tries to close the gap between scientific findings and the attempt to implement those results into day-to-day military training.

As for the German perspective, the Bundeswehr defines itself in the line of tradition with the Prussian reformers (Epkenhans, 2018). Already here it shows that the question of resilience has played a role in the thinking of military leaders. Clausewitz (*1. July 1780 as Carl Philipp Gottlieb Clauswitz; †16. November 1831) writes extensively about the mental aspects of warfare in his book on the nature of war (Clausewitz, *On War*, 1832).

He characterizes the concept of "Gewohnheit," which is best translated as accustomation. "Accustomation strengthens the body in great efforts, the soul in great danger, and the verdict against the first impression (of war). Everywhere a precious prudence is gained through them, which reaches from the hussar and shooter to the general of the division and makes the general's actions easier" (Clausewitz, 1832).

In a broader sense, adapted to today's terminology, one could equate accustomation to the concept of hardiness and resilience. From that point of view, in addition to the actual operational experience, realistic exercise and training are the keys to establish the resilience of the individual as well as of the whole army.

"It is infinitely important that the soldier, high or low, at whatever level he stands, should not first see for the first time in war the phenomena of war which, for the first time, put him in astonishment and embarrassment" (Clausewitz, 1832).

To build up hardiness and resilience, soldiers should therefore be confronted with the threatening environment of warfare as realistically as possible. One of the critical problems of this approach is balancing the increasing level of realism while mitigating the resulting increase in danger and keeping the soldiers mentally and physically out of harm's way.

A crucial improvement in this context could be the utilization of immersive virtual reality training simulators. With the ever-advancing technological progress, training simulators play an increasing role in military training. While they have become a fundamental component of training in some fields, such as the training of aviators and crewmen, simulators are less common in others, for example, dismounted infantry training. Now that the recent development in technology provides portable virtual reality devices, which can display a complex scenario, there will be the question to what extent training in a virtual environment can be implemented. And if so, how effective is it? Is it possible to create and implement simulated training for resilience training to the same extent it is already used within other fields, generating the same training and cost effectiveness?

Developing and implementing high-resolution realistic simulations can be a very cost-intensive process. Furthermore, there has been controversy in the pertinent literature as to the role of fidelity, amount of detail and originality in simulated training. Is it really necessary, for example, to use absolutely high-end graphics and a 3D sound system in order to get immersed in virtual reality? Or does less technological effort suffice to enthrall the soldiers? What is the impact of the soldiers' "immersion" into a virtual reality on the effectiveness of training projects using such virtual reality?

In conclusion, it can be said that the contribution of applied military psychology in building resilience and hardiness in military personnel contains several factors: from an academic approach by sharpening the conception of human behavior for leaders, to hands-on methods to cope with individual stress, psychology already has a big footprint in military training. The next milestone that has to be achieved will be integrating the advantages of this training in a virtual environment, being aware of and finding answers to the questions that will arise.

## References


Annen, H., Niederhauser, M., & Zueger, R. (2018). Swiss Army Resilience Training – Current State of Work and Initial Findings. Presentation at 53rd IAMPS, Bucharest.

Bartone, P. T. (2008). *What is Hardiness?* Retrieved from http://www.hardiness-resilience.com

Bartone, P. T. & Hystad, S. W. (2010). Increasing mental hardiness for stress resilience in operational settings. In P. T. Bartone, B. H. Johnsen, J. Eid, J. Violanti, & J. C. Laberg (Eds.), *Enhancing Human Performance in Security Operations* (pp. 257–274). Springfield: Charles C. Thomas.

Beckmann, J. & Elbe, A. M. (2007). Motivation und Expertise [Motivation and expertise]. In N. Hagemann, M. Tietjens, & B. Strauss (Eds.), *Psychologie der sportlichen Hoechstleistungen [Psychology of Peak Performance in Sports]* (pp. 140–155). Goettingen: Hogrefe.

Castro, C. A. & Adler, A. (2011). Military mental health training: Building resilience. In: Southwick, S. M., Litz, B. T., Charney, D., & Friedman, M. J. (Eds.), *Resilience and Mental Health: Challenges across the Lifespan* (pp. 323–339). Cambridge: Cambridge University Press.

Clausewitz, C. (1832). *On War.* Princeton: Princeton University Press. Howard, M., & Paret, P. (Editors and translators) 1989.

Eberspaecher, H. (2011). *Gut sein, wenn's drauf ankommt: Von Top-Leistern lernen [Being Good When It Matters: Learning from Peak Performers].* Munich: Carl Hanser Verlag.

Eid, J., Brun, W., Laberg, J. C., & Johnsen, B. H. (1998). Participating in a Prisoner-of-War Exercise: Stress Reactions and Coping. Paper presented at the 34th International Applied Military Psychology Symposium, May 25–28, 1998, Paris, France.

Epkenhans, M. (2018). Saeulen der Tradition [Pillars of tradition]. In *Zeitschrift fuer Innere Fuehrung [Magazine for Leadership Development and Civic Education] 62,* (pp. 28–35).

Escolas, S. M., Pitts, B. L., Safer, M. A., & Bartone, P. T. (2013). The protective value of hardiness on military posttraumatic stress symptoms. *Military Psychology, 25*(2), 116–123.

Explosive Ordnance Disposal (EOD) Training Unit. (2018a). Information Briefing. Stetten am kalten Markt.

Explosive Ordnance Disposal (EOD) Training Unit. (2018b). Records. Stetten am kalten Markt.

Farrow, D. & Abernethy, B. 2007. Wahrnehmung von Expertinnen und Experten im Sport: Einige Kernfragen und -probleme [Perception of experts in sports: Some core questions]. In N. Hagemann, M. Tietjens, & B. Strauss (Eds.), *Psychologie der sportlichen Hoechstleistungen [Psychology of Peak Performance in Sports]* (pp. 71–92). Goettingen: Hogrefe.

Frankl, E.V. (2006). *Man's Search for Meaning.* Boston: Beacon Press.

Greene, C. H. & Staal, M. A. (2017). Resilience in US Special Operations Forces. In S. Bowles & P. T. Bartone (Eds.), *Handbook of Military Psychology* (pp. 177–192). Cham, Switzerland: Springer International Publishing.

Kabatt-Zinn, J. (2011). *Gesund durch Meditation [Healthy through Meditation].* Munich: Knaur/MensSana.

Kaluza, Gerd (2011). *Stressbewaeltigung. Trainingsmanual zur psychologischen Gesundheitsfoerderung [Stress managment. A training manual for health promotion].* Berlin: Springer.

Knapik, J. J., Reynolds, K. L., & Harman, E. (2004). Soldier load carriage: Historical, physiological, biomechanical, and medical aspects. *Military Medicine, 169*(1), 45–56.

Knapik, J. J., Sharp, M. A., Darakjy, S., Jones, S. B., Hauret, K. G., & Jones, B. H. (2006). Temporal changes in the physical fitness of US Army recruits. *Sports Medicine, 36*(7), 613–634.

Knapik, J. J., Sharp, M. A., & Steelman, R. A. (2017). Secular trends in the physical fitness of United States army recruits on entry to service, 1975–2013. *Journal of Strength and Conditioning Research, 31*(7), 2030–2052.

Kobasa, S. (1979). Stressful life events, personality, and health: An inquiry into hardiness. *Journal of Personality and Social Psychology, 37*(1), 1–11.

Krueckel, O. (2018). Chaos driven situations management retrieval system (CHARLY). Presentation at 53rd IAMPS, Bucharest.

Laberg, Ch., Johnsen, H. B., Eriksen B. S., & Zachariassen, K. K. (2000). Coping with interrogations. In C. McCann & R. Pigeau (Eds.), *The Human in Command: Exploring the Modern Military Experience* (pp. 333–344). New York: Kluwer Academic.

Locke, E. A. & Latham, G. P. (1990). *A Theory of Goal Setting and Task Performance*. Englewood Cliffs, NJ: Prentice-Hall.

Maddi, S. R. (1967). The existential neurosis. *Journal of Abnormal Psychology, 72*(4), 311–325.

Moran, A., Toner, J., & Campbell, M. (2018). Attention and concentration. In A. Mugford & J.G. Cremades (Eds.), *Sport, Exercise and Performance Psychology. Theories and Applications* (pp. 233–250). New York: Routledge.

O'Grady, S. (n.d.) AZQuotes.com. Retrieved August 16, 2019, from AZQuotes.com. Website: https://www.azquotes.com/author/26095-Scott_O_Grady

Salas, E., Priest, H. A., Wilson, K. A., & Burke, C. S. (2006). Scenario-based training: Improving military mission performance and adaptability. In A. B. Adler, C. A. Castro, & T. W. Britt (Eds.), *Operational Stress. Military Life: The Psychology of Serving in Peace and Combat: Operational Stress* (pp. 32–53). Westport: Praeger Security International.

Seligman, M. E. & Fowler, R. D. (2011). Comprehensive soldier fitness and the future of psychology. *American Psychologist, 66*(1), 82–86.

Southwick, S. & Charney, D. (2012). Facing fear. An adaptive response. In S. Southwick & D. Charney (Eds.), *Resilience: The Science of Mastering Life's Greatest Challenges* (pp. 46–63). Cambridge: Cambridge University Press.

Tenenbaum, G. (2003). Expert athletes: An integrated approach to decision-making. In J. L. Starkes & K. A. Ericsson (Eds.), *Expert Performance in Sport: Advances in Research on Sport Expertise* (pp. 191–218). Champaign: Human Kinetics.

Tenenbaum, G. & Sacks, D.N. (2007). Die emotionale Seite der Hoechstleistung [The emotional side of peak performance]. In N. Hagemann, M. Tietjens, & B. Strauss (Eds.), *Psychologie der sportlichen Hoechstleistungen [Psychology of Peak Performance in Sports]* (pp. 118–139). Goettingen: Hogrefe.

University of Michigan. 1962. *The Fraternal Monitor, 73–74*. Research & Review Service of America Incorporated.

Wagstaff, C. & Leach, J. (2015). The value of strength-based approaches in SERE and sport psychology. *Military Psychology, 27*, 65–84.

Werner, E. E. (1971). *The Children of Kauai: A Longitudinal Study from the Prenatal Period to Age Ten*. Honolulu: University of Hawaii Press.

Wesemann, U., Kowalski, J., Zimmermann, P., Rau, H., Muschner, P., Lorenz, S., … Willmund, D. (2016). From hero to Pro – change in attitude towards mental illness in deployed soldiers using the preventive computer program CHARLY. *Wehrmedizinische Monatsschrift, 60*(1), 2–7.

Wyss, T. & Annen, H. (2012). Influence of progressively increased physical strain, sports, and leadership style on fitness, injuries, dropout rates, military performance capability, stress, and motivation in Swiss recruits. Magglingen/Birmensdorf: Internal Study. BASPO/MILAK.

Yates, F. J. & Tschirhart, M.D. (2006). Decision-making expertise. In K.A. Ericsson, N. Charness, P. J. Feltovich, & R. R. Hoffman (Eds.), *The Cambridge Handbook of Expertise and Expert Performance* (pp. 421–438). Cambridge: University Press.

Yerkes, R.M. & Dodson, J.D. (1908). The relation of strength of stimulus to rapidity of habit-formation. *Journal of Comparative Neurology and Psychology, 18*, 459–482.

# 12

# SUSTAINABLE TEAM LEADERSHIP

## Social identity and collective leadership for military and society

*António Palma Rosinha, Hermes de Andrade Jr, and Marcos Aguiar de Souza*

A prompt, timely, and appropriate response that guarantees collective security requires fine cooperation between different stakeholders, people, and nature. Leaders must be prepared to transition rapidly from brutal environments to situations of close interaction with different realities and, ultimately, to maintain a high level of emotional self-control and efficiency. Volatility, uncertainty, complexity, and ambiguity (VUCA) foster a hazardous climate that places new demands on leadership and challenges leaders' ability to adapt. The complexity and prevalence of the challenges that potentially undermine national and international security, such as demographic shocks, the increase in terrorism and transnational crime, maritime piracy, cyber threats, regional conflicts, and resource disputes (Borges & Rodrigues, 2016) have forced teams, organizations, and states to take a broader look at leadership. Thus, although most definitions of leadership acknowledge that leaders are responsible for the leadership process, we should not assume that this role is fixed and predetermined—it can be contingent because events can be highly unpredictable and volatile (Rosinha, Matias & Souza, 2017). Furthermore, formal leaders do not act alone but interact with all those around them. The influence process is not unidirectional: quite the contrary; followers play a decisive role in influencing both leaders and outcomes. Trait leadership theories and behavioural, situational, transactional, transformational, and visionary leadership theories have been an important step towards viewing individuals as stakeholders in the leadership process. However, these theories are self-focused and their scope is narrow because modern organizations have become increasingly complex and are built on non-linear network structures that employ highly skilled teams (Daft, 2014). Traditional leadership models are too restrictive and too focused on the leader-subordinate dyad; therefore, a multilevel approach to leadership is required. Both research on leadership and a theoretical framework for leadership are still in the early stages.

Thus, leadership should be understood as a complex and dynamic process where different organizational roles are performed by several individuals (Gronn, 2002) and where those roles can be switched among work teams (Day, Gronn & Salas, 2004). This paper is based on the following assumptions: (i) leadership is a developmental process; (ii) leadership emerges when proximity to the in-group prototype is oriented toward common goals and expectations by affirming the group identity that gives it legitimacy; (iii) what distinguishes collective leadership from traditional models is the fact that by sharing leadership, influence processes go beyond the traditional top-down influence exercised exclusively and unilaterally by formal leaders, recognizing individual expertise and allowing leadership, in its purest form, to be widely distributed among several individuals; (iv) leadership should be sustainable, ensuring that more people and their descendants have access to resources and opportunities, a democratic path, and a democratic future; and (v) sustainable leadership is a factor of positive participation that is essentially democratic (and not only discursive) and cannot be otherwise. Its gregariousness is a natural antidote to abuse and misuse of authority by generating trust and, of course, by promoting transparency. The key issue addressed in this study is whether preparing individuals and teams for sustainable leadership (collective, participatory, effectively shared, and virtuous, laying out a path for future generations in a turbulent world) will be rewarded by a social identity that has trust in authority and a truly representative governance that presents solutions to bring about democratic change, with emphasis on the participatory process. Thus, we strongly question the paradigm of leadership as an individual phenomenon and propose that the construction of leadership identity is a social process, given its nature and the requirements of communication and media scenarios; we point out the reductionism of the current studies on leadership, which focus only on the figure of the formal leader, neglecting the development of leadership as an inherently multilevel and longitudinal social process; we emphasize the participatory phenomenon of sustainable leadership, arguing that its goal is to aggregate values and virtues and to transform society into a democratic environment, as opposed to simply a group of people who have no commitment to the truth, and criticize the lack of professionalism and transparency that make a future in the shadow of authoritarianism even more uncertain.

## Assumption (i): Leadership is a developmental process

### *The complexity of leadership*

In its quest to obtain a more accurate picture of the true nature of leadership, the extant research on leaders has ignored the most appropriate ontological locus for leadership, which rests in the relationship between the leader-team dyad and the processes that explain how it emerges (Antonakis, Fenley & Liechti, 2011). Leadership can be defined as the ability of an individual to influence, motivate, and enable others to contribute toward the effectiveness and success of the organizations of which they are members (House et al., 1999). Vieira (2002) argues that leadership is defined as a process of influencing human behaviour that goes beyond the use of official authority, in order to achieve the purposes, goals, and objectives defined and provided by the appointed organizational leader. Most concepts define leadership as an influence process or ability, which one member of a group exerts over other members to achieve certain goals by eliciting motivation. Therefore, leaders are the main source of influence for processes and tasks, dominating the decision-making process and playing a key role in defining the strategy, dynamics, motivation and identity of a work group. This single-leader perspective views leadership as a specialized function that cannot be shared without jeopardizing group effectiveness (Ensley, Pearson & Pearce, 2013). A dynamic, interdisciplinary, and inclusive process, leadership cannot be separated from the core functions of organizations. Reality may have forced leaders to adapt to the changes brought about by globalization, but leadership theories have been slower to do so. Recent research based on previous studies stresses the need to redefine leadership. The traditional image of the leader as hero is slowly being replaced by a different image, that of a collective group of people who exert influence through formal and informal relationships between individuals (Chrobot-Mason, Gerbasi & Cullen-Lester, 2016; Yukl, 1999).

### *The developmental nature of leadership*

Leadership development must focus less on leadership theories and more on developmental science. Murphy and Johnson (2011) examined the seeds of leader development and concluded that those seeds germinate in the stages before adulthood. The most important developmental experiences that occur during the sensitive periods of childhood and adolescence, when personality is formed and consolidated, influence the development of leadership into adulthood. Identity development and self-regulation are initial factors that relate to future experiences of leadership development and leadership effectiveness. Thus, early learning experiences of leadership, which are part of a cadet's basic training, combined with education and physical and military training, play an important role in the development of future leaders. Designing individual leadership development programs is part of a broader process of adult development (Day, Harrison & Halpin, 2005). This is in agreement with Kegan's (1994) view of leader development as a progression through stages of moral growth and individuation.

By looking at leadership development as part of human development, we can implement a longitudinal multilevel approach in stages, where prior knowledge can be absorbed and transferred to more complex levels that require new knowledge, skills, and attitudes.

Examining this development involves mapping and understanding the patterns of change of individuals, groups, and teams over time (Day, 2000). These patterns of change reflect transitional phases between stages that represent changes in thinking based on mental schemes through which individuals intellectually adapt and organize their environment. Piaget (1976) believed that building knowledge requires balancing, that is, assimilating new knowledge and accommodating new structures of thought. Examining this development involves mapping and understanding the patterns of change of individuals, groups, and teams over time (Day, 2000). These patterns of change reflect transitional phases between stages that represent changes in thinking based on mental schemes through which individuals intellectually adapt and organize their environment. Piaget (1976) believed that building knowledge requires balancing, that is, assimilating new knowledge and accommodating new structures of thought.

## Assumption (ii): Leadership emerges when the proximity to the in-group prototype is oriented toward common goals and expectations by affirming the group identity that gives it legitimacy

### *Building leadership identity as a social process: Formation of psychological groups*

What distinguishes a set of individuals from a group? Is it cohesion, camaraderie, team spirit, unity, or a sense of belonging? There are several terms to describe the force that binds the members of a group, and the concept of attraction is common to all. The theories of self-categorization (Turner, Hogg, Oakes, Reicher & Wetherell, 1987) and social identity (Tajfel & Turner, 1986) help explain how attraction emerges. Participation in and adherence to a psychological group (that is, an individual believes they are a member of the group, regardless of how third parties would classify them) explain in part how the underlying processes in the social psychology of leadership unfold. When social identity becomes salient, people experience depersonalization at the cognitive level, stop perceiving themselves as unique individuals, and begin seeing themselves as members of the group (Turner et al., 1987). Hence, they begin acting according to the norms and values of the group (Reicher, Spears & Postmes, 1995). The psychological process of depersonalization makes it possible to develop group processes that extend to cooperation, to the definition of rules, and, more importantly, to social influence and leadership. People must understand and accept themselves as members of a specific group before group-based processes such as the emergence of leadership can occur (Reicher, Haslam & Hopkins, 2015). According to self-categorization theory, self-concepts consist of cognitive self-categorizations in which members see themselves as similar to one another at various levels of inclusion. At one end of the spectrum, people are seen as unique individuals who are different from other individuals (i.e., personal identity). At the other end, they are seen as part of humanity as a whole (that is, collective identity). In-group and out-group categorizations fall between these two extremes. A person's relative in-group prototypicality, idea, or behaviour is a key factor that affects that person's degree of influence within a group (McGarty, 1999). Prototypicality is the cognitive criterion that determines how people or entities are placed together in a group and separated from others. Because prototypicality is an elastic entity, individuals are able to compare themselves to the prototype at different times of their membership to the group (i.e., self-prototypicality), to compare people who share the same prototype and, finally, to compare themselves to other people. The derivation of a theoretical personal identity accentuates the loss of personal identity in favour of group identity. The key factor that determines leadership is proximity to the in-group prototype, which in turn intensifies influence over the group (Turner, 1991). To the extent that there is variability among group members regarding the prototypicality of the in-group, there is a gradient of relative influence within that group. At any given moment, and in any inter-group context, the member of the group closest to the prototype spontaneously emerges in a social system and may become its leader (Yukl, 2002). In sum, the more an individual is seen as an integral part of the group, and the more that individual represents the group and advocates for it, thereby reinforcing a sense of belonging and emphasizing group identity, the more likely that individual is to be seen as leader of that group. The leader might not interact with and influence all the elements in a group, but only with the people who legitimize him or her. The leader's legitimacy stems from a process of mutual influence which results in joint action that is oriented towards common goals and expectations.

## *The role of social identity in building relationships*

Dyadic leadership relationships are the basic building blocks of leadership networks (Kalish, 2013). Identity is an important part of self-concept because it helps answer the questions, who am I? and who do other people know me to be? (Ibarra, Wittman, Petriglieri & Day, 2014). Earlier research on identity and leadership initially focused on how people see themselves as leaders, emphasizing the socially constructed process by which individuals internalize leader identity as part of their self-concept (De Rue & Ashford, 2010). However, the construction of this identity stems from the work groups and organizations to which people belong. Role-based identity theory suggests that leader is only one of many roles with which individuals can identify. A leader's identity should not be seen as static or unidirectional, but rather as a process of mutual influence in which social interactions between individuals, as well as contextual factors, are responsible for that identity (Chrobot-Mason, Gerbasi & Cullen-Lester, 2016). Recent studies by DeRue, Nahrgang, and Ashford (2015) suggest that when individuals perceive the group as being warmer, their identification with the group is stronger and several people contribute to leadership. In short, individuals who have a stronger collective identity are more motivated to engage in leadership behaviours on behalf of the group. Research in the area of leadership development suggests that, in addition to developing a personal identity, successful leaders also develop a strong sense of collective identity (Ibarra et al., 2014). Inexperienced leaders emphasize individual identities (I) and more experienced leaders emphasize a more collective identity (we), one which strongly identifies with the values and priorities of the team or organization as a whole (Day & Harrison, 2007). Along the same line, other researchers

(Van Knippenberg, Van Knippenberg, De Cremer & Hogg, 2004) show that would-be leaders receive stronger support when using self-oriented group behaviours rather than self-oriented to their own personality. When someone is considered a leader by favouring the in-group when distributing resources, it proves the theory that leadership is about "having it done by us," which involves the principle of social identity. Thus, leadership is a psychological group process in which the members of a group elect one of their own as their leader. At least two key aspects improve the odds of a group member emerging as leader: engaging in "doing it for us" behaviours and actively building and rebuilding the very meaning of "us" (Chrobot-Mason, Gerbasi & Cullen-Lester, 2016). The greatest challenge for formal leaders is meeting the needs and expectations of the group while also meeting the needs and expectations of the organization.

## Assumption (iii): What distinguishes collective leadership from traditional models is the fact that, by sharing leadership, influence processes go beyond the traditional top-down influence exercised exclusively and unilaterally by formal leaders, recognizing individual expertise and allowing leadership, in its purest form, to be broadly distributed among several individuals

### *Understanding collective leadership*

We are faced with a paradigm shift in leadership, which is becoming the purview of the collective rather than the individual both in formal and informal relationships (Carter, DeChurch, Braun, & Contractor et al., 2015). These new leadership practices are referred to in different ways, such as distributed leadership, shared leadership, relational leadership, polarized leadership, and collective leadership. These approaches acknowledge that, as collective systems, organizations are complex relational systems and that leadership stems from interdependent interactions and connections that intertwine to form larger structural patterns. Although leader-follower relationships are already addressed in traditional and contemporary leadership approaches (for example, transformational leadership), what distinguishes this approach is the acknowledgement that teams, organizations, coalitions, communities, networks, and systems carry out their leadership roles and activities through distributed processes and collective social behaviours that change over time. The macro component of an organization's leadership can be depicted as a tripartite representation, according to variables that describe collective leadership as concentration (people), roles, and time (Contractor, DeChurch, Carson, Carter & Keegan, 2012). Role concentration is the degree of leadership distribution from one individual to multiple individuals, with each role corresponding to a function—Navigator—Engineer—Social Integrator—Liaison (Carson & Tesluk, 2007). Role modelling refers to a person's ability to lead by example, displaying effective ethical and behavioural patterns through social learning processes and becoming an ideal prototype for all members of an organization, who is capable of activating collective identities and goals (De Cremer & van Knippenberg, 2002). Delegation is defined as the transfer of power and competencies and is expanded through the practice of shared leadership (Pearce, 2004).

In this networked environment, influence involves using techniques that appeal to reason, emotions, or values to gain the commitment of the group and the support of supervisors, as well as to negotiate and interact effectively with the members of the local population. The dispersed and fluid nature of operations requires the ability to analyze problems systematically and to act decisively without outside guidance or advice. Decision-making involves detecting, perceiving, and understanding the elements of the environment and projecting future states or events. Finally, support consists of displaying respect, sociability, individual consideration, and empathy (Hannah,

Jennings & Nobel, 2010). An example of its applicability to the field of communication is how reality and realism remake the tense environment of data collection in loco of the report. Concern over the repercussions of the news represents a challenge for communication professionals, who are easily criticized for decisions and attitudes that are perceived as wrong and unethical or that are uncomfortable to some audiences, who are increasingly aware of their identity. As cognitive and affective attributes demanded by the agent of communication, various forms of courage arise, such as physical, moral, and social courage (confronting a superior or colleague) and psychological courage (the risk of experiencing a new skill where it may fail; Hannah, Jennings & Nobel, 2010). We have also considered the cognitive styles associated with resilience, which is defined as the ability to find meaning and purpose in potentially stressful events, turning them into opportunities for personal growth. We have also encountered concepts related to psychological states, such as hope, optimism, resilience, and confidence. Self-confidence in the ability to perform leadership tasks was found to be a key factor. Positive emotional demonstration, or positive affectivity, as opposed to negative affectivity, is crucial for leaders who seek to elicit energetic and positive emotional responses, provide stress-relief resources, and rally the team (Rosinha et al., 2017). Similarly, a high level of emotional self-regulation is considered a key attribute in maintaining a positive attitude under stress. Several self-regulation attributes have been identified, such as the ability to process large amounts of information, which implies "thinking about thinking" or monitoring and controlling one's own cognitive processes. Another critical factor of self-regulation is that, in order to remain effective, leaders must have a "genuine and visceral motivation to lead" (Hannah, Jennings, & Nobel, 2010, p. 436). The motivation to lead others can come from multiple sources, such as seeing oneself as a leader (identity-based), social or normative motives, or the result of a calculated cost-benefit analysis.

## Assumption (iv): Leadership should be sustainable, ensuring access to resources and opportunities for more people and their descendants, and a democratic way and democratic future

### *Sustainable of what?*

The term *sustainability* is widely used but little explained. It is conceptually misunderstood and regarded as a fashion accessory (Hasna, 2010) or a matter of common sense (Moldan, Janouaková, & Hák, 2012). The concept is inconsistently interpreted and ambiguously applied and reveals an incomplete perception of the issues of poverty, environmental degradation, and the role of economic growth (Mori & Christodoulou, 2012).

What we now call sustainable development (SD) has evolved as an integrating concept, an umbrella under which a set of interrelated issues is uniquely organized. It is a variable process of change with sustainability as its ultimate goal. In the same context, sustainability is the ability of a human, natural, or mixed system to resist or adapt to endogenous or exogenous change for an indefinite period, represented as a goal or end point (Hove, 2004). Therefore, to achieve sustainability, SD is usually required. The SD concept remains contested because of the different positions taken in relation to what can be considered fair (Todorov & Marinova, 2011).

It does not explicitly embrace future thinking, and almost all published definitions of the SD concept are based on principles of sustainability, e.g., long-term perspective, the key importance of local conditions, and understanding the nonlinear evolution of environmental and human systems (Moldan et al., 2012) because the term *sustainability* comes from renewable resources and has become the main motto of the ecological movement.

This sense of the concept refers to the existence of the ecological conditions required to support human life at a specific level of welfare through future generations, which is ecological sustainability, not sustainable development. According to Ayres (2008), sustainability is a normative concept about how humans should act in relation to nature and how they are responsible for each other and for future generations.

As defined in this study, the sustainability paradigm also postulates the preservation of diversity in its broadest sense—sociodiversity—in addition to biodiversity, that is, that the system of values, practices, and symbols of identity that enables the reproduction of the social fabric and ensures national integration is maintained over time.

The political basis for sustainability is linked to the process of deepening democracy and building citizenship. At the macro level, this translates into the democratization of society, and, at the micro level, into the democratization of the state. The first of these goals requires strengthening social and community organizations, redistributing assets and information to the subordinate sectors, increasing their organizations' analysis and decision-making capabilities. The second goal is achieved by opening the state apparatus to citizen control, by reforming political parties and electoral processes and by incorporating the concept of political responsibility into public activity (Viana, Silva & Diniz, 2001).

Without directly referring to sustainability, Fritjof Capra describes this concept in his book *The Turning Point*, which was also published in the 1980s. In the work, Capra presents his concept or systemic view of the world from an ecological perspective that contrasts with the fragmentary system that supported the modern world of the late twentieth and early twenty-first centuries. For Capra, any concept of sustainability must be directly linked to the complex web of interdependencies that have the capacity (inherent in nature) to transform human culture.

Being sustainable means acknowledging that reality is a continuous, systemic, non-linear process in which relationships (social, natural, and socio environmental) occur both simultaneously and subsequently and that each party inevitably has responsibilities in the process, as the pattern in nature is a network of interdependencies between beings and resources coexisting in a given environment. Other definitions of sustenance, such as to bear, to nourish, or to fortify, also mean to maintain, provided they are being used correctly.

This conceptual approach inspired us to conceptualize sustainable leadership for democracy. Therefore, our definition of sustainable leadership is a collective leadership with an ideal, inspiring, and consensual nature, which includes delegation of powers, sharing of trustworthy actions, and commitment to a future for all.

## Assumption (v): Sustainable leadership is a vector of positive participation with democracy in its essence (not in speech) and does not allow it to be otherwise. Its gregarious sense becomes a natural antidote to abuse and misuse of authority by generating trust and of course inducing transparency

This section will demonstrate how sustainable leadership helps legitimize participation.

Sustainable leadership is ethical, shared, participatory, and distributive; is based on the best human traits and virtues; and works under the assumption of trust based on prior conduct; it effectively bets on its own perpetuity and on its multiplier effect over space and time. Over space because it is an intercultural and international chain of leadership that many adhere to when provided with a sound role model, and over time because it ensures that future generations have the opportunities that are denied by authoritarian and individualistic capitalist alienation.

The importance of social participation is recognized by scholars and rulers because representative democracy cannot adequately meet the demands of society. The concept of participation has a highly symbolic charge and is emblematic within the ideological context and the social structure that perceives it. The extant literature seems to agree that "engaging" generally means "being involved, taking part in or influencing processes, decisions and activities in a particular context or field of action" (Nirenberg, 2006, p. 121).

However, the risk of widespread participation in the democratic environment is trivializing the process, with people having an idea of what it is they are participating in and confirming that they want to participate but in which their participation is not characterized (Carpentier, 2012) and becomes a fad that cannot yield the expected benefits.

Broadening the discussion, Sirvent (2004) refers to two forms of participation: real participation, in which members of an institution or group effectively influence all processes of institutional life, including decision-making, and are able to change the power structures, and symbolic participation, which involves "actions that have little or no influence over institutional politics and management and which generate in individuals and groups the illusion of a nonexistent power" (Sirvent, 2004, p. 129).

We see this as the problem of participation, as its symbolic activity, which is present in symbolic interactionism (Carvalho, Borges & Rêgo, 2010) or symbolic power (Wacquant, 2013), is not significant enough to implement solutions and cannot raise the awareness required for it to take effect.

In symbolic interactionism, because people's actions only acquire meaning through the actions of others and during a social act, the objects of the perceived environment are defined and redefined. These dynamics consist of symbolic interactions, which are not a direct reaction to the actions and gestures of the other, but an interpretation of these actions or gestures based on the meaning attributed to them (Carvalho, Borges & Rêgo, 2010). That is, leaders should set the example by ensuring transparency in their attitudes and by participating as citizens, motivating other citizens to replicate their participatory behaviour and becoming inspiring leaders. Without this inductive, gestural, and active effect caused by this "other" that we see in the mirror, the process of participation has little hope of transforming reality.

Symbolic power emerges from the symbolic mechanisms of group formation and domination and is experienced through recognition/false recognition, that is, a constant and diversified effort to inculcate and impose categories of perception that help form social reality, shaping its representation (Wacquant, 2013) and portraying figures of authority as worthy of distrust. Effective participation is linked to commitment and trust: trust in the authorities and in the quality of their leadership, which must transcend discourse (negative symbolism).

Committed people and organizations will move away from the symbolic and imaginary dimension (collective) and towards the micro reality dimension, in which participation is required because the state has flaws that must be corrected.

To ascertain if this participatory activity is sustainable, we examined a literature review aimed at improving the quality of public services. The review's authors (Struecker & Hoffmann, 2017) used as benchmarks a set of 56 articles published from 2005 to 2015 in high-impact journals and on consolidated metadatabases.

Kluvers and Pillay (2009) point out that the interest in a more participative model of decision-making results from the growing lack of trust in democratic institutions. In this regard, social participation may be the answer to reestablish the bonds between citizens, military, and governments by lending credibility to public actions (Mattia & Zappellini, 2014; Eriksson 2012; Karkin & Calhan, 2012; Kloby, 2009; Kluvers & Pillay, 2009; Lawton & Macaulay, 2014; Yang & Pandey, 2011).

Likewise, the literature suggests that increasing citizen participation confers legitimacy to decisions (Karkin & Calhan, 2012; Lawton & Macaulay, 2014), provided that the process is genuinely collaborative and does not lend itself to particular interests (Boin et al., 2011) or to the capture of the state by elites due to power asymmetries and social exclusion (Aiyar, 2010).

Kloby (2009) also argues that citizen engagement improves the quality of decision-making (Lawton & Macaulay, 2014), which represents the will of taxpayers, and leads to greater efficiency and savings in most public actions (Neshkova, 2014). Along the same lines, Aiyar (2010) highlights the momentum that participation can bring to the performance of the state, making it more agile and responsible; dialogue and articulation between the state and its citizens ensure that local demands are heard, while at the same time enabling effective decision-making.

The desire of public authorities to improve service performance and to promote social participation must necessarily be accompanied by new forms of accountability and concern for the legitimacy of decisions (Simmons & Birchall, 2005). Thus, social participation is an opportunity to bring information to citizens, contributing to transparency and accountability (Aiyar, 2010; Eriksson, 2012).

However, some scholars suggest that accountability must be achieved through citizens' participation in decision making and service delivery assessment (Kloby, 2009). The learning acquired in the participation process empowers participants to exercise citizenship effectively, allowing them to contribute their knowledge and experience to achieve the best solutions (Eriksson, 2012; Lawton & Macaulay, 2014; Neshkova, 2014).

Citizens can even offer innovative solutions to collective problems because they are not constrained by the traditional model of decision-making generally adopted by the public administration (Neshkova, 2014), as well as proposing new forms of public service delivery (Osborne & Strokosch, 2013).

Despite the advantages pointed out in the literature, social participation has its critics. The articles identified in the systematic review are unanimous about the benefits of citizen engagement; however, they also refer to negative aspects found in other studies, such as: the risk of losing control of the process, which can be dominated by interest groups (Neshkova, 2014; Osborne & Strokosch, 2013), and the population's lack of expertise and lack of time to make good decisions (Eriksson, 2012), as many individuals are incapable of constructive criticism and thrive on alienation and conflict (Musso et al., 2011 cit. by Struecker & Hoffmann, 2017).

The need for transformative leadership and cultural change is one of the recommendations of a reviewing on social participation (from 2005 to 2015), as both are important factors that foster sustainability (Struecker & Hoffmann, 2017, p. 379). Therefore, we will now organize the elements and process of our proposal on sustainable leadership.

So far, we have seen that the reductionism of the studies that focus on formal leadership has ignored that the development of leadership is intrinsically a multilevel and longitudinal social process. The evidence shows that leaders emerge when the proximity to the in-group prototype is oriented towards common goals and expectations by affirming the group identity that gives them legitimacy.

When we consider the contributions of the studies that address multilevel phenomena, the empirical limitations of reductionism become clear. That which is true at one level of analysis (the individual) cannot be directly extrapolated to other levels (the collective of individuals) through a cause-and-effect correlation.

Reductionism in the study of leadership has been shown to lead to discontinuity of effects between levels of analysis. Higher-level phenomena are more than the product of lower-level processes. Any attempts to transpose dyadic, group, or organizational level aspects to the individual level run the risk of producing distorted ideas about how leadership emerges (Fisher & To, 2012).

Developing the collective identity of the organization leads to the emergence of more plural forms of leadership. Leadership is a group psychological process that involves the principle of social identity, reinforcing behaviours that emphasize "doing it for us," in the sense of constructing and actively rebuilding the very meaning of "us." The more individuals identify with their organization and coworkers, the more likely they are to consider others as sources of leadership and to be seen as potential leaders, even if they are not formal leaders.

Examining how influence emerges within the group and the multiple formal and informal relationships that are established is a new approach to the leadership process. Thus, shared leadership is required in order to face the challenges posed by the new global challenges.

The following modelling dimensions (Figure 12.1) have been considered inputs for sustainable leadership: (a) Spiritual, which involves the factors that motivate long-term and life projects; (b) Emotional, which encompasses affective attributes and emotional intelligence; (c) Cognitive, which incorporates knowledge about the process in community and personal limitations (self-knowledge, mental structuring); (d) Values, which sustain life in society and harmony with others; (e) Culture, which includes all the specificities of the individual and collective context and their representations; (f) Beliefs, in the sense of degree of involvement/commitment; (g) Attitude, which is manifested in the direct direction of transformation vis-à-vis the other and that which is new; (h) Will, which involves firmness of purpose and endurance; (i) Empathy, which involves the decision to live and share with others; and (j) Technology, which uses resources, knowledge, and experience combined with competence to achieve sustainability.

The transformation process that occurs in the formation of sustainable leadership has several stages (Figure 12.2).

First, the individual becomes aware of his or her situation as an individual leader and reflects on the dimensional elements described above.

In the second stage, there is the realization that one's individual condition can be expanded by the condition of the other (by example).

In the third stage, after the decision has been made to expand the process, the real and symbolic interactions and interrelationships between the individual and the other become

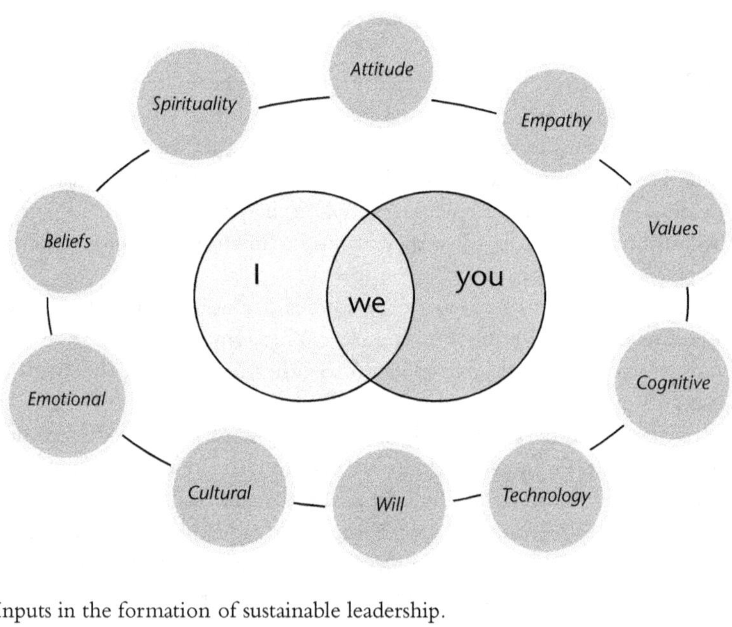


*Figure 12.1* Inputs in the formation of sustainable leadership.

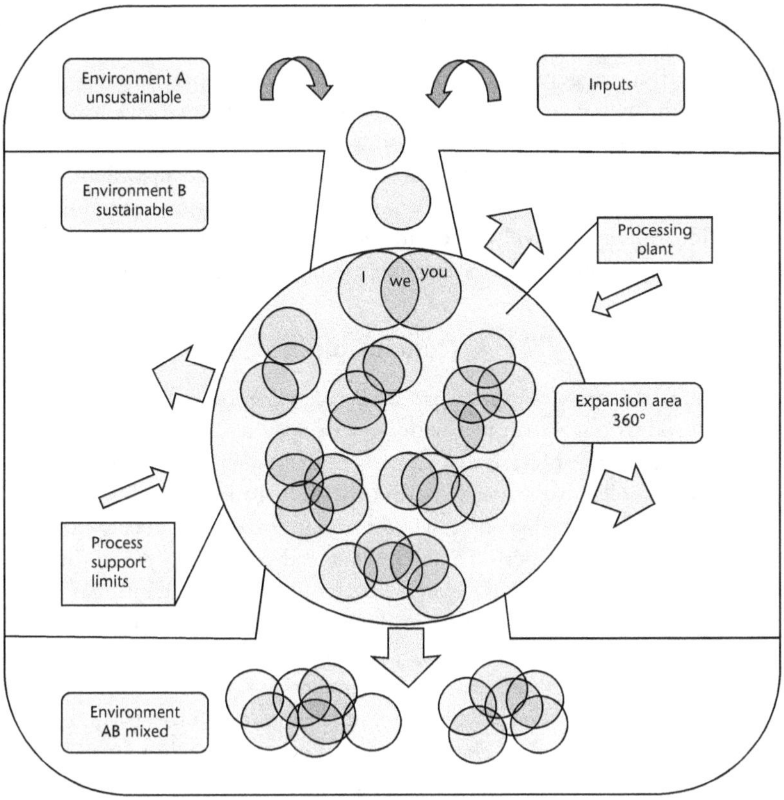


*Figure 12.2* Sustainable leadership formation process.

*Source:* Proposed by the authors

significant for both, which is a feature of collective leadership. From that moment on, decisions will be made together.

The fourth stage is the moment of multiplication or aggregation, since several individuals acting in the same propitious environment (environment *b*) will tend to exchange references, experiences, and meanings. The moment occurs when several individual conditions become references for one another and vice versa.

The fifth stage is that of interactivity, in which the benefits of synergy for sustainability are achieved.

Environment *b*, in which ecosystems interact with people, is a sustained and controlled environment because it is preserved.

Those who participate in the process of leadership formation give back to the environment by being sustainable, collectively responsible for their actions, and aware of the consequences for environment *b*. Leaders are formed with multiple skills and types of intelligence and can elicit pro-sustainable attitudes in the team. Therefore, these pro-sustainability attitudes maintain the environment's habitability conditions over time. This synergistic mechanism is facilitated within the limits of the sustainability of environment *b*.

The result of this synergy is an important artefact, that is, a team that can be described as self-managed due to the meaningful exchanges and trust generated in the process.

When sustained teams form in this way, the environment of origin (environment *a*) receives them again, but with a new configuration, which is directed from the individual to the other and leads to the transformation and formation of leaders, a fundamental part of the process of societal transformation.

The chain reaction triggered by the emergence of multiple sustainable teams in the environment of origin will characterize it as an *ab* environment, that is, a hybrid environment capable of changing the status quo and the existing power configurations.

Modelling dimensions, considered as inputs for sustainable leadership, will be recognized from individual and personal references until their collective construction: a moment in which the context of individual insertions will be considered and progressivity will be achieved through the process of transformation that passes through the different stages mentioned.

## Conclusion

Leadership is one of the most complex and multifaceted organizational phenomena; thus, it cannot be understood through a simple or unitary perspective. Leadership theories have been slow to incorporate innovative notions that emphasize the synergistic strength of joint leadership.

What distinguishes collective leadership from traditional models is the fact that, by sharing leadership, influence processes go beyond the traditional influence unilaterally exercised by formal leaders, recognizing individual expertise and allowing leadership, in its purest form, to be widely distributed among several individuals. Shared leadership prevents traditional leadership from collapsing because, in practical terms, a single leader is not capable of dealing with multiple factors that require diversified knowledge, which can only be achieved through the communication, information sharing, and social interaction that characterize formal and informal networks.

Leadership can emerge anywhere, at any time, and is not limited to a particular configuration or person. This has been reflected in changes in the literature, which now addresses leadership as a relational property rather than as an individual entity. The integrative framework of the network-centred organizational context shows how a collective engages in the social process of leadership, examining whether the relationships that foster direction, alignment, and commitment among employees are present, as well as their possible predictors.

All leadership is about self-identity, self-leadership, and shared and collective leadership. Communicators must manage themselves strategically, becoming more aware of their thoughts, feelings, and actions, and be able to acquire wider and more complex contact and relationship networks, fostering change at all levels of the organization through collective leadership. Only by reaching these deeper levels of consciousness can one arrive at the process of dematerialization of the self, eliminating narcissism and promoting more inclusive forms that recognize the potential and expertise of others, allowing the influence processes that occur in organizational networks to happen naturally. The principles of collective leadership recognize diversity of competencies, the need for dynamic distribution, and exchange of information as a result of a set of interactions that requires a broader awareness of teamwork, and as fundamental pillars in this construction.

This paper has immediate practical implications, the need to consolidate a leadership model that is sufficiently parsimonious, that consists of more than a typology of competencies, and that can address to the systemic complexity of global phenomena. It would be useful to establish the foundations for leadership programs, which include self-leadership, team leadership, and collective leadership, as theoretical orientations that enable the creation of mental schemes, which are vital to the stages of leadership development. It should also be emphasized that implementing collective leadership requires institutions to be aware of the need for networking and the notion of sharing, as well as recognize individual competencies.

Thus, we strongly question the paradigm of leadership as an individual phenomenon and propose that the construction of leadership identity is a social process, as well as a way to meet the increasing need to adapt to a network-centric environment dominated by volatility, uncertainty, complexity, and ambiguity, given its nature and the requirements of communication and media scenarios; we point out the reductionism of the current studies on leadership, which focus only on the figure of the formal leader, neglecting the development of leadership as an inherently multilevel and longitudinal social process; we emphasize the participatory phenomenon of sustainable leadership, arguing that its goal is to aggregate values and virtues and to transform society into a democratic environment, as opposed to simply a group of people who have no commitment to the truth, and criticize the lack of professionalism and transparency that make a future in the shadow of authoritarianism even more uncertain.

This study can contribute to leadership research, not only to our understanding of the process of collective leadership, but also by transforming organizations and individuals and implementing leadership development programs. This investigation is a strategic requirement if we wish to develop a generation of leaders who effectively care for a future for all, in the sense that it is worth participating in their project.

Sustainable, collective, shared and participatory leadership is essentially democratic, and not only discursive, and cannot be otherwise. Its gregariousness, based on shared power, fosters the subject's identity and identification with the organization, and is a natural antidote to abuse and misuse of authority by generating trust and, of course, by inducing transparency.

## References


Aiyar, Y. (2010). Invited spaces invited participation: Effects of greater participation on accountability in service delivery. *India Review, 9*(2), 204–229.

Antonakis J., Fenley, M., & Liechti, S. (2011). Teaching leadership—can charisma be taught? Tests of two interventions. *Academy of Management Learning & Education, 10*(3), 374–396.

Ayres, R.U. (2008). Sustainability economics: Where do we stand? *Ecological Economics, 67*(2), 281–310.

Boin, A., Ansell, C., Lodge, M., Parrado, S., Wegrich, K., & Xue, L. (2011). A time for public administration. *Public Administration, 89*(2), 221–225.

Borges, J.V., & Rodrigues, T. R. (Eds.). (2016). Ameaças e Riscos Transnacionais no Novo Mundo Global, Fronteira do Caos. ISBN: 9789898647603.

Capra, F. (1884). *The Turning Point: Science, Society, and the Rising Culture*. Bantam Books.

Carpentier, N. (2012). O conceito de participação. Se eles têm acesso e interagem, eles participam de fato? *Revista Fronteiras – estudos midiáticos, 14*(2), 164–177.

Carson, J. B., & Tesluk, P. E. (2007). Leadership from within: A look at leadership roles in teams. Paper presented at the 67th Annual Meeting of the Academy of Management, Philadelphia.

Carter, D. R., DeChurch, L. A., Braun, M. T., & Contractor, N. S. (2015). Social network approaches to leadership: An integrative conceptual review. *Journal of Applied Psychology, 100*(3), 597–622.

Carvalho, V., Borges, L., & Rêgo, D. (2010) Interacionismo simbólico: Origens, pressupostos e contribuições aos estudos em Psicologia Social. *Psicologia: ciência e profissão, 30*(1), 146–161.

Chrobot-Mason, D., Gerbasi, A., & Cullen-Lester, K.L. (2016). Predicting leadership relationships: The importance of collective identity. *The Leadership Quarterly, 27*, 298–311.

Contractor, N., DeChurch, L., Carson, J., Carter, D., & Keegan, B. (2012). The topology of collective leadership. *The Leadership Quarterly, 23*, 994–1011.

Daft, R. (2014). *The Leadership Experience* (6th Ed.). Mason, OH: Cengage Learning.

Day, D.V. (2000). Leadership development: A review in context. *The Leadership Quarterly, 11*(4), 581–613.

Day, D.V., Gronn, P., & Salas, E. (2004). Leadership capacity in teams. *The Leadership Quarterly, 15*(6), 857–880.

Day, D. V., & Harrison, M. M. (2007). A multilevel identity-based approach to leadership development. *Human Resource Management Review, 17*, 360–373.

Day, D.V., Harrison, M. M., & Halpin, S. M. (2005). Building an integrative, adult lifespan theory of Army leader development. Paper presented at the 20th Annual Conference of the Society for Industrial and Organizational Psychology, Inc., Los Angeles, CA.

De Cremer, D., & van Knippenberg, D. (2002). How do leaders promote cooperation? The effects of charisma and procedural fairness. *Journal of Applied Psychology, 87*, 858–866.

De Rue, D. S., & Ashford, S. J. (2010). Who will lead and who will follow? A social process of leadership identity construction in organizations. *Academy of Management Review, 35*, 627–647.

De Rue, D. S., Nahrgang, J. D., & Ashford, S. J. (2015). Interpersonal perceptions and the emergence of leadership structures in groups: A network perspective. *Organization Science, 26*, 1192–1209.

Ensley, M. D., Pearson, A., & Pearce, C. L. (2013). Top management team process, shared leadership, and new venture performance: A theoretical model and research agenda. *Human Resource Management Review, 13*(2), 329–346.

Eriksson, K. (2012). Self-service society: Participative politics and new forms of governance. *Public Administration, 90*(3), 685–698.

Fisher, C., & To, M. (2012). Using experience sampling methodology in organizational behavior. *Journal of Organizational Behavior, 33*(7), 865–877.

Gronn, P. (2002). Distributed leadership. In K. Leithwood, P. Hallinger, K. Seashore-Louis, G. Furman-Brown, P. Gronn, W. Mulford & K. Riley (Eds.) *Second International Handbook of Educational Leadership and Administration* (pp. 653–696). Dordrecht, the Netherlands: Kluwer.

Hannah, S. T., Jennings, P., & Nobel, O. B. (2010). Tactical military leader requisite complexity: Toward a referent structure. *Military Psychology, 22*, 412–449.

Hasna, A.M. (2010). Sustainability classifications in engineering: Discipline and approach. *International Journal of Sustainable Engineering, 3*(4), 258–276.

House, R. J., Hanges, P. J., Ruiz-Quintanilla, S. A., Dorfman, P. W., Falkus, S. A., & Ashkanasy, N. M. (1999). Cultural influences on leadership and organizations: Project GLOBE. In W. H. Mobley, M. J. Gessner, & V. Arnold (Eds.), *Advances in Global Leadership* (pp. 171–233). Bingley, UK: Emerald Group Publishing Ltd.

Hove, H. (2004). Critiquing sustainable development: A meaningful way of mediating the development impasse? *Undercurrent, 1*, 1.

Ibarra, H., Wittman, S., Petriglieri, G., & Day, D. (2014). Leadership and identity: An examination of three theories and new research directions. In D. V. Day (Ed.), *The Oxford Handbook of Leadership and Organizations* (pp. 285–300). USA: Oxford University Press.

Kalish, Y. (2013). Harnessing the power of social network analysis to explain organizational phenomena. In J. M. Cortina, & R. S. Landis (Eds.), *Modern Research Methods for the Study of Behavior in Organizations* (pp. 99–136). NY: Routledge.

Karkin, N., & Calhan, H. (2012). An interactive e-participation model for the public administration system in Turkey: SIBIYO. *Ege Academic Review, 12*(1), 107–125.

Kegan, R. (1994). *In Over our Heads: The Mental Demands of Modern Life*. Cambridge, MA: Harvard University.

Kloby, K. (2009). Less is more: Exploring citizen-based financial reporting in local government. *Journal of Public Budgeting, Accounting & Financial Management, 21*(3), 368–392.

Kluvers, R., & Pillay, S. (2009). Participation in the budgetary process in local government. *Australian Journal of Public Administration, 68*(2), 220–230.

Lawton, A., & Macaulay, M. (2014). Localism in practice: Investigating citizen participation and good governance in local government standards of conduct. *Public Administration Review, 74*(1), 75–83.

Mattia, C. D., & Zappellini, M. (2014). Ética e co-produção de serviços públicos: uma fundamentação a partir de Habermas/Ethics and coproduction of public services: A rationale from Habermas. *Cadernos EBAPE BR, 12*(3), 573–589.

McGarty, C. (1999). *Categorization in Social Psychology*. London: SAGE.

Moldan, B., Janouaková, S., & Hák, T. (2012). How to understand and measure environmental sustainability: Indicators and targets. *Ecological Indicators, 17*, 4–13.

Mori, K., & Christodoulou, A. (2012). Review of sustainability indices and indicators: Towards a new City Sustainability Index (CSI). *Environmental Impact Assessment Review, 32*(1), 94–106.

Murphy, S. A., & Johnson, S.K. (2011). The benefits of a long-lens approach to leader development: Understanding the seeds of leadership. *Leadership Quarterly, 22*, 459–470.

Musso, J., Weare, C., Bryer, T., & Coope, T., (2011). Toward strong democracy in global cities? Social capital building, theory driven reform, and the Los Angeles Neighborhood Council experience. *Public Administration Review, 71*(1), 102–111.

Neshkova, M. (2014). Does agency autonomy foster public participation? *Public Administration Review, 74*(1), 64–74.

Nirenberg, O. (2006). *Participación de adolescentes em proyectos sociales: aportes conceptuales y pautas para su evaluatión*. Buenos Aires: Paidós.

Osborne, S., & Strokosch, K. (2013). It takes two to tango? Understanding the coproduction of public services by integrating the services management and public administration perspectives. *British Journal of Management, 24*(1), 31–47.
Pearce, C. L. (2004). The future of leadership: Combining vertical and shared leadership to transform knowledge work. *Academy of Management Executive, 18*, 47–57.
Piaget, J. (1976). *A equilibração das estruturas cognitivas.* Rio de Janeiro: Zahar.
Reicher, S. D., Spears, R., & Postmes, T. (1995). A social identity model of individuation phenomena. *European Review of Social Psychology, 6*(1), 161–198.
Reicher, S. D., Haslam, A., & Hopkins, N. (2015). Social identity and the dynamics of leadership: Leaders and followers as collaborative agents in the transformation of social reality. *Leadership Quarterly, 16*, 547–568.
Rosinha, A. P., Matias, L. J. S., & de Souza, M. A. (2017). Leadership in extreme conditions and under severe stress: Case study analysis. In M. Holenweger, M. K. Jager & F. Kernic (Eds.), *Leadership in Extreme Situations* (pp. 93–112). Cham, Switzerland: Springer International Publishing.
Simmons, R., & Birchall, J. (2005). A joined-up approach to user participation in public services: Strengthening the participation chain. *Social Policy & Administration, 39*(3), 260–283.
Sirvent, M. (2004). *Cultura popular y participación social: uma investigación en el barrio de Mataderos (Buenos Aires).* Buenos Aires: Miño y Dávila.
Struecker, D., & Hoffmann, M. (2017). Participação social nos serviços públicos: Caracterização do estado da arte por meio da bibliometria e da revisão sistemática. *REGE - Revista de Gestão, 24*, 371–380.
Tajfel, H., & Turner, J. C. (1986). The social identity theory of intergroup behaviour. In S. Worchel & L. W. Austin (Eds.), *Psychology of Intergroup Relations* (pp. 7–24). Chicago: Nelson-Hall.
Todorov, V., & Marinova, D. (2011). Modelling sustainability. *Mathematics and Computers in Simulation, 1*(7), 1397–1408.
Turner, J. C. (1991). *Social Influence.* Milton Keynes: Open University Press.
Turner, J. C., Hogg, M. A., Oakes, P. J., Reicher, S. D., & Wetherell, M. S. (1987). *Rediscovering the Social Group: A Self-Categorisation Theory.* Oxford, England: Blackwell.
Van Knippenberg, D., Van Knippenberg, B., De Cremer, D., & Hogg, M. A. (2004). Leadership, self, and identity: A review and research agenda. *The Leadership Quarterly, 15*(6), 825–856.
Viana, G., Silva, M., & Diniz, N. (2001). *O desafio da sustentabilidade.* São Paulo: Fundação Perseu Abramo.
Vieira, B. (2002). *Liderança militar.* Lisboa: Editora Atena, Academia Militar.
Wacquant, L. (2013). Poder simbólico e fabricação de grupos: como Bourdieu reformula a questão das classes. *Novos estudos CEBRAP, 96*, 87–103.
Yang, K., & Pandey, S. (2011). Further dissecting the black box of citizen participation: When does citizen involvement lead to good outcomes? *Public Administration Review, 71*(6), 880–892.
Yukl, G. (1999). An evaluative essay on current conceptions of effective leadership. *European Journal of Work and Organizational Psychology, 8*, 33–48.
Yukl, G. A. (2002). *Leadership in Organizations, Prentice Hall* (5th Ed.).

# 13

# THE PHYSICAL BRAVERY STUDY

## Findings and implications for small, innovative research studies with military/veteran populations

*Kristen J. Vescera, Jacie Brown, Catherine Hausman, and Bruce Bongar*

Increasingly, researchers are attempting to gain new understandings on serious problems such as suicide and self-destructive behaviors and trauma in military, veteran, and civilian populations. A promising area of research focuses not just on traditional risk factors but also on strengths and resilience. These concepts shift the conversation in line with the positive psychology movement, which asserts that focusing on personal strengths may have the potential to identify and increase awareness around values and qualities in the interest of a person and serve as protective factors (Peterson & Seligman, 2004). This chapter focuses on the concept of physical bravery as a potential strength that could reveal information about resilience, born and learned traits, and altruism. Studying physical bravery could lead to addressing problems of stigma around suicide and trauma not only with military and veteran populations but could be generalize to other populations and the civilian population in the United States as a whole.

Concerns about the rising rates of suicide completion are prevalent in the United States despite growing efforts in research and prevention (Centers for Disease Control [CDC], 2016). In 2016, the CDC found a 24% rise in suicide rates from 1999 to 2014, and rates are currently increasing. In this direction, such risks in the military and veteran population are increasing at an even more alarming rate. Current military and veteran populations make up about 1% of the overall population in the United States, with about 20 veterans completing suicide per day and the rate rising higher than overall national numbers (Department of Veterans Affairs, 2016). In 2016, the suicide rate for veterans was found to be 1.5 times greater than non-veterans after age and gender adjustments were made (Department of Veterans Affairs, 2018). Additionally, suicide deaths with veterans between the ages of 18 and 34 increased from 40.4 suicide completions per 100,000 in 2015 to 45 per 100,000 in 2016 (Department of Veterans Affairs, 2018).

While the VA is working toward gathering information on the number of veteran suicides, neither active nor reserve components have a streamlined suicide tracking system that captures concise data between all branches of currently serving service members. However, based on quarterly Department of Defense (DoD) reports, there is still a significant amount of suicide completion within each active military branch. Overall, some of the most current known data shows that active component suicide rates alone increased from 254 to 273 service members from 2013 to 2014 (Franklin, 2015), and data from the United States Army have found that active-duty service members have a predominate high risk for suicide completion (Greene-Shortridge, Britt, & Castro, 2007; Rüsch, Corrigan, Todd, & Bodenhausen, 2010).

This elevated suicide risk in military and veterans could be potentially due to the adverse effects of trauma, including posttraumatic stress disorder (PTSD). Hoge, Auchterloine, and Milliken (2006) found that Operation Iraqi Freedom (OIF) veterans were diagnosed with PTSD after deployment at a rate between 12% and 20%. Further research by Ramchand and colleagues (2010) found that close to half of treatment-seeking veterans who served in the current Iraq and Afghanistan conflicts have a PTSD diagnosis. Other estimates suggest that up to half of treatment-seeking veterans of the Iraq and Afghanistan conflicts are diagnosed with PTSD (Ramchand et al., 2010). Suicide risk may also be influenced by veteran fear of stigma as a barrier to treatment-seeking behavior (Hoge et al., 2004). Thus, more work is needed to understand problems with suicide and trauma-based disorders in military and veteran populations.

Currently the Department of Veteran Affairs (VA) is using evidence-based protocols (EBPs) that include treatment interventions like prolonged exposure (PE) and cognitive processing therapy (CPT), as well as a nationwide suicide prevention protocol. PE is a treatment intervention that assists in teaching veterans and others who have suffered trauma to gradually face previously avoided traumatic memories, emotions, and situations (Foa, Hembree, & Rothbaum, 2007). CPT is a more directed form of cognitive behavioral therapy (CBT) in which it is used to specifically treat PTSD or other trauma-related disorders (Resick, Monson, & Chard, 2017). CPT is used to challenge the maladaptive beliefs that a person can develop as a result of trauma (Resick, Monson, & Chard, 2017). Essentially, the administration of CPT and PE focuses on reducing symptoms that veterans are currently facing. While these treatments are effective for the majority of patients, many with PTSD will not benefit from these treatments, either due to a lack of access (this could be due to not seeking treatment due to stigma or difficulty navigating the VA system) or a lack of response to the treatments (i.e., symptoms are not reduced).

Research on acts of physical bravery has the potential to reveal information regarding biological factors, military training, resilience, and other protective factors. Physical bravery is a construct of interest for several key reasons. First, by examining physical bravery as a strength, researchers have the potential to contribute to the literature on suicide prevention efforts while avoiding the negative stigma that can be a barrier for working with both military and veteran populations (Hoge et al., 2004). Next, physical bravery could be particularly relevant to the concept of military culture as the PBS was designed around the military and veteran populations' potential interest on this topic. Because the military uses training that makes service members prone to overcome fear and help others, such as remaining calm in the face of distress (Bryan, Jennings, Jobes, & Bradley, 2012), there seem to be cultural underpinnings of later mental health challenges upon reintegration to civilian life. The ability to suppress emotions and avoid certain experiences during training and deployment may put the veteran at risk for suicidal behaviors and other trauma-related mental health issues (Bryan et al., 2012). This understanding could help provide cultural competence in military populations that can assist both clinically and in research. Last, understanding innate and learned traits of physical bravery can inform the military's selection methods of future service members as well as ongoing training to create a more mentally fit military force. Kugel (2014) suggested that those who receive sufficient training may be more likely to act in a brave manner.

## Construct definitions

Variability has been found in literature between related constructs to bravery such as "courage" and "heroism." Noting that there are differences in definitions, meaning, and measurement may help researchers avoid confusion and establish clear definitions and distinguishing characteristics to these constructs.

### *Courage*

Rate, Clarke, Lindsay, and Sternberg (2007) encountered 29 different definitions of courage and were ultimately unable to delineate a clear operational definition of courage as a construct. They ultimately identified three factors that define courage to include "self-sacrifice/risk for others," "non-physical/social-oriented acts for noble ends," and "self-focused perseverance despite fear" (p. 89). Simply stated, Asarian (1981) theoretically defined courage as an intentional and laborious behavior in the face of conflict and fear. His writing explored courage through a lens of holding onto one's values in the face of struggle. Asarian's (1981) work, however, sometimes used the word "hero" while describing courageous people, and while some believe that heroism and courage are interchangeable, others infer that courage and heroism are related but different and should be distinguished appropriately (Weinstein, 2012). Alternatively, Kugel (2014) simply described courage as a persistence in action despite fear or threat to life. This definition alone does not necessitate that the courageous act is done to aid another person as it had above. For the present chapter, however, Kugel's definition is used in order to parsimoniously understand the concept of physical bravery.

### *Heroism*

Pury and Lopez (2010) stated that heroism is one of the highest levels of courage as heroism requires action despite risk and without understandable personal gain (Franco, Blau, & Zimbardo, 2011). As a community, we normally do not think of the average person as a "hero." An act of heroism is defined as putting someone else's safety and well-being above one's own (Black, 2015). This action to help separates the definition of courage from heroism in this chapter. The major distinction between the construct of heroism and physical bravery as it is described in this chapter is that heroism denotes an involvement of a conscious thought process, whereas individuals who act physically brave in a specific event may or may not take conscious efforts to risk their lives for another. Further, heroism is an overall higher level and complex tendency than acting physically brave during a specific event (Franco et al., 2011; Weinstein, 2012). Therefore, heroism can involve large amounts of courageous and brave acts (Hummel, 2018).

For example, in 2018, Virgil Smith was recognized by the White House for receiving the "Young Hero Award" (Ordinary Americans Are, 2018). This award was granted due to his brave actions in saving 17 of his neighbors stranded after Hurricane Katrina with only the use of his air mattress. If this was a conscious effort, this may be an example of heroism, but there are things that happen every day that can warrant an act of bravery, even from the most "typical" civilian. The previous year (2017), Katelyn Ibarra was granted the "Young Hero Award" for performing first-aid to car accident victims before first responders arrived (Ordinary Americans Are, 2017). Although these acts are outstanding examples of young people going above and beyond, their behavior could be seen as reckless and not necessarily conscious. Aristotle theorized that bravery may be the gear that connects fear and recklessness (Pury & Lopez, 2010).

### *The hero complex*

Some researchers have reported thinking that instead of this reckless type of act being that of "selfless bravery," it is out of that individual's "hero complex." In other words, the negative definition of recklessness conceals their genuine intentions (Pury & Lopez, 2010). Others have said that a brave act puts the individual in the spotlight to be regarded as important and selfless instead of the act actually being selfless. When we think of service members, we normally think of the hero whose job it is to act and actually be brave. As previously mentioned, service members are not the only ones who perform acts of physical bravery. Even if it is done for selfish reasons and not as a selfless act, it could still be considered an act of bravery. The only person who would know the motives behind the act is the person performing the act or anyone they decide to tell. What we can objectively see is the act itself. Readers may find themselves reflecting on whether the motives behind the act of bravery matter.

### *Altruistic suicide*

In respect to the topic of heroism, altruistic suicide may be considered heroism despite the risk to one's life. In Weinstein's (2012) study, participants reported that they experienced fear during heroic events. Altruistic suicide can be described as intentionally pursuing an act that results in the end of life for purposes of helping other individuals, groups, or society as a whole (Stack, 2004). There is an apparent link in the ideas that these individuals could be acting to save someone else despite consciously understanding that they could die and the intention behind suicide in which a person consciously understands that they may die (Bongar & Sullivan, 2013). In 1897, Emile Durkheim was the first to define altruistic suicide and described that altruistic and heroic acts as well as terrorist acts are under the umbrella of altruistic suicide. Differentiation may be necessary when looking at either act's value upon society (lauded or condemned, respectively; Leenaars, 2004). This would understandably make this concept vague and potentially controversial (Leenaars, 2004).

## ***Physical bravery***

When a layperson thinks of bravery, he or she may think of fictional superheroes (e.g., Superman and Wonder Woman) that have been portrayed in comic books and action movies. Others may consider those in the military forms of tangible heroes that are glorified by the media. In actuality, the United States Armed Forces provide the country with the protection it requires to function as a free nation. As of November 2018, there are 1,281,900 active-duty service members in the United States military (Defense Manpower Data Center, 2018). However, no two service members normally share the same military career or life history. They also have had differing reasons that have factored into their decision of enlisting into the military (Hummel, 2018). Most, if not all, of these men and women may have been put into a situation where they have had to make the choice of putting themselves in harm's way in order to protect someone else, whether that is another service member or a civilian.

Weinstein (2012) described heroism, courage, and bravery through differing action levels. In this case, heroism is the most difficult accomplishment and is glorified like the superheroes noted above, while courage and bravery follow respectively. Conceptualizing through this lens, bravery can be viewed as incorporating less altruism and risk-taking than heroism. There are differences in types of bravery, however: physical and moral (Hummel, 2018). Hummel (2018) stated that physical bravery includes objective experience or risk of physical harm to the person conducting a brave act, while moral bravery involves other objective risks like job loss, imprisonment, or other consequences.

Because of the overabundance of definitions for courage, heroism, and bravery already discussed, Hummel's (2018) definition for physical bravery will be used as it relates most specifically to the Physical Bravery Survey (PBS) discussed below. As with all constructs with different definitions, the caveat to this definition may be inconsistent with descriptions in other literature. Therefore, the definition used throughout this chapter distinguishes physical bravery as an action in which one saves the life of another while freely risking his or her life (Hummel, 2018). Physical bravery in this chapter is regarded under specific situations that individuals have experienced.

## Physical bravery as an innate trait

Researchers have not definitively found whether physical bravery is a learned or innate (genetic) trait. According to current research, there does not seem to be a scientific test to identify the presence of this specific trait within an individual. For example, if a family member has shown to actively engage in physical bravery, it may not automatically indicate that their children will also perform a brave act. However, it has been shown throughout research that certain characteristics can be heredity. An example of this is alcoholism, as alcoholism itself cannot be passed down through generations, but the addictive gene has been shown to be passed on to an alcoholic's children (Pombo, Ferreira, Bicho, & Levy, 2017). This does not guarantee that the child will become addicted to anything related to alcohol or any other substance, just that they are at greater risk for developing a dependency.

## Training and physical bravery as a learned behavior

Military service members are well trained and have unique abilities like keeping calm in the face of pain that others may not (Bryan, et al. 2012). The military partakes in extensive and rigorous training in order to be able to perform at their best if they were to ever encounter combat situations. For example, the strenuous training military personnel go through prepares them for high-stress, intense situations are unmatched in many other civil servant positions. Also, many authorities see the United States military in ways that demonstrate a high level of duty and patriotism. This training may make service members especially likely to be able to overcome fear and help others who are in peril. As Kugel (2014) pointed out, those who receive ample training may be more likely to act courageously. Because military receive special training, they may be a population especially likely to act bravely. Thus, examining physical bravery in this population is especially timely and relevant.

However, elements of military training are not just relevant for combat scenarios. For example, in 2009, Chelsey Sullenberger (Sully) landed his airplane safely in the Hudson River after both engines were disabled. Thanks to Sully, all 155 souls on board survived the "crash" (Sullenberger, 2018). It should be considered whether Sully's act of bravery was due either to his Air Force training or another factor. Many civilians have a sense of duty and patriotism but never receive the training to prepare them for dangerous situations. At most, if civilians work for a large corporation, they are provided with training that attempts to prepare them for a "active shooter" scenario. However, civilians have shown that they too can perform acts of physical bravery without the rigorous training of the military. Filjak and Pavlina (2000) posit that bravery can be instilled in military personnel through training, motivation, and leadership and can occur under certain specific circumstances. This chapter supports the concept of bravery as a learned characteristic over innate, which is consistent with the stance of Kugel (2014) yet contrasts with findings in studies from both Hummel (2018) and Weinstein (2012) that training was not necessarily related to reported experiences of acting bravely.

In particular, Weinstein (2012) found a higher frequency of having done heroic acts reported by undergraduates than for military cadets. In this study, training did not appear related to having committed heroic acts (Weinstein, 2012). In studies by both Hummel (2018) and Kugel (2014) on active-duty military and veterans, participants were more likely to state the personal belief that physical courage is learned and not innate. For Hummel (2018), participants appeared to believe that training was influential, whereas the actual experience of training was not significantly associated with likelihood to have acted bravely. In contrast, in the study by Kugel (2014), participants who reported more than two acts of physical courage also indicated greater influence of instinct, level of training, and sense of duty in the likelihood to commit a heroic act than those who did fewer courageous acts.

It is possible that those who are in the military for longer, given that they are continuing their training and have a strong sense of duty, are likely to rise to leadership positions given their exemplification of courage. Also, considering the inconsistent findings regarding training, it could be that training is necessary to have the capability to act bravely in certain situations (i.e., for military personnel), yet it may be the person's innate characteristics (i.e., personality factors) that determine whether the training will be implemented in a real-world, real-time situation. Thus, it is entirely possible that *both* innate and learned aspects are involved with physical bravery. At this time, the relationship between training and the likelihood to act bravely remains unclear and warrants further study. The relationship between training, leadership, and physical bravery needs to be examined, particularly to discern any inconsistencies between peoples' beliefs and actuality.

There may be a relationship between a person's tendency to be resilient and their likelihood of having more training or of being in a leadership role. Freedman (2004) found that training and comradery were related to resilience in 9/11 first responders. These first responders were able to act efficiently and in accordance with their training without experiencing psychological numbing. A sense of comradery was noted as key in these moments. Freedman (2004) further commented that these individuals demonstrated significant resilience in the face of an extreme situation.

## Physical bravery and resilience

The Merriam-Webster Collegiate Dictionary defines resilience as the ability to overcome hardship when faced with extreme adversity (Mish, 2012). Hardship in this instance most often refers to a past traumatic event. The development of resiliency within individuals and how resiliency can serve as a protective factor against some mental disorders has gained attention within the academic community following publicized mass public shootings and large-scale school shootings. Following a traumatic event, an individual attempts to make sense of what the experience was and how it has affected them as a person. During this time, the individual who has experienced trauma is also trying to effectively provide themselves with barriers against the adverse effects of the traumatic experience. Some similar personality characteristics are found in people who have demonstrated resiliency (Hummel, 2018). For example, the level of social support can greatly influence a person's attainable level of resiliency (Hummel, 2018).

Communities can experience trauma in a similar matter to that of an individual (i.e., a sexual assault victim). The military in itself is a tight-knit community, and communities have been shown to have their own way of building resiliency. One of the ways is to confide in and support one another through such traumas while alienating the people around them who "do not belong" to that particular community (Day, Lawson, & Burge, 2017). A more well-known phenomenon related to this topic is the concepts of "thin blue line" or "blue curtain of silence" related to law enforcement. Though this behavior is portrayed more by military and paramilitary groups, this type of behavior is not mutually exclusive to those communities that are faced with daily threats of trauma. When the Virginia Tech shooting occurred, students who sought campus counseling instead of reaching out for mental help outside of the community showed a larger decrease in symptoms (Day et al., 2017). It was hypothesized that those mental health providers who also went through the same trauma were better providers for those students (Day et al., 2017). A research study has shown that those in law enforcement and the military who reach out for support from their comrades and receive it are more likely to develop resiliency and a healthier overall emotional state. There is currently research being designed to gain more evidence of community resilience (Bonanno, Romero, & Klein, 2015). More research is needed in the interest on this population (and similar populations; first responders, police officers, firefighters, etc.) that has historically exhibited distrust and resistance to seeking mental health.

In a study conducted by Kimhi, Eshel, Leykin, & Lahad (2017), researchers measured the resiliency of Jewish Israeli nationals who experienced terror attacks. Results from this study showed that a sense of coherence was one of the best predictors of community resiliency (Kimhi et al., 2017). The military in itself has their own code of honor and at some points their own language. Banding together in times of crisis could change the course of their mental health. There is also a stigma associated with seeking mental health professionals within this tight-knit community. Unfortunately, in more severe cases of depression and PTSD, seeking help from fellow comrades is sometimes not enough to support them long term.

Physical bravery has been hypothesized to act as strength against individual suicidal ideation and mental health disorders. Some researchers believe the greater the number of physical bravery acts a person has committed, the less likely they are to attempt or complete suicide. However, the current research does not currently strongly support this hypothesis. The research is not available, yet if it is not being studied or being considered as a topic of study, it will be in the future. In addition, physical bravery itself is still in the infancy of research development. As a science, physical bravery needs to have a more concrete foundation before then branching out to include other areas of interest such as physical bravery and how it can impact suicidal ideation. Implications for measurement of physical bravery should be identified and discussed more fully to continue professional discussion in the field.

## Measuring physical bravery

Physical bravery within military, police, and first responder populations has historically been difficult to measure due to the closed social culture that these populations normally exhibit. In 2012, in lieu of this gap in research, the Clinical Crises and Emergencies research group at Palo Alto University created the PBS as a tool to evaluate physical bravery within these populations as well the civilian populations as a partial control (Hummel, 2018). A comprehensive literature review and consultations with authorities in the field was conducted to maximize face and content validity. In Hummel's (2018) study, the PBS consisted of qualitative- and quantitative-related questions from participants who engaged in physically brave acts and consisted of 47 items (includes a range of items, namely Likert-scale, multiple choice option, and open-ended response-types). The survey takes about 20–25 minutes for full completion.

The PBS aims to understand the different demographics that can impact an individual's decision to perform an act of physical bravery. Also, it focuses on the endorsement of different characteristics of bravery (i.e., training, heritability, etc.) and how these demographics and characteristics could influence one another in performing an act of physical bravery. Samples of the six-point Likert-type (from 1 - strongly disagree to 6 - strongly agree) questions asked within the PBS are: "during my childhood, I had role models (parents or others) who showed brave behavior," "do you believe that people who risk their own physical safety to save others are born that way or life makes them that way?", "In the situation where I was the most physically at risk and I was attempting to save another human being/s from harm or possible death; I believe that my behavior was due to: Instincts, Training, Experience, Duty, Fear of repercussions, Protect or save someone I deeply cared for; Concern for another human being; Peer pressure or a group momentum towards action." The latter questions asked for participants to individually rate each factor (Hummel, 2018).

Some results of the PBS have shown that there are no significant differences in the number of acts of physical bravery when comparing active-duty military to a civilian population (Hummel, 2018). These results may seem surprising due to military personnel being perceived to perform more brave acts than the average person.

Everyday people like Virgil Smith, who have no affiliation or training with the military, perform acts of physical bravery. Researchers found that there are many different individual factors that can influence a person's decision on whether to commit an act of physical bravery. For example, based on the PBS administered to veterans and active-duty service members, higher levels of adventurousness, humor, and cognitive flexibility have been found in people who reportedly exhibit acts of physical bravery (Hummel, 2018). In another study conducted by Rachman (1990), results showed that self-confidence was a characteristic that was high in individuals who committed acts of physical bravery, although Hummel (2018) found in her analysis that self-confidence was not a factor. The major difference in these studies was the population studied. Hummel (2018) hypothesized that self-confidence was not a significant contributing factor in military personnel who committed physical bravery acts due to the similar personalities that military and paramilitary personnel tend to exhibit.

## Challenges in research methodology with military populations

There is a dearth of information regarding research methodology on certain topics for those studying military populations. Aspects such as how to go about building relationships for access to participants, developing military culture-sensitive survey items, and other considerations are critical to consider for this line of research. Further, these issues may be particularly challenging to navigate for smaller (i.e., outside of major VA- or DoD-funded research) researchers to navigate. Yet, it is these same types of researchers, with fresh perspectives, that may provide crucial innovation for the field. The present section aims to lay out the key considerations for new research groups seeking to develop research projects with military and veteran populations. External factors, like collecting data within the military structure, may be just as salient as internal factors such as service member mental health stigma.

### *External factors*

There are multiple challenges in initiating the collection of data for research. The National Research Council (2014) found that the plethora of demographic survey data collected under the DoD is not maintained to be conducive for outside research programs. This issue with availability of DoD data for outside researchers may be due to several problems. First, that demographic information collected at the start of a service member's career is not extensive. Second, there is a lack of data preservation to include a lack of a central storage system in which various amounts of demographic are not consolidated in a clear and concise manner. Last, there are unclear procedures for ease of use for data requests from outside agencies (National Research Council, 2014). Further, this Council suggests that a gold standard for research with this population is longitudinal research that follows individuals throughout their military careers.

In 2011, the DoD Health Related Behavior Survey was conducted as the largest data collection at the time for all five branches of the military; the Army, Navy, Marine Corps, Air Force, and Coast Guard (Barlas, Higgins, Pflieger, & Diecker, 2013). A web-based format was used for the first time with this population and showed several benefits, including reduced research costs and a broadened range of respondents across the United States beyond specific military bases. This was also found to decrease the burden on the unit leadership of these locations. A shortened and condensed questionnaire along with considerations for sample design and incorporation of assessment for health issues also influenced an overall DoD and Coast Guard response rate of 22% and 37%, respectively (n = 39,877; Barlas, Higgins, Pflieger, & Diecker, 2013).

These surveys above have proven successful when thinking about largely funded projects. For independent researchers, however, it may be more difficult to understand the "red tape" to navigate the data collection process. Past researchers have shown to be creative when targeting military and veteran populations. For example, Warner, Appenzeller, Mullen, Warner, and Grieger (2008)'s methodology for collecting data from deploying Army soldiers involved obtaining approval from a unit's Brigade and Division commanders as well as an approved institutional review board prior to participation in the survey. Alternatively, Bryan et al., (2015) and Currier, Holland, and Malott (2015) recruited students across colleges in the United States that had a current or prior military service status through the use of Student Veteran Coordinators. These examples highlight some factors to consider when designing an independent research study that does not involve national sampling as noted above.

### *Internal factors*

Greene-Shortridge, et al. (2007) found evidence that the general public generally hold negative stereotypes toward those with psychological problems, which then leads to discriminations and internalization of these negative beliefs by the individual. Further, they found that because military members feel as though their unit holds them accountable for their psychological problems (like the rest of the population), treatment seeking is largely inhibited both in public and governmental domains due to stigma. Taken further, the phenomenon "healthy soldier effect" (p. 99) was coined in earlier wars to describe the difference in suicide rates between veterans and the overall civilian population (Kang et al., 2015).

Before recent wars in Iraq and Afghanistan, this concept was based on the assumption that recently separated active-duty military members were healthier than the general public based on initial entry requirements and access to care during and after military service. The increase in recent numbers is inconsistent with this effect (Kang et al., 2015). In consideration of physical bravery as a strength, contributions to literature regarding suicide completion and efforts to avoid stigma may be positively impacted by this construct.

## Implications

As noted above, within military populations, the strain of combat and other potentially traumatic situations tends to be quite prevalent. Physical bravery has just started being investigated with respect to potential implications regarding the mental health and overall mindset of our military, as well as civilians who have engaged in acts of selfless bravery. Higher levels of resiliency have been present in military personnel that were measured as more likely to engage in physical bravery (Hummel, 2018). In similar terms, it has been shown that the more physical bravery acts a person has committed, the higher their resiliency. This could mean resiliency acts as a protective factor in not developing severe mental health concerns, including but not limited to major depressive disorder (MDD) and PTSD. Both diagnoses on average tend to carry an elevated mortality rate.

As mental health professionals, we do not want to imply that people who have been through trauma do not need to see licensed mental health professionals. PTSD treatment alone costs the United States $2029 per person, much of which goes to the treatment of service members (Painter, Fortney, Austen, & Pyne, 2017). Physical bravery has the potential to act as a protective factor against severe suicidal ideation, and possibly help military men and women feel more comfortable coming forward about their experiences. This could challenge (1) the stigma of reaching out for help, (2) arguments against the stigma that they are weak to ask for help, and (3) against the stigma that no one can help except for those who have experienced the same type of traumas. Cigrang et al. (2017) studied different treatments that have the potential to decrease PTSD symptom severity while also limiting the stigma surrounded by seeking mental help. Participants were active-duty military and randomized to either receive Prolonged Exposure for Primary Care (PE-PC) as an immediate intervention or were put into a delayed treatment control condition (Cigrang et al., 2017). PE-PC consists of four 30-minute sessions; instead of referring to these sessions as mental health treatment, they are referred to as primary care appointments, therefore decreasing the outward stigma associated with receiving professional mental health (Cigrang et al., 2017). Researchers found using PE-PC resulted in a larger reduction of PTSD symptoms than the delayed control intervention (Cigrang et al., 2017). This treatment, coupled with strength-based approaches, could be helpful to further bridge the gap between stigma and treatment-seeking behavior.

PE-PC appears to be an effective intervention once troops have returned stateside, but there is little evidence on its effectiveness with actively deployed military personnel. In combat environments, there are not always mental health professionals available, nor are there primary care professionals. Research has shown that suicidal ideation peaked at deployment month 2, 6, and 12 (Warner et al., 2011). In 2007, suicide rates for deployed military were 24/100,000 (Warner et al., 2011). That may seem low, but the active-duty military population had the lowest suicide rates until 2004, when they started to grow 4–5 times over a relatively short number of years. In response, the military implemented suicide prevention training throughout initial military training as well as continuing annual training sessions (Warner et al., 2011). These new trainings included developing positive life coping skills, help-seeking behavior, and raising awareness (Warner et al., 2011). Since the implementation, the annual suicide rate has dropped from 24/100,000 to 16/100,000 (Warner et al., 2011). It appears that the suicidal prevention program that focused on strength-based approaches may decrease rates of suicide in a combat environment. It is suggested that promoting training in physical bravery may further assist resilience for these service members.

Over 50% of Operation Enduring Freedom/Operation Iraqi Freedom (OEF/OIF) veterans have described reintegration back into civilian life as a hardship (Pease, Billera, & Gerard, 2016). When discussing military personnel's reintegration back into civilian life, it is hard to know what they have experienced if you yourself have not had the same experience. However, this does not mean as mental health professionals that we give up on helping these individuals. The military culture is one that unintentionally can install a core belief that seeking help makes them look week and/or vulnerable (Pease et al., 2016). Through this belief, they then view themselves as a burden to those around them (Pease, Billera, & Gerard, 2016). This can include their family and friends when they return home. Returning military may have difficulties connecting with their families and friends initially due to culture shock (Pease, Billera, & Gerard, 2016). Mental health professionals' awareness of these potential issues is the first step in maintaining the therapeutic alliance with a veteran seeking treatment (Pease, Billera, & Gerard, 2016). Here, awareness of innate and learned physical bravery characteristics may provide helpful insights to clinicians and service members as they reintegrate in their home environments.

## Conclusion

The PBS survey was initially developed for the measurement and analysis of physical bravery within military populations and civilians. Since physical bravery is a relatively new and under-researched phenomenon, the potential growth for this topic has can be exponential. Similar to studies discussed previously, it would be beneficial to know how service members measure and portray physical bravery while still in combat zones. While this type of research may be difficult to obtain, it would be beneficial to learn if there is a significant difference in perceptions of physical bravery being a learned or innate trait based on the environment where the participant is given the PBS.

Recently, researchers have been steering the research in a new direction to include more than military populations and community college students. These populations include law enforcement, emergency medical services (EMS), firefighters, and other first responders. These particular populations have been chosen as possible future subject groups to further the research due to the increased probability that they will be in the unique position of routinely risking their own lives to protect and save another's (Eyre, 2017). These occupations tend to attract a certain personality type, and there could be similarities correlated between the type of person who commits brave acts and the type of person who goes into these occupational fields. For example, police officers and firefighters tend to exhibit higher scores in excitement-seeking than a normative population (Salters-Pedneault, Ruef, & Orr, 2010). Therefore, these questions that aim to understand physical bravery with the military and veterans as well as other populations are especially warranted for future research.

## References


Asarian, R. D. (1981). *The psychology of courage: A human scientific investigation*. Ph.D. dissertation, Duquesne University, United States – Pennsylvania. Retrieved from Dissertations & Theses: Full Text. (Publication No. AAT 8121943).

Barlas, F. M., Higgins, W. B., Pflieger, J. C., & Dieker, K. (2013). *2011 Health Related Behaviors Survey of Active Duty Military Personnel*. Retrieved from https://apps.dtic.mil/dtic/tr/fulltext/u2/a582287.pdf

Black, L. (2015). *Are heroes born or made? An analysis of heroic behavior based on U.S. military personnel and veterans*. (Unpublished doctoral dissertation). Pacific Graduate School of Psychology, Palo Alto University, Palo Alto, CA.

Bonanno, G. A., Romero, S. A., & Klein, S. I. (2015). The temporal elements of psychological resilience: An integrative framework for the study of individuals, families, and communities. *Psychological Inquiry*, *26*(2), 139–169.

Bongar, B., & Sullivan, G. (2013). *The Suicidal Patient: Clinical and Legal Standards of Care* (3rd ed.). Washington, D.C.: American Psychological Association.

Bryan, C. J., Jennings, K. W., Jobes, D. A., & Bradley, J. C. (2012). Understanding and preventing military suicide. *Archives of Suicide Research*, *16*(2), 95–110.

Bryan, C. J., Roberge, E., Bryan, A. O., Ray-Sannerud, B., Morrow, C. E., & Etienne, N. (2015). Guilt as a mediator of the relationship between depression and posttraumatic stress with suicide ideation in two samples of military personnel and veterans. *International Journal of Cognitive Therapy*, *8*(2), 143–155.

Centers for Disease Control and Prevention (CDC). (2016). *Increase in suicide in the United States, 1999–2014*. Retrieved from http://www.cdc.gov/nchs/products/databriefs/db241.htm

Cigrang, J. A., Rauch, S. A., Mintz, J., Brundige, A. R., Mitchell, J. A., Najera, E., … Peterson, A. L. (2017). Moving effective treatment for posttraumatic stress disorder to primary care: A randomized controlled trial with active duty military. *Families, Systems & Health*, *35*(4), 450–462.

Currier, J. M., Holland, J. M., & Malott, J. (2015). Moral injury, meaning making, and mental health in returning veterans. *Journal of Clinical Psychology, 71*(3), 229–240.

Day, K. W., Lawson, G., & Burge, P. (2017). Clinicians' experiences of shared trauma after the shootings at Virginia Tech. *Journal of Counseling & Development, 95*(3), 269–278.

Defense Manpower Data Center (2018). Retrieved from https://www.dmdc.osd.mil/appj/dwp/index.jsp

Department of Veterans Affairs, Office of Mental Health and Suicide Prevention (2018). *VA national suicide data report 2005–2016.* Retrieved from https://www.mentalhealth.va.gov/docs/data-sheets/OMHSP_National_Suicide_Data_Report_2005-2016_508.pdf

Department of Veterans Affairs, Office of Suicide Prevention (2016). *Suicide among veterans and other Americans 2001–2014.* Retrieved from https://www.mentalhealth.va.gov/docs/2016suicidedatareport.pdf

Eyre, A. (2017). The making of a hero: An exploration of heroism in disasters and implications for the emergency services. In P. Murphy and K. Greehalgh (Eds.), *Fire and Rescue Services: Leadership and management perspectives* (pp. 113–129). Cham, Switzerland: Springer International publishing.

Filjak, T., & Pavlina, Ž. (2000). Psychological Model of Soldier Bravery. Office of Naval Research International Field Office, Grant N00014-00-1-1086, 108.

Foa, E. B., Hembree, E. A., & Rothbaum, B. O. (2007). *Prolonged Exposure Therapy for PTSD: Therapist Guide: Emotional Processing of Traumatic Experiences.* New York: Oxford University Press.

Franco, Z. E., Blau, K., & Zimbardo, P. G. (2011). Heroism: A conceptual analysis and differentiation between heroic action and altruism. *Review of General Psychology, 15*(2), 99–113.

Franklin, K. (2015). *Department of Defense quarterly suicide report calendar 2015 2nd quarter.* Retrieved from Defense Suicide Prevention Office website: http://www.dspo.mil/Portals/113/Documents/DoD%20Quarterly%20Suicide%20Report%20CY2015%20Q2.pdf.

Freedman, T. G. 2004. Voices of 9/11first responders: Patterns of collective resilience. *Clinical Social Work Journal, 32*(4), 377–393.

Greene- Shortridge, T. M., Britt, T. W., & Castro, C.A. (2007). The stigma of mental health problems in the military. *Military Medicine, 172*(2), 157–161.

Hoge, C. W., Auchterlonie, J. L., & Milliken, C. S. (2006). Mental health problems, use of mental health services, and attrition from military service after returning from deployment to Iraq or Afghanistan. *Journal of the American Medical Association, 295*(9), 1023–1032.

Hoge, C. W., Castro, C. A., Messer, S. C., McGurk, D., Cotting, D. I., & Koffman, R. L. (2004). Combat duty in Iraq and Afghanistan, mental health problems, and barriers to care. *The New England Journal of Medicine, 351*(1), 13–22.

Hummel, K. L. (2018). *Physical Bravery among United States Special Operations Unit Members. Dissertation Abstracts International: Section B: The Sciences and Engineering. ProQuest Information & Learning.* Retrieved from https://paloaltou.idm.oclc.org/login?url=https://search.ebscohost.com/login.aspx?direct=true&db=psyh&AN=2018-34220-007

Kang, H. K., Bullman, T. A., Smolenski, D. J., Skopp, N. A., Gahm, G. A., & Reger, M. A. (2015). Suicide risk among 1.3 million veterans who were on active duty during the Iraq and Afghanistan wars. *Annals of Epidemiology, 25*(2), 96–100.

Kimhi, S., Eshel, Y., Leykin, D. & Lahad, M. (2017). Individual, community, and national resilience in peace time and in the face of terror: A longitudinal study. *Journal of Loss and Trauma, 22*(8), 698–713.

Kugel, U. (2014). *Physical courage among military personnel and veterans.* (Unpublished doctoral dissertation). Pacific Graduate School of Psychology, Palo Alto University, Palo Alto, CA.

Leenaars, A. A. (2004). Altruistic suicide: A few reflections. *Archives of Suicide Research, 8*(1), 1–7.

Mish, F. C. (2012). *Resilience. In Merriam-Webster's Collegiate Dictionary.* Springfield, MA: Merriam-Webster.

National Research Council. (2014). *The Context of Military Environments: An Agenda for Basic Research on Social and Organizational Factors Relevant to Small Units.* Washington, DC: The National Academies Press.

Ordinary Americans Are. (2017). Retrieved from https://themedalofhonor.com/citizen-honors

Ordinary Americans Are. (2018). Retrieved from https://themedalofhonor.com/citizen-honors

Painter, J. T., Fortney, J. C., Austen, M. A., & Pyne, J. M. (2017). Cost-effectiveness of telemedicine-based collaborative care for posttraumatic stress disorder. *Psychiatric Services, 68*(11), 1157–1163.

Pease, J. L., Billera, M., & Gerard, G. (2016). Military culture and the transition to civilian life: Suicide risk and other considerations. *Social Work, 61*(1), 83–86.

Peterson, C., & Seligman, M. E. P. (2004). *Character Strengths and Virtues: A Handbook and Classification.* Oxford: Oxford University Press.

Pombo, S., Ferreira, J., Bicho, M., & Levy, P. Q. (2017). Is there a genetic support for the Cloninger (type I/II) clinical classification of alcohol addiction? *Psychiatry Research, 258*, 621–623.

Pury, C. L. S., & Lopez, S. J. (2010). *The Psychology of Courage: Modern Research on an Ancient Virtue.* Washington, D.C: American Psychological Association.

Rachman, S. J. (1990). *Fear and Courage*, 2nd Ed. New York: W. H. Freeman and Company.

Ramchand, R., Schell, T. L., Karney, B. R., Osilla, K. C., Burns, R. M., & Caldarone, L. B. (2010). Disparate prevalence estimates of PTSD among service members who served in Iraq and Afghanistan: Possible explanations. *Journal of Traumatic Stress, 23*(1), 59–68.

Rate, C. R., Clarke, J. A., Lindsay, D. R., & Sternberg, R. J. (2007). Implicit theories of courage. *The Journal of Positive Psychology, 2*(2), 80–98.

Resick, P. A., Monson, C. M., & Chard, K. M. (2017). *Cognitive Processing Therapy for PTSD: A Comprehensive Manual.* New York, N.Y.: The Guilford Press.

Rüsch, N., Corrigan, P. W., Todd, A. R., & Bodenhausen, G. V. (2010). Implicit self-stigma in people with mental illness. *The Journal of Nervous and Mental Disease, 198*(2), 150–153.

Salters-Pedneault, K., Ruef, A. M., & Orr, S. P. (2010). Personality and psychophysiological profiles of police officer and firefighter recruits. *Personality and Individual Differences, 49*(3), 210–215.

Stack, S. (2004). Emile Durkheim and altruistic suicide. *Archives of Suicide Research, 8*(1), 9–22.

Sullenberger, C. B. (2018). *We saved 155 lives on the Hudson. Now let's vote for leaders who'll protect us all.* Retrieved from https://www.washingtonpost.com/opinions/we-saved-155-lives-on-the-hudson-now-lets-vote-for-leaders-wholl-protect-us-all/2018/10/29/554fd0e6-d87c-11e8-a10f-b51546b10756_story.html?noredirect=on&utm_term=.68f83a5c48e9

Warner, C. H., Appenzeller, G. N., Mullen, K., Warner, C. M., & Grieger, T. (2008). Soldier attitudes toward mental health screening and seeking care upon return from combat. *Military Medicine, 173*(6), 563–569.

Warner, C. H., Appenzeller, G. N., Parker, J. R., Warner, C., Diebold, C. J., & Grieger, T. (2011). Suicide prevention in a deployed military unit. *Psychiatry: Interpersonal & Biological Processes, 74*(2), 127–141.

Weinstein, H. (2012). *Beyond courage: The psychology of heroism.* (Unpublished doctoral dissertation). Pacific Graduate School of Psychology, Palo Alto University, Palo Alto, CA.

# 14

# MILITARY RECRUITING IN THE UNITED STATES

## Selection, assessment, training, well-being, and performance coaching

*Stephen V. Bowles, Bettina Schmid, Laurel K. Cofell Rashti, Susan J. Scapperotti, Tracy D. Smith, Paul T. Bartone, and Peter Mikoski*

### Military recruiting in the United States: Recruiter characteristics, lifestyle, well-being, and performance coaching

#### *Introduction and history*

Maintaining critical force strength and composition is imperative to the national security of the United States. World events continue to demand a strong focus on recruiting personnel for all military services, which can be challenging when the economy is good and it is difficult to recruit applicants. Each military service has a recruiting command organization funded to focus specifically on the recruiting mission. These commands are responsible for sourcing highly qualified applicants to become Airmen, Marines, Sailors, and Soldiers, thus maintaining military readiness for each service.

Historically, the United States relied heavily on conscription, that is, compulsory enlistment. In 1973, the United States transitioned to an all-volunteer force. In the years following, as the United States recovered from the Vietnam War, the military forces were also recovering, and they had to reassess their role, culture, and identity. What eventually emerged was a military force focused on professionalism (Kitfield, 1995). This shift elevated the importance of recruiting, which promotes the importance of serving one's country through recruiting high-quality youth who will make up the emerging professional force (Rostker, 2006).

The U. S. Congress mandates personnel strength requirements for each component (both active and reserve) of the Army, Navy, Air Force, and Marine Corps (National Defense Authorization Act for Fiscal Year 2018, 2017). Each recruiting command organization headquarters determines the accession goals for subordinate units, down to the lowest levels (e.g., recruiting stations). Requirements are based on data that take into account a number of factors for each geographical area, including demographics, desired military end strength, job specialty, state of the economy, education and income levels, political leanings, and propensity to serve. Recruiters experience the daily pressures of recruiting, a mission they have little input in determining and one that ultimately is the standard by which they measure their success as recruiters.

The U.S. Army was the first service to assign psychologists to support recruiting. The Office of the Command Psychologist in U.S. Army Recruiting Command (USAREC) was established in the early 2000s. At that time a recruiter selection instrument was developed and implemented along with an assessment process at the recruiter school along with a professional coaching program that was provided to some USAREC leaders in mid-level management (Borman et al. (2004); Bowles, Cunningham, De La Rosa, & Picano, 2007; Bowles & Picano, 2006; White, Borman, & Bowles, 2001). Today, the U.S. Army continues to be the only service that has assigned psychologists, though all the services continue to have service-specific screening requirements for recruiter applicants. In 2018, the USAREC Office of the Command Psychologist was awarded the Society of Military Psychology's prestigious Julius E. Uhlaner Award for their innovative recruiter assessment and selection process providing better candidates (American Psychological Association, 2018). This chapter provides an overview of the U.S. military recruiter and the role of psychology in recruiting. It describes the characteristics and screening of recruiters, initial and ongoing recruiter training, lifestyle and challenges, motivation and positives of recruiting, characteristics of failing recruiters, the healthy workplace, and coaching in recruiting.

## Future force characteristics of the recruiter

### *Service branch-specific screening requirements*

All services screen for the best non-commissioned officers (typically enlisted E-5 and E-6) for both active and reserve military recruiting. The methods for selecting active and reserve component recruiters vary by service. In the past, selection had little to do with the actual practice of military recruiting. The selection criteria included good performance evaluations, passing scores on the physical fitness test, and a good credit history. However, changes in the demand characteristics of the environment and advances in personnel selection practices have resulted in enhanced recruiter selection procedures to ensure candidates have the necessary maturity and skills to serve in this role.

Currently, selection of recruiters is a multistep process drawing from several sources. Information regarding a candidate's personal background, legal history, finances, medical and psychological functioning, and character are taken into account. All branches require the recommendation of the candidate's unit commander who may also be asked to identify any current or past legal, disciplinary, or financial concerns; comment on current hardships or character-related concerns; and describe the individual's occupational and physical performance. In addition, potential recruiters may also undergo formal background checks from various agencies. Military justice databases and personnel records are searched to ensure that any past instances of misconduct are identified. These steps occur simultaneously and effectively winnow down the pool of eligible recruiter candidates (Department of the Navy, Headquarters, U.S. Marine Corps, 2012; U.S. Navy Personnel Command, 2018; U.S. Department of the Air Force, 2018; U.S. Department of the Army, 2016).

### *Behavioral health mental status exam*

Medical and psychological screenings are part of the recruiter selection process for all branches, with varying degrees of stringency. Psychological suitability for recruiting duty is determined in various ways. For example, prospective Marine recruiters must have a medical examination within 12 months of the recruiting class reporting date; disqualifying conditions may include "problems with stress or psychological dysfunction" (Department of the Navy, Headquarters, U.S. Marine Corps, 2012). Similarly, Navy recruiter candidates undergo a medical records screening by a medical provider and may be disqualified for recruiting duty if they have "a history of stress-related problems such as depression, suicidal ideation, nervous breakdowns, and anger management issues" (U.S. Navy Personnel Command, 2018). The policy governing Air Force recruiting selection specifies a "Mental Health review will be evaluated … No record of emotional instability, personality disorder, sexual misconduct or other unresolved mental health problems that will interfere with the ability to perform Recruiting duty (sic)" (U.S. Department of the Air Force, 2018).

U.S. Army recruiter candidates undergo a stringent psychological screening process. Per Army Regulation 601-1 (U.S. Department of the Army, 2016), recruiter candidates are required to have a behavioral health suitability assessment. This assessment is conducted by a trained clinician and consists of a face-to-face evaluation and review of available medical records, and may include objective psychological measures (Gomez, Agnor, Jenkins, McClenen, & Schmid, 2015). The clinician reviews psychiatric history, substance use history, medical concerns (including pain conditions), social history, current psychosocial stressors, education and work history, legal and disciplinary history, and financial status. Individuals may be deemed unsuitable for recruiting duty if they have a diagnosis or condition that causes emotional instability. Individuals may be temporarily or permanently screened out for current and past psychological concerns as well as a history of problematic behavior.

While some recruiter candidates volunteer for the duty, the majority are chosen by supervisors and may have varying degrees of motivation. Recruiting duty differs from other job application processes in that many of the candidates may not even want the job, and in extreme cases may seek to amplify disqualifying information. Similarly, among those who volunteer to be recruiters, there may be a tendency to minimize disqualifying information. As such, it is important that the screening process address individuals who may try to falsify or overstate positive or negative information.

## *Recruiter aptitude*

The effective recruiter should possess certain personality traits, interpersonal skills, and technological abilities. Past research indicates there are personality characteristics that successful Army military recruiters appear to share. These include agreeableness, hardiness, interpersonal skills, self-esteem, and social perceptiveness (Borman et al., 2003; Borman et al., 2002). In the past, the U.S. Air Force used the EQ-I emotional intelligence instrument, and found the characteristics most likely to relate to success on the job were assertiveness, empathy, happiness, problem solving, and self-awareness (as cited in Stein & Book, 2011). In the U.S. Army, a series of studies was conducted from 2000 to 2005 with the Noncommissioned Officer Leadership Skills Inventory (NLSI) (Borman et al., 2004; White, Borman, & Bowles, 2001). The NLSI measures characteristics related to recruiter performance for interpersonal skills, leadership capability, and work orientation. In 2008 the instrument was renamed the Warrior Attributes Inventory (WAI), which, like the NLSI, was implemented and administered at Digital Training Facilities worldwide. Since then, a new instrument has evolved out of the WAI called the Noncommissioned Officer Special Assignment Battery (NSAB) (Horgen et al., 2013); however, it has not yet been fully implemented.

Many studies have investigated the relationship between sales performance and personality traits. Researchers have found positive correlations between sales performance and openness and conscientiousness (Furham & Fudge, 2008); a general factor of personality including altruism, emotional stability, agreeableness, conscientiousness, and extraversion (Sitser, Van Der Linden, & Marise, 2013); and multi-tasking (Conte & Gintoff, 2005). Further, Furham and Fudge (2008) found a negative correlation between agreeableness and sales performance, speculating that high agreeableness is negatively correlated with competitive drive in team sales settings. While many studies, such as the one conducted by Furham and Fudge (2008), have failed to find significant linear relationships between extraversion and sales performance, Grant (2013) found that ambiverts outperformed both introverts and extraverts in sales performance. Such studies can be used to inform selection of military recruiters as well as training recruiters to key skills. Additionally, a meta-analysis found five key drivers of sales performance to be selling-related knowledge, degree of adaptiveness, role ambiguity (negatively correlated), cognitive aptitude, and work engagement (Verbeke, Dietz, & Verwaal, 2011). As a psychologist it is important to understand the training and skills of the recruiting occupation.

## Recruiter training in the Air Force recruiting school

Each service has similar training and approaches in meeting the recruiting mission. As an example, we will use the U.S. Air Force to illustrate the process of training new recruiters. The Air Force recruiter falls under one of the three components: Regular Air Force (Reg AF), Air Force Reserve (AFRes), or Air National Guard (ANG). Air Force recruiter training is conducted at Lackland Air Force Base, Texas, and it is controlled by the Air Education and Training Command. The courses are accredited by the Community College of the Air Force, and the credits can be applied towards an Associates of Applied Science in Human Resources degree (Department of the Air Force, 2017).

Although students may have achieved expertise in their previous career field, they often find recruiting to be a new and unfamiliar challenge. The curriculum is demanding and students often find it easier if they connect with fellow classmates. Certified instructors train the students on a civilian sales model that is contracted to the Air Force. Currently, under guidance of the Secretary of the Air Force, the three Air Force components are partnering to merge the three separate recruiting courses into a single Total Force course (Department of USAF, 2017).

The scope of recruiter training consists of time and stress management, sales presentations, sales telephone techniques, persuasive presentation, eligibility requirements for military service, and applicant processing procedures. This training includes basic computer applications and advanced computer technology. Students are taught the sales model by the instructor through lecture and demonstration.

The Air Force recruiting school has a unique approach to student evaluation. First, students must pass a written test. Second, students take a performance test in which they must demonstrate proficiency in accession of a simulated prospect from start to finish. This is accomplished in six phases including the following tasks:

1. Give a speech to fellow classmates (simulating high school or college students) to develop public speaking skills as well as teach fellow recruiters how to prospect for applicants.
2. Collect and prioritize potential leads following the speech.
3. Complete a telephone call to the highest qualified lead (played by the instructor). The student must determine if the applicant tentatively meets eligibility requirements as well as properly demonstrate sales techniques.
4. Meet with the applicant in person (also played by the instructor).
5. Brief the applicant for the military entrance physical exam and complete required documents.
6. Performance of component-specific training and evaluation.

## On-boarding training for regular Air Force recruiting

After graduation, RegAF recruiters report to their new assignment where they spend their first year working on certification in recruiting. They begin their training under the supervision of the Flight Chief (e.g., Station Commander in the Army, Leading Petty Officer in the Navy, and Sub Station Commander in the Marine Corp). Training is tracked through the Air Force Training Record (AFTR) database. The supervisor documents the training on the Job Qualification Standards (JQS), which outlines training tasks and timelines for completion. The Flight Chief conducts a 14-day shadowing program that includes demonstration, observation, and feedback. The new recruiter creates an annual, quarterly, monthly, and daily plan of tasks which is reviewed by the supervisor.

Over the course of the first year, the recruiter continues to work with guidance from his/her supervisor. Within the first quarter, a squadron trainer (from a higher echelon) schedules a visit of at least 2 days for training which includes demonstration, observation, and feedback on any training deficiencies. In addition to figuring out the new job and its flow, the recruiter is also assigned a recruiting quota known as their goal. The goal for a rookie recruiter is gradually increased as he/she gains proficiency.

The Squadron Production Superintendent personally reviews the recruiter's training and performance for the first year, and the individual receives certification if all goals and expectations have been met. The recruiter will then join the remaining members within the unit to receive routine professional training. For production, recruiters who miss their monthly assigned goal are placed on a remedial plan which includes close supervision and guidance to help them improve.

## The inside life of the (Air Force) recruiter

Recruiting can be a stressful job, but most recruiters will agree that it can also be very fulfilling. Recruiters develop trusting relationships with applicants, providing career guidance, mentorship, and general support in addition to helping them enlist in the military. In the recruiter-applicant relationship, each party has roles and responsibilities. The common expectations of an applicant include being mentally and physically fit for military service, being honest, following recruiter directions, being on time for appointments, and serving as a referral for other potential applicants. Expectations of the recruiter in this relationship include maintaining regular communication with applicants, timely and expedient processing, friendly and courteous behavior, being knowledgeable about the Air Force, and being honest and forthcoming.

Each recruiter is assigned a territory called a "zone." This zone specifies a population of Americans within the age range and demographics that the military strives to recruit. The recruiter makes an annual, quarterly, and daily plan of how he/she will meet the monthly enlistment goal for the assigned zone. The following sections describe core tasks that are typically incorporated into a recruiter's plan, including planning, school visitation, zone advertising, marketing, social media, telephone prospecting, recruiter generated mail, sales appointments, applicant perpetuation, applicant processing, and administrative action. If recruiters can properly plan and act on these core tasks, they should find success within their jobs.

### *Planning*

Planning is the foundation for a recruiter's job. Thoughtful scheduling will ensure recruiters are working ahead of the timelines. Plans must be feasible and flexible.

## School visitation

Unlike recruiters from other branches of the military, Air Force recruiters have large territorial zones that they cover independently. Recruiters' zones generally consist of 6 to 10 schools (e.g., high schools, community colleges, universities, and vocational schools) they are expected to visit. Prioritizing the schools ensures that recruiters focus their time and effort on the institutions where the students are more likely to have interest in the military. In addition to prospecting for potential applicants, recruiters also focus on influencers within the school. Influencers can include guidance counselors, teachers, coaches, principals, secretaries, and any official within the school who may facilitate contact with the student population. The public school systems generally allow recruiters to access students through career fairs, lunchroom setups, and guest speaking and provide student informational lists to recruiters.

## Zone advertising

Recruiters are given a budget to advertise within their zones. Advertising is targeted to the positions and demographics for their component's needs. If resources are low, recruiters will normally set up take-one boxes or hang up flyers within their zones to promote awareness. Schools usually designate a location that allows military recruiters to leave information for the students.

## Marketing

Joint Advertising Market Research Studies (JAMRS) aids and assists the branches of the military with market research. This agency ensures the campaigns for each of the branches are based on empirical data. As Air Force recruiting moves toward the Total Force model, the marketing strategy will adjust accordingly. The budget and campaign strategies vary from year to year at the national level.

## Social media

Social media has become a new, inexpensive tool that allows recruiters to attract a larger pool of applicants, connect with younger generations, and target specific audiences. In order to make this tool effective, the recruiter looks at social media trends and peak times for posting and strategically incorporates these into the recruiting plan.

## Telephone prospecting

Telephone prospecting is a way in which recruiters can identify people who may be interested in military service, with the goal of the call being to schedule an appointment with a qualified applicant. Recruiters collect lists of leads obtained from various sources such as schools and ASVAB testing lists.

## Recruiter-generated mail

Recruiter-generated mail consists of mailing out fliers or brochures advertising the Air Force to potential applicants. However, it is becoming an obsolete tactic of recruiting.

## *Sales appointments*

By meeting face-to-face, the recruiter attempts to help an individual meet their career and or life goals through joining the service. Recruiters focus on getting to know their applicants and understanding their needs and circumstances. This information guides the discussion on the benefits of military service. Recruiters are also trained on overcoming objections. Applicants may have personal concerns or misguided information, and recruiters help them work through those concerns. Recruiters establish trust with the applicant, determine their eligibility, and help determine the next step in the process towards joining the military.

## *Applicant perpetuation*

This is the process of asking satisfied customers to provide referrals to the recruiter. Recruiters are taught to perpetuate with anyone they encounter. They perpetuate through applicant appointments; applicants who have already enlisted; handing out business cards; and asking school guidance counselors, teachers, friends, and family members if they know of anyone interested in hearing or learning more about the Air Force.

## *Applicant processing*

Applicant processing is mandatory for all applicants of the Armed Forces. This includes taking the Armed Services Vocational Aptitude Battery (ASVAB), a timed multi-aptitude test (may be completed at various locations such as a local National Guard Armory), and completing a thorough medical examination at a Military Entrance Processing Station (MEPS).

## *Administrative action*

Recruiters have many administrative responsibilities. Attention to detail by recruiters is crucial to ensure that the applicants' forms are completed correctly the first time to avoid delays and errors.

A week of recruiter activity will normally consist of the following:

1. Two to three planned school visits.
2. Two to three appointments per day (1 hour per appointment).
3. Telephone calls (approximately 100 per week).
4. One to two hours of social media advertising/posting per week.
5. Ten hours of Administrative Action per week.
6. One to two hours of zone posting/canvassing per week. Zone canvassing is considered any type of act or action that promotes the U.S. Air Forces with the goal of generating leads and brand awareness. Zone posting is when the recruiting posts information all around the community to aid in recruiting efforts.
7. MEPS processing (e.g., transporting applicants to and from MEPS at least once per week).

Recruiting duty can be rewarding, and the connection between applicant and recruiter can ultimately make the difference in an applicant's life. The job can be demanding and stressful at times, but those who stay positive and strive to improve their skills can find success.

## Challenges of recruiting

Irrespective of service, most new recruiters suddenly enter a world of selling, far different from their previous military career occupation. This can be challenging for many recruiters, as they may have to make more than 10,000 contacts to produce 50 to 60 recruits that successfully complete basic training (Randolph, 2012). To gain more insight on the lives of recruiters, a survey was conducted at the U.S. Army Recruiting and Retention School with 26 recruiters and 11 junior-level recruiting managers. These 37 soldiers participated in focus groups to discuss the three-year cycle of six specific areas in recruiting (Table 14.1).

In conjunction with the results of this survey, we have used extensive observations, experience, and discussions with recruiters to illuminate many of the challenges that recruiters typically face. To begin, recruiters usually move to an environment miles away from any military installation and are not afforded the typical support systems to which they have become accustomed. The recruiter has to adapt to a new home, possibly fewer hours with his or her family, and life outside of a military installation. Many recruiters have neither the interest nor intention of making this duty into a permanent career. They often have less time for themselves and their family due to the long hours demanded on the job.

For the recruiter, working a 60-hour week is not uncommon. In addition to the long hours, this is one of the toughest military jobs outside of combat deployment. Once recruiters have accomplished their monthly mission, the pressure begins again to find the next applicants for the upcoming month.

*Table 14.1* Survey conducted at U.S. Army Recruiting and Retention School, Fort Jackson, SC 2004

| | *Year 1 recruiter* | *Year 2 recruiter* | *Year 3 recruiter* |
|---|---|---|---|
| What are the hardest aspects of recruiting to stay motivated in? | Time management<br>Locating quality prospects<br>Salesmanship | Time management<br>Locating quality prospects<br>Delayed entry maintenance | Time management<br>Delayed entry maintenance<br>Work hours/time off |
| Which areas of your training enabled you to be the most effective? | Salesmanship<br>Time management<br>Product knowledge<br>Locating quality prospects | Salesmanship<br>Time management<br>Closing | Delayed entry maintenance<br>Product knowledge<br>Locating (maintaining) quality prospects |
| What aspects of your job are most problematic? | Workload<br>Time management<br>Locating quality prospects | Time management<br>Mission completion<br>Workload | Time management<br>Mission completion<br>Workload |
| Which aspects of your life, other than your job, are the most problematic? | Health lifestyle<br>Time management<br>Education opportunities | Health lifestyle<br>Recreation opportunities<br>Time management | Health lifestyle<br>Time management<br>Spouse/relationships |
| What are the most stressful aspects of your life? | Work<br>Time management<br>Spouse/relationships | Time management<br>Work<br>Spouse/relationships | Work<br>Time management<br>Spouse/relationships |
| What best helps you to cope with the stresses of your life? | Spouse/significant other<br>Exercise<br>Talking to friends | Spouse/significant other<br>Exercise<br>Talking to family | Spouse/significant other<br>Exercise<br>Talking to friends |

With the pressures of the job environment, the recruiter needs a supportive family. Most spouses, parents, and siblings are supportive of the recruiter; however, there are challenges. Families of recruiters are constantly adjusting to new environments, as the service member changes assignments and duty locations every few years. The long hours of the job mean less time at home, which may become a source of marital conflict and cause children to act out. It is not uncommon for a recruiter to miss eating meals with the family on a regular basis. Due to the high cost of living in some areas, housing may be less affordable and recruiters may live in less desirable areas.

Recruiters are often left with little time for themselves, and may experience a decline in healthy lifestyle habits. Some recruiters report sleep problems due to the stress of the job. Generally, they report being in worse physical condition than before recruiting and often gain weight. Most do not maintain a healthy diet. Often, they eat fast food out of their car since they are usually travelling between appointments. Such time constraints are worsened when many feel pressured to work while on vacation. Recruiter production cycle over three years may be low as they start their jobs and low as they end their recruiting duty and begin planning to move back into their previous occupations.

## Recruiter motivation and personal resources

Though many aspects of recruiting are stressful, it can also be extremely rewarding and motivating. Many acknowledge their time in recruiting was a period of personal growth. They develop business skills, including marketing, advertising, and sales techniques, and further develop human relations skills. Many recruiters also grow in cultural awareness when moving to a different social environment (e.g., military to civilian).

One of the most rewarding aspects of recruiting comes from the satisfaction of helping a young recruit find direction. Recruiters stated that once a new recruit returns from basic training, they display significant growth, with a greater sense of discipline and new values. They have been given an opportunity and a means to meet their goals and to develop as individuals.

This change reflects positively on the military as well. When a community witnesses the dramatic transformation of one of its youth, the reputation and image of the military is enhanced. Recruiters also find satisfaction in promoting the military while helping out in the community. Recruiters are the ambassadors of the military services for the civilian sector, serving as great role models and positive forces for our youth and communities.

Staying motivated is necessary for recruiters to be effective and successful in their jobs. In the previously mentioned survey and focus group, participants were asked to identify the factors that motivate them to be successful at work. The most common responses were the feeling of success (i.e., achieving recruiting goals), positive feelings associated with helping others (they saw themselves helping applicants find solid career paths and ways to serve their country), time off with family and friends, and providing a good quality of life for themselves and their families. These factors were inter-related in that recruiters placed importance on having balance in their lives. The success at work provided time off and the opportunity to spend time with loved ones, and this gave them a chance to recharge and have energy for the next week.

Taking a different approach, Dertouzos and Garber (2006) looked at recruiter motivation and production from a human resource management perspective. Their study of Army recruiting suggested the combined effect of implementing multiple policies could have a significant effect on production. One key implication of this study is that having an understanding of individual recruiter characteristics associated with production (and the interaction with local markets) can be used to inform decisions about selection of personnel for recruiting and where to place them (e.g., recruiters are more productive when assigned to a station in their home state). This study also confirmed that setting performance goals is associated with increased effort and productivity. The authors also found that completing a recruiting assignment was beneficial to a soldier's career, and high-performing recruiters were promoted ahead of their peers, which may be a motivating factor for these soldiers. In a subsequent study, Dertouzos and Garber (2008) looked at models of recruiter performance evaluation, noting that the standards by which recruiters are evaluated must be perceived as fair in order to have a motivating effect.

## Psychologists work with failing recruiters and work settings

With new recruiters, the transition into sales may be difficult. Recruiters were once confident high performers in their previous military jobs. However, in recruiting, they are dealing with the civilian world, having a harder time completing their mission, and may begin to question themselves. It can also be hard, especially in times of complicated social and world issues, to work outside of a normal military structure; some may find it difficult to work with civilians when they feel that they do not offer much gratitude or support.

There is also constant pressure to perform at high standards. Individuals may notice that other recruiters in their station display symptoms of "cracking under the pressures" of recruiting. They may be consistently late for work, show lower performance, engage in unsafe driving, avoid going home, gain or lose weight, experience sleep problems, increase their dependence on alcohol/drugs, show signs of burn out, and demonstrate changes in their interpersonal behavior (angry outbursts or negative attitude). Recruiters coming back from war settings may also exhibit anxiety or depressive symptoms described above after experiencing the stresses of recruiting. They may seek help in performance enhancement training to get back on the right track, and at this time, the trainer and/or supervisor may recognize more help is needed. Work stresses, along with a lack of family time (and the guilt associated with it), may result in a supervisor or self-referral by recruiters and self-referral by family members for counseling to address family needs with those of military operations.

### *Resources for struggling military recruiters*

Struggling military recruiters may engage in treatment with civilian or nearby military healthcare resources, but may also seek help from resources like Military One Source, a contracted service that provides a broad range of consultation and mental health services, or Reserve chaplains or chaplains from nearby installations. Given the stigma still associated with seeking mental health services, recruiters may turn to resources with strict confidentiality and may not tell their leaders that they are seeking help.

Recruiters with significant problems may require coordination of care with their healthcare provider and command. This serves the purpose of ensuring the service members receive the needed care and leaders are able to modify the individual's work requirements appropriately. In some cases, based on the mental health recommendation, struggling recruiters may need to have their duties limited or modified to allow time for treatment. If a recruiter has been away for a period of time for treatment, the process of returning to work will vary depending on the circumstances (e.g., seriousness of their condition, time away from work). If the recruiter is not able to integrate back into recruiting, he/she may be assigned less stressful administrative recruiting tasks. In some instances, a return to production recruiting may be difficult and not fit into the member's life circumstances, resulting in a transfer back to another military assignment. In other more serious cases where a member is not fit for duty, further medical and administrative actions may occur. Healthcare providers can assist command by helping leaders understand which duties (described "Inside the Life a Recruiter" section) the recruiter is likely to be able to successfully execute.

## Promoting recruiter well-being

Many recruiters do not have easy access to the usual resources located on U.S. military installations. Managing personal time has to include an investment in exercise, getting adequate sleep, and eating nutritious food. Managers have a vested interest in ensuring that recruiters maintain physical well-being, in part because a physically fit recruiter provides a better representation of their of service. In addition, recruiters who are not engaging in basic self-care are placing themselves at risk for unsafe driving, safety mishaps, and decreased performance (Litwiller, Snyder, Taylor, & Steele, 2017).

Maintaining family support is also critically important for recruiting communities. Recruiters may work long hours and be away from home, and the support network typically present at military installations may not be available. If family members are not inclined to get involved in "non-military" social or community avenues of engagement, recruiting duty may be a very lonely assignment. Commander efforts to bolster family support may include regular gatherings at family-friendly venues during a time that is generally convenient for all and does not require extra time at work. Ultimately, however, recruiters and their families should be encouraged to engage in community events and be given the time to do so. These efforts not only improve family support and engagement but may also provide prospects through community contact, such as religious or community organizations. Leaders may encourage recruiters' families to visit the recruiter during certain events, such as fairs and other public engagements. This approach also supports a "family-friendly" military image that may appeal to potential recruits.

## Promoting an ethical recruiting environment

Just as senior leaders can impact the psychological health of the workplace and employees, senior leaders and managers can also foster a workplace culture that encourages unethical or dangerous conduct, also known as behavioral drift. In their review of the literature, Appelbaum, Iaconi, and Matousek (2007) identified several common factors that may contribute to workplace deviance, some of which may occur in recruiting environments. First, organizational culture factors, such as a "bottom line" focus on numbers and quotas, may tempt individuals to cut corners to meet mission requirements. A real-life example of this may be falsifying documents to push forward an unqualified prospect.

Conversely, a sense of shared well-being and organizational concern with individual welfare may decrease politically deviant behaviors such as showing favoritism and undue criticism, whereas job-related frustration is likely to result in interpersonal deviant behaviors. Organizations that uphold high standards of behavior and hold individuals accountable for maintaining organizational policies have fewer instances of property deviance, such as stealing from the organization.

Ensuring leaders model ethical values is also critically important. If individuals in roles of authority are allowed to act in a deviant manner, the behavior they embody becomes acceptable, and over time the degree of deviance may increase, resulting in a significant shift in organizational norms. Appelbaum et al. (2007) also identified several factors that can encourage "positive deviant behaviors," such as innovation. By setting up environments in which employees are engaged, feel a personal stake in the organizational outcomes and a sense of shared mission, and are rewarded for process improvement and innovation, leaders can encourage positive innovation. Finally, leaders can encourage their service members to intervene if they become concerned about the behavior or well-being of a peer by providing bystander intervention training. As Elliman, Shannahoff, Metzler, and Toblin (2018) have demonstrated, peers are often the ones observing deviant or concerning behavior and may be most able to intervene at an early stage. Because recruiters are usually placed in small units and have significant autonomy and minimal oversight, it is critical that organizational culture encourage ethical behavior.

## *Creating a healthy workplace*

Leaders are encouraged to develop policies that require subordinate leaders to follow or develop programs that promote recruiter well-being. Some research suggested important areas of well-being related to production were realistic work demands, co-worker support, spirituality, friendship support, and medical services support (Bowles, Bartone, Cooke & Swisher, 2018). Hardiness has also been found to predict performance in recruiting (Bowles & Bartone 2017). Further, Bartone & Bowles (2018) have shown that personality hardiness interacts with active, problem-focused coping to predict improved recruiter performance. One model that has demonstrated success in corporate America is the American Psychological Association Center for Organizational Excellence's "psychologically healthy workplace."

Psychologists can facilitate a "psychologically healthy workplace" by encouraging recruiting organizations to adopt practices in five different domains: employee involvement, work-life balance, employee growth and development, health and safety, and employee recognition. Recruiting commands can incorporate each of these domains while still meeting and in many cases facilitating organizational goals. Supervisors can improve work-life balance by ensuring that the recruiter has projected their vacation time throughout the year in their annual planning. They may also limit the number of hours worked per day, ensure that time off is predictable, and allow family participation when appropriate. Ensuring that recruiters get adequate time for rest and leisure is another way to promote workplace health and safety, as fatigue is a significant contributor to workplace mishaps and decreased productivity (Hafner, Stepanek, Taylor, Troxel, & Van Stolk, 2017). Leaders can promote a safe and healthy work environment by ensuring that the recruiter has budgeted their time with exercise and wellness check-ups as well as provide a physically and psychologically safe and secure workplace environment.

Organizational leaders can also recognize individual employee strengths and interests and allow employees to "self-sort" into particular tasks or missions, thereby improving employee involvement and self-determination in taking ownership of their duty schedule and responsibilities. Further, leaders can promote growth by allowing the recruiter to become an office trainer to newly assigned recruiters and or offer development opportunities such as coaching or attendance at career-enhancing military training or schools. Finally, leaders may use existing means of recognition, such as awards, or may develop their own processes for recognition at lower levels of the organization, potentially using incentives such as time off. These strategies, which may contribute to a psychologically healthy workplace, not only promote employee satisfaction, morale, and stress management, but also result in improved work quality, performance, and productivity (Grawitch, Gottschalk, & Munz, 2006).

## Coaching

Research since the mid-2000s found mixed results for non-psychologist coaching of leaders in military recruiting for production. Two studies on coaching military leaders have directly investigated the impact of coaching in military recruiting. Using a within-subject design, Bowles and Picano (2006) found that coaching enhanced the quality of life for First Sergeants in the U.S. Army Recruiting Command. In an effort to determine the utility of coaching in this military environment, the researchers measured participant changes in goal attainment, quality of life, and productivity as a result of coaching interventions. Nineteen participants attended 8 to 12 hours of instruction (leadership and goal setting) and agreed to participate in telephone coaching sessions over the next six months, conducted by a coach highly experienced in Army recruiting. After controlling for the age of the recruiter, results revealed a significant correlation between adherence and work satisfaction such that individuals who closely followed the advice of the coach reported higher work satisfaction than those who did not adhere to the advice of the coach. Further, a relationship between life satisfaction and adherence approached significance, indicating that life satisfaction might increase alongside adherence to coaching. Perhaps due to the small sample size and shorter coaching time period, there was not a significant relationship between coaching and productivity.

In a subsequent study, Bowles, Cunningham, De La Rosa, and Picano (2007) attempted to address some of these methodological concerns. Thirty Station Commanders (middle managers) and 29 executive managers (Company Commanders and First Sergeants) were recruited to participate in both an 8 to 10-hour leadership, energy management, goal setting training, and a series of coaching sessions over the course of one year. As in the previous study, the non-psychology coaches were individuals with extensive backgrounds in military recruiting, supervised by a psychologist. Using both within- and between-subject designs, the investigators found that participants both outperformed their un-coached more experienced incumbent peers and developed stronger leadership competencies. The results indicated that coached participants developed stronger skills in the areas of task-goal planning, time management, daily progress reporting, analysis, and general leadership. Further, the benefits of coaching were strongest for middle managers when compared to executive managers. This limited number of studies found in our review provides support for the value of coaching for either greater work satisfaction or productivity for these leaders in the recruiting environment.

In turning to the civilian literature, the coaching articles focused on other valuable areas, but none were related to sales and performance. Grant, Curtayne, and Burton (2009), using a randomized controlled waitlist design, assigned 41 high-level managers to receive 10 weeks of professional coaching (four coaching sessions administered every two weeks). Participants both scored higher in measures of goal attainment, resilience, and well-being and scored lower on measures of depression. Similarly, MacKie (2014) investigated the impact of strength-based coaching on senior-level managers. Using a non-equivalent group waitlist design, MacKie assigned 37 Australian senior managers to either an intervention or a waitlist control group (who received coaching after the intervention group had finished). The study utilized 11 professional executive coaches to provide goal oriented coaching over six 90-minute sessions. Results indicated that members of the intervention condition reported higher levels of transformational leadership, satisfaction, and willingness to put in extra effort than members of the control condition.

As indicated by Grant and colleagues and MacKie, professional coaches provide effective, targeted coaching that leads to increased well-being and productivity. In a 2016 meta-analysis of coaching research, Jones, Woods, and Guillaume found that results were significantly higher when coaching was provided by an internal member of an organization rather than an outsider. Coaching presents a promising way to effectively enhance military recruiters' well-being and performance, though further research needs to be conducted in military recruiting. Finally, some best practices identified by successful coaching that may be applied in the recruiting community that are used in business are goal setting, consultative feedback, and cognitive restructuring of ideas (Vandaveer, Lowman, Pearlman, & Brannick, 2016).

## Summary

Recruiters are the face of the U.S. services they represent, tasked with attracting and hiring high-quality men and women to meet military personnel end strength and readiness goals. Recruiting is a military career field that is both challenging and rewarding. Psychologists have had an important role in studying, advising, and supporting military recruiters and leaders on behavioral health matters inherent in selection, training, operations, resilience and well-being, and professional growth.

Researchers and leaders in recruiting have gradually gained a better understanding and refined the selection and screening process to bring on board service members who will best represent the military and have the requisite skills for success, such as communication skills, assertiveness, optimism, self-confidence, emotional intelligence, well-being, adaptability, self-awareness, and stress resilience. In recruiting school, psychologists have provided consultation to leaders and faculty on various aspects of training, as recruiters obtain skills such as product knowledge, time management, marketing, locating prospects, selling, communicating through social media, and processing recruits.

Today's recruiters face a fast-paced, goal-oriented lifestyle that includes a varied schedule. Recruiters often spend long hours focusing on production, which leaves less time for attention to personal health and fitness, family, and recreational pursuits. New recruiters often struggle with learning how to organize and manage their time, as well as learn key skills specific to this field such as locating prospects in a shrinking pool of propensed individuals and influencing people about the value of military service. The resulting imbalance can take a toll on the individual recruiter, as well as his or her family. The spouse or significant other can be one of the most important sources of support, but this relationship is also vulnerable to the strain of the recruiter's demanding job. Research from surveys, focus groups, and observation indicates that it is important, though, for recruiters to schedule time for activities that promote resilience and well-being such as exercise, connecting with family and friends, and spirituality. In some cases, professional psychological intervention is needed for those who have become overwhelmed and show signs such as interpersonal work problems, domestic difficulties, substance abuse, misconduct, or health problems. Further study is needed on the topic of sustainment of recruiters.

Despite the stresses, there are many positives to recruiting. Recruiters improve their interpersonal and business skills while offering exciting opportunities to young citizens. Areas that motivate recruiters include spending time with family members, providing quality of life to their families, accomplishing mission goals, and the positive feeling of putting a young person in the military and watching them excel. Additionally, a recruiting assignment can help to accelerate one's career.

Leaders in recruiting can have a great deal of influence over the lives of recruiters. A poor leadership culture can be a factor in behavioral drift, in which the workplace becomes a negative environment. Positive leadership, in which proper standards are upheld and healthy workplace practices are in place, can provide a productive and positive work environment. These can be furthered by involving psychologists and other professionals for training, consultation, and coaching.

There are three recommendations we would make to sustain recruitment: First, senior leaders at the highest levels can incentivize "psychologically healthy recruiting centers" by recognizing and rewarding recruiting units and centers that are especially adept at managing dimensions of workplace well-being. Second, in order to ensure adequate attention is paid to sustaining recruiter well-being, we recommend that wellness initiatives be codified in regulations and policies put forth by higher-level recruiting commands. Third, recruiting commands should provide coaching by psychologists to help recruiting units that struggle with meeting mission goals or desire to improve their organizational culture.

The field of psychology has positively influenced recruiting in multiple ways and still has much to contribute in supporting those who serve as military ambassadors to communities throughout the United States and inspiring young citizens to serve their country by enlisting in the armed forces.

**Note:** The conclusions and opinions expressed in this chapter are those of the authors. They do not reflect the official position of the Uniformed Services University of the Health Sciences, National Defense University, United States Air Force, United States Army, Department of Defense, or the United States Government.

## References


American Psychological Association. (2018). *Society for Military Psychology Julius E. Uhlaner Award*. Retrieved from https://www.apa.org/about/awards/div-19-uhlaner.aspx?tab=3.

Appelbaum, S. H., Iaconi, G. D., & Matousek, A. (2007). Positive and negative deviant workplace behaviors: Causes, impacts, and solutions. *Corporate Governance: The International Journal of Business in Society*, 7(5), 586–598.

Bartone, P. T. & Bowles, S. (2018). Hardiness and active coping as resilience factors in military recruiting. Draft manuscript in preparation, National Defense University.

Borman, W. C., White, L. A., Bowles, S.V., Horgen, K. E., Kubisiak, U. C., & Penney, L. M. (2003). U.S. Army recruiter selection research: An update. Paper Presented at the Annual Conference of the International Military Testing Association, Pensacola, FL.

Borman, W. C., White, L. A., Horgen, K. E., Penney, L. M., Kubisiak, U. C., Kaufman, J. D., & Bowles, S. V. (2002). U.S. army recruiting research: An update. Paper presented at the 38th International Applied Military Psychology Symposium, Netherlands.

Borman, W. C., White, L. A., Young, M. C., Kubisiak, U. C., Horgen, K E., Penney, L. M., & Connell, P. W. (2004). U.S. Army Recruiter Selection Research: Development of the Noncommissioned Officer Leadership Skills Inventory (NLSI). Paper presented at the 46th Annual Conference of the International Military Testing Association, Brussels, Belgium.

Bowles, S. & Bartone, P. (2017). Hardiness and problem focused coping resilience factors for military recruiters. Poster session presented at the American Psychological Association Convention, Washington, DC.

Bowles, S. Bartone, P., Cooke, F., & Swisher, S. (2018). Important dimensions for well-being in military recruiters. Poster session presented at the American Psychological Association Convention, San Francisco, CA.

Bowles, S., Cunningham, C. J., De La Rosa, G. M., & Picano, J. (2007). Coaching leaders in middle and executive management: Goals, performance, buy-in. *Leadership & Organization Development Journal*, *28*(5), 388–408.

Bowles, S. V., & Picano, J. J. (2006). Dimensions of coaching related to productivity and quality of life. *Consulting Psychology Journal: Practice and Research*, *58*(4), 232.

Conte, J. M. & Gintoft, J. N. (2005). Polychronicity, big five personality dimensions, and sales performance. *Human Performance*, *18*(4), 427–444.

Dertouzos, J. N. & Garber, S. (2006). *Human Resource Management and Army Recruiting: Analysis of Policy Options*. Santa Monica, CA: Rand.

Dertouzos, J. N. & Garber, S. (2008). *Performance Evaluation and Army Recruiting*. Santa Monica, CA: Rand.

Elliman, T. D., Shannahoff, M. E., Metzler, J. N., & Toblin, R. L. (2018). Prevalence of bystander intervention opportunities and behaviors among US Army soldiers. *Health Education & Behavior*, *45*(5), 741–747.

Furham, A. & Fudge, C. (2008). The Five Factor model of personality and sales performance. *Journal of Individual Differences*, *29*(1), 11–16.

Gomez, C. L., Agnor, C. J., Jenkins, C. M., McClenen R., & Schmid, B. (2015). Enhancing evaluations for recruiting duty by addressing competency in behavioral health providers. Poster Presented at the Annual Meeting of the American Psychological Association, Toronto, Canada.

Grant, A. M., Curtayne, L., & Burton, G. (2009). Executive coaching enhances goal attainment, resilience and workplace well-being: A randomised controlled study. *The Journal of Positive Psychology*, *4*(5), 396–407.

Grant, A. M. (2013). Rethinking the extraverted sales ideal: The ambivert advantage. *Psychological Science*, *24*(6), 1024–1030.

Grawitch, M. J., Gottschalk, M., & Munz, D. C. (2006). The path to a healthy workplace: A critical review linking healthy workplace practices, employee well-being, and organizational improvements. *Consulting Psychology Journal: Practice and Research*, *58*(3), 129–147.

Hafner, M., Stepanek, M., Taylor, J., Troxel, W. M., & Van Stolk, C. (2017). *Why sleep matters (RB-9962-VH).* Retrieved from https://www.rand.org/pubs/research_briefs/RB9962.html(LC).

Horgen, K. E., Nye, C. D., White, L. A., LaPort, K. A., Hoffman III, R. R., Drasgow, F., & Conway, J. S. (2013). Validation of the noncommissioned officer special assignment battery (No. ARI-TR-1328). Personnel Decisions Research INST (PDRI) INC, Tampa, FL.

Jones, R. J., Woods, S. A., & Guillaume, Y. R. (2016). The effectiveness of workplace coaching: A metaanalysis of learning and performance outcomes from coaching. *Journal of Occupational and Organizational Psychology, 89*(2), 249–277.

Kitfield, J. (1995). *Prodigal Soldiers: How the Generation of Officers Born of Vietnam Revolutionized the American Style of War.* Washington, DC: Brassey's.

Litwiller, B., Snyder, L. A., Taylor, W. D., & Steele, L. M. (2017). The relationship between sleep and work: A meta-analysis. *Journal of Applied Psychology, 102*(4), 682–699.

MacKie, D. (2014). The effectiveness of strength-based executive coaching in enhancing full range leadership development: A controlled study. *Consulting Psychology Journal: Practice and Research, 66*(2), 118–137.

National Defense Authorization Act for Fiscal Year 2018, 4 U.S.C §§ 401-402. (2017). Retrieved from https://www.congress.gov/115/crpt/hrpt404/CRPT-115hrpt404.pdf.

Randolph, J. G. (2012). *Selecting the right Marines for recruiting duty: The important role of the commander.* Retrieved from http://www.dtic.mil/dtic/tr/fulltext/u2/a600539.pdf.

Rostker, B. (2006). *I Want You!: The Evolution of the All-Volunteer Force.* Santa Monica, CA: Rand.

Sitser, T., van der Linden, D., & Born, M. P. (2013). Predicting sales performance criteria with personality measures: The use of the general factor of personality, the Big Five and narrow traits. *Human Performance, 26*(2), 126–149.

Stein, S. J. & Book, H. E. (2011). *The EQ Edge: Emotional Intelligence and Your Success.* John Wiley & Sons.

U.S. Department of the Air Force. (2018). *Special Duty Catalog (SPECAT).* Retrieved from http://static.e-publishing.af.mil/production/1/afrs/publication/afrsi84-101/afrsi84-101.pdf.

U.S. Department of the Air Force. (2017). *Air Force Instruction 36-2002.* Retrieved from http://static.e-publishing.af.mil/production/1/af_a1/publication/afi36-2002/afi36-2002.pdf.

U.S. Department of the Air Force. (2018). *Air University Community College of the Air Force 2017-2019 General Catalog Number 21.* Retrieved from page 65, https://www.my.af.mil/gcss-af/USAF/AFP40/d/s6925EC13447C0FB5E044080020E329A9/Publications/2017-2019%20CCAF%20General%20Catalog.pdf.

U.S. Department of the Army. (2016). *Assignment of Enlisted Personnel to the U.S. Army Recruiting Command. Army Regulation 601-1. Washington, DC.* Retrieved from http://www.eis.army.mil/uploaded_files/AEPARMY_RC_HQDA_11OCT2016.pdf.

U.S. Navy Personnel Command. (2018). *Recruiting Duty (MILPERSMAN 1306-964).* Retrieved from https://www.public.navy.mil/bupers-npc/reference/milpersman/1000/1300Assignment/Documents/1306-964.pdf/Documents/1306-964.pdf P4.

Vandaveer, V. V., Lowman, R. L., Pearlman, K., & Brannick, J. P. (2016). A practice analysis of coaching psychology: Toward a foundational competency model. *Consulting Psychology Journal: Practice and Research, 68*(2), 118–142.

Verbeke, W., Dietz, B., & Verwaal, E. (2011). Drivers of sales performance: A contemporary meta-analysis. Have salespeople become knowledge brokers? *Journal of the Academy of Marketing Science, 39*(3), 407–428.

White, L. A., Borman, W. C., & Bowles, S. (2001). New tools for selecting high potential Army recruiters. Paper presented at the *109th Annual Convention of the American Psychological Association*, San Francisco, CA.

# 15
# VALIDITY ASSESSMENT IN MILITARY PSYCHOLOGY

*Noah K. Kaufman and Shane S. Bush*

Military personnel are at risk for cognitive, emotional, and behavioral problems. Whether in training, routine daily activities, or combat environments, the potential exists for physical or psychological trauma. Psychological and neuropsychological evaluations, therefore, serve an essential role in understanding the possible impact of such events on the functioning and wellbeing of military personnel. For the collected data to reflect the constructs of interest (e.g., anxiety, memory), test-takers must respond accurately to questions about their backgrounds, put forth good effort on cognitive ability tests, and respond in a consistent and honest manner to inventories of psychological functioning. Failure to respond in a valid manner limits the usefulness of the results for the purposes for which they were intended, except when the purpose of the evaluation is to assess for the presence of deceptive behavior (e.g., malingering).

Military personnel may generate invalid data for multiple reasons; for example, they may minimize problems in order to remain with or return to their units, or they may exaggerate or fabricate problems to avoid unpleasant or dangerous duties or to obtain benefits. Accordingly, a multi-method approach to assessing the validity of symptoms and performance is an essential component of psychological and neuropsychological evaluations of military personnel, as well as veterans.

The purpose of the present chapter is to provide an evidence-based review of validity assessment, including discussion of common biasing factors operating *during* and *after* data collection; symptom, performance, and response validity; and misconceptions about validity assessment. Because the availability and use of standardized assessment measures varies within and among countries and cultures, an effort has been made to emphasize principles and procedures that are broadly relevant and therefore more readily useful to practitioners across a range of specific contexts.

## Psychological assessment in the military

In the United States, the field of psychological assessment got much of its impetus from World War I (Yoakum & Yerkes, 1920) because it was necessary "to point out the feeble-minded and those incapable of military service because of mental deficiency and … to find those of unusual or special ability" (p. 10). Test development and validation during this early period included uncommonly large sample sizes (N = 1,726,966), and test-taking instructions were reflective of the times (Yoakum & Yerkes, 1920):

> Attention! The purpose of this examination is to see how well you can remember, think, and carry out what you are told to do. We are not looking for crazy people. The aim is to help find out what you are best fitted to do in the Army. The grade you make in this examination will be put on your qualification card and will also go to your company commander. Some of the things you are told to do will be very easy. Some you may find hard. You are not expected to make a perfect grade, but do the very best you can. (p. 53)

At present, aptitude testing remains an integral part of psychological assessment in the military with new recruits. However, it is also currently important for mental health professionals to accurately characterize mental health problems among military personnel and veterans. Because both types of assessment, cognitive and general mental health, routinely take place in the military; validity assessment is also routinely needed.

## Deception and validity assessment in the military

Deception can be an understandable, and effective, strategy to improve one's circumstances. Animals use deception for this reason (Stone & Boone, 2007), as do humans, including very young children (Peterson & Peterson, 2015). The implication here is that one need not be particularly experienced or sophisticated to intentionally deceive.

In his discussion of this topic, Frederick (2012) emphasized that intentional deception to achieve a desired outcome is not inherently bad and that it can sometimes be understandably adaptive: "It is therefore usually not useful to consider people 'bad' who malinger. In the right context, malingering can be considered heroic behavior—a way to deceive enemies. King David, for example, once escaped from his enemies by feigning mental illness" (p. 229). That said, there are times where the intentional use of deception is not heroic, in which case strategies to correctly identify deceivers are needed. For example, deception by criminal defendants to improve their legal circumstances can be a cowardly act to evade punitive servitude.

Another form of servitude, although not of a penal nature, is military service. In the military, deception through exacerbation or fabrication of problems is used to avoid service and to obtain medical discharges and/or disability-based benefits. However, the percentage, or base rate, of deceivers in the military has not been well established. To illustrate, in a recent large-scale study with a sample size of 28,065,568 military healthcare visits, clinicians only diagnosed 1,074 individuals as malingering or having factitious disorder (Lande & Williams, 2013). Other estimates put the base rate at about 0.5% (Armed Forces Health Surveillance Center [AFHSC], 2013). If one accepts these numbers, deception to improve one's circumstances in the military is practically non-existent.

In contrast, Denning and Shura (2017) found that 33% to 52% of disability-seeking military veterans were malingering mild traumatic brain injury (mTBI), which does not take into consideration other forms of malingering. Armistead-Jehle (2010) reported very high rates (58%) of inadequate effort output among veteran examinees referred for mTBI-related evaluations, a theme supported by data from Wall, Graver, and Shurak (2013), who found that over 50% failed validity testing in their sample of 1,178 active-duty soldiers, military retirees, and military beneficiaries. Meanwhile, Whitney, Shepard, Williams, Davis, and Adams (2009) reported lower rates (17%) of veterans from either Operation Iraqi Freedom or Operation Enduring Freedom scoring below cut-offs on validity testing. Similarly, over half of American Vietnam veterans in a study by Freeman, Powell, and Kimbrell (2008) exaggerated symptoms of posttraumatic stress disorder (PTSD).

Because studies examining exaggeration of problems do not cover military personnel who deny or minimize symptomatology for the purpose of returning to, or remaining with, their unit or for other psychological reasons, the full extent of deception in the military remains unknown. Moreover, because currently reported estimates of intentionally deceptive behavior in the military vary so radically, it is premature to conclude that a solid understanding as to the base rate for this behavior exists.

It is not premature, however, to appreciate that estimates of malingering in the military are influenced by two factors: (1) the negative consequences of diagnosing malingering *in the military* and (2) the methods used to identify malingerers. Regarding the first factor, those identified as malingering may be tried in military court (i.e., court-martialed) under Article 115 of the Uniform Code of Military Justice (UCMJ). Although there are exceptions in non-military settings (e.g., *U.S. v. Greer, 1998*), non-military malingering generally does not trigger, or enhance, legal consequences. Regarding the second factor, many healthcare clinicians working in a military context do not use scientific methods to detect malingering, which is problematic given what has been known about the superiority of actuarial methods, as opposed to unstructured clinical judgment, since the 1950s (Meehl, 1954) and that methods for measuring deception have existed for decades (Pankratz, 1979; Rey, 1941). While the field of neuropsychology fully embraces the importance of validity assessment (Bush et al., 2005), the same does not hold for other healthcare fields.

In short, some mental health clinicians working with military populations are timid about labeling soldiers and veterans as malingerers because of harsher consequences, which affects base rate estimates. Similarly, mental health clinicians often do not go beyond unstructured clinical judgment to detect deception, also skewing base rate estimates.

## *Biasing factors influencing psychological and neuropsychological test scores*

Psychological and neuropsychological assessment in any context involves the collection of data. These data are vulnerable to many biasing influences, as conveyed in Figure 15.1. It is, therefore, always necessary for psychologists to consider these influences and, sometimes, account for their impact on the data.

The influence of demographic variables such as sex, education level, racial group, and age can be statistically controlled for with many neuropsychological tests, the importance of which has been articulated in the neuropsychological literature. In contrast, other tests (e.g., most measures of IQ, academic achievement, personality) tend to use stratified general U.S. population norms (Strauss, Sherman, & Spreen, 2006), meaning the examinee's score is computed in comparison to others their age and what is normal for all others in the entire United States, across levels of sex, education level, racial group, and socioeconomic status.

Calibrating scores based on multiple demographic factors is not always necessary because some referral questions call for a comparison of the examinee to what is normal in the general population. Although it is typically necessary to control for age, it is not always necessary to control for other demographic variables, especially if the demographic variable does not confound the neuropsychological data point of interest (e.g., memory, attention, executive functioning). If the demographic variable *does* confound the neuropsychological data point, it is then necessary to consider the nature of the relationship. To illustrate, age, sex, and racial group status can have a *unidirectional* relationship with neuropsychological data points, whereas education level has a *bidirectional* relationship with neuropsychological data points (Boone, 2013a), meaning that

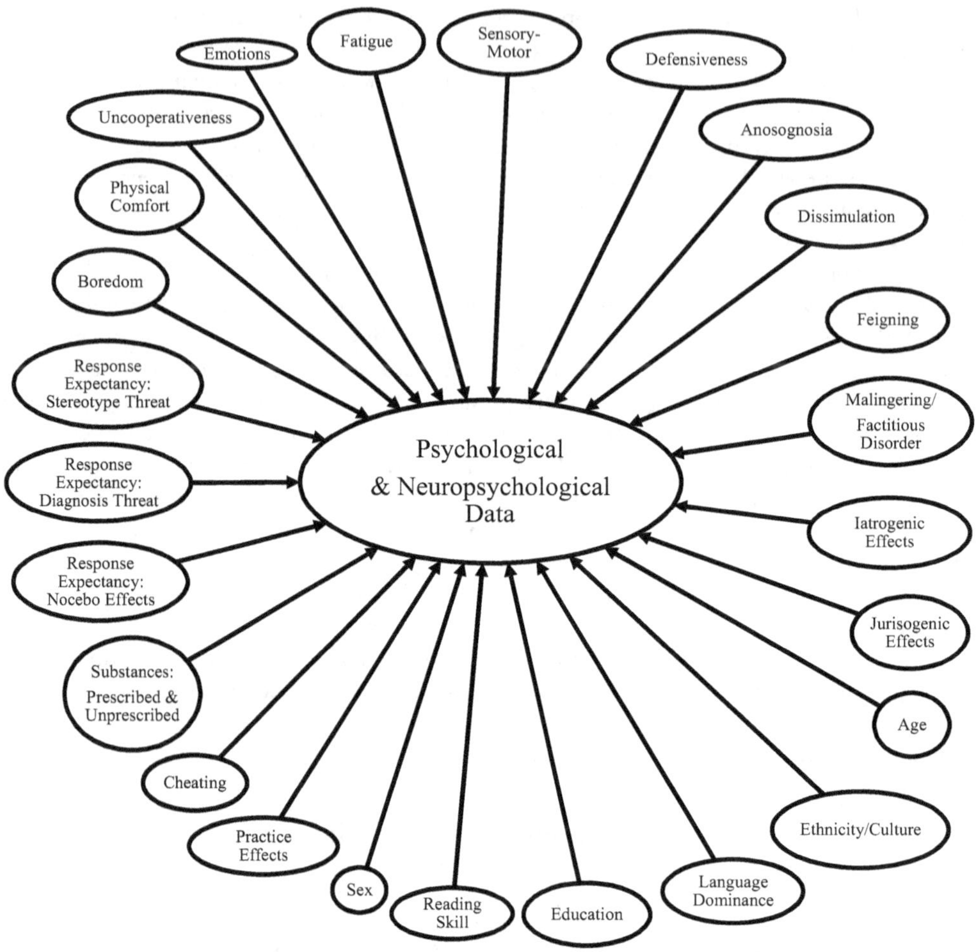


*Figure 15.1* Biasing factors influencing psychological and neuropsychological data.

cognitive ability influences how much education a person gets, and how much education a person gets influences their cognitive abilities.

Regarding the manner in which examinees respond to questions from psychological and neuropsychological tests (i.e., their *response style*), different terminology is encountered. Rogers and Bender (2013) provided helpful definitions, the following of which are included in Figure 15.1:

> *malingering* ("deliberate fabrication or gross exaggeration of psychological or physical symptoms for the fulfillment of an external goal"); *feigning* ("deliberate fabrication or gross exaggeration of psychological or physical symptoms without any assumptions about goals [for this behavior]"; *dissimulation* ("general term to describe an inaccurate portrayal of symptoms ... used when more precise terms [do not clearly apply]"); and *defensiveness* ("the polar opposite of malingering; it is the deliberate denial or gross minimization of symptoms in the service of an external goals"). (p. 518)

Other response styles in Figure 15.1 are *uncooperativeness* (Vore, 2007), which is when an examinee is unwilling to go through with all or part of the evaluation, and *factitious disorder*, which is fabricated or exaggerated health problems for the purpose of receiving attention from healthcare providers. Other terms not included in Figure 15.1 include social desirability, impression management, suboptimal effort, over-reporting, and secondary gain, all of which have been described by Rogers (2008) as lacking in specificity. Arguably, defensiveness encapsulates both social desirability and impression management, while the other three terms are more specifically characterized using malingering, factitious disorder, feigning, or defensiveness.

### *Biasing factors influencing psychologist's interpretation of test scores*

In addition to threats to score validity during the data collection process, threats to score validity exist *after* the data have been collected (see Figure 15.2). Evaluators can be influenced, without their awareness, by positive and negative qualities of the examinee.

For example, a well-groomed, polite, higher-status examinee (e.g., military officer) might be perceived by the evaluator in a more positive light, positively skewing data interpretation. This

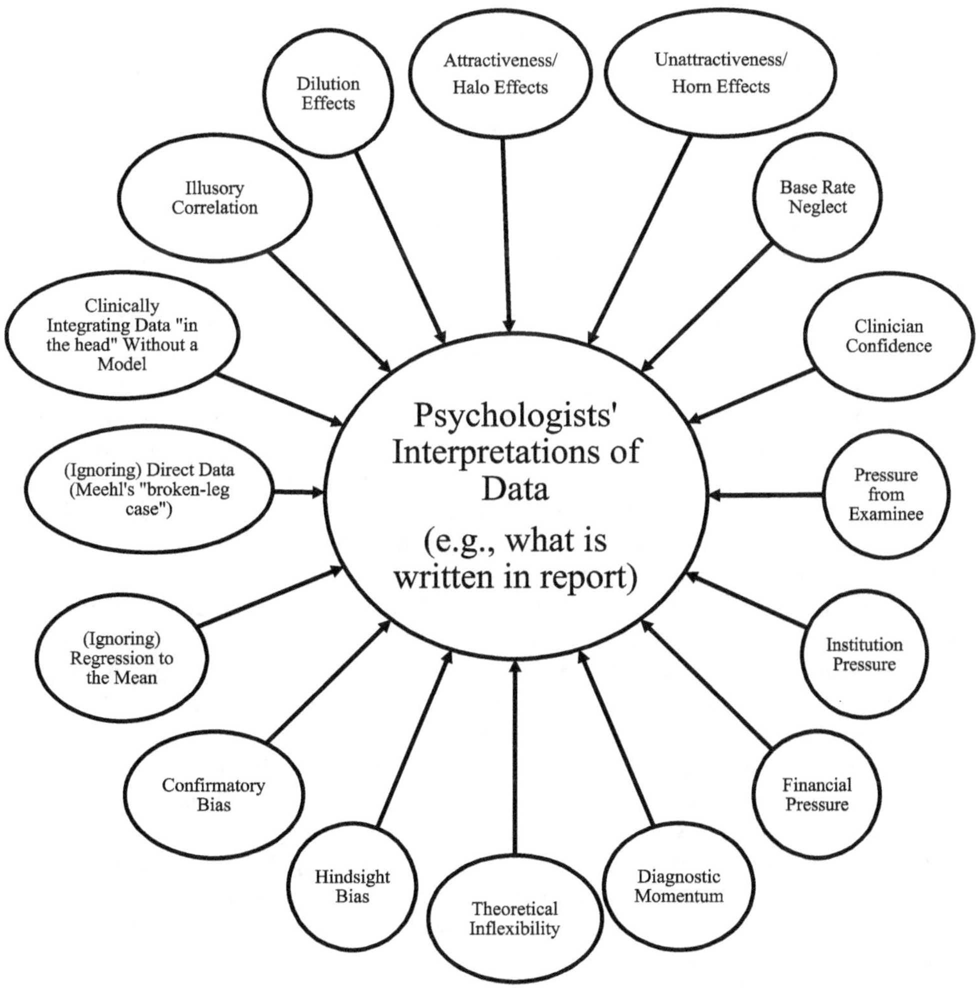


*Figure 15.2* Biasing factors influencing psychologists' interpretations of data.

is the *attractiveness/halo effect*. In contrast, evaluators may develop subtle biases against examinees with unattractive characteristics, which is the *unattractiveness/horn effect*. In both scenarios, factors of tangential importance (i.e., examinee attractiveness) create bias in data interpretation.

A similar biasing effect can occur when other factors of tangential importance dilute the impact of valid variables during data interpretation. As an example, Whitehead (1986) reported that diagnostic decision-making accuracy declined among clinicians given Rorschach scores. In other words, diagnostic accuracy was better *without* the Rorschach scores. To further appreciate the *dilution effect*, consider the following from Faust, Bridges, and Ahern (2009):

> A ceiling on predictive accuracy is commonly approached or reached once the three to five most valid, and least redundant, variables have been identified. Beyond this point, additional information tends to increase confidence, not accuracy. More problematically, at times continuing to add information may decrease accuracy because weaker predictors may dilute the stronger predictors. (p. 14)

As the foregoing quote asserts, clinician over-confidence can erode accuracy, because confidence and accuracy do not consistently correlate: "confidence accounts for 2% of variance in judgment accuracy" (Miller, Spengler, & Spengler, 2015; p. 9). Hence, clinicians are encouraged to proceed humbly when interpreting data, to minimize the likelihood that overconfidence will skew data interpretation, thereby resulting in less valid conclusions.

When interpreting psychological and neuropsychological data, it is essential to consider the *base rate* of occurrence for the event or phenomenon because extremely uncommon events are harder to correctly identify, while extremely common events are easier to correctly identify. With specific respect to validity assessment, two examples are worth describing. First, it is known that examinees with healthy nervous systems score low on neuropsychological tests at a certain base rate (Binder, Iverson, & Brooks, 2009). When this rate is not considered, evaluators are more likely to over-pathologize. Second, failure of validity tests also occurs at a rate, although, as described in this chapter, that rate varies by population. When clinicians apply the wrong failure rate to examinees who do not "pass" validity tests, they are likely to make incorrect conclusions about score validity. Rogers and Bender (2013) called this the "fall-through-the-ice mentality" (p. 529). Like a hapless ice skater who drowns after falling through the surface of a frozen pond, evaluators cease any further, or nuanced, explanation of validity data after failure of a single validity test, and the examinee is immediately labeled with a pejorative term (e.g., malingerer, dissimulator), ending any further consideration of the validity data.

Base rates are also important when interpreting symptoms that are suggestive of a disorder. Just as clinicians need to consider how often non-deceivers fail validity tests, they need to consider how often non-disordered members of a population present with symptoms suggestive of a disorder. Failure to do this can result in the clinician buying into what is called an *illusory correlation* (Borum, Otto, & Golding, 1993; Faust & Ahern, 2012).

Psychological and neuropsychological tests are not perfectly reliable, meaning that different scores can be obtained under different circumstances. The more unreliable a test, the more vulnerable test scores are to what is called *regression to the mean* (Campbell & Kenny, 1999), which is when extreme scores do not replicate upon subsequent testing. Instead, scores regress back toward the most likely score, which is the mean. This happens when individuals get an unusually high, or low, score. Therefore, clinicians should consider the likelihood that an extreme score is artifactual, rather than accurate, and that it may not replicate upon future testing.

Validity assessment boils down to decision-making precision by the clinician, which can be carried out "in the head" of the clinician, using clinical/professional judgment, or using more structured methods, including actuarial models, which are mathematical formulas developed through research to optimize decision-making accuracy. If the question of interest can be answered with an actuarial model, it is then the preferred decision-making method, given the vast literature on the superiority of actuarial methods, compared with clinical/professional judgment (Dawes, Faust, & Meehl, 1989; Grove, 2005; Grove & Meehl, 1996; Meehl, 1954). In reality, however, actuarial models are not available for every important decision, including whether a particular examinee, in a particular set of circumstances, has produced valid psychological or neuropsychological data. Not having an actuarial model to judge score validity puts clinicians in a difficult position, because they must render a decision about the validity of an examinee's behavior without the help of a mathematical model that does all the hard thinking. Therefore, it is important for clinicians making this judgment call to acknowledge the limitations of their "in-the-head" decision-making abilities, especially since researchers have shown that clinical judgment about the validity of psychological and neuropsychological data is questionable (Faust, Hart, Guilmette, & Arkes, 1988; Heaton, Smith, Lehman, & Vogt, 1978; Rosenhan, 1973; Trueblood & Binder, 1997; Walters, White, & Greene, 1988).

Rigid, inflexible adherence to decision rules can be just as problematic as confidently making clinical judgments about the validity of psychological and neuropsychological data "in the head." Perhaps the most compelling example of this was provided by the strongest advocate for actuarial models, Paul Meehl (1954), who described a hypothetical example involving an actuarial model that could predict, with a 0.90 probability, that a professor would go to the movies on a particular night. As the example goes, the professor breaks his leg, throwing off the accuracy of the actuarial model. Because the professor breaks his leg (a rare and unexpected circumstance *not* built into the actuarial model), the 0.90 accuracy rate no long holds. The important takeaway from Meehl's "broken-leg" case is that clinical judgment is still important. Hence, choosing between actuarial models and clinical judgment is, perhaps, a false dichotomy. Both methods for integrating psychological and neuropsychological data have a place.

While not an actuarial model, per se, Slick, Sherman, and Iverson (1999) developed what is now called the "Slick criteria" for Malingered Neurocognitive Dysfunction (MND). Originally, these authors proposed three categories of MND: (1) *possible* MND (i.e., examinee has an external incentive to deceive and there is evidence of discrepancy in their self-report or they fail a symptom validity test); (2) *probable* MND (i.e., examinee has an external incentive to deceive and there is stronger evidence of deception in the data [e.g., two or more failed forced-choice validity tests, but *not* below chance-level scoring]); and (3) *definite* MND (i.e., incentive to deceive and below chance-level scoring on forced-choice validity testing). Slick and Sherman (2013) revised and extended their criteria to include definite and probable *Primary* MND, *Secondary* MND (i.e., when MND criteria are met among those with legitimate neurodevelopmental, psychiatric, or neurological disorders of moderate or severe severity), and MND *by Proxy* (i.e., when examinee deception is influenced by another individual).

Presently, attempts are underway to operationally define deception in a manner that falls short of an actuarial model but which goes beyond "in-the-head" clinical judgment. What is *not* apparent, however, is the extent to which this more structured approach improves decision-making about the validity of psychological and neuropsychological data, especially in all populations and circumstances. Moreover, structured approaches like the "Slick criteria" may fuel unjustified confidence about appraising validity (Trueblood & Binder, 1997), a liability that is not specific to mental health professions (Curci, Lanciano, Battista, Guaragno, & Ribatti, 2019; Gillis & Moran, 1981; Meyer, Payne, Meeks, Rao, & Singh, 2013).

Another category of biasing influence, operating after psychological and neuropsychological data have been collected, involves outside pressure, from the examinee (or their supporters), from the Department of Veterans Affairs (VA), or from another party paying the evaluator. Evaluators commonly feel pressure to make clinical decisions that examinees will embrace, including *not* calling them out as deceptive. Meanwhile, there can be pressure on contracted evaluators *not* to question the motives and intentions of military personnel and veterans. Poyner (2010), for example, effectively articulated her disappointing experience as an independently hired psychologist "to conduct compensation and pension evaluations with military veterans seeking VA disability benefits for PTSD and/or Traumatic Brain Injury (TBI)" (p. 130). Despite recommendations to assess for deception with veterans, Dr. Poyner was released from her VA contract because she "used malingering instruments" in her assessments.

## *Symptom, performance, and response validity*

Larrabee (2012) published an important article distinguishing between *symptom validity tests* (SVTs) and *performance validity tests* (PVTs) and underscored the need to use these specific terms, based on "strong diagnostic discrimination for PVTs and SVTs, with a particular emphasis on minimizing false positive errors" (p. 1). Symptom validity is defined as the validity of self-reported symptoms, while performance validity is the validity of performances on tests that require the examinee to mentally exert themselves (e.g., IQ, neuropsychological, and academic achievement tests). Before this publication, it was common to refer to all validity testing, both symptom and performance, as SVTs and/or to use various other terms with less definitional clarity (e.g., "poor or inadequate effort" or "biased responding"). Of note, Larrabee's SVT/PVT terminology is not mutually exclusive with the Rogers and Bender (2013) terminology described earlier.

In 2014, Bush (2014) proposed an additional term, *response validity*, which describes "the accuracy of information provided by examinees during clinical interviews" (p. 434). While little has been written about this newer validity term, it can be operationalized by comparing historical self-report to other data points with potentially more validity. Among soldiers, response validity can be of particular relevance with regard to historical substance use, both prescribed and recreational substance use, given the motivation to underreport problematic substance-use patterns (Harrison, 1997). Hence, comparing self-report to pharmacy and legal (e.g., DWI) records can help identify discrepancies.

Many of the biasing factors in Figure 15.1 can be ruled out, or in, using SVTs and PVTs. Defensiveness, for example, can be measured using validity scales from questionnaires like the Personality Assessment Inventory (PAI) (Morey, 2007) and Minnesota Multiphasic Personality Inventory-2, Restructured Form (MMPI-2-RF) (Ben-Porath, 2012, 2013; Ben, Porath, & Tellegen, 2008). The same holds for intentional over-reporting of symptoms (i.e., dissimulation, feigning, malingering, and factitious disorder). However, it is important to emphasize that these SVTs are not "malingering" tests, per se, since malingering is just one possible explanation for SVT failure.

Reading skill, boredom, fatigue, and uncooperativeness can also affect how examinees report symptoms, which can be detected using SVTs. For example, the Inconsistency (ICN) scale on the PAI "reflects the consistency with which the respondent completed items with similar content" (p. 27). Likewise, the Infrequency (INF) scale on the PAI "is useful in identification of people who complete the PAI in an atypical way because of carelessness, confusion, reading difficulties, or other sources of random responding" (p. 28). Failure on SVTs can be a result of deception (e.g., malingering), but also reading skill, boredom, fatigue, and so forth. Therefore, clinicians must critically think through SVT failures. That said, SVTs can be invaluable in the assessment of valid self-reporting.

Like SVTs, PVTs are vital in the assessment of valid performances on tests involving mental effort. Accordingly, PVT results can help detect deception (e.g., malingering) involving low scores on cognitive tests.

## *Misconceptions about validity assessment*

Although not an exhaustive list, the following five misconceptions about validity assessment are worth fleshing out: (1) examinees who intentionally deceive are "bad"; (2) the base rate for examinee deception is definitively known; (3) examinee deception is of trivial significance; (4) examinee deception occurs in a straightforward, predictable manner (e.g., within the same evaluation and/or across settings); and (5) cutoff scores on validity tests are extremely accurate and work equally well with all populations. Each of these misconceptions will be unpacked sequentially.

Stripped to its core, a mental health assessment is an attempt to identify and explain problematic, if not also embarrassing, behavior, such as excessive anxiety, inattention, forgetfulness, substance-use problems, depression, impulsivity, and so forth. Viewed this way, it is understandable why examinees would be motivated to hide their problematic behavior. It is also understandable why examinees would use deception in an attempt to "get" something, such as money for having a disability or justification for avoiding a difficult, or life-threatening, experience (e.g., active combat). When examinee deception is put on a finer scale of measurement, it becomes easier to appreciate that it is not a dichotomous, "good versus bad" phenomenon. This continuum of invalid responding has special relevance to the military, where malingerers are not just scorned, but court-martialed. In short, automatically concluding that examinee deception equates to "bad" behavior oversimplifies an important issue, which precludes an opportunity to understand the examinee on a more nuanced level.

As discussed earlier in relation to examinees with healthy nervous systems scoring low on neuropsychological tests at a known rate (Binder et al., 2009), familiarity with the base rate of occurrence of an important behavior is crucial because accurate measurement of the important behavior hinges on how often it is likely to be encountered in a given population. Unfortunately, it is not uncommon for mental health evaluators to both under- and overestimate the base rate of deception. The former mistake leads to false-negative conclusions about deception, whereas the latter mistake causes false-positive conclusions about deception.

The obvious question, then, pertains to the base rate for deception. Unfortunately, there is no easy answer to this question because the base rate for deception seems to vary by population and circumstance, and accurate estimates of deception have not been established. Beyond divergent *military-specific* deception base rate estimates (AFHSC, 2013; Armistead-Jehle, 2010; Denning & Shura, 2017; Freeman et al., 2008; Lande & Williams, 2013; Wall et al., 2013; Whitney et al., 2009), the following *general* (i.e., not military-specific) estimates of deception also vary dramatically: 15% (Rogers, Harrell, & Liff, 1993), 15% to 48% (Youngjohn, Burrows, & Erdal, 1995), 38.5% (Mittenberg, Patton, Canyock, & Condit, 2002), 40% (Larrabee, 2003), 1% to 20% (median values) (Sharland & Gfeller, 2007), 5% to 30% (Slick, Tan, Strauss, & Hultsch, 2004), 16% (Rogers, Sewell, & Goldstein, 1994), 15% to 17% (Frederick, 2000), 3.4% (Ruff, Klopfer, & Blank, 2016), and 0% to 30% (Young, 2015).

Clinicians embracing the responsibility of judging the motives for others' problematic behavior leads to awareness about what happens when examinee deception is not identified and when examinee deception is incorrectly identified. With the first scenario, failure to detect deception, examinees are likely to receive disability-based support that is extremely expensive or to avoid unpleasant duty when there is no justifiable medical or psychological reason. For example, Frueh, Grubaugh, Elhai, and Buckley (2007) reported a 79.5% increase in payments to veterans with PTSD between 1999 and 2004, in contrast to a 12.2% increase in payments made to veterans with other disabilities during this same time period. Viewed through a financial lens, PTSD-related payments to veterans from 1999 to 2004 increased by 148.8% (i.e., $4.3 billion each year), compared to an increase of 41.7% in payments for other disabilities. In the second set of circumstances, incorrectly identifying deception, great harm can come to the examinee, including severe legal consequences under Article 115 of the UCMJ. Therefore, the consequential significance of accurate, and inaccurate, decision-making about examinee deception cannot be overstated; it is huge.

Deception occurs in unpredictable ways, sometimes within the same evaluation, but also across different settings (Rogers, Salekin, Sewell, Goldstein, & Leonard, 1998). Accordingly, a clinician might encounter powerful evidence of deception earlier in an assessment process, before the examinee has developed a feel for the demands of the circumstances, but later find less, or no, evidence of deception. Likewise, deception may occur as a function of other factors, such as the quality of the informed consent process (e.g., forewarning examinees about the importance of full participation and engagement, versus failing to do so) or the referral question (Rogers & Salekin, 1998). Hence, mental health evaluators should operate with humility about their ability to correctly identify deception.

Rogers and Bender (2013) emphasized the lack of "laser accuracy of cut scores" (p. 519) on tests of validity, meaning that there is a margin of error associated with any given validity test cut score. Accordingly, they argued that rigid application of a cut score to a given examinee can result in spurious findings, the "fall-through-the-ice mentality" described previously. To minimize this mistake, Rogers and Gillard (2011) recommend using the standard error of measurement (SEM) associated with the test to minimize error. Namely, they assert that validity scores within 1 SEM should be considered *indeterminate*, rather than making more definitive conclusions about the presence or absence of deception. Additionally, the cutoff scores reported in test manuals are sometimes found to be less accurate than scores later established through more extensive research after publication of the test. For this reason, clinicians should remain abreast of ongoing research on validity assessment measures.

Another strategy to avoid false-positive conclusions based on a single failed validity test is to establish a threshold for score invalidity based on *multiple* failed validity tests, not just one. Schutte and Axelrod (2013) described the benefits and liabilities with this approach, including the increased potential for false-positive conclusions when validity tests are strongly correlated (i.e., are multicollinear). In her discussion of neurocognitive symptom validity, Boone (2013b) applied this multiple-cut-score approach to those with dementia:

> [I]f a patient fails 85% of the [PVTs] administered, this would be comparable to the performance of patients with advanced dementia … and such a test taker would not be able to live independently, drive, handle finances, or manage other important ADLs. Thus, if a patient failing 85% of [PVTs] does not demonstrate markedly impaired function in ADLs, the [PVT] failure can be relied upon as evidence of noncredible neurocognitive performance. (p. 47)

Two things should be evident with regard to validity test cut-scores: (1) a rigid, or cavalier, application of cut-scores is likely to increase decision-making error, especially when assessing those with legitimate neurocognitive and/or intellectual deficits (Graue, 2006; Shandera et al., 2010), and, as emphasized previously, (2) evaluators should not use a one-size-fits-all approach to cut-score use.

## Conclusions

Psychological assessment in the United States flourished as a result of World War I because it was necessary to accurately identify those who should be excluded from military service, as well as those of uncommon aptitude. While aptitude testing remains important, mental health professionals working with military personnel must also accurately characterize mental health problems. To effectively accomplish these tasks, evaluators should endeavor to increase the quality of data *collection* and *interpretation*. A vital step in data collection is to properly utilize SVTs and PVTs. It is equally important to stay aware of limitations imposed by "in-the-head" clinical judgment about the presence of deception, as well as misconceptions about validity assessment. While actuarial models are not available for every challenging decision faced by the clinician, other methods (e.g., the revised and extended "Slick criteria" for MND) offer clinicians a more structured way of identifying data among military personnel that are of questionable validity due to examinee deception. Thus, a multi-method approach to validity assessment is ideal. Because accurate identification and understanding of invalid psychological and neuropsychological data among military personnel remains a challenging and nuanced process, it should be carried out thoughtfully and with an awareness of available research, which continues to evolve.

## References

Armed Forces Health Surveillance Center. (2013). Malingering and factitious disorders and illnesses, active component, US Armed Forces, 1998–2012). *Medical Surveillance Monthly Report, 20*(7), 20–24.

Armistead-Jehle, P. (2010). Symptom validity test performance in US veterans referred for evaluation of mild TBI. *Applied Neuropsychology, 17*(1), 52–59.

Ben-Porath, Y. S. (2012). *Interpreting the MMPI-2-RF.* Minneapolis, MN: University of Minnesota Press.

Ben-Porath, Y. S. (2013). *Using the MMPI-2-RF validity scales in forensic assessments.* Retrieved from http://www.apadivisions.org/division-18/publications/newsletters/gavel/2013/07/forensic-assessments.aspx

Ben-Porath, Y. S., & Tellegen, A. (2008). *MMPI-2-RF (Minnesota Multiphasic Personality Inventory-2 Restructured Form): Manual for Administration, Scoring, and Interpretation.* Minneapolis: University of Minnesota Press.

Binder, L. M., Iverson, G. L., & Brooks, B. L. (2009). To err is human: "abnormal" neuropsychological scores and variability are common in healthy adults. *Archives of Clinical Neuropsychology, 24*(1), 31–46.

Boone, K. B. (2013a). *Clinical Practice of Forensic Neuropsychology: An Evidence-Based Approach.* New York, NY: Guilford Press.

Boone, K. B. (2013b). A reconsideration of the Slick et al. 1999 criteria for malingered neurocognitive dysfunction. In K. B. Boone (Ed.), *Assessment of Feigned Cognitive Impairment* (pp. 29–49). New York, NY: The Guilford Press.

Borum, R., Otto, R., & Golding, S. (1993). Improving clinical judgment and decision making in forensic evaluation. *The Journal of Psychiatry & Law, 21*(1), 35–76.

Bush, S. S. (2014). Assessment of symptom and performance validity in veterans. In S. S. Bush (Ed.), *Psychological Assessment of Veterans* (pp. 434–452). New York, NY: Oxford University Press.

Bush, S. S., Ruff, R. M., Tröster, A. I., Barth, J. T., Koffler, S. P., Pliskin, N. H., .... Silver, C. H. (2005). Symptom validity assessment: Practice issues and medical necessity: NAN Policy & Planning Committee. *Archives of Clinical Neuropsychology, 20*(4), 419–426.

Campbell, D. T., & Kenny, D. A. (1999). *A Primer on Regression Artifacts.* New York, NY: Guilford Publications.

Curci, A., Lanciano, T., Battista, F., Guaragno, S., & Ribatti, R. M. (2019). Accuracy, confidence, and experiential criteria for lie detection through a videotaped interview. *Frontiers in Psychiatry, 9*, 1–14.

Dawes, R. M., Faust, D., & Meehl, P. E. (1989). Clinical versus actuarial judgment. *Science, 243*(4899), 1668–1674.

Denning, J. H., & Shura, R. D. (2017). Cost of malingering mild traumatic brain injury-related cognitive deficits during compensation and pension evaluations in the veterans benefits administration. *Applied Neuropsychology: Adult, 26*(1), 1–16.

Faust, D., & Ahern, D. C. (2012). Clinical judgment and prediction. In D. Faust (Ed.), *Coping with Psychiatric and Psychological Testimony* (6 ed., pp. 147–208). New York, NY: Oxford University Press.

Faust, D., Bridges, A. J., & Ahern, D. C. (2009). Methods for the identification of sexually abused children: Issues and needed features for abuse indicators. In K. Kuehnle & M. Connell (Eds.), *The Evaluation of Child Sexual Abuse Allegations: A Comprehensive Guide to Assessment and Testimony* (pp. 3–19). New York: Wiley.

Faust, D., Hart, K. J., Guilmette, T. J., & Arkes, H. R. (1988). Neuropsychologists' capacity to detect adolescent malingerers. *Professional Psychology: Research and Practice, 19*(5), 508–515.

Frederick, R. I. (2000). Mixed group validation: A method to address the limitations of criterion group validation in research on malingering detection. *Behavioral Sciences & the Law, 18*(6), 693–718.

Frederick, R. I. (2012). Malingering/cooperation/effort. In D. Faust (Ed.), *Coping with Psychiatric and Psychological Testimony* (6 ed., pp. 229–247). New York, NY: Oxford University Press.

Freeman, T., Powell, M., & Kimbrell, T. (2008). Measuring symptom exaggeration in veterans with chronic posttraumatic stress disorder. *Psychiatry Research, 158*(3), 374–380.

Frueh, B. C., Grubaugh, A. L., Elhai, J. D., & Buckley, T. C. (2007). US Department of Veterans Affairs disability policies for posttraumatic stress disorder: Administrative trends and implications for treatment, rehabilitation, and research. *American Journal of Public Health, 97*(12), 2143–2145.

Gillis, J. S., & Moran, T. J. (1981). An analysis of drug decisions in a state psychiatric hospital. *Journal of Clinical Psychology, 37*(1), 32–42.

Graue, L. O. (2006). *Detection of malingered mental retardation.*

Grove, W. M. (2005). Clinical versus statistical prediction: The contribution of Paul E. Meehl. *Journal of Clinical Psychology, 61*(10), 1233–1243.

Grove, W. M., & Meehl, P. E. (1996). Comparative efficiency of informal (subjective, impressionistic) and formal (mechanical, algorithmic) prediction procedures: The clinical–statistical controversy. *Psychology, Public Policy, and Law, 2*(2), 293–323.

Harrison, L. (1997). The validity of self-reported drug use in survey research: An overview and critique of research methods. *National Institute on Drug Abuse Research Monograph, 167*, 17–36.

Heaton, R. K., Smith, H. H., Lehman, R. A., & Vogt, A. T. (1978). Prospects for faking believable deficits on neuropsychological testing. *Journal of Consulting and Clinical Psychology, 46*(5), 892–900.

Lande, R. G., & Williams, L. B. (2013). Prevalence and characteristics of military malingering. *Military Medicine, 178*(1), 50–54.

Larrabee, G. J. (2003). Detection of malingering using atypical performance patterns on standard neuropsychological tests. *The Clinical Neuropsychologist, 17*(3), 410–425.

Larrabee, G. J. (2012). Performance validity and symptom validity in neuropsychological assessment. *Journal of the International Neuropsychological Society, 18*(04), 625–630.

Meehl, P. E. (1954). *Clinical versus Statistical Prediction: A Theoretical Analysis and a Review of the Evidence.* Minneapolis, MN: University of Minnesota Press.

Meyer, A. N., Payne, V. L., Meeks, D. W., Rao, R., & Singh, H. (2013). Physicians' diagnostic accuracy, confidence, and resource requests: A vignette study. *JAMA Internal Medicine, 173*(21), 1952–1958.

Miller, D. J., Spengler, E. S., & Spengler, P. M. (2015). A meta-analysis of confidence and judgment accuracy in clinical decision making. *Journal of Counseling Psychology, 62*(4), 553–567.

Mittenberg, W., Patton, C., Canyock, E. M., & Condit, D. C. (2002). Base rates of malingering and symptom exaggeration. *Journal of Clinical and Experimental Neuropsychology, 24*(8), 1094–1102.

Morey, L. C. (2007). *Personality Assessment Inventory Professional Manual.* Lutz, FL: Psychological Assessment Resources.

Pankratz, L. (1979). Symptom validity testing and symptom retraining: Procedures for the assessment and treatment of functional sensory deficits. *Journal of Consulting and Clinical Psychology, 47*(2), 409–410.

Peterson, E., & Peterson, R. (2015). Understanding deception from a developmental perspective. In M. Kirkwood (Ed.), *Validity Testing in Child and Adolescent Assessment: Evaluating Exaggeration, Feigning, and Noncredible Effort* (pp. 42–61). New York, NY: The Guilford Press.

Poyner, G. (2010). Psychological evaluations of veterans claiming PTSD disability with the Department of Veterans Affairs: A clinician's viewpoint. *Psychological Injury and Law, 3*(2), 130–132.

Rey, A. (1941). L'examen psychologique dans les cas d'encephalopathie traumatique. *Archives de Psychologie, 28*, 286–340.

Rogers, R. (2008). An introduction to response styles. In R. Rogers (Ed.), *Clinical Assessment of Malingering and Deception* (3 ed., pp. 3–13). New York, NY: Guilford Press.

Rogers, R., & Bender, S. D. (2013). Evaluation of malingering and related response styles. In R. K. Otto (Ed.), *Handbook of Psychology* (2 ed., pp. 517–540). Hoboken, NJ: John Wiley & Sons, Inc.

Rogers, R., & Gillard, N. D. (2011). Research methods for the assessment of malingering. In B. Rosenfeld & S. Penrod (Eds.), *Research Methods in Forensic Psychology* (pp. 174–188). Hoboken, NJ: John Wiley & Sons Inc.

Rogers, R., Harrell, E. H., & Liff, C. D. (1993). Feigning neuropsychological impairment: A critical review of methodological and clinical considerations. *Clinical Psychology Review, 13*(3), 255–274.

Rogers, R., & Salekin, R. T. (1998). Beguiled by Bayes: A re-analysis of Mossman and Hart's estimates of malingering. *Behavioral Sciences & the Law, 16*(1), 147–153.

Rogers, R., Salekin, R. T., Sewell, K. W., Goldstein, A., & Leonard, K. (1998). A comparison of forensic and nonforensic malingerers: A prototypical analysis of explanatory models. *Law and Human Behavior, 22*(4), 353–367.

Rogers, R., Sewell, K. W., & Goldstein, A. M. (1994). Explanatory models of malingering: A prototypical analysis. *Law and Human Behavior, 18*(5), 543–552.

Rosenhan, D. L. (1973). On being sane in insane places. *Science, 179*(4070), 250–258.

Ruff, R. M., Klopfer, J., & Blank, J. (2016). Homogenous base rates for malingering in neuropsychological examination of litigants. *Psychological Injury and Law, 9*(2), 143–153.

Schutte, C., & Axelrod, B. N. (2013). Use of embedded cognitive symptom validity measures in mild traumatic brain injury cases. In D. A. Carone & S. S. Bush (Eds.), *Mild Traumatic Brain Injury: Symptom Validity Assessment and Malingering* (pp. 159–181). New York, NY: Springer Publishing Company.

Shandera, A. L., Berry, D. T., Clark, J. A., Schipper, L. J., Graue, L. O., & Harp, J. P. (2010). Detection of malingered mental retardation. *Psychological Assessment, 22*(1), 50–56.

Sharland, M. J., & Gfeller, J. D. (2007). A survey of neuropsychologists' beliefs and practices with respect to the assessment of effort. *Archives of Clinical Neuropsychology, 22*(2), 213–223.

Slick, D. J., & Sherman, E. M. (2013). Differential diagnosis of malingering. In D. A. Carone & S. S. Bush (Eds), *Mild Traumatic Brain Injury: Symptom Validity Assessment and Malingering* (pp. 57–72). New York, NY: Springer Publishing Company.

Slick, D. J., Sherman, E. M., & Iverson, G. L. (1999). Diagnostic criteria for malingered neurocognitive dysfunction: Proposed standards for clinical practice and research. *The Clinical Neuropsychologist, 13*(4), 545–561.

Slick, D. J., Tan, J. E., Strauss, E. H., & Hultsch, D. F. (2004). Detecting malingering: A survey of experts' practices. *Archives of Clinical Neuropsychology, 19*(4), 465–473.

Stone, D. C., & Boone, K. B. (2007). Feigning of physical, psychiatric, and cognitive symptoms: Examples from history, the arts, and animal behavior. In K. B. Boone (Eds.), *Assessment of Feigned Cognitive Impairment* (pp. 3–12). New York, NY: Guilford Press.

Strauss, E., Sherman, E., & Spreen, O. (2006). *A Compendium of Neuropsychological Tests.* Administration, Norms, and Commentary. New York: Oxford University Press.

Trueblood, W., & Binder, L. M. (1997). Psychologists' accuracy in identifying neuropsychological test protocols of clinical malingerers. *Archives of Clinical Neuropsychology, 12*(1), 13–27.

*U.S. v. Greer*, No. 96–11443, 158 F.3d 228 (Court of Appeals, 5th Circuit 1998). Retrieved from https://casetext.com/case/us-v-greer-18

Vore, D. (2007). The disability psychological independent medical evaluation: Case law, ethical issues, and procedures. In A. M. Goldstein (Ed.), *Forensic Psychology: Emerging Topics and Expanding Roles* (pp. 489–510). Hoboken, NJ: John Wiley & Sons, Inc.

Wall, A. M., Graver, C. J., & Shurak, N. R. (2013). Base rates of suboptimal effort in a broad clinical population. *The Clinical Neuropsychologist, 27*(4), 607–608.

Walters, G. D., White, T. W., & Greene, R. L. (1988). Use of the MMPI to identify malingering and exaggeration of psychiatric symptomatology in male prison inmates. *Journal of Consulting and Clinical Psychology, 56*(1), 111–117.

Whitehead, W. C. (1986). *Clinical decision making on the basis of Rorschach, MMPI, and automated MMPI report data.* University of Texas Health Science Center at Dallas, *Unpublished doctoral dissertation.*

Whitney, K. A., Shepard, P. H., Williams, A. L., Davis, J. J., & Adams, K. M. (2009). The medical symptom validity test in the evaluation of operation Iraqi freedom/operation enduring freedom soldiers: A preliminary study. *Archives of Clinical Neuropsychology, 24*(2), 145–152.

Yoakum, C. S., & Yerkes, R. M. (1920). *Army Mental Tests.* New York, NY: Henry Holt & Company.

Young, G. (2015). Malingering in forensic disability-related assessments: Prevalence 15±15%. *Psychological Injury and Law, 8*(3), 188–199.

Youngjohn, J. R., Burrows, L., & Erdal, K. (1995). Brain damage or compensation neurosis? The controversial post-concussion syndrome. *The Clinical Neuropsychologist, 9*(2), 112–123.

# 16

# REINTEGRATION AND MILITARY FAMILY HEALTH

## Military training and its relationship to post-deployment role conflict in intimate partner relationships

*E. Ann Jeschke, Jessica M. LaCroix, Amber M. Fox, Laura A. Novak, and Marjan Ghahramanlou Holloway*

### Background

Transitioning from the deployment environment to the home environment is considered by researchers and service members alike to be one of the most stressful and demanding aspects of military service (Cameron Kelly, Howe-Baksdale, & Gitelson, 2011; Glynn, 2012), making it difficult for everyone involved (Sayers, 2011; Marlantes, 2011). Reintegration is an experience shrouded in unfamiliar emotions, behaviors, and adjustments. A successful homecoming requires service members and their families to adapt to the changes that have occurred during deployment. The family unit's critical role in the health and combat readiness of the service member necessitates that helping professionals focus on family health (Cabrera, Figley, Yarvis, & Cox, 2012; Cooper, Pasquina, & Drach, 2011; Duncan, 1976). Researchers studying the post-deployment reintegration process have shown that maintaining a cohesive support network, especially within the family structure, reduces the likelihood that service members will experience mental health conditions (Borus, 1973; D. W. King, King, Foy, Keane, & Fairbank, 1999; L. A. King, King, Fairbank, Keane, & Adams, 1998; Koenen, Stellman, Stellman, & Sommer Jr, 2003). For this reason, policy makers and researchers have "taken an avid interest in developing tools to make the transition process easier for service members and their families" in order to reduce the potential for long-term psycho-social difficulties stemming from post-deployment reintegration (Garber & Zamorski, 2012, p. 397).

### Theoretical objectives

We theoretically engage the broad question of how post-deployment reintegration impacts family health by focusing on the service member's intimate partner relationship within the larger military life-cycle. Our specific goal is to construct a meaningful frame for post-deployment challenges that are experienced across the military and impact the well-being of each partner and the intimate partner relationship. This frame can serve as the foundation from which other more specific post-deployment challenges can be explored. We present two different ways in which reintegration challenges in the intimate partnership have been framed; next, we attempt to integrate cogent insights from both frames to articulate a more nuanced foundation upon which a broader set of post-deployment challenges impacting intimate partner relationship health can be based.

## Key definitions

We use the term *intimate partner relationship* to include any current or previous pair-bonded partnership wherein both parties are/were dependent on each other to achieve certain outcomes, needs, or goals in life (Finkel, Simpson, & Eastwick, 2017). Types of relationships included within our definition are: current spouses (including common law), dating partners (including first date), former marital partners (due to separation or divorce), and former dating partners (Saltzman, Fanslow, McMahon, & Shelley, 1999). We rely on the Centers for Disease Control (Saltzman et al., 1999) categorizations of violence as: physical, sexual, threats of physical or sexual, and emotional (including coercive tactics). The word *universal* is also important to our analysis: we want to achieve the most good for the most people by capturing challenges most service members and intimate partners experience, regardless of age, gender, sexual orientation, military branch, rank, and/or career specialty.

## Theoretical orientation

We want to avoid too narrowly framing reintegration challenges for several reasons. First, a narrow frame risks seeing a partial image as the whole picture. This picture could potentially cause the needs of many service members and their intimate partners to go unexplored and unmet. Second, if reintegration challenges go unmet, they could, in turn, lead to chronic behavioral health concerns and a possible reduction in combat readiness. Third, we are working within a logic like that of worker's compensation. Namely, if the job causes certain negative health outcomes, then there is a responsibility for the employer to provide resources to counteract the consequences of degraded health.

## Outline of analysis

Current research on military family health focusing on intimate partner relationships tends to fall into two general categories. One body of literature examines the relationship between reintegration challenges and intimate partner violence (IPV). In contrast, an emerging body of literature looks at reintegration challenges as role conflict relative to role adjustment after periods of separation. We provide an overview of each set of literature to evaluate how insights from both work together to address reintegration challenges faced by most service members and their intimate partners. The first two sections of our analysis focus on the IPV frame. Specifically, we critically review literature linking military training with post-deployment IPV. Next, we consider the role of military training in the creation and maintenance of the warrior ethos and rigid masculine identity, which may provide a pathway for IPV. The third section expands on the notion of role conflict found in male gender role conflict (MGRC) to broaden the IPV frame. To address universal post-deployment reintegration challenges, we identify enhanced cognitive flexibility as one strategy for ameliorating rigid role identification.

## Section I: Military training and interpersonal violence

We begin with literature that has the potential to achieve our goals because it is based on military training. Within this frame, popular press and research connect military training to IPV. Some even suggest that when it comes to issues of family health, not enough time has been spent evaluating the connection between military training and IPV (Gimbel & Booth, 1994).

In *Odysseus in America*, Jonathan Shay (2002), a clinical psychologist best known for his work with Vietnam veterans, suggests that military training and combat exposure lead service members to engage in uncontrolled fits of passion upon coming home from deployment. Shay claims that the effective service member is trained by the military to excel at the art of deception, disregarding rules, responding with instant aggression, and suppressing emotions, particularly compassion, horror, guilt, tenderness, and grief. Shay illustrates how military training develops a warrior spirit in service members that is nourished by bitterness, violence, and anger, and suggests that one of the ugliest outcomes can be seen post-deployment, when male service members express "hostility and habitual disrespect toward women" (p. 65). Shay argues that "rage and the inability to have authority over anger is the single most destructive" aspect of service members' post-deployment life (Shay, 2002). To Shay, these post-deployment challenges are rooted in military training that benefits combat missions but presents in the civilian world as aggressive, even criminal, behavior. Shay concludes that what the military prizes in most service members during times of war is precisely that which makes service members least suitable for domestic life post-deployment.

More recent research also illustrates IPV as a significant concern for service members post-deployment. Williston and colleagues report that "military-related calls into the National Domestic Violence hotline increased from 457 in 2006 to over 1,100 in 2010, and 61% of these calls reported physical abuse" (Williston, Taft, & VanHaasteren, 2015, p. 55). The authors suggest that the etiology of IPV in these situations is in the service members' military training, therefore setting them apart from their civilian counterparts. It is the mission-driven violence and aggressive nature of military training imparted on service members that make this population more prone to IPV upon coming home. The authors also posit that IPV may be more prevalent among the cohort of service members who have been deployed in support of the Global War on Terror (GWOT) due to the occupational stress of ambiguous missions and multiple deployments.

We turn to a military narrative that illustrates the very personal reality of post-deployment IPV faced by service members and their intimate partners. In the book *Thank You for Your Service*, David Finkel (2013) includes the filed report stating the nature of Corporal Jessie Robinson's violent attacks against his wife. In the report, Kristy, Jessie's wife, describes how Jessie repeatedly shoved her, would not let her leave the house, and taunted her. Jessie said that she was lucky she had not made him mad enough to beat her more. As the previous research supports, Kristy assumes that her husband's violent behavior is caused by his training and combat exposure. She concludes that there is no other explanation for why her normally funny, charming husband would be so aggressive.

There is no doubt IPV can be a reality of post-deployment reintegration. Having said that, two key issues need to be acknowledged. First, the connection between military training and IPV is not inevitable in the case of every service member. Second, while theoretically, a causal link between military training and IPV has been postulated, an empirical basis for this assertion is non-existent—scientifically, we must acknowledge that correlation does not mean causation. Relying on this insight and respecting the reality of IPV in the military population post-deployment, we move forward to evaluate whether this initial set of literature achieves our goal of framing a universal set of reintegration challenges that are experienced across all service members and their intimate partners.

## *Assumptions and logical inconsistencies of military training as etiology of intimate partner violence*

Connecting IPV to military training provides a broad way to explore reintegration challenges for intimate partnerships. Ostensibly, all service members receive military training. However, there are conceptual problems with the way press and researchers have connected military training and IPV. The first problem is that much of the literature speaks about military training in generic terms, without making any distinction between the many types of military training. The conflation of military training with combat training, as well as the conflation of deployment with combat exposure, narrows the focus to a select group of service members—predominantly male, enlisted, and in combat arms. Furthermore, it focuses on the male service member as the perpetrator of violence against the intimate partner. As it stands, this perspective excludes female service members and service members who function in support roles. The latter category of service members are much more in number than those serving in direct combat roles. More importantly, a recent meta-analysis established that service members experience a high level of IPV victimization (Sparrow et al., 2018). On the logic provided by the research above, IPV does not immediately appear to meet our objective of framing reintegration challenges experienced by most service members and their intimate partners. We will return to a new logic in the next section to continue our analysis of IPV relative to military training. We turn now to explore our second point concerning workers' compensation.

If combat training and/or combat exposure is at cause for IPV, the military would need to research and develop resources regardless of our utilitarian analysis. In evaluating the workers' compensation logic, we come upon a second logical inconsistency when attempting to connect combat training and/or combat exposure to IPV; that is, the implicit logic assumes a contradictory notion of adaptability. In the literature above, combat training and/or exposure, to the exclusion of any other type of experience, is given pre-eminent power to form a rigid service member. According to those who forward this perspective, upon returning home, service members are rigid and violent. The implicit logic appears to be that the psychic skills formed by combat training are forged in the crucible of combat, hardening service members such that they are inflexible upon coming home. However, if it is true that combat training takes a pacific civilian and creates an aggressive, yet highly adaptive service member, is it not logically consistent to assume that the same person, when given proper reintegration resources, would return to a state of tranquil domesticity? However, the narrative frame does not discuss reintegration training as an option, only that combat training and/or combat exposure is at cause for criminal behaviors in the service member.

The final assumption we want to challenge is that all violence is of the same nature. In other words, the previously mentioned literature assumes that the purposeful application and management of violence on the battlefield is equivalent to violence expressed as outbursts of uncontrollable rage in domestic settings. While there is certain "common-sense" logic in this assumption, it leads to the superficial conclusion that IPV is an inevitable consequence of combat training and exposure. Clearly, not all service members who experience combat training and/or exposure engage in IPV, and there are many service members—as well as civilians—who have experienced no combat training and/or exposure but nonetheless perpetrate IPV.

Violence is a complex human phenomenon. Many conceptually sophisticated thinkers have struggled to taxonomize the various nuanced forms of violence. Adriane Raine's (2013) book *The*

*Anatomy of Violence* drives home the aforementioned point. The insights Raine gained from tests performed with incarcerated perpetrators of domestic violence established that there can be different forms of aggression. Raine's experience highlights the distinction between *reactive* and *proactive* aggression. The former is a result of an individual spontaneously responding without forethought to a provocation, whereas the latter is planned and premeditated. Measuring for reactive and proactive aggression, he found that IPV was "strongly characterized by reactive aggression" (p. 85). Although we do not wish to draw exact comparisons between Raine's experiments and combat training, it seems fair to say that combat training is based on planned, non-reactive aggression. In fact, just war theory, the ethical framework that is meant to guide service member actions in combat, is based on the notion of restraint where the moral use of aggression is proportional to the operational context. Practically speaking, service members in combat arms are trained to engage in a highly controlled escalation of force that moves from shout, to shove, to shoot. Additionally, these service members are taught to recognize and respond appropriately when *de-escalation* of force is warranted, i.e., when to move from shove to shout, rather than shoot.

While it is likely that combat training is connected to IPV in some instances, the strength of that association is hard to evaluate given the complex nature of IPV. Considering the logical inconsistencies discussed, we cannot conclude that combat training and/or combat exposure is a direct cause of IPV in all cases. However, the previous literature raises an important question that must be answered to thoroughly evaluate whether IPV is a wide enough frame to capture universal reintegration challenges; that is, is there a general form of military training that all service members experience that might lead to post-deployment IPV? We continue in the next section by exploring the extent to which military training as the formation of a masculine warrior ethos may be connected to IPV.

## Section II: Military training as formation of masculine warrior ethos

Realizing that IPV is a reality for some service members post-deployment and that military training may be one of the notable causes for this form of aggression, this section continues to explore whether military training due to its contribution to the formation of the warrior ethos engenders a rigid masculine identity that may subsequently contribute to IPV. Since all service members are encouraged explicitly and implicitly to form a warrior ethos across the military life-cycle, this literature allows us to consider the IPV frame in relationship to all service members.

*Warrior ethos* is a term that is often heard in relation to military values. It has been variously captured across times through implicit or explicit warrior codes related to the profession of arms and war fighting cultures. While there is no one explicit and delineated definition of warrior ethos according to which all service members are uniformly trained, it is a general notion that permeates the values, habits, traditions, actions, and identities of all service members. As the warrior ethos applies to all military branches, career specialties, ranks, and genders, we rely on this construct to establish a way to explore a general military training that could potentially be connected to IPV. We begin by looking at ways in which the warrior ethos manifests as masculinity and move forward with research on masculine gender role conflict (MGRC) to establish if IPV is sufficient to capture the most pervasive forms of post-deployment reintegration challenges.

Attempting to define the warrior ethos as masculine is difficult because both the warrior ethos and masculinity function as amorphous, although potently meaningful, ideals that are not specifically delineated by exact content. These culturally normative standards constitute and guide interpretation of value and meaning through symbols, language, and embodied action. Although it is difficult to reduce the content of the warrior ethos and masculinity to a specific set of qualities, signs, empirical perceptions, or behaviors, by structuring meaning and behavior they serve as a medium of reflexive self-interpretation, self-enactment, and enactment with others in dyadic and group relationships. In spite of the complexities of attempting to concretize these concepts, we rely on Leo Braudy's (2003) meticulous analysis to establish that the warrior ethos is (and likely always has been) intertwined with socio-cultural concepts of masculinity.

In his book *From Chivalry to Terrorism: War and the Changing Nature of Masculinity*, Braudy (2003) evaluates how men of war become men of peace by performing a historical investigation of ideal masculinity relative to military norms. His analysis establishes that prior to Napoleon Bonaparte, military training was not connected to a masculine warrior ethos. Instead, the warrior ethos was connected to a man's genealogy, body type, or aristocratic standing. With the development of nation-states and the need for standing armies, Napoleon grounded warrior ethos in a form of military training that was able to take any common man and transform him into a warrior. By the end of World War I, military training was seen as the method for forming not only combat service members, but also masculine warrior ethos both inside and outside of the military.

In the history of the word "man," even the very structure of Indo-European languages has been heavily influenced by "wartime masculinity … and the language of social class" (Braudy, 2003, p. 4). Braudy cogently argues that military masculinity defines being a man in Western society, and that war is a social institution that binds together different masculine rites and traditions. Even though what is coded as masculine constantly changes across cultures and contexts, what has been deemed as man, manly, or masculine is coupled with a set of martial qualities that are attributed to the warrior ethos. Braudy explains that the larger enterprise of war—not just combat—emphasizes "the physical prowess of men," distills "national identity into the abrupt contrast between winning and losing," and "enforces an extreme version of male behavior as the ideal model" of manhood (p. xvi). In times of war, the effort to systematically and uniformly engender a masculine ideal across the military "becomes a top-down and bottom-up effort"(p. xvi). As such, the warrior ethos, especially in war-time "ensures a belief in a purified, totemistic, and one-sided masculinity as a way of ensuring victory" (p. 54).

Considering how many women serve in the military, the reader might be thinking that the creation and maintenance of the warrior ethos does not apply to roughly 15% of the military population (Office of the Deputy Assistant Secretary of Defense for Military Community and Family Policy, 2016). Although not the focus of his book, Braudy (2003) notes that women have always been engaged in the profession of arms, although rates have fluctuated across history. His insight is that women shape the masculine warrior ethos by altering what is labeled as masculine, not by de-coupling masculinity from warrior ethos. As such, female service members are still expected to develop and comport to a masculine warrior ethos to succeed in the profession of arms. One of the most basic forms of training that instills this warrior ethos is drill. At the beginning of the seventeenth century when Napoleon was attempting to mold average men into noble warriors, drill became the basic form of military training, to promote cohesion and a warrior identity.

In his book *Keeping Together in Time: Dance and Drill in Human History*, William H. McNeil (2008) confirms Braudy's previous point through an analysis of synchronized group movement in military drill practice. In this book, McNeil shows how drill practice consolidates military groups, instills identity, alters human feelings, and orients behavior around a common unified goal. After describing his experience in Army basic training where he drilled for hours, McNeil concludes that what seemed to be pointless marching was the means of creating and establishing his identity as a good warrior. Everyone in the military experiences drill as the foundation of all other training. Drill imprints a holistic military identity through movement, symbols, language, habits, and tradition, allowing service members to embody a warrior ethos.

Unlike the medieval period wherein codes of conduct for both masculine warriors and female maidens were explicitly articulated, modern warfare maintains more implicit expectations around masculine and feminine behavior in relationship to the warrior ethos and intimate partnership. In the book *Women and War*, Jean Bethke Elshtain (1995) provides a detailed analysis of how men and women, as combatant or noncombat warriors, are allowed participatory action when functioning in relationship to masculine warrior ethos. Like Braudy's analysis, Elshtain (1995) views the warrior ethos in dialectic tension with the broader socio-political discourse. While recognizing that women have taken part in war, she notes that even if they wore a military uniform, most women have only served as witnesses to men's courage and the results of violent action. Rarely are women who serve as "the ferocious few" allowed to take center stage. Dominant social discourse and contemporary popular understanding have developed symbols, language, and structures that delineate men as fighters playing the role of just warrior and women as non-combatants playing the role of "beautiful soul," even if these women serve in the military. Elshtain (1995) suggests that among the parts women have played in the social narratives of war, they have always been given the socially sanctioned position to act as sexual and maternal partners to men. Until recently, women were not allowed to engage in combat positions. This change in social expectations of gender roles came with mixed reception. Disagreement persists when discussing women serving in combat positions (Sparkjoy Studios, 2009; Swick & Moore, 2018). Breaking out of gender roles, regardless of a person's relationship to the military, is difficult. Implicit ways of being, written into social expectations and norms, serve as moral codes that permeate our collective social consciousness in times of war and peace (Braudy, 2003; Elshtain, 1995). As Braudy explains, the specific content concerning "what male and female represent often doesn't matter as much as their opposition to each other" (Braudy, 2003, p. 213).

At this point, we have established that the warrior ethos is masculine, that all service members are trained in this warrior ethos, and that this ethos both implicitly and explicitly shapes expectations of intimate partner relationships. We now turn to a body of literature on the topic of male gender role conflict (MGRC). Emerging from gender theory and developed in parallel with social role theory (Eagly, 2013), this area of psychological research evaluates how patriarchal social structures not only negatively impact the well-being of women, but also men through gender role conflict. Since the 1960s, there has been growing interest in the way gender identity shapes the experience of being men and women in the world. First, researchers evaluated the subordination of women within hierarchical male structures known as patriarchy. In the late 1970s, researchers also became interested in exploring how restrictive gender roles are harmful to men (Hancock, 2001).

MGRC is based on the assumption that men and women are socialized through rewards and punishment as well as behavioral modeling to conform to gender stereotyped roles (Eagly, 2013; O'Neil, Helms, Gable, David, & Wrightsman, 1986; Pleck, 1983). Gender identity sets implicit expectations of behavior for oneself and others. As such, gender identity guides intra- and interpersonal behavior. Individuals may experience gender in more flexible or rigid adherence to socially sanctioned expressions of gender. MGRC suggests that when men rigidly identify with a confined understanding of "gender roles, and the obligations inherent in these roles," this rigidity "can serve as gender-specific stressors" (Mussap, 2008, p. 72). Rigid identification and adherence to stereotyped male roles has become known as hegemonic, conventional, or traditional masculinity (Barrett, 1996; Cowan & Mills, 2004; Gallagher & Parrott, 2011; Mussap, 2008; O'Neil, 2008). In other words, gender role conflict emerges when one person's rigid adherence to a traditional gender identity conflicts with his/her reflexive behavior towards him-/herself or interpersonal behavior with another person. Similarly, gender role conflict can emerge when someone else's behavior conflicts with a person's rigid gender bound expectation of others' behavior. Restrictive and rigid attachment to gender roles can have negative consequences not only for men but also women, their careers, family, and interpersonal relationships (O'Neil, 2008).

Having established that a masculine warrior ethos shapes the gender identity and conformity of both male and female service members, we assume both men and women service members experience MGRC because their warrior identity is formed, enacted, and judged against a masculine ideal. Unfortunately, O'Neil and others do not provide a definition of traditional masculinity that can link a masculine warrior ethos to MGRC. Therefore, we turn to psychological scales that measure MGRC to gain a clearer picture of the operant definition of masculinity against which gender role conflict is being evaluated. MGRC scales capture a set of prescriptive and proscriptive social norms, known as masculine ideologies that regulate men and women's behavior in relationship to gender ideals. The following constructs are aggregated from these 16 measures: *independence* (self-reliance), *avoiding femininity* (anti-femininity), *achievement/status driven*, *providing for family* (responsibility), *rationality*, *physical toughness* (aggression/violence), *restrictive emotionality* (stoicism), *sexuality* (anti-homosexuality, entitlement to sexual dominance, entitlement to offspring), *control*, and *risk-taking* (adventure) (Thompson Jr & Bennett, 2015).

To link the masculine warrior ethos to IPV requires that we first establish a link between the masculine contents in MGRC and those the military seeks to form when developing a masculine warrior ethos. In other words, we are seeking to understand how MGRC might mediate the relationship between warrior ethos and IPV. Colonel Theodore Scott Westhusing (2003) addressed the topic of the formation of warrior ethos within American war fighting. Many of the personality attributes assigned to the scales of MGRC are present in Westhusing's analysis of warrior ethos, such as toughness, responsibility, competition, aggressiveness, focus, achievement, risk-taking, and rationality. It seems there is at least some overlap when considering the contents of masculinity found in MGRC and the military warrior ethos.

Within the MGRC literature, there is a specific interest in relating traditional masculine behavior and gender rigidity to interpersonally violent behavior, especially in relationship to same-sex relationships, intimate partner relationships, and children. Research examining hostility posits that interpersonal aggression is common in instances where people do not adhere or conform to traditional gender roles (D'Augelli, Grossman, & Starks, 2006; Wirtz, Poteat, Malik, & Glass, 2018). Indeed, the recent controversy regarding transgender people joining and serving in the military illustrates the extent to which rejection of gender nonconforming individuals is socially and institutionally sanctioned (Philipps, 2018). Interpersonal aggression is an outgrowth of gender dissatisfaction and gives rise to a need to maintain control and exert power over others. There is a significant correlation between gender role conflict (GRC) and positive attitudes toward different instances of sexual violence toward women, including sexual harassment, rape myths, hostile sexism, and self-reported sexual and dating violence toward women, suggesting that restrictive gender roles are associated with potential mental health issues across genders (O'Neil, 2008). If MGRC is one possible byproduct of the formation of a masculine warrior ethos, this would potentiate troubling incidences of IPV and might serve as our frame for common reintegration challenges.

## *Obstacles to linking masculine warrior ethos to intimate partner violence through male gender role conflict*

While the general contents assigned to masculinity by the MGRC literature seems to comport well to aspects of masculine warrior ethos, the challenges of connecting this to IPV are many.

Unfortunately, O'Neil (2008) does not provide specific details or statistical information that allow us to evaluate the significance of connections between MGRC and IPV. The specific articles captured in O'Neil's meta-analysis, as well as subsequent publications attempting to assess the relationship between MGRC and IPV, focus solely on men as perpetrators against women and rely on college-aged men or prison inmates to test hypotheses (Cowan & Mills, 2004; Gallagher & Parrott, 2011; Jakupcak, Lisak, & Roemer, 2002; Jakupcak, Tull, & Roemer, 2005; Mussap, 2008). These results are not necessarily generalizable to a military population because military training and its contribution to the formation of the warrior ethos includes both competitive and collaborative attributes. In Westhusing's research, which relies heavily on just war theory, the author articulates a set of collaborative virtues that serve to moderate violence and allow for service members to flourish in the civilian life they seek to protect. He states that the military achieves its mission of decisive victory by subjecting members to "codes of behavior in accord with the common life (*sensus communis*) which they protect. The explicit and implicit norms within such 'codes' are expressed not only by the force of custom, but are also expressed in the moral and legal structure of the military" (Westhusing, 2006, pp. 5–6). In a detailed analysis, Westhusing articulates how virtues such as benevolence and justice function within the warrior ethos; however, for the purposes of this book chapter, it is enough to point out that military training seeks to form a masculine warrior ethos that can contain aggression and counteract rigidity in service members.

Another obstacle to connecting a masculine warrior ethos to IPV through MGRC is that there has been no justification given for choosing the content that was included in traditional masculinity. Instead, MGRC is based on a static and historically concrete definition of traditional masculinity. Originally, O'Neil (1981) relies on a definition of traditional masculinity that was defined from popular press, commercials, and social journals from the 1970s to capture normative social opinion on manhood. While Braudy (2003) and Elshtain (1995) rely on popular culture to explicate the shifts and contents of gender, they are both careful to avoid linking exact behavior to outcomes and provide meticulous justification for every connection they make. Furthermore, Braudy's (2003) analysis shows the complex dialectical tension that links social change to a constantly adapting masculine warrior ethos to interpreted categories such as gender and violence. Relying on a static definition of gender, O'Neil (2008) links traditional masculinity to violent behaviors against women. Thereafter, he develops measurements that test for a connection between masculinity and violence. This measurement then proves his assumption that traditional masculinity leads to violence. In so doing, MGRC inadvertently reifies the very patriarchy it is attempting to counteract.

Furthermore, to give the military singular credit for inculcating a masculine warrior ethos which leads to IPV would contradict the central claim of all the literature upon which we rely. Braudy (2003), Elshtain (1995), and O'Neil (2008) all claim that masculinity, whether traditional or as a warrior ethos, are in dialectical tension with broader social dynamics. Connecting masculine warrior ethos to IPV through MGRC does not provide a means to sift out what aspects of traditional masculinity have been formed before joining the military.

If all service members are formed in a masculine warrior ethos and masculinity is directly linked to IPV, then there would be more cases of IPV across the military. Based on meta-analytic methods, Marshall, Panuzio, and Taft (2005) found that rates of IPV experienced in the military and veteran populations ranged anywhere from 13% to 58%. While Marshall, Panuzio, and Taft (2005) are seriously concerned with the prevalence of IPV in military and veteran populations, they are cautious of making any strong statement about linking military formation directly to IPV due to an inordinate number of confounding variables. Without ignoring the impact and reality of IPV in the military, forming service members in a masculine warrior ethos does not necessarily entail reactive violence amongst intimate partners. Additionally, the previously discussed research attempting to connect MGRC to IPV focuses only on two forms of interpersonal violence: first, men's perpetration of violence against women intimate partners and second, men's perpetration of violence against men who fail to achieve standards of traditional masculinity. IPV in the military remains a serious issue that needs to be addressed. However, given the difficulties directly linking military training with IPV, and similar issues linking military training with IPV via MGRC, IPV alone cannot provide a wide enough frame to create a foundation for universal reintegration challenges.

Evaluating military training, MGRC, and IPV gave us two critical insights with which we will proceed: first, that rigid identification with a role can lead to interpersonal conflict in intimate relationships, and second, that individuals are highly adaptive and able to take on new roles or role-bound meaning when given the proper training.

## Section III: Role conflict as the frame for universal reintegration challenges

Without dismissing IPV as a very important reality in the military, this section explores a body of literature related to natural role conflict that emerges after periods of separation. To widen the notion of role conflict beyond that contained in IPV, we turn to empirical literature on intimate partner relationships.

When evaluating post-deployment intimate partner relationship challenges, it is important to understand that both partners in the intimate partnership as well as the relationship itself are constantly adapting and changing across the military life-cycle. As the service member changes, so does the intimate partner and the intimate relationship. Finkel, Simpson, and Eastwich (2017) established 14 core aspects of relationships that together provide a comprehensive relationship theory. This comprehensive relationship theory articulates what a relationship is, how a relationship operates, what tendencies people bring to a relationship, and how context affects a relationship. Since we are evaluating common reintegration challenges within the context of the broader military life-cycle, relying on this integrated model is helpful because it views the relationship in the context of broader social structures as the partners and the relationship change across time.

Relationships depend on specific qualities of each partner as well as the unique patterns that develop within the relationship (Finkel et al., 2017). People bring basic personalities, temperaments, standards, and goals to a relationship that influence the dynamics of each partner and the well-being of the relationship. As interdependence between partners increases, they merge to share similar thoughts, feelings, intentions, and patterns of behavior. Shared experiences also provide the opportunity for each person to establish his/her unique role within the intimate partner relationship. Intimate partner relationships are embedded in a social milieu that shapes the trajectory of both partners and the relationship and establishes contextual dynamics within which the relationship ebbs and flows. Overtime, everyday events impact the development of relationships, for better or worse. As such, partners constantly evaluate each other against an ever-shifting set of negative and positive constructs and perceptions. Conflicts surrounding miscommunication and/or misperceived expectations challenge the stability of relationships over time. However, high external demands on the relationship, such as long periods of separation, perceiving other relationships as more valuable, death, and disaster, predict a worse relationship quality when the demands exceed partners' coping resources. The way people evaluate their partner, cope with stress, and resolve conflict impacts the relationship. Relationships wherein each partner perceives the other as having good intentions, and wherein each partner resolves conflict in a mutually amicable way, generally have more stability across time.

When looking at the military life-cycle, service member intimate partnerships experience both chronic challenges and acute stressors. An understanding of intimate partner relationship development helps to establish that long periods of separation, due to deployment or other causes, could be detrimental to the well-being of the intimate partner relationship. During periods of long separation, service members and their intimate partners forego consistent and everyday opportunities to merge through shared cognitive, affective, motivational, and behavioral interactions. Separation hinders the ability of shared goals and trajectories to establish daily routines. Consequently, each partner must establish a new set of individual goals and trajectories relative to daily tasks that must be accomplished in his or her unique environment, without the physical presence of the other partner. While the intimate partner relationship may still exist during a deployment separation, the thoughts, perceptions, expectations, feelings, intentions, and behaviors that define each role continue to shift on an individual level rather than a dyadic level, limiting the extent to which each partner's role evolves synergistically with the other.

As our introduction noted, all service members come home from war changed. However, often missed during discussions of reintegration challenges are the difficulties faced by intimate partners, who also change during and after deployment-related separation periods. Intimate partners, regardless of gender, take on expanded responsibilities such as care giving, household duties, handling finances, employment, and managing a long-distance relationship with the deployed service member. One of the most important and demanding aspects of reintegration is a reallocation of these roles (Bommarito, Sherman, Rudi, Mikal, & Borden, 2017). One study on post-deployment reintegration found that one in five service members and their intimate partners experienced significant difficulty with reintegration challenges related to: uncertainty about domestic roles, adjusting to new routines, no longer feeling needed by the intimate partner, and re-establishing joint decision-making. Over half of those studied experienced moderate difficulties in one or more of these areas. As intimate partners engage the process of becoming reacquainted, sharing experiences that occurred during separation, restructuring daily routines, redistributing chores, managing strong emotions, and renegotiating responsibilities, military intimate partners report insecurity, numbness, stress conceding autonomy, and awkwardness surrounding sexual intimacy. These stressors lead to problems reconnecting, difficulty communicating, and heightened conflict (Knobloch & Theiss, 2012).

During deployment separation, each partner's role changes because the intimate relationship is no longer the center of gravity defining the roles of each intimate partner. Considering role adaptations that occur during a period of separation, it seems that role conflict in the intimate relationship is likely to emerge as a natural aspect of post-deployment reintegration because each partner will need to redefine boundaries of their own and their partner's role to re-engage in a collaborative intimate partnership. Some degree of role conflict during reintegration and subsequent role adjustment is to be expected in all military intimate partnerships. However, just because something is normal does not mean it does not present salient challenges to intimate relationships. If unaddressed, role conflict could lead to confusion within the partnership that might prompt or reinforce rigid role conformity in either or both partners, leading to a disintegration of well-being in the intimate partnership. Furthermore, role conflict, as established by the MGRC literature, can emerge as an outgrowth of entrenched and rigid role conformity. In other words, a service member or intimate partner might experience heightened role conflict if either or both partners attempt to rigidly conform to the roles that emerged during general military or specific deployment related separation where intimate partners are away from each other in different locations for an extended period of time.

## Role conflict and implications for future research

Relying on the concept of role conflict in relationships after periods of separation seems to expand upon some of the insights learned from the IPV frame to include other post-deployment challenges without necessarily eliminating the significance of IPV when discussing military intimate partner relationship well-being. First, role conflict views the post-deployment reintegration as one aspect of relationship development within the larger military life-cycle. Second, role conflict focuses on both intimate partners as well as the intimate relationship. Third, role conflict applies to any period of separation that occurs across the military life-cycle and is not dependent on rank, military career specialty, deployment type, or gender. Fourth, role conflict allows for challenges to emerge organically as part of the natural reality of reintegration. Finally, role conflict does not exclude the possibility that rigid role conformity may contribute to more extreme versions of interpersonal conflict that manifest as IPV. As such, role conflict facilitates discussion about: (1) how military training may lead to rigid identification with aggression or combat trauma, (2) power dynamics illustrated in the MGRC literature, and (3) IPV as a highly prevalent reality among service members and their romantic partners. In fact, role conflict and subsequent difficulties with role adjustment can provide nuanced insight into how to continue researching IPV within the military community. Consequently, it seems that role conflict as a natural or chronic consequence of separation across the military life-cycle is well suited to achieve our normative goal of finding a way to frame common post-deployment reintegration challenges that serves the most people across the military.

The vulnerability-stress-adaptation (VSA) model of relationship dysfunction fits into the broader theory of relationships on which we are relying and confirms the common challenges of role conflict as a normal part of role adjustment following reintegration, as well as concerns expressed by the IPV frame (Karney & Bradbury, 1995). VSA imports a diathesis-stress model to relationship dynamics to evaluate the interaction between predisposed vulnerabilities each partner brings to the intimate relationship and the impact that life stressors have on intimate relationship well-being over time. According to VSA, vulnerabilities of each partner and the strength of the intimate partnership is tested when partners respond to stress through positive or negative behavioral activities to manage conflict. The theory assumes that links between predisposed vulnerabilities, stressors, and behavior patterns manifest in both adaptive and maladaptive ways, leading to changes in relationship stability over time. The theory seeks to explain how intimate relationships develop, succeed, and fail over time and what adaptive qualities lead to stable partnerships.

Langer, Lawrence, and Barry (2008) utilized the VSA model to study 103 newlyweds for three years to determine the extent to which pre-existing vulnerabilities as personality traits contribute to physical aggression when mixed with the stress of adapting to marriage. Results indicated that individuals under increased stress had more difficulties employing adaptive behaviors, such as active listening, when attempting to engage in relationship-supporting behaviors. Further, personality traits, such as trait impulsivity and trait aggressiveness, were associated with perpetration of physical aggression for both men and women. Further, husbands' trait aggressiveness was associated with wives' experiences of stress as well as wives' perpetration of physical aggression. As such, when applying the VSA model to role adjustment after periods of separation, we see that role conflict is a common reintegration challenge that can potentially lead to more profound problems depending on the interactions between pre-existing vulnerabilities and chronic stress on both the individual and dyadic levels.

Having achieved our normative analysis, we wish to briefly suggest potential directions for future research. A number of recent studies and policy papers have started to investigate the common reintegration challenges discussed above and concluded that resilience training in the form of developing life-skills for both partners is essential (Bommarito et al., 2017; Danish & Antonides, 2013; Marek & D'Aniello, 2014; Park, 2011; Sayers, 2011). However, these studies do not provide specific suggestions for what type of resilience training might work best when addressing common reintegration challenges such as role conflict. We suggest that researchers focus on developing and piloting reintegration programs that seek to counteract the rigid role identification that contributes to interpersonal conflict.

## Cognitive agility

Research on cognitive agility largely stems from organizational and decision-making literature. *Cognitive agility* is defined as the ability to deliberately adapt cognitive processing strategies in accordance with dynamic shifts in situational and environmental demands (Good & Yeganeh, 2012). Cognitive agility is composed of three core components: the ability to *focus* on relevant information, *openness* to new information, and *cognitive flexibility* to move between the two perceptual lenses. Focused attention refers to one's ability to oppose and effectively manage incoming distraction. Openness refers to one's ability to observe and to seek new information in any environment. Finally, cognitive flexibility refers to one's ability to switch mental activity in favor of what is most appropriate given the situation and environmental demands, i.e., the ability to go back and forth effectively between focused attention and openness.

Cognitive agility is a tool for dynamic decision-making (Brehmer, 1992). Dynamic decision-making has four distinct characteristics: (1) a series of decisions is required; (2) decisions are interdependent; (3) the state of the decision problem changes, both as a consequence of the individual's actions as well as the actions of other people; and (4) decisions must be made in real time (Brehmer & Allard, 1991). Dynamic decision-making has been studied in the context of military operations (Brehmer, 2012), but as outlined above, role adjustment following reintegration also requires dynamic decision-making. To the extent that partners become rigid in the thoughts, feelings, and behaviors prescribed by the roles that have evolved during separation, the dynamic role adjustment process becomes more difficult. The service member and his or her partner struggle with dynamic decision-making within their relationship due to their cognitive rigidity.

Cognitive agility training, with its emphasis on enhancing focus, openness, and especially cognitive flexibility, may improve dynamic decision-making not only in operational settings, but also within intimate relationships. Cognitive flexibility, while identified as one component of cognitive agility, has also been explored independently of the focus and openness constructs. Dennis and Vander Wal (2010), developers of the Cognitive Flexibility Inventory, state the following: "there is no consensus within the literature about precisely how to define or measure this construct. Generally speaking, the ability to switch cognitive sets to adapt to changing environmental stimuli appears to be the core component for most operational definitions of cognitive flexibility" (p. 242). In addition to the ability to disengage, switch, and revise a mental set, cognitive flexibility is marked by the ability to inhibit habitual responses (Rende, 2000), for instance, responses habituated from rigid role identification. Cognitive flexibility has been linked to several beneficial outcomes, including greater overall psychological well-being, better stress management, reduced overall stress, long-term post-trauma adjustment, and higher life satisfaction (Keith, Velezmoro, & O'Brien, 2015). In contrast, cognitive *inflexibility* is the inability to adapt cognitive processes to new situations and ultimately contributes to difficulty identifying problems and effective solutions (Schotte & Clum, 1987). Thus, cognitive agility training—with its focus on enhancing cognitive flexibility—may benefit service members and their intimate partners during reintegration and subsequent role adjustment.

Reintegration programs for military intimate partner relationships targeting cognitive agility developed within a VSA framework could be an important asset to the military because such programs focus on counteracting role rigidity and enhance adaptive, flexible thinking. Enhancing cognitive agility would enable each partner to learn to adjust to new roles when navigating through novel situations unfolding in uncertain, dynamic environments. The capacity for adaptive thinking is a central element of operational resilience in the face of unexpected challenges and could also become a central element of intimate relationship resilience if targeted resources are developed to meet the needs of the intimate partnership during separation and reintegration. In this way, improving cognitive agility could enhance combat readiness both in times of war and peace by equipping service member and their intimate partners with skills needed to adapt to ever-changing individual and relational demands. Military leadership training programs have already been exploring cognitive flexibility training for service members in the context of virtual game-based learning (Morrison & Fletcher, 2002; Raybourn, Deagle, Mendini, & Heneghan, 2005). For example, a program at Fort Leavenworth that provides training in cultural competencies for service members incorporates cognitive flexibility training to promote adaptability to combat zones (Griffin & McClary, 2015). Further, the United States Special Operations Command (USSOCOM) Preservation of the Force and Family (POTFF) has partnered with educators and scientists at the Uniformed Services University of the Health Sciences (USUHS) to develop complementary cognitive agility programs for Special Operations Forces (SOF) members and their spouses. In addition to enhancing overall mission readiness, programs such as those supported by USSOCOM also have the potential to reduce role conflict in intimate partnerships during the post-deployment reintegration process. Future research should explore the feasibility and acceptability of cognitive agility programs among military personnel and their intimate partners, as well as establish their effectiveness in improving relationship functioning.

## Conclusion

Transitioning from deployment to home is a stressful experience within the military life-cycle. In this chapter, we focused on the intimate partner relationship to broadly understand ways in which post-deployment reintegration impacts family health. Applying a utilitarian justification to our analysis, we first thoroughly explored IPV and concluded that logical inconsistencies make it difficult to uniformly apply IPV across the military as a frame for understanding common reintegration challenges. We also evaluated the extent to which military training contributes to IPV, as it is a serious concern that must be addressed. However, literature supporting this link tends to conflate combat training with general military training and creates a narrow research agenda, one that potentially overlooks subtler, yet broader, psychological concerns. Without dismissing IPV as a very important reality in the military, we illustrated that role conflict as a likely universal aspect of post-deployment role adjustment could serve as a foundational frame to capture a broad spectrum of conflict that ranges from simple misunderstandings to extreme IPV. We concluded by suggesting that expanding cognitive flexibility literature and developing training aimed to reduce rigid role identification among service members and their intimate partners could help reduce role conflict.

**Note:** The opinions and assertions expressed herein are those of the author(s) and do not necessarily reflect the official policy or position of the Uniformed Services University or the Department of Defense.

## References

Barrett, F. J. (1996). The organizational construction of hegemonic masculinity: The case of the US Navy. *Gender, Work & Organization*, *3*(3), 129–142.

Bommarito, R. K., Sherman, M. D., Rudi, J. H., Mikal, J. P., & Borden, L. M. (2017). Challenges facing military spouses during postdeployment reintegration: A review of the literature and current supports. *Military Behavioral Health*, *5*(1), 51–63.

Borus, J. F. (1973). Reentry: I. Adjustment issues facing the Vietnam returnee. *Archives of General Psychiatry*, *28*(4), 501–506.

Braudy, L. (2003). *From Chivalry to Terrorism: War and the Changing Nature of Masculinity*. New York: Random House.

Brehmer, B. (1992). Dynamic decision making: Human control of complex systems. *Acta Psychologica*, *81*(3), 211–241.

Brehmer, B., & Allard, R. (1991). Real time, dynamic decision making: The effects of complexity and feedback delays. In J. Rasmussen, B. Brehmer, & J. Leplat (Eds.), *Distributed Decision Making. Cognitive Models for Cooperative Work* (pp. 319–334). New York: John Wiley and Sons.

Brehmer, B. (2012). Dynamic decision making in command and control. In C. McCann & R. Pigeau (Eds.), *The Human in Command: Exploring the Modern Military Experience* (pp. 233–248). New York: Springer Science & Business Media.

Cabrera, D., Figley, C., Yarvis, J., & Cox, A. (2012). The warrior's family. In N. D. Ainspan & W. Penk (Eds.), *When the Warrior Returns: Making the Transition at Home* (pp. 1–14). Annapolis, MD: Naval Inst Press.

Cameron Kelly, D., Howe-Baksdale, S., & Gitelson, D. (2011). *Treating Young Veterans: Promoting Resilience Through Practice and Advocacy* (1st ed.). New York: Springer Publishing Company.

Cooper, R. A., Pasquina, P. F., & Drach, R. (2011). *Warrior Transition Leader Medical Rehabilitation Handbook*. Fort Detrick, MD: Borden Institute.

Cowan, G., & Mills, R. D. (2004). Personal inadequacy and intimacy predictors of men's hostility toward women. *Sex Roles*, *51*(1–2), 67–78.

Danish, S. J., & Antonides, B. J. (2013). The challenges of reintegration for service members and their families. *American Journal of Orthopsychiatry*, *83*(4), 550–558.

D'Augelli, A. R., Grossman, A. H., & Starks, M. T. (2006). Childhood gender atypicality, victimization, and PTSD among lesbian, gay, and bisexual youth. *Journal of Interpersonal Violence*, *21*(11), 1462–1482.

Dennis, J. P. & Vander Wal, J. S. (2010). The Cognitive Flexibility Inventory: Instrument development and estimates of reliability and validity. *Cognitive Therapy and Research*, *34*, 241–253.

Duncan, M. D. (1976). The military family: Its future in the all-volunteer context. In N. L. Goldman & D. R. Segal (Eds.), *The Social Psychology of Military Service* (pp. 135–149). Beverly Hills, California: Sage Publications, Inc.

Eagly, A. H. (2013). *Sex Differences in Social Behavior: A Social-Role Interpretation*. New York: Psychology Press.

Elshtain, J. B. (1995). *Women and War*. Chicago: University of Chicago Press.

Finkel, D. (2013). *Thank You for Your Service*. New York, NY: Sarah Crichton Books.

Finkel, E. J., Simpson, J. A., & Eastwick, P. W. (2017). The psychology of close relationships: Fourteen core principles. *Annual Review of Psychology*, *68*, 383–411.

Gallagher, K. E., & Parrott, D. J. (2011). What accounts for men's hostile attitudes toward women? The influence of hegemonic male role norms and masculine gender role stress. *Violence against Women*, *17*(5), 568–583.

Garber, B. G., & Zamorski, M. A. (2012). Evaluation of a third-location decompression program for Canadian Forces members returning from Afghanistan. *Military Medicine*, *177*(4), 397–403.

Gimbel, C., & Booth, A. (1994). Why does military combat experience adversely affect marital relations? *Journal of Marriage and the Family*, *56*(3), 691–703.

Glynn, S. M. (2012). The immediate post-deployment period: Deployments are hard, but homecoming can be even harder. In N. D. Ainspan & W. Penk (Eds.), *When the Warrior Returns: Making the Transition at Home* (pp. 15–28). Annapolis, MD: Naval Institute Press.

Good, D., & Yeganeh, B. (2012). Cognitive agility: Adapting to real-time decision making at work. *Organizational Development Practitioner*, *44*(2), 13–17.

Griffin, M. B., & McClary, R. B. (2015). A way to teach critical thinking skills so learners will continue using them in Operations. *Military Review*, *95*(6), 108–118. Retrieved from https://www.armyupress.army.mil/Portals/7/military-review/Archives/English/MilitaryReview_20151231_art018.pdf.

Hancock, T. (2001). *The influence of male gender role conflict on life satisfaction*. Retrieved from https://ro.ecu.edu.au/theses/1072.

Jakupcak, M., Lisak, D., & Roemer, L. (2002). The role of masculine ideology and masculine gender role stress in men's perpetration of relationship violence. *Psychology of Men & Masculinity*, *3*(2), 97–106.

Jakupcak, M., Tull, M. T., & Roemer, L. (2005). Masculinity, shame, and fear of emotions as predictors of men's expressions of anger and hostility. *Psychology of Men & Masculinity*, *6*(4), 275–284.

Karney, B. R., & Bradbury, T. N. (1995). The longitudinal course of marital quality and stability: A review of theory, methods, and research. *Psychological Bulletin*, *118*(1), 3–34.

Keith, J., Velezmoro, R., & O'Brien, C. (2015). Correlates of cognitive flexibility in veterans seeking treatment for posttraumatic stress disorder. *The Journal of Nervous and Mental Disease*, *203*(4), 287–293.

King, D. W., King, L. A., Foy, D. W., Keane, T. M., & Fairbank, J. A. (1999). Posttraumatic stress disorder in a national sample of female and male Vietnam veterans: Risk factors, war-zone stressors, and resilience-recovery variables. *Journal of Abnormal Psychology*, *108*(1), 164–170.

King, L. A., King, D. W., Fairbank, J. A., Keane, T. M., & Adams, G. A. (1998). Resilience–recovery factors in post-traumatic stress disorder among female and male Vietnam veterans: Hardiness, postwar social support, and additional stressful life events. *Journal of Personality and Social Psychology*, *74*(2), 420–434.

Knobloch, L. K., & Theiss, J. A. (2012). Experiences of US military couples during the post-deployment transition: Applying the relational turbulence model. *Journal of Social and Personal Relationships*, *29*(4), 423–450.

Koenen, K. C., Stellman, J. M., Stellman, S. D., & Sommer Jr, J. F. (2003). Risk factors for course of posttraumatic stress disorder among Vietnam veterans: A 14-year follow-up of American Legionnaires. *Journal of Consulting and Clinical Psychology*, *71*(6), 980–986.

Langer, A., Lawrence, E., & Barry, R. A. (2008). Using a vulnerability-stress-adaptation framework to predict physical aggression trajectories in newlywed marriage. *Journal of Consulting and Clinical Psychology*, *76*(5), 756–768.

Marek, L. I., & D'Aniello, C. (2014). Reintegration stress and family mental health: Implications for therapists working with reintegrating military families. *Contemporary Family Therapy*, *36*(4), 443–451.

Marlantes, K. (2011). *What It Is Like to Go to War*. New York, NY: Atlantic Monthly Press.

Marshall, A. D., Panuzio, J., & Taft, C. T. (2005). Intimate partner violence among military Veterans and active duty servicemen. *Clinical Psychology Review*, *25*(7), 862–876.

McNeill, W. H. (2008). *Keeping Together in Time: Dance and Drill in Human History*. New York, USA: ACLS Humanities E-Book.

Morrison, J. E., & Fletcher, J. D. (2002). *Cognitive Readiness*. Alexandria, VA: Institute for Defense Analyses.

Mussap, A. J. (2008). Masculine gender role stress and the pursuit of muscularity. *International Journal of Men's Health*, 7(1), 72–89.

Office of the Deputy Assistant Secretary of Defense for Military Community and Family Policy. (2016). *Profile of the military community (Demographics). Washington D.C.* Retrieved from http://download.militaryonesource.mil/12038/MOS/Reports/2016-Demographics-Report.pdf.

O'Neil, J. M. (1981). Male sex role conflicts, sexism, and masculinity: Psychological implications for men, women, and the counseling psychologist. *The Counseling Psychologist*, *9*(2), 61–80.

O'Neil, J. M. (2008). Summarizing 25 years of research on men's gender role conflict using the Gender Role Conflict Scale: New research paradigms and clinical implications. *The Counseling Psychologist*, *36*(3), 358–445.

O'Neil, J. M., Helms, B. J., Gable, R. K., David, L., & Wrightsman, L. S. (1986). Gender-role conflict scale: College men's fear of femininity. *Sex Roles*, *14*(5–6), 335–350.

Park, N. (2011). Military children and families: Strengths and challenges during peace and war. *American Psychologist*, *66*(1), 65–72.

Philipps, D. (2018). Ban was lifted, but transgender recruits still can't join up [Newspaper]. Retrieved October 20, 2018, from https://www.nytimes.com/2018/07/05/us/military-transgender-recruits.html.

Pleck, J. H. (1983). *The Myth of Masculinity*. Cambridge: MIT Press.

Raine, A. (2013). *The Anatomy of Violence: The Biological Roots of Crime*. New York: Random House.

Raybourn, E. M., Deagle, E., Mendini, K., & Heneghan, J. (2005). Adaptive thinking & leadership simulation game training for special forces officers. *I/ITSEC 2005 Proceedings, Interservice/Industry Training, Simulation and Education Conference Proceedings*, November 28–December 1, Orlando, Florida, USA.

Rende, B. (2000). Cognitive flexibility: Theory, assessment, and treatment. *Seminars in Speech and Language*, *21*(2), 121–132.

Saltzman, L. E., Fanslow, J. L., McMahon, P. M., & Shelley, G. A. (1999). *Intimate Partner Violence Surveillance: Uniform Definitions and Recommended Data Elements. Version 1.0.* Atlanta (GA): National Center for Injury Prevention and Control, Centers for Disease Control and Prevention.

Sayers, S. L. (2011). Family reintegration difficulties and couples therapy for military veterans and their spouses. *Cognitive and Behavioral Practice, 18*(1), 108–119.

Schotte, D. E., & Clum, G. A. (1987). Problem-solving skills in suicidal psychiatric patients. *Journal of Consulting and Clinical Psychology, 55*(1), 49–54.

Shay, J. (2002). *Odysseus in America: Combat Trauma and the Trials of Homecoming*. New York, NY: Scribner.

Sparkjoy Studios. (2009). *Women in combat pros and cons—Sisters in arms*. Retrieved from https://sistersinarms.ca/history/women-in-combat-pros-and-cons/.

Sparrow, K., Dickson, H., Kwan, J., Howard, L., Fear, N., & MacManus, D. (2018). Prevalence of self-reported intimate partner violence victimization among military personnel: A systematic review and meta-analysis. *Trauma, Violence, & Abuse*. Epub ahead of print. doi.org/10.1177/1524838018782206.

Swick, A., & Moore, E. (2018). *The (mostly) good news on women in combat*. Retrieved from https://www.cnas.org/publications/reports/an-update-on-the-status-of-women-in-combat.

Thompson Jr, E. H., & Bennett, K. M. (2015). Measurement of masculinity ideologies: A (critical) review. *Psychology of Men & Masculinity, 16*(2), 115–133.

Westhusing, T. S. (2003). *The Competitive and Cooperative Aretai within the American Warfighting Ethos*. Atlanta: Emory University.

Westhusing, T. S. (2006). Equality within military organizations. *Journal of Military Ethics, 5*(1), 5–11.

Williston, S. K., Taft, C. T., & VanHaasteren, K. O. (2015). Military veteran perpetrators of intimate partner violence: Challenges and barriers to coordinated intervention. *Aggression and Violent Behavior, 21*, 55–60.

Wirtz, A. L., Poteat, T. C., Malik, M., & Glass, N. (2018). Gender-based violence against transgender people in the United States: A call for research and programming. *Trauma, Violence, & Abuse*, Epub ahead of print. doi.org/10.1177/1524838018757749.

# 17
# LONE WOLF TERRORISM

*Bruce Bongar, Anna Feinman, and Renata Sargon*

The goal of terroristic acts is for the terroristic group to ultimately gain power and control of the region they are attacking in order to enforce laws and a lifestyle consistent with their own beliefs, which they feel are better than existing laws (Loza, 2007). The psychology behind terrorism is that these groups use violence, intimidation, shock, and destruction in order to make citizens begin to doubt and mistrust the government that is intended to protect them (Loza, 2007). If the group is successful at this, then the idea is that the citizens of that region may become attracted to joining the terrorist group, may be fearful enough to concede to the rule of this group, or at minimum place severe doubts regarding the government's ability to control or protect the area (Loza, 2007). Terroristic acts are a phenomenon that may seem like a recent development—especially to Americans whose first main exposure was the attacks on 9/11. However, terrorism has been occurring in many forms since the 1880s (Post, McGinnis, & Moody, 2014).

Terrorism has come in four main waves since the 1880s, and some scholars consider that now, with the rise of technology and communication abilities, we are in the fifth wave of terrorism known as lone wolf terrorism (Post et al., 2014). The first wave of terrorism is known as the Anarchist wave, and it occurred in the 1880s, starting in Russia and eventually moving to Europe, the Americas, and Asia (Post et al., 2014). The second wave, known as the Anti-Colonial wave, happened as a result of people's dissidence in conjunction with the dissolution of colonial powers after World War I, lasting until World War II (Post et al., 2014). Similarly, the third wave of terrorism occurred after the Vietnam war as a result of people rebelling against their parents' generation who were loyal to the government, and it is known as the New Left wave (Post et al., 2014). In 1979, the Religious wave of terrorism began that is still occurring and developing today with the suicide bombing of the United States Embassy in Tehran by Hezbollah, an Islamic-based terrorist group (Post et al., 2014). The rationalization of suicide bombings by Hezbollah's leader, Sheikh Fadlallah, set a dangerous and lasting precedent for terror attacks that also set the stage for lone wolf attacks.

## Suicide terrorism

Sheikh Fadlallah preached that suicide bombings are the ultimate act of martyrdom and began to lay the foundation of many terrorist tactics that are still being utilized in the present day, including the use of communications and media to spread propaganda, martyrdom as a righteous and holy act, female terrorists, and training of specially selected individuals to carry out these suicide attacks (Bruce, 2013; Post et al., 2014). Ironically, although suicide is considered a sin in Islam, the reframing of suicide bombings as an act of martyrdom makes it acceptable for those who become radicalized (Post et al., 2014).

Islamic people from the Middle East historically have been prone to becoming radicalized because of the long-standing political, social, and economic issues, including the interference of foreign non-Islamic countries, throughout much of the region. The instability in these areas allows terrorist leaders to spread propaganda about how other religions and ways of life are threatening, unholy, and detrimental to the people of the Middle East (Loza, 2007). The people then begin to feel that they have been treated unfairly and become angry and morally outraged (Mastors & Siers, 2014).

Radicalization is posited to be a process involving three steps (Kruglanski et al., 2014). It begins with individuals who feel that they have lost meaning and significance in their lives and begin to search for ways to find meaning and feel significant (Kruglanski et al., 2014). Secondly, violence is seen as not only the most effective but also the only way to attain significance (Kruglanski et al., 2014). Lastly, the concept of being significant supersedes any other goal the individual may have, so it becomes the main focus and it is seen as only achievable through violent methods (Kruglanski et al., 2014).

An individual's worldview is also a contributing factor to whether they are prone to radicalization. Borum (2014) posited that three specific worldviews, authoritarianism, dogmatism, and fundamentalism, are linked to one's likelihood to become radicalized. Authoritarianism is linked to extremism and the denial of civil and human rights, dogmatism involves intolerance because of the inability to consider other points of view, and fundamentalism indicates black-and-white thinking as well as an apocalyptic orientation (Borum, 2014). If an individual has one or more of these worldviews, they are more prone to radicalization.

The radicalization and use of suicide bombers is intended to cause a sense of shock and awe that makes a community more susceptible to mistrust the government and perhaps even side with the terror group. As stated earlier, suicide is reframed as a glorious and righteous act done in the name of martyrdom for the cause, and an individual who does this act is praised and romanticized (Bruce, 2013; Hafez, 2006; Post, 2009). Religion is even used to reward this act, with religious leaders promising that Allah sees these individuals as special and that they will be immediately accepted into and rewarded in heaven with all the pleasures of life, including sexual gratification from 72 virgins (Loza, 2007).

Terroristic groups that utilize suicide bombers have specific identifying factors they search for when recruiting people. First, they search for individuals that are willing to make the ultimate sacrifice who tend to have weak personalities and low self-esteem (Bruce, 2013). They are careful to select people who do not have severe psychiatric disorders; however, due to the denunciation of suicide in the Qur'an, they are unable to specifically pick those who have suicidal ideation, as this would be a sin (Bruce, 2013; Lankford, 2011). Individuals susceptible to becoming suicide bombers also tend to feel isolated and lack a sense of belonging (Spaaij, 2010).

The terror group leaders encourage martyrdom and take advantage of the economic unrest of the region by promising to care for the family and loved ones of the suicide bomber (Bruce, 2013; Lankford, 2011). This individual not only feels a sense of duty spurred on by the events of the region, but then also begins to feel a sense of duty to the terror group to carry out this act after all the tactics used to radicalize them and the promises for a rewarding afterlife for them and better lifestyle for their families (Bruce, 2013).

## Lone wolf terrorism

As stated in the introduction, it is posited that we are now living in the fifth wave of terrorism known as Lone Wolf terrorism, which has arisen due to the advent of the internet and the accessibility of the internet and social media across the world (Post et al., 2014). Some scholars trace lone wolf terrorism back to the anarchist movement in the 19th century, but it took much longer for this term to be utilized and recognized (Pitcavage, 2015). For example, it became much more prominent in the United States in 1995 with the Oklahoma City bombing, culminating in much more attention to the topic around 2009 where there were a number of lone wolf attacks which mobilized law enforcement and academia into studying this specific phenomenon (Pitcavage, 2015).

Lone wolf terrorists are individuals that conduct terrorist attacks on their own accord, sometimes without even officially belonging to an identified terrorist group (Post et al., 2014). The propaganda of terror groups is now able to be spread through social media platforms and reach people worldwide. In fact, terror groups specifically recruit individuals that are technologically savvy in order to become more efficient and effective and disseminating information (Post et al., 2014). ISIS (Islamic State of Iraq and Syria), also known as ISIL (Islamic State of Iraq and Levant), has been especially effective at using the internet to recruit members, sometimes becoming so successful that individuals are carrying out attacks in their name without even technically belonging to the group (Post et al., 2014). Although many people associate the term "terrorism" with Islamic terror groups, it is important to recognize that many white supremacist groups are considered terroristic groups because of the one definition that specifies that the terror acts that are committed are driven by a specific ideology (Pitcavage, 2015). Pitcavage (2015) stated that lone wolf terrorists fall into seven "broad categories: (1) anarchists/left-wing extremists; (2) domestic Muslim extremists; and various categories of right-wing extremists; (3) sovereign citizens; (4) tax protestors; (5) other antigovernment extremists; (6) anti-abortion extremists; and (7) white supremacists" (p. 1666). In the sample examined in this study, 86% of lone wolf terrorists belonged to a right-wing extremist movement, with white supremacists making up the majority of this category (49%) (Pitcavage, 2015).

Due to the definitional, conceptual, and methodological issues as well as the nature of terrorism, it has been difficult to conduct research on lone wolf terrorism (Becker, 2014). There has been some debate about the definition of what a lone wolf terrorist is. Many of the existing definitions define the term differently based on some key points. For example, law enforcement believes that the term "lone wolf" may add some element of glamour and prefer the term "terrorists acting alone," a term also used by the United Nations Counter-Terrorism Committee Executive Directorate (CTED) (Spaaij & Hamm, 2015). The term "lone wolf" comes from the media and of the perpetrators themselves and is not a definition given by social sciences (Spaaij & Hamm, 2015). This also varies by country, as in some countries a lone wolf is one who performs terroristic acts because of an affiliation with a terrorist group, whereas in other countries this term is used to mean the exact opposite – someone who acts alone without a direct affiliation to a terrorist group (Spaaij & Hamm, 2015). There has also been some disagreement as to whether or not a couple or trio of individuals can be defined as lone wolf terrorists or if it must be one unaffiliated individual. Some researchers may call individuals in this situation "lone wolf packs," which is still a point of contention because on the one hand, they are at minimally affiliated with each other, but on the other hand lack the structure of an identified terrorist group (Pitcavage, 2015; Spaaij & Hamm, 2015). Most researchers agree that when there are at least four individuals, it is then considered a terrorist group (Pitcavage, 2015). Borum, Fein, and Vossekuil (2012) used a three-dimensional definition of lone wolf terrorism, which consisted of "loneness, or independence of activity; direction, or autonomy of decision making; and motivation, or clarity of causation/purpose."

There are two main perspectives on what motivates lone wolf terrorists presented by Pitcavage (2015). The first perspective is that these individuals suffer from psychological maladjustment and that this and not political ideation is what motivates them to kill others. They desire to be part of a group but are unable to fit in due to personality deficits and do have some sort of mental illness and problems with aggression and anger (Pitcavage, 2015). The second perspective combines this first perspective with the idea that because of their anger, mental instability, and desire to belong (in addition to the individual feeling unsafe and threatened), they become drawn to terrorist groups and commit acts in the name of such groups without being directed to act by the actual group (Pitcavage, 2015).

Individuals living in democratic states and the states themselves ironically tend to become most susceptible to lone wolf radicalization and attacks—due to the fact that individuals there have more of a right to privacy and because the government is not as intrusive regarding their personal lives as they are in other countries (Weissmann, Busch, & Schouten, 2014). For example, the United States has had more lone wolf terror attacks than any other country (Spaaij, 2010). Lone wolf terrorists select targets based on their ideology and locations for attacks based on their familiarity with the area, and they tend to use firearms as a main method of attack; however, in recent years explosives have been used as well (Becker, 2014). There is increasing concern that lone wolf terrorists act impulsively, which would make it more difficult to thwart the attack. However, there is also evidence that even in these cases, there is some element of planning (Meloy & Pollard, 2017). It seems to be the case that although the attack begins with a plan, a triggering event leads to an attack shortly after said event where only a few elements of the plan are used (Meloy & Pollard, 2017).

Lone wolf terrorists tend to share may similar risk factors as suicide bombers, so they are recruited in similar ways—based upon the characteristics that terror group leaders look for, and because of the internet, terror group leaders are able to cast a wider net to attract even more individuals willing to give up their lives (Bruce, 2013). Again, these leaders look for individuals that have low self-esteem, weak personalities in terms of willingness to follow a strong leader, and a lack of psychiatric issues who also are searching for meaning and significance in their lives and wish to feel a sense of belonging rather than isolation (Bruce, 2013; Hafez, 2006; Lankford, 2011; Post, 2009; Spaaij, 2010). By committing terrorist attacks, this individual not only finds meaning and a sense of belonging, but also gains notoriety and glory within their organization, which is especially appealing to them (Cottee & Hayward, 2011). Corner and Gill (2015) found that a lone wolf terrorist was 13.49 times more likely than an individual belonging to a terror group to have a mental illness and was more likely to be experiencing a major life change, be a recent victim of prejudice, and experience chronic stress. This data can be used to identify and determine methods of treating and deradicalizing lone wolf terrorists should they be caught.

## Counterterrorism approaches

### *Challenges specific to lone-wolf actors*

The literature overwhelmingly supports the contention that lone-wolf actors are difficult to detect, categorize, or systematize in any meaningful way (Bakker & De Graaf, 2011; Gill, Horgan, & Deckert, 2014; McCauley, Moskalenko, & Van Son, 2013). Moreover, despite the fact that a history of trauma, personality disorder traits, and substance abuse have been advanced as possible common links among these individuals, striking differences between actors have been observed in background, age, and motivation (McCauley et al., 2013). Although instances of lone-wolf terrorist acts have increased in recent years, one of the biggest difficulties for both researchers and counterterrorism organizations is the relatively low number of lone-actor perpetrated acts (Bakker & De Graaf, 2011; Spaaij, 2010). Additionally, these attacks continue to be one of the more unpredictable forms of terrorism—making prevention difficult for counterterrorism practitioners, intelligence agencies, and law enforcement.

The lone wolf's preference for isolation and lack of ties to a formal organization make identification, monitoring, and detention extremely difficult. Compared to hierarchically organized terrorist groups, these individuals are better able to avoid early detection, as their plans are rarely communicated to others (Bakker & De Graaf, 2011). Lone actors display a range of backgrounds with significant variation in motivations and ideologies, as well as utilizing a wide spectrum of means to send their message. These can range from lengthy rhetoric posted on social media to sending hate mail to specific targets (Bakker & De Graaf, 2011). This lack of discernible recurring methodology adds to the difficulty of creating prevention strategies for these attacks.

There is literature to suggest that a number of attackers post rhetoric, plans of attack, and even intended targets on social media prior to perpetrating their attacks (Merari, 2010). However, the increasing use of social media as a means of global communication makes it difficult to sift through the information to differentiate between individuals expressing extremist ideologies (Pantucci, Ellis, & Chaplais, 2015). Mainstream social media as well as more under-the-radar forums and chat rooms are increasingly being used to attract, groom, and launch potential lone actors (Weimann, 2012). In the United States and many other Western countries, attempts to limit freedom of speech have been met with substantial pushback, limiting the power of counterterrorism units to investigate unless a documented act of violence has occurred (Bakker & de Graaf, 2011).

Of particular concern is the social contagion aspect of lone-wolf terrorism: a relatively large number of lone-actor manifestos continue to circulate on the World Wide Web written by individuals such as Ted Kaczynski, Mohammed Bouyeri, and Elliot Rodger. Copycat behavior is inspired not only by perceived common ground and shared feelings of alienation, but also the near-religious reverence imparted onto these "success stories" (Pantucci et al., 2015). Methods of attack such as shootings, bombings, or anthrax letters continue to cycle though rising and falling in popularity depending on effectiveness of recent attacks (Bakker & de Graaf, 2011).

Despite often lacking means and encountering difficulties associated with the absence of organizational support of terrorist networks, lone-wolf attacks have often proven to be devastating.

## *Traditional counterterrorism tactics*

Law enforcement, intelligence agencies, counterterrorism organizations, and military efforts have increasingly focused their efforts on fighting the lone-wolf terrorist threat. Several approaches are described within the literature produced in the last two decades, coinciding with the rise of this type of perpetrator.

Initial efforts have focused on the application of tactics that have been previously successful against traditional terrorist organizations as follows.

Confidential informants have been previously used with some success as a source of information. Their contributions have led to the disruption of plans as well as the incarceration of suspected terrorists. Confidential informants have been individuals who experienced a change of heart or members of law enforcement placed within cells undercover (Doosje et al., 2016).

Counter-radicalization has been a central strategy employed by the U.S. government, with officials attempting to engage with communities most likely to be affected by terrorism. These programs include outreach to community leaders, education on controversial policies, and establishing trust within relevant neighborhoods. The efficacy of these programs continues to be evaluated, and addressing terrorist proclivities prior to violent escalation has become the key to decreasing the occurrence of terrorist acts. In the case of traditional terrorism, involving the community increases the cooperation between law enforcement and the general population, leading to more effective protection strategies (Barnes, 2012; Lakhani, 2012).

Agencies have employed electronic and physical surveillance as a means of monitoring, studying, and preventing a variety of terrorist activity. This tactic alone does not seem to increase the efficacy of finding potential terrorist actors, and limited success has been achieved when surveillance has been used in conjunction with other means as a strategy for detection and disruption (Barnes, 2012).

Last, physical denial of access to both information and firearms has been considered as a possible counter-terrorism tactic (Barnes, 2012). This has proven to be an incredibly difficult approach as specialized knowledge including but not limited to instructions on: bomb-making, weapons training, and publicly available schematics and architectural plans for potential soft targets is readily available on the internet (Barnes, 2012). Restricting access to firearms is unlikely to be successful as constitutionally protected rights are often difficult to circumvent without tangible proof (Rapoport, 2013).

## *Counterterrorism tactics as applied to lone-wolf actors*

These aforementioned strategies have proven to be difficult to employ in the case of lone-wolf terrorism for a variety of reasons. Confidential informants are unlikely to be able to extract useful information, as these individuals tend towards isolation and have difficulty with social interactions (Merari, Diamant, Bibi, Broshi, & Zakin, 2009; Gill et al., 2014). Despite these tendencies, lone-wolf actors exist as a part of a broader community and often interact with members of this community in a way that increases awareness of possible planned attacks (Pantucci et al., 2015). Indeed, members of the community are often aware about presenting extremist ideology and/or plan to engage in violent activities expressed by the lone-wolf actor. Lone-wolf terrorists will often show signs of impending plans via identification with and activities related to a particular social movement, which is clearly visible to community members such as neighbors, doctors, or even librarians (Gill et al., 2014). It is imperative to seek the help of influential leaders in solidifying community engagement by promoting aversion to terrorist activities through both passive and active means (Bakker & De Graaf, 2011). These efforts can be reasonably expected to increase the likelihood of detection and disruption of planned lone-wolf attacks by various agencies (Pantucci et al., 2015).

Scholars in the field of counterterrorism stress the need to understand how attacks happen rather than the traditional approach of examining motivations behind the attacks when attempting to locate potential lone-wolf actors (Bakker & De Graaf, 2011). Shone (2010) emphasized the importance of understanding the process of radicalization as it applies to lone-wolf actors. Factors in radicalization can include feelings of insignificance, a sense of humiliation, a lack of career prospects, and personal uncertainty. These factors can contribute to the desire to identify strongly with an ideology that promotes black-and-white thinking and decreases uncertainty. The potential lone actor's social environment or lack thereof may also contribute to radicalization (Doosje et al., 2016). Successful counterterrorism organizations need to be vigilant about gaining insight into any signals that a would-be attacker may give off, by effectively utilizing intelligence projects, capturing data, and, most importantly, and working in concert with other agencies (Shone, 2010).

Researchers contend that despite the seemingly spontaneous nature of the violent lone-wolf attacks, it behooves experts in the field to look for a catalyzing event. Moreover, because lone actors perpetrate terror outside of a hierarchical structure, other extremists often inspire them. With this in mind, an effective counterterrorism strategy will include the circulation of a counter-narrative that aims to delegitimize previous attempts and attacks (Bakker & De Graaf, 2011). As part of this strategy, informational and educational programs for schools and universities as well as awareness campaigns for parents and guardians are encouraged.

Bakker and De Graaf (2011) caution counterterrorism organizations to carefully toe the line between communicating potential threats to pertinent audiences and giving lone actors the public space that they inherently crave. Media coverage, which often increases with the level of human tragedy achieved, increases the payoff for these actors (Phillips, 2011). This level of public visibility is likely to encourage other such would-be perpetrators and significantly exacerbate the problem.

Scholars in the field encourage counterterrorism actors to utilize the Internet as an instrument of early detection. Citing isolationism in their offline life, researchers contend that lone actors derive social interaction from their online connections (Brynielsson et al., 2012). Agencies that are able to monitor radicalization efforts by analyzing linguistic patterns and tracking movements on social media can potentially access important information about these lone actors prior to a planned attack (Cohen, Johansson, Kaati, & Mork, 2014). Focusing on and analyzing online interactions can lead to the detection of planned lone-wolf attacks. However, Striegher (2013) warns that although online intelligence has previously been utilized, there is no research to indicate that this has been successful.

A final recommendation is one that has become the subject of much debate among policy makers, law enforcement and military, and the general public. A look at perpetrators of recent shooting sprees across the world indicates that the perpetrators were able to easily access semi-automatic firearms. Members of the counterterrorism community continue to voice reservations about the relatively easy access to these dangerous weapons. Researchers encourage agencies to pay careful attention to those applying for a permit to carry a weapon or seeking membership in shooting clubs (Bakker & De Graaf, 2011).

The recommendations discussed above shed some light on potential counterterrorism tactics to be used as the threat of lone-wolf terrorism continues to grow. Overall, this list of possible approaches demonstrates there is still much work to be done in understanding, analyzing, and preventing lone-wolf terrorist acts.

## Goals

Deradicalization occurs when an individual moves away from extremist views towards more moderate beliefs (Kruglanski et al., 2014). Unfortunately, there is a paucity of research when it comes to the best and most effective methods for deradicalization. There are some known differences seen in deradicalization programs in the Middle East and South Asia versus those seen in Europe (Dalgaard-Nielsen, 2013). In the Middle East and South Asia, deradicalization programs focus on a less extreme and more accurate interpretation of the Qur'an with the hopes that religious individuals will conclude that suicide is in fact a sin and not the ultimate sacrifice one can make (Dalgaard-Nielsen, 2013). The programs in Europe focus on increasing the individual's vocational, social, and financial skills (Dalgaard-Nielsen, 2013) in order to help combat the despair an individual may feel due to the political, social, and economic unrest that is prevalent in the Middle East, while at the same time combating psychological factors of the individual (Dalgaard-Nielsen, 2013; Lankford, 2011). All of the aforementioned programs use a variety of religious scholars, former terrorists, and experts to establish rapport and build trust, and success is measured by the individual's lack of criminal and terroristic activities—although there is no long-term data to support the efficacy of either type of program (Dalgaard-Nielsen, 2013).

Some believe that there is no way to deradicalize an individual and that their actions must also be met with violence and war (Loza, 2007). However, this may prove to be counterproductive, as it may validate the fact that these individuals feel under attack and that their entire way of life is threatened. Additionally, due to the advent of the Internet and social media platforms, experts must begin to look at also using internet-based interventions to combat the use of propaganda disseminated by terror groups—preferably by combining facets of programming from the Middle East, South Asia, and Europe in order to ensure all angles of the complex factors influencing radicalization are addressed.

## What the Israel Prison Service can teach us about lone-wolf actors

The Israel Prison Service (IPS) holds the highest number of incarcerated terrorists including the largest number of non-local inmates in the world. Additionally, the system's unique population includes lone-wolf actors and a number of inmates often referred to as "would-be suicides," or captured individuals who, for a variety of reasons, failed to complete the planned terrorist act (Merari, 2010). Professor Ariel Merari, a scholar of suicide terrorism who has been tracking the phenomenon for nearly 30 years, has conducted a number of studies aimed at better understanding the background and motivation of suicide bombers in general and more recently lone actors incarcerated within the Service (Merari, 2010).

In a study looking at "would-be" suicide bombers, Merari and colleagues narrowed their approach by reasoning that the most accurately representative individuals for study are those that attempted to detonate their device but failed due to a malfunction (Merari, 2010). The researchers conducted a number of clinical interviews with three distinct groups which included "would-be" suicide bombers, group leaders who organized suicide attacks, and a control group of non-suicide terrorists. Findings showed that over half of the "would-be" actors were not involved in any terrorist activities prior to their attempt and none had been arrested prior to their attempt compared to a quarter of the controls (Merari et al., 2009). Despite previous literature that contends that suicide terrorists lack shared personality characteristics (Silke, 2003; Sageman, 2004), Merari and colleagues found that the majority of "would-be" actors in their sample displayed dependent and avoidant personality characteristics. The researchers noted that no psychopathic tendencies were observed among the "would-be" bombers (Merari et al., 2009). Overall, profiles compiled by these researchers described marginalized, shy individuals with a history of failed social interactions, difficulties at school, and perceived parental disappointment. A number of the "would-be" actors also presented with suicidal tendencies (Merari et al., 2009). Despite the admittedly small sample size, these findings provided a launching point for future research.

A more recent study conducted by Ganor and Merari (2018) presented at the International Homeland Security Forum was the first of its kind and included data collected from 700 attackers and interviews with 45 imprisoned lone-wolf actors with the aim of defining motivations, creating profiles, and formulating prevention techniques (Ganor & Merari, 2018). Motivations for planned attacks ranged from pre-existing psychological conditions and ideological (nationalistic and religious) factors to trigger events including geopolitical changes, traumatic experiences, and social contagion. Over half of the lone-wolf actors interviewed presented with mental disorders, specifically psychosis and/or suicidal ideation, and a similar number of interviewees expressed regret at surviving the attack. Although the number of female attackers was low, this subgroup displayed a notably high incidence of reported familial problems. Of particular concern is the finding that over 10% of these actors utilized social media prior to their attacks, with over half of those individuals expressing intent (Ganor & Merari, 2018).

The data collected by Merari and colleagues has allowed experts in the field to develop a better understanding of the motivations of lone-wolf actors. Moreover, it has established a possible pattern by which multidisciplinary counterterrorism teams could potentially locate and immobilize would-be threats. A similar approach to studying lone actors in the United States could potentially yield further information on how to deal with this growing threat.

## A multidisciplinary approach to lone-wolf terrorism

The study of lone-wolf terrorism is deeply rooted in the multidisciplinary approach, yet there are differing views and limited research on the subject. Scholars in the field of counterterrorism often emphasize the need for a multidisciplinary approach, but rarely operationalize what this step would look like in practice (De Roy van Zuijdewijn & Bakker, 2016). The following section will address multidisciplinary approach to the global fight against traditional terrorism on the macro level, and then explore the potential for a multi-pronged community-based approach on the micro level.

Schmid (2005) looked at a possible multi-pronged approach to battle traditional terrorism. Much of what he found can be relevant to fighting lone-wolf actors as well. On a macro scale, several actors are instrumental in counterterrorism efforts, with one such actor being the government whose policies are often the target of the lone actor's anger. Schmid (2005) recommends four "pillars" of government in order to prevent terrorist actions: good governance, democracy, rule of law, and social justice. This authority contends that corruption, political violence, and desperate actions (such as those of a lone-wolf actor) are to be expected in a society with long-standing injustices and inequalities (Schmid, 2005). Schmid further emphasizes the importance of economic and social measures, stating that suppressing the flow of funds to terrorist organizations is key to their demise. On the other side, the researcher contends that although the correlation between poverty and the existence of terrorism is low, the correlation between respect for human rights and absence of terrorism is significant. Another economic motivation for terrorism can be unemployment and the resulting feelings of hopelessness in younger individuals. All of these factors contribute to the importance of social and economic measures in terrorism prevention (Schmid, 2005).

Schmid, along with a number of his colleagues, encourages the use of psychological-communication-educational measures to decrease the susceptibility of easily influenced groups. These groups include individuals already identifying as part of a terrorist organization, unaffiliated members of the community, and other vulnerable populations. Researchers propose a series of offensive and defensive psychological operations, including promotion of desired perceptions among both members of the community and potential members of terrorist organizations (Crelinsten & Schmid, 1992). These efforts are meant to dissuade potential future actors from embracing terrorism, deny means for carrying out an attack, and increase international cooperation in the battle against the global terrorist threat (Schmid, 2005).

On the legal front, a number of agencies have acted to aid countries with few resources of their own in counterterrorism efforts. Furthermore, agencies such as the Terrorism Prevention Branch provide legal assistance to the global community by training judges and lawyers, aiding in the drafting of legislation, and expediting extradition. Over 70% of countries belonging to the UN have requested such aid, and policy makers and researchers are optimistic that this organization will continue to be a resource (Schmid, 2005).

Lastly, Schmid (2005) encourages counterterrorism organizations to continue to utilize good intelligence as a means for disrupting terrorist activity. The researcher contends that although intelligence often contains accurate, up-to-date information on terrorist activities, the information fails to be utilized in a timely manner due to information-sharing constraints. Bureaucratic issues and fear of revealing sources continue be problems that prevent full cooperation across all actors (Schmid, 2005). Each of the major actors described above is involved in the global fight against terrorism and is crucial to counterterrorism efforts across the board; however, the lone-wolf terrorist is far more likely to be thwarted at the community level.

On a micro community level, a possible approach includes forming teams consisting of members from mental and public health, educators, religious leaders, law enforcement personnel, and other representatives of the legal system (Weine, Cohen, & Brannegan, 2015). Required training for the teams would focus on threat and risk assessment and best practices, and the team would play a role in building awareness of the current threat to the community through its leaders and advocates. Researchers encourage potential teams to hold multidisciplinary conferences with the hope of disseminating accurate and crucial information to relevant stakeholders (Weine et al., 2015).

An important consideration in multidisciplinary work is the balance between increasing cooperation and the exchange of information among agencies without compromising the privacy and right to confidentiality of individuals being assessed. Based on research that looks at personal characteristics of lone-wolf actors, De Roy van Zuijdewijn and Bakker (2016) encourage researchers and counterterrorism agencies alike to focus their attention on individuals with a known history of mental health issues who are attempting to obtain a permit to carry a gun. This type of information can only be gleaned if mental health organizations and law enforcement or other agencies involved in issuing gun permits are able to work together. Notably, this approach works best with individuals who are already predetermined to be a potential threat (De Roy van Zuijdewijn & Bakker, 2016).

Another approach to multidisciplinary cooperation provides a framework for the role of government officials, law enforcement, and mental health organizations in formulating preventative activities for vulnerable populations, specifically younger individuals. A plan like this was implemented in the United States in 2011; however, some concerns were raised that this program was geared more towards gathering intelligence than actual preventative care (Weine et al., 2017). With this in mind, researchers encourage actors looking to implement similar programs to seek a buy-in from the community, prioritize collaboration, and emphasize the protection of civil liberties (Weine et al., 2017).

Increasingly, mental health professionals are being called to the forefront of the struggle with lone-wolf terrorism. Mental health practitioners are encouraged to provide aid within their traditional scope of practice, including providing a safe space for clients to express a variety of views, build trust within the community, and seek consultations in cultural competency related to the context of their clients' experiences. Additionally, mental health workers are invited to become actively involved in the multidisciplinary approach to counterterrorism by attending threat assessment training, building relationships with law enforcement and community leaders, and consulting with a professional peer group to discuss violence and extremism (Weine et al., 2017).

As the study of lone-wolf terrorism and counterterrorism tactics continues to grow, it becomes clear that an interdisciplinary approach and cooperation will be the key to fighting the lone-wolf terrorist threat.

## Conclusion

Due to the rise and accessibility of the Internet and use of social media around the globe, the current geo-political stage is set to facilitate the rise of lone-wolf terrorism. Unfortunately, due to nature of lone-wolf terrorism as well as a number of methodological issues, research on the matter has been difficult to conduct. For this reason, relatively little is known about motivations, risk factors, and commonalities of this population. The current perspective is that the motivation for the actions of these individuals is likely the psychological maladjustment that they suffer from, rather than political motivation. Secondary motivating factors are feelings of intense anger, the desire for recognition and belonging, and mental instability (Pitcavage, 2015). The work of Merari and colleagues in this field has been promising as well as informative, and it would behoove practitioners of counter-terrorism, law enforcement, and other relevant actors to continue to pay attention to future literature coming out of Israel.

Prevention of lone-wolf attacks is incredibly difficult for counterterrorism practitioners around the world, as this continues to be one of the most unpredictable forms of terrorism. Monitoring is difficult due to the lone actor's tendency towards isolation. Several common tactics for prevention have been attempted, including the use of confidential informants, counter-radicalization, electronic and physical surveillance, and physical denial of access to information and firearms, all with limited to varying success. The literature suggests utilizing the Internet as an instrument of early detection as well as paying careful attention to individuals applying to carry a weapon.

Due to the complex nature of lone-wolf terrorism, a multidisciplinary approach is key. Cooperation across law enforcement agencies; counterterrorism practitioners; scholars in the fields of politics, psychology and economics; and other relevant actors is crucial to the containment of this threat. Further and more extensive research is needed in order to better understand and eventually eliminate lone-wolf terrorism across the globe.

## References


Bakker, E., & De Graaf, B. (2011). Preventing lone wolf terrorism: Some CT approaches addressed. *Perspectives on Terrorism, 5*(5–6), 43–50.

Barnes, B. D. (2012). Confronting the one-man wolf pack: Adapting law enforcement and prosecution responses to the threat of lone wolf terrorism. *Boston University Law Review, 92*, 1613–1662.

Becker, M. (2014). Explaining lone wolf terror target selection in the United States. *Studies in Conflict & Terrorism, 37*(11), 959–978.

Borum, R. (2014). Psychological vulnerabilities and propensities for involvement in violent extremism. *Behavioral Sciences & the Law, 32*(3), 286–305.

Borum, R., Fein, R., & Vossekuil, B. (2012). A dimensional approach to analyzing lone offender terrorism. *Aggression and Violent Behavior, 17*(5), 389–396.

Bruce, G. (2013). Intrinsic and external factors and influences on the motivation of suicide attackers. *Journal of Military and Veterans' Health, 21*(3), 27–33.

Brynielsson, J., Horndahl, A., Johansson, F., Kaati, L., Martenson, C., & Svenson, P. (2012). Analysis of weak signals for detecting lone wolf terrorists. *IEEE European Intelligence and Security Informatics Conference* (pp. 197–204).

Cohen, K., Johansson, F., Kaati, L., & Mork, J. C. (2014). Detecting linguistic markers for radical violence in social media. *Terrorism and Political Violence, 26*(1), 246–256.

Corner, E., & Gill, P. (2015). A false dichotomy? Mental illness and lone-actor terrorism. *Law and Human Behavior, 39*(1), 23–34.

Cottee, S., & Hayward, K. (2011). Terrorist (e) motives: The existential attractions of terrorism. *Studies in Conflict & Terrorism, 34*(12), 963–986.

Crelinsten, R. D., & Schmid, A. P. (1992). Western responses to terrorism: A twenty-five year balance sheet. *Terrorism and Political Violence, 4*(4), 307–340.

Dalgaard-Nielsen, A. (2013). Promoting exit from violent extremism: Themes and approaches. *Studies in Conflict & Terrorism, 36*(2), 99–115.

De Roy van Zuijdewijn, J., & Bakker, E. (2016). Analysing personal characteristics of lone-actor terrorists: Research findings and recommendations. *Perspectives on Terrorism, 10*(2), 42–49.

Doosje, B., Moghaddam, F. M., Kruglanski, A. W., De Wolf, A., Mann, L., & Feddes, A. R. (2016). Terrorism, radicalization and de-radicalization. *Current Opinion in Psychology, 11*, 79–84.

Ganor, B., & Merari, A. (2018). The art of counter-terrorism and the lone wolf phenomenon. Presented at International Homeland Security Forum, Jerusalem, Israel.

Gill, P., Horgan, J., & Deckert, P. (2014). Bombing alone: Tracing the motivations and antecedent behaviors of lone-actor terrorists. *Journal of Forensic Sciences, 59*(2), 425–435.

Hafez, M. M. (2006). Rationality, culture, and structure in the making of suicide bombers: A preliminary theoretical synthesis and illustrative case study. *Studies in Conflict & Terrorism, 29*(2), 165–185.

Kruglanski, A. W., Gelfand, M. J., Bélanger, J. J., Sheveland, A., Hetiarachchi, M., & Gunaratna, R. (2014). The psychology of radicalization and deradicalization: How significance quest impacts violent extremism. *Political Psychology, 35*(S1), 69–93.

Lakhani, S. (2012). Preventing violent extremism: Perceptions of policy from grassroots and communities. *The Howard Journal of Criminal Justice, 51*(2), 190–206.

Lankford, A. (2011). Could suicide terrorists actually be suicidal? *Studies in Conflict & Terrorism, 34*, 337–366.

Loza, W. (2007). The psychology of extremism and terrorism: A Middle-Eastern perspective. *Aggression and Violent Behavior, 12*(2), 141–155.

Mastors, E., & Siers, R. (2014). Omar al-Hammami: A case study in radicalization. *Behavioral Sciences & the Law, 32*(3), 377–388.

McCauley, C., Moskalenko, S., & Van Son, B. (2013). Characteristics of lone-wolf violent offenders: A comparison of assassins and school attackers. *Perspectives on Terrorism*, 7(1), 4–24.

Meloy, J. R., & Pollard, J. W. (2017). Lone-actor terrorism and impulsivity. *Journal of Forensic Sciences, 62*(6), 1643–1646.

Merari, A. (2010). *Driven to Death: Psychological and Social Aspects of Suicide Terrorism*. New York, NY, US: Oxford University Press.

Merari, A., Diamant, I., Bibi, A., Broshi, Y., & Zakin, G. (2009). Personality characteristics of "self martyrs"/"suicide bombers" and organizers of suicide attacks. *Terrorism and Political Violence, 22*(1), 87–101.

Pantucci, R., Ellis, C., & Chaplais, L. (2015). *Lone-Actor Terrorism*. Occasional Papers.

Phillips, P. J. (2011). Lone wolf terrorism. *Peace Economics, Peace Science and Public Policy, 17*(1), 1–29.

Pitcavage, M. (2015). Cerberus unleashed: The three faces of the lone wolf terrorist. *American Behavioral Scientist, 59*(13), 1655–1680.

Post, J. M. (2009). Reframing of martyrdom and jihad and the socialization of suicide terrorists. *Political Psychology, 30*(3), 381–385.

Post, J. M., McGinnis, C., & Moody, K. (2014). The changing face of terrorism in the 21st century: The communications revolution and the virtual community of hatred. *Behavioral Sciences & The Law, 32*(3), 306–334.

Rapoport, D. C. (2013). The four waves of modern terrorism: David C. Rapoport. In J. G. Horgan & K. Braddock (Eds.), *Terrorism Studies* (pp. 63–82). London: Routledge.

Sageman, M. (2004). *Understanding Terror Networks*. University of Pennsylvania Press.

Schmid, A. P. (2005). Prevention of terrorism: Towards a multi-pronged approach. In T. Bjorgo (Ed.), *Root Causes of Terrorism: Myths, Reality and Ways Forward* (pp. 223–240). London & New York: Routledge.

Shone, A. (2010). *Countering Lone Wolf Terrorism: Sustaining the CONTEST Vision*. Henry Jackson Society.

Silke, A. (2003). *Terrorists, Victims and Society: Psychological Perspectives on Terrorism and Its Consequences*. London: John Wiley & Sons.

Spaaij, R. (2010). The enigma of lone wolf terrorism: An assessment. *Studies in Conflict & Terrorism, 33*(9), 854–870.

Spaaij, R., & Hamm, M.S. (2015). Key issues and research agendas in lone wolf terrorism. *Studies in Conflict & Terrorism, 38*(3), 167–178.

Strieger, J. L. (2013). Early detection of the lone wolf: Advancement of counter-terrorism investigations with an absence or abundance of information and intelligence. *Journal of Policing, Intelligence and Counter Terrorism, 8*(1), 35–53.

Weimann, G. (2012). Lone wolves in cyberspace. *Journal of Terrorism Research, 3*(2), 75–90.

Weine, S., Cohen, J., & Brannegan, D. (2015). Moving beyond motive-based categories of targeted violence (No. ANL/GSS-15/6). Argonne National Lab. (ANL), Argonne, IL (United States).

Weine, S. M., Stone, A., Saeed, A., Shanfield, S., Beahrs, J., Gutman, A., & Mihajlovic, A. (2017). Violent extremism, community-based violence prevention, and mental health professionals. *The Journal of Nervous and Mental Disease, 205*(1), 54–57.

Weissman, S.H., Busch, K.G., & Schouten, R.S. (2014). Introduction to this issue: The evolution of terrorism from 1914 to 2014. *Behavioral Sciences and the Law, 32*(3), 259–262.

# 18

# WHY DO THEY LEAVE? A CONCEPTUAL MODEL OF MILITARY TURNOVER

*H. Canan Sümer and İpek Mete*

## Overview

Understanding the process within which voluntary turnover unfolds is vital for all organizations due to the tangible and intangible costs involved in losing personnel that the organizations would have desired to keep. Voluntary turnover is even more challenging for military organizations because of the size of the workforce as well as the context-specific need for preserving morale, motivation, and unit readiness. However, most models of voluntary turnover are developed for civilian organizations and fail to capture the unique aspects of the process in the military context (e.g., Schwerin, Kline, Olmsted, & Wilcove, 2006). Furthermore, these models have conventionally focused on the attitude-intentions-turnover linkages and largely ignored factors such as dispositions, job and organizational characteristics, quality of life concerns, and unexpected life events that may be particularly critical in understanding the route that leads to turnover for military members. The military turnover model we propose in this chapter is an attempt to fill this void in the literature. It represents a revised version of the model developed by H. Canan Sümer and Cyril van de Ven (2007) as part of their work in a NATO task group on military recruitment and retention. The original model has been updated based on more recent empirical and theoretical evidence and advancements in the turnover literature.

## An overview of the turnover literature and models

Employee turnover has been a topic of empirical interest for about a century now (Hom, Lee, Shaw, & Hausknecht, 2017). Attesting the importance of the topic, a web of science literature search for the period 1945–2018 revealed 3506 articles in which employee turnover or its derivative concepts (voluntary turnover, turnover intentions, intentions to quit, quit intentions, or employee retention) was a part of the title. Turnover that is voluntary, dysfunctional, and organizationally avoidable (Griffeth & Hom, 2001) can be very costly for work organizations. The cost of recruiting, selecting, and developing replacements is estimated to be more than 200% of annual pay of the employee (Allen, Bryant, & Vardaman, 2010). In addition to the tangible costs, there are intangible or hidden costs, which may include loss of social capital; adverse publicity and reduced customer satisfaction; loss of morale, motivation, and efficiency in the stayers; and turnover contagion. Retaining qualified personnel is a main concern for military organizations because of both the scale of investments made in attracting, selecting, and training members as well as the context-specific need for preserving morale, motivation, and unit readiness. Hence identifying critical organizational, job, and individual factors in the turnover process has obvious utility implications for military organizations.

Early models of turnover, mostly capturing civilian turnover process, were relatively simple, relating turnover directly to job attitudes like satisfaction and commitment (Mobley, 1977; Mobley, Horner, & Hollingsworth, 1978; Newman, 1974). Later models were more complex, yet still attitude centered (Griffeth & Hom, 2001; Hom & Griffeth, 1991; Hom, Griffeth, & Sellaro, 1984). Griffeth and Hom (2001) proposed satisfaction and organizational commitment as the major mediators in the turnover process. In their model, job satisfaction, which is influenced by job (e.g., job complexity and group cohesion), individual (e.g., negative affectivity), and labor market–related factors, affects turnover through perceptions of the costs and benefits of job seeking and turnover, job search, and finally evaluation of alternatives. Organizational commitment, on the other hand, is influenced by job- and organization-related (e.g., procedural justice and job security) and individual-related (e.g., commitment propensity) factors. Effects of commitment on turnover behavior are mediated by thoughts of quitting.

In a recent all-inclusive review of the turnover literature, Rubenstein, Eberly, Lee, and Mitchell (2015, 2018) meta-analytically examined a long list of frequently studied constructs in the turnover literature for both their predictive strength and generalizability and tested a meta-analytic path model of voluntary turnover. They reported that among the four exit/turnover route alternatives, the route with the strongest support was the one linking job attitudes or perceptions about the job to turnover via withdrawal cognitions. Furthermore, withdrawal behaviors (i.e., lateness and absenteeism) failed to mediate the relationship between withdrawal cognitions on actual turnover, suggesting that quitters quit without necessarily withdrawing from work. Finally, although this meta-analytic study introduced some new variables critical in the turnover process (e.g., job engagement, emotional stability, fit, and rewards), it still yielded strong support for the attitude-intention-behavior–centered nature of the turnover process.

Some researchers have argued against the idea that accumulated job dissatisfaction is the major and immediate cause of voluntary turnover (Holtom, Goldberg, Allen, & Clark, 2017; Holtom, Mitchell, Lee, & Inderrieden, 2005; Lee & Mitchell, 1994; Zimmerman, 2008). Holtom et al. state, "organizational leaders … must develop clear strategies for attracting and retaining good employees. However, these plans must move beyond methods to combat job dissatisfaction if they expect to be effective. They also must systematically address shocks and the critical role of these shocks in the voluntary turnover process" (2005, p. 337–338). Lee, Mitchell, and colleagues proposed the "unfolding model," which basically states that unexpected life events, or shocks, cause voluntary turnover more often than accumulated job dissatisfaction. The model proposes four potential courses of actions or paths leading to turnover. Three of these paths involve some type of shock, positive or negative, leading to a quitting decision, whereas the last path involves job dissatisfaction as the main cause of turnover. Therefore, the model does not deny the role of job attitudes altogether in the voluntary turnover process, yet it acknowledges external events as one of the critical factors that influence employees' decision to leave.

Although job attitudes are still an important predictor of turnover intentions in the military context (Knapp, McCloy, & DiFazio, 1993; Smith, Holtom, & Mitchell, 2011), there seems to be a growing convergence on the idea that most models developed for civilians fail to capture the essence of the process for military members (Capon, Chernyshenko, & Stark, 2007). Accordingly, a number of researchers attempted to explore how turnover unfolds in the military taking the unique aspects of the context into account. For example, both Kerce (1992) and Wilcove, Schwerin, and Wollosin (2003) proposed models where quality of life perceptions have

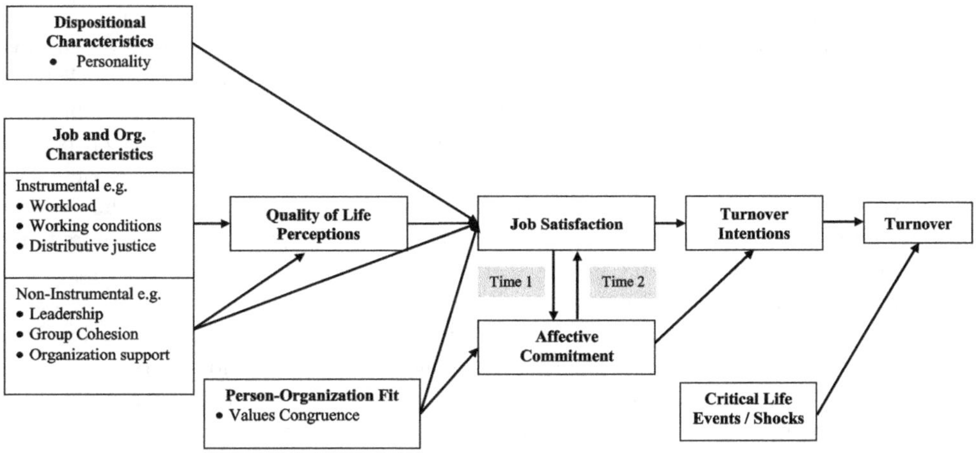


*Figure 18.1* Structural model of military turnover.

precedence over factors frequently explored in the civilian models. Yet the literature on both civilian and military voluntary turnover is still largely descriptive in nature and fails to offer decision-makers the longed-for prediction power concerning a member's decision to stay or leave. Our purpose is to propose a model that establishes the temporal nature of the turnover phenomenon in the military context.

The proposed model classifies the factors expected to play a role in military turnover under three categories: distal (i.e., *dispositional* and *job and organizational characteristics*), intermediate (i.e., *person-organization fit, quality of life perceptions, job satisfaction*, and *affective commitment*), and proximal factors (i.e., *turnover intentions* and *shocks*). Figure 18.1 depicts how these factors are proposed to be related to one another in the military withdrawal process.

## Distal factors

### *Dispositional characteristics*

Although not directly included in most models, personal dispositions are acknowledged in the literature as critical factors in the process of turnover. For instance, Griffeth and Hom (2001) included negative affectivity, the tendency to perceive oneself and environment negatively, among the factors contributing to job dissatisfaction in employee turnover. Similarly, in his conceptual model of quality of life (QoL) outcomes, Dowden (2000) proposed that personal dispositions are likely to influence three organizational outcomes (retention, absenteeism, and individual performance) through their effects on attitudinal factors, morale, motivation, and perceived stress.

Criticizing the dominance of attitudinally based models, Zimmerman (2008) argued that "researchers need to consider whether some individuals have a propensity to quit regardless of having a work environment designed to increase their job satisfaction, whereas other employees may be more likely to stay even under less-than-ideal circumstances" (p. 310). In his meta-analysis, he examined correlations of Big Five personality factors and trait affect with both intentions to quit and actual turnover. Furthermore, he tested a theoretically developed dispositional model of turnover. He found that, while negative affect ($\rho = 0.31$) and emotional stability ($\rho = -0.29$) were stronger predictors of intentions to quit, conscientiousness ($\rho = -0.22$) and agreeableness ($\rho = -0.27$) were stronger predictors of actual turnover. Personality traits had direct effects on intentions to quit (emotional stability) and turnover (conscientiousness and agreeableness) along with their indirect effects through job satisfaction. Results of this meta-analysis simply showed that employees may develop turnover intentions or may quit their jobs without necessarily being dissatisfied with them. Zimmerman's findings also indirectly support the role of impulsivity in unplanned quitting as described in the unfolding model (Lee & Mitchell, 1994) because conscientiousness and agreeableness are argued to represent impulsivity as the higher-order trait (Clark & Watson, 1999). Further support for the role of personal dispositions in the turnover process comes from a recent meta-analysis by Rubenstein et al. (2018). These authors found that emotional stability ($\rho = -0.19$), conscientiousness ($\rho = -0.16$), internal locus of control ($\rho = -0.10$), and internal motivation ($\rho = -0.16$) were predictive of turnover.

Recently, a narrower personality attribute, namely *grit*, emerged as a noteworthy and functional predictor of turnover. Grit refers to the extent to which an individual shows persistent interest, passion, and effort towards a long-term goal, and it distinguishes outstanding performers in many occupations (Duckworth, Peterson, Matthews, & Kelly, 2007). Empirical evidence suggests that grit is associated with higher educational attainment and retention (Credé, Tynan, & Harms, 2017; Duckworth & Quinn, 2009). Duckworth et al. (2007) conducted six consecutive studies to establish both a partial nomological network of the construct as well as to examine its predictive ability. They found that although grit showed some conceptual overlap with conscientiousness, it appeared to be conceptually distinct from it. They also found that military academy cadets with a grit score one standard deviation above the mean were more than 60% likely to finish the three-month summer training. Similarly, Eskreis-Winkler, Shulman, Beal, and Duckworth (2014) reported that *gritty* trainees were more likely to complete a 24-day Army Special Operations Forces selection course. These results offer grit as a critical individual difference variable in military retention.

Among the reviewed dispositional factors, the three Big Five dimensions (conscientiousness, agreeableness, and emotional stability) emerge to be critical for the attitude-centered turnover process and/or unplanned quitting that may result from unexpected turn of events. In addition to these relatively broad personality factors, grit, a narrower trait, seems to be important in the military turnover process. Hence, in the proposed model conscientiousness, agreeableness, emotional stability, and grit play a role in the development of military turnover intentions directly and through their influence over job satisfaction.

## *Job and organizational characteristics*

It needs to be emphasized at the outset that in the proposed model, job and organizational characteristics refer to the perceptions of the person rather than the actual characteristics. Job and organizational characteristics cover a broad range of factors. A distinction can be made between instrumental (e.g., workload) and non-instrumental (e.g., work group cohesion) job and organizational characteristics relevant for the military turnover. Instrumental characteristics refer to observable, quantifiable, and measurable aspects of conditions of employment, whereas non-instrumental characteristics refer to intangible, unquantifiable, and mostly relational aspects of one's work environment.

### *Instrumental characteristics*

There exists empirical and/or theoretical evidence concerning the role of instrumental factors in the development of negative work attitudes, which are well-established antecedents of turnover in both general and military contexts. Our literature review suggested criticality of three groups of instrumental factors: workload related factors (Dunn & Morrow, 2002; Sanchez, Bray, Vincus, & Bann,

2004; Stetz, Castro, & Bliese, 2007), working conditions or tempo (e.g., Castro & Adler, 1999; Dunn & Morrow, 2002; Huffman, Adler, Castro, & Dolan, 2000, Huffman, Adler, Dolan, & Castro, 2005), and distributive justice (McIntyre, Bartle, Landis, & Dansby, 2002, Suurd Ralph & Holmvall, 2016).

Using qualitative methods, Dunn and Morrow (2002) examined members' reasons for leaving the Canadian Forces (CF). Workload as it related to perstempo[1] and opstempo[2] was identified to be the second most commonly mentioned reason for leaving the CF. In another study conducted with 288 U.S. soldiers deployed in Europe, it was found that the decision to leave the military was influenced by perceptions of role overload (Huffman et al., 2005). Using a comprehensive database collected from 24,881 members of the Active Duty and Reserve/Guard components, Sanchez et al. (2004) reported job pressure as the most important predictor of job satisfaction. Another study with the U.S. Army reservists demonstrated that higher workload perceptions were linked with lower psychological well-being and higher intentions to leave (Stetz et al., 2007).

Although related to workload, working conditions seem to have a relatively independent impact on the process of work withdrawal. Studies suggest that frequent and long deployments, geographical isolation from family, overnight duty, long work hours, and high tempo, typical of most military jobs (Dunn & Morrow, 2002; Sanchez et al., 2004), play a role in the process of turnover. Demerouti, Geurts, Bakker, and Euwema (2004) examined the effects of rotation and timing of shifts on job attitudes, turnover intentions, work-home conflict, health, and absenteeism among Dutch military police officers. They found that rotating shift workers held less positive attitudes toward their jobs. Those with fixed shifts, however, reported higher levels of job satisfaction and professional efficacy and lower turnover intentions and cynicism. Interestingly, Carboon et al. (2009) reported that increased number of deployments was associated with lower separation risk. However, the authors also noted that the relationship between frequency of deployments and turnover is more complex, such that the nature (e.g., hostile) or length of deployments might influence the direction of this link.

*Distributive justice*, which basically refers to "fairness of outcome allocation" (Tepper, 2000, p. 179), is the third group of instrumental factors expected to be critical in the proposed model. Using three large random samples from the U.S. Military Equal Opportunity Climate Survey database, McIntyre et al. (2002) found that distributive justice perceptions affected job satisfaction through work group efficacy. In another military study in the CF, Suurd-Ralph and Holmvall (2016) reported that distributive justice, along with procedural and interpersonal justice, predicted overall justice perception, which in turn influenced turnover intentions via psychological strain. Furthermore, empirical evidence and the conceptual work on QoL (Dowden, 2000; Kerce, 1992; Schwerin et al., 2006), reviewed below, suggest that instrumental characteristics influence job satisfaction through their effects on QoL perceptions/concerns of military members.

## *Non-instrumental characteristics*

Among non-instrumental characteristics deemed relevant in the military turnover process are leadership and/or leader-member relations (Britt, Davison, Bliese, & Castro, 2004; Eberly, Bluhm, Guarana, Avolio, & Hannah, 2017; Pohl, Bertrand, & Ergen, 2016; Sanchez et al., 2004), group cohesion (Ahronson & Cameron, 2007; Griffeth & Hom, 2001), and organizational support (Dupré & Day, 2007; Jordan, Gabriel, Teasley, Walker, & Schraeder, 2015).

1 *Perstempo*: The pace at which personnel are deployed or are sent to temporary duty.

2 *Opstempo*: The pace of operations as indicated by the number and length of deployments, training exercises, temporary duty assignments, and the like (see Huffman et al., 2005).

Leadership/supervision is critical not only for boosting motivation and performance, but also for improving general adaptation of military members (Britt et al., 2004). Rubenstein et al.'s (2018) meta-analysis suggests a significant negative association between positive leadership and voluntary turnover ($\rho = -0.24$). In a longitudinal study with U.S. Army officers, Payne and Huffman (2005) reported that when received from a supervisor, mentoring significantly predicted affective commitment a year later, which in turn affected voluntary turnover after 10 years. Consistently, Pohl et al.'s (2016) study with 337 Belgian soldiers reported that perceived supervisor support predicted affective commitment, normative commitment, and job satisfaction. Effective leadership was also shown to have a buffer effect against stressful conditions in the military (Britt et al., 2004; Eberly et al., 2017). A recent multilevel study with U.S. soldiers deployed to an active war zone reported that transformational leadership indirectly reduced turnover intentions via on-the-job embeddedness when the soldiers' units were highly exposed to extreme contexts (Eberly et al., 2017).

Another non-instrumental factor that is proposed to be effective in military turnover is group cohesion. Meta-analytic findings by Rubenstein et al. (2018) suggest a significant association between peer/group relations (e.g., cohesion, coworker support) and voluntary turnover ($\rho = -0.14$). In their study with 447 CF employees, Ahronson and Cameron (2007) found that task-related dimensions of group cohesion, namely group integration and individual attraction to the group, significantly predicted job satisfaction.

Organizational support is another non-instrumental characteristic that has a distal effect on military turnover. Rubenstein et al.'s (2018) meta-analytic study reported that organizational support had a negative association with turnover ($\rho = -0.19$). In a study conducted with 450 military personnel, Dupré and Day (2007) reported that both organizational and supervisory support had direct associations as well as indirect relationships with turnover intentions via job satisfaction. Furthermore, Jordan et al. (2015) found that perceived organizational support positively influenced cadets' decision to sign a contract to become officers in the U.S. military.

To summarize, based on the reviewed military and nonmilitary literatures, it is proposed that non-instrumental characteristics influence job satisfaction and affective component of organizational commitment directly. Furthermore, as explained below, they have an indirect effect on job satisfaction through QoL concerns.

## Intermediate factors

### *Person-organization fit*

The fit between person and environment can be expressed as the degree of adaptation an individual exhibits with respect to his/her vocational niche (Mumford & Stokes, as cited in Gustafson & Mumford, 1995). Among different forms of P-E fit, person-organization (P-O) fit refers to the extent to which an employee's personal values and the employing organization's values are congruent or compatible (Kristof, 1996; Lauver & Kristof-Brown, 2001).

Several researchers reported P-O fit as being predictive of important organizational outcomes, such as job satisfaction, organizational commitment, and turnover intentions (Arthur, Bell, Villado, & Doverspike, 2006; Kristof-Brown, Zimmerman, & Johnson, 2005; Lauver & Kristof-Brown, 2001). In their multiple meta-analyses on different types of fit, Kristof-Brown et al. (2005) found that P-O fit had substantial correlations with job satisfaction ($\rho = 0.44$), organizational commitment ($\rho = 0.51$), and turnover intentions ($\rho = -0.35$). Arthur et al.'s (2006) meta-analysis also reported similar associations where turnover was more strongly associated with value congruence ($\rho = 0.38$) than other types of congruence ($\rho = 0.26$). Arthur et al. also reported a partially mediated relationship between P-O fit and turnover via work attitudes (i.e., job satisfaction, organizational commitment, and turnover intentions). In short, evidence concerning the primacy of values congruence as a predictor of work attitudes and turnover intentions over the other measures is quite convincing. Hence, in the model, we expect them to influence turnover intentions via job attitudes.

## *Quality of life perceptions*

QoL perceptions could be defined as an individual's global sense of well-being nourished by his/her feelings about various life domains, such as standard of living, job itself, leisure and recreation, health, marriage/intimate relationship, and relations with children (Dowden, 2000). In models attempting to capture QoL in the military, perceptions of global QoL are assumed to have significant effects on critical outcome variables such as performance, commitment, and retention either directly or indirectly (Dowden, 2000; Kerce, 1992; Schwerin, 2006; Wilcove et al., 2003). Dowden's QoL model assumes that: (1) QoL domains (e.g., job characteristics, intimate relations, and friendship) are strong contributors of global QoL perceptions; (2) dispositions influence global QoL perceptions; (3) enhancing global QoL perceptions positively influences satisfaction, commitment, and motivation; and (4) global QoL perceptions influence important organizational outcomes (retention, absenteeism, and performance) via satisfaction, commitment, and motivation. The reason QoL appears to be more relevant in a conceptual framework of military turnover is simply that being in the military has an overarching effect on both work and non-work lives of the members (Ryan & Burrell, 2012; Schwerin et al., 2006). According to Morrow (2007), in the CF, QoL variables influence people's retention intentions such that they account for 10%–20% of the variance in stay–leave decisions over and above the influence of other variables.

Wilcove et al. (2003) categorized QoL under two domains: non-work (e.g., marriage, relationship with child, standard of living) and work (e.g., professional development, shipboard life satisfaction). In their model, global QoL perceptions loaded onto the work domain. Furthermore, non-work factors were directly linked with reenlistment intentions, whereas work factors (including global QoL perceptions) were indirectly linked with reenlistment intentions via organizational commitment. When analyzed individually, global QoL perceptions had strong correlations with organizational commitment (0.46), intention to stay at the military at the next decision point (0.31), and intention to stay at the military until retirement (0.28). Similarly, Hindelang and colleagues (2004), who validated Wilcove et al.'s (2003) model in the U.S. Marine Corps, found that work domain factor of QoL was associated with reenlistment intentions via organizational commitment. Global QoL perceptions again strongly correlated with intention to stay for both married (0.38) and unmarried (0.36) Marines. To sum, evidence suggests that QoL perceptions mediate the relationship of environmental and dispositional factors with organizational outcomes. Consistently, in the proposed model, QoL factors mediate the effects of dispositional, job, and organizational characteristics on job satisfaction.

## *Work attitudes: Job satisfaction and affective commitment*

In most models of turnover, job satisfaction is assumed to influence turnover behavior largely through turnover thoughts and intentions (Hom et al., 1984; Mobley, 1977; Rubenstein et al., 2015). However, despite this well-theorized and supported indirect effect on turnover, job satisfaction has also been shown to directly influence turnover. For instance, Rubenstein and colleagues' (2018) meta-analysis reported a relatively strong negative association between job satisfaction and turnover ($\rho = -0.28$).

However, meta-analytic findings as well as theoretical arguments suggest that because of contractual obligations, satisfaction may have a weaker (yet still significant) influence on withdrawal cognitions and actual turnover for military samples than for civilian samples (Carsten & Spector, 1987; Farkas & Tetrick, 1989; Hom, Caranikas-Walker, Prussia, & Griffeth, 1992). Carsten and Spector (1987), for example, found that predictability of turnover especially by satisfaction decreased with time, and this decrease was more evident in the military samples. Furthermore, an unfolding model of voluntary turnover suggests that a considerable portion of voluntary turnover is not induced by job dissatisfaction; shocks or significant life events seem to have more explanatory power in the experience of turnover (Holtom et al., 2005; Lee & Mitchell, 1994). Yet, as stated by Holtom and colleagues (2005) themselves, the concept of shocks does not replace job dissatisfaction as an antecedent of voluntary turnover. Even in studies supporting the role of shocks in the turnover experience, dissatisfaction was the major predictor of turnover for some participants. Job satisfaction was found to strongly correlate (0.63) with turnover intentions in a recent study conducted with Swedish Army personnel (Österberg & Rydstedt, 2018). Hence, in the proposed model, job dissatisfaction (along with or irrespective of shocks) is expected to play a critical role in the process of withdrawal.

Organizational commitment, which refers to a relatively stable and more global attitude toward the employing organization, is a frequently studied variable in relation to employee withdrawal. In a meta-analysis, Griffeth, Hom, and Gaertner (2000) reported that organizational commitment predicted turnover ($\rho = -0.23$) better than did overall satisfaction ($\rho = -0.19$). The predictive power of commitment was even larger for military samples ($\rho = -0.28$). As Lytell and Drasgow suggest, this difference may stem from the unique nature of the military context, where "physical risks, lengthy time commitments, and separations from home may require a strong commitment to the military in order for attrition to be avoided" (2009, p. 347).

Meyer and Allen's (1997) conceptualization of commitment as a three-component structure (i.e., affective commitment – AC, continuance commitment – CC, and normative commitment – NC) offers a framework to explore the nature of the relationships between satisfaction, commitment, and turnover in the military turnover process. Within the military context, AC refers to a soldier's emotional attachment to, identification with, and involvement in the military service or unit. CC refers to perceptions of the costs associated with leaving the military and taps into perceptions of both available job alternatives and the personal sacrifices to be made by leaving the organization. Finally, NC refers to a soldier's felt moral obligation to stay with the military (Gade, 2003). In a longitudinal study with U.S. Air Force Academy cadets, Holtom and colleagues (2014) reported that AC was significantly associated with actual turnover after two years, whereas NC and CC were not significant predictors. Similarly, in their study with 462 male noncommissioned members at the CF, Godlewski and Kline (2012) reported that AC ($\beta = -0.50, p < 0.01$) was a stronger predictor of turnover intentions than NC ($\beta = -0.17, p < 0.01$). Turnover intentions predicted actual turnover in turn ($\beta = 0.31, p < 0.01$). Further empirical evidence from the U.S. (Langkamer & Ervin, 2008) and Canadian (Charbonneau & Wood, 2018) military samples suggest that AC is negatively associated with turnover intentions.

Although both satisfaction and commitment have been identified as critical variables in the turnover process, there seems to be a lack of agreement concerning the nature of the relationship between these variables in relation to employee withdrawal. Some studies suggest that commitment mediates the relationship between satisfaction and turnover intentions (e.g., Heffner & Gade, 2003; Williams & Hazer, 1986). In an old but noteworthy study, Farkas and Tetrick (1989) argued that the relationship between commitment and satisfaction was much more complicated than a simple unidirectional relationship. Results of their longitudinal study on a sample of Navy enlisted personnel suggested existence of a cyclical relationship between the two variables. Mathieu's (1991) study with 588 Army and Navy Reserve Officer Training Cadets also supported a recursive link between organizational commitment and job satisfaction. Similarly, Tremble, Payne, Finch, and Bullis (2003) emphasized the benefits of longer tracking periods in fully capturing the development of organizational commitment. Van Maanen's study, cited by Tremble et al. (2003), using a sample of police officers, indicated that it was not before the first 30 months of employment that organizational commitment stabilized. Exploring the causal nature of the relationship between job satisfaction and commitment, Huang and Hsiao (2007) found stronger support for the model depicting a reciprocal relationship between the two variables. More recent reviews on turnover also suggest that turnover should be considered a dynamic process and scholars should employ longitudinal methods to understand this process (Holtom, Mitchell, Lee, & Eberly, 2008; Hom et al., 2017; Lee, Hom, Eberly, Junchao, & Mitchell, 2017).

We propose that both AC and job satisfaction contribute to turnover intentions, and the nature of the relationship between these two variables is likely to be cyclical in nature. That is, satisfaction is expected to play a role in the development of AC, but, once established, AC can be expected to influence satisfaction.

## Proximal factors

### *Turnover intentions*

In almost all models of turnover (Mobley, Griffeth, Hand, & Meglino, 1979; Mobley et al., 1978), turnover intentions are the most proximal predictors of turnover behavior and they mediate the relationship between job attitudes, mainly satisfaction, and turnover behavior. Rubenstein et al.'s (2018) meta-analysis suggests a strong association between withdrawal cognitions (including turnover intentions, job search intentions, expected utility of withdrawal) and actual voluntary turnover ($\rho = 0.56$).

In an earlier attempt to understand the process of military turnover, Knapp et al. (1993) reported that although predictive ability of satisfaction concerning turnover behavior was weaker, the association between turnover intentions and turnover seemed stronger in the military context. It seems plausible to argue that once turnover intentions grow, the path to turnover seems more definite in the military, most likely because of the contractual nature of the jobs. Accordingly, in the proposed model, turnover intentions, which are directly influenced by satisfaction and affective commitment, are treated as one of the direct determinants of turnover behavior.

### *Critical life events/shocks*

Recently there has been an increased recognition that turnover is not always a "slow burn," deliberative process (Kammeyer-Mueller, Wanberg, Glomb, & Ahlburg, 2005, p. 646). As mentioned before, according to the "unfolding model" of voluntary turnover (Lee & Mitchell, 1994), unexpected life events, or shocks, cause voluntary turnover more often than accumulated job dissatisfaction. Shocks are jarring, usually unexpected, events that prompt thoughts of quitting. They can be positive (e.g., an unsolicited job offer), neutral (e.g., change in supervisor), or negative (e.g., harassment), and also they can be personal, work-related, or professional.

Three of the four paths suggested by the unfolding model involve a shock as the triggering event for turnover. In Path 1, a shocking event (e.g., an irresistible job offer) triggers the realization of a preexisting plan of action, and the person leaves without considering alternatives or his/her commitment to the organization. In this path, job dissatisfaction is basically irrelevant in turnover. In Path 2, the shocking event (e.g., after being passed over for promotion) is negative and leads to a rather quick decision without a search for alternatives. In Path 3, the triggering event (e.g., a serious job offer which makes the person compare and contrast it with the current job and other alternatives) can be positive, negative, or neutral, and unlike the first two paths, resulting dissatisfaction initiates a search for alternatives. Although relative dissatisfaction is a possibility in Path 3, the person may be quite satisfied with the current job before the shock. Compared to the first two paths, Path 3 requires considerable deliberation. In Path 4, the turnover process is not initiated by a shock; the decision to quit evolves rather gradually. In this path to turnover, lack of a compatible fit with the organization results in dissatisfaction and reduced commitment, and this in turn leads to quitting with or without searching for alternatives (Donnelly & Quirin, 2006).

Empirical evidence has in general yielded support for the role of shocks in turnover (Holtom et al., 2005, 2017; Kammeyer-Mueller et al., 2005; Morrell, Loan-Clarke, & Wilkinson, 2004). For example, Holtom et al.'s (2005) study indicated that in a sizeable proportion of decisions to quit, shocks, but not job dissatisfaction, played a critical role and that Path 3 represented the most widely experienced shock-based pathway to turnover. Kulik, Treuren, and Bordia (2012) conducted exit interviews with 228 employees in an Australian organization. Sixty-eight percent of leavers reported some sort of shock, whereas the remaining 32% reported a history of dissatisfaction with their job prior to leaving. Military studies also support the influence of shocks in the turnover process. Holt, Rehg, Lin, and Miller (2007) tested the unfolding model with a sample of 182 Air Force Officers who voluntarily quit the army. Sixty-two percent of these officers stated that they experienced a type of shock before leaving, and the model accurately classified 47% of the participants' decisions to leave. Based on the reviewed literature, shocks are expected to play a role in the military turnover process, either along with or independently of the progression from dissatisfaction and reduced commitment.

## Conclusion

Reviewed evidence suggests unique dynamics of the withdrawal process in the military context. The proposed military turnover model contributes to the existing literature in several ways. First, it represents one of the rare attempts to capture the complex decision-making processes involved in military turnover by focusing mainly on the psychological processes and the individual's subjective experiences. Second, departing from the descriptive approaches conventionally used to understand military turnover, the proposed model presents a comprehensive causal framework that allows for prediction. Third, unlike typical attitude-centered models, this model acknowledges value congruence and especially four context-relevant personality factors (conscientiousness, agreeableness, emotional stability, and grit), as being important in the process of military turnover. Finally, by integrating shocks, this model aims to account for turnover that is not necessarily attitude based.

Overall, we believe that the proposed model is a small step in the right direction. However, we also believe that the model presented here may benefit from some further refinements. Refinement efforts may focus on three issues. First, the proposed model implicitly focuses on late turnover, and factors playing a role in early turnover (e.g., turnover occurring during or right after initial training) are not specifically addressed or separated from the factors critical in late turnover. So, efforts may be directed at identifying antecedents of early turnover and, perhaps, linking them to both recruitment and late turnover processes.

Second, although empirical evidence still supports precedence of both job satisfaction and commitment in the process of turnover, there also exists evidence pointing relevance of other job attitudes, especially job engagement, in this process. Closely connected to the eudemonic tradition of well-being, engagement is defined as "a relatively enduring state of mind referring to the simultaneous investment of personal energies in the experience or performance of work" (Christian, Garza, & Slaughter, 2011, p. 95). In a study conducted in the military, Alarcon, Lyons, and Tartaglia (2010) found that work engagement was a significant predictor of turnover intentions and it was predicted by work, work-group, and organizational factors directly and leadership indirectly. Rubenstein et al.'s (2018) meta-analysis also indicated that engagement ($\rho = -0.20$) was a substantial predictor of turnover in its own. Hence, it may be wise to include this more relevant and encompassing attitudinal variable in the model, and examine its contribution above and beyond job satisfaction and commitment.

Third, in the proposed model, demographic variables, such as gender and ethnicity, are not directly addressed; they are assumed to have an influence on work attitudes especially through QoL perceptions. However, Rubenstein et al.'s (2018) meta-analysis suggested that demographic factors such as age ($\rho = -0.21$) and having children ($\rho = -0.20$) had relatively strong effects on turnover. Hence, there is a need for a more thorough examination of demographic variables critical in military turnover and the mechanisms through which these demographic variables contribute to employee withdrawal.

We believe, following the finalization of the conceptual framework, the model should be subjected to empirical testing, preferably using a longitudinal approach. Since the proposed model is a generic one, the fit of the model in varying military contexts should be tested and compared. Result of this empirical testing is expected to both contribute to existing knowledge on turnover and have implications for military recruitment, selection, and training practices.

## References


Ahronson, A., & Cameron, J. E. (2007). The nature and consequences of group cohesion in a military sample. *Military Psychology, 19*(1), 9–25.

Alarcon, G., Lyons, J. B., & Tartaglia, F. (2010). Understanding predictors of engagement within the military. *Military Psychology, 22*(3), 301–310.

Allen, D. G., Bryant, P. C., & Vardaman, J. M. (2010). Retaining talent: Replacing misconceptions with evidence-based strategies. *Academy of Management Perspectives, 24*(2), 48–64.

Arthur Jr, W., Bell, S. T., Villado, A. J., & Doverspike, D. (2006). The use of person-organization fit in employment decision making: An assessment of its criterion-related validity. *Journal of Applied Psychology, 91*(4), 786–801.

Britt, T. W., Davison, J., Bliese, P. D., & Castro, C. A. (2004). How leaders can influence the impact that stressors have on soldiers. *Military Medicine, 169*(7), 541–545.

Capon, J., Chernyshenko, O. S., & Stark, S. (2007). Applicability of civilian retention theory in the New Zealand military. *New Zealand Journal of Psychology, 36*(1), 50–56.

Carboon, I., Creamer, M., Forbes, A. B., McKenzie, D. P., McFarlane, A. C., & Kelsall, H. L. (2009). The relationship between deployment and turnover in Australian navy personnel. *Military Psychology, 21*(2), 233–240.

Carsten, J. M., & Spector, P. E. (1987). Unemployment, job satisfaction, and employee turnover: A meta-analytic test of the Muchinsky model. *Journal of Applied Psychology, 72*(3), 374–381.

Castro, C. A., & Adler, A. B. (1999). The impact of operations tempo on soldier and unit readiness. *Parameters, 29*, 86–95.

Charbonneau, D., & Wood, V. M. (2018). Antecedents and outcomes of unit cohesion and affective commitment to the Army. *Military Psychology, 30*(1), 43–53.

Christian, M. S., Garza, A. S., & Slaughter, J. E. (2011). Work engagement: A quantitative review and test of its relations with task and contextual performance. *Personnel Psychology, 64*(1), 89–136.

Clark, L. A., & Watson, D. (1999). Temperament: A new paradigm for trait psychology. In L. A. Pervin & O. P. John (Eds.), *Handbook of Personality: Theory and Research* (2nd ed., pp. 399–423). New York: Guilford Press.

Credé, M., Tynan, M. C., & Harms, P. D. (2017). Much ado about grit: A meta-analytic synthesis of the grit literature. *Journal of Personality and Social Psychology, 113*(3), 492–511.

Demerouti, E., Geurts, S. A. E., Bakker, A. B., & Euwema, M. (2004). The impact of shift-work on work-home conflict, job attitudes and health. *Ergonomics, 47*(9), 987–1002.

Donnelly, D. P., & Quirin, J. J. (2006). An extension of Lee and Mitchell's unfolding model of voluntary turnover. *Journal of Organizational Behavior, 27*(1), 59–77.

Dowden, C. (2000). Quality of life in the Canadian Forces: Conceptualization and measurement of the QOL construct. *Sponsor Research Report* No.00-05. Director Human Resources Research and Evaluation, National Defense Headquarters, Ottawa, Canada.

Duckworth, A. L., Peterson, C., Matthews, M. D., & Kelly, D. R. (2007). Grit: Perseverance and passion for long-term goals. *Journal of Personality and Social Psychology, 92*(6), 1087–1101.

Duckworth, A. L., & Quinn, P. D. (2009). Development and validation of the Short Grit Scale (GRIT–S). *Journal of Personality Assessment, 91*(2), 166–174.

Dunn, J. R., & Morrow, R.O. (2002). *Should I Stay or Should I Go?: Attrition Questionnaire Revision Project – Phase 1 Findings*. SRR 2002-09. Directorate of Human Resources Research and Evaluation, Ottawa, Canada.

Dupré, K. E., & Day, A. L. (2007). The effects of supportive management and job quality on the turnover intentions and health of military personnel. *Human Resource Management, 46*(2), 185–201.

Eberly, M. B., Bluhm, D. J., Guarana, C., Avolio, B. J., & Hannah, S. T. (2017). Staying after the storm: How transformational leadership relates to follower turnover intentions in extreme contexts. *Journal of Vocational Behavior, 102*, 72–85.

Eskreis-Winkler, L., Shulman, E. P., Beal, S., & Duckworth, A. L. (2014). The grit effect: Predicting retention in the military, the workplace, school and marriage. *Frontiers in Psychology, 5*, 36.

Farkas, A. J., & Tetrick, L. E. (1989). A three-wave longitudinal analysis of the causal ordering of satisfaction and commitment on turnover decisions. *Journal of Applied Psychology, 74*(6), 855–868.

Gade, P. A. (2003). Organizational commitment in the military: An overview. *Military Psychology, 15*(3), 163–166.

Godlewski, R., & Kline, T. (2012). A model of voluntary turnover in male Canadian Forces recruits. *Military Psychology, 24*(3), 251–265.

Griffeth, R. W., Hom, P. W., & Gaertner, S. (2000). A meta-analysis of antecedents and correlates of employee turnover: Update, moderator tests, and research implications for the next millennium. *Journal of Management, 26*(3), 463–488.

Griffeth, R. W., & Hom, P. W. (2001). *Retaining Valued Employees*. Thousand Oaks: Sage.

Gustafson, S. B., & Mumford, M. D. (1995). Personal style and person-environment fit: A pattern approach. *Journal of Vocational Behavior, 46*(2), 163–188.

Heffner, T. S., & Gade, P. A. (2003). Commitment to nested collectives in special operations forces. *Military Psychology, 15*(3), 209–224.

Hindelang, R. L., Schwerin, M. J., & Farmer, W. L. (2004). Quality of life (QOL) in the US Marine Corps: The validation of a QOL model for predicting reenlistment intentions. *Military Psychology, 16*(2), 115–134.

Holt, D. T., Rehg, M. T., Lin, J. H., & Miller, J. (2007). An application of the unfolding model to explain turnover in a sample of military officers. *Human Resource Management, 46*(1), 35–49.

Holtom, B. C., Goldberg, C. B., Allen, D. G., & Clark, M. A. (2017). How today's shocks predict tomorrow's leaving. *Journal of Business and Psychology, 32*(1), 59–71.

Holtom, B. C., Mitchell, T. R., Lee, T. W., & Eberly, M. B. (2008). Turnover and retention research: A glance at the past, a closer review of the present, and a venture into the future. *The Academy of Management Annals, 2*(1), 231–274.

Holtom, B. C., Mitchell, T. R., Lee, T. W., & Inderrieden, E. J. (2005). Shocks as causes of turnover: What they are and how organizations can manage them. *Human Resource Management, 44*(3), 337–352.

Holtom, B. C., Smith, D. R., Lindsay, D. R., & Burton, J. P. (2014). The relative strength of job attitudes and job embeddedness in predicting turnover in a US military academy. *Military Psychology, 26*(5–6), 397–408.

Hom, P. W., Caranikas-Walker, F., Prussia, G. E., & Griffeth, R. W. (1992). A meta-analytical structural equations analysis of a model of employee turnover. *Journal of Applied Psychology, 77*(6), 890–909.

Hom, P. W., & Griffeth, R. W. (1991). Structural equations modeling test of a turnover theory: Cross-sectional and longitudinal analyses. *Journal of Applied Psychology, 76*(3), 350–366.

Hom, P. W., Griffeth, R. W., & Sellaro, C. L. (1984). The validity of Mobley's (1977) model of employee turnover. *Organizational Behavior and Human Performance, 34*(2), 141–174.

Hom, P. W., Lee, T. W., Shaw, J. D., & Hausknecht, J. P. (2017). One hundred years of employee turnover theory and research. *Journal of Applied Psychology, 102*(3), 530–545.

Huang, T. C., & Hsiao, W. J. (2007). The causal relationship between job satisfaction and organizational commitment. *Social Behavior and Personality: an International Journal, 35*(9), 1265–1276.

Huffman, A. H., Adler, A., Castro, C., & Dolan, C. (2000, November). *Retention and the U.S. Army Officer in Europe*. Paper presented at the 42nd Annual Conference of the International Military Testing Association, Edinburgh, UK.

Huffman, A. H., Adler, A. B., Dolan, C. A., & Castro, C. A. (2005). The impact of operations tempo on turnover intentions of Army personnel. *Military Psychology, 17*(3), 175–202.

Jordan, M. H., Gabriel, T. J., Teasley, R., Walker, W. J., & Schraeder, M. (2015). An integrative approach to identifying factors related to long-term career commitments: A military example. *Career Development International, 20*(2), 163–178.

Kammeyer-Mueller, J. D., Wanberg, C. R., Glomb, T. M., & Ahlburg, D. (2005). The role of temporal shifts in turnover processes: It's about time. *Journal of Applied Psychology, 90*(4), 644–658.

Kerce, E. W. (1992). *Quality of Life: Meaning, Measurement, and Models* (NPRDC-TR-92-15). San Diego: Navy Personnel Research and Development Center.

Knapp, D. J., McCloy, R. A., & DiFazio, A. S. (1993, August 22). *Prediction of military turnover using intentions, satisfaction, and performance*. Paper presented at the 101st Annual Convention of the American Psychological Association, Toronto, Canada.

Kristof-Brown, A. L., Zimmerman, R. D., & Johnson, E. C. (2005). Consequences of individuals' fit at work: A meta-analysis of person-job, person-organization, person-group, and person-supervisor fit. *Personnel Psychology, 58*(2), 281–342.

Kristof, A. L. (1996). Person-organization fit: An integrative review of its conceptualizations, measurement, and implications. *Personnel Psychology, 49*(1), 1–49.

Kulik, C. T., Treuren, G., & Bordia, P. (2012). Shocks and final straws: Using exit-interview data to examine the unfolding model's decision paths. *Human Resource Management, 51*(1), 25–46.

Langkamer, K. L., & Ervin, K. S. (2008). Psychological climate, organizational commitment and morale: Implications for Army captains' career intent. *Military Psychology, 20*(4), 219–236.

Lauver, K. J., & Kristof-Brown, A. (2001). Distinguishing between employees' perceptions of person-job and person-organization fit. *Journal of Vocational Behavior, 59*(3), 454–470.

Lee, T. W., Hom, P. W., Eberly, M. B., Junchao (Jason) Li, & Mitchell, T. R. (2017). On the next decade of research in voluntary employee turnover. *Academy of Management Perspectives, 31*(3), 201–221.

Lee, T. W., & Mitchell, T. R. (1994). An alternative approach: The unfolding model of voluntary employee turnover. *Academy of Management Review, 19*(1), 51–89.

Lytell, M. C., & Drasgow, F. (2009). "Timely" methods: Examining turnover rates in the US military. *Military Psychology, 21*(3), 334–350.

Mathieu, J. E. (1991). A cross-level nonrecursive model of the antecedents of organizational commitment and satisfaction. *Journal of Applied Psychology, 76*(5), 607–618.

McIntyre, R. M., Bartle, S. A., Landis, D., & Dansby, M. R. (2002). The effects of equal opportunity fairness attitudes on job satisfaction, organizational commitment, and perceived work group efficacy. *Military Psychology, 14*(4), 299–319.

Meyer, J. P., & Allen, N. J. (1997). *Commitment in the Workplace*. Thousand Oaks, CA: Sage.

Mobley, W. H. (1977). Intermediate linkages in the relationship between job satisfaction and employee turnover. *Journal of Applied Psychology, 62*(2), 237–240.

Mobley, W. H., Griffeth, R. W., Hand, H. H., & Meglino, B. M. (1979). Review and conceptual analysis of the employee turnover process. *Psychological Bulletin, 86*(3), 493–522.

Mobley, W. H., Horner, S. O., & Hollingsworth, A. T. (1978). An evaluation of precursors of hospital employee turnover. *Journal of Applied Psychology, 63*(4), 408–414.

Morrell, K., Loan-Clarke, J., & Wilkinson, A. (2004). The role of shocks in employee turnover. *British Journal of Management, 15*(4), 335–349.

Morrow, R. (2007). Perstempo/opstempo and quality of life. *Final Report of Research Task Group on Recruitment and Retention of Military Personnel, HFM, 107*, 3I1–3I16.

Newman, J. (1974). Predicting absenteeism and turnover: A field comparison of Fishbein's model and traditional job attitude measures. *Journal of Applied Psychology, 59*(5), 610–615.

Österberg, J., & Rydstedt, L. (2018). Job satisfaction among Swedish soldiers: Applying the Job Characteristics Model to newly recruited military personnel. *Military Psychology, 30*(4), 302–310.

Payne, S. C., & Huffman, A. H. (2005). A longitudinal examination of the influence of mentoring on organizational commitment and turnover. *Academy of Management Journal, 48*(1), 158–168.

Pohl, S., Bertrand, F., & Ergen, C. (2016). Psychological contracts and their implications for job outcomes: A social exchange view. *Military Psychology, 28*(6), 406–417.

Rubenstein, A. L., Eberly, M. B., Lee, T., & Mitchell, T. R. (2015). Looking beyond the trees: A meta-analysis and integration of voluntary turnover research. In *Academy of Management Proceedings* (*Vol. 2015*, No. 1, p. 12779). Briarcliff Manor, NY: Academy of Management.

Rubenstein, A. L., Eberly, M. B., Lee, T. W., & Mitchell, T. R. (2018). Surveying the forest: A meta-analysis, moderator investigation, and future-oriented discussion of the antecedents of voluntary employee turnover. *Personnel Psychology*, *71*(1), 23–65.

Ryan, D., & Burrell, M. (2012). What they deserve: quality of life in the US military. In J. H. Laurence & M. D. Matthews (Eds.), *The Oxford Handbook of Military Psychology* (pp. 381–399). New York: Oxford University Press.

Sanchez, R. P., Bray, R. M., Vincus, A. A., & Bann, C. M. (2004). Predictors of job satisfaction among Active Duty and Reserve/Guard Personnel in the U.S. Military. *Military Psychology*, *16*(1), 19–35.

Schwerin, M. J. (2006). Quality of life and subjective well-being among military personnel: An organizational response to the challenges of military life. In T. W. Britt, A. B. Adler, & C. A. Castro (Eds.), *Military Culture. Military Life: The Psychology of Serving in Peace and Combat: Military Culture* (pp. 145–179). Westport, CT: Praeger Security International.

Schwerin, M. J., Kline, T. L., Olmsted, M. G., & Wilcove, G. L. (2006, May 20). *Validation of Navy quality of life retention models*. Paper presented at the American Association for Public Opinion Research 61st Annual Conference, Montréal, Canada.

Smith, D. R., Holtom, B. C., & Mitchell, T. R. (2011). Enhancing precision in the prediction of voluntary turnover and retirement. *Journal of Vocational Behavior*, *79*(1), 290–302.

Stetz, M. C., Castro, C. A., & Bliese, P. D. (2007). The impact of deactivation uncertainty, workload, and organizational constraints on reservists' psychological well-being and turnover intentions. *Military Medicine*, *172*(6), 576–580.

Sümer, H. C., & van de Ven, C. A. (2007). A proposed model of military turnover. *Final Report of Research Task Group on Recruitment and Retention of Military Personnel, HFM*, *107*, 4B1–4B18.

Suurd Ralph, C. D., & Holmvall, C. M. (2016). Examining the relationships between the justice facets and turnover intent: The mediating roles of overall justice and psychological strain. *Military Psychology*, *28*(4), 251–270.

Tepper, B. J. (2000). Consequences of abusive supervision. *Academy of Management Journal*, *43*(2), 178–190.

Tremble, T. R., Payne, S. C., Finch, J. F., & Bullis, R. C. (2003). Opening organizational archives to research: Analog measures of organizational commitment. *Military Psychology*, *15*(3), 167–190.

Wilcove, G. L., Schwerin, M. J., & Wolosin, D. G. (2003). An exploratory model of quality of life in the US Navy. *Military Psychology*, *15*(2), 133–152.

Williams, L. J., & Hazer, J. T. (1986). Antecedents and consequences of satisfaction and commitment in turnover models: A reanalysis using latent variable structural equation models. *Journal of Applied Psychology*, *71*(2), 219–231.

Zimmerman, R. D. (2008). Understanding the impact of personality traits on individuals' turnover decisions: A meta-analytic path model. *Personnel Psychology*, *61*(2), 309–348.

# 19

# TRANSITION FROM MILITARY TO CIVILIAN LIFE

*Harprit Kaur and Swati*

Transition is part and parcel of military life; it might be in terms of a new duty station, an overseas deployment, a field posting or a change of command. An important transition which veterans have to go through is returning to civilian life. Transitions that take place within the military setup are managed by providing assistance through various programmes or courses; however, veterans remain emotionally underprepared to manage transition to civilian employment (Baruch & Quick, 2007).

Transition from military to civilian life is a challenging task and proves quite difficult for some veterans (Rogers, 1944; Blackburn, 2016). In a survey conducted on Canadian armed forces veterans pertaining to life after service, 27% of veterans reported it to be difficult to very difficult, 17% rated it as neither easy nor difficult and 56% said it was easy (Thompson, Vantil, Zamorski, Garber, Dursun, Fikeretoglu & Pedlar, 2016). There has been an agreement on the negative effect of transition on the well-being of retired military personnel and their families (Rogers, 1944; Adler, Zamorski & Britt, 2011; Demers, 2011; Thompson et al., 2016). Even after years of research, the process of military to civilian transition is not clear; researchers have only been able to unpack issues pertaining to it (Cooper, Caddick, Godier, Cooper & Fossey, 2018). McNeil, Lecca and Wright (1983) asserted that using the term **'retirement' is not appropriate for veterans**; it would be more accurate to describe it as a change of career. Military personnel who retire after 20 years of service still have 15 career-oriented years with them.

While going through this midlife career transition from military to civilian life, veterans face various adjustment issues along with anxieties. A core aspect of these anxieties is a veteran's **inseparable identification with the military**. Researchers have advocated that with passage of time, the military is becoming more civilianized because of convergence with non-military jobs (Biderman & Sharp 1968; Biderman 1964; Moskos, 1977), thus making it easy for veterans to adjust; however, the military is an institution. Moskos (1988) reported that members of an institution consider or perceive themselves as different from the rest of society. Biderman and Sharp (1968) referred to the military as a totally isolated institution where work, sleeping, play, worship, eating, nursing, burial and education all take place within the institution along with colleagues. This arrangement of interactions actually leads to **isolation from the rest of society**. Standford (1971) has called midlife military to civilian transition as re-socialization into civilian society by concluding one career and initiating another. The military has always been viewed as the epitome of security (Kilpatrick & Kilpatrick, 1979), providing a secure social environment. Leaving **the secure environment of the military** where one has been socialized, getting into civilian life and going through re-socialization is stressful and at times labelled a crisis (Giffen & McNeil, 1967; Ogburn, Bellino, Williams and Gordon, 1969; Doherty, 1983).

Losing a job after investing years of your life can be traumatic, and when such a loss occurs at a particular age (mid-life), it can be catastrophic (Parnes & King, 1977). McDonald (1977) compared military retirement to the cutting of an umbilical cord. To explain the transition, a corollary from the process of change can be brought in where first an individual encounters the inevitable change, breaks up with the old life and finally commits to a new life.

## *Human adaptation to transition*

Schlossberg (1981) gave a transition theory to explain the transition process in adults. It provides an insight into the help required during transition. Schlossberg's theory assimilated the transition process into 4Ss—self, situation, support and strategies. In the **self** part, each individual has their own set of psychological resources that determine how they will react to transition and the adjustment it calls for. Along with this, an individual's perception also affects their assets. The second aspect, **situation**, pertains to what led to transition, its trigger, the time of transition and so on. The third aspect is **support** which refers to what actually helps one deal with the stress related to transition and can include friends, family, colleagues and so on. Fourth is **strategies**; Schlossberg talks about a response that leads to modification of the situation, by and large here Lazarus and Folkman's (1984) coping modes were used, i.e., taking direct action, inhibiting action and seeking more information regarding the problem.

Adults keep experiencing transition in their lives; reaction to this transition depends upon the kind of transition, an individual's perception regarding it, the background of transition and its effect on their lives. There is no end to the transition, rather, **transition is a process** with different phases, i.e., going in, assimilating, moving through and coming out (Anderson, Goodman & Schlossberg, 2012). The authors of the model have also emphasized that preparedness for the difficulties helps the individual to go through the process smoothly. Moving out of transition refers to the stage where the individual is done with one transition and is in search of something new.

## *Facing the inevitable change*

### *Psychological adjustment*

A service personnel's return to civilian life does not involve any basic training or a transition period. One day you are a part of the organization, and the next day you are no longer associated with it. Some veterans may encounter problems during their pre-retirement time, while others experience difficulties after retirement (Drier, 1995). Bellino (1970) explained retirement as a process and not merely a status; it is "a transition from one period of life to another, leading to new interpersonal and social adjustment" (pp 580–583). Various authors have described problems, anxieties, needs and stressors of veterans during the transition and adjustment process. Kilpatrick and Kilpatrick (1979) described retirement from military as a psychological loss that leads to grief and mourning. According to them, reaction to this loss includes four stages, initially **protest**, then a feeling of **despair**, after which an individual **detaches** himself and finally **recovers**. All this needs to be dealt with like **'grief work'** so as to place it in a proper perspective. McNeil and Giffen (1967) provided the term **'retirement syndrome'** describing pre-retirement symptoms among military personnel. They found these symptoms centred on the gastrointestinal tract and cardiovascular system. Not all military officers retire at the same age or rank. Rank at the time of retirement, reason for retirement and age at the time of retirement all play a significant role in post-retirement adjustment. Berkey (1972) found that those military officers who retire after failing to make it to the next rank consider themselves inadequate; whatever success they achieved previously does not lend sufficient self-pride to ameliorate the feeling of degradation of missing out on promotion. These results were found to be true for all rank officers who missed out on their promotion goal. Failing to achieve a rank in the military is one aspect; there are other individuals who take retirement as a second opportunity to achieve something big in life. Garber (1971) also noted that a civilian career post-retirement is another opportunity for advancement. Adjustment and satisfaction post-retirement are also determined if the retirement is **voluntary or involuntary** in nature. Knippa (1979) found that voluntarily retired military officer were significantly higher on life satisfaction than involuntarily retired military officer. A similar kind of study was conducted with a civilian population, and the results repeated. In addition, researchers found that early retirees, be it voluntary or involuntary, were less satisfied than on-time retirees (Price, Walker & Kimmel, 1979).

Stanford (1968) and McClure (1993) explained that voluntary retirement is not always voluntary; at times it is due to promotion pass over, not meeting fitness standards or undesirable transfers. Another way of looking at voluntary retirement as involuntary is that only a few higher ranks work until superannuation; others have to retire early. So the **circumstances** under which the individual retires make a lot of difference to his psychological well-being.

## *Loss of identity*

Identity lies on a continuum of personal to social, where personal identity refers to how one describes oneself as a unique individual. It is our attitude, behaviour and interest that make us different from other individuals (Cruwys & Gunaseelan, 2016). Social identity is part of an individual's self-concept. It is an insight to an individual that he belongs to a particular group (Stets & Burke 2000). It assists an individual to keep his interest, behaviour and attitude in sync with the other members of the group he belongs to and demarcate himself from members of the outer groups (Cruwys & Gunaseelan, 2016). Identities are dynamic and keep evolving as per personal, cultural and social factors (Lawler, 2008). Erikson's (1959) model helps us to understand the process of identity formation; however, Para (2008) explained that in today's sociocultural environment, the identity formation process starts in late adolescence or early adulthood as people delay committing themselves to a specific set of values and beliefs. Late adolescence and early adulthood is a time when individuals join the military and form a **military identity** which is different from civilian identity.

Major transitions of life take place when an individual shifts from one social situation to another; this shift requires a shift in identity as well, for instance, from single to married, student to employed, adolescent to adult, military personnel to veteran. A transition is called successful when both old and new identities negotiate and integrate (Areul & Castro, 2018). The identity which has worked in one situation might not be suitable for a new situation, leading to discord in social situations. An individual can function well only after reshaping his identity (Kidwell & Dunham, 1995).

Military-to-civilian transition is one of the major life transitions. Identity issued during this phase have been considered of paramount importance to deal with adjustment challenges (Coll & Weiss, 2013; Castro & Kintzle, 2014; Cooper et al., 2018; Orazem et al., 2017; Robinson et al., 2017). This issue is most pertinent for those who are **forced to leave the military** because of health problems, joined the service early before even forming a civilian identity or are undergoing midlife aging crisis (Smith & True, 2014; Cooper et al., 2018).

Experts emphasize that the feelings a veteran experiences while undergoing the transition to civilian life, such as **fatigue, anger, sadness or irritability**, are due to identity challenges and are quite normal (Castro & Kintzle, 2014). As the area of military civilian transition is being explored these days, there has been increasing consensus that identity-related disorders need not be pathologized, as distress is common in major life transitions (Cooper et al., 2018).

There are fewer **females in the military**; hence, there are limited studies on women veterans. Demers (2013) reported that women who joined the service gained confidence and became more capable and strong; however, when they went back to civilian life, their transition was quite challenging as their military identity was a mis-fit to the traditional civilian female identity and, as a result, they felt **lonely**.

McNeil and Giffen (1965) found that not only veterans but their wives and children also experienced **loss of prestige** after retirement. Bellino (1970) also confirmed that **families of military personnel** hold a social position per the military rank, and retirement makes them move into a different culture altogether. Frank (1993) reported that a few military wives experience grief over the loss of social status post-retirement.

A study conducted in 2017 by Praharso, Tear and Cruwys found that social identities work as psychological resources that help in the maintenance of well-being. This evidence came from a longitudinal study in which participants who experienced significant **life stressors** were protected from depression and reported significantly greater life satisfaction if they retained their membership of a group. It clearly indicates that even if an individual lacks social support before a stressful life event, if he becomes part of the group, this can prevent degradation of their psychological health.

### *Stereotypes*

Fiske and Taylor (2013) explained stereotyping as **'casual bias'** common in day-to-day life. Every group has certain characteristics which ultimately evolve as group stereotypes. These stereotypes work as an **in-group/out-group boundary**. Stereotypes have both positive and negative implications; they can either facilitate or hinder transition to civilian life. A positive stereotype associated with veterans is that they are viewed as protectors. A negative stereotype associated with veterans, especially combat veterans, is that they suffer from PTSD. Stereotypes can cause prejudice and stigma (Allport, 1954). Stereotypes also lead to perceived discrimination and stigma, causing conflict between groups. Stereotyping lowers the physical and mental well-being of the person who faces the stereotype (Fiske & Taylor, 2013). For instance, veterans feel distressed if they feel they are unable to live up to the **heroic stereotype** (Smith & True, 2014).

Soldiers generally perceive civilians as a homogenous group, while 90% of the UK civilian public perceived veterans to be physically or mentally damaged (Ashcroft, 2016). Veterans who suffer from chronic physical or mental health issues are **stereotyped as disabled** (Gerber, 2003). Minority groups with common minority identities can utilize various techniques to tackle stereotype threats (Fiske & Taylor, 2013). One example of this is veterans with similar health impairment such as blindness or amputation; they bond together to enhance the veterans' benefits (Gerber, 2003). Dealing with stereotype threat is important, as veterans who develop a **stigmatized identity** face difficulty in their transition (Amiot, Sablonniere, Terry & Smith, 2007).

## ***Breaking with old life***

### *Revisiting past life*

Military life involves many turning points and transitions. The experience of remembering these turning points and transitions is generally intense and vivid (Enz & Talarico, 2016). Remembering them at times leads to heightened emotional response, resulting in **watershed moments**.

More active members and veterans in service at times make efforts to **memorialize** their time spent at a particular posting by carrying some insignia of that place; their home itself is full of such insignias, revealing the story of their life or the kind of life they led during their service tenure. For those individuals who face grief after leaving the service up to a subclinical level, memorialization of this kind promotes their psychological health, as it helps in positive remembrance (Watkins, Cole & Weidemann, 2010).

Revisiting memories at time when a person feels alone in the new setup might help in providing relief; however, if an individual remains in constant mode of reminiscence, it can lead to adjustment problems. **Reminiscence bump**, which has been defined as innumerable memories in middle-age adulthood and older age from one's adolescence and early adulthood (Koppel & Berntsen, 2014; Rubin, Wetzler & Nebes, 1986), is a distinctive and defining feature of autobiographical memory (Koppel & Berntsen, 2014; Eysenck & Keane, 2010; Goldstein, 2015; Rathbone, Moulin, Conway & Holmes, 2012). A U.S. military veteran–associated suicide prevention program in the year 2016 revealed that in 2014, 65% of veterans who committed suicide were 50 years old or older, when reminiscence bump is the most common occurrence (Mobbs & Bonanno, 2018).

Remembering one's past might pave way to a **purpose of life**. Research states that it can facilitate individual development. Erikson (1959) has appreciated the importance of reviewing one's past life, especially in his last stage of development where an individual either gains integrity or is lost in despair. When one reviews his past, he might end up feeling that some crucial part of the self went unlived. The purpose is to integrate these conflicts into one's life and gain wisdom through the review.

Butler (1963) was one of the first to talk about the adaptive role in reviewing personal past. He coined the term 'life review' and, according to him, it is a process to consciously recall one's past so as to integrate it into one's life story. People of all ages use recall from the past as a source of guidance and work on unresolved conflicts to have a meaningful picture of life (Butler, 1963). It is up to an individual how he utilizes his memories while going through the transition phase.

### *Service-related nostalgia*

Nostalgia is a **sentimental yearning** for the past. Negative emotions can trigger nostalgia (Wildschut, Sedikides, Amdt & Routledge, 2006). Transition from military to civilian life is a major life event which might at times lead to fear, dissatisfaction, insecurity and worry that causes nostalgia (Davis, 1979). The positive aspects of nostalgia are a heightened feeling of social connectedness (Wilschut et al., 2006; Wildschut, Sedikides, Routledge, Arndt & Cordaro, 2010; Zhou, Sedikides, Wildschut & Gao, 2008), escalation of defences against existential threat (Juhl, Routledge, Amdt, Sedikides & Wildschut, 2010; Routledge, Amdt, Wildschut, Sedikides, Hart, Juhl & Schlotz, 2011) and heightened positive self-regard (Wildschut et al., 2006). Veterans might utilize these nostalgic memories of their service time as a means of thwarting the civilian transition struggle.

### *Civilian military divide*

Military life is quite challenging and one who chooses this life has to undergo rigorous training. The entry-level training is actually meant to do away with traces of civilian life and imbibe in the trainee **military values**, transforming them from civilians to soldiers. During their tenure in the military, soldiers face innumerable instances which could prove morally injurious, such as taking someone's life (Maguen et al., 2010, 2011), failing to prevent the death of a comrade (Grunnert, Smucker, Weis

& Rusch, 2003) and so on. Though a great amount of emphasis is laid on taking a soldier out of any major traumatic event, the moral and ethical implications still receive minimal attention (Litz et al., 2009). Lack of attention to moral aspects leads to guilt and shame. This guilt and shame is exaggerated once veterans enter civilian life due to lack of understanding and no shared experience. There is no substantial evidence on the disparity between military and civilian life; however, there are various articles on the web and numerous books focussing on the adjustment to civilian life.

One of the transition stressors while entering civilian life is the **socialized masculinity** of the military. While serving in the military, an individual undergoes rigorous training and spends their tenure in the fields sculpting their personality as per war norms, which are masculine in nature. Stoicism refers to a stereotypical masculine role (Cheng, 1999; Jansz, 2000; Murray et al., 2008), and today's soldier's mindset is similar to many of its facets. Veterans undergoing transition from a military environment that promotes stoicism to a civilian life which does not encourage such an attitude can cause **internal dissonance** (Mobbs & Bonanno, 2018).

Women officers undergo almost the same kind of socialization. The U.S. Department of Veterans Affairs in 2016 found that from 2001 to 2014, **women veterans' suicide** rate doubled in comparison to their civilian counterparts. The military environment's emphasis on masculinity and a striking divergence from that in civilian life could be the potential source of transition stress for women veterans (Mobbs & Bonanno, 2018).

There are those veterans who join **new work, post-retirement**. It is important to understand that career progression in the military is altogether different than in civilian employment. Career progression in the military is based on annual evaluation with the chain of command and little involvement from service personnel. In the military, **competition** is not obvious; rather it is to attain personal pride by accomplishing goals rather than seeking a personal advantage over colleagues. Trust and faith in colleagues, supporting each other and going beyond the call of duty to accomplish tasks is the military way of life. Civilian life is quite different, where job security might depend upon achieving individual targets and career advancement is attained by competing with colleagues on daily basis. All this can be a surprise for veterans upon entering into the civilian world of work.

## *Building and committing to new life*

### *Planning*

Improved and increased pre-retirement assistance has been recommended from time to time. Even if there is no assistance from the military regarding pre-retirement planning, a soldier is well aware of the time when he becomes eligible to retire, so he has an opportunity to plan for subsequent civilian employment (Wolpert, 2000). McNeil (1976) reported that an efficient pre-retirement planning begins from the day an individual joins the military. Though the individual is certain that he has to move into civilian society post-retirement and the exact timing of retirement eligibility is known, only a few individuals end up planning ahead.

Sharp and Biderman (1966, 1967) explained that the cause of low rates of pre-retirement planning is **high optimism that transition to civilian life is easy**. McClure (1993) gave two factors for the low rate of pre-retirement planning. First, a competing need to keep one's self-image positive by giving first priority to work over plans for future. Second is rationalization, by focussing on familiar tasks and providing a justifiable reason for postponing planning for the unfamiliar scenario.

Henry (1978) reported that preparing and planning for retirement result in a favourable factor during midlife military-to-civilian transition. McClure (1993) found that those military officers who planned well are those who had military experience which could be utilized once they moved to civilian employment. When a military officer is unsure how to transfer his experience into a civilian job, planning ahead becomes difficult. Standford (1971) explored the factors that facilitate reintegration into civilian life. His study consisted of those army personnel who were about to retire and were undergoing an anticipation phase as they were in middle of pre-retirement planning and adjustment. He found that high-ranking personnel viewed retirement favourably. Fuller and Redfering (1976) studied various factors such as number of years in service, rank and pre-retirement planning effect on post-retirement adjustment. Pre-retirement planning came out to be the only significant factor affecting post-retirement adjustment.

## *Re-employment*

All military retirees do not have the **career maturity required for a smooth transition** into civilian life with a new job. Biderman (1969) found that generally military men plan to replicate their service life work style into civilian setup rather than opting for something completely new. Positions occupied by veterans in civilian jobs involved **less command** of men and materials than military jobs (Biderman, 1973). A majority of veterans are able to find satisfactory jobs in a civilian setup though at lower pay scale and skill utilization (Biderman & Sharp, 1968).

Reemployment provides a boost to a **veteran's healthy adjustment**. Attempts made before retiring to prepare veterans for employment in civilian environments would make their transition a little smoother. Henry (1978) found that employment opportunities for veterans can be improved before retirement by focussing on their education, training programmes for civilian occupation, actually finding a civilian job even before leaving military and finally by developing skills that can be utilized in a civilian job.

Key to a successful reemployment of veterans is **transferability of skill to the new job**. Garber (1971) also reported transferable skills of veterans into the civilian labour market are their biggest asset. Standford (1971) found that transition of veterans is smoother into new civilian jobs if they continue to work in a field similar to their military job. Garrett (1961), too, reported a similar finding where he found a significant relationship between military skill transfer and a similar civilian job. Even in the financial aspect, veterans who have been found to work on a similar kind of civilian work as their military field are better off than ones who are engaged in different work (Cooper, 1981).

Transferability of skill helps in smoother transition to the new job and lack of transferability of skills leads to frustration (Biderman, 1969; Biderman & Sharp, 1968; Ostertag, 1976). More than frustration: it can also be a cause of stress and strain. In one study on Navy personnel, it was found that there are two chief strains in the workplace related to low transferability, i.e., low self-esteem and job dissatisfaction (French, Doehrman, Davis-Sacks & Vinokur, 1983). In fact, low transferability of skills is an eminent reason that leads to low satisfaction among veterans (Biderman & Sharp, 1968).

There has been a positive correlation between rank in military and level of education (Dunning & Biderman, 1973). A better education leads to good chances of serving in a civilian job (Biderman, 1969; Lenz, 1967; Collings, 1969). Collings (1969) found that it is only education that affects veteran's salary, particularly starting salary.

To have a **meaningful role in society** is not only essential for a person's identity but also helps in gratification of his need to have sense of purpose. Work or any other meaningful activity is important to form identity and gives feedback regarding an individual's skill, capacity and value in society. Failure to find meaningful work creates disturbance in an individual's life (Kroger, 2007).

## *Peer support*

For centuries, veterans have been **helping each other adjust** to the civilian environment (Gerber, 2003). This tradition is also being followed today, where the veteran community helps other veterans. Peer groups are those groups who have a similar social identity and interests. People are keener to help those who have a similar social identity than a stranger. Research conducted by Haslam, Jetten, Postmes & Haslam (2009) found that mental functioning and well-being of a vulnerable group can be improved through interventions. One of the interventions used for people suffering from mental illness and stigma was peer group support; this intervention reinforced in them the feeling of having a similar social identity and assisted them in rejecting negative stereotypes. Not only does peer support provide assurance during the transition phase, it also gives veterans an opportunity to share and relive their military life experience as people in a civilian environment might not be able to relate to it. There is a program being run in Canada named 'Sol 2 Lead'. In this program, one veteran who has established himself in a particular field acts as a mentor to veterans who want to succeed in that field (Rehman, 2017). Brown (2016) revealed that he and his fellow veterans who had served in U.S. forces during the Iraq war bonded together and helped each other go through transition when they returned to university. Later, they started working independently.

There is increasing confirmation that identification with a social group enhances and protects well-being. *The Social Cure: Identity, Health and Well-Being*, edited by Jetten, Haslam and Haslam (2012), explains how group membership and an individual's social identity affect his well-being. In the same book, the authors explained how social isolation negatively affects health and psychological well-being on one side and becoming a part of a social network leads to benefits.

## *Social recognition*

Veterans, after leaving the military, face identity issues as they are required to part with their military identity and formulate a civilian identity. For formulating this new identity, support from culture and one's social environment act as reinforcers. When a person is recognized, he feels valued and **his identity is validated**. This validation ultimately helps in formulating a well-integrated post-retirement identity in veterans (Thompson et al., 2016). Recognition gives them a sense that they have been seen as they actually are; however, not all types of recognition from civilians makes sense to veterans. For instance, Norwegian veterans posted in Afghanistan desired only to be respected and understood rather than being honoured on a personal level (Gustavsen, 2015).

It has been reported that military personnel and veterans want to be seen in a specific military role and experience. Canadian and Norwegian veterans and military personnel felt disappointed when civilians could not comprehend their deployments (Gustavsen, 2015; Dentry-Travis, 2013); they got limited media coverage and absent or low reporting of causalities, and this was perceived as how society supports their armed personnel (Dentry-Travis, 2014).

Ten years ago, a concern was raised in the United Kingdom regarding the lack of awareness among civilians regarding military duties. As a result in 2008, the United Kingdom's prime minister invited a study to improve the understanding and appreciation of UK armed forces in the eyes of the British public (Davies, Clark & Sharp, 2008).

**A strong hierarchy system** is followed in the armed forces, and this hierarchy of military identities is deeply rooted in veterans. It makes a lot of difference to the veterans who were low in the hierarchy that they are appreciated for supporting their comrades, wearing the uniform and ultimately helping in making the mission a success (Gustavsen, 2015). Military personnel and veterans feel marginalized, as they perceive they are not being understood by civilians (Davies et al., 2008; Smith & True, 2014). For civilians, the military experience or knowledge is through the media and movies. Various medals, honours and badges have a little meaning in civilian life.

Though it has been seen that recognition for military personnel and veterans is important, how society can make them feel recognized is not yet clear.

## Conclusion

The transitional journey from military to civilian life is a complex and multifaceted task. Soldiers, by virtue of their selection and training, are resilient; however, they are not superhuman beings. The transition process and reintegrating into civilian life are stressful for them and can lead to psychological issues. This is a theoretical framework to bring into the limelight various aspects pertaining to the transition from military to civilian life. The goal is to find the means to ease this journey of transition and assist veterans in adjusting well to civilian life. We hope this chapter will be fruitful in paving the way for studies on veterans' transition and reintegration into civilian life.

## References


Adler, A. B., Zamorksi, M., & Britt, T. W. (2011). The psychology of transition: Adapting to home after deployment. In A. B. Adler, P. D. Bliese & C. A. Castro (Eds.), *Deployment Psychology: Evidence-Based Strategies to Promote Mental Health in the Military* (pp. 153–174). Washington, DC: American Psychological Association.

Allport, G. W. (1954). *The Nature of Prejudice*. New York, NY: Addison-Wesley Publishing Company Inc.

Amiot, C. E., Sablonniere, R., Terry, D. J., & Smith, J. R. (2007). Integration of social identities in the self: Toward a cognitive-developmental model. *Personality and Social Psychology Review*, *11*(4), 364–388.

Anderson, M. L., Goodman, J., & Schlossberg, N. K. (2012). *Counselling Adults in Transition: Linking Schlossberg's Theory with Practice in a Diverse World* (4th ed.). New York: Springer Publishing Company, LLC.

Areul, H. R., & Castro, A. C. (2018). Military transition process and veteran identity. In E. L. Weiss & C. A. Castro (Eds.), *American Military Life in 21st Century: Social, Cultural, and Economic Issues* (pp. 485–496). California: ABC-CLIO, LLC.

Ashcroft, M. (2016). *The Veterans' Transition Review–Second Follow-up Report*. London, U.K. Retrieved from http://www.veteranstransition.co.uk/.

Baruch, Y., & Quick, J. C. (2007). Understanding second careers: Lessons from a study of U.S. Navy admirals. *Human Resource Management*, *46*(4), 471–491.

Bellino, R. (1970). Perspectives of military and civilian retirement. *Mental Hygiene*, *54*(4), 580–583.

Berkey, B. R. (1972). Military retirement syndrome: The civilian side. *Military Medicine*, *137*(7), 278–281.

Biderman, A. D. (1964). Sequels to a military career: The retired military professional. In M. Janowitz (Ed.), *The New Military: Changing Patterns of Organization* (pp. 287–336). New York, NY: Russell Sage Foundation.

Biderman, A. D. (1969). Relationships between active duty and post-retirement careers. In D. A. B. Wilson (Ed.), *Manpower Research in a Defence Context* (pp. 427–441). New York, NY: American Elsevier Publishing Company, Inc.

Biderman, A. D. (1973). Where do they go from here?–Retired military in America. *Annals of the American Academy of Political and Social Science*, *406*(1), 146–161.

Biderman, A. D., & Sharp, L. M. (1968). The convergence of military and civilian occupational structures: Evidence from studies of military retired employment. *American Journal of Sociology*, *73*(4), 381–399.

Blackburn, D. (2016). Transitioning from military to civilian life: Examining the final step in a military career. *Canadian Military Journal*, *16*(4), 53–61.

Brown, N. (2016). Combat veterans: Adaptation to transition. *Psychiatric Times*, *33*(6), 1–3.

Butler, R. N. (1963). The life review: An interpretation of reminiscence in the aged. *Psychiatry, Journal for the Study of Interpersonal and Biological Processes*, *26*(1), 65–76.

Castro, C. A., & Kintzle, S. (2014). Suicides in the military: The post-modern combat veteran and the Hemingway effect. *Current Psychiatry Reports*, *16*(8), 460–469.

Cheng, C. (1999). Marginalized masculinities and hegemonic masculinity: An introduction. *The Journal of Men's Studies*, 7(3), 295–315.

Coll, J. E., & Weiss, E. L. (2013). Transitioning veterans into civilian life. In A. Rubin, E. L. Weiss, & J. E. Coll (Eds.), *Handbook of Military Social Work* (pp. 281–297). Hoboken, NJ: John Wiley and Sons Inc.

Collings, K. J. (1969). Employment of middle-aged job applicants, with special emphasis on retired military personnel. Unpublished doctoral dissertation, University of Washington, Seattle.

Cooper, L., Caddick, N., Godier, L., Cooper, A., & Fossey, M. (2018). Transition from the military into civilian life: An exploration of cultural competence. *Armed Forces & Society*, *44*(1), 156–177.

Cooper, R. V. L. (1981). *Military Retirees' Post-Service Earnings and Employment*. Santa Monica, CA: Rand Corporation.

Cruwys, T., Gunaseelan, S. (2016). "Depression is who I am": Mental illness identity, stigma and wellbeing. *Journal of Affective Disorders*, *189*, 36–42.

Davis, F. (1979). *Yearning for Yesterday: A Sociology of Nostalgia*. New York: Free Press.

Davies, Q., Clark, B., & Sharp, M. (2008). *Report of inquiry into national recognition of our armed forces. A report to the prime minister*. Retrieved from http://www.ppu.org.uk/militarism/recognition_of_our_armed_forces.pdf.

Demers, A. L. (2011). When veterans return: The role of community in reintegration. *Journal of Loss and Trauma*, *16*(2), 160–179.

Demers, A. L. (2013). From death to life: Female veterans, identity negotiation, and reintegration into society. *Journal of Humanistic Psychology*, *53*(4), 489–515.

Dentry-Travis, S. J. (2013). Canadian soldiers' constructions of their role-sets. *Personal Construct Theory and Practice*, *10*, 28–39.

Dentry-Travis, S. J. (2014). Soldiers within society: A comparison of Canada, the United Kingdom and France. Ph.D. Thesis, University of Reading, UK.

Doherty, T. F. (1983). Patterns of adaptation to military retirement: An exploratory study. Unpublished doctoral dissertation, Smith College School for Social Work, Northampton, MA.

Drier, H. N. (1995). *Out of Uniform: A Career Transition Guide for Ex-Military Personnel*. Lincolnwood, IL: NTC Publishing Group.

Dunning, B. B., & Biderman, A. D. (1973). The case of military "retirement". *Industrial Gerontology*, *17*(1), 18–37.

Enz, K. F., & Talarico, J. M. (2016). Forks in the road: Memories of turning points and transitions. *Applied Cognitive Psychology*, *30*(2), 188–195.

Erikson, E. H. (1959). Identity and the life cycle. *Psychological Issues*, *1*, 1–171.

Eysenck, M. W., & Keane, M. T. (2010). *Cognition and Emotion. Cognitive Psychology* (pp. 571–605). (6th ed.). Hove and New York: Psychology Press.

Fiske, S.T., & Taylor, S.E. (2013). *Social Cognition–From Brains to Culture*. Washington: Sage.

Frank, R. A. (1993). Military retirement in the post-cold war era. In F. W. Kaslow (Ed.). *The Military Family in Peace and War* (pp. 214–240). New York, NY: Springer.

French, R. P. Jr., Doehrman, S. R., Davis-Sacks, M. L., & Vinokur, A. (1983). *Career Change in Midlife: Stress, Social Support, and Adjustment*. Ann Arbor, MI: Institute for Social Research.

Fuller, R. L., & Redfering, D.L. (1976). Effects of preretirement planning on the retirement adjustment of military personnel. *Sociology of Work and Occupations*, *3*(4), 479–487.

Garber, D. L. (1971). Retired soldiers in second careers: Self-assessed change, reference group salience, and psychological well being. Unpublished doctoral dissertation, University of Southern California, Los Angeles.

Garrett, D. G. (1961). Retirement experiences and employment status of United States Air Force retired enlisted personnel. Unpublished masters thesis, University of Wisconsin, Madison.

Gerber, D. A. (2003). Disabled veterans, the state and the experience of disability in Western societies, 1914–1950. *Journal of Social History*, *36*(4), 899–916.

Giffen, M. B., & McNeil, J. S. (1967). Effect of military retirement on dependents. *Archives of General Psychiatry*, *17*(6), 717–722.

Goldstein, E. B. (2015). *Cognitive Psychology: Connecting Mind, Research, and Everyday Experience* (4th ed.). Stamford, CT: Cengage Learning.

Grunnert, B. K., Smucker, M. R., Weis, J. M., & Rusch, M. D. (2003). When prolonged exposure fails: Adding an imagery-based cognitive restructuring component in the treatment of industrial accident victims suffering from PTSD. *Cognitive and Behavioral Practice*, *10*(4), 333–346.

Gustavsen, E. (2015). The construction of meaning among Norwegian Afghanistan veterans. *International Sociology*, *31*(1), 21–36.

Haslam, S. A., Jetten, J., Postmes, T., & Haslam, C. (2009). Social identity, health and well-being: An emerging agenda for applied psychology. *Applied Psychology, 58*(1), 1–23.

Henry, W. H., Jr. (1978). Second careers for retired military officers; the transfer of educational attainments, skills and experience to civilian occupations. Unpublished doctoral dissertation, Rutgers The State University of New Jersey, New Brunswick.

Jansz, J. (2000). Masculine identity and restrictive emotionality. In A. H. Fischer (Ed.), *Gender and Emotion: Social Psychological Perspectives* (pp. 166–186). New York: Cambridge University Press.

Jetten, J., Haslam, C., & Haslam, S. A. (2012). *The Social Cure: Identity, Health and Well-Being*. New York, NY, US: Psychology Press.

Juhl, J., Routledge, C., Arndt, J., Sedikides, C., & Wildschut, T. (2010). Fighting the future with the past: Nostalgia buffers existential threat. *Journal of Research in Personality, 44*(3), 309–314.

Kidwell, J. S., Dunham, R. M., Bacho, R. A., & Portes, P. R. (1995). Adolescent identity exploration: A test of Erikson's theory of transitional crisis. *Adolescence, 30*(120), 785–794.

Kilpatrick, A. C., & Kilpatrick, E. G., Jr. (1979). Retirement from the military: Problems of adjustment. *Social Casework, 60*(5), 282–288.

Knippa, W. B. (1979). The relationship of antecedent and personality variables to the life satisfaction of retired military officers. Unpublished doctoral dissertation, The University of Texas, Austin.

Koppel, J. M., & Berntsen, D. (2014). Does everything happen when you are young? Introducing the youth bias. *The Quarterly Journal of Experimental Psychology, 67*(3), 417–423.

Kroger, J. (2007). *Identity Development: Adolescence through Adulthood* (2nd ed.). London, UK: Sage Publications.

Lawler, S. (2008). *Identity: Sociological Perspectives*. Cambridge: Polity Press.

Lazarus, R., & Folkman, S. (1984). *Stress, Appraisal, and Coping*. New York: Springer.

Lenz, A. J. (1967). *Military Retirement and Income Maximization: An Examination of the Economic Incentives to Extended Military Service*. Springfield, VA: Clearinghouse for Federal Scientific and Technical Information.

Litz, B. T., Stein, N., Delaney, E., Lebowitz, L., Nash, W. P., Silva, C., & Maguen, S. (2009). Moral injury and moral repair in war veterans: A preliminary model and intervention strategy. *Clinical Psychology Review, 29*(8), 695–706.

Maguen, S., Lucenko, B. A., Reger, M. A., Gahm, G. A., Litz, B. T., Seal, K. H., … Marmar, C. R. (2010). The impact of reported direct and indirect killing on mental health symptoms in Iraq war veterans. *Journal of Traumatic Stress, 23*(1), 86–90.

Maguen, S., Vogt, D. S., King, L. A., King, D. W., Litz, B. T., Knight, S. J., & Marmar, C. R. (2011). The impact of killing on mental health symptoms in Gulf War veterans. *Psychological Trauma: Theory, Research, Practice, and Policy, 3*(1), 21–26.

McClure, H. M. (1993). Alienated patriots: A sociological portrait of military retirees. Unpublished doctoral dissertation, University of California, Berkeley.

McDonald, T. F. (1977 (July, 11)). Early retirement I: Few are ready for the "silver bullet". *Army Times*, 1–33.

McNeil, J. S. (1976). Individual and family problems related to retirement from the military service. In H. F. McCubbin, B. B. Dahl, & E. J. Hunter (Eds.), *Families in the Military System* (pp. 237–257). Beverly Hills, CA: Sage Publications.

McNeil, J. S., & Giffen, M. B. (1965). The social impact of military retirement. *Social Casework, 46*(4), 203–207.

McNeil, J. S., & Giffen, M. D. (1967). Military retirement: The retirement syndrome. *American Journal of Psychiatry, 123*(7), 848–854.

McNeil, J. S., Lecca, P. J., & Wright, Jr. R. (1983). *Military Retirement: Social Economic and Mental Health Dilemmas*. Totowa, NJ: Rowman and Allanheld.

Mobbs, M. C., & Bonanno, G. A. (2018). Beyond war and PTSD: The crucial role of transition stress in the lives of military veterans. *Clinical Psychology Review, 59*, 1–8.

Moskos, C. C. (1977). From institution to occupation: Trends in military organization. *Armed Forces and Society, 4*(1), 41–50.

Moskos, C. C. (1988). Institutional and occupational trends in armed forces. In C. C. Moskos & F. R. Wood (Eds.), *The Military; More Than Just a Job?* (pp. 15–26). Washington, DC: Pergamon-Brassey's International Defense Publishers, Inc.

Murray, G., Judd, F., Jackson, H., Fraser, C., Komiti, A., Pattison, P., … Robins, G. (2008). Big boys don't cry: An investigation of stoicism and its mental health outcomes. *Personality and Individual Differences, 44*(6), 1369–1381.

Ogburn, B., Bellino, R., Williams, R., & Gordon, R. (1969). Problems of the military retiree. *The Journal of the Florida Medical Association, 56*(4), 245–248.

Orazem, R. J., Frazier, P. A., Schnurr, P. P., Oleson, H. E., Carlson, K. F., Litz, B. T., & Sayer, N. A. (2017). Identity adjustment among Afghanistan and Iraq War veterans with reintegration difficulty. *Psychological Trauma: Theory, Research, Practice, and Policy*, *9*(Suppl 1), 4–11.

Ostertag, R. H. (1976). The relationship of adjustment to the concept of dogmatism in former members of the military services. Unpublished doctoral dissertation, United States International University, San Diego, CA.

Para, E. A. (2008). The role of social support in identity formation: A literature review. *Graduate Journal of Counseling Psychology*, *1*(1), 1–9.

Parnes, H. S., & King, R. (1977). Middle aged job losers. *Industrial Gerontology*, *4*, 77–95.

Praharso, N. F., Tear, M. J., & Cruwys, T. (2017). Stressful life transitions and wellbeing: A comparison of the stress buffering hypothesis and the social identity model of identity change. *Psychiatry Research*, *247*, 265–275.

Price, K. F., Walker, J. W., & Kimmel, D. C. (1979). Retirement timing and retirement satisfaction. *Aging and Work: A Journal on Age, Work and Retirement*, *2*(4), 235–245.

Rathbone, C. J., Moulin, A., Conway, M. A., & Holmes, E. A. (2012). Autobiographical memory and the self. In N. Braisby, & A. Gellatly (Eds.). *Cognitive Psychology* (pp. 546–576). (2nd ed.). Oxford, England: Oxford University Press.

Rehman, M. (2017). *Mentoring transitioning CAF personnel program seeks CAF veteran mentors*. Canadian Military Families. Retrieved from http://cmfmag.ca/progs_services/mentoring-transitioning-caf-personnel-program-seeks-caf-veteran-mentors/?platform=hootsuite.

Robinson, J., Littlefield, P., & Schleuning, A. (2017). *Transforming Veterans' Experiences during Military-to-Civilian Transition: Gaps and Opportunities*. Washington, DC: Center for Innovation, U.S. Veterans Administration.

Rogers, C. R. (1944). Psychological adjustments of discharged service personnel. *Psychological Bulletin*, *41*(10), 689–696.

Routledge, C., Arndt, J., Wildschut, T., Sedikides, C., Hart, C. M., Juhl, J., Vingerhoets, J. J. M., & Schlotz, W. (2011). The past makes the present meaningful: Nostalgia as an existential resource. *Journal of Personality and Social Psychology*, *101*(3), 638–652.

Rubin, D. C., Wetzler, S. E., & Nebes, R. D. (1986). Autobiographical memory across the life span. In D. C. Rubin (Ed.). *Autobiographical Memory* (pp. 202–221). New York, NY: Cambridge University Press.

Schlossberg, N. K. (1981). A model for analyzing understanding human adaptation to transition. *The Counselling Psychologist*, *9*(2), 2–18.

Sharp, L. M., & Biderman, A. D. (1966) *The Employment of Retired Military Personnel*. Washington, DC: Bureau of Social Science Research, Inc.

Sharp, L. M., & Biderman, A. D. (1967). Out of uniform: The employment experience of retired servicemen who seek a second career. *Monthly Labor Review*, *90*, 15–21.

Smith, R. T., & True, G. (2014). Warring identities, identity conflict and the mental distress of American veterans of the wars in Iraq and Afghanistan. *Society and Mental Health*, *4*(2), 147–161.

Stanford, E. P. (1968). Anticipation of retirement by military personnel. Unpublished doctoral dissertation, Iowa State University, Ames.

Stanford, E. P. (1971). Retirement anticipation in the military. *The Gerontologist*, *11*(1), 37–42.

Stets, J. E., & Burke, P. J. (2000). Identity theory and social identity theory. *Social Psychology Quarterly*, *63*(3), 224–237.

Thompson, J. M., VanTil, L. D., Zamorski, M. A., Garber, B., Dursun, S., Fikretoglu, D., Pedlar, D. J. (2016). Mental health of Canadian Armed Forces veterans—Review of population studies. *Journal of Military, Veteran and Family Health*, *2*(1), 70–86.

Watkins, N., Cole, F., & Weidemann, S. (2010). The war memorial as healing environment: The psychological effect of the Vietnam Veterans Memorial on Vietnam War combat veterans' posttraumatic stress disorder symptoms. *Environment and Behavior*, *42*(3), 351–375.

Wildschut, T., Sedikides, C., Arndt, J., & Routledge, C. D. (2006). Nostalgia: Content, triggers, functions. *Journal of Personality and Social Psychology*, *91*(5), 975–993.

Wildschut, T., Sedikides, C., Routledge, C., Arndt, J., & Cordaro, F. (2010). Nostalgia as a repository of social connectedness: The role of attachment-related avoidance. *Journal of Personality and Social Psychology*, *98*(4), 573–586.

Wolpert, D. S. (2000). Military retirement and the transition to civilian life. In J. A. Martin, L. N. Rosen, & L. R. Sparacino (Eds.). *The Military Family: A Practice Guide for Human Service Providers* (pp. 103–119). Westport, CT: Praeger.

Zhou, X., Sedikides, C., Wildschut, T., & Gao, D. G. (2008). Counteracting loneliness: On the restorative function of nostalgia. *Psychological Science*, *19*(10), 1023–1029.

# PART III

# Making a choice: Mental health issues and prospects in the military

# 20

# MILITARY BURNOUT AND WORK ENGAGEMENT

## A qualitative systematic literature review

*Maria José Chambel, Sílvia Lopes, Filipa Castanheira, and Carolina Rodrigues-Silveira*

The military context has been acknowledged as a setting with different types of working conditions that influence military well-being. However, what do we actually know about military well-being and its development over time? Previous reviews on the topic have tended to be limited and geared towards analyzing specific situations or/and samples. For example, Buckman et al. (2011) conducted a systematic review of studies examining the effects of deployment length and a 'mismatch' between the expected and actual length of deployments on the health and well-being of military personnel. Bauer, Newbury-Birch, Robalino, Ferguson, and Wigham (2018) conducted a systematic review of the effectiveness of well-being interventions for military personnel adjusting to civilian life. Smith-MacDonald, Norris, Raffin-Bouchal and Sinclair (2017) conducted a systematic review to analyze the relationship between spirituality and mental well-being in combat veterans. No systematic review has been conducted on the burnout and work engagement of military personnel. Given that burnout and work engagement are probably the most disseminated indicators of work well-being, to which much of occupational health research is dedicated (Marques-Pinto & Chambel, 2008), an integration of the research on this topic will contribute to further data-based knowledge and will facilitate future research by shedding light upon current gaps in the literature.

Burnout, which has a longer history, was identified in the 1970s by Freudenberger and Maslach as a state of emotional exhaustion and loss of motivation and commitment, affecting aid-providing professionals as a response to limitations (overly high expectations, bureaucracy, absence of financial resources and support, loss of authority) in their work context (Schaufeli & Enzmann, 1998). Later on, Maslach and colleagues redefined the concept of burnout as a crisis in the relationship with work, affecting workers of different professional areas, not only aid-providing professionals, characterized by three dimensions: exhaustion – general reactions to stress, such as emotional and physical fatigue, depression, psychosomatic complaints and anxiety; cynicism (depersonalization) – an attitude of indifference or mental detachment, divestment towards work; the loss of professional efficacy (loss of personal fulfillment) – decreased feelings of self-efficacy at work (Demerouti, Bakker, Nachreiner & Schaufeli, 2001; Maslach, Jackson, & Leiter, 1996).

In line with a positive psychology trend, the concept of work engagement emerged more recently, characterized by high levels of energy, involvement and motivation in the fulfillment of one's work (Schaufeli & Buunk, 2003). The concept was introduced by Kahn (1990), as an individual's strong identification with his/her work which, in a dynamic relationship, leads to the individual investing high energy levels in the performance of his/her tasks, consequently giving rise to high involvement in this performance. It should be noted that in its initial formulation, this positive psychological state of engagement was regarded as the opposite pole to burnout, and it was only necessary to invert the level of the latter to identify the level of engagement. In the meantime, Schaufeli and colleagues (Schaufeli & Bakker, 2004; Schaufeli, Salanova, González-Romá & Bakker, 2002) developed a new perspective which considers work engagement as a positive cognitive-affective state, characterized by vigor, dedication and absorption. Vigor is conveyed through high levels of energy and mental resilience and the desire and ability to invest efforts in one's work; dedication is characterized as a feeling of relevance, enthusiasm, inspiration, pride and challenge towards one's work; and absorption is similar to a state of persistent flow (Csikszentmihalyi, 1990), in which the individual is immersed in concentration, losing his/her sense of the passage of time, and happy in his/her work involvement. This psychological state of engagement is a positive indicator of work well-being, which presents a negative and relatively weak relationship with burnout. Thus, having work engagement is more than simply not having burnout, and its evaluation should therefore be specific. In fact, in order to analyze both burnout and engagement, two separate processes need to be examined, since the determinants that trigger well-being in individuals are not necessarily the absence of the determinants that contribute to ill-being and pathology (Seligman & Csikszentmihalyi, 2000).

The conservation of resources (COR) theory has been used to explain the development of burnout and work engagement (Hobfoll, 2002). This theory considers that individuals wish to obtain and retain valued resources, namely objects (e.g. tools), personal characteristics (e.g. emotional stability), conditions (e.g. social support) and energies (e.g. money) that they value as a means to acquire and strengthen other resources. Thus, they will seek to maximize resource gains while minimizing resource losses and avoiding potential threats. The threat of loss, actual loss or lack of a gain of resources after investment is conceptualized as demands. That is, as stress mounts, the individual must increasingly divert psychological resources to combat its negative effects until those resources are exhausted and the individual feels overwhelmed and no longer able to cope with work. Thus, the prolonged experience of low resources and high demands leads to an erosion of other resources such as energy, identification and perceived efficacy, which is the burnout process (Hobfoll, Freedy, Lane, & Geller, 1990). In fact, in the work context, burnout may be frequent, particularly because work demands typically enclose a higher rate of resources used by employees, in comparison with the rate of resources the employees are able to replenish (Freedy & Hobfoll, 1994). However, high resources in a high-demand environment should lead to optimal functioning, leading to a reinvestment of resources such as time and energy into the work environment. Thus, engagement can be conceptualized as a result of continuous resource gain, since initial gain begets further gain. Individuals who possess strong resource pools experience spirals of resource gain, and that resource surplus promotes engagement. In accordance with the COR, the relationship between job demands and strain (burnout) depends on the investment of resources, and the relationship between job resources and well-being (engagement) is reliant upon the accumulation of more resources. In fact, this theory assumes that employees' burnout and engagement will increase or decrease over time in response to changes in resources, while engagement occurs either as a "gain/loss cycle" and burnout either as a "downward/upward spiral."

# Method

## *Procedure*

To provide an indication of the potential size and nature of the available literature examining the work engagement and burnout of military personnel, a qualitative systematic review was conducted (Paré, Trudel, Jaana, & Kitsiou, 2015). We followed the recommendations of Daudt, van Mossel, and Scott (2013) and the Preferred Reporting Items for Systematic Reviews and Meta-Analyses (PRISMA; Liberati et al., 2009).

The first step of the systematic review consisted of using two databases, namely ISI Web of Science and MEDLINE. Titles and abstracts were identified using two queries. The first query used the combination ["work engagement" AND "military"]. The second query used the combination ["burnout" AND "military"]. A total of 374 papers were found. An initial screening was then conducted in order to remove duplicates and non-English language papers, and a total of 315 papers remained for the next step of the systematic review.

In order to eliminate papers that did not address the scope of the present review, inclusion and exclusion criteria were established (Daudt et al., 2013). The inclusion criteria were: (1) studies conducted with military personnel sample, (2) empirical qualitative and quantitative studies, (3) evaluating work engagement or/and burnout and (4) being published in a scientific journal. As exclusion criteria, the following were not included: (1) literature reviews and meta-analyses, including an indirect analysis of work engagement and burnout; (2) studies with a mixed sample composed not only of military personnel but also civilians and (3) studies published in Congress proceedings.

On the basis of the inclusion and exclusion criteria, two reviewers critically and independently appraised the quality of 315 papers by assessing each paper's title and correspondent abstract. As a result of the titles and abstracts' appraisal, a total of 62 papers were considered to meet the inclusion criteria. A third reviewer then checked the remaining papers and analyzed all 62 full-text articles. Following the full-text articles' analysis, 11 papers were excluded, due to the fact that they did not evaluate work engagement and/or burnout, and 2 had been published in Congress proceedings. Hence, 49 papers met the inclusion criteria and were used to perform the qualitative systematic review. In Figure 20.1, it is possible to observe the flow diagram of the systematic selection of studies undertaken in the present study.

# Analysis

Table 9.1 summarizes the reviewed studies, providing information on (a) area of publication (i.e. *Psychology, Social Sciences* vs. *Medicine, Health, Mental Health, Nursing*), (b) country, (c) participants (i.e. sample size), (d) study design (i.e. cross-sectional, longitudinal, intervention), (e) indicator studied (i.e. burnout, engagement, both), (f) levels of burnout/work engagement, (g) antecedents/consequences of the studied military burnout and engagement (i.e. individual vs. contextual).

# Results

## *When, where and what is investigated?*

The review of the articles (see Table 20.1) revealed that publications began in 1986 but were scarce (i.e. zero, one or two studies per year) up to 2009, while in 2015 and 2016, more published studies were observed, eight and seven, respectively. Regarding the choice of journal publications, this review pointed to a high diversity, and the majority of journals had published only one or

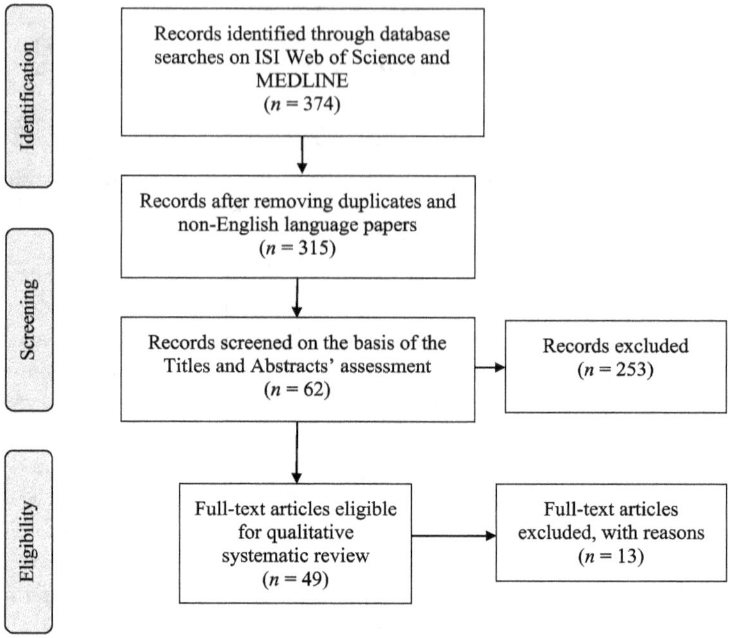


*Figure 20.1* Flow diagram of the systematic selection of studies.

*Source:* Adapted from Liberati et al. 2009.

two studies, with the exception of *Military Psychology* (n = 8) and *Military Medicine* (n = 11). As for the scientific area of these publications, the review revealed that around half (n = 26) had been published in *Psychology* and *Social Sciences* and the other (n = 23) in *Medicine, Health, Mental Health and Nursing.* The majority of the studies had been conducted in the United States (n = 30) and Europe (n = 12), and most had focused on the negative side of military well-being: over half of the analyzed articles (n = 34) concentrated on burnout, work engagement was investigated in 11, and both burnout and work engagement in 4 studies.

## Who was investigated, how and for what purpose?

The sample size used varied from 32 to 320,503, although the typical sample size varied mostly due to the nature of the studies. Indeed, it was common to find samples with hundreds of military personnel in specific studies of militaries that had been deployed in peacekeeping missions (Boermans, Kamphuis, Delahaij, van den Berg, & Euwema, 2014; Chambel & Oliveira-Cruz, 2010), whereas other studies presented smaller samples only with dozens of participants focusing on trainees from specific courses (Morgan et al., 2011) or occupations (mostly health care providers such as nurses: Bartz & Maloney, 1986; van Wijk, 1997; physicians: Simons et al., 2016; or dentists: Shelley & Wong, 1991), while others focused on more general larger samples with thousands of participants, for example the population of active workers for the National Army (Bryan, Goodman, Chappelle, Prince & Thompson, 2018; Ivey, Blanc, & Mantler, 2015; Jiang et al., 2015; Vie, Scheier, Lester, & Seligman, 2016). Of the analyzed articles, 19 used occupational/organizational specific samples, whereas heterogeneous samples, that is, samples comprising participants with various occupations, were used in 30 of the studies. Health services (i.e. nurses, psychiatrists) were the most frequently investigated.

*Table 20.1* Study characteristics

| *Authors (year)* | *Journal (area)* | *Country* | *Study design* | *Participants* | *(Scale) level(s) of well-being [rating scale/range score]* | *Aim of study* |
|---|---|---|---|---|---|---|
| Adler et al. (2017) | Military Medicine (Medicine, Health, Mental Health, Nurse) | USA | Quantitative, cross-sectional, correlation and regression analysis. | 344 U.S. military medical staff members. | (MBI-HSS) Emotional Exhaustion = 13.4 [0–18 points]; Depersonalization = 11.7 [0–18 points] | Antecedents of burnout: self-care, team care and health-promotion leadership as independent variables. |
| Afonso & Gomes (2009) | Psychology/Psicologia Reflexão e Crítica (Psychology, Social Sciences) | Portugal | Quantitative, cross-sectional, descriptive statistics and regression models. | 95 Portuguese military agents of the Republican National Guard. | (MBI-GS) emotional exhaustion = 1.68 [0–6 points]; cynicism = 1.82 [0–6 points]; personal accomplishment = 4.69 [0–6 points] | Antecedents of burnout: global stress, proactive coping, desire to leave the job/profession, organizational commitment and satisfaction with life. |
| Alessandri et al. (2018) | Journal of Occupational and Organizational Psychology (Psychology, Social Sciences) | Italy | Quantitative, longitudinal, two-wave study, 1 year time-lag. | 416 new military cadets of an Italian military academy. | (MBI-GS) Mean values for burnout were not provided | Antecedents of burnout: Emotional stability as an independent variable and self-efficacy beliefs in managing negative emotions at work as a mediator variable. |
| Alves, Bendassolli, & Gondim (2017) | Avances en Psicología Latinoamericana (Psychology, Social Sciences) | Brazil | Quantitative, cross-sectional, validity study. | 525 Brazilian police officers. | (MBI-HSS) Emotional Exhaustion = 31.3 [0–54 points] | Antecedents of burnout: emotional labor as an independent variable. |

*(Continued)*

*Table 20.1* (Continued)

| *Authors (year)* | *Journal (area)* | *Country* | *Study design* | *Participants* | *(Scale) level(s) of well-being[rating scale/range score]* | *Aim of study* |
|---|---|---|---|---|---|---|
| Bakir, Ozer, Ozcan, Cetin, & Fedai (2010) | Bulletin of Clinical Psychopharmacology (Medicine, Health, Mental Health, Nurse) | Turkey | Quantitative, cross-sectional, descriptive statistics. | 377 Turkish military nurses in Ankara. | (MBI – Turkish version) emotional exhaustion = 24.8 for <29 years, and 25.3 for >30 years [0–36 points]; depersonalization = 10.7 for <29 years, and 10.6 for > 30 years [0–20 points]; personal accomplishment = 20.8 for < 29 years, and 20.7 for > 30 years [0–40 points] | Antecedents of burnout: Depressive symptoms as the independent variable. |
| Ballenger-Browning et al. (2011) | Military Medicine (Medicine, Health, Mental Health, Nurse) | USA | Quantitative, cross-sectional, descriptive statistics and regression models. | 97 mental health providers at military facilities. | (MBI-HSS) emotional exhaustion = 16.6 [0–54 points]; depersonalization = 4.3 [0–30 points]; personal accomplishment = 39.7 [0–54 points] | Antecedents of burnout: provider demographics, social support, institutional factors and beliefs about psychotherapy and medication used as independent variables. |

*(Continued)*

*Table 20.1* (Continued)

| *Authors (year)* | *Journal (area)* | *Country* | *Study design* | *Participants* | *(Scale) level(s) of well-being[rating scale/range score]* | *Aim of study* |
|---|---|---|---|---|---|---|
| Bartz & Maloney (1986) | Research in Nursing & Health (Medicine, Health, Mental Health, Nurse) | USA | Quantitative, cross-sectional, descriptive analysis and regression models. | 89 full-time registered intensive care nurses. | (MBI-HSS) emotional exhaustion = 21.49 [0–54 points]; depersonalization = 6.60 [0–30 points]; personal accomplishment = 37.11 [0–48 points] | Antecedents of burnout: demographic variables as independent variables. |
| Boermans, Kamphuis, Delahaij, van den Berg, & Euwema (2014) | Stress and Health (Psychology, Social Sciences) | Netherlands | Quantitative, longitudinal, two-wave study, 6 month time-lag. | 971 Dutch Armed Forces peacekeepers. | (UWES) Teamwork Engagement = 4.28 [0–6 points] | Antecedents and consequents of teamwork engagement: Organizational constraints during a military operation as an independent variable and fatigue symptoms as the dependent variable. |
| Britt & Bliese (2003) | Journal of Personality (Psychology, Social Sciences) | USA | Quantitative, cross-sectional, multilevel analysis. | 1,181 U.S. soldiers deployed on a peacekeeping mission. | (a) Self-engagement = 4.20 [1–5 points] | Antecedents and consequents of self-engagement: stressors (lower amounts of sleep, work stress, family stress) as independent variables and psychological distress as dependent variable. |

*(Continued)*

*Table 20.1* (Continued)

| *Authors (year)* | *Journal (area)* | *Country* | *Study design* | *Participants* | *(Scale) level(s) of well-being[rating scale/range score]* | *Aim of study* |
|---|---|---|---|---|---|---|
| Britt, Adler, & Bartone (2001) | Journal of Occupational Health Psychology (Psychology, Social Sciences) | USA | Quantitative, longitudinal, two-wave study, 4–5 month time-lag. | 161 soldiers. | (b) Engagement = 3.93 [1–5 points] | Antecedents and consequents of work engagement: personality hardiness as the independent variable and deriving benefits from the deployment as the dependent variable. |
| Britt, Castro, & Adler (2005) | Personality and Social Psychology Bulletin (Psychology, Social Sciences) | USA | Quantitative, longitudinal, two-wave study, 3–4 month time-lag. | 176 soldiers currently working at their home station. | (a) Self-engagement = 3.72 [1–5 points] | Antecedents and consequents of work engagement: work demands (days training, work hours and subjective work overload) as independent variables and physical symptoms as the dependent variable. |
| Britt, Dickinson, Moore, Castro, & Adler (2007) | Journal of Occupational Health Psychology (Psychology, Social Sciences) | USA | Quantitative, longitudinal, two-wave study, 6 month time-lag. | 1,685 U.S. soldiers on a peacekeeping mission to Kosovo. | (a) Mean value for job engagement was not provided [1–5 points] | Consequents of work engagement: morale and benefits of deploying as dependent variables. |

*(Continued)*

*Table 20.1* (Continued)

| *Authors (year)* | *Journal (area)* | *Country* | *Study design* | *Participants* | *(Scale) level(s) of well-being[rating scale/range score]* | *Aim of study* |
|---|---|---|---|---|---|---|
| Bryan, Goodman, Chappelle, Prince & Thompson (2018) | Military Psychology (Psychology, Social Sciences) | USA | Quantitative, cross-sectional, latent profile analysis. | 7,550 U.S. Air Force remote warriors. | (MBI-GS) Exhaustion = 13.4 [0–30 points]; Cynicism = 11.7 [0–30 points]; Professional Efficacy = 24.4 [0–36 points]. | Antecedents of burnout: demographics and Air Force remote warriors' career as independent variables. |
| Carvalho & Chambel (2017) | Armed Forces & Society (Psychology, Social Sciences) | Portugal | Quantitative, cross-sectional, structural equation modeling. | 175 military employees working in three units of the Portuguese Marine Corps: Naval School, Marine Rifle School and Marine Technology School. | (MBI-GS &UWES) Exhaustion = 2.53 [1–7 points]; Cynicism = 1.92 [1–7 points]; Vigor = 4.53 [1–7 points]; Dedication = 4.61 [1–7 points] | Antecedents of burnout and work engagement: job demands–control–support (JDCS) as an independent variable and work-family conflict (WFC) and family-work conflict (FWC) as mediator variables. |
| Castanheira, Chambel, Lopes & Oliveira- Cruz (2016) | Military Psychology (Psychology, Social Sciences) | Portugal | Quantitative, cross-sectional, structural equation modeling. | 322 officers and sergeants and 1,045 soldiers of the Portuguese Army. | (UWES) Officers and Sergeants' Work Engagement = 5.43 [0–6 points]; Soldiers' Work Engagement = 4.37 [0–6 points] | Antecedents of work engagement: perceived social impact and social worth as independent variable and prosocial motivation as mediator variable. |

*(Continued)*

*Table 20.1* (Continued)

| *Authors (year)* | *Journal (area)* | *Country* | *Study design* | *Participants* | *(Scale) level(s) of well-being[rating scale/range score]* | *Aim of study* |
|---|---|---|---|---|---|---|
| Chambel & Oliveira-Cruz (2010) | Military Psychology (Psychology, Social Sciences) | Portugal | Quantitative, longitudinal, three-wave study, 6 month time-lag. | 387 soldiers on a peacekeeping mission. | (MBI-GS & UWES) emotional exhaustion = 1.57 (T1), 1.62 (T2), and 1.68 (T3) [0–6 points]; cynicism = 1.57 (T1), 1.69 (T2), and 1.85 (T3) [0–6 points]; Vigor = 5.14 (T1), 5.00 (T2), and 4.78 (T3) [0–6 points]; Dedication = 5.40 (T1), 5.17 (T2), and 4.85 (T3) [0–6 points] | Antecedents of burnout and work engagement: psychological contract breach as the independent variable. |
| Chambel, Castanheira, Oliveira-Cruz & Lopes (2015) | Military Psychology (Psychology, Social Sciences) | Portugal | Quantitative, cross-sectional, structural equation modeling. | 1,045 Portuguese soldiers. | (MBI-GS & UWES) emotional exhaustion = 4.06 [1–7 points]; cynicism = 3.81 [1–7 points]; Work engagement = 4.35 [1–7 points] | Antecedents of burnout and work engagement: perceived organizational support and leader-member exchange as independent variables, and autonomous motivation and controlled motivation as mediator variables. |

*(Continued)*

*Table 20.1* (Continued)

| *Authors (year)* | *Journal (area)* | *Country* | *Study design* | *Participants* | *(Scale) level(s) of well-being[rating scale/range score]* | *Aim of study* |
|---|---|---|---|---|---|---|
| Chappelle et al. (2014) | Military Psychology (Psychology, Social Sciences) | USA | Quantitative, cross-sectional, descriptive statistics. | 1,094 U.S. Air Force drone operators. | (MBI-GS) emotional exhaustion = 11.99 [0–30 points]; cynicism = 9.08 [0–30 points]; personal accomplishment = 26.12 [0–36 points] | Antecedents of burnout: demographics and career as independent variables. |
| Cragun, April, & Thaxton (2016) | Military Medicine (Medicine, Health, Mental Health, Nurse) | USA | Quantitative, cross-sectional, t-tests and multiple regression models. | 105 respondents (nurses, technicians and physicians) from a Military Emergency Department in USA. | (ProQOL V) Never Deployed providers' Burnout = 21.8 [0–50 points]; Previously Deployed providers' Burnout = 22.3 [0–50 points] | Antecedents of burnout: population professions and deployment status as independent variables. |
| Delahaij, Kamphuis, & van den Berg (2016) | Military Psychology (Psychology, Social Sciences) | Netherlands | Quantitative, longitudinal, two-wave study, 6 month time-lag. | 123 service members from the Police Training Group and 41 from the Air Task Force of the Dutch Armed Forces deployed in NATO mission ISAF. | (MBI-GS & UWES) Exhaustion = 1.00 [1–5 points]; Work Engagement = 4.11 [1–5 points] | Antecedents of burnout and work engagement: self-efficacy as an independent variable and threat situations and family support as moderator variables. |

*(Continued)*

*Table 20.1* (Continued)

| *Authors (year)* | *Journal (area)* | *Country* | *Study design* | *Participants* | *(Scale) level(s) of well-being[rating scale/range score]* | *Aim of study* |
|---|---|---|---|---|---|---|
| Demerouti, Geurts, Bakker, & Euwema (2004) | Ergonomics (Medicine, Health, Mental Health, Nurse) | Netherlands | Quantitative, cross-sectional, descriptive statistics. | 3,122 Dutch Military Police. | (MBI-GS) Fixed shifts: emotional exhaustion = 1.71 and 1.75 [0–6 points]; cynicism = 1.62 and 1.50 [0–6 points]; personal accomplishment = 4.35 and 4.47 [0–6 points]; Rotating shifts: emotional exhaustion = 1.51 and 1.43 [0–6 points]; cynicism = 2.03 and 2.10 [0–6 points]; personal accomplishment = 4.20 and 4.11 [0–6 points] | Antecedents of burnout: shiftwork as the independent variable and work-home conflict as the mediator variable. |
| Espinoza-Parra, Molero, & Fuster-Ruizdeapodaca (2015) | Revista de Psicología Social (Psychology, Social Sciences) | Chile | Quantitative, cross-sectional, structural equation modeling. | 985 Chilean military police officer candidates. | (UWES) Work Engagement = 4.18 [Information about the range of the rating scale was not provided] | Antecedents and consequents of work engagement: Transformational leadership as an independent variable and job satisfaction as the dependent variable. |

*(Continued)*

*Table 20.1* (Continued)

| *Authors (year)* | *Journal (area)* | *Country* | *Study design* | *Participants* | *(Scale) level(s) of well-being[rating scale/range score]* | *Aim of study* |
|---|---|---|---|---|---|---|
| Harrington, Bean, Pintello, & Mathews (2001) | Administration in Social Work (Medicine, Health, Mental Health, Nurse) | USA | Quantitative, cross-sectional, descriptive statistics and regression models. | 139 respondents of the Air Force Family Advocacy Program. | (MBI-GS) Several occupations compared: emotional exhaustion = 25.6, 19.8, 18.4, 20.6, and 15.5 [0–54 points]; cynicism = 7.2, 4, 3.5, 2.1, and 2.4 [0–30 points]; personal accomplishment = 39.3, 41.3, 39, 40.8, and 36.3 [0–48 points] | Consequents of burnout: intention to leave as the dependent variable. |
| Ivey, Blanc, & Mantler (2015) | Journal of Occupational Health Psychology (Psychology, Social Sciences) | Canada | Quantitative, cross-sectional, structural equation modeling. | 1,224 Canadian Armed Forces personnel. | (UWES) Work Engagement = 4.56 [Information about the range of the rating scale was not provided] | Antecedents of work engagement: trust in team work and job significance as independent variables, and willingness to deploy on operations, turnover intentions and psychological distress as mediator variables. |

*(Continued)*

*Table 20.1* (Continued)

| *Authors (year)* | *Journal (area)* | *Country* | *Study design* | *Participants* | *(Scale) level(s) of well-being[rating scale/range score]* | *Aim of study* |
|---|---|---|---|---|---|---|
| Jiang et al. (2015) | Human Resource Management (Psychology, Social Sciences) | USA | Quantitative, cross-sectional, structural equation modeling. | 13,182 U.S. Armed Forces personnel from the Army, Navy, Marines, Air Force, and the Coast Guard. | (Gallup Organization Q-12 scale) Work Engagement = 3.56 [1–5 points] | Antecedents and consequents of work engagement: anti-sexual harassment practices and sexual harassment incidents as independent variables and affective commitment and intentions to stay as dependent variables. |
| Kok, Herrell, Grossman, West, & Wilk (2016) | Psychiatric Services (Medicine, Health, Mental Health, Nurse) | USA | Quantitative, cross-sectional, descriptive statistics and regression models. | 488 Military Mental Health Service providers | (c) 21% (N = 103) of respondents reported elevated levels of job-related burnout | Descriptive study: determine the burnout level of Military Mental Health Service providers. |
| Lang, Patrician, & Steele (2012) | Journal of Nursing Scholarship (Medicine, Health, Mental Health, Nurse) | USA | Quantitative, cross-sectional, descriptive statistics and regression models. | 257 Army nursing personnel. | (MBI-HSS) emotional exhaustion = 23 [0–48 points]; depersonalization = 8 [0–30 points]; personal accomplishment = 38 [0–54 points] | Antecedents of burnout: deployment and perceptions of the practice environment as independent variables. |

*(Continued)*

*Table 20.1* (Continued)

| *Authors (year)* | *Journal (area)* | *Country* | *Study design* | *Participants* | *(Scale) level(s) of well-being[rating scale/range score]* | *Aim of study* |
|---|---|---|---|---|---|---|
| Lang, Pfister, & Siemens (2010) | Military Medicine (Medicine, Health, Mental Health, Nurse) | USA | Quantitative, cross-sectional, descriptive statistics. | 187 U.S. army nurses assigned to a large military treatment facility. | (MBI-HSS) emotional exhaustion = 25 [0–54 points]; depersonalization = 8 [0–30 points]; personal accomplishment = 39 [0–48 points] | Descriptive study: determine the burnout level of nurses and comparing civilian career nurses versus military career nurses. |
| Matthew et al. (2015) | Military Medicine (Medicine, Health, Mental Health, Nurse) | United Kingdom | Quantitative, intervention study, descriptive statistics. | 40 Resistance Instructors from the British Military personnel (n = 17 permanent and 23 external). | (MBI-GS) Permanent Resistance Instructors: emotional exhaustion = 5.71 [0–30 points]; cynicism = 5.24 [0–30 points]; personal accomplishment = 29.82 [0–36 points]; External Resistance Instructors: emotional exhaustion = 9.17 [0–30 points]; cynicism = 8.61 [0–30 points]; personal accomplishment = 28.04 [0–36 points] | Antecedents of burnout: analyze the level of burnout of military instructors in a training program. |

*(Continued)*

*Table 20.1* (Continued)

| *Authors (year)* | *Journal (area)* | *Country* | *Study design* | *Participants* | *(Scale) level(s) of well-being[rating scale/range score]* | *Aim of study* |
|---|---|---|---|---|---|---|
| Moran et al. (2008) | Medicine & Science in Sports & Exercise (Medicine, Health, Mental Health, Nurse) | Israel | Quantitative, cross-sectional, physiological measures. | 227 female soldiers. | (d) Burnout feeling = 3.87 for the stress fracture group and 3.39 for the non-stress factor group [1–7 points]; Strain feeling = 4.24 for the stress fracture group and 4.23 for the non-stress factor group. | Consequents of burnout: stress fracture during basic training as the dependent variable. |
| Morgan et al. (2011) | Journal of the International Neuropsychological Society (Psychology, Social Sciences) | USA | Quantitative, intervention study, descriptive statistics. | 32 U.S. Military Special Operations personnel. | (MBI-GS) emotional exhaustion = 6.8 [0–54 points]; cynicism = 4.2 [0–30 points]; personal accomplishment = 31.15 [0–54 points] | Consequents of burnout: visuospatial executive function as the dependent variable. |
| Morgan, Cho, Hazlett, Coric, & Morgan (2002) | The Yale Journal of Biology and Medicine (Medicine, Health, Mental Health, Nurse) | USA | Quantitative, cross-sectional, physiological measures. | 41 soldiers. | (MBI-HSS) emotional exhaustion = 14 [0–54 points]; depersonalization = 7 [0–30 points]; personal accomplishment = 37 [0–48 points] | Consequents of burnout: physiological changes as the dependent variable. |
| Patrician, Shang, & Lake (2010) | Research in Nursing & Health (Medicine, Health, Mental Health, Nurse) | USA | Quantitative, cross-sectional, descriptive statistics and regression models. | 357 Army Medical Department Registered Nurses. | (MBI-HSS) 40% of the total sample had a score of 27 or above indicating high emotional exhaustion [0–54 points] | Antecedents of burnout: demographics and organizational characteristics as independent variables. |

*(Continued)*

*Table 20.1* (Continued)

| *Authors (year)* | *Journal (area)* | *Country* | *Study design* | *Participants* | *(Scale) level(s) of well-being[rating scale/range score]* | *Aim of study* |
|---|---|---|---|---|---|---|
| Quick, Joplin, Nelson, Mangelsdorff, & Fiedler (1996) | Military Psychology (Psychology, Social Sciences) | USA | Quantitative, cross-sectional, descriptive statistics and regression models. | 1,497 basic military trainees. | (e) Four sub-groups compared: burnout = 33.54, 37.97, 33.31, 37.97 [5–50 points] | Antecedents of burnout: self-reliant as the independent variable. |
| Sargent et al. (2016) | Military Medicine (Medicine, Health, Mental Health, Nurse) | USA | Quantitative, cross-sectional, descriptive statistics and regression models. | 523 health care providers in a United States Military Medical Center. | (MBI-HSS) emotional exhaustion = 19.99 [0–52 points]; depersonalization = 4.84 [0–26 points]; personal accomplishment = 40.56 [0–48 points] | Antecedents of burnout: demographics and areas of greatest frustration with work (seven options: computers, parking, administrative support, collateral duties, inadequate staffing, supplies and life/work balance). |
| Serec, Bajec, Petek, Švab, & Selič (2012) | ZdravniskiVestnik - Slovenian Medical Journal (Medicine, Health, Mental Health, Nurse) | Slovenia | Quantitative, cross-sectional, structural equation modeling. | 390 soldiers from the Slovenian Armed Forces. | (MBI-HSS) emotional exhaustion = 14.46 [0–54 points]; depersonalization = 8.71 [0–30 points]; personal accomplishment = 30.37 [0–54 points] | Antecedents of burnout: personality traits and coping as independent variables. |

*(Continued)*

*Table 20.1* (Continued)

| *Authors (year)* | *Journal (area)* | *Country* | *Study design* | *Participants* | *(Scale) level(s) of well-being[rating scale/range score]* | *Aim of study* |
|---|---|---|---|---|---|---|
| Shelley & Wong (1991) | Military Medicine (Medicine, Health, Mental Health, Nurse) | USA | Quantitative, cross-sectional, descriptive analysis. | 77 Military Dentists. | (MBI-HSS) emotional exhaustion = 20.35 [0–54 points]; depersonalization = 9.58 [0–30 points]; personal accomplishment = 40.15 [0–48 points] | Antecedents of burnout: demographic variables as independent variables. |
| Simons et al. (2016) | Military Medicine (Medicine, Health, Mental Health, Nurse) | USA | Quantitative, cross-sectional, descriptive statistics. | 27 military orthopaedic residents and 12 military staff surgeons. | (MBI-HSS) military orthopaedic residents: emotional exhaustion = 23.37 [0–54 points]; depersonalization = 8.41 [0–30 points]; personal accomplishment = 37.56 [0–48 points]; military staff surgeons: emotional exhaustion = 19.75 [0–54 points]; depersonalization = 7.08 [0–30 points]; personal accomplishment = 37.75 [0–48 points] | Descriptive study: determine the prevalence rate of burnout among orthopaedic surgery residents and staff surgeons in a military orthopaedic residency. |

*(Continued)*

*Table 20.1* (Continued)

| *Authors (year)* | *Journal (area)* | *Country* | *Study design* | *Participants* | *(Scale) level(s) of well-being[rating scale/range score]* | *Aim of study* |
|---|---|---|---|---|---|---|
| Souza, Torres, Barbosa, Lima, & Souza (2015) | Psychology/Psicologia Reflexão e Crítica (Psychology, Social Sciences) | Brazil | Quantitative, cross-sectional, multiple regression models. | 228 cadets at a military academy in Brazil. | (MBI for students) emotional exhaustion = 3.34 [1–5 points]; depersonalization = 2.41 [1–5 points]; personal accomplishment = 2.86 [1–5 points] | Antecedents of burnout: subjective well-being as an independent variable and self-efficacy as a mediator variable. |
| Taghva, Imani, Kazemi, & Shiralinia (2015) | Archives of Psychiatry and Psychotherapy (Medicine, Health, Mental Health, Nurse) | Iran | Quantitative, cross-sectional, structural equation modeling. | 215 active duty personnel of the Islamic Republic of Iran Army. | (MBI-HSS) Mean values for emotional exhaustion and depersonalization were not provided. | Consequents of burnout: depression as a mediator variable and self-destructive behaviors as a dependent variable. |
| van Wijk (1997) | Military Medicine (Medicine, Health, Mental Health, Nurse) | South Africa | Quantitative, cross-sectional, descriptive analysis. | 46 Military nurses. | (f) Mean values for burnout were not provided. | Antecedents of burnout: demographic and contextual variables as independent variables. |
| Varner & Foutch (2014) | Journal of the American Academy of Physician Assistants (Medicine, Health, Mental Health, Nurse) | USA | Quantitative, cross-sectional, descriptive statistics. | 150 family medicine providers on active duty in the U.S. Air Force. | (MBI-HSS) emotional exhaustion = 16.04 [0–52 points]; depersonalization = 12.67 [0–26 points]; personal accomplishment = 26.85 [0–48 points] | Descriptive study: determine the level of burnout of air force family medicine providers. |

*(Continued)*

*Table 20.1* (Continued)

| *Authors (year)* | *Journal (area)* | *Country* | *Study design* | *Participants* | *(Scale) level(s) of well-being[rating scale/range score]* | *Aim of study* |
|---|---|---|---|---|---|---|
| Vie, Scheier, Lester, & Seligman (2016) | Military Psychology (Psychology, Social Sciences) | USA | Quantitative, cross-sectional, factorial validity and reliability analysis. | 320,503 U.S. Active Duty soldiers. | (g) Mean values for work engagement were not provided. | Validation of the U.S. Army Global Assessment Tool, which includes the assessment of work engagement. |
| Vinokur, Pierce, & Lewandowski-Romps (2009) | Stress and Health (Psychology, Social Sciences) | USA | Quantitative, longitudinal, two-wave study, 1 year time-lag. | 1,009 Air Force personnel | (SMBM) emotional exhaustion = 2.85 (T1) and 2.88 (T2) [1–7 point]; physical fatigue = 3.73 (T1) and 3.68 (T2) [1–7 point]; cognitive weariness = 2.44 (T1) and 2.46 (T2) [1–7 point]; job burnout = 2.94 (T1) and 2.95 (T2) [1–7 point]. | Consequents of burnout: perceived health as the dependent variable. |
| Vinokur, Pierce, Lewandowski-Romps, Hobfoll, & Galea (2011) | Journal of Occupational Health Psychology (Psychology, Social Sciences) | USA | Quantitative, longitudinal, two-wave study, 1 year time-lag. | 1,009 U.S. Air Force personnel deployed to wars. | (SMBM) Job Burnout = 2.88 (T1) and 2.82 (T2) [1–7 points] | Antecedents of burnout: effects of war exposure, post-traumatic stress (PTS) symptoms and resource losses as independent variables. |

*(Continued)*

*Table 20.1* (Continued)

| *Authors (year)* | *Journal (area)* | *Country* | *Study design* | *Participants* | *(Scale) level(s) of well-being[rating scale/range score]* | *Aim of study* |
|---|---|---|---|---|---|---|
| Vogelgesang, Leroy, & Avolio (2013) | The Leadership Quarterly (Psychology, Social Sciences) | USA | Quantitative, longitudinal, three-wave study, 3 weeks time-lag from Time 1 to Time 2, and 6 weeks time-lag from T2 and T3. | 451 cadets from a United States military academy. | (h) Work Engagement = 3.45 [1–5 points] | Antecedents and consequents of work engagement: leaders' transparent communication as an independent variable and performance as a dependent variable. |
| Walters, Matthews, & Dailey (2014) | Military Medicine (Medicine, Health, Mental Health, Nurse) | USA | Quantitative, cross-sectional, descriptive statistics. | 53 active duty Army health care providers, including combat medics, physician assistants, and physicians. | (MBI-HSS) emotional exhaustion = 30.17 [0–54 points]; depersonalization = 15.92 [0–30 points]; personal accomplishment = 34.34 [0–48 points] | Antecedents of burnout: deployment analyzed and number of leave days accrued as independent variables. |

*(Continued)*

*Table 20.1* (Continued)

| *Authors (year)* | *Journal (area)* | *Country* | *Study design* | *Participants* | *(Scale) level(s) of well-being[rating scale/range score]* | *Aim of study* |
|---|---|---|---|---|---|---|
| Whealin et al. (2007) | Military Medicine (Medicine, Health, Mental Health, Nurse) | USA | Quantitative, cross-sectional, descriptive and regression analysis. | 253 U.S. Army, Army Reserve Units, U.S. Air Force, and U.S. Navy personnel. | (MBI-HSS) Enlisted: emotional exhaustion = 17.87 [0–54 points]; depersonalization = 19.46 [0–30 points]; personal accomplishment = 34.75 [0–48 points]; Officers: emotional exhaustion = 16.6 [0–54 points]; depersonalization = 18.3 [0–30 points]; personal accomplishment = 36.18 [0–48 points]; | Antecedents of burnout: exposure to traumatic events as the independent variable. |
| Zheng et al. (2015) | The Leadership Quarterly (Psychology, Social Sciences) | USA | Quantitative, cross-sectional, structural equation modeling. | 338 military personnel deployed in combat zones overseas. | (MBI-GS) Exhaustion = 3.01 [0–6 points] | Antecedents of burnout: ethical leadership as an independent variable, team cohesion as a mediator variable and conscientiousness as a moderator variable. |

*Note*: MBI-HSS = Maslach Burnout Inventory – Human Services Survey; MBI-GS = Maslach Burnout Inventory – General Survey; SMBM = Shirom–Melamed Burnout Measure; UWES = Utrecht Work Engagement Scale; (a) Self- Engagement: four-item scale that was a modified version of a measure used in past research (Britt & Bliese, 2003; Britt et al., 2001; Britt, Castro, & Adler, 2005); (b) Engagement in Meaningful Work; (c) Participants rated their level of professional burnout. Response options were very low, low, medium, high, and very high. Burnout status was categorized into elevated levels (responses of high and very high) and non-elevated levels (responses of very low, low and medium); (d) At the method section, the authors mentioned the assessment of burnout and exhaustion; however, the scale used was not identified by the authors; (e) Bodensteiner, Gerloff, and Quick (1989) 10-item burnout scale; (f) At the method section, the author mentioned the assessment of burnout and exhaustion; however, the scale used was not identified by the author; (g) Four items adapted from Wrzesniewski, McCauley, Rozin and Schwartz (1997); (h) May, Gilson and Harter (2004) scale of engagement; ProQOLV = Professional Quality of Life V survey instrument.

The vast majority (n = 38) of studies were cross-sectional, only nine had a longitudinal design (7 with two waves and 2 with three waves), and two studies reported interventions.

Concerning the aim of the studies reviewed, we observed that more than half analyzed well-being (i.e. burnout or engagement) predictors (n = 28); seven analyzed these well-being indicators as mediators or moderators; seven analyzed the consequences of engagement/burnout; six described the prevalence of well-being (i.e. burnout) among military groups; one reported the validation of a well-being measure that included a work engagement dimension. Of the studies that researched antecedents, the majority (n = 17 out of 28) analyzed situational variables, 8 analyzed individual variables and 3 a mixture of situational and individual variables.

## *What are the burnout and engagement levels among military personnel?*

The majority of studies adopted the Maslach Burnout Inventory (MBI; n = 31 out of 38 studies) to measure burnout and the Utrecht Work Engagement Scale (UWES) to measure engagement (n = 8 out of 15 studies).

Although the most widely used instrument to measure burnout is the Maslach Burnout Inventory (MBI), 13 studies adopted the MBI – General Survey (MBI GS), whereas 16 used the MBI Health Services Survey (MBI HSS), and 2 used specific versions (e.g., MBI student version and MBI Turkish version). Furthermore, some studies only focused on the measurement of exhaustion (n = 4), a few assumed exhaustion and cynicism/depersonalization as the core dimensions (n = 6), and the vast majority treated the burnout syndrome as being composed of three dimensions (n = 21). In addition, the scales' range changed across the studies, using the same version of the same instrument. Indeed, some studies assumed that scores were calculated by the sum of the items' scores, whereas others assumed the average score across items' scores. Therefore, scores that vary within a 7-point scale (0–6 or 1–7) may be found for the same scale, whereas others assume much broader scales (0–54, for example). The possibility of comparing the scores was further jeopardized, as several studies were also found to use the short versions of the original scale (probably to improve the scale's reliability, items had sometimes been eliminated), and this is also reflected in a stronger variation of the limits of the scale.

As for Engagement, 8 out of the 15 studies used the Utrecht Work Engagement Scale (UWES); however, some adopted a "single" dimension approach (n = 5) and some used vigor and dedication as core dimensions (n = 2). One study (Boermans et al., 2014) aggregated the engagement scores at the team level, referring to team work engagement. Despite this variability, all the studies calculated the average score on a 7-point Likert scale (ranging from 0 to 6 or from 1 to 7). The remaining 7 studies adopted other measures of engagement. Out of these 7, 4 studies measured the concept of "self-engagement" or "job engagement," which is a different concept and refers to how responsible an individual feels for his/her job performance and how much job performance matters to the individual.

## Discussion

The aim of this chapter was to review, summarize and evaluate the research on military burnout and engagement. These psychological states were conceptualized by the conservation resource theory model (Hobfoll, 2002) as negative and positive indicators of workplace well-being. Along with several inclusion criteria, our review focused on 49 studies. Our qualitative systematic literature review revealed that military burnout and engagement had been researched by *two different lenses*: psychology and medicine. This emerged as a factor influencing the studies' theoretical framework and aim: for psychology, it has been crucial to understand what explains burnout and engagement or its consequences, whereas for medicine, pinpointing the level of well-being, particularly the level of burnout, has taken priority. The review further indicated that the majority of studies were cross-sectional, and military burnout levels were found to be relatively high to moderate in the vast majority of the reviewed studies. Work engagement did not fall below average levels.

## *Burnout more frequently studied than work engagement*

A major trend that is salient when revisiting the literature on military burnout and engagement is a prevailing negative bias of studies addressing mental illness rather than mental "wellness." This is illustrated by the fact that the number of publications on burnout exceeds that of engagement by a ratio of 2, 6:1.

This is not exclusive to the studies on military personnel. Indeed, Schaufeli and Salanova (2007) listed the papers published on workplace well-being in the *Journal of Occupational Health Psychology* between 1996 and 2005 and found that out of 233 manuscripts, only 14 referred exclusively to positive indicators. The authors found a ratio of 14:1 in favor of papers focusing on mental illness. In the same vein, Macik-Frey, Quick and Nelson (2007) analyzed the literature on occupational health in from 1990 to 2005 and confirmed that stress and burnout were the predominant trends.

It is important to create a shift in this tendency towards illness, as failing to capture the positive aspects of work and the inherent resources are inappropriate and incomplete. As argued by Turner, Barling, and Zacharatos (2002, p. 715), "... it is time to extend our research focus and explore more fully the positive sides, so as to gain full understanding of the meaning and effects of working." Indeed, most organizations expect their workers to take responsibility and initiative, to be committed to the organization and the team, to be involved with the job and feel accountable for high quality performance. Thus, they need employees who feel engaged, i.e. people who feel energetic and dedicated, and who are absorbed by their work (Bakker & Schaufeli, 2008).

Hence, we need studies that address the positive sides of work and the effective functioning of individuals. The literature on burnout and engagement has demonstrated that these indicators can be explained by different variables (threat of loss or availability of resources; Hobfoll, 2002), and we now have extensive evidence showing how the loss of resources is associated with feelings of burnout. However, we know much less about the resources that are fundamental to promote engagement. Carvalho and Chambel (2017) sampled 175 military employees from three units of the Portuguese Marine Corps and found that autonomy and the supervisory support were related to engagement, and that this relationship was mediated by work–family enrichment. Delahaij, Kamphuis, and van den Berg (2016) found that strong self-efficacy helped service members deal with exposure to threatening situations during deployment, leading to more work engagement. These authors further demonstrated that service members with low self-efficacy benefited from family support when threat exposure was high, whereas service members with high self-efficacy benefited from family support when threat exposure was low. Espinoza-Parra, Molero, and Fuster-Ruizdeapodaca (2015) studied a sample of Chilean military police officer candidates and found a positive association between transformational leadership and the recruits' level of engagement. Ivey et al. (2015) sampled Canada Armed Forces personnel and found that trust in team work and job significance were predictors of engagement.

Furthermore, there is evidence demonstrating that engaged workers perform better than non-engaged workers (Motyka, 2018). Studies among military personnel have supported this relationship. Indeed, studies have found that engaged military personnel report more affective commitment to their individual branch of the service – Army, Navy, Marines, Air Force and

Coast Guard (Jiang et al., 2015), have better performance (third-party ratings; Vogelgesang, Leroy & Avolio, 2013) and have fewer turnover intentions (Ivey et al., 2015; Jiang et al., 2015).

As explained by Bakker, Schaufeli, Leiter and Taris (2008), engaged employees perform better since they often experience positive emotions which broaden people's thought-action repertoire (Fredrickson, 2003); experience better psychological and physical health; create their own job and personal resources (e.g., support from others), increasing their ability to deal with job demands; and transfer their engagement to others, which is particularly important when performance is the result of the combined effort of individual employees. Boermans et al. (2014) sampled 971 Dutch peacekeepers within 93 teams who were deployed and found that within teams there was an adequate level of agreement to justify aggregation of engagement at the collective level. These authors found that team members reported fewer fatigue symptoms after deployment if they were part of highly engaged teams during deployment, demonstrating that investing in team work engagement is important for those working in highly demanding jobs.

Having argued the need to increase the focus on well-being and good functioning, we also believe it is important to continue to study burnout, as it has implications for workers' health, also in the military context. For example, Taghva, Imani, Kazemi, & Shiralinia (2015) studied 215 active duty personnel of the Islamic Republic of Iran Army and found that burnout influenced depression and had an effect on self-destructive behavior. Morgan et al. (2011) sampled 32 U.S. Military Special Operations personnel and concluded that the assessment of burnout symptoms may help identify military personnel at risk for stress-related executive dysfunction.

Hence, research on military personnel's well-being would benefit from more studies simultaneously accounting for burnout and engagement levels. Indeed, some studies have highlighted that engagement can serve as a protector from highly demanding situations (Britt, Castro & Adler, 2005; Delahaij, et al., 2016). Unfortunately, the very few studies examining burnout and engagement together (Carvalho & Chambel, 2017; Chambel, Castanheira, Oliveira-Cruz, & Lopes, 2015; Chambel & Oliveira-Cruz, 2010; Delahaij, Kamphuis & van den Berg, 2016) focused on the antecedents of burnout and engagement and treated them as two independent parallel indicators. Therefore, there is still much to explore on the potential interaction effects between these two workplace well-being indicators.

## *The measure of burnout and engagement*

The majority of studies adopt the same instrument to measure burnout and engagement. Although there is an apparent consensus on how to measure these well-being indicators, closer examination points to a considerable variability of versions, structure and range of the scale, even within the same instrument. This heterogeneity compromises the ability to make comparisons across studies and to reach a conclusion on the burnout or engagement levels of military personnel.

Indeed, the MBI GS and MBI HSS versions have important differences that should be taken into account. On the one hand, the scales' content changes significantly across the versions. For example, the MBI GS measures cynicism, which reflects a detached attitude towards work, whereas the MBI HSS measures depersonalization, which reflects a detached attitude towards people; another example is that the MBI GS measures exhaustion, whereas the MBI HSS measures emotional exhaustion. Although these changes reflect the authors' attempt to better adapt the instrument to the different contexts, when making comparisons across studies, one must take into account that these are not the same concepts. Furthermore, the scales' structure is different across versions (for example, exhaustion is measured with 5 items in the MBI GS, and emotional exhaustion has 9 items in the MBI HSS), and even adjustments within the same version were found in some studies with the same version of the MBI. Depending on how the scale's score is calculated, this can have a strong impact on the scale's limit, the consequence being that it is possible to find the "same" dimension (e.g. exhaustion) being scored within different ranges within and across MBI versions.

On a final note, some studies translated burnout scale scores into low, medium and high burnout, using an established normative range for health care workers (Maslach, Jackson & Leiter, 1996). Although the reference to a threshold to interpret the meaning of MBI scores as low/moderate/high may be appealing, the fact that the scale's range frequently changes across studies to assume a threshold may be misleading. Furthermore, to our knowledge, there are no normalization studies of MBI among military personnel. Therefore, given the specific characteristics of this population, any attempt to use cutoff points identified in other contexts may result in serious misinterpretations.

## The studies' design

Our systematic review reveals that research on military burnout and engagement has been dominated by cross-sectional studies, with a minority representing longitudinal studies. Moreover, if we use the Kelloway and Francis (2013) criteria that consider a longitudinal study to be one that employs three or more measures, this review only identifies two studies. Vogelgesang, Leroy and Avolio (2013) used a military cadet sample and researched how leader behavioral integrity relates to individual follower work engagement and how that relationship, in turn, connects to performance. Chambel and Oliveira-Cruz (2010) investigated peacekeeping militaries and analyzed the impact of the noncompliance with Army obligations (psychological contract breach) in this mission on the burnout and engagement of military personnel during and at the end of the mission. Although this predominance of cross-sectional studies has been identified as a characteristic in occupational health literature (e.g. Häusser, Mojzisch, Niesel & Schulz-Hardt, 2010; Luchman & Gonzalez-Morales, 2013; Taris & Schaufeli, 2016), it is a significant limitation. Cross-sectional studies only enable an understanding of whether the relationship between variables is significant but do not permit the establishment of unidirectional causation (e.g. De Lange, Taris, Kompier, Houtman, & Bongers, 2003; Zapf, Dormann, & Frese, 1996). Thus, a cross-sectional approach is inappropriate to show, for example, that some personal, job or organizational characteristics lead military personnel to lose or gain resources and develop burnout or engagement. With this design, it is also possible that a reverse causation may exist and military burnout and engagement influence their occupational situation perceptions as a loss or gain of resources. Future research with longitudinal designs that can address such patterns of causation (e.g., cross-lagged panel designs) would be valuable in underlying processes associated with employee attempts to garner, conserve and utilize resources in the context of military burnout and engagement development. In these longitudinal studies, it is important to consider the change in military well-being. Mäkikangas, Kinnunen, Feldt and Schaufeli's (2016) systematic review showed that a change of well-being, namely of burnout and work engagement, is more frequent than stability. We need more evidence on the specific features of the military context (i.e. training, war deployment, peacekeeping deployment) that may influence a change in military burnout and work engagement.

This systemic review also shows that intervention studies are rare in the literature on military burnout and engagement (Matthew et al., 2015; Morgan et al., 2011). The intervention studies not only provide valuable information on the causal relationships but also enable a systematic examination of the organizational actions (e.g. training, simulations, recruitment) and their efficacy to promote military well-being. Thus, researchers should be encouraged to employ more intervention studies in order to increase the ecological validity of military actions and consequently increase their evidence-based nature.

## *Antecedents of military burnout and work engagement*

The causes of burnout and engagement are generally divided into situational and individual causes (Aronsson et al., 2017). Our systematic review also reveals that the individual and situational antecedents have been considered in military research. As far as individual antecedents are concerned, we observed that personality factors (i.e., self-efficacy) are important in explaining military burnout. This is exemplified by Alessandri et al. (2018), who showed in their longitudinal study with cadets that emotional self-efficacy beliefs in managing negative emotions at work are a key mechanism that contributes to mediating the negative relationship between emotional stability and job burnout. In line with the COR theory (Hobfoll, 1989), some personal characteristics (i.e. personality) should be considered resources that the individual should use in order to deal with threatening conditions and prevent negative outcomes, namely burnout. This assumption has been empirically supported in prior meta-analysis studies (Alarcon, Eschleman, & Bowling, 2009; Swider & Zimmerman, 2010), which found that the Big Five factors are consistently related to each of the three dimensions of burnout. Furthermore, Alarcon et al. (2009) also found that lower-order personality factors (e.g., self-esteem, self-efficacy, locus of control, positive affectivity, negative affectivity, optimism, proactive personality and hardiness) had a significant negative relationship with burnout. Thus, it would be interesting in future research with militaries to include the Big Five and other lower-order personality factors as predictors of burnout development, particularly in high-demand situations.

Moreover, in line with the COR theory (Hobfoll, 2002), as individuals not only strive to protect their resources, but also to accumulate them, some personality factors should be considered personal resources that tend to generate other resources, which may result in positive outcomes like work engagement. In fact, the review of literature conducted by Mäkikangas et al. (2013) showed that Big Five factors (e.g., extraversion and neuroticism) and lower-order individual factors (e.g., self-efficacy) were consistently related to higher work engagement. For instance, employees with high extraversion and low neuroticism showed higher work engagement. In addition, self-efficacy was found to be related to higher absorption levels of employees (i.e., one dimension of work engagement). This systematic review reveals that the relationship between personality and military engagement has not yet been investigated; thus, we suggest that future studies analyze not only the effect of military personality on burnout but also on engagement.

Concerning the situational factors, this systematic review reveals that different job and organizational characteristics contribute to explaining military well-being. However, related to the explanation of military burnout, this review makes it possible to distinguish different study groups. The first group includes studies published on the scientific areas of medicine, health, mental health and nursing which seek to describe the participants' burnout level. The assumption is that military personnel confront high job demands and consequently have high levels of stress, namely high levels of burnout. The study of Simons et al. (2016), describing the levels of orthopedic residents and staff surgeons' burnout in an army medical center, and the study of Varner and Foutch (2014) that describes the burnout of Air Force family medicine providers are examples of this. These studies supported the idea that these military personnel are vulnerable to burnout but did not contribute to explaining why and how the development of this chronic stress indicator occurs. In fact, these studies did not analyze the professional experiences of these military personnel and, as such, did not bring to light the levels of demands they encounter or whether the demands they may encounter are solely related to the fact that they are health professionals or are also due to the military context.

The second group that this systematic review identifies includes studies that were also published in medicine, health, mental health and nursing and compared the burnout of military personnel with a deployment experience with those who had not had this experience. The assumption is that a deployment implies high demands that contribute to the emergence of burnout. However, the results of these studies were not conclusive, and some of the studies did not observe significant differences (i.e., Cragun, April & Thaxton [2016], which included health care providers in a military emergency department) or observed better results in some burnout dimensions in militaries with experiences of deployment (e.g., Lang, Patrician & Steele [2012] observed that nurses who had participated in a deployment in Iraq had lower exhaustion than nurses of a large army teaching hospital in the United States). We believe that these inconclusive results may arise due to the fact that both the personnel who had been deployed and those that had not may have had enough resources to meet the demands they encountered; those without a deployment experience may encounter higher demands and lower resources than personnel with such experience. In fact, the studies included in this second group did not analyze whether the deployment experiences involved threat or loss of resources and were not conclusive as to the influence of deployment on burnout.

In addition to the limitations of studies in the two aforementioned groups, our systematic review encountered a third group of studies that analyzed the specific demands of a deployment and its influence on the development of burnout. This is exemplified by Vinokur, Pierce, Lewandowski-Romps, Hobfol and Galea (2011), who showed in their longitudinal study with a large sample of U.S. Air Force personnel deployed to the war that exposure to traumatic experiences during deployment predicted a loss of resources, which in turn promoted the burnout of returning veterans.

Finally, the fourth group of studies that the systematic review identified includes studies pointing to the fact that some job demands and/or job resources that were salient to explain the burnout of civil professionals were also valuable to explain military burnout. For example, Demerouti, Geurts, Bakker and Euwema (2004) showed that shift rotations in military jobs were related to cynicism and professional efficacy (both burnout dimensions). Carvalho and Chambel (2017) observed that high job demands (i.e., workload) and low supervisory support were related to work-family conflict, which in turn was related to military burnout.

The studies identified in these third and fourth groups of this systematic review show that the military personnel who perceived high demands and low resources displayed higher levels of burnout. This observation is in line with the COR assumption that burnout occurs when a person perceives high demands that threaten something of value to him/her. The threats or losses will tax or exhaust the resources the personnel have available to confront the situation (Hobfoll, 1989). Furthermore, these results are also in line with the meta-analysis of Lee and Ashforth (1996), showing that job demands were important predictors of burnout; the meta-analysis conducted by Alarcon (2011), which observed that higher demands and lower resources were associated with burnout and the systematic review of Seidler et al. (2014), pointing to high job demands, low possibility to exert control and non-supportive workplaces being important to explain the development of burnout. Thus, we suggest that future studies continue to analyze different situations with high demands and low resources that explain the development of burnout in the military.

Related to the explanation of military engagement, this systematic review enables us to conclude that this positive psychological state has been exclusively researched by studies on psychology and reveals that some job and organizational characteristics are important resources that predict it. For example, the study by Ivey et al. (2015) with a large sample of Canadian Armed Forces showed that trust in team and job significance predicted work engagement which, in turn, predicted military personnel's willingness to deploy an operation, their psychological distress and their turnover intention. In the same vein, the longitudinal study of Vogelgesang et al. (2013) with military cadets showed that leader behavior integrity mediated the positive influence of leader transparent communication on military personnel engagement, which in turn influenced their performance.

These systematic review results are in line with the COR theory (Hobfoll, 2002), as previously mentioned, and presuppose that engagement occurs as a "gain/loss cycle" since resources tend to generate other resources, which may result in positive outcomes like the positive psychological state. The meta-analyses by Halbesleben (2010) and by Christian et al. (2011) confirmed this assumption that job resources are the most important predictors of employee engagement. Furthermore, in an attempt to more clearly delineate the nature of resources, Hobfoll (1988) underlined social support and argued that it can both widen one's pool of available resources and replace or reinforce other resources that have been lacking. In fact, this systematic review reveals that social support, particularly leader support, is an important resource to explain military engagement. Not only does the previously referred to study of Vogelgesang, et al. (2013) underline the role of the leader to explain cadets' engagement, but, for example, Espinoza-Parra et al. (2015) also observed that leadership (i.e. transformational style) was related to military engagement which, in turn, was related to work satisfaction. Reinforcing the salient leader role in the military context (Hannah & Sowde, 2012), this systematic review reveals the promotion of military engagement through the leader. Moreover, as leaders can make followers more certain of the social and material resources available to them and have an important role in the supply of resources that aid subordinates to cope with job demands and allow them to deploy their resources more effectively, their relationship with subordinates should be an important situational characteristic to explain not only military engagement but also burnout. In the same vein, the meta-analysis by Harms, Bai, and Han (2016) showed that transformational leadership and the leader-member exchange (LMX) were negatively related to subordinates' burnout. This systematic review only identified one study (Zheng et al., 2015) that associated leadership with subordinate emotional exhaustion (i.e. a burnout dimension). The authors researched a sample of military personnel deployed in combat zones and observed that team cohesion partially mediated the relationship between ethical leadership and emotional exhaustion, particularly when personnel had high conscientiousness. Thus, we recommend that future studies analyze the effects of leadership on the development of military burnout and engagement.

In addition, this systematic review also reveals that military engagement has a role in buffering the relationship between demands and strain, namely burnout. For example, in a longitudinal multilevel study, Britt and Bliese (2003), with a sample of personnel deployed on a peacekeeping mission to Bosnia, verified that in situations with high stressors, the engaged personnel reported lower psychological distress than the disengaged personnel. This role of engagement is not presupposed in the COR theory. However, in line with Schaufeli and Taris (2014), the development process of engagement and that of burnout should not be independent. Thus, future research should deepen this line of research and analyze both processes jointly, namely the role of engagement in burnout development and vice versa.

## Consequences of burnout and work engagement

Compared to the analysis of antecedents, this systematic review shows that military well-being (i.e. burnout and work engagement) consequences have received less interest from researchers.

Moreover, this review also reveals that studies only analyze individual consequences of military well-being, and the majority analyze attitudinal consequences such as work satisfaction, organizational affective commitment and turnover intentions. However, this review shows that studies on burnout preferred to analyze the direct consequences of this stress syndrome, and the majority of studies that analyzed engagement consequences chose mediation models, and this positive psychological state played the role of mediator. For example, Vinokur, Pierce and Lewandowski-Romps (2009) verified in a longitudinal study with a large sample of Air Force personnel deployed to the wars in Iraq, Afghanistan and supporting locations that burnout and health perceptions had reciprocal effects. Jiang et al. (2015), with large samples of U.S. Armed Forces from the Army, Navy, Marines, Air Force, and Coast Guard, observed that engagement and psychological distress mediated the relationship between anti–sexual harassment practices/sexual harassment incidents and affective commitment and intention to stay. These observations are in line with previous studies that demonstrated that burnout and engagement have consequences for health and job attitudes (e.g., job satisfaction and organizational commitment) (Maslach, Schaufeli, & Leiter, 2001; Taris, 2006; Taris & Schaufeli, 2016). However, previous meta-analytical studies have also demonstrated that these indicators of workers' well-being are important performance predictors underlining the practical relevance of burnout and engagement research for organizational results. Taris's (2006) meta-analysis confirmed the relationship between burnout, namely exhaustion, and in-role behavior, organizational citizenship behavior and customer satisfaction, and Swider and Zimmerman's (2010) meta-analysis also confirmed that job burnout had a significant relationship with job performance. As for engagement, Harter, Schmidt, and Hayes (2002) showed that work engagement was related to higher profitability and customer satisfaction/loyalty, and Christian et al.'s (2011) meta-analysis found that work engagement predicted both the task (in-role) and the contextual (extra-role) performance. Unfortunately, this systematic review only includes one study (Vogelgesang et al., 2013) that analyzed the relationship between burnout and performance and confirmed the negative consequence of this indicator of military chronic stress. Future studies should analyze the effects of burnout and engagement on military performance and consequently help military institutions to better understand personnel behaviors, namely those that contribute to the efficiency and effectiveness of these institutions.

## Conclusions

By means of the present systematic review, it was possible to elaborate on several theoretical, practical considerations. From the theoretical point of view, we noted that to date, both burnout and work engagement are being researched through two different lenses: psychology and medicine that resulted in a difference in the approach and objectives, such as the theoretical framework and the aim of the study. However, regardless of the lens used (i.e., psychological or medical), the prevalence of studies focusing on mental illness, or rather mental "wellness," is remarkable, and, as previously noted, relying on the conceptualization of the conservation of resources theory, studies analyzing the resources promoting military engagement are needed.

From the practical point of view, we are also able to elaborate on some final concluding remarks. For instance, in a methodological scope, this systematic literature review clearly showed some limitations of previous studies that should be overcome in future research. Firstly, the cross-sectional predominant design of previous studies, which does not allow inferences about the causal relationships among the variables and does not allow one to get a "picture" of the burnout and work engagement evolution levels across time. Secondly, although the majority of studies adopt the same instrument to measure burnout and work engagement, the systematic review revealed a considerable variability of versions, structure, and range of the scale used, even when the same instrument was used. In such a fashion, this limitation compromised the ability to make comparisons across studies and to make inferences about military burnout and work engagement levels. Finally, still from a practical point of view, the present review unveiled some antecedents of military burnout and work engagement, which can be classified into individual (e.g., personality) or situational antecedents (e.g., job demands), and some future pathways were pointed out. To mention a few, more research is needed in order to inspect: (1) the nature of the relationship between personality and military engagement, (2) how different situations with high demands and low resources contributes to explain the development of burnout in military personnel, (3) the effects of leadership on the development of military burnout and engagement and (4) the buffering role of work engagement on the relationship between job demands and burnout. Regarding the consequences of burnout and work engagement, compared to the analysis of antecedents, this systematic review allowed verifying that military burnout and work engagement consequences have been receiving less interest from researchers. In addition, when consequences are studied, the majority of studies focused only on observing the relationship between individual attitudinal characteristics (e.g., work satisfaction) and military well-being. As such, we suggested that future studies should observe the effects of burnout and work engagement on military members' behavior, such as their performance.

To summarize, the current work contributed to sum up all the empirical evidence obtained up to date on military burnout and work engagement, and by doing so, it highlighted future pathways for the research of work well-being in a military setting.

## References

Adler, A. B., Adrian, A. L., Hemphill, M., Scaro, N. H., Sipos, M. L., & Thomas, J. L. (2017). Professional stress and burnout in U.S. military medical personnel deployed to Afghanistan. *Military Medicine, 182*(3), e1669, 2017.

Afonso, J. M. P., & Gomes, A. R. (2009). Stress ocupacional em profissionais de segurança pública: Um estudo com militares da Guarda Nacional Republicana. [Occupational stress in professionals of public security: A study with Military Agents of the Republican National Guard]. *Psicologia: Reflexão e Crítica, 22*(2), 294–303.

Alarcon, G. (2011). A meta-analysis of burnout with job demands, resources, and attitudes. *Journal of Vocational Behavior, 79*, 549–562.

Alarcon, G., Eschleman, K. J., & Bowling, N. A. (2009). Relationships between personality variables and burnout: A meta-analysis. *Work Stress, 23*, 244–263.

Alessandri, G., Perinelli, E., De Longis, E., Schaufeli, W. B., Theodorou, A., Borgogni, L., … Cinque, L. (2018). Job burnout: The contribution of emotional stability and emotional self-efficacy beliefs. *Journal of Occupational and Organizational Psychology, 91*, 823–851.

Alves, J. S. C., Bendassolli, P. F., & Gondim, S. M. G. (2017). Trabalho emocional e burnout: Um estudo com policiais militares. [Emotional labor and burnout: A study with the Military Police]. *Avances en Psicología Latino-americana, 35*(3), 459–472.

Aronsson, G., Theorell, T., Grape, T., Hammarstrom, A., Hogstedt, C., Marteinsdottir, I. … Hall, C. (2017). A systematic review including meta-analysis of work environment and burnout symptoms. *BMC Public Health, 17*(1), 264.

Bakir, B., Ozer, M., Ozcan, C. T., Cetin, M., & Fedai, T. (2010). The association between burnout, and depressive symptoms in a Turkish Military Nurse sample. *Klinik PsikofarmakolojiBülteni-Bulletin of Clinical Psychopharmacology, 20*(2), 160–163.

Bakker, A. B., & Schaufeli, W. B. (2008). Positive organizational behavior: Engaged employees in flourishing organizations. *Journal of Organizational Behavior, 29*, 147–154.

Bakker, A. B., Schaufeli, W., Leiter, M. P., & Taris T. (2008). Work engagement: An emerging concept in occupational health psychology. *Work & Stress, 22*(3), 187–200.

Ballenger-Browning, K. K., Schmitz, K. J., Rothacker, J. A., Hammer, P. S., Webb-Murphy, J. A., & Johnson, D. C. (2001). Predictors of burnout among military mental health providers. *Military Medicine, 176*(3), 253–260.

Bartz, C., & Maloney, J. P. (1986). Burnout among intensive care nurses. *Research in Nursing & Health, 9*, 147–153.

Bauer, A., Newbury-Birch, D., Robalino, S., Ferguson, J., & Wigham, S. (2018). Is prevention better than cure? A systematic review of the effectiveness of well-being interventions for military personnel adjusting to civilian life. *PLOS ONE 13*(5), e0190144.

Bodensteiner, W. D., Gerloff, E. A., & Quick, J. C. (1989). Uncertainty and stress in an R&D project environment. *R&D Management, 19*(4), 309–322.

Boermans, S. M., Kamphuis, W., Delahaij, R., van den Berg, C., & Euwema, M. C. (2014). Team spirit makes the difference: The interactive effects of team work engagement and organizational constraints during a military operation on psychological outcomes afterwards. *Stress and Health, 30*, 386–396.

Britt, T. W., Adler, A. B., & Bartone, P. T. (2001). Deriving benefits from stressful events: The role of engagement in meaningful work and hardiness. *Journal of Occupational Health Psychology, 6*, 53–63.

Britt, T. W., & Bliese, P. D. (2003). Testing the stress-buffering effects of self-engagement among soldiers on a military operation. *Journal of Personality, 71*, 245–266.

Britt, T. W., Castro, C. A., & Adler, A. B. (2005). Self-engagement, stressors, and health: A longitudinal study. *Personality and Social Psychology Bulletin, 31*, 1475–1486.

Britt, T. W., Dickinson, J. M., Moore, D. M., Castro, C. A., & Adler, A. B. (2007). Correlates and consequences of morale versus depression under stressful conditions. *Journal of Occupational Health Psychology, 12*, 34–47.

Bryan, C. J., Goodman, T., Chappelle, W., Prince, L., & Thompson, W. (2018). Subtypes of severe psychological distress among US Air Force remote warriors: A latent class analysis. *Military Psychology, 30*(5), 437–448.

Buckman, J. E. J., Sundin, J., Greene, T., Fear, N. T., Dandeker, C., Greenberg, N., & Wessely, S. (2011). The impact of deployment length on the health and well- being of military personnel: A systematic review of the literature. *Occupational & Environmental Medicine, 68*, 69–76.

Carvalho, V. S., & Chambel, M. J. (2017). Work–family conflict and enrichment mediates the relationship between job characteristics and well-being at work with Portuguese Marine Corps. *Armed Forces & Society, 44*(2), 301–321.

Castanheira, F., Chambel, M. J., Lopes, S., & Oliveira-Cruz, F. (2016). Relational job characteristics and work engagement: Mediation by prosocial motivation. *Military Psychology, 28*(4), 226–240.

Chambel, M. J., Castanheira, F., Oliveira-Cruz, F., & Lopes, S. (2015). Work context support and Portuguese soldiers' well-being: The mediating role of autonomous motivation. *Military Psychology, 27*(5), 297–310.

Chambel, M. J., & Oliveira-Cruz, F. (2010). Breach of psychological contract and the development of burnout and engagement: A longitudinal study among soldiers on a peacekeeping mission. *Military Psychology, 22*, 110–127.

Chappelle, W., McDonald, K., Prince, L., Goodman, T., Ray-Sannerud, B. N., & Thompson, W. (2014). Assessment of occupational burnout in United States Air Force Predator/Reaper "Drone" Operators. *Military Psychology, 26*(5–6), 376–385.

Christian, M. S., Garza, A. S., & Slaughter, J. E. (2011). Work engagement: A quantitative review and test of its relations with task and contextual performance. *Personnel Psychology, 64*, 89–136.

Cragun, J. N., April, M. D., & Thaxton, R. E. (2016). The impact of combat deployment on health care provider burnout in a military emergency department: A cross-sectional Professional Quality of Life Scale V Survey study. *Military Medicine, 181*(8), 730–734.

Csikszentmihalyi, M. (1990). *Flow, the Psychology of Optimal Experience*. New York: Global Learning Communities.

Daudt, H. M., van Mossel, C., Scott, S. J. (2013). Enhancing the scoping study methodology: A large, inter-professional team's experience with Arksey and O'Malley's framework. *BMC Medical Research Methodology, 13*(1), 48.

Delahaij, R., Kamphuis, W., & van den Berg, C. E. (2016). Keeping engaged during deployment: The interplay between self-efficacy, family support, and threat exposure. *Military Psychology, 28*(2), 78–88.

De Lange, A. H., Taris, T. W., Kompier, M. A. J., Houtman, I. L. D., & Bongers, P. M. (2003). The very best of the millennium: Longitudinal research and the demand- control-(support) model. *Journal of Occupational Health Psychology, 8*(4), 282–305.

Demerouti, E., Bakker, A. B., Nachreiner, F., & Schaufeli, W. B. (2001). The job demands-resources model of burnout. *Journal of Applied Psychology, 86*, 499–512.

Demerouti, E., Geurts, S. A. E., Bakker, A. B., & Euwema, M. (2004). The impact of shiftwork on work-home interference, job attitudes and health. *Ergonomics, 47*, 987–1002.

Espinoza-Parra, S., Molero, F., & Fuster-Ruizdeapodaca, M. J. (2015). Transformational leadership and job satisfaction of police officers (carabineros) in Chile: The mediating effects of group identification and work engagement. *Revista de Psicología Social, 30*(3), 439–467.

Fredrickson, B. L. (2003). Positive emotions and upward spirals in organizations. In K. Cameron, J. Dutton & R. Quinn (Eds.), *Positive Organizational Scholarship* (pp. 163–175). San Francisco: Berrett Koehler.
Freedy, J. R., & Hobfoll, S. E. (1994). Stress inoculation for reduction of burnout: A conservation of resources approach. *Anxiety, Stress, & Coping, 6*, 311–325.
Halbesleben, J. R. B. (2010). A meta-analysis of work engagement: Relationships with burnout, demands, resources, and consequences. In A. B. Bakker and M. P., Leiter (Eds.), *Work Engagement: A Handbook of Essential Theory and Research*, (pp. 102–111). New York: Psychological Press.
Hannah, S. T., & Sowde, W. J. (2012). Leadership in the profession of arms. In M. Rumsey (Ed.), *Oxford Handbook of Leadership* (pp. 291–310). Oxford University Press: New York.
Harrington, D., Bean, N., Pintello, D., & Mathews, D. (2001). Job satisfaction and burnout: Predictors of intentions to leave a job in a military setting. *Administration in Social Work, 25*(3), 1–16.
Harter, J. K., Schmidt, F. L., & Hayes, T. L. (2002). Business-unit-level relationships between employee satisfaction, employee engagement, and business outcomes: A meta-analysis. *Journal of Applied Psychology, 87*, 268–279.
Harms, P. D., Bai, Y., & Han, G. (2016). How leader and follower attachment styles are mediated by trust. *Human Relations, 69*, 1853–1876.
Häusser, J. A., Mojzisch, A., Niesel, M., & Schulz-Hardt, S. (2010). Ten years on: A review of recent research on the job demand–control (-support) model and psychological well-being. *Work & Stress, 24*, 1–35.
Hobfoll, S. E. (1988). *The Ecology of Stress*. Washington, DC: Hemisphere.
Hobfoll, S. E. (1989). Conservation of resources: A new attempt at conceptualizing stress. *American Psychologist, 44*, 513–524.
Hobfoll, S. E. (2002). Social and psychological resources and adaptation. *Review of General Psychology, 6*(4), 307–332.
Hobfoll, S. E., Freedy, J., Lane, C., & Geller, P. (1990). Conservation of social resources: Social support resource theory. *Journal of Social and Personal Relationships*, 7, 465–478.
Ivey, G. W., Blanc, J.-R., & Mantler, J. (2015). An assessment of the overlap between morale and work engagement in a nonoperational military sample. *Journal of Occupational Health Psychology, 20*(3), 338–347.
Jiang, K., Hong, Y., McKay, P. F., Avery, D. R., Wilson, D. C., & Volpone, S. D. (2015). Retaining employees through anti–sexual harassment practices: Exploring the mediating role of psychological distress and employee engagement. *Human Resource Management, 54*, 1–21.
Kahn, W. A. (1990). Psychological conditions of personal engagement and disengagement at work. *Academy of Management Journal, 33*, 692–724.
Kelloway, E. K., & Francis, L. (2013). Longitudinal research and data analysis. In L. E. Tetrick, M. Wang, & R. R. Sinclair (Eds.), *Research Methods in Occupational Health Psychology: Measurement, Design, and Data Analysis* (pp. 374–393). New York, NY: Routledge.
Kok, B. C., Herrell, R. K., Grossman, S. H., West, J. C., & Wilk, J. E. (2016). Prevalence of professional burnout among military mental health service providers. *Psychiatric Services, 67*, 137–140.
Lang, G. M., Patrician, P., & Steele, N. (2012). Comparison of nurse burnout across Army Hospital practice environments. *Journal of Nursing Scholarship, 44*(3), 274–283.
Lang, G. M., Pfister, E. A., & Siemens, M. J. (2010). Nursing burnout: Cross-sectional study at a large army hospital. *Military Medicine, 175*(6), 435–441.
Lee, R. L., & Ashforth, B. (1996). A meta-analytic examination of the correlates of the three dimensions of job burnout. *Journal of Applied Psychology, 81*(2), 123–133.
Liberati, A., Altman, D. G., Tetzlaff, J., Mulrow, C., Gøtzsche, P. C., Ioannidis, J. P., … Moher, D. (2009). The PRISMA statement for reporting systematic reviews and meta-analyses of studies that evaluate health care interventions: Explanation and elaboration. *PLoS Medicine, 6*(7), e1000100.
Luchman, J. N., & González-Morales, M. G. (2013). Demands, control, and support: A meta-analytic review of work characteristics interrelationships. *Journal of Occupational Health Psychology, 18*(1), 37–52.
Macik-Frey, M., Quick, J. C., & Nelson, D. L. (2007). Advances in occupational health: From a stressful beginning to a positive future. *Journal of Management, 33*, 809–840.
Mäkikangas, A., Feldt, T., Kinnunen, U., & Mauno, S. (2013). Does personality matter? Research on individual differences in occupational well-being. In A. B. Bakker (Ed.), *Advances in Positive Organizational Psychology* (*Vol. 1*, pp. 107–143). Bingley, UK: Emerald.
Mäkikangas, A., Kinnunen, U., Feldt, T., & Schaufeli, W. (2016). The longitudinal development of employee well-being: A systematic review. *Work & Stress, 30*, 46–70.
Marques-Pinto, A., & Chambel, M. J. (2008). *Burnout e engagement em context organizacional. Estudos com amostras portuguesas. [Burnout and engagement in organizational context. Studies with Portuguese samples]*. Lisbon, Portugal: Livros Horizonte.

Maslach, C., Jackson, S. E., & Leiter, M. P. (1996). *Maslach Burnout Inventory Manual* (3rd edition). Palo Alto, CA: Consulting Psychology Press.

Maslach, C., Schaufeli, W. B., & Leiter, M. P. (2001). Job burnout. *Annual Review of Psychology, 52*, 397–422.

Matthew, G., Meek, D., Gibbs, T., Sawford, H., Wessely, S., & Greenberg, N. (2015). What are the psychological effects of delivering and receiving high-risk survival resistance training? *Military Medicine, 180*(2), 168–177.

May, D. R., Gilson, R. L., & Harter, L. M. (2004). The psychological conditions of meaningfulness, safety and availability and the engagement of human spirit at work. *Journal of Occupational and Organizational Psychology*, 77(1), 11–37.

Moran, D. S., Israeli, E., Evans, R. K., Yanovich, R., Constantini, N., Shabshin, N., ... Finestone, A. (2008). Prediction model for stress fracture in young female recruits during basic training. *Medicine & Science in Sports & Exercise, 40*(11 Suppl), S636–S644.

Morgan, C. A., Cho, T., Hazlett, G., Coric, V., & Morgan, J. (2002). The impact of burnout on human physiology and on operational performance: A prospective study of soldiers enrolled in the combat diver qualification course. *Yale Journal of Biology and Medicine, 75*, 199–205.

Morgan, C., Russell, B., McNeil, J., Maxwell, J., Snyder, P., Southwick, S., & Pietrzak, R. (2011). Baseline burnout symptoms predict visuospatial executive function during survival school training in special operations military personnel. *Journal of the International Neuropsychological Society, 17*(3), 494–501.

Motyka, B. (2018). Employee engagement and performance: A systematic literature review. *International Journal of Management and Economics, 54*(3), 227–244.

Paré, G., Trudel, M.-C., Jaana, M., & Kitsiou, S. (2015). Synthesizing information systems knowledge: A typology of literature reviews. *Information & Management, 52*, 183–199.

Patrician, P. A., Shang, J., & Lake, E. (2010). Organizational determinates of work outcomes and quality care ratings among Army Medical Department registered nurses. *Research in Nursing and Health, 33*(2), 99–110.

Quick, J. C., Joplin, J. R., Nelson, D. L., Mangelsdorff, A. D., & Fiedler, E. (1996). Self- reliance and military service training outcomes. *Military Psychology, 8*(4), 279–293.

Sargent, P., Millegan, J., Delaney, E., Roesch, S., Sanders, M., Mak, H., Mallahan, L., Raducha, S., & Webb-Murphy, J. (2016). Health care provider burnout in a United States military medical center during a period of war. *Military Medicine, 181*(2), 136–142.

Schaufeli, W., & Salanova, M. (2007). Work engagement: An emerging psychological concept and its implications for organizations. In S. W. Gilliland, D. D. Steiner, & D. P. Skarlicki (Eds.), *Research in Social Issues in Management (Vol. 5): Managing Social and Ethical Issues in Organizations* (pp. 135–177). Greenwich, CT: Information Age Publishers.

Schaufeli, W., Salanova, M., González-Romá, V., & Bakker, A. B. (2002). The measurement of engagement and burnout: A two sample confirmatory factor analytic approach. *Journal of Happiness Studies, 3*, 71–92.

Schaufeli, W. B., & Bakker, A. B. (2004). Job demands, job resources and their relationship with burnout and engagement: A multi-sample study. *Journal of Organizational Behavior, 25*, 293–315.

Schaufeli, W. B., & Buunk, B. P. (2003). Burnout: An overview of 25 years of research and theorizing. In M. J. Schabracq, J. A. M. Winnubst, & C. L. Cooper (Eds.), *Handbook of Work and Health Psychology* (pp. 383–425). Chichester: Wiley.

Schaufeli, W. B., & Enzmann, D. (1998). *The Burnout Companion to Study and Practice: A Critical Analysis*. Philadelphia: Taylor & Francis.

Schaufeli, W. B., & Taris, T. W. (2014). A critical review of the job demands-resources model: Implications for improving work and health. In G. Bauer & O. Hämmig (Eds), *Bridging Occupational, Organizational and Public Health: A Transdisciplinary Approach* (pp. 43–68). Dordrecht: Springer.

Seidler, A., Thinschmidt, M., Deckert, S., Then, F., Hegewald, J., Nieuwenhuijsen, K., & Riedel-Heller, S. G. (2014). The role of psychosocial working conditions on burnout and its core component emotional exhaustion: A systematic review. *Journal of Occupational Medicine and Toxicology, 9*, 10.

Seligman, M. E., & Csikszentmihalyi, M. (2000). Positive psychology. An introduction. *American Psychologist, 55*(1), 5–14.

Serec, M., Bajec, B., Petek, D., Švab, I., & Selič, P. (2012). A structural model of burnout syndrome, coping behavior and personality traits in professional soldiers of the Slovene armed forces. *Zdravniski Vestnik - Slovenian Medical Journal, 81*, 326–336.

Shelley, J. J., & Wong, M. (1991). Prevalence of burnout among military dentists. *Military Medicine, 156*(3), 113–118.

Simons, B. S., Foltz, P. A., Chalupa, R. L., Hylden, C. M., Dowd, T. C., & Johnson, A. E. (2016). Burnout in U.S. military orthopaedic residents and staff physicians. *Military Medicine, 181*(8), 835–839.

Souza, L. A. S., Torres, A. R. R., Barbosa, G. A., Lima, T. J. S., & Souza, L. E. C. (2015). Bem-estar subjetivo e burnout em cadetes militares: O papel mediador da autoeficácia. [Subjective well-being and burnout in military cadets: The mediating role of self-efficacy]. *Psychology/PsicologiaReflexão e Crítica, 28*(4), 744–752.

Smith-MacDonald, L., Norris, J. L., Raffin-Bouchal, S., & Sinclair, S. (2017). Spirituality and mental well-being in combat veterans: A systematic review. *Military Medicine, 182*, 11/12:e1920.

Swider, B. W., & Zimmerman, R. D. (2010). Born to burnout: A meta-analytic path model of personality, job burnout, and work outcomes. *Journal of Vocational Behavior, 76*(3), 487–506.

Taghva, A., Imani, M., Kazemi, M. R., & Shiralinia, K. (2015). The role of burnout and depression in self-destructive behavior of the Islamic Republic of Iran Army personnel. *Archives of Psychiatry and Psychotherapy, 2*, 41–47.

Taris, T. W. (2006). Is there a relationship between burnout and objective performance? A critical review of 16 studies. *Work & Stress, 20*(4), 316–334.

Taris, T. W., & Schaufeli, W. B. (2016). The Job Demands-Resources model. In S. Clarke, T. M. Probst, F. Guldenmund & J. Passmore (Eds.), *The Wiley Blackwell Handbook of the Psychology of Occupational Safety and Workplace Health* (pp. 157–180). Chichester: John Wiley.

Turner, N., Barling, J., & Zacharatos, A. (2002). Positive psychology at work. In C. R. Snyder & S. J. Lopez (Eds.), *Handbook of Positive Psychology* (pp. 715–728). New York: Oxford University Press.

van Wijk, C. (1997). Factors influencing burnout and job stress among military nurses. *Military Medicine, 162*(10), 707–710.

Varner, D. F., & Foutch, B. K. (2014). Depression and burnout symptoms among Air Force family medicine providers. *Journal of the American Academy of Physician Assistants, 27*(5), 42–46.

Vie, L. L., Scheier, L. M., Lester, P. B., & Seligman, M. E. P. (2016). Initial validation of the U.S. army global assessment tool. *Military Psychology, 28*(6), 468–487.

Vinokur, A. D., Pierce, P. F., & Lewandowski-Romps, L. (2009). Disentangling the relationships between job burnout and perceived health in a military sample. *Stress and Health, 25*, 355–363.

Vinokur, A. D., Pierce, P. F., Lewandowski-Romps, L., Hobfoll, S. E., & Galea, S. (2011). Effects of war exposure on air force personnel's mental health, job burnout and other organizational related outcomes. *Journal of Occupational Health Psychology, 16*(1), 3–17.

Vogelgesang, G. R., Leroy, H., & Avolio, B. J. (2013). The mediating effects of leader integrity with transparency in communication and work engagement/performance. *The Leadership Quarterly, 24*(3), 405–413.

Walters, T. A., Matthews, E. P., & Dailey, J. I. (2014). Burnout in Army health care providers. *Military Medicine, 179*(9), 1006–1012.

Whealin, J. M., Batzer, W. B., Morgan, C. A., Detwiler, H. F., Schnurr, P. P., & Friedman M. J. (2007). Cohesion, burnout, and past-trauma in tri-service medical and support personnel. *Military Medicine, 172*(3), 266–272.

Wrzesniewski, A., McCauley, C., Rozin, P., & Schwartz, B. (1997). Jobs, Careers, and Callings: People's relations to their work. *Journal of Research in Personality, 31*(1), 21–33.

Zapf, D., Dormann, C., & Frese, M. (1996). Longitudinal studies in organizational stress research: A review of the literature with reference to methodological issues. *Journal of Occupational Health Psychology, 1*, 145–169.

Zheng, D., Witt, L. A., Waite, E., David, E. M., van Driel, M., McDonald, D. P., … Crepeau, L. J. (2015). Effects of ethical leadership on emotional exhaustion in high moral intensity situations. *The Leadership Quarterly, 26*(5), 732–748.

# 21

# STRESS, BURNOUT AND COPING IN THE MILITARY ENVIRONMENT

*Yonel Ricardo de Souza and Fabio Biasotto Feitosa*

In 1936 Hans Selye (1936) was the first researcher to work with stress and defined it as "a non-specific organic response to stressful situations to the organism," conceptualizing this response as General Adaptation Syndrome. Rolhfs et al. (2005) define stress as the individual's perception of the imbalance between the physical and psychological demands and the resources to face them in an activity considered important. Selye's view was that stress was a defense of the body, perfected by the evolutionary process, enabling humans to adapt to menacing events. McEwen (2008) corroborates with Selye and adds that stress is "the pressure that life exerts on us and the way this pressure makes us feel." Arantes and Vieira (2002) also observed that the main characteristic of this mechanism is the fact that stress is linked to a specific agent to be produced and always combined with some external agent environment. Bianchi (2001) understands that stress represents the response of the organism, and its causative is defined as a stimulus or stressor agent.

The reaction to stress triggers the process of struggle or flight. McEwen (2008) says, then, stress is the organic reaction that occurs in the face of the stressors that stops being normal when the reactions to these events become chronic, and can generate illnesses. In this way, the mechanism of stress can be described from the perceived reactions of the organism to an emergency. Lehmann, Foster, Dickhuth and Gastmann (1998) show that, in the state generated by increased perception of stimuli, stress generates emotional excitation that, by altering homeostasis, triggers an adaptation process characterized by increased secretion of adrenaline, noradrenaline, cortisol and other hormones, leading to various systemic changes with physiological and psychological disorders. Thus, for a life event to be stressful, it needs to be perceived by the body as threatening or challenging (Ballone & Ortolani, 2007; Cohen, Kamarck & Mermelstein, 1983). To Folkman et al. (1986), it is the interaction between the environment and the individual who defines stress; that is, the individual feels stress when the demands from the environment exceed the resources that the individual has.

Bianchi (2001) warns that although the stress reaction may result in physiological or psychological disturbances, this reaction is what makes possible the survival and adaptation of living beings to the environmental stimuli to which they are constantly exposed. The stress reaction is part of a unified mind-body system; that is, the organic responses are not isolated but are closely related between the different systems, and for this reason prolonged exposure to the stressor stimulus can generate a major problem, causing illness of the organism (Arantes & Vieira, 2002). Rio (1995) divides stress into physical; psychic; by overload; by monotony and chronic stress, which the current literature conceptualizes as burnout.

Hans Selye (1951), in defining General Adaptation Syndrome, divided it into three phases: alert, resistance and exhaustion. The alert phase is considered the positive phase of stress (eustress), when the human being feels energized through increased production of adrenaline and usually reaches a feeling of fullness. At this stage, all bodily responses enter into a state of general readiness; that is, every organism is mobilized without specific or exclusive involvement of any particular organ. In the second phase (resistance), stress persists and the person automatically tries to deal with stressors in order to maintain their internal homeostasis. At this stage the body begins to become accustomed to the stimuli that cause stress. During this stage, the body adapts its reactions and its metabolism to support for a greater period of time. In this state the reaction can be channeled to a specific organ or to a certain system (cardiac, respiratory), skin, muscles and so on (Ballone & Ortolani, 2007).

If stressors persist in frequency or intensity, there is a breakdown in the person's resistance and it goes into the phase of distress when there is a marked drop in our adaptive capacity (Lipp, Pereira & Sadir, 2005). That is when adaptation mechanisms begin to fail and deficits in energy reserves arise. This phase is serious and may lead to death of some organisms. Most somatic and psychosomatic symptoms are most noticeable in this third phase. The state of exhaustion is the sum of the non-specific general reactions that develop as a result of prolonged exposure to stressors to which adaptation has been developed and, subsequently, the organism can't maintain it. At this stage, the biological modifications that appear resemble those of the alarm reaction (shock phase). However, the organism is no longer able to balance itself, leading to adaptive failure. The effects of chronic stress on the individual make up the pathophysiological substrate of psychosomatic diseases.

In the military field, several studies have addressed the incidence of stress phenomena from the perspective of work psychology (Leiter, Clark & Durup, 1994; Johnsen, Laberg & Eid, 1998; Deahl et al., 2000; Day & Livingstone, 2001). Wilcox, 1994 points out that stress is an integral part of military training. Van Heerden and De Beer (2014) add that the physical and psychological demands of the military profession are higher than most civilian professions, and that the military context is particularly favorable to high-stress stressors such as physical exhaustion, prolonged exposure to climate change, abstention from the home and exposure to dangerous situations.

It highlights, then, the fundamental importance of the environment and its interaction in the triggering of stress and its consequences in the military. The purpose of this chapter is therefore to discuss how the environment can influence the manifestations of stress and burnout, as well as to show which coping strategies seem to be more efficient in dealing with such demands. For this, the chapter is divided into three parts: the first seeks to address the relationship between physical stress and stress perceived in strenuous military environments, leading the reader to understand that their manifestations are not necessarily proportional and should have independent interpretations. The second part seeks to show that environmental factors tend to be more or less significant in the incidence of burnout in the military universe when exposed to an inhospitable environment. The third part tries to present which strategies of coping seem to be more effective in the fight against chronic stress in environments under strong psychological pressure, giving the reader access to the best psychological strategies to deal with such situations.

## Differences between physical and perceived stress

As discussed previously, stress is integral to everyday military life. The fundamental destination of the use of force is combat and, since it involves an imminent risk of death, constitutes one of the greatest stressors in existence. Therefore, even in basic training, the military is subjected to situations of potential stress. Basic training is designed to place the cadets in various forms of stress, both physical and psychological. The purpose of stress exposure in training is to replicate a similar situation of actual combat, where the military must act in a balanced way under strong psychological pressure (Wilcox, 1994;Van Heerden & De Beer 2014).Therefore, the most reliable observations in the literature regarding psychological (perceived) or physical stress in the military are from studies in strenuous combat or training environments.While some students are in better physical condition than others, there are mechanisms for everyone to experience stress, such as activity uniformity, hunger, sleep deprivation and thirst.Aiming to deepen research in this area of knowledge, several researches have been carried out.

A very broad, consistent, and general study of perceived military stress was that of Pflanz and Sonnek (2002). In this study, the prevalence and sources of occupational stress for military personnel and the relationship between work stress and emotional health in the military population were observed. The sample consisted of 472 active personnel from the U.S. Air Force who completed a survey of 65 items involving reported life events, perceptions about occupational stress and perceptions about the relationship between work stress and emotional health. As a result, the authors pointed out that military personnel were significantly more prone to suffering from stress than civilian workers; that 26% of the sample reported suffering from significant stress at work; that for 15%, stress at work was causing significant emotional distress and 8% reported that stress at work was severe enough to impair their emotional health. The authors concluded that generic stressors were more involved than specific stressors of the military.

It should be noted that the study of Pflanz and Sonnek (2002) did not involve real training or combat situations, where the exposed environment tends to raise stress levels considerably, as described by Smith et al. (2008) in their study on the appearance and persistence of symptoms of self-reported posttraumatic stress disorder in 50,814 U.S. military personnel who fought in the wars in Iraq and Afghanistan between 2001 and 2006. The findings indicated a threefold increase in the appearance of new symptoms or diagnoses of posttraumatic stress disorder after exposure in combat.

However, any relationship one might try to make between the psychological and physical stresses tends to be complex, first as to the approach: literature treats physical stress as a consequence, a biological view, focused on a non-specific physiological response, i.e. as a syndrome that consists of all the physiological changes that occur in the biological system when it is affected by a stimulus or by an excessive or harmful load. The expression nonspecific response means that the organism responds in a stereotyped manner to a wide variety of different stimuli or agents such as intoxication, nervous tension, heat, cold or muscle fatigue (Ribeiro & Marques, 2009). The perceived stress approach is process focused. It is the eminently psychological approach, focused on the dynamic interaction between the individual and the environment and on the subjective assessment of the stress perceived by the individual.

The main publications on the measurement of physical stress include the use of biochemical markers, such as cortisol (salivary or serum), creatine, kinase and lactate dehydrogenase, while the application of specific inventories, such as POLMS, BRUNEL-BRUMS or PS-14, is most often used in the measurement of perceived stress. It is very common to find studies that evaluate the associations between physical activity and (perceived) stress, such as an excessive reading of cortisol associated with psychological fatigue, converging the great majority of these works for significant positive relations between them, as the studies of Nguyen-Michel, Unger, Hamilton, and Spruijt-Metz (2006) and VanKim and Nelson (2013) do. However, comparison between these two stresses is rare, especially in the military environment.

In addition, studies show that perceived or psychological stress is associated with increased vulnerability to physical illness (Pedersen, Bovbjerg, Zachariae & Piscitelli, 2010), depression and physical symptoms (Cohen, 1986), reinforcing the view that the psychological and somatic dimensions of stress are interconnected, although they are independent. Córdova and Navas

(2000) cite behavioral responses associated with physical stress, such as mood changes, apathy, lack of motivation, sleep disturbances, irritability and depression. Thus, physical and perceived stress would influence military performance in situations under strong psychological pressure, such as strenuous training and actual combat operations, and affect the mental health of the military member and therefore of the family members.

When the focus is specifically on the relationships between perceived and physical stress, the most comprehensive research that addresses this relationship pertains to the psychology of sport. High-performance sport works with both demands and any imbalance between them can lead to loss of performance or illness of the athlete (Lehmann et al., 1998; Mannrich, 2007; Freitas, Miranda & Bara Filho, 2009; VanKim & Nelson, 2013).

The physical condition of the athlete (or military) in strenuous training is a preponderant factor for success. The physical requirements in this situation lead the subject not only to physical but behavioral and psychological limits, because they force him to control his emotional balance in the search for better results (Mannrich, 2007). Reviewing the literature, Freitas et al. (2009), when observing the relationship between exercise and psychological condition, concluded that intense exercise negatively affects mood. The accumulation of fatigue during intense periods of training combined with periods of inadequate recovery can lead to overtraining (Mannrich, 2007).

Garrett and Kirkendall (2003) understand that excessive training is due to the accumulation of stress that can be directly provoked by training and results in decreased performance, usually accompanied by psychological or physiological signals. The same authors cite that excess training is a process that can lead to persistence to fatigue, change in mood and frequent illnesses. Mannrich (2007) in his master's dissertation sought to relate biochemical markers and the psychological state of professional soccer athletes of the first division of Brazil in a longitudinal study with possibilities of injury. For this, he used the creatine kinase, lactate dehydrogenase, alanine transferase and aspartate transferase markers and the BRUM psychological scale. The author verified that the incidence of lesions had a significant and positive relationship with the increase in creatine and lactate values, as well as increased tension and fatigue and decreased vigor, concluding that the BRUM scale was a good psychological predictor of injuries (physical stress).

However, if we take stress as an imbalance between the physical and/or psychological demands and the resources to face them in an activity considered important, we can arrive at a common denominator from different approaches. It is known that stress can be physical, psychological, by overload or by monotony and chronic stress, and whatever the classification, prolonged exposure to the stressor stimulus can generate psychosomatic problems, causing illness of the organism.

The study by Souza and Feitosa (2017) is quite relevant in order to make a relation between the perceived and physical stresses in a strenuous military activity in the Brazilian Amazon region. In this study, 48 volunteer military personnel from the Brazilian Army were submitted to the PSS-14 inventory and blood samples were collected before and immediately at the end of the adaptation period to life in the jungle. Correlations were detected between all biochemical variables and significant and positive differences between pre- and post-collections. The mean CK post was outside the parameters of normality and consequent significant difference pre/post, which points to the manifestation of physical stress during the stage. No correlation was found between perceived pre- or post-stress and any biochemical variables related to physical stress. The results found in this study suggest that the perception of control and emotional stability alone can't be taken into account in resource management and decision-making in situations of evident wear.

The synthesis of what was presented in this first part highlights the influence of physical and perceived stress on the exercise of the military profession. The prevalence of both in all contexts, especially in combat training and employment situations, imposes on the masters of fractions the knowledge of the mechanisms of action stressors and how these can act differently from individual to individual. It is also necessary to understand that, although interdependent, physical and perceived stresses do not necessarily have a significant positive relation; that is, changing one factor will not necessarily compromise another depending on the subjectivity of each individual. Military training is an excellent tool for knowing the military's individuality and exercising resilience factors to stress, discussed in the third part of this chapter.

## Incidence of burnout in the military universe

One of the consequences of stress, or reactions arising in the presence of stress, is burnout. Burnout is understood as a psychological syndrome resulting from chronic emotional tension at work. It is a subjective internal experience that generates negative feelings and attitudes in the individual's relationship with their work (dissatisfaction, attrition, loss of commitment), undermining their professional performance and bringing undesirable consequences to the organization (absenteeism, abandonment of employment, low productivity). Burnout is characterized by emotional exhaustion, depersonalization and decreased personal achievement (Tamayo & Tróccoli, 2002). The same authors characterize the emotional exhaustion by the symptoms of fatigue, emotional imbalance and negativism. According to the same authors, depersonalization is manifested by the coldness and hardness in dealing with clients and co-workers. On the other hand, the decrease in professional achievement occurs in the form of feelings of incompetence and/or perceived unsatisfactory performance.

The causes of burnout are not yet precisely known, but according to Prins et al. (2007), there are three triggering conclusions:

- It results from chronic stress in the work environment and the lack of resources (psychological) to deal with the situation – emphasizes pressure and society.
- It results from the individual's inadequacy in the organizational environment (idealistic and highly motivated individuals): people who, realizing that their efforts are not enough to achieve the goals, give up or suffer from the situation – emphasizes personal and individual issues.
- It results from the person's inadequacy with the work environment – emphasizes interactions between the individual and the social environment.

Malach, Schaufeli and Leiter (2001) point out that organizations that maintain a well-established hierarchy level, which are very rule centered, are more susceptible to the development of burnout amongst their employees. It has already been emphasized in this chapter that the military context is more favorable to stressors of high demand, such as physical exhaustion, exposure to climatic changes, prolonged abstention from the home and exposure to dangerous situations (Van Heerden & De Beer, 2014). Different authors (Alarcon, Lyons & Tartaglia, 2010; Chambel, & Oliveira-Cruz, 2010; Ivie & Garland, 2011) agree that the military has a dangerous and stressful activity level, a factor that is improving the studies on burnout and engagement in this group. In the case of the manifestation of burnout in a military environment, the literature is broad and many studies address the issue.

Leiter et al. (1994), in their study of 473 Canadian military personnel, found considerable differences in levels of burnout, organizational impairment and psychosomatic symptoms between men and women, indicating that women are more susceptible to diseases and reactions of impotence when confronted with organizational problems. Deahl et al. (2000) studied the relationship between participation in armed conflict and psychiatric morbidity in 106 British soldiers when they returned from Yugoslavia, concluding the high incidence of psychiatric morbidity is not an inevitable consequence of armed conflict.

Quite representative is the study by Chambel and Oliveira-Cruz (2010), a longitudinal study of 387 individuals on a peace mission, which analyzed the impact of Army missions on burnout levels and soldiers' engagement. The authors found that during the mission, burnout levels increased, while levels of professional engagement decreased significantly. The finding showed that there may be an association between the type of professional function exercised, the overload of work activities and the levels of burnout in the military.

A detailed evaluation of the most prevalent factors in the military context in burnout is the work of Souza, Feitosa and Bezerra (in press) on Brazilian career soldiers who served in the Brazilian Amazon region. This study used a sample of 122 volunteer officers, sergeants and sergeants who served in the garrison of Porto Velho who filled out a questionnaire with socio-demographic, labor, physical and social information and a Burnout Inventory (MBI). The results showed that the military in the region had medium to high levels of burnout, according to the classification of the international evaluation protocol. It was verified that the main socio-demographic, labor, physical and social conditions that in the present study were associated with a greater perception of burnout were: shorter service time, sedentary lifestyle, extra work, lower hierarchical rank and living away from relatives. Due to the fact that it has not yet been published, the tables of the study by Souza et al. (in press) with the most significant indexes in the incidence of burnout follow (Tables 21.1–21.4).

All of the above studies are focused on the process approach. Reinforcing the information presented in the first part of this chapter, this approach is eminently psychological, focused on the dynamic interaction between the individual and the environment and on the subjective evaluation of the stress perceived by the individual. For Lazarus and Folkman (1984), it is the

*Table 21.1* Burnout dimensions within the objective variables

| *Variable* | *Emotional exhaustion* | | *Job satisfaction* | | *Depersonalization* | |
|---|---|---|---|---|---|---|
| | *Mean/SD* | *Rating* | *Mean/SD* | *Rating* | *Mean/SD* | *Rating* |
| General Sample | 25,09 ± 6,71 | Medium | 29.08 ± 5.29 | High | 11.07 ± 3.84 | Medium |
| Operational Personnel[a] | 25.54 ± 6.59 | Medium | 28.04 ± 5.66 | High | 11.33 ± 4.26 | Medium |
| Administrative Personnel[a] | 24.76 ± 6.85 | Medium | 29.86 ± 4.91 | High | 10.87 ± 3.53 | Medium |
| Officers[b] | 23.73 ± 5.80 | Medium | 31.27 ± 4.44* | High | 10.33 ± 3.63 | Medium |
| Sublieutenants/sergeants[b] | 25.53 ± 6.89 | Medium | 28.37 ± 5.27* | High | 11.30 ± 3.91 | Medium |
| Male[c] | 24.95 ± 6.84 | Medium | 30.04 ± 4.95 | High | 10.98 ± 3.50 | Medium |
| Female[c] | 22.20 ± 7.85 | Medium | 27.40 ± 4.57 | High | 9.40 ± 2.83 | Medium |
| Married/stable union[d] | 24.61 ± 6.62 | Medium | 29.38 ± 4.97 | High | 10.90 ± 3.71 | Medium |
| Single/widow(er)[d] | 27.83 ± 6.93 | High | 27.33 ± 6.12 | High | 12.00 ± 4.30 | Medium |
| Relatives live nearby[e] | 22.75 ± 6.65* | Medium | 29.46 ± 5.40 | High | 9.79 ± 3.24 | Medium |
| Relatives live far[e] | 25.66 ± 6.66* | Medium | 28.99 ± 5.28 | High | 11.38 ± 3.94 | Medium |
| Spouse shares the expenses[f] | 23.60 ± 6.82 | Medium | 29.71 ± 5.58 | High | 11.02 ± 3.71 | Medium |
| Spouse does not share the expenses[f] | 25.29 ± 6.41 | Medium | 29.21 ± 4.62 | High | 10.75 ± 3.79 | Medium |

*Note*: *Significant difference ($p < 0.05$) from the comparison group. Indexes of significance:
a *Emotional Exhaustion* ($p = 0.462$); *Job Satisfaction* ($p = 0.113$); *Depersonalization* ($p = 0.536$).
b *Emotional Exhaustion* ($p = 0.198$); *Job Satisfaction* ($p = 0.011$)*; *Depersonalization* ($p = 0.291$).
c Too reduced number of women to proceed to statistical analyses.
d *Emotional Exhaustion* ($p = 0.080$); *Job Satisfaction* ($p = 0.107$); *Depersonalization* ($p = 0.285$).
e *Emotional Exhaustion* ($p = 0.036$)*; *Job Satisfaction* ($p = 0.525$); *Depersonalization* ($p = 0.068$).
f *Emotional Exhaustion* ($p = 0.123$); *Job Satisfaction* ($p = 0.371$); *Depersonalization* ($p = 0.681$).

*Table 21.2* Burnout dimensions for years of service

| *Years of service* | *Emotional exhaustion* | | *Job satisfaction* | | *Depersonalization* | |
|---|---|---|---|---|---|---|
| | *Mean/SD* | *Rating* | *Mean/SD* | *Rating* | *Mean/SD* | *Rating* |
| Up to 5 years | 24.71 ± 8.11 | Medium | 28.00 ± 6.01* | High | 11.88 ± 4.39 | Medium |
| 6–10 years | 28.44 ± 4.69* | High | 27.19 ± 4.45 | High | 12.85 ± 3.87* | High |
| 11–15 years | 26.58 ± 6.77 | High | 27.83 ± 5.92 | High | 12.25 ± 4.18 | High |
| 16–20 years | 25.18 ± 5.03 | Medium | 28.64 ± 4.08 | High | 12.27 ± 2.67 | High |
| 21–25 years | 22.89 ± 6.81* | Medium | 29.19 ± 5.22 | High | 9.81 ± 3.35* | Medium |
| 26–30 years | 25.32 ± 7.72 | Medium | 30.53 ± 4.92 | High | 9.58 ± 2.87* | Medium |
| Over 30 years | 19.78 ± 5.26* | Medium | 35.67 ± 2.23* | Medium | 8.00 ± 3.89* | Medium |

*Note*: *Significant difference ($p < 0.05$) from the comparison group. Indexes of significance: *Emotional Exhaustion* ($p = 0.010$)*; *Job Satisfaction* ($p = 0.004$)*; *Depersonalization* ($p = 0.002$)*.

*Table 21.3* Burnout dimensions according to frequency of physical exercises

| *Frequency* | *Emotional exhaustion* | | *Job satisfaction* | | *Depersonalization* | |
|---|---|---|---|---|---|---|
| | *Mean/SD* | *Rating* | *Mean/SD* | *Rating* | *Mean/SD* | *Rating* |
| Sometimes | 28.33 ± 5.34* | High | 24.53 ± 5.12* | High | 13.73 ± 2.87* | High |
| Frequently | 26.00 ± 7.11 | Medium | 29.26 ± 5.26 | High | 11.43 ± 4.01 | Medium |
| Very often | 24.48 ± 6.79 | Medium | 28.76 ± 4.91 | High | 10.96 ± 3.67 | Medium |
| Always | 23.28 ± 6.43* | Medium | 30.80 ± 4.76* | High | 9.61 ± 3.47* | Medium |

*Note*: *Significant difference ($p < 0.05$) from the comparison group. Indexes of significance: *Emotional Exhaustion* ($p = 0.043$)*; *Job Satisfaction* ($p = 0.001$)*; *Depersonalization* ($p = 0.005$)*.

*Table 21.4* Burnout dimensions according to extra work hours

| *Frequency* | *Emotional exhaustion* | | *Job satisfaction* | | *Depersonalization* | |
|---|---|---|---|---|---|---|
| | *Mean/SD* | *Rating* | *Mean/SD* | *Rating* | *Mean/SD* | *Rating* |
| Never | 20.50 ± 6.05* | Medium | 31.07 ± 5.35 | Medium | 8.92 ± 3.90* | Medium |
| Sometimes | 24.12 ± 6.35* | Medium | 28.99 ± 4.75 | High | 10.49 ± 3.63* | Medium |
| Frequently | 29.46 ± 4.80* | High | 27.50 ± 5.58 | High | 13.08 ± 3.50* | High |
| Very often/always | 26.13 ± 7.61 | Medium | 30.04 ± 6.30 | High | 12.13 ± 4.05 | High |

*Note*: * Significant difference ($p < 0.05$) from the comparison group. Indexes of significance: *Emotional Exhaustion* ($p = 0.000$)*; *Job Satisfaction* ($p = 0.254$); *Depersonalization* ($p = 0.003$)*.

interaction between the environment and the individual that defines stress. This perspective also fits for burnout. Because of this, psychological, environmental or social stressors only have the ability to trigger a reaction to stress, even if it becomes chronic. It will, however, be mediated through the coping capacity that the person has learned to employ during his or her life history to the way each individual has created the conditions to handle new or adverse situations.

## Coping in military environments

Coping was first defined by Lazarus and Folkman (1984) as a set of cognitive and behavioral efforts used by individuals to deal with specific internal or external demands that arise in stress situations and are assessed as overloading or exceeding your personal resources. It is assumed that coping strategies are deliberate actions that can be learned, used and discarded. Therefore, unconscious and unintentional defense mechanisms, such as negation, displacement and regression, cannot be considered coping strategies. In addition, according to Ryan-Wenger (1992), somatization, domination and competence are seen as results of coping efforts rather than as strategies.

Antoniazzi, Dell'Aglio and Bandeira (1998) cite that the coping model of Lazarus and Folkman (1984) is based on four concepts:

1. Coping is a process that takes place between the individual and the environment.
2. The coping function is the management of the stress situation, not the domain of the stress.
3. The coping processes presuppose evaluation, that is, perception, interpretation and cognitive representation.
4. The process is a mobilization of effort in order to manage the internal and external demands that arise from interaction with the environment.

Although the conceptual difference of coping styles and coping strategies is not consensual, Antoniazzi et al. (1998) make a distinction between the two: coping styles are related to personality traits or results and are reflected in tendencies to use a coping reaction to a greater or lesser degree in the face of a stressful situation. Coping strategies are cognitive or behavioral actions taken during a stress episode, linked to situational factors, and reflect actions, behaviors or thoughts used to deal with a stressor.

Lazarus and Folkman (1984) also point out that coping can be understood from two types of strategies: problem-focused strategies and strategies focused on emotion. The first type aims to analyze and define the situation, considering its costs and benefits and seeking alternatives to solve it. This type includes strategies that affect the environment and the subject. Strategies that affect the environment try to modify the stressor through changes in external pressures, obstacles, resources, procedures etc. The strategies that affect the subject include changes in the individual's aspirations, reduction of the participation of the self, search for alternative participation channels, development of new patterns of behavior and learning new procedures and resources.

Emotion-centered strategies refer to the cognitive processes responsible for reducing the emotional distress generated by a stressful situation. These strategies are used by individuals when they realize that stressors cannot be modified and that it is necessary to continue interacting with them. The purpose is to maintain hope and optimism, to deny both the situation and its consequences or to act as if the circumstances do not matter. These strategies include avoidance, guilt, escape, detachment, selective attention, positive comparisons and the extraction of positive aspects from negative events (Lazarus & Folkman, 1984).

In response to different types of stress or different moments in time, the influence of the coping strategy being problem focused or emotion focused may vary. According to Carver and Scheier (1994), at first these two categories are easily perceived, but their effects are not so clear. Emotion-focused coping can facilitate problem-focused coping by removing tension, and, similarly, problem-focused coping can lessen the threat by reducing emotional tension. However, the way in which the individual uses coping is determined in part by his resources, which include health and energy, existential beliefs, problem-solving skills, social skills, social support and material resources. Coping can also be determined by variables that reduce the use of personal resources, such as environmental (Savóia, Santana & Mejias, 1996).

Rudolph, Dennig and Weisz (1995) understand that coping strategies are part of a process, influenced by many variables. In this process, two concepts are involved: the moderators and the mediators. Moderators are pre-existing variables that affect the direction or intensity of the relationship between independent and dependent variables. In general, moderators reflect the characteristics of the individual, such as level of development, gender, experience and temperament; of the context, as paternal influence and the interaction between them. Mediators are mechanisms through which the independent variable is able to influence the dependent variable, such as cognitive evaluation and development of attention. Beresford (1994) classifies these variables into physical and psychological ones that include physical, moral, ideological beliefs, previous coping experiences, intelligence and other personal characteristics.

Lazarus and Folkman (1984) divide coping strategies into eight factors:

1. Confrontation, when the individual decides to face the stressful situation, not moving away from the situations of risk
2. Removal, when the individual distances himself from the possible stressful situations at the moment in which they occur
3. Self-control, when energy is focused on maintaining emotional balance until the stressful situation passes
4. Social support, defined as a resource where the individual seeks external affective support, such as friends, colleagues or family
5. Acceptance of responsibility, when the individual understands that the stressful situation is the result of their actions and accepts its consequences
6. Escape-avoidance, when the individual avoids the appearance of possible risk situations
7. Problem solving, which consists of the application of rational techniques to solve the stressful situation
8. Positive reappraisal, that is, taking advantage of the lessons provided by the stressful situation that the subject is experiencing at that moment

After the classification in 1984, Folkman and Lazarus (1988) separate emotion-focused strategies: self-control, social support, acceptance of responsibility and positive reassessment, from strategies focused on the problem: confrontation, withdrawal, escape-avoidance and problem solving. On the other hand, Tap, Costa and Alves (2005) adopt six coping strategies: focusing (focusing on the problem and how to solve it), social support (desire or need for help), retraining (rupture of social interactions, withdrawal), conversion (change in behaviors, cognitive positions or values), control (coordination of behaviors, containment of emotions) and refusal (inability to perceive and accept the reality of the situation). These same authors present two categories of coping: coping perceived as positive (e.g. control, social support and focus) and coping perceived as negative (e.g. withdrawal and refusal).

In the military field, the literature is quite comprehensive on coping strategies. The reason is obvious: coping seems to be the solution to the acute and chronic stress problems explored so far in this chapter during the exercise of the military profession. However, as coping strategy and style also depend on life experience, there are no ready-made formulas for each individual in the context of the same stressful situation.

Johnsen et al. (1998) studied the effects of different coping strategies and mental health problems in a context of radical changes in a military organization in Norway, pointing out that subjects who focused on the coping strategy presented greater symptoms of mental health problems. Day and Livingstone (2001) examined the positive and negative impact of different coping styles on acute and chronic stress in 521 Canadian military personnel based on self-reported health symptoms. After finding that the items ambiguity, overload and lack of stimulation at work were the only ones associated with increased symptoms, it was verified that only negative coping styles (withdrawal and refusal) were directly related to the associated symptoms.

Deahl et al. (2000) studied the relationship between participation in armed conflict and psychiatric morbidity in 106 British soldiers when they returned from Yugoslavia, where they all received, before the mission, an operational stress training package. High scores of the assessment instruments pointed to significant alcohol abuse and withdrawal behavior (negative coping). However, the study concluded that the high incidence of psychiatric morbidity is not an inevitable consequence of armed conflict.

Overdale and Gardner (2012) examined how coping adaptability and social support strategy influence responses of recruits in training in 2,293 New Zealand Defense Force personnel. Correlating perceived social support, adaptability, coping, perceived difficulty, self-reported performance and sense of integration to the military corporation, and results pointed out that the support provided by the instructors was related to less difficulty in coping and greater performance and feeling of belonging to the Force. Support from external sources was related to greater adaptability to coping, performance and feeling of belonging to the Force and, paradoxically, increased difficulty in developing coping strategies. Adaptation to coping is provided by the beneficial effects of external support. In short, coping employability and social support viability are key tools for success in military training.

Motivated by the challenge posed to the candidate at high physical and mental levels during extremely severe stress conditions, Van Heerden and De Beer (2014) studied motivational and coping responses in the process of selecting 73 military personnel for the Special Forces course in South Africa for four weeks. To do this, they compared the constructs of: sense of consistency, hardiness, control assignment and self-efficacy between selected and unselected ones and explored what was considered important to success in the selection process. The researchers did not find significant differences between the two groups, although all the candidates presented high levels of sense of coherence and self-efficacy and medium levels of hardiness and attribution of control. In addition to the results, the authors pointed out that the use of coping strategies contributes to the management and improvement of military well-being, especially in severe stress conditions, where aspects such as hardiness and physical and mental resilience are preponderant.

Souza and Feitosa (2015) carried out a study with students of the Course of Operations in the Jungle of the Brazilian Army, whose objective was to investigate which coping strategies can predict success or failure. To that end, the sample consisted of 36 volunteer soldiers (18 dropouts and 18 graduates) out of a total of 63 candidates (57.14%), all men, officers and career sergeants. Everyone in the "graduating" group had scores above the "dropouts" group. However, significant differences were observed only in the strategies of self-control and positive re-evaluation. In general, emotion-focused strategies were prominent in relation to problem-focused strategies. The findings of this study suggest that, in situations of intensive military training, coping strategies focused on emotion, mainly self-control and positive reassessment, may act positively on the individual's performance, possibly making him more resilient, aiding in the success of the training.

Several studies support the significant difference of self-control strategies (Clemons, 1996; Everly, McCormack & Strouse, 2012) and positive reassessment (Meredith, Sherbourne & Gaillot, 2011; Gerzina & Porfeli, 2012), which were more emphasized in the previous studies, indicating that such strategies contribute to building the military's resilience. In turn, resilience is a fundamental virtue in the training of the military, it is trainable, assists in the stabilization of psychological stress levels and is directly related to success in military training and operations (Escolas, Pitts, Safer & Bartone, 2013; McGeary, 2011; Meredith et al., 2011; Nash, Steenkamp, & Conoscenti, 2011).

Riolli and Savicki (2010) evaluated the coping strategies of U.S. soldiers who had recently arrived in Iraq and found that individuals whose strategies were more focused on emotion had lower psychological symptoms such as depression. Kumar, Parkash and Mandal (2013) cite that the military should develop coping strategies focused on emotion in order to avoid frustration, cognitive disorientation and aggravation of stress levels that decisively influence performance. The studies of Riolli and Savicki (2010), Kumar et al. (2013) and de Souza and Feitosa (2015) bring up an interesting aspect: the greater efficiency of emotion-oriented strategies when compared to problem-oriented strategies. It seems that the focus on internal issues of the psyche, such as self-control, are most relevant in situations where the focus of the problem cannot be manipulated or avoided, limiting the use of some strategies focused on the problem, such as withdrawal or escape.

## Final considerations

Once considered fundamental in the survival of the species, stress has been prevalent in the military profession and, when it becomes chronic, can bring problems for the mental health of the military and family. The purpose of this chapter was to introduce the reader to an understanding of the mechanisms of the action of stress and its prevalence in the military environment. The reader can also understand that although there are more stressful environments than others, it is the interaction of the subject with the environment that is the main stress-triggering factor. In the military milieu, physical and psychological stresses are distinct, although interdependent as to the psychosomatic responses in the individual. Caution is advised in evaluating the military's stress, always taking into consideration the individuality of each one.

From the beginnings of humanity, stress has been an important factor for the survival of the species. Nowadays, especially in the exercise of the military profession, stress is still crucial, especially in real situations of the use of troops. The chronification (third stage) of stress, widely known in the literature as burnout, can lead to illness of military personnel and loss of functional performance. The most obvious signs that lead to burnout were presented in this chapter, as well as coping strategies that can be developed and trained in simulations of troop employment to protect the military from the harmful effects of burnout. The most efficient strategies seem to be those focused on emotion, so the advice of professionals specialized in the development of such strategies seems to be a valid investment in preserving the health of the military public.

## References

Alarcon, G., Lyons, J. B., & Tartaglia, F. (2010). Understanding predictors of engagement within the military. *Military Psychology, 22*(3), 301–310.

Antoniazzi, A. S., Dell'Aglio, D. D., & Bandeira, D. R. (1998). O conceito de coping: uma revisão teórica (The concept of coping: A theoretical revision). *Estudos de psicologia, 3*(2), 273–294.

Arantes, M. A. D. A., Vieira, M. J. F. (2002). *Estresse (Stress)*. São Paulo: Casa do Psicólogo.

Ballone, G. J., & Ortolani, I. V. (2007). *Da emoção à lesão: um guia de medicina psicossomática (From emotion to injury: A guide to psychosomatic medicine)*. São Paulo: Manole.

Beresford, B. A. (1994). Resources and strategies: How parents cope with the care of a disabled child. *Journal of Child Psychology and Psychiatry, 35*(1), 171–209.

Bianchi, E. R. F. (2001). Conceito de stress: evolução histórica (Concept of stress: Historical evolution). *Nursing, 4*(39), 16–19.

Carver, C. S., & Scheier, M. F. (1994). Situational coping and coping dispositions in a stressful transaction. *Journal of Personality and Social Psychology, 66*(1), 184–195.

Chambel, M. J., & Oliveira-Cruz, F. (2010). Breach of psychological contract and the development of burnout and engagement: A longitudinal study among soldiers on a peacekeeping mission. *Military Psychology, 22*(2), 110–127.

Clemons, E. P. (1996). Monitoring anxiety levels and coping skills among military recruits. *Military Medicine, 161*(1), 18–21.

Cohen, S. (1986). Contrasting the hassles scale and the perceived stress scale: Who's really measuring appraised stress? *American Psychologist, 41*(6), 716–718.

Cohen, S., Kamarck, T., & Mermelstein, R. (1983). A global measure of perceived stress. *Journal of Health and Social Behavior, 24*(4), 385–396.

Córdova, A., & Navas, F. J. (2000). Os radicais livres e o dano muscular produzido pelo exercício: papel dos antioxidantes (Free radicals and muscle damage produced by exercise: The role of antioxidants). *Revista Brasileira de Medicina do Esporte*, *6*(5), 204–208.

Day, A. L., & Livingstone, H. A. (2001). Chronic and acute stressors among military personnel: Do coping styles buffer their negative impact on health? *Journal of Occupational Health Psychology*, *6*(4), 348–360.

Deahl, M., Srinivasan, M., Jones, N., Thomas, J., Neblett, C., & Jolly, A. (2000). Preventing psychological trauma in soldiers: The role of operational stress training and psychological debriefing. *British Journalof Medical Psychology*, *73*(1), 77–85.

Escolas, S. M., Pitts, B. L., Safer, M. A., & Bartone, P. T. (2013). The protective value of hardiness on military posttraumatic stress symptoms. *Military Psychology*, *25*(2), 116–123.

Everly Jr, G. S., McCormack, D. K., & Strouse, D. A. (2012). Seven characteristics of highly resilient people: Insights from Navy SEALs to the "Greatest Generation". *International Journal of Emergency Mental Health*, *14*, 87–93.

Folkman, S., & Lazarus, R. S. (1988). Coping as a mediator of emotion. *Journal of Personality and Social Psychology*, *54*(3), 466–475.

Folkman, S., Lazarus, R. S., Gruen, R. J., & DeLongis, A. (1986). Appraisal, coping, health status, and psychological symptoms. *Journal of Personality and Social Psychology*, *50*(3), 571–579.

Freitas, D. S. D., Miranda, R., & Bara Filho, M. (2009). Marcadores psicológico, fisiológico e bioquímico para determinação dos efeitos da carga de treino e do overtraining (Psychological, physiological and biochemical markers for determining the effects of training load and over training). *Revista Brasileira de Cineantropometria e Desempenho Humano*, *11*(4), 457–465.

Garrett, W. E., & Kirkendall, D. T. (2003). *A ciência do exercício e dos esportes (The Science of Exercise and Sports)*. Porto Alegre: Artmed.

Gerzina, H. A., & Porfeli, E. J. (2012). Mindfulness as a predictor of positive reappraisal and burnout in standardized patients. *Teaching and Learning in Medicine*, *24*(4), 309–314.

Ivie, D., & Garland, B. (2011). Stress and burnout in policing: Does military experience matter? *Policing: An International Journal of Police Strategies & Management*, *34*(1), 49–66.

Johnsen, B. H., Laberg, J. C., & Eid, J. (1998). Coping strategies and mental health problems in a military unit. *Military Medicine*, *163*(9), 599–602.

Kumar, U., Parkash, V., & Mandal, M. K. (2013). Stress in extreme conditions: A military perspective. In D. M. Pestonjee & S. Pandey (Eds.), *Stress and Work: Perspectives on Understanding and Managing Stress* (pp. 101–126). New Delhi: Sage.

Lazarus, R. S., & Folkman, S. (1984). *Stress, Appraisal, and Coping*. New York: Springer publishing company.

Lehmann, M., Foster, C., Dickhuth, H. H., & Gastmann, U. (1998). Autonomic imbalance hypothesis and overtraining syndrome. *Medicine and Science in Sports and Exercise*, *30*(7), 1140–1145.

Leiter, M. P., Clark, D., & Durup, J. (1994). Distinct models of burnout and commitment among men and women in the military. *The Journal of Applied Behavioral Science*, *30*(1), 63–82.

Lipp, M. E. N., Pereira, M. B., & Sadir, M. A. (2005). Crenças irracionais como fontes internas de stress emocional (Irrational beliefs as internal sources of emotional stress). *Revista Brasileira de Terapias Cognitivas*, *1*(1), 29–34.

Mannrich, G. (2007). *Perfil dos marcadores bioquímicos de lesões músculo esquelética, relacionado ao estado psicológico, em atletas profissionais de futebol*. Master's Dissertation. University of the State of Santa Catarina, Florianópolis, Santa Catarina, Brazil.

Maslach, C., Schaufeli, W. B., & Leiter, M. P. (2001). Job burnout. *Annual Review of Psychology*, *52*(1), 397–422.

McEwen, B. S. (2008). Central effects of stress hormones in health and disease: Understanding the protective and damaging effects of stress and stress mediators. *European Journal of Pharmacology*, *583*(2–3), 174–185.

McGeary, D. D. (2011). Making sense of resilience. *Military Medicine*, *176*(6), 603–604.

Meredith, L. S., Sherbourne, C. D., & Gaillot, S. J. (2011). *Promoting Psychological Resilience in the US Military*. Rand Corporation.

Nash, W. P., Steenkamp, M., Conoscenti, L., & Litz, B. T. (2011). The stress continuum model: A military organizational approach to resilience. In *Resilience and Mental Health: Challenges Across the Lifespan*. New York: Cambridge University Press.

Nguyen-Michel, S. T., Unger, J. B., Hamilton, J., & Spruijt-Metz, D. (2006). Associations between physical activity and perceived stress/hassles in college students. *Stress and Health: Journal of the International Society for the Investigation of Stress*, *22*(3), 179–188.

Overdale, S., & Gardner, D. (2012). Social support and coping adaptability in initial military training. *Military Psychology*, *24*(3), 312–330.

Pedersen, A. F., Bovbjerg, D. H., Zachariae, R., & Piscitelli, R. (2010). Stress and susceptibility to infectious disease. In R. J. Contrada & A. Baum (Eds.), *The Handbook of Stress Science: Biology, Psychology and Health* (pp. 425–446). New York, NY, US: Springer Publishing Company.

Pflanz, S., & Sonnek, S. (2002). Work stress in the military: Prevalence, causes, and relationship to emotional health. *Military medicine, 167*(11), 877–882.

Prins, J. T., Gazendam-Donofrio, S. M., Tubben, B. J., Van Der Heijden, F. M., Van De Wiel, H. B., & Hoekstra-Weebers, J. E. (2007). Burnout in medical residents: A review. *Medical Education, 41*(8), 788–800.

Ribeiro, J., & Marques, T. M. P. F. S. (2009). A avaliação do stresse: A propósito de um estudo de adaptação da escala de percepção de stresse (The evaluation of stress: The purpose of a study of adaptation of the scale of perception of stress). *Psicologia, Saúde & Doenças, 10*(2), 237–248.

Rio, R. P. (1995). *O fascínio do stress (The Fascination of Stress)*. Belo Horizonte: Del Rey.

Riolli, L., & Savicki, V. (2010). Coping effectiveness and coping diversity under traumatic stress. *International Journal of Stress Management, 17*(2), 97–113.

Rudolph, K. D., Dennig, M. D., & Weisz, J. R. (1995). Determinants and consequences of children's coping in the medical setting: Conceptualization, review, and critique. *Psychological Bulletin, 118*(3), 328–357.

Ryan-Wenger, N. M. (1992). A taxonomy of children's coping strategies: A step toward theory development. *American Journal of Orthopsychiatry, 62*(2), 256–263.

Savóia, M. G., Santana, P. R., & Mejias, N. P. (1996). Adaptação do inventário de Estratégias de Coping de Folkman e Lazarus para o português (Adapting the Folkman and Lazarus Coping Strategies Inventory to Portuguese). *Psicologia usp*, 7(1–2), 183–201.

Selye, H. (1936). A syndrome produced by diverse nocuous agents. *Nature, 1383479*, 32–32.

Selye, H. (1951). The general-adaptation-syndrome. *Annual Review of Medicine*, 2(1), 327–342.

Smith, T. C., Ryan, M. A., Wingard, D. L., Slymen, D. J., Sallis, J. F., & Kritz-Silverstein, D. (2008). New onset and persistent symptoms of post-traumatic stress disorder self reported after deployment and combat exposures: Prospective population based US military cohort study. *British Medical Journal, 3367640*, 366–371.

Souza, Y. R., & Feitosa, F. B. (2015). The impact of coping strategies on the operations in the jungle course of the Brazilian army. *Military psychology, 27*(6), 348–353.

SouzaY. R., & Feitosa, F. B. (2017). Risk of single appraisal in complex environment: Physical and perceived stress in military training on Amazon jungle. *Journal of Sports and Physical Education*, 4(3), 13–17.

Souza, Y. R., Feitosa, F. B., & Bezerra, G. L. in press. Burnout syndrome in Brazilian Army Officers in the Amazon region. *Administration and Policy in Mental Health and Mental Health Services Research*

Tamayo, M. R., & Tróccoli, B. T. (2002). *Burnoutno trabalho (Burnout at Work)*. Brasília: UnB/Finatec.

Tap, P., Costa, E. S., & Alves, M. N. (2005). Escala Toulousiana de Coping (ETC): estudo de adaptação à população portuguesa (Coping Toulousiana Scale (ETC): Study of adaptation to the Portuguese population). *Psicologia, Saúde & Doenças*, 6(1), 47–56.

Van Heerden, A., & De Beer, M. (2014). Exploring the role of motivational and coping resources in a Special Forces selection process. *SA Journalof Industrial Psychology, 40*(1), 1–13.

VanKim, N. A., & Nelson, T. F. (2013). Vigorous physical activity, mental health, perceived stress, and socializing among college students. *American Journalof Health Promotion, 28*(1), 7–15.

Wilcox, V. L. (1994). Burnout in military personnel. In F. D. Jones, L. R. Sparacino, V. L. Wilcox, & J. M. Rothberg (Eds.), *Military Psychiatry: Preparing in Peace for War* (pp. 31–50). Washington DC, USA: TMM Publications.

# 22

# STRESS EXPERIENCES AND ABILITIES TO COPE

## Civil population versus military personnel

*Vijay Parkash*

Work stress is becoming a priority issue for all organizations today due to its economic and human consequences. There is a growing belief in all spheres of public and private sectors that the experience of stress at work has undesirable consequences for both employees and organizations. This belief has been reflected both in public and media interest and in increasing concern voiced by the trade unions and professional and scientific bodies. This chapter gives an overview of how changing work environments have led to increasingly stressful job characteristics and how these characteristics have psychological implications for the military population. It is proposed that while current stress models present fruitful frameworks for stress research, their application in the military setting varies considerably. Many of the major theoretical models that depict the stress process are described, with particular attention paid to the most influential.

Although the stress response is an individualized experience, there are, without a doubt, a wide variety of pressures in today's work settings with the potential to generate adverse stress reactions (distress) for many personnel. Excessive demands associated with multiple roles can lead to role strain, characterized by the person feeling stressed or physically and emotionally drained (Barnett & Brennan, 1995). Adverse work experiences, particularly experiences that are uncertain or outside the employee's control, result in unpleasant emotional and physiological states (Judge & Colquitt, 2004). The most vulnerable to these adverse experiences are military personnel, who most of the time are deployed in situations that place high demands on them. Though stress during war or counterinsurgency operations is understandable, it is becoming apparent that service life is stressful even in the absence of such situations. Based on data from the United States military population, a prominent researcher in this field, Pflanz (2001) asserted that military personnel suffer from occupational stress; a majority reported suffering from significant work stress and more than 40% considered work stress a significant contributor to the onset of mental illness. Though various models and researches exist on the impact of stress on general populations across continents, there are very few studies highlighting the impact of stress on military personnel. The present article attempts to encapsulate the current stress models and the existing framework of research on stress in understanding the impact of stress on military personnel. Further, a new approach that combines many of the features of existing models is proposed, which includes strong roles for military training, organizational support, and subjective perceptions. It is suggested that the proposed model could be a useful new direction for stress research in military studies.

## Defining stress

The definition of stress is not simply a question of semantics; it is crucial that there exists a consensus on what can be understood as stress and the differentiation in the nature of stress for military and civil population. A lack of such agreement would seriously hamper research in this direction and the subsequent development of effective stress management strategies for military personnel. A lack of knowledge of the relevant scientific literature on stress has led to an unfortunate and popular misconception that a certain amount of stress is necessary for optimal performance. This belief has been widely used to justify poor management practices for decades and belies a lack of intensive research, as it presents stress as an indefinable and immeasurable construct. For instance, the landmark study by Selye (1976) differentiated between eustress and distress. Eustress is the positive and pleasant side of stress. On the other hand, distress refers to a destructive type of stress that depletes one's energy reserves and taxes the maintenance and defence of the bodily systems, potentially causing harm to both physical and psychological health. However, experts have not been able to set boundary conditions on when a particular amount of stress will continue being positive and when it will go beyond the threshold of positive stress and transform into negative stress. More so for a military population which is selected and trained to undergo acute stress on a daily basis, it is literally impossible to delineate boundary conditions for what will be considered eustress and/or distress. Consequently, a common-sense view of stress is weighted to the negative side.

Despite all the hyperbole surrounding various definitions of stress, it has been concluded in several different reviews that essentially and mainly there are three different types of conceptions of the nature of stress. The first approach is the 'engineering approach' where stress is seen as a stimulus or characteristic of the environment in the form of level of demand. Here occupational stress is an aversive characteristic of the work environment and is treated as an independent variable and environmental cause disrupting the well-being of employees. The second is the 'physiological approach' where the definition of stress is based upon the physiological or biological changes that occur in the person when they are in a stress state, that is, as a dependent variable and a physiological response to a threatening or damaging environment. The third approach conceptualizes work stress in terms of the dynamic interaction between the person and their work environment. This approach has been termed the 'psychological approach'.

The first two approaches are surrounded by specific criticisms at the empirical and conceptual level. Both engineering and physiological models do not adequately account for the existing data; that is, they ignore the mediation of strong cognitive as well as situational (context) factors in the overall stress process. Secondly, on a conceptual level, engineering and physiological models of stress follow a relatively simple stimulus-response paradigm and largely ignore individual differences of a psychological nature and the perceptual and cognitive processes that might underpin them. Therefore, both these approaches treat the person as a passive vehicle for translating the stimulus characteristics of the environment into psychological and physiological response parameters. They generally ignore the interactions between the person and their various environments, which are an essential part of systems-based approaches to biology and psychology. However, it is only the psychological approach which pays special attention to environmental factors and, in particular, to the psychosocial and organizational contexts of work stress. In this approach, stress is either inferred from the existence of problematic person-environment interactions or measured in terms of the cognitive processes and emotional reactions which underpin those interactions. The psychological viewpoint is perhaps the most popular conceptualization today and is considered superior (Cox & Griffiths, 1995), as it treats the individual as an active agent and takes into account the effects of cognitive and situational factors on performance and well-being.

This viewpoint has special relevance for the military population, as their operating environments undergo continuous change and their cognitive reengineering during training alters the way they perceive and cope with stress. To some extent, psychological models on stress attempt to overcome criticisms of earlier approaches (Mark & Smith, 2008).

Most contemporary stress theories are variants of the psychological approach, and the two most distinct among them are interactional and transactional. The interactional models focus on structural features of the person's interaction with their work environment, whereas the transactional models focus on the psychological mechanisms, such as a cognitive appraisal of stress and coping with stress, behind these person-environment interactions. It is the overall psychological approach which brings researchers together to a consensus on defining stress. Moreover, psychological models attempt to provide suitable answers to stress issues in both civilian and military population, but in the case of the latter, the propositions are barely empirically tested and therefore the similar applicability of principles still remains questionable. Taking a lead from various stress models integratively reviewed by Mark and Smith (2008), the following sections outline a selection of key frameworks important in guiding research and practice in the area to discern how application of various approaches to stress varies for civil and military population.

## Interactional approaches

The interactive approach features how structural characteristics of the person's interaction with their work environment lead to stress.

### *Person-environment fit model (French, 1973)*

Professor John French (1973) formulated a theory of work stress based on the explicit concept of person-environment interaction, which is an extension of Lewin's (1951) concept suggesting that an individual's personal characteristics interact with their work environment to determine strain and consequently behaviour and health. Two basic facets of fit were identified, first the degree to which an employee's attitudes and abilities meet the demands of the job and second the extent to which the job environment fulfils the employee's needs, for example, to what extent an individual is encouraged and permitted to use their knowledge and skill set at work. Lack of fit in either of these domains can cause problems; the greater the gap or mis-fit (either subjective or objective) between environmental variables (E) and person variables (P), the greater the strain as demands exceed abilities and need exceeds supply (Sonnentag & Frese, 2013). With reference to the application of this model for military population, the individual characteristics and work environment determine strain; however, job environment not fitting employees' needs does not create strain for the military population, as they have an exclusively trained sense of following their leaders under any circumstances and are guided by patriotism and motivation to serve their country. Moreover, according to Lazarus (1991), the P-E fit model represented an advance in thinking; however, the concept of fit between the person and environment is too mechanical, with emphasis on stable relationships rather than giving consideration to individual differences and subjective well-being factors that may vary personal or situational experiences causing mis-fit. Though subsequently Buunk, de Jong, Ybema and de Wolff (1998) stated that empirical support for the theory is limited, this model provides a good explanation for the factors that create stress at the workplace. Application of this model can be extended to the military population if subjective experiences and environmental exceptions are included in the model with due consideration of certain characteristic psychological aspects of military environment and their all-time united sense of purpose.

## The job characteristics model and the Vitamin model

As delineated by Mark and Smith (2008), job characteristics such as task identity, task significance, skill variety, autonomy, and feedback are at the centre of the job-characteristics model of stress. The model suggests that the significant psychological states of experiencing responsibility, its meaningfulness, and knowledge of outcomes are created by various job characteristics. It is proposed that motivation, satisfaction, absenteeism, stress, and so forth are the cognitive and behavioural outcomes of various mental states caused by positive or negative work attributes (Hackman & Oldham, 1980, as cited by Mark & Smith, 2008). These outcome variables of the mentioned core job characteristics have been widely used in the literature (Kompier, 2003). Nevertheless, the model received criticisms, as only a few psychological states were considered in analyzing positive and negative characteristics. This model completely ignores the role of environmental factors and focuses only on the outcome variable. A general environmental constancy may probably be assumed for civil populations, but since continuous change in the environment is an unavoidable condition for work in the military, this model in its current form cannot explain stress outcomes in the military environment, as on one hand, the military as an organization is much more structured, and on the other, the environmental characteristics have the potential for rapid and uncertain change. However, the inclusion of specific environmental factors can widen the scope of this model not just for the military population but also for the civilian population.

Extending the job characteristics model further, Warr's (1987) Vitamin model considers certain job characteristics as analogous to the vitamins working in human body, which by virtue of their varied functioning have an effect on the mental health of the individual. In other words, some job characteristics such as task significance, safety, increase in salary, and so on can affect stress experiences and coping, but only up to a threshold, after which there is no positive or negative effect (Buunk et al., 1998). On the other side, some job characteristics such as job demands, autonomy, skill utilisation, task feedback, and social support have an additional decrement effect, such that moderate levels are the most beneficial, but too much or too little can have negative health effects (van Veldhoven, de Jonge, Broersen, Kompier, & Meijman, 2002). Notwithstanding the interesting premise of the Vitamin model, researchers state that the model is mixed and inconclusive (Buunk et al., 1998; Sonnentag & Frese, 2013) and lacks empirical evidence. The constant effects explanation in this model provides answers to how an increase in salary, safety, and task significance can only drive the military population to a certain extent, but along with standard fixedness of these aspects, it is the other job characteristics of the military environment which affect motivation to deal with combat stress. Moreover, given the inherent variations and standard policy-driven governance of various characteristics in military, the applications would significantly vary from that of the civil environment. Empirical support and expansion of the process of stress as it prevails in the military can make this model potent for application in military contexts, which does not seem to be a part of it in its current form.

## The role stress approach

The role stress approach, also known as social environment model (Mark & Smith, 2008), underscores the individual's own subjective perceptions of stressors. According to Kompier (2003), role issues (such as role conflict, role ambiguity, and role expectations) are treated as central stressors in this approach, while personality variables, demographics, and social support moderate subjectively perceived environmental stressors (such as role ambiguity, conflict, lack of participation, job security, workload, lack of challenge). Though the inclusion of individual differences and environmental stressors makes this model appropriate for the military population, difficulty in empirically testing this model highlights its complexity. Inclusion of too many factors makes it difficult to test this model on any population. Moreover, the role issues considered in this model are generally constant or well-defined in the military, whereas the same may not be the case for civil population. Not the exact present form but only a mindful selection of factors relevant in the military setup can make this model suitable for application in the military population.

### *Demand control support model*

The original demand-control model of Karasek (1979) focused on job demand and job control as two important job characteristics. Job control is sometimes called 'decision latitude', which may be referred to as the working individual's potential control over his tasks and conduct during the working day, and job demands are the psychological stressors involved in accomplishing the workload (Karasek, 1979). Karasek posits that work characteristics may not be linearly associated with health and stress as proposed by earlier models. He demonstrated that employees with low decision latitude and high job demands were found to be stressful and have low satisfaction at work. Later on, Johnson and Hall (1988) added social support as one of the job characteristics. Subsequent research evidence has suggested that support may act as a buffer in high demand situations (Cooper, Dewe, & O'Driscoll, 2001; Karasek & Theorell, 1990; Lim, 1996). Despite the expansion of the model with inclusion of social support, the model is still considered limited in the number of job characteristics in explaining the various stressful aspect of the job. Though it provides an interesting take on decision latitude and job demands which successfully explains day-to-day stress in military organizations, the model, however, is not able to include the role of individual differences in perceiving stress, as it fails to explain why the same levels of demand and control in two individuals may give rise to different behavioural or health outcomes (Perrewe & Zellars, 1999). Furthermore, seeing the inherent effect of military-specific regular training and the exclusively trained sense of having heightened control over tasks in the military, its applications would significantly vary for military personnel in comparison with civilians. If individual difference and subjective experiences specific to the military context are added as factors, this model could probably be relatively potent in explaining the process of stress in the military population.

## Transactional approaches

The transactional approach underscores the role of subjective perceptions of the environment and acknowledges the role of individual differences (coping, appraisal, personality, locus of control) in understanding workplace stress. Most transactional theories of stress focus on the cognitive processes and effective behaviour underpinning the person's interaction with their environment. They attempt to address the possible imbalance between demands and ability or competence to complete assigned tasks at the workplace.

### *Effort-reward imbalance model*

Siegrist (1996) proposed that the experience of chronic stress can be best defined in terms of an imbalance between high costs spent and low gains received. In simple language, according to this model, low returns yielded from high efforts invested at work lead to stress. The key feature of this model is its emphasis on subjective perceptions of the environment which, rather than being variable, is generally a standard trained aspect of the military population wherein the environment is objectively and uniformly seen by all personnel. The sources of effort invested at work are distinguished as extrinsic sources such as demands of the job and intrinsic sources such as the motivation of the individual worker in a demanding situation. Subsequently, the rewards expected from these efforts include financial gratification, socio-emotional reward, and status control (e.g., promotion prospects and better deployments). According to Peter and Siegrist (1999), the focal point of the model is reciprocity, where effort at work should be compensated by suitable rewards, and a mismatch between these will lead to stressful experiences. As compared to the civil population, the concept of reciprocity has a very low potential of application in the military, as patriotism takes a higher seat. Although the imbalance in the model is not an actual mismatch but the subjective perception of the employees with reference to their efforts made and rewards granted; in light of a greater sense of purpose and dutifulness to the nation, this aspect of imbalance takes a backseat in causing stress among military personnel. The application of this model is slightly complicated for a military organization due to the heavy emphasis on reciprocity. The reciprocal explanation does not apply in the military, as personnel's motivation towards their line of duty is not binding on what they receive from the organization. Military personnel work endlessly towards the service of the nation even when their expectations are not fulfilled. Stress due to unfulfilled expectations may be applicable for civilian employees but not for military personnel.

## *The cognitive theory of psychological stress and coping*

This theory primarily considers the coexistence of an individual and environment in a dynamic relationship. The theory has two key features: appraisal and coping. Here, an appraisal is the evaluative process that gives these person-environment transactions their meaning (Holroyd & Lazarus, 1982). Stress is experienced when an individual perceives that the demands of the situation are beyond their coping capabilities. The later refinement of this theory suggests both primary and secondary components of the appraisal process. The first stage is where encounters are subjectively evaluated to see what is at stake in terms of potential risk (Perrewe & Zellars, 1999). The secondary appraisal is contingent upon the recognition of the problem that exists and involves detailed analysis and the generation of possible coping strategies. These conceptualizations may hold true in the case of civil populations, but the case in military populations is altogether different, as military training and the tasking approach barely leave any scope for consideration of any problem beyond one's efforts. Similarly, the coping strategies and approaches of military personnel are immediate and qualitatively different than those of civilians.

Folkman and Lazarus (1980) described many types of coping behaviours and suggested that they could be aggregated into two major categories. First is problem-focused coping, which attempts to cope using more rational problem-solving approaches, and second is emotion-focused coping, which is emotion oriented and attempts to cope by using emotional responses per the requirements of the different situations. Emotion-focused coping does not alter or solve the stressful situation but only emotionally prepares the individual to bear consequences or escaping from it. Again, these coping conceptualizations lack similar application in military settings, as military personnel are regularly trained for focusing on problem solving and high emotional control is practiced even under the gravest circumstances. The model primarily focuses on the job situation, subjective perceptions, and the potential influence of various factors of individual differences. Though the model successfully incorporates all aspects of stress, its complexity and variations in its conceptual applications in the military make it difficult to empirically evaluate this model. Briner, Harris, and Daniels (2004) have suggested that the conception of primary and secondary appraisal is too simplistic and doesn't include individuals' histories and anticipated futures. Despite some of the major odds, this theory comes closest to explaining stress in the military organization. Firstly, the dynamic nature of the coexistence of stress and environment relates most to the dynamic nature of the job in the military. Secondly, problem-focused and emotion-focused coping attempt to explain how military personnel endure continued stress at the workplace. However, the process of how stress affects the individual is not explained in this model. This model can be applied in a military scenario if it elucidates how stress affects military personnel assuming the inherent task approach differences and regular training of keeping up high motivation and self-confidence and remaining problem focused and relatively immune to emotional encounters.

### *Cox's transactional model of occupational stress*

Cox's transactional model has a clarified structure and emphasizes individual differences in understanding stress (Cox & Ferguson, 1991). The stress here is defined as an individual phenomenon resulting from a transaction between individual and situation. Cox's framework has five stages. The first stage includes 'the sources of demand relating to the person' and forms part of the individual's environment. These demands are either derived from the environment or an individual's psychological needs. The second stage is an individual's perception of these demands and his own capability to cope with them. Stress is experienced when an individual realizes he cannot cope with the demands of the situation given his present capability. The third stage is associated with response to stress: emotional changes occur while an individual tries to cope with stress. The fourth stage includes the consequences of coping with stress. The fifth stage revolves around general feedback that occurs in relation to all other stages of the model. Cox and Ferguson (1991) state that this framework consists of primary and secondary appraisal, where the primary appraisal concerns monitoring and secondary appraisal involves decision-making grounded in a problem-solving context. They also suggest that individual differences (locus of control, hardiness, and coping resources) exert mediating and moderating effects determining stress. While the model attempts to incorporate most of the missing features in earlier suggested models, it becomes difficult to extend its application to military given the aspects of military approach, as explained in the previous section. More aptly, the problem is about explaining the aspect that generally military personnel do not reach the level of experiencing heightened stress, as their capabilities and positive perceptions enable them to cope successfully. Though this model is comprehensive in its approach and provides a stepwise explanation of stress at the workplace, this model may better help in understanding stress in military personnel, as it factors in all possible variables affecting an individual.

### *Cybernetic theory of organizational stress*

Derived from concepts of cybernetics or systems control, the theory states that behaviour is directed at reducing deviations from a specific goal-state. The theory is based on application of cybernetics to living systems, explaining living systems maintain themselves in steady states or homeostasis (Cummings & Cooper, 1998). Stress is experienced when this homeostasis is disrupted from either inside or outside the system by forces and an attempt is made to restore the original balance. This theory is criticized for being too simplistic and not taking into account the role of individual differences and their context specificity. Thus the model per se cannot be extended to a military organization, as stress in the military is more complicated than a disturbed state of homeostasis. Moreover, human behaviour in an organization like the military and the functional contexts of military personnel cannot be understood just on the basis of maintaining homeostasis.

Based on the important features of the earlier models, many new models have been propounded to explain the stress experiences in variety of occupations and different situations. One such model is the job demands-resources model by Demerouti, Bakker, Nachreiner and Schaufeli (2001) that takes the lead from various approaches and categories of psychosocial factors into job demands and job resources. Job demands refer to physical or social aspects of a job that require effort and have physical and mental costs. Job resources refer to organizational aspects that help with the achievement of work goals and reduce demands. These two affect the conditions which influence key processes leading to health impairment and stress. Researchers suggest that this model is a heuristic-based model which can be applicable to any occupational setting regardless of the particular demands or resources involved (Llorens, Bakker, Schaufeli, & Salanova, 2006); whereas, few researchers assert that this has focused only on the characteristics of the work environment (Xanthopoulou, Bakker, Demerouti & Schaufeli, 2007) and the individual characteristics remain out of its coverage, which contributes to huge differences between civilians and trained military personnel.

The demand-induced strain compensation model (de Jonge & Dormann, 2003) is another model which was developed keeping in mind stress in service jobs. Its central assumption is that emotional, cognitive, and physical resources are matched as per the emotional, cognitive, and physical demands in the environment (van Veldhoven, Taris, de Jonge, & Broersen, 2005), which can be called the triple match principle. Compared to other models, if explained in specific and not a civil population–generalized perspective, the application of the triple match principle in military organizations can provide a glimpse into the effect of the emotional, cognitive, and physical demands of combat and non-combat military tasks on personnel. According to the model, cognitive demands are met by cognitive capabilities, emotional demands are met by emotional capabilities, and physical demands are met by the physical capabilities of an employee. However, this compartmentalized understanding of demands and resources is not adept at understanding stress in military personnel. On a superficial or civil population level, this compartmentalization seems efficient, but how these resources are used to attend to these different demands in a military environment is not clear. This distinction of the triple match principle is not appropriate for a military organization; however, a more relaxed approach to understanding these demands and resources can widen the scope of application in the military environment. The demand skill support model (van Veldhoven et al., 2005) is another such model with the objective of explaining the process of stress using a minimum number of factors across different situations and occupations. This model proposes that quantitative and qualitative demands are more likely to relate to stress, and skill utilization and social support are more likely to relate to attitudinal outcomes and well-being. This model is an example of a blend between interactive and transactional theories, but it is too narrow in its approach to explain the process of stress and coping, especially in a military setup where stress is a continuous process and varies in its intensity throughout service. The very difference in the characteristics of civilian and military populations and barely any research on military samples make it evident that these models cannot similarly explain the stress experiences in the military environment.

## General adaptation syndrome

One of the most popular theories of stress deserves to be approached separately from other models. Famously known as the father of stress research, Hans Selye referred to non-specific physiological response to harmful stimuli as general adaptation syndrome (GAS). According to Selye (1956), all kind of stressors, including pathogens, physical stressors, and psychosocial stressors, elicit a common pattern of physiological response which proceeds in the characteristic three stages of alarm (including immediate flight or fight), resistance, and exhaustion. During the first stage of GAS, the alarm stage, the organism's physiological changes reflect the initial reactions necessary to meet the demands made by the stressor agent. This approach asserts that when an individual is faced with excessive stimulation or any stressor, the hormone output from the adrenal cortex increases rapidly. After the initial stage of alarm reactions, the second stage of resistance comes into play with an intention to get consequent improvement or disappearance of symptoms by means of resisting or having a complete adaptation to the stressor. For the obvious reason of resistance or adaptation, the output of corticosteroids (cortisol) remains high but stable during the resistance stage. Finally, if the stressors acting on the individual are so severe and prolonged that individual's resources and somatic defences get depleted, the third stage of exhaustion occurs. Since the secretion of corticosteroid hormones has been constantly high during the resistance stage, due to over-prolonged exposure, the anterior pituitary and the adrenal cortex lose their capacity to secrete hormones, and the organism can no longer adapt to the stressor. Hence, stress symptoms reappear, and, if the stress response continues at the same degree without subsiding, vulnerable organs as determined by the individual's genetic and surrounding environmental factors are likely to break down. Selye's paradigm of GAS claims that any stressor or noxious agent, physical or psychosocial in nature, would mobilize a similar GAS response (Selye, 1974).

Though Selye's model had huge momentum, critiques of Selye's model disagree with the presence of GAS and suggest that each stressor elicits its own distinct physiological reactions (Lazarus, 1977; Mason, 1975). Though these researchers agree that there is a nonspecific physiological response to stressors, they emphasize the appraisal of any stimulation as stressful and claim that the physiological response is concomitant to the emotional reaction that occurs when situations are appraised as stressful. In his later works, Selye (1974, 1980) acknowledged that there are both specific as well as general (nonspecific) factors in the physiological response to a stressor. He also suggested that the GAS does not occur or is at least not destructive in response to all kinds of stressors; however, he maintained that the nonspecific response is not always mediated psychologically (Cohen, Kessler, & Gordon, 1995). Based on these aspects, many new ways were proposed to explain stress, some prominent ones of which have been briefly highlighted in previous sections. The specific separate focus on Selye's model here is an attempt to delineate the difference in its applications in military contexts.

In most organizations across continents, Selye's model of stress is being used for the formulation of training modules. This model is followed in the military as well, but largely due to lack of an adapted model for stress in the military. This dilemma is due to the paucity of the distinction between stress in the military and general population. To be able to formulate a model for military organizations, researchers need to pin down the differentiation between the two. The following subsection attempts to elaborate on stressors in the military organization. The subsection includes various forms of stress and different environments which are sources of these stressors.

## Stresses for military populations

In the context of military settings, much attention has been devoted to the relationship between combat and non-combat stress. The aspects of combat activities have been clearly established as precipitants of psychological stress. Research has established that combat, exposure to heavy casualties, deployment of units in a war-type zone, and unexpected mobilizations of units are all correlated with higher levels of psychological distress (Pflanz & Sonnek, 2002). It seems to be assumed that the stress of military life is primarily attributable to such things as unstable deployments, exposure to combat, and the threat of bodily harm: getting killed. The periodic change of posting station, the stationing of personnel overseas, and lack of control over duty assignments are just a few examples of the more mundane aspects of military life that may cause stress and affect the mental health of military personnel (Pflanz & Sonnek, 2002).

Going beyond combat and wartime assignments, researchers have found that work stress is a significant source of distress for military personnel during routine peacetime assignments as well. Manning and colleagues (1981) found that military mental health patients identified work-related problems as the primary contributor to their emotional problems. Another study examined the prevalence of reported job stress in military mental health patients and concluded that work stress significantly leads to emotional distress and in some cases to mental illnesses (Pflanz, 2001). The most common work stressors reported were a change in work responsibilities and change in work hours. Hence, it may be inferred that job stress in the military may have little to do with the fact that military personnel deal with the difficult business of war and may stem from more subtle aspects of military culture that foster work stress.

## Psychological stressors in the military

Military deployments, in simpler terms, may be called Harsh, Ambiguous, Risky, and Distant (HARD; Pflanz & Sonnek, 2002). The personnel deployed in the field live in austere conditions. The deployment environments can be uncomfortable, harsh, and severe. The troops have limited basic facilities and recreational options available with them. The temperature and climate can range from very cold to very hot. In addition to harsh work environments, ambiguity and uncertainty are, to a great extent, a part of military deployments. The troops being deployed are often not aware of when they are deploying, where they are going, or when they are returning. The variability and unpredictability of deployments add to the experience of their stress. Also, military deployments are risky to a great extent. Especially today, deploying military personnel are heading out to counterinsurgency and low-intensity conflict areas. In these dangerous environments, the troops are exposed to the threat of enemy fire and terrorist attacks. These challenging hazardous conditions create stress, irritability, anxiety, and depression among the troops. Lastly, military deployments are distant, and military personnel are far away from home. They remain disconnected and separated from their family life (see Kumar, Parkash, & Mandal, 2013).

## Combat stress and stressors in counterinsurgency

In pursuit of national security, military personnel are always combat ready and under a great deal of combat stress. Combat stress is the mental, emotional, or physical tension, strain, or distress resulting from exposure to combat and combat-related conditions. It is important to emphasize that combat stress is not restricted only to combat, but may also arise from combat-like conditions present during military operations other than war. In an operational area characterized by continuous action and high danger, forces may experience high rates of combat stress casualties unless timely interventions are applied to manage this stress. Combat stress reactions have the potential to disable the most courageous soldiers and influence the success or failure of a unit in accomplishing its mission. As my colleagues and I have put forth in our earlier works (Kumar, Parkash, & Mandal, 2013), the scenario that the military personnel are facing these days is not in the form of conventional warfare; rather, it is nonconventional warfare that more prominently involves counterinsurgency operations. Having limited human potential and striving for the security of the motherland, the troops deployed in such areas become vulnerable to heightened stress experience that has the potential to adversely affect their mental health. Troops trained in conventional warfare experience significant stress in such counterinsurgency operations (Goel, 1998; Ray, 1997).

In conventional operations of war, limited periods of intense stress followed by adequate recovery phases do not significantly sap the psychological resources of the soldier unless the operations are unduly prolonged. The situation in counterinsurgency is diametrically opposite. Prolonged spells of stress punctuated by quantitatively and qualitatively inadequate opportunities for rest and relaxation impose immense and often unbearable demands on even otherwise robust soldiers. This may result in psychological distress, combat stress disorder, or post-traumatic stress disorder (Badrinath, 2003; Haas, 2003). Several operational factors such as harsh weather conditions, unavailability of rest time, fatigue, unpredictability of threat, zero error syndromes, extended tenures of stay, absence of recreational avenues, lack of adequate social support in terms of healthy behaviour by superiors, domestic worries, problems related to leave, and so on increase the level of frustration (Goel, 1998; Ray, 1997, as cited by Chaudhury, Goel, & Singh, 2006). Troops in long-term postings in the counterinsurgency environment are subjected to high-risk operations, forcing them to remain more vigilant round the clock. Extended and prolonged exposure to such a stressful environment adversely affects the physical and mental state of the soldiers. In threatful counterinsurgency situations, the fear and possibility of getting killed anytime coupled with long duty hours without rest put tremendous mental pressure on the troops that exceeds their cognitive capacities to resist and which may ultimately lead to maladaptive psychological functions and aberrant behaviour (Kumar, Parkash, & Mandal, 2013). It is hardly a matter of dispute that just like in the civil population, stress is a significant determinant of performance, emotional well-being, and other work outcomes in military settings (Kavanagh, 2005). What is not known is how these stressors function differently for military personnel. The nature of these stressors has been explained by various frameworks of soldier well-being. One of the most used models consists of risk factors or stressors, such as combat exposure and deployment length, which lead to a variety of negative outcomes, including decreased emotional well-being, poorer performance, posttraumatic stress disorder, and suicide (Harms, Krasikova, Vanhove, Herian, & Lester, 2013). This relationship is moderated by a number of protective factors such as personality characteristics, sources of personal support such as leaders or spouses, and training. The difficulty lies in gathering accurate information concerning the emotional and mental well-being of military personnel. They keep constantly rotating in and out of combat and even hesitate to acknowledge or report such problems while actively serving (Hoge et al., 2004). Another issue with getting an accurate estimate from the military population is that stress often manifests long after combat has ended (Milliken, Auchterlonie, & Hoge, 2007). Given the critical nature of the military profession, a balanced mind and attitude are prerequisites for the job.

## A few military-specific propositions

On the basis of the issues presented above, a new stress framework is suggested for the military population. The model acknowledges the importance of the role played by organizational support and prior training in dealing with stressors in the military setting. This framework aims to represent key aspects of the stress process without getting entangled in complexities of innumerable factors and mental processes. Many aspects of this model are supported by both interactional and transactional theories, notably from Selye's (1956, 1974) general adaptation syndrome. The proposed model given in Figure 22.1 encapsulates environmental and individual factors in coping with stress in military personnel and can be called 'military model of alarm, resistance, and coping'.

*Figure 22.1* Proposed military model of alarm, resistance, and coping.

Adapting from Selye's GAS and considering the military-specific factors of training, motivation, and fight readiness, it can be asserted that when faced with stressors, military personnel, by the virtue of their regularly continued training, do not use the flight path and stressors are dealt only with fight mode of responding. During the alarm stage, an inherent capacity to fight the immediate threat and stress is activated. Military personnel are trained at a young age to fight every threat in the workplace. Their orientation is developed in such a way that they instinctively react to engage with any stimuli posing threat to them or their counterparts. Vigorous training and skill development towards positive engagement with stress make them physiologically and psychologically mechanized to deal with stress. Hence, the first stage consists of recognition and acceptance of the nature of stress. A meticulous analysis of situational demands is conducted and the military personnel chalk out a course of actions based on their cognitive mapping suggesting various alternatives with a problem-solving orientation. The alarm stage ends when the individual has acknowledged the situation at hand and is ready with a course of action in hand.

The second stage of resistance begins when the individual starts collecting available resources to sustain resistance for as long as the situation demands. The first resource available to the individual is his own training. The greatest strength of the military personnel is his cognitive capability, which has been well tested prior to their induction and functioning in the field. Military training, with the help of cognitive science principles, restructures the mental orientation of personnel in such a way that their perspective on stress transforms and they are cognitively reengineered to process mental workloads efficiently. The cognitive potential is very closely related to their self-regulation and self-efficacy to deal with stress. The strength and efficiency of utilizing cognitive capacity vary for individuals and therefore act as moderating variable in this process. Due to stringent training to deal with different work as well as personal problems, there are barely any significant individual differences in the approaches adopted by different personnel. Therefore, regardless of minor individual differences, it is the well-trained sense of fight readiness which fosters the capability to resist stressors. Organizational support is the second resource available to military personnel. The resistance stage is supplemented with organizational support such as reinforcements, buddy support, medical care, leave, social security, and so forth. This stage is where the environmental factors exert the highest influence. The camaraderie and brotherhood in the military setup make it easier for them to seek organizational support and persevere against stressors. Even though the military organization is hierarchical in nature, the respect for brothers in arms and family well-being runs through all levels. Supported by all this, unlike Selye's GAS model, there is no exhaustion stage for military personnel while dealing with stressors. They perceive resistance as a part of the job training and simultaneously keep assimilating newly learned skills and supportive elements to deal with stress. Hence, the third stage that in general arrives for majority of military personnel is that of adaptive coping, the individual restoring back to normal functioning.

Although exceptions lie in every field, a generalizably significant strength of military personnel perceive stress and unexpected threat as an opportunity to acquire a new set of skills and test their training. They are focused on solving the problem and that too in such a way that the same stress demand becomes a mundane task in the future. Moreover, it emotionally strengthens them with positive emotions as it confirms their faith in their training and organizational support. This exclusive endurance to stress by military personnel is part of their adaptive coping nature, as they are able to internalize the skills acquired and positively appraise their own capability to deal with stress effectively. In this stage, the job satisfaction is highest and psychological distress is least. Against this backdrop, another conceptual mediation-moderation model of stress for the military population may be hypothesized (Figure 22.2). The model proposes that organizational support mediates the relationship between stressors at work and resistance capability of military personnel. Likewise, military training moderates the relationship between the stressor and organizational support, such that when military training is positively perceived, the extent of organizational support and resulting resistance to stress is high and when military training is negatively perceived, the dependence on organizational support and resulting resistance is low, the former being more prevalent in general and the latter being an exception.

This model is not intended at this early stage to be a predictive model; however, it proposes a theoretical framework which can be tested through future research and to which more relevant variables can be introduced. Many aspects of this proposed model were supported by recent research (Kumar, Parkash, & Mandal, 2013). The nature of mental processes that takes place when dealing with stress, that is, rational procedures and analysis of the situation, is not hypothesized in the proposed model in order to keep it easily comprehensible and to avoid complexities. This framework is suggestive of ways of illustrating which variables and factors may relate to stress and coping experiences in military personnel. The aspects of how and why they relate may be explored empirically through future research. As far as coping with stress is concerned, it is posited that as a general approach, emotion-focused, maladaptive, avoidant, or escape types of coping approaches seldom exist among military personnel, and problem-focused, rational, and adaptive coping dominates their behaviour due to the specialized training that they undergo to deal with different problems, more refined abilities that they possess as prerequisites of induction, and their exclusively trained sense to best utilize the abilities in somewhat concretely varied manners as compared to the civil population.

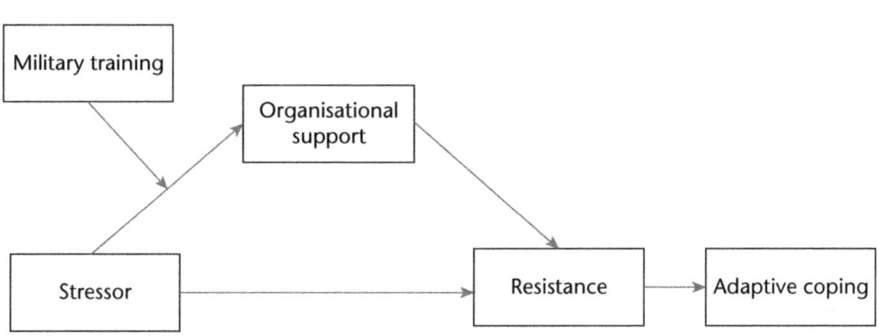


*Figure 22.2* Proposed conceptual model depicting moderated mediation of stress and coping in military setting.

## Summing up

The increase of work-related stress in our society highlights the need to address the economic and social costs of work-related stress and review existing research paradigms for the same. Models that explain the stress experiences of the civil population may not be applicable similarly to the military population due to its various inherently different characteristics. An overview of various models relating to work-related stress formulated for the civilian population suggests that many of them leave many unanswered questions to include a role of subjective perceptions, individual differences, and environmental context or are otherwise complex, hard to support, and lack predictive validity. Keeping in mind the distinct work environment and nature of work of the military population, most models fail to tap into key features in the military context. With few attempts to explain stress experiences specific to the military, a slightly different line of thought has been proposed to combine features of existing models and to represent the differences and complexities of the stress process in military populations. The proposed models attempt to strike a balance between an integration of aspects of interactive and transactional models while still maintaining a balance between simplicity and complexity. It cannot be denied that the chapter posits more questions to the reader rather than providing answers, but it would ignite a relatively different line of thought specific to military populations among critiques and fellow researchers.

## References

Badrinath, P. (2003). Psychological impact of protracted service in low intensity conflict operations on Armed Forces personnel: Causes and remedies. *Journal of United Service Institution of India, 83*(1), 38–58.

Barnett, R. C., & Brennan, R. T. (1995). The relationship between job experiences and psychological distress: A structural equation approach. *Journal of Organizational Behaviour, 16*(3), 259–276.

Briner, R. B., Harris, C., & Daniels, K. (2004). How do work stress and coping work? Toward a fundamental theoretical reappraisal. *British Journal of Guidance and Counselling, 32*(2), 223–234.

Buunk, B. P., de Jong, J., Ybema, J. F., & Wolff, C. J. (1998). Psychosocial aspects of occupational stress. In P. J. D. Drenth, H. Thierry, & C. J. de Wolff (Eds.), *Handbook of Work and Organizational Psychology* (pp. 145–182). Hove, UK: Psychology Press.

Caplan, R. D., Cobb, S., French, J. R., Harrison, R. D., & Pinneau, S. R. (1975). *Job Demands and Worker Health: Main Effects and Occupational Differences.* Washington: U.S. Government Printing Office.

Chaudhury, S., Goel, D. S., & Singh, H. (2006). Psychological effects of low intensity conflict operations. *Indian Journal of Psychiatry, 48*(4), 223–231.

Cohen, S., Kessler, R. C., & Gordon, L. U. (1995). Strategies for measuring stress in studies of psychiatric and physical disorder. In S. Cohen, R. C. Kessler, & L. U. Gordon (Eds.), *Measuring Stress: A Guide for Health and Social Scientists* (pp. 3–26). New York: Oxford University Press.

Cooper, C. L., Dewe, P. J., & O'Driscoll, M. P. (2001). *Organizational Stress: A Review and Critique of Theory, Research, and Applications.* London: Sage Publications.

Cox, T. (1978). *Stress.* London: Macmillan.

Cox, T., & Ferguson, E. (1991). Individual differences, stress and coping. In C. L. Cooper & R. Payne (Eds.), *Wiley Series on Studies in Occupational Stress. Personality and Stress: Individual Differences in the Stress Process* (pp. 7–30). Oxford, England: John Wiley & Sons.

Cox, T., & Griffiths, A. (1995). The nature and measurement of work stress: Theory and practice. In J. R. Wilson & E. N. Corlett (Eds.), *Evaluation of Human Work: A Practical Ergonomics Methodology* (pp. 783–803). London: Taylor & Francis.

Cox, T., & Mackay, C. J. (1981). A Transactional approach to occupational stress. In E. N. Corlett & J. Richardson (Eds.), *Stress, Work Design and Productivity* (pp. 429–454). Chichester: Wiley & Sons.

Cummings, T. G., & Cooper, C. L. (1998). A cybernetic theory of organizational stress. In C. L. Cooper (Ed.), *Theories of Organizational Stress* (pp. 101–121). New York: Oxford University Press.

de Jonge, J., & Dormann, C. (2003). The DISC model: Demand-induced strain compensation mechanisms in job stress. In M. Dollard, H. R. Winefield, & A. H. Winefield (Eds.), *Occupational Stress in the Service Professions* (pp. 57–88). London: Taylor & Francis.

Demerouti, E., Bakker, A. B., Nachreiner, F., & Schaufeli, W. B. (2001). The job demands-resources model of burnout. *Journal of Applied Psychology, 86*(3), 499–512.

Folkman, S., & Lazarus, R. S. (1980). An analysis of coping in a middle-aged community sample. *Journal of Health and Social Behaviour, 21*(3), 219–239.

French, J. R. P. Jr. (1973). Person-role fit. *Occupational Mental Health, 3*(1), 15–20.

Goel, D. S. (1998). Psychological aspects of counter-insurgency operations. *Combat, 27*(1), 43–48.

Haas, K. L. (2003). Stress and mental health support to Australian Defence Health Service personnel on deployment: A pilot study. *Australia Defense Force Health, 4*(1), 19–22.

Hackman, J. R., & Oldham, G. R. (1980). *Work Redesign*. Reading, MA: Addison-Wesley.

Harms, P. D., Krasikova, D.V., Vanhove, A. J., Herian, M. N., & Lester, P. B. (2013). Stress and emotional well-being in military organizations. In P. L. Perrewe, C. C. Rosen, & J. R. B. Halbesleben (Eds.), *The Role of Emotion and Emotion Regulation in Job Stress and Well Being* (pp. 103–132). Bingley: Emerald Group Publishing.

Hoge, C.W., Castro, C.A., Messer, S. C., McGurk, D., Cotting, D. I., & Koffman, R. L. (2004). Combat duty in Iraq and Afghanistan, mental health problems, and barriers to care. *New England Journal of Medicine, 351*(1), 13–22.

Holroyd, K. A., & Lazarus, R. S. (1982). Stress, coping and somatic adaptation. In L. Goldberger, & S. Breznitz (Eds.), *Handbook of Stress: Theoretical and Clinical Aspects* (pp. 21–35). New York: Free Press.

Johnson, J.V., & Hall, E. M. (1988). Job strain, workplace social support and cardiovascular disease: A cross-sectional study of a random sample of Swedish working population. *American Journal of Public Health, 78*(10), 1336–1342.

Judge, T. A., & Colquitt, J. A. (2004). Organizational justice and stress: The mediating role of work-family conflict. *Journal of Applied Psychology, 89*(3), 395–404.

Karasek, R. (1979). Job demands, job decision latitude and mental strain: Implications for job redesign. *Administrative Science Quarterly, 24*(2), 285–306.

Karasek, R., & Theorell, T. (1990). *Healthy Work: Stress, Productivity and the Reconstruction of Working Life*. New York: Basic Books.

Kavanagh, J. (2005). *Stress and Performance A Review of the Literature and its Applicability to the Military*. Santa Monica, CA: RAND Corporation.

Kompier, M. (2003). Job design and well-being. In M. Schabracq, J. Winnubst, & C. L. Cooper (Eds.), *Handbook of Work and Health Psychology* (pp. 429–454). Chichester, West Sussex: John Wiley and Sons.

Kumar, U., Parkash, V., & Mandal, M. K. (2013). Stress in extreme conditions: A military perspective. In D. M. Pestonjee, & S. Pandey (Eds.), *Stress and Work: Perspectives on Understanding and Managing Stress* (pp. 101–126). New Delhi: Sage.

Lazarus, R. S. (1977). Cognitive and coping processes in emotion. In A. Monat, & R. S. Lazarus (Eds.), *Stress and Coping* (pp. 145–158). New York: Columbia University Press.

Lazarus, R. S. (1991). Psychological stress in the workplace. *Journal of Social Behaviour and Personality, 6*(7), 1–13.

Lewin, K. (1951). *Field Theory in Social Science; Selected Theoretical Papers*. New York: Harper & Row.

Lim, V. K. G. (1996). Job insecurity and its outcomes-moderating effects of work-based and non-work based social support. *Human Relations, 49*(2), 171–194.

Llorens, S., Bakker, A. B., Schaufeli, W., & Salanova, M. (2006). Testing the robustness of the job demands-resources model. *International Journal of Stress Management, 13*(3), 378–391.

Manning, F. J., Kukura, F. C., DeRouin, E. M., McCarroll, J. E, Zych, K. A., & Edwards, F. (1981). Outpatient mental health facilities in the U.S. Army. Europe: Patient characteristics, complaints and dispositions at three sites. *Military Medicine, 38*(1), 7–13.

Mark, G. M., & Smith, A. P. (2008). Stress models: A review and suggested new direction. In J. Houdmont, & S. Leka (Eds.), *Occupational Health Psychology* (*Vol. 3*, pp. 111–144). Nottingham: Nottingham University Press.

Mason, J. W. (1975). A historical view of stress field, Part-II. *Journal of Human Stress, 1*(1), 22–36.

Milliken, C. S., Auchterlonie, J. L., & Hoge, C.W. (2007). Longitudinal assessment of mental health problems among active and reserve component soldiers returning from the Iraq war. *The Journal of the American Medical Association, 298*(18), 2141–2148.

Perrewe, P. L., & Zellars, K. L. (1999). An examination of attributions and emotions in the transactional approach to the organizational stress process. *Journal of Organizational Behaviour, 20*(5), 739–752.

Peter, R., & Siegrist J. (1999). Chronic psychosocial stress at work and cardiovascular disease: The role of effort–reward imbalance. *International Journal of Law and Psychiatry, 22*(5–6), 441–449.

Pflanz, S. E. (2001). Occupational stress and psychiatric illness in the military: Investigation of the relationship between occupational stress and mental illness amongst military mental health patients. *Military Medicine, 166*(6), 457–462.

Pflanz, S. E., & Sonnek, S. (2002). Work stress in the military: Prevalence, causes, and relationship to emotional health. *Military Medicine, 167*(11), 877–882.

Ray, A. (1997). *Kashmir Diary. Psychology of Militancy*. Delhi: Manas Publications.

Selye, H. (1956). *The Stress of Life*. New York: McGraw-Hill.

Selye, H. (1974). *Stress without Distress*. London: Hodder & Stoughton.

Selye, H. (1976). *Stress in Health and Disease*. Boston, MA: Butterworth.

Selye, H. (1980). Stress, aging, and retirement. *Journal of Mind & Behaviour, 1*(1), 93–110.

Siegrist, J. (1996). Adverse health effects of high-effort/low-reward conditions. *Journal of Occupational Health Psychology, 1*(1), 27–41.

Sonnentag, S., & Frese, M. (2013). Stress in organisations. In I. B. Weiner (Ed.), *Handbook of Psychology* (2nd ed., pp. 560–592). Hoboken, NJ: John Wiley & Sons.

van Veldhoven, M., de Jonge, J., Broersen, S., Kompier, M., & Meijman, T. (2002). Specific relationships between psychosocial job conditions and job-related stress: A three level analytical approach. *Work and Stress, 16*(3), 207–228.

van Veldhoven, M., Taris, T. W., de Jonge, J., & Broersen, S. (2005). The relationship between work characteristics and employee health and well-being: How much complexity do we really need? *International Journal of Stress Management, 12*(1), 3–28.

Warr, P. B. (1987). *Work, Unemployment, and Mental Health*. Oxford: Clarendon Press.

Xanthopoulou, D., Bakker, A. B., Demerouti, E., & Schaufeli, W. B. (2007). The role of personal resources in the job demands-resources model. *International Journal of Stress Management, 14*(2), 121–141.

# 23

# MILITARY-RELATED MENTAL HEALTH MORBIDITIES

## A neurobiological approach

*Shobit Garg and Jyoti Mishra*

The nature of military-related morbidities has changed over time considerably. Preventive measures like improved protective gear have been effective in reducing military morbidity and mortality. Moreover, survival rates of post-battle physical injuries have improved due to better evacuation and tertiary care facilities. Therefore, recently psychological morbidities as an effect of war have started to outnumber physical injuries. Behavioral neuroscience has grown leaps and bounds studying normal and abnormal psychology of humans. Studying abnormal psychological phenotypes, that is, military psychological morbidity, from a behavioral neuroscience perspective is pertinent to understand the etiology, progression, presentation, treatment and prevention of illnesses. A simplistic tool to execute this is neurobiology.

What is neurobiology? Neurobiology is a branch of biology that deals with nervous system functions and structures. This branch focuses on how cellular level functions interact to model the various circuits involved in execution of normal and abnormal behavior. A neurobiological approach is the understanding of the neurochemical and neuro-electrical dynamics brought upon by genetic underpinnings or environmental conditions. In the following sections, we will deal with important military-related mental health morbidities like suicide, post-traumatic stress disorder (PTSD), sleep disorder, substance use disorders (SUDs) and traumatic brain injuries from a neurobiological perspective.

### Suicidal and related issues

Suicide is a significant public health issue within the Department of Defense. It currently remains the leading cause of mortality in the West (Ghahramanlou-Holloway, Baer, Neely, Koltko, & Nielsen, 2017; Armed Forces Health Surveillance Center (AFHSC), 2014). The unadjusted suicide rates per Western data for the four services have been reported as 23.8 (Army), 18.5 (Air Force), 17.9 (Marine Corps), and 16.3 (Navy) per 100,000 (Ghahramanlou-Holloway et al., 2017).

There are several risk factors reported, mostly via cross sectional or retrospective studies. These factors are demotion within the last two years, early military separation, dishonorable discharge from duty, relationship issues, and access to means (e.g., firearms). Specific subpopulations within the military such as infantrymen and combat engineers are also at higher risk for suicide. Military personnel suffering from physical pain, injury, or sleep problems, as well as increased rates of both

Axis I and Axis II disorders, significantly elevate suicide risk (Ghahramanlou-Holloway et al., 2017). Recent studies have found no correlation between deployment status and suicidal rates, but if a correlation exists, it has been found to be mediated by the development of depression and PTSD.

Factors protecting against suicide include adequate social support, higher unit cohesion and good workplace relationships, training and preparation, religious involvement, spouse's preparedness for deployment, and individual resilience (Ghahramanlou-Holloway et al., 2017). Suicide is a heterogeneous condition. Research pertaining to the neurobiology of suicide is mainly from postmortem studies of nonmilitary populations. Differences from the civilian population are largely because of epigenetic interactions citing various risk and protective factors as mentioned above. Many explanatory and predictive models of suicidal behavior have been hypothesized and a few are explained below (Nizamie & Garg, 2012).

## Unified model of suicide

Suicidal behavior is conceptualized as the outcome of the balance between an individual's threshold for suicidal acts and triggers of suicidal behavior. Threshold refers to the propensity for manifesting suicidal behavior and may be considered trait related. In contrast, triggers are precipitant contributing to the probability of suicidal acts and are considered state related. Various risk factors for suicidal behavior have been categorized according to whether they affect the threshold or the trigger. Risk factors affecting threshold could be genetic (e.g., family history), biological (e.g., chronic combat stress), neurochemical (e.g., low 5-HIAA in CSF), and so on, whereas factors affecting triggers for suicidal behavior could be social, financial, and family crisis in a military context. This model functions in the line of biopsychosocial theory (Mann & Arango, 1992).

## A four-pathway clinical-biochemical model

There are four hypothesized pathways to suicide in terms of clinical features and biochemical variables. Of these pathways, the elevated CRH-anxiety agitation pathway appears to be an acute/state pathway. Hypothesized low serotonin function (correlated with low serum cholesterol) associated with impulsiveness could be a state or trait pathway. Hopelessness appears to be a trait pathway, although its severity is certainly triggered to the severity of clinical depression. Anhedonia severity appeared as an acute suicide risk factor, suggesting a state variable, and it may be a secondary variable affected by the severity of anxiety or other depression-related symptoms (Fawcett, Busch, Jacobs, Kravitz, & Fogg, 1997).

## Stress-diathesis model

The relation between risk factors can be described in explanatory models of suicide, such as the stress-diathesis model. Acute psychosocial crises and psychiatric disorders such as depression are common proximal stressors leading to suicidal behavior, while pessimism or hopelessness and aggression or impulsivity are components of the diathesis for suicidal behavior. Familial or genetic factors, childhood experiences, and other factors, including cholesterol concentrations, could influence the diathesis (Hawton & Heeringen, 2009).

Postmortems of suicidal probands have shown greater serotonin neurons, more tryptophan hydroxylase 2 gene expression in the brainstem with reduction in dentate gyrus volume, and immature neuronal architecture. Molecular studies have demonstrated that in people who have attempted suicide, prefrontal localized hypo-function and impaired serotonergic responsivity was found to be proportional to the lethality of suicide attempts (Heeringen & Mann, 2014).

Neurobiological perspectives shed light on the treatment options available for suicidal risk. Electroconvulsive therapy has short-term life-saving effects in high suicidal risk. More recently, ketamine in subanesthetic doses was reported to have an acute antisuicide effect that lasted for about one week (Kellner et al., 2005; Larkin & Beautrais, 2011). Most antidepressant drugs enhance serotonergic levels, leading to a putative antisuicidal effect. A single randomized clinical trial of suicidal patients showed that a selective serotonin-reuptake inhibitors had a better antidepressant effect and greater reduction in suicidal ideation than did a noradrenergic–dopaminergic drug (Grunebaum et al., 2012; Heeringen & Mann, 2014).

## Combat stress reaction

During World War I (WWI), combat stress was referred to as shell shock, effort syndrome, war neurosis, Da Costa's syndrome, or irritable heart syndrome. The U.S. Department of Defense (DoD) describes **combat stress reaction** as the "expected, predictable, emotional, intellectual, physical, and/or behavioral reactions of service members who have been exposed to stressful events in combat or military operations other than war." One of the most commonly studied postcombat disorders is PTSD (Campise, Geller, & Campise, 2006).

PTSD is a significant issue that ought to be addressed and monitored in the military. PTSD can occur after someone experiences or witnesses a traumatic event. Examples can be combat, a terrorist attack, sexual or physical assault, a serious accident, or a natural disaster. PTSD can be a significant condition, especially when symptoms continue for more than one month after exposure to a trauma and cause significant distress or impairment in social, occupational, or other important areas of functioning. There are four main categories of symptoms: intrusions (e.g., nightmares), avoidance, negative alterations in cognitions and mood (e.g., guilt), and alterations in arousal and activity (e.g., hyperarousal). The vast majority of people recover quickly and have no long-term effects. The incidence rate of PTSD for active duty personnel is 0.6%. These rates could be artificially low due to stigma, minimization of symptoms, and mistrust (or misunderstanding) of mental health professionals. Post-deployment PTSD rates rise to 10% of the respective population (Mehlum & Weisaeth, 2002; Johnston, Robinson, Earles, Via & Delaney, 2017). The Vietnam War raised the awareness of PTSD and the decades that followed taught that "PTSD is not a constant or static condition, but a disorder that may actually wax and wane throughout a lifetime" (Falk, Hersen, & Van Hasselt, 1994; Johnston et al., 2017).

### *Neurological model of post-traumatic stress disorder*

Various cognitive aspects of PTSD have recently gained attention, including attentional biasing, reduced recall of both traumatic and nontraumatic material, and impaired executive skills. Several brain areas are implicated in PTSD (Scaglione & Lockwood, 2014):

*Over-activation of the amygdala*: Hyper-reactivity of the amygdala is a robust finding in PTSD. In PTSD the amygdala also has exaggerated responses to non-trauma-related affective material, such as fearful facial expressions as well as attentional tasks. The degree of amygdala activation is also correlated with PTSD severity.

*Underactivation and cell loss of the hippocampus*: In PTSD, diminished activation of the hippocampus is often found. A meta-analysis of MRI volumetric studies of PTSD by Smith (2005) indicated a 6% loss in hippocampal volume. Cortisol neurotoxicity (due to chronic stress) is a neuropathological correlate of these findings. The hippocampus has some inhibitory influence on the amygdala under physiological conditions. In PTSD, the amygdala-hippocampal circuits related to attentional biasing, encoding, and memory processes are overactive, possibly due to increased innervation from the original traumatic incident and a decrease in neuronal volume (Mahan & Ressler, 2012; Morey et al., 2012).

*Dysregulation of frontal-subcortical circuit*: Normal brain functioning requires complex feedback circuitry between subcortical and cortical systems. The frontal lobes function as a cortical-subcortical network. All **frontal-subcortical circuits** are closed-loop "final effector mechanisms" which require integration of external and internal/limbic information and the initiation of appropriate responses. There is a dynamic balance maintained between bottom-up survival-oriented activity and top-down directing of goal-oriented planning behavior which is found to be disrupted in PTSD, particularly anterior cingulated **frontal-subcortical circuits** and the medial division of the orbitofrontal systems which involve connections to the amygdala. This dysregulation could account for a number of neurocognitive processes in PTSD, such as fear conditioning, habituation, and extinction, cognitive–emotional interactions, and emotional processing (Scaglione & Lockwood, 2014).

## Treatment

Cognitive-behavioral therapy is the most empirically supported treatment for PTSD. It is effective in only about 50% of PTSD patients. No straightforward pharmacotherapy is effective in PTSD either. Cognitive rehabilitation of executive functions has also shown to be useful in PTSD, especially in veterans returning from combat with head injuries in addition to PTSD (Scaglione & Lockwood, 2014). Prolonged exposure (PE) therapy is also an evidence-based therapeutic option in PTSD (IOM, 2012).

## Substance use disorders

Historically, the worst substance problems were evident in 1971 in the Vietnam War, where 34% of soldiers admitted to marijuana use and 50% to the use of heroin. More service members were medically evacuated for drug use than for war wounds. Alcohol abuse and other substance use have been the 5th and 10th most commonly occurring disqualifiers for the Navy personnel. Those who are heavy drinkers (five or more drinks at least once per week) are more likely to be late to work, to leave work early, to exhibit decreased job performance, and to suffer more on-site injuries (Fisher, Hoffman, Austin-Lane, & Kao, 2000). The military lifestyle has numerous challenges like deployment to danger zones and loss of personal freedom not encountered by civilian population. When challenged, those with inadequate coping styles would consume alcohol in harmful patterns (25% of both military men and women). Substance use disorder is a common endemic affecting the young, adult male population. One five-year longitudinal study found that 75% of U.S. Navy recruits used alcohol prior to enlistment and 31% had used illegal drugs (Kennedy, Jones, & Grayson, 2006). One of the core features of SUDs and major factors of relapse is craving, which has several neurobiological underpinnings explained below as psychobiological models (Garg, Dharmadhikari, & Sinha, 2012).

## Psychobiological model

These models are directly influenced by biological neural systems (neural circuitry, reward systems, and neuroanatomical systems) and are as follows.

### *Neuroadaptive model*

There is hyper-sensitization of the dopamine neural transmitter system (implicated in reward pathways) in substance abuse. This in turn increased the incentive salience of drugs. "Incentive salience"' makes stimuli more attractive and turns ordinary wanting into excessive drug craving (Robinson & Berridge, 1993). Drugs cause neurobiological adaptations to maintain homeostasis, and if drug consumption ceases, an imbalance in cerebral activity results in craving. In the early withdrawal state, craving occurs to alleviate the imbalance. In a later recovery state, the altered brain functions return to their original state, but cravings may appear suddenly. This reactivation can be triggered by stress, which may activate the reward memory (Anton, 1999). The reward memories may also reactivate neurochemical processes associated with past experiences of drug use and cause craving.

### *Theory of neural opponent motivation*

Chronic drug exposure causes allostasis, defined as "a state of chronic deviation of the regulatory systems with establishment of a new set point." An allostatic state involves a feed-forward mechanism rather than the negative feedback mechanism of homeostasis. Over-activation of the brain reward system triggers the brain stress circuit or anti-reward system in order to limit the reward. Long-term drug intake dysregulates the underlying neurochemical functions, resulting in an allostatic state. An anti-reward system and neurotransmitter changes create powerful negative reinforcement. Craving arises from the action of memory of the rewarding effects of drug use superimposed on a negative emotional state (Koob & Moal, 2008).

### *Temporal-difference reinforcement learning model*

In a temporal-difference reinforcement learning model (TDRL), actions are selected to maximize future rewards which are based on the strength of a value signal, defined as the expected future reward discounted by the expected time to the reward. This value signal is carried by dopamine and produces temporal-difference learning in the normal brain. TDRL is based on assumptions about cocaine, which produce a phasic increase in dopamine directly and push a person towards irrational behavior (Redish, 2004).

### *Model of interoceptive dysregulation*

Interoception refers to the sensations that originate from the interior of the body. The interoceptive state is mediated by the anterior insular cortex which has bidirectional connections to the amygdala and the ventral striatum. Alteration in interoceptive processing is due to an altered "prediction error," which refers to the difference between the value of the anticipated sensation (i.e., the hoped-for result) and the value of the current interoceptive state. Dysregulation of the insular cortex minimizes the body prediction error, causing non-adaptive adjustment of the body prediction error, which causes craving in the addiction process (Paulus, Tapert & Schulteis, 2009).

## Treatment of substance use disorders

Although treatment review in detail is beyond the scope of this chapter, few salient points like principles of treatment, tobacco cessation program, and preventive strategies are discussed. Military personnel with substance abuse diagnoses warrant outpatient treatment (Level I). Similarly, a substance dependence diagnosis generally warrants either intensive outpatient treatment (Level II) or residential treatment (Level III). Exceptions might be those who previously completed treatment for substance dependence and remain sober for a significant period of time but then had a brief relapse. If they want to remain sober and demonstrate motivation to follow a recovery plan, they may be best served by an outpatient treatment (Kennedy et al., 2006).

### *Tobacco cessation*

Smoking is the single most important health risk in military personnel. Smoking affects personnel readiness through lower levels of physical fitness, increased injuries, and more sick days. Again, increasing tobacco use is due to stressful circumstances like deployment and high workload. The prevalence of smoking in the military (any smoking in the past 12 months) is as follows: Army, 35.6%; Navy, 36.0%; Marine Corps, 38.7%; Air Force, 27.0%. Almost every military installation in the West offers programs for tobacco cessation. Most of these programs use a combination of a behaviorally based program combined with nicotine replacement therapy (NRT) such as patches and gum and buproprion hydrochloride. Abstinence rates with these tobacco cessation program remain high (Peterson, 2006).

## Prevention of substance use disorders

Various military policies around the globe mandate prevention training for 100% of new military members, and annual training is required for all troops, in addition to random urine drug testing. The basic objectives of promoting mission readiness and the health and wellness of troops through the prevention of substance abuse are promoted via these policies. The *Navy's program* is a good example of how to disseminate best-practice information on alcohol abuse prevention (navdweb.spawar.navy.mil). The "three Rs" (relationship, relevance, and responsibility) are identified as a core program: a positive mentoring relationship, the relevance of everyone's role in the success of the mission, and the responsibility to learn and integrate expectations, as well as leadership's responsibility to provide information and facilitate the prevention program (Kennedy et al., 2006).

## Sleep issues in military psychology

According to a survey, around 48% of the military population suffers from significant sleep problems with no significant differences between deployment status and high and low combat exposure. Moreover, 31% of respondents were in "extreme short sleeper category" (five or fewer hours of sleep per night) (Campbell et al., 2017).

### *Sleep regulation*

There is two-process model of sleep regulation regulating many aspects of sleep including its timing, duration, stages, and quality. The first biological process, called sleep homeostasis, is a drive for sleep that increases with time awake and attenuates with time asleep. The sleep homeostatic drive would continuously increase with sleep deprivation such as on a typical day of 18 hours of being awake. The second biological process of sleep regulation is circadian rhythm, which is a 24-hour biological clock that synchronizes the timing of sleep and wakefulness. Various hormones like melatonin and core body temperature are directed by the circadian clock. These two processes are physiologically distinct; when aligned, they interact to promote consolidated periods of sleep during darkness at night and wakefulness during the day. Regulated sleep is known to have vital implications for healthy cognition and behavior (Campbell et al., 2017).

## Health consequences

The American Academy of Sleep Medicine and Sleep Research Society recently recommended that adults obtain 7–9 hrs of sleep per night (Watson et al., 2015). Sleep durations below this recommended amount, or poor sleep quality, have been associated with negative health consequences like weight gain, insulin resistance, increased risk of heart failure and coronary artery disease (due to insomnia), and immune dysfunction, to name a few (Campbell et al., 2017). Per the most comprehensive study by Mysliwiec et al. (2013) in military personnel who underwent polysomnography, the most commonly occurring sleep disorders were: mild obstructive sleep apnea (OSA; 27.2%), insomnia (24.7%), moderate-to-severe OSA (24%), behaviorally induced insufficient sleep syndrome (8.9%), snoring (5.3%), and paradoxical insomnia (5.1%). The most prevalent psychiatric co-morbidities were depression (22.6%), anxiety (16.8%), and PTSD (13.2%). Sleep problems could precede or co-occur or be a consequence of psychiatric co-morbidities. Moreover, deployment status is associated with poor quality of sleep (around 70%) but mediated by mental health morbidities (Campbell et al., 2017). Fatigue remains a challenge to high performance, and sleep disruption is one of the several contributors to fatigue. Fatigue is especially relevant in light of continuous operations and sustained operations due to related performance and safety decrements and possible adverse outcomes. The Aerospace Medical Association has released a position statement as countermeasures for aviation fatigue (Caldwell et al., 2009):

1. Fatigue is a physiological problem that cannot be overcome by motivation, training, or will power.
2. People cannot reliably self-judge their own level of fatigue-related impairment.
3. There are wide individual differences in fatigue susceptibility that must be taken into account but which presently cannot be reliably predicted.
4. There is no one-size-fits-all "magic bullet" (other than adequate sleep) that can counter fatigue for every person in every situation.
5. Valid counter-fatigue strategies will enhance safety and productivity, but only when they are correctly applied.

### *Maritime operations*

Shift work is an unavoidable reality in maritime operations. Shift work that is not aligned with the day/night, wake/sleep cycle is a threat to psychological and physical health, resulting in loss of sleep homeostasis and a pathological condition known as shift work disorder (SWD). Of particular concern is the common 5/10 schedule in which sailors are on-duty for 5 h, followed by 10 h off (15-hour day). The 5/10 schedule results in different clock hours throughout the day–night cycle over a 72-hour period, resulting in de-synchrony between the sleep homeostatic drive and the internal circadian clock. The 5/10 shift schedule is associated with low psychological resilience and psychomotor performance (Campbell et al., 2017).

### *Treatment implication*

Molecules like benzodiazepines, Z drugs like zolpidem (for the short term), and like ramelteon (for chronic usage) can be used for insomnia as per the recommendations. Nonpharmacological interventions, specifically cognitive behavioral therapy for insomnia (CBT-I), is the recommended first-line treatment for insomnia (Schutte-Rodin, Broch, Buysse, Dorsey & Sateia, 2008).

## Neuropsychological issues in military

Neuropsychology is the science and study of brain-behavior relationships and the clinical application of that knowledge. The primary goal of neuropsychology in armed forces is to help service members who have experienced neurological disorders or injuries by providing assessment, diagnosis, and treatment plans to foster effective recovery (Green, Jacobson, Waggoner & Armistead-Jehle, 2017).

*Traumatic brain injury* (TBI) is one of the most commonly occurring neurological conditions. Other neurological morbidity includes seizure disorders, cerebral vascular accidents, neoplasms, neurodegenerative conditions, hypoxia, etc. TBI can result in a variety of cognitive, emotional, behavioral, and physical sequelae, depending on the severity and the location of the cerebral damage. The cognitive deficits often involve attention and concentration, executive functioning, memory, and expressive language. Emotional and behavioral problems include apathy, irritability, disinhibition, depression, anxiety, and mood lability. The physical symptoms include dizziness, balance problems, vision changes, hearing changes, and headaches (Ryan, Zazeckis, French & Harvey, 2006; Green et al., 2017).

There have been two types of TBI: blunt force and blast wave (bTBI). Few of the neurobiological mechanisms have been elucidated to explain blast wave TBI (Courtney & Courtney, 2015):

1. *Acceleration mechanism*: Rotational accelerations of the brain caused by exposure to a blast wave may result in bTBI.
2. *Direct cranial entry*: A pressure transient traverses the skull and directly injures brain tissue. The pressure transient may result from direct transmission due to bulk motion of the skull. Localized peak pressures may result from constructive interference of internally reflected waves.
3. *Thoracic mechanism*: Combination of a pressure transient reaching the brain via the thorax and a vagally mediated reflex result in bTBI. Transient increased intracranial pressures in the cerebral vasculature may result from high speed propagation of a pressure transient without significant bulk motion.

*Concussion injury.* Concussion or mTBI has been called the "signature injury" in military circles. Approximately 80% of concussions occur in garrison, calling into question whether it is truly a deployment-related issue that has a higher representation than other battle-related injuries. Concussive symptoms such as headache, photophobia, phonophobia, sleep difficulties, dizziness, and psychiatric conditions after three months of recovery is known as persistent postconcussive symptoms (PPCS) and is prevalent in around 15% of individuals. PPCS needs detailed evaluation which includes neuropsychological testing, a neurological examination, an audiology examination, a psychiatric interview, an EEG, and neuroimaging to look for evidence of diffuse axonal injury or hemorrhage (Ryan et al., 2006; Green et al., 2017).

*Gulf War Syndrome.* Gulf War syndrome constitutes a wide range of physical (e.g., fatigue, pain, sleep disturbance, fever, rashes, tremor, and sexual dysfunction), cognitive (e.g., attention and memory problems), and psychological (e.g., depression and anxiety) symptoms (Hom, Haley & Kurt, 1997; Ryan et al., 2006). Various theories of the origin of Gulf War syndrome exist. These range from the influence of preexisting conditions or exposure to toxins or inoculations to psychological problems such as posttraumatic stress disorder or depression.

*Malingering.* World War II saw the publication of multiple articles on malingering in order to avoid military service or discipline. Common methods of malingering purported at this time were consuming alcohol, epinephrine, sugar, and cathartics; claims of pain or other sensory problems (e.g., blindness); claims of motor dysfunction; feigning of insanity; self-mutilation or exaggeration of real symptoms; etc.

Malingering in a military context (as per Uniform Code of Military Justice) refers to any service member who for the express purpose of avoiding a military duty feigns illness, physical disability, or a mental health issue or intentionally inflicts an injury to oneself (Joint Service Committee on Military Justice, 2005). The malingerer may simulate depressive symptoms, make suicidal comments, or adopt nonexistent psychotic symptoms to avoid a specific military duty, to be relieved from a specific duty (e.g., deployment), or to attempt to leave military service prematurely. Malingering could pose difficult diagnostic dilemmas for a military psychologist. There is no litmus test for malingering. The following red flags should raise the suspicion: the member keeps adding symptoms during the interview, the symptoms do not make diagnostic sense, there seems to be clear secondary gain in avoiding duty, elevations on psychological testing suggest faking a bad profile, and/or a resolution of symptoms occurs immediately after the desired outcome (Ryan et al., 2006).

The phenomenon of denial of actual symptoms which are present is known as reverse malingering, faking good, or dissimulation. Individual are evaluated for depression, suicidal or homicidal ideation, an alcohol incident, or domestic abuse. The secondary gain is to avoid the stigma of a mental health diagnosis or a negative career impact. In cases of suspected simulation and dissimulation, information must be collected from at least four sources. The first source is companions or supervisors. The second source is provided by the clinical interview. The third source is the medical record, and the fourth is psychological testing (Ryan et al., 2006).

## Conclusion and future directions

A neurobiological approach or perspective has important implications for military-related mental health morbidities. Studying neurobiology via the help of neuroimaging, either structural or functional, would help us to better delineate brain regions and networks, e.g., for those involved in suicide risk, and could be a way to track the effectiveness of interventions targeting such specific neural networks and brain regions. Future imaging–genetic approaches which combine neuroimaging and genomic study would help us to study the association between circuitry-related changes and gene (with epigenetic modification) function and behavioral neuroscience. Neurobiological models using state-of-the-art high-throughput resting-state functional MRI could help us to enumerate quantitative phenotypes for molecular genetic studies and biomarkers of pathological processes in the brain. These biomarkers in future would encourage more sophisticated and personalized treatments like transcranial magnetic stimulation in the military context.

## References

Anton, R. (1999). What is craving: Models and implications for treatment. *Alcohol Research and Health, 23*, 165–173.

Armed Forces Health Surveillance Center (AFHSC). (2014). Suicides and suicide attempts among active component members of the U.S. Armed Forces, 2010–2012: Methods of self-harm vary by major geographic region of assessment. *Medical Surveillance Monthly Report, 21*, 2–5.

Caldwell, J. A., Mallis, M. M., Caldwell, J. L., Paul, M. A., Miller, J. C., & Neri, D. F. (2009). Fatigue countermeasures in aviation. *Aviation, Space, and Environmental Medicine, 80*, 29–59.

Campbell, J. S., Markwald, R., Chinoy, E. D., Germain, A., Griese, R. E., Lim, I., & Bowles, S.V. (2017). A sleep primer for military psychologists. In S.V. Bowles, & P.T. Bartone (Eds.), *Handbook of Military Psychology* (pp. 239–260). Washington, DC, USA, Springer.

Campise, R. L., Geller, S. K., & Campise, M. E. (2006). Combat stress. In C.H. Kennedy & E.A. Zillmer (Eds.). *Military Psychology* (pp. 215–240). New York: The Guilford Press.

Courtney, A., & Courtney, M. (2015). The complexity of biomechanics causing primary blast-induced traumatic brain injury: A review of potential mechanisms. *Frontiers in Neurology*, *6*, 1–12.

Falk, B., Hersen, H., & Van Hasselt, V. B. (1994). Assessment of post-traumatic stress disorder in older adults: A critical review. *Clinical Psychology Review*, *14*, 383–416.

Fawcett, J., Busch, K. A., Jacobs, D., Kravitz, H. M., & Fogg, L. (1997). Suicide: A four-pathway clinical biochemical model. *Annals of the New York Academy of Sciences*, *836*, 288–301.

Fisher, C. A., Hoffman, K. J., Austin-Lane, J., & Kao, T. (2000). The relationship between heavy alcohol use and work productivity loss in active duty military personnel: A secondary analysis of the 1995 Department of Defense worldwide survey. *Military Medicine*, *165*, 355–361.

Garg, S., Dharmadhikari, A. S., & Sinha, V. K. (2012) Craving in substance use disorders. *Indian Journal of Social Psychiatry*, *28*(1–2), 43–52.

Ghahramanlou-Holloway, M., Baer, M. M., Neely, L.L., Koltko, V., & Nielsen, M. K. (2017). Suicide prevention in the United States military. In S.V. Bowles, P. T. Bartone (Eds.). *Handbook of Military Psychology* (pp. 73–88). Washington, DC, USA, Springer.

Green, R. R., Jacobson, D. A., Waggoner, J. W., & Armistead-Jehle, P. (2017). Neuropsychology in the Military. In S.V. Bowles & P.T. Bartone (Eds.). *Handbook of Military Psychology* (pp. 137–156). Washington, DC, USA, Springer.

Grunebaum, M. F., Ellis, S. P., Duan, N., Burke, A. K., Oquendo, M. A., & Mann, J. (2012). Pilot randomized clinical trial of an SSRI vs bupropion: Effects on suicidal behavior, ideation, and mood in major depression. *Neuropsychopharmacology*, *37*, 697–706.

Hawton, K., & Heeringen, K.V. (2009). Suicide. *Lancet*, *373*, 1372–1381.

Heeringen, K., & Mann, J. J. (2014). The neurobiology of suicide. *Lancet Psychiatry*, *1*, 63–72.

Hom, J., Haley, R. W., & Kurt, T. L. (1997). Neuropsychological correlates of Gulf War syndrome. *Archives of Clinical Neuropsychology*, *12*, 531–544.

IOM (Institute of Medicine). (2012). *Treatment for Posttraumatic Stress Disorder in Military and Veteran Populations: Initial Assessment*. Washington, DC: The National Academies Press. ISBN: 978-0-309-25421-2, 396 pages. Retrieved from http://www.nap.edu/catalog.php?record_id=13364

Johnston, S. L., Robinson, C., Earles, J. E., Via, J., & Delaney, E. M. (2017). State of psychology in the US armed forces. In S.V. Bowles & P. T. Bartone (Eds.), *Handbook of Military Psychology* (pp. 1–19). Washington, DC, USA, Springer.

Joint Service Committee on Military Justice. (2005). *Manual for Courts-Martial*. Washington, DC: U.S. Government Printing Office.

Kellner, C. H., Fink, M., Knapp, R., Petrides, G., Husain, M., Rummans, T., … Malur, C. (2005). Relief of expressed suicidal intent by ECT: A consortium for research in ECT study. *American Journal of Psychiatry*, *162*, 977–982.

Kennedy, C. H., Jones, D. E., & Grayson, R. (2006). Substance abuse services and gambling treatment in the military. In C. H. Kennedy & E. A. Zillmer (Eds.), *Military Psychology* (pp. 163–192). New York, London: The Guilford Press.

Koob, G. F., & Le Moal, M. (2008). Addiction and the brain antireward system. *Annual Review of Psychology*, *59*, 29–53.

Larkin, G. L., & Beautrais, A. L. (2011). A preliminary naturalistic study of low-dose ketamine for depression and suicide ideation in the emergency department. *International Journal of Neuropsychopharmacology*, *14*, 1127–1131.

Mahan, A. L., & Ressler, K. J. (2012). Fear conditioning, synaptic plasticity and the amygdala: Implications for posttraumatic stress disorder. *Trends in Neurosciences*, *35*, 24–35.

Mann, J. J., & Arango, V. (1992) Integration of neurobiology and psychopathology in a unified model of suicidal behavior. *Journal of Clinical Psychopharmacology*, *12*, 2S–7S.

Mehlum, L., & Weisaeth, L. (2002). Predictors of posttraumatic stress reactions in Norwegian U.N. peacekeepers 7 years after service. *Journal of Trauma and Stress*, *15*(1), 17–26.

Morey, R. A., Gold, A. L., LaBar, K. S., Beall, S. K., Brown, V. M., Haswell, C. C.,… McCarthy, G. (2012). Amygdala volume changes in posttraumatic stress disorder in a large case-controlled veterans group. *Journal of the American Medical Association: Psychiatry*, *69*, 1169–1178.

Mysliwiec, V., Gill, J., Lee, H., Baxter, T., Pierce, R., Barr, T.L.,… Roth, B. J. (2013). Sleep disorders in US military personnel: A high rate of comorbid insomnia and obstructive sleep apnea. *Chest*, *144*(2), 549–557.

Nizamie, S. H., & Garg, S. (2012). Depression and suicide. In N. G. Desai (Ed.), *Dealing with Depression in Medically Ill Patients* (pp. 50–73). New Delhi, India: Elsevier.
Paulus, M. P., Tapert, S. F., & Schulteis, G. (2009) The role of interoception and alliesthesia in addiction. *Pharmacology, Biochemistry and Behaviour, 94*, 1–7.
Peterson, A.L. (2006). Clinical health psychology and behavioral medicine in military healthcare settings. In C. H. Kennedy & E. A. Zillmer (Eds.), *Military Psychology* (pp. 74–104). New York, London: The Guilford Press.
Redish, A. D. (2004). Addiction as a computational process gone awry. *Science, 306*, 1944–1947.
Robinson, T., & Berridge, K. (1993). The neural basis of craving: An incentive-sensitization theory of addiction. *The Behavioural and Brain Sciences, 18*, 247–291.
Ryan, L. M., Zazeckis, T. M., French, L.M., & Harvey, S. (2006). Neuropsychological practice in the military. In C. H. Kennedy & E. A. Zillmer (Eds.), *Military Psychology: Clinical and Operational Applications* (pp. 105–129). New York, London: The Guilford Press.
Scaglione C., & Lockwood, P. (2014). Application of neuroscience research to the understanding and treatment of posttraumatic stress disorder (PTSD). *International Journal of Applied Science and Technology, 4*, 35–45.
Schutte-Rodin, S., Broch, L., Buysse, D., Dorsey, C., & Sateia, M. (2008). Clinical guideline for the evaluation and management of chronic insomnia in adults. *Journal of Clinical Sleep Medicine, 4*(5), 487–504.
Smith, M. E. (2005). Bilateral hippocampal volume reduction in adults with post-traumatic stress disorder: A meta-analysis of structural MRI studies. *Hippocampus, 15*, 798–807.
Watson, N. F., Badr, M. S., Belenky, G., Bliwise, D. L., Buxton, O., Buysse, D., … Tasali, E. (2015). Joint consensus statement of the American Academy of Sleep Medicine and Sleep Research Society on the recommended amount of sleep for a healthy adult: Methodology and discussion. *Journal of Clinical Sleep Medicine, 11*, 931–952.

# 24

# CLINICAL HEALTH PSYCHOLOGY APPLICATIONS IN MILITARY SETTINGS

*Ryaja Johnson and Larry C. James*

The health of the U.S. Armed Forces is paramount for the defense of the United States. More than 2.8 million service members have been deployed in support of wars in Afghanistan and Iraq since 2001 (Meadows et al., 2018). Riggs and Mallonee (2017) estimated that 15%–35% of service members returning from the wars experienced substantial psychological distress and reintegration problems. The most common injuries sustained by service members from almost 20 years of continuous war are, posttraumatic stress disorder (PTSD; 4%–25%), mild traumatic brain injury (mTBI; 8%–20%), and depression (5%–15%; RAND, 2016; Riggs & Mallonee, 2017). It is noteworthy to mention that service members report a variety of other problems that are psychological or impact psychological functioning; these include anxiety, sleep disturbances, diabetes, chronic pain, asthma, relationship difficulties, substance use problems, and suicidal ideations (Hunter et al., 2018; Ogbeide, Landoll, Nielsen, & Kanzler, 2018; Riggs & Mallonee, 2017).

The widely-accepted definition of "health" was authored by the World Health Organization in the Preamble to the Constitution: "Health is a state of complete physical, mental and social well-being and not merely the absence of disease or infirmity" (1948). The U.S. Department of Defense (DoD) has taken considerable steps to improve the population health and mission readiness of the 1.37 million active duty service members and an additional 170,000 Guard and Reserves members seen within the Military Health System (MHS; Military Health System, 2019). One major step taken by the DoD in most military primary care treatment facilities was the integration of behavioral health personnel, mainly clinical health psychologists, as full-time team members in what is referred to as "Patient-Centered Medical Homes" (PCMH; Hunter, Goodie, Dobmeyer, & Dorrance, 2014).

To maintain the health of service members, MHS adopted Quadruple Aim as their ultimate goal. Quadruple Aim consists of the following four objectives:

- Readiness: Ensuring that the total military force is medically ready to deploy and that the medical force is ready to deliver health care anytime, anywhere in support of the full range of military operations, including humanitarian missions.
- Population Health: Improving the health of a population by encouraging healthy behaviors and reducing the likelihood of illness through focused prevention and the development of increased resilience.

- Experience of Care: Providing a care experience that is patient and family centered, compassionate, convenient, equitable, safe, and always of the highest quality.
- Per Capita Cost: Creating value by focusing on quality, eliminating waste, and reducing unwarranted variation; considering the total cost of care over time, not just the cost of an individual health care activity (MHS, 2013).

The role of a clinical health psychologist naturally complements the DoD and MHS aims to maintain a healthy fighting force. The primary objective of a clinical health psychologist is to apply scientific knowledge of the interconnectedness of behavioral, emotional, cognitive, social, and biological components in health and disease to promote and maintain health. Clinical health psychologists accomplish their objective from a psychological perspective of prevention, treatment, and rehabilitation of illnesses and the improvement of the health care system (Council of Clinical Health Psychology Training Program, 2019).

Clinical health psychologists understand common medical problems and treatments and take a biopsychosocial approach to screening, assessment, and intervention (Hunter et al., 2014). Targeting behavioral factors that contribute to a patient's physical health is the distinct focus in clinical settings (Clinical Health Psychology Specialty Council, 2017; Council of Clinical Health Psychology Training Program, 2019). Clinical health psychologist can employ numerous evidence-based interventions designed to target the diverse array of problem presentations common in primary care. Interventions primarily consist of cognitive–behavioral approaches (Haibach et al., 2017). A clinical health psychologist is typically referred to as Behavioral Health Consultant (BHC) when integrated into a primary care setting because there is an emphasis on consultation, with the BHC providing feedback to the primary care physician on a unified treatment plan for a patient (Corso et al, 2012; Dobmeyer, 2013; Landoll, Nielsen, & Waggoner, 2017; Ogbeide et al., 2018).

## Unique clinical applications for military veterans suffering from mental health and/or medical disorders

Service members experience a variety of mental and medical concerns during their military careers. Common complaints are PTSD, mTBI, depression, anxiety, insomnia and nightmares, diabetes, chronic pain, asthma, relationship difficulties, substance use problems, and suicidal ideations (Dobmeyer, 2013; Hunter et al., 2018; Ogbeide et al., 2018; Parker et al., 2017; RAND, 2016; Riggs & Mallonee, 2017). Additionally, service members compared to civilians are more likely to smoke cigarettes and misuse alcohol (DoD, 2013). Weight management has also become a great concern in recent years (Haibach et al., 2017; Meadows et al., 2018; Parker et al. 2017). In 2015, the prevalence of obesity amongst service members ranged from 6.4% in the Marine Corps to 18.0% in the Army (Meadows et al., 2018).

When service members returning from deployment suffer from mental health disorders or preventable chronic diseases, their readiness for duty is greatly impeded. Additionally, service members' ability to effectively reintegrate with families and their civilian communities is negatively impacted after deployment (Riggs & Mallonee, 2017). Given the large number of service members who have been deployed and the subsequent high psychological burden associated with combat, it is of no surprise that the DoD has devoted considerable attention and resources to behavioral health support (Russell et al., 2014).

The DoD has recognized that the health and ultimate readiness of the Armed Forces depend upon psychosocial well-being. Since the late 1990s, the DoD has been implementing mental health-related programs to increase readiness (Hunter et al., 2014). One of the major evidence-based transformations within DoD was the integration of biomedical and psychosocial care into the primary care setting (Haibach et al., 2017; Mauksch, Peek, & Fogarty, 2017). The integration of clinical health psychologists within primary care clinics in military medical facilities is known as "Primary Care Mental Health" (PCMH); which is largely modeled after the civilian consultative model of integration known as "Primary Care Behavioral Health" (PCBH; Dobmeyer, 2013; Landoll et al., 2017).

PCBH is a population-based health care service delivery model within primary care in which the medical staff and BHC work as a team to manage behavioral health problems and biopsychosocially influenced health conditions (Hunter et al., 2014; Ogbeide et al., 2018; Reiter, Dobmeyer, & Hunter, 2018). PCBH was designed to resemble the primary care model in structure and function with an emphasis on evidence-based assessment, intervention, and consultation for common mental health concerns, health risk behaviors, and chronic medical conditions (Corso et al., 2012; Haibach et al., 2017; Hunter et al., 2014). The BHC is a team-based generalist, providing fast-paced, same-day accessible services, consulting with providers and educating patients, inquiring about patient's clinical progress, and arranging follow-up care or referral to mental health specialists (Dobmeyer, 2013; Mauksch et al., 2017; Ogbeide et al., 2018; Reiter et al., 2018).

PCBH assessment and intervention are typically 15- to 30-minute appointments and are not considered psychotherapy due to patients visiting the BHC only once or twice (Corso et al., 2012; Haibach et al., 2017; Hunter et al., 2014). Interventions are largely empirically supported cognitive–behavioral treatments, e.g., behavioral activation, cognitive restructuring, goal-setting, problem-solving, relaxation training, self-monitoring, and stimulus control (Corso et al., 2012; Haibach et al., 2017; Hunter et al., 2014). Interventions are designed to improve patients' everyday functioning (Reiter et al., 2018).

Using PCBH as a guide, in 2010 the DoD and MHS developed PCMH with the intent of integrating clinical health psychologists within primary care clinics to serve as BHCs to the medical team (Hunter et al., 2014). The goal was to improve the detection, diagnosis, and treatment of psychosocial health issues in service members seeking medical care from their primary care physician (Hunter et al., 2014; Ray-Sannerud et al., 2012). Within the PCMH model, the primary care physician retains full responsibility for service members' care decisions (Ray-Sannerud et al., 2012).

The BHC receives a "warm handoff" of the patient from the primary care physician. A warm handoff is a transfer of care between two medical team members that occurs in front of the patient (Agency for Healthcare Research & Quality, 2017). A central expectation of the BHC by the medical team is a swift response to the primary care physician (Mauksch et al., 2017). The BHC's time with the patient is limited (typically one or two sessions, 15 to 30 minutes per session) and solution focused (Ogbeide et al., 2018; Ray-Sannerud et al., 2012). Within this brief window of time, the BHC is required to make rapid and accurate assessment and treatment decisions in support of the primary care physician's treatment plan (Corso et al., 2012). Given the time constraints in the primary care setting, the clinical health psychologist tends to be highly directive, ask closed-ended questions, ask the patient to generate specific examples of problems, educate patients about diagnoses, teach behavioral skills, and encourage skill practice between appointments (Corso et al., 2012). The BHC provides recommendations and feedback to the primary care physician to engage in collaborative decision-making regarding the overall care of the patient (Ray-Sannerud et al., 2012). If necessary, the BHC refers the patient to traditional psychotherapy (specialty mental health care) when the patient's symptoms or functioning has not improved after several BHC visits (Ogbeide et al., 2018). It is estimated that between 16% and 54% of patients with mild to moderate symptoms who are treated by BHCs are referred to specialty mental health providers (Mauksch et al., 2017).

The PCMH model of care is known by different names in the Air Force, Army, and Navy. Within each department, the implementation of PCMH varies as well.

## *Air Force*

The Air Force implemented the Behavioral Health Optimization Program (BHOP) circa 2000 (Dobmeyer, 2013; Hunter et al., 2014). The intent of BHOP was to improve clinical preventive services while meeting the needs of service members who reported low interest in receiving behavioral health care in spite of experiencing significant behavioral health concerns (Funderburk, Dobmeyer, Hunter, Walsh, & Maisto, 2013). The Air Force integrated clinical health psychologist into primary care settings in an attempt to achieve the following BHOP goals: (a) offer acceptable services to patients and their primary care providers, (b) include improved shared decision-making with patients, (c) increase access to behavioral health services, (d) have collaborative team-based health care management, (e) broaden the range of behavioral health services within primary care, and (f) be prevention-focused with early identification and treatment of behavioral health conditions (Hunter et al., 2014).

The BHC's fast pace mirrors the rapid pace of the primary care clinic. Typically, BHCs have open slots in their schedules for same-day appointments, which allows for a larger number of airmen to receive behavioral health services (Funderburk et al., 2013; Hunter et al., 2014). BHOP is managed centrally and has demonstrated the ability to be deployed across a variety of treatment environments within the Air Force (Landoll et al., 2017).

Ray-Sannerud et al. (2012) conducted a longitudinal study on patients who received behavioral health care at an Air Force Base integrated family medicine clinic. The results of their study suggest that approximately two years after the final appointment with a BHC, patients maintained global mental health functioning improvements, gained over the course of the BHC's intervention. The findings in the Ray-Sannerud study align with the growing body of research that supports the claim of symptom reduction and functional improvement across the course of integrated primary care treatment in general.

## *Army*

The Army launched the Re-Engineering Systems of Primary Care Treatment of Depression and PTSD in the Military (RESPECT-Mil) in 2004 (Hunter et al., 2014). RESPECT-Mil uses a care management model to enhance the identification, assessment, and treatment of the two most common behavioral health concerns that present in military primary care settings: PTSD and depression. RESPECT-Mil integrates clinical health psychologists into the primary care setting, providing soldiers with education, screening tools, and treatment. The Army is seeking to implement RESPECT-Mil into additional primary care settings in an effort to reduce the stigma associated with seeking mental health care (Army Medical Services Corps, n.d.). The narrow focus of RESPECT-Mil is in contrast to the PCBH model which was designed to target all behavioral health problems that present in a primary care setting (Hunter et al., 2014).

RAND conducted an independent implementation evaluation of RESPECT-Mil in 2015. The findings suggest that RESPECT-Mil is identifying a considerable number of service members who report depression and PTSD symptoms (Wong et al., 2015). Of the 13% of soldiers who screened positive, 65% resulted in a probable diagnosis of a mental health disorder (Wong et al., 2015). For the soldiers in the "depression prominent" category who received treatment, 42% experienced at least a 50% reduction in depression symptoms from baseline to the last follow-up assessment. Similarly, 33% of the soldiers in the "PTSD prominent" category who received treatment experienced at least a 50% decrease in symptoms (Wong et al., 2015).

### *Navy*

The Navy initially launched Behavioral Health Integration Program (BHIP) in 2003 to encourage collaborative primary care physicians and BHC care and management of psychosocial problems in the primary care settings (Hunter et al., 2014). In 2010, the U.S. Navy Bureau of Medicine and Surgery mandated that the Navy focus on readily accessible, comprehensive, and patient-centered care (Navy and Marine Corps Public Health Center Public Affairs, 2016). Out of that mandate, Medical Home Port (MHP) was developed—the Navy's version of the PCMH model (Navy and Marine Corps Public Health Center Public Affairs, 2016; Patient-Centered Primary Care Collaborative, 2019).

MHP aims to ensure that the medical care received by sailors and marines is all-inclusive and integrated with other care provided within the health care system (Patient-Centered Primary Care Collaborative, 2019). Presently, BHIP is the program being used by the Navy to seamlessly integrate mental health care into MHP. BHIP-MHP is increasing same-day access to mental health care and health-behavior services by integrating clinical health psychologists (referred to as Internal Behavioral Health Consultants; IBHCs) in the primary care setting (Navy and Marine Corps Public Health Center Public Affairs, 2016). Additionally, IBHCs improve triage to specialty mental health for patients who require specialty services (Hunter et al., 2014).

## Delivery of innovative clinical health psychology applications in a military setting

There is a growing body of research that confirms that integrated behavioral health models of service delivery can be effective and cost efficient, can improve primary care physician and patient satisfaction with care and patients' access to behavioral health services, and can lead to better patient health outcomes due to improved treatment adherence (Corso et al., 2012; Funderburk et al., 2013; Landoll et al., 2017; Ogbeide et al., 2018). Additionally, integrated behavioral health models reach populations that otherwise may not receive behavioral health services (Ogbeide et al., 2018). Patients receive more comprehensive care and validation of the importance of their behavioral health care needs (Shearer, 2012). Furthermore, patients are significantly more likely to engage in specialty mental health care after contact with a BHC (Ogbeide et al., 2018).

Integrating clinical health psychologists into primary care is now becoming a core element of standardized care across all branches of service (Parker et al., 2017). In a military medical setting, clinical health psychologists are tasked with not only providing clinical care for the common concerns that present in primary care settings but also providing care across a broad range of behavioral health concerns related to deployment (Riggs & Mallonee, 2017). Treatment engagement, insomnia and nightmares, stress management, physical and social environments, and spiritual practice have important physical and mental implications to service members' readiness (Haibach et al., 2017; Parker et al. 2017).

The United States DoD has taken significant steps to integrate clinical health psychology applications: how it evaluates, mitigates, and treats psychological injuries associated with deployment (Hunter et al., 2014; Riggs & Mallonee, 2017). Steps taken include deployment of mental health providers into theater with service members; consistent screening of service members for psychological complaints; education of leaders and troops regarding the signs, symptoms, and treatment of common psychological health problems; and efforts to reduce stigma associated with seeking psychological health care (Riggs & Mallonee, 2017).

James, Folen, Porter, and Kellar (1999) as well as others (James & Folen, 1999) pioneered innovative clinical health psychology applications in military settings. These authors embedded military health psychologists directly into primary care and specialty clinics more than two decades ago and pioneered the application of Neuro EEG biofeedback, pharmacological intervention for smoking cessation prescribed by health psychologists. Also, in 1999, James and Folen (1999) designed a clinical psychopharmacology training program for military clinical health psychologists. The program led to military clinical health psychologists being credentialed to prescribe psychotropic medications, order and interpret medical labs, and independently perform health assessments on medical patients at their military hospital.

Delivery of behavioral health care in the military is partially grounded on the assumption that service members are most comfortable receiving care from familiar service providers. It is important that the same behavioral health provider be stabilized within an assigned unit to provide service members with continuity of care, foster trusting relationships, and simplify behavioral health care access and utilization (Russell et al., 2014). Thus, an innovative approach is to increase the number of clinical health psychologists embedded within units. A clinical health psychologist within a unit will be best positioned to detect health risk behaviors, diagnose mental health problems, and provide behavioral health interventions for chronic medical conditions.

Each branch of service is currently at various stages of embedding military psychologists within operational units. These positions offer psychologists the opportunity to interact directly with the troops and leaders on a consistent basis (Riggs & Mallonee, 2017). Due to the nature of primary care settings, clinical health psychologists tend to see patients after they have developed medical concerns. If clinical health psychologists were embedded in military units with regular contact with leaders and troops, they could individually engage with a service member to rapidly identify and address issues before they become pathological (Riggs & Mallonee, 2017).

Given the high ops tempo of the military, there is a significant need for preventative care and brief interventions for service members with functional complaints that can potentially adversely impact individual readiness as well as unit readiness. Preventive interventions may take the form of educational briefings during commander's calls. Briefings can be specifically tailored for the unit's concerns or generalized for an overview of healthy behavioral strategies (Johnston, Robinson, Earles, Via, & Delaney, 2017). Unit leaders and troops may be encouraged to view the clinical health psychologist as working with the unit to achieve their goals, rather than working to disrupt their goals. The clinical health psychologist should strive to improve unit performance by improving physical health rather than treating mental disorders (Riggs & Mallonee, 2017).

Consultation with commanders has been a key role for military behavioral health providers in all service branches for many years (Johnston et al., 2017). Military units have well-recognized leadership hierarchies and are typically close-knit teams. The attitudes and viewpoints of unit members often reflect those of the unit leaders (Riggs & Mallonee, 2017). Educating unit commanders and leaders about what clinical health psychologists can do to further the mission of the unit can be valuable (Riggs & Mallonee, 2017). By working with commanders, clinical health psychologists may make psychological interventions be seen as part of an overall effort to improve the readiness and physical performance of the unit (Johnston et al., 2017; Riggs & Mallonee, 2017). Having a clinical health psychologist on site may enhance a commander's understanding of the connection between body and mind, thereby enabling unit leadership to detect when their troops are experiencing behavioral health issues. Commanders can have troops meet with the clinical health psychologist early enough to prevent medical conditions from degrading performance both professionally and personally (Johnston et al, 2017; Riggs & Mallonee, 2017). Finally, the embedded model may decrease negative stigma about receiving mental health treatment by encouraging early help seeking (Johnston et al, 2017).

The primary rationale for the DoD's inclusion of behavioral health providers within military units is to increase service members' access to care and promote early prevention and intervention (Russell et al., 2014). An embedded clinical health psychologist can provide on-the-spot consultation to commanders and risk assessment and crisis intervention to service members, as well as referrals to medical providers and/or specialty mental health providers. With a clinical health psychologist in the unit, a biopsychosocial assessment can be made to view troops with medical concerns in an occupational context (Mauksch et al., 2017). Potential behavioral health patients receive their first touch more quickly, and service members have better access with less stigma. Rather than waiting for outside referrals or overcoming resistance to go to a specialty clinic, the barriers are lowered for service members when clinical health psychologists are embedded in their units (Mauksch et al., 2017). Clinical health psychologists embedded into units enhance service member's operational effectiveness, increase psychological readiness to prevent negative mission impact, and reduce mental health stigma (Johnston et al., 2017).

For those clinical health psychologists who remain integrated into primary care clinics, it is still important to visit units periodically to interact with leaders and troops. While in a deployed environment, as part of or alongside operational units, clinical health psychologists are better able to interact informally with the troops. Through these interactions, they are able to evaluate any changes in behavior that might become pathological somewhere down the line (Riggs & Mallonee, 2017). While in a stateside environment, periodic visits to units will allow clinical health psychologists to identify behaviors that might warrant attention or treatment, detect early any psychological adjustment issues allowing for quick and brief interventions, and offer informal counseling to encourage positive health changes without documenting the service member's medical record (Riggs & Mallonee, 2017). Regular interaction with service members improves the chances that they will receive specialty mental health care when it is needed (Ogbeide et al., 2018; Riggs & Mallonee, 2017).

Clinical health psychologists in the military are best represented by the manner in which the specialty continues to leverage the biopsychosocial model to anticipate and meet the needs of the DoD, the MHS, and service members (Parker et al., 2017). The embedding of clinical health psychologists into operational units is a natural outgrowth of the biopsychosocial model of treatment. Unit embedding provides the type of continuity and access envisioned in the concept of the MHS's Quadruple Aim.

## Overview of military clinical health psychology graduate training programs

It is important to note that military psychologists and clinical health psychologists working in a military setting are not the same. As a specialized branch of psychology, military psychology applies psychological knowledge and practice to promote the overall readiness of individual service members and the military as a whole (Johnston et al., 2017). The application of psychological research to address military problems requires military psychologists to fully grasp military policies, procedures, and operations (Johnston et al., 2017).

Most military psychologists incorporate into their clinical work the practice and concepts of clinical health psychologists within general behavioral health clinics and/or primary care settings, but they are not considered clinical health psychologists (Parker et al., 2017). Individuals graduating from military-specialized clinical health psychology postdoctoral fellowship programs or those who obtain board certification in clinical health psychology are considered clinical health psychologists in the military (Parker et al., 2017). Military settings account for one-third of all American Psychological Association (APA)-accredited postdoctoral fellowships in clinical health psychology (Parker et al., 2017). The Air Force, Army, and Navy each offer APA-accredited training in clinical health psychology.

### *Air Force*

The Air Force offers an intensive predoctoral internship rotation and a two-year postdoctoral fellowship in clinical health psychology. Interns and fellows are exposed to a broad range of clinical health psychology-related clinical problems. These include but are not limited to the following: cardiac and pulmonary rehabilitation, diabetes education program, insomnia management, pain management, primary care behavioral health consultant service, sleep clinic, tobacco cessation, and weight management (59th Medical Wing, 2018; Parker et al. 2017). Interns learn evidence-based biopsychosocial assessment and treatment strategies (e.g., biofeedback and relaxation) to target a broad spectrum of clinical concerns (Parker et al. 2017).

The curriculum of the Air Force's Clinical Health Psychology Fellowship is designed to teach the application of psychological principles and behavioral strategies for management and prevention of chronic disease using a scientist-practitioner model of training (59th Medical Wing, 2018). Fellows are trained to be outstanding clinicians, researchers, teachers, supervisors, and consultants in the field of clinical health psychology and become uniquely equipped to address the complex presenting problems faced by airmen (59th Medical Wing, 2018).

### *Army*

The Army offers intensive predoctoral internship rotation and a two-year postdoctoral fellowship in clinical health psychology. Depending on the location of the training, the experience interns receive may vary. Interns serve in integrated primary care settings and are exposed to biopsychosocial assessment and treatment experiences as well as learning skills such as mindfulness-based stress reduction (Parker et al. 2017).

The curriculum of the Army's Clinical Health Psychology Postdoctoral Fellowship is designed to provide integrated training using a practitioner-scholar model to develop skills related to the unique application of clinical health psychology in the Army. Fellows are trained to develop and carry out treatment protocols, create and direct clinical programs, consult and collaborate with other providers in patient care, and educate other health care professionals about clinical health psychology contributions to patient care (San Antonio Military Medical Center/Brooke Army Medical Center, 2018). Fellows receive a breadth of experiences in pain clinic, amputee center, GI clinic, health lifestyle programs, oncology, cardiothoracic, and bariatric surgery (Parker et al. 2017; San Antonio Military Medical Center/Brooke Army Medical Center, 2018).

### *Navy*

The Navy offers a predoctoral internship rotation in clinical health psychology. The main objectives in the internship program are to ensure interns leave with experience and knowledge in the management of patients with medical complaints and have effective tools to offer sailors and marines in an operational setting (Parker et al. 2017). Skills taught are similar to those highlighted in the other military services' programs with a focus on biopsychosocial assessment to view patients in occupational, social, and personal contexts (Parker et al. 2017). The Navy does not offer a postdoctoral fellowship in clinical health psychology.

## Recommendations for civilian university training programs to enhance their curricula

Interest in military-related health psychology has undoubtedly risen due to ongoing war for almost 20 years. With the increasing prevalence of PTSD, mTBI, and behavioral health issues among service members, there is a full-scale push to increase the number of mental health providers and resources to behavioral health support within the DoD (Barry & Barry, 2017; Russell et al., 2014). This increasing need for mental health providers has made civilian university training programs an integral part in the training and development of future clinical health psychologists working in military settings.

The APA's taxonomy for education and training in professional psychology health service specialties provides an aspirational educational and training sequence for professional psychology health services specialties training programs (APA, 2012). Clinical health psychology specialty education and training begins at the doctoral level, continues during internship with experiential training, and usually concludes with intense postdoctoral training but can also include post-licensure training (Council of Specialties in Professional Psychology, 2019). The taxonomy highlights four stages of education and training (doctoral, internship, postdoctoral, and post-licensure training). Within each of the four stages of education and training, there are four levels of education and training opportunities (exposure, experience, emphasis, and major study). Each level of education and training opportunity represents an increased intensity in the education and training activities, *exposure* being the lowest intensity and *major study* being the highest intensity (APA, 2012).

At the doctoral level, civilian universities have opportunities to enhance their training programs to include an emphasis in military-related health psychology training at each level of education and training opportunity. Figure 24.1 is an example of how civilian universities can provide students with opportunities in coursework and clinical experience to learn about clinical health psychology in military settings.

The DoD requires their mental health providers to have training in evidence-based practices (EBP; Barry & Barry, 2017). The majority of the psychologists who are primarily being hired by the DoD to fill BHC positions have a cognitive-behavioral or behavioral-theoretical orientation (Funderburk et al., 2013). Ninety-six percent of Air Force BHP in the Funderburk et al. (2013) study reported using cognitive–behavioral approaches as at least one form of intervention. It is imperative that civilian training programs offer variety of EBP in their curriculum and externship/practicum sites to greatly increase students' ability to secure relevant internships and otherwise prepare for a career as a clinical health psychologist in a military setting (Barry & Barry, 2017).

For a psychologist to be culturally competent is as necessary in a military setting as in any other culture. To optimally perform as a clinical health psychologist in military settings, one must understand military norms as they pertain to culture, behaviors, and expectations, as well as other aspects that directly or indirectly impact mental health care efforts (Johnston et al., 2017; Riggs & Mallonee, 2017). Clinical health psychologists must be able to understand and speak the language of their particular branch of service and avoid psychological jargon when interacting with service members (Johnston et al., 2017). Taking military psychology classes will enable students to gain specialized knowledge and approaches to working with the diverse psychological needs of service members. Students will also become familiar with military culture, military and mental health policy, and the psychological and physiological effects of military duty.

Some graduate programs offer courses and externship/practicum opportunities for clinical and research training with military populations at DoD facilities, enabling students to gain

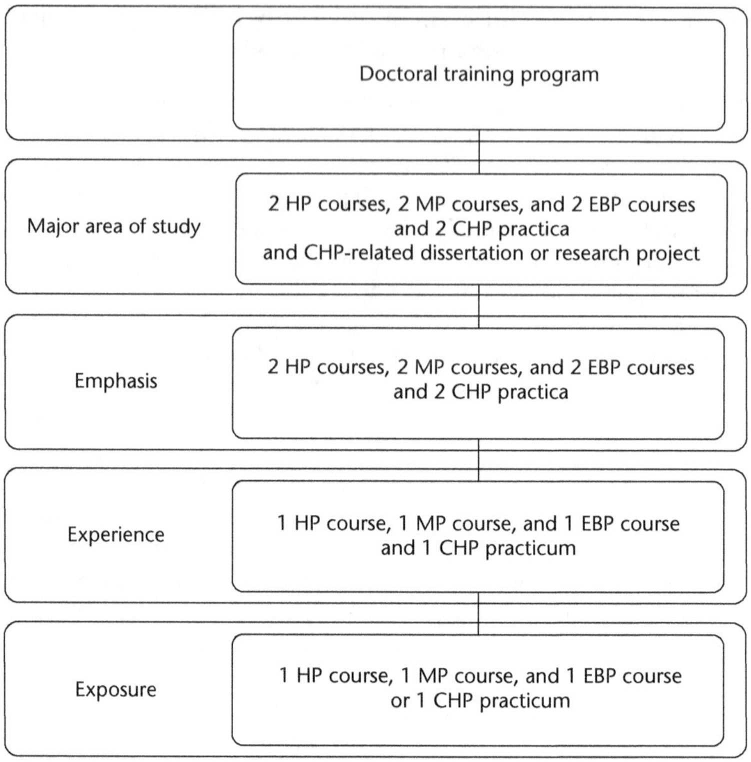


*Figure 24.1* Note: Clinical Health Psychology (CHP), Health Psychology (HP), Military Psychology (MP), Evidence-Based Practices (EBP).

*Source*: Adapted from Taxonomy for Clinical Health Psychology (Council of Specialties in Professional Psychology, 2019).

clinical experience directly with service members (Barry & Barry, 2017). Having clinical health psychology rotations within DoD externship/practicum experience will allow students the opportunity to promptly apply what they learn in the classroom as well as expose students to the rigors of BHC work prior to internship or postdoctoral experience.

## Challenges and innovations for the future

### *Challenges*

Military culture and values and the attitudes held by service members may inhibit treatment-seeking even if clinical health psychologists are embedded within units. The core values of each branch of service embody mental, moral, and physical strength. These values can contribute to maladaptive denial of pain and distress and therefore a reluctance to acknowledge the need for help (Riggs & Mallonee, 2017). In a study exploring barriers to seeking mental health treatment, researchers found that 65% of the service members did not seek treatment due to beliefs that they should be able to handle their problems on their own or because they did not want to believe they had a problem (Kim, Britt, Klocko, Riviere, & Adler, 2011).

Stigma surrounding behavioral health problems and biopsychosocially influenced health conditions is complex, as there are perceived external and internal negative consequences of acknowledging psychological problems. Service members in all branches of service express concerns that seeking mental health care will negatively impact their careers or security clearances (Johnston et al., 2017; Riggs & Mallonee, 2017). The chief stigma-related concern expressed by service members is worrying that their commanders or members of their unit will negatively respond to the presence of psychological health problems and/or seeking mental health care (Riggs & Mallonee, 2017).

A way internalized stigma manifests is reluctance to acknowledge to others, including care providers, behavioral health concerns (Riggs & Mallonee, 2017). This stigma is a significant cause of symptom under-reporting and has been well-documented in research exploring service members' behavioral health treatment-seeking decisions (Britt, Wright, Sipos, & McGurk, 2018; Riggs & Mallonee, 2017). Additionally, service members' negative attitudes toward mental health treatment are associated with lower interest in care; this is in spite of recognition of significant problems (Britt et al., 2018).

Another major challenge the DoD faces is the cost and logistics associated with training and employing hundreds of additional clinical health psychologists to embed into units. Currently, the Air Force trains and employs one or two clinical health psychologists each year (59th Medical Wing, 2018). The Army has one clinical health psychologist training position open per year (San Antonio Military Medical Center/Brooke Army Medical Center, 2018). According to Parker et al., there were less than 2% of active duty Naval psychologists trained as clinical health psychologists (2017). Currently, military settings account for one-third of all APA-accredited postdoctoral fellowships in clinical health psychology (Parker et al., 2017). Military and civilian institutions will have to increase the annual training of clinical health psychologists to meet the demand.

## Innovation

Attempts to reduce the stigma associated with receiving behavioral health services have focused efforts on normalizing psychological reactions to deployment and educating service members about symptoms of PTSD and other behavioral health problems (Riggs & Mallonee, 2017). Embedding clinical health psychologists into units increases accessibility to behavioral health care and early detection of behavioral health concerns, as well as reducing the stigma associated with being referred to mental health specialty clinics (Johnston et al., 2017; Mauksch et al., 2017; Riggs & Mallonee, 2017; Russell et al., 2014). To alleviate service members' fear that seeking mental health treatment will adversely impact their careers, clinical health psychologists can offer informal counseling to encourage positive behavioral change without documenting the counseling on the service member's medical records (Riggs & Mallonee, 2017).

In most clinical settings, due to time constraints and concerns about patient reactions, primary care physicians may find it difficult to speak with patients about behavioral health services (Landoll et al., 2017). Non-physicians spend more time with patients and may be in a better position to be a conduit to behavioral health services (Landoll et al., 2017). Clinical health psychologists can train medical personnel and military unit leaders on how to communicate the benefits of seeking BHC services. Training may increase positive perceptions of the usefulness of BHC for service members as well as medical personnel and unit leadership.

Within the PCMH model, the Air Force has recently instituted approaches to improve team members' understanding of the consultative model. The Air Force has developed clinical pathways for anxiety, depression, obesity, and sleep difficulties. These pathways enable the BHP to develop an initial treatment plan in collaboration with the rest of the patient's primary care team (Landoll et al., 2017). Team members become more accustomed to consulting with the BHP when there is a clearer understanding of how behavioral health contributes to the overall health of each service member.

The training conducted at the unit level can provide a realistic description of behavioral health problems and treatments as well as focus on increased access to a BHC and the effectiveness of the short-term, solution-focused interventions. Emphasis on unit readiness and improvement of physical performance may mitigate service members' reluctance to seek care, alleviate negative attitudes towards behavioral health care, and reduce stigma concerns within the unit (Britt et al., 2018).

## Conclusion

The health of the U.S. Armed Forces is paramount for the defense of the United States. The DoD has taken considerable steps to improve the population health and mission readiness of its active duty, Guard, and Reserves members. The role of a clinical health psychologist naturally complements the DoD and MHS objectives to maintain a healthy fighting force. A clinical health psychologist within a unit will be best positioned to detect health risk behaviors, diagnose mental health problems, and provide behavioral health interventions for chronic medical conditions. By working with commanders, clinical health psychologists may make psychological interventions be seen as part of an overall effort to improve the readiness and physical performance of the unit (Johnston et al., 2017; Riggs & Mallonee, 2017). Clinical health psychologists embedded into units enhance service members' operational effectiveness, increase psychological readiness to prevent negative mission impact, and reduce mental health stigma (Johnston et al., 2017).

Military culture and values and the attitudes held by service members may inhibit treatment-seeking even if clinical health psychologists are embedded within units. Stigma surrounding behavioral health problems and biopsychosocially influenced health conditions are complex, as there are perceived external and internal negative consequences of acknowledging psychological problems. To alleviate service members' fear that seeking mental health treatment will adversely impact their careers, clinical health psychologists can offer informal counseling to encourage positive behavioral change without documenting the counseling on the service member's medical records (Riggs & Mallonee, 2017). Embedding clinical health psychologists into military units increases accessibility to behavioral health care and early detection of behavioral health concerns, as well as reducing the stigma associated with being referred to mental health specialty clinics (Johnston et al., 2017; Mauksch et al., 2017; Riggs & Mallonee, 2017; Russell et al., 2014).

## References

59th Medical Wing. (2018). *Clinical health psychology fellowship*. Retrieved from https://www.59mdw.af.mil/About/Education-Training/Clinical-Health-Psychology-Fellowship/.

Agency for Healthcare Research and Quality. (2017). *Warm handoff: Intervention*. Retrieved from http://www.ahrq.gov/professionals/quality-patient-safety/patient-family-engagement/pfeprimarycare/interventions/warmhandoff.html.

American Psychological Association. (2012). *Education and training guidelines: A taxonomy for education and training in professional psychology health service specialties*. Retrieved from http://www.apa.org/ed/graduate/specialize/taxonomy.pdf.

Army Medical Services Corps. (n.d.). *Army clinical psychologists*. Retrieved from https://www.goarmy.com/content/dam/goarmy/downloaded_assets/pdfs/amedd/amedd-clinical-psychologist-success.pdf.

Barry, J. A., & Barry, D. M. (2017). Military psychology students: Contributions, pathways, and opportunities. In S.V. Bowles, & P.T. Bartone, (Eds.), *Handbook of Military Psychology: Clinical and Organizational Practice* (pp. 383–397). Cham, Switzerland: Springer.

Britt, T.W., Wright, K. M., Sipos, M. L., & McGurk, D. (2018). Testing the effects of using a behavioral health provider to deliver a routine post-deployment assessment for soldiers returning from combat. *Psychological Services*. Epub ahead of print. doi: 10.1037/ser0000266.

Clinical Health Psychology Specialty Council. (2017). *Petition for the recognition of a specialty in professional psychology*. Retrieved from https://www.apa.org/ed/graduate/specialize/crsppp.aspx.
Corso, K. A., Bryan, C. J., Corso, M. L., Kanzler, K. E., Houghton, D. C., Ray-Sannerud, B., & Morrow, C. E. (2012). Therapeutic alliance and treatment outcome in the primary care behavioral health model. *Families, Systems, & Health, 30*(2), 87–100.
Council of Clinical Health Psychology Training Program. (2019). *What is clinical health psychology?* Retrieved from http://www.cchptp.org/.
Council of Specialties in Professional Psychology. (2019). *CHP 2018 education and training guidelines*. Retrieved from https://www.cospp.org/clinical-health-psychology.
Department of Defense (DoD). (2013). *2011 health-related behavior (HRB) survey of active duty military personnel*. Retrieved from https://www.health.mil/Military-Health-Topics/Access-Cost-Quality-and-Safety/Health-Care-Program-Evaluation/Survey-of-Health-Related-Behaviors/2011-Health-Related-Behavior-Survey-Active-Duty.
Dobmeyer, A. C. (2013). Primary care behavioral health: Ethical issues in military settings. *Families, Systems, & Health, 31*(1), 60–68.
Funderburk, J. S., Dobmeyer, A. C., Hunter, C. L., Walsh, C. O., & Maisto, S. A. (2013). Provider practices in the primary care behavioral health (PCBH) model: An initial examination in the Veterans Health Administration and United States Air Force. *Families, Systems, & Health, 31*(4), 341–353.
Haibach, J. P., Haibach, M. A., Hall, K. S., Masheb, R. M., Little, M. A., Shepardson, R. L., … Hausmann, L. R. (2017). Military and veteran health behavior research and practice: Challenges and opportunities. *Journal of Behavioral Medicine, 40*(1), 175–193.
Hunter, C. L., Funderburk, J. S., Polaha, J., Bauman, D., Goodie, J. L., & Hunter, C. M. (2018). Primary care behavioral health (PCBH) model research: Current state of the science and a call to action. *Journal of Clinical Psychology in Medical Settings, 25*(2), 127–156.
Hunter, C. L., Goodie, J. L., Dobmeyer, A. C., & Dorrance, K. A. (2014). Tipping points in the Department of Defense's experience with psychologists in primary care. *American Psychologist, 69*(4), 388–398.
James, L. C., & Folen, R. A. (1999). A paradigm shift in the scope of practice for health psychologists: Training health psychologists to be primary care case managers. *Professional Psychology: Research & Practice, 30*(4), 352–356.
James, L. C., Folen, R. A., Porter, R. I., & Kellar, M. A. (1999). A conceptual overview of a proactive health psychology service: The Tripler health psychology model. *Military Medicine: An International Journal, 164*(6), 396–400.
Johnston, S. L., Robinson, C., Earles, J. E., Via, J., & Delaney, E. M. (2017). State of psychology in the US armed forces. In S.V. Bowles, & P. T. Bartone, (Eds.), *Handbook of Military Psychology: Clinical and Organizational Practice* (pp. 1–15). Cham, Switzerland: Springer.
Kim, P. Y., Britt, T. W., Klocko, R. P., Riviere, L. A., & Adler, A. B. (2011). Stigma, negative attitudes about treatment, and utilization of mental health care among soldiers. *Military Psychology, 23*, 65–81.
Landoll, R. R., Nielsen, M. K., & Waggoner, K. K. (2017). US Air Force behavioral health optimization program: Team members' satisfaction and barriers to care. *Family Practice, 34*(1), 71–76.
Mauksch, L., Peek, C. J., & Fogarty, C. T. (2017). Seeking a wider lens for scientific rigor in emerging fields: The case of the primary care behavioral health model. *Families, Systems, & Health, 35*(3), 251–256.
Meadows, S. O., Engel, C. C., Collins, R. L., Beckman, R. L., Cefalu, M., Hawes-Dawson, J., … Williams, K. M. (2018). *2015 Department of Defense Health Related Behaviors Survey (HRBS)*. Santa Monica, CA: RAND Corporation.
Military Health System. (2013). *MHS Quadruple Aim*. Retrieved from https://health.mil/Reference-Center/Glossary-Terms/2013/04/09/MHS-Quadruple-Aim.
Military Health System. (2019). *Patients by beneficiary category*. Retrieved from https://www.health.mil/I-Am-A/Media/Media-Center/Patient-Population-Statistics/Patients-by-Beneficiary-Category.
Navy and Marine Corps Public Health Center Public Affairs. (2016). *Ongoing collaboration results in improvements to mental health care for beneficiaries*. Retrieved from https://www.navy.mil/submit/display.asp?story_id=95218.
Ogbeide, S. A., Landoll, R. R., Nielsen, M. K., & Kanzler, K. E. (2018). To go or not go: Patient preference in seeking specialty mental health versus behavioral consultation within the primary care behavioral health consultation model. *Families, Systems, & Health, 36*(4), 513–517.
Parker, J., Afanador, J.H., Goodie, J. L., Porter, S. J., Weits, G. I., & Cassidy, D. G. (2017). Training and practice in military specialty psychology. In S.V. Bowles, & P. T. Bartone, (Eds.), *Handbook of Military Psychology: Clinical and Organizational Practice* (pp. 65–72). Cham, Switzerland: Springer.

Patient-Centered Primary Care Collaborative. (2019). *U.S. Navy medical home port.* Retrieved from https://www.pcpcc.org/initiative/us-navy-medical-home-port.

RAND Corporation. (2016). *Understanding Treatment of Mild Traumatic Brain Injury in the Military Health System.* Santa Monica, CA: RAND Corporation.

Ray-Sannerud, B. N., Dolan, D. C., Morrow, C. E., Corso, K. A., Kanzler, K. E., Corso, M. L., & Bryan, C. J. (2012). Longitudinal outcomes after brief behavioral health intervention in an integrated primary care clinic. *Families, Systems, & Health, 30*(1), 60–71.

Reiter, J. T., Dobmeyer, A. C., & Hunter, C. L. (2018). The primary care behavioral health (PCBH) model: An overview and operational definition. *Journal of Clinical Psychology in Medical Settings, 25*(2), 109–126.

Riggs D. S., & Mallonee S. (2017). Barriers to care for the complex presentation of post-traumatic stress disorder and other post-combat psychological injuries. In S.V. Bowles, & P.T. Bartone, (Eds.), *Handbook of Military Psychology: Clinical and Organizational Practice* (pp. 33–44). Cham, Switzerland: Springer.

Russell, D. W., Whalen, R. J., Riviere, L. A., Clarke-Walper, K., Bliese, P. D., Keller, D. D., … Thomas, J. L. (2014). Embedded behavioral health providers: An assessment with the Army National Guard. *Psychological Services, 11*(3), 265–272.

San Antonio Military Medical Center/Brooke Army Medical Center. (2018). *Clinical health psychology postdoctoral fellowship.* Retrieved from https://www.appic.org/Postdocs/Universal-Psychology-Postdoctoral-Directory-UPPD/Detail/id/2513.

Shearer, D. S. (2012). Civilian primary care prescribing psychologist in an army medical center. *Journal of Clinical Psychology in Medical Settings, 19*(4), 430–433.

Wong, E. C., Jaycox, L. H., Ayer, L., Batka, C., Harris, R., Naftel, S., & Paddock, S. M. (2015). *Evaluating the Implementation of the Re-Engineering Systems of Primary Care Treatment in the Military (RESPECT-Mil).* Santa Monica, CA: RAND Corporation.

World Health Organization. (1948). *Constitution of WHO: Principles.* Retrieved from https://www.who.int/about/mission/en/.

# 25

# SUICIDE PREVENTION STRATEGIES IN MILITARY POPULATIONS

*Marissa N. Eusebio, Abigale Brady, and Bruce Bongar*

Death by suicide among military populations has captured the attention and concern of leaders within the military, policy-makers, and the general public (Ramchand, Acosta, Burns, Jaycox, & Pernin, 2011). The increase in prevalence of psychopathology after war is a testament to the amount of stress that military members can experience. While deployed, military members may be exposed to environmental, physiological, and emotional stressors (Kennedy & Zillmer, 2012). These stressors can have numerous psychological consequences, including posttraumatic stress disorder (PTSD), substance abuse, depression, and suicidality (Bush et al., 2013; Kennedy & Zillmer, 2012). Likely the increased exposure to stressful and traumatic events combines with mental health stigma to create a "tough-it-out" mentality. Furthermore, this avoidance and suppression can lead to increased psychopathology, which can put the individual at risk for developing suicidal ideation.

In 2009, military suicide rates rose above civilian rates for the first time in 28 years, which is often attributed to the wars in Iraq and Afghanistan (Kennedy & Zillmer, 2012). Within this trend, and still today, nine years later, the Army has the highest suicide rate of all of the branches, with the Marine Corps coming in a somewhat distant second (Kennedy & Zillmer, 2012). The list of possible suicide risk factors within military branches is extensive, including interpersonal difficulties/lack of social support, substance use, and depressed mood. Protective factors can also come in many forms; however, a central theme is that of social support (Kennedy & Zillmer, 2012). Furthermore, 61% of deaths by suicide involve the use of a firearm; of these, 17% used a military-issue firearm (Bush et al., 2013). Risk and protective factors as well as individual characteristics combine with systemic factors to influence military suicide rates.

Among different military components, suicide rates vary in size and growth trends. According to the Department of Defense's Suicide Event Report (DoDSER), active duty military saw an increase in rate of suicides since 2013, beginning at 18.7 deaths by suicide for every 100,000 service members, increasing to 21.1 for every 100,000 in 2016 (Department of Defense, 2012–2016). Typically, the rates of suicide increase in active duty populations once deployment has concluded. It has been suggested that Operation Iraqi Freedom (OIF) and Operation Enduring Freedom (OEF) have exposed service members to stress during active duty to the extent that suicidal ideation develops despite historical trends, though it is unclear if this is a result of this specific deployment or mental health history (Harmon, Cooper, Nugent, & Butcher, 2016). Other risk factors for active duty military members include decreased fear associated with death and a desire to escape both psychological and environmental stress (Harmon et al., 2016). The reserves and National Guard components of the military have overall higher rates of suicides than the active duty component but have either fluctuated without a trending direction or remained stable (Department of Defense, 2012–2016).

## Comparison of military branches

Definitive statements about the different suicide rates among active duty and reserve and National Guard components cannot be made. Likewise, certain interbranch trends of suicidal behaviors can be observed. This can result due to accumulated stressors and stigmas associated with each branch's individual culture. For example, one Navy survey found that sailors were experiencing stress related to chain of command and deployment changes. A mental health provider recommended taking time ashore when possible, which resulted in stigma related to abandoning the crew (Millegan, Delaney, & Klam, 2016). As such, under-reporting is a reasonable assumption to make when considering that the Navy has the lowest suicide rate of the four main branches (Department of Defense, 2016). Additionally, in 1978, the Navy established its Special Psychiatric Rapid Intervention Teams (SPRINT) model. SPRINT typically consists of one mental health provider and a medical corpsman that are deployed upon request to evaluate mental health both at the individual and at the unit level (Milligan et al., 2016). While stigma still existed among crew members, the increased access to mental healthcare could be related to the Navy's low suicide rates. Across all branches, limitations of suicide rate trends should be considered, for example, lack of standardization of data collection procedures across branches and the absence of adjusting suicide rates based on enlistment numbers (i.e. higher suicide rates among officers in a branch with fewer officers than other branches) (Kennedy & Zillmer, 2012).

While trends can be difficult to directly attribute to specific factors, there are some relationships regarding military suicide and risk factors that can be reliably studied. Differences in suicide rates between branches can also be affected by deployment setting. A positive relationship has been observed between exposure to violence and suicidal ideation. One theory behind this relationship is the normalization of death, which in addition to suicidal ideation can lead to risk-taking behaviors (Bryan, Kanzler, Durham, West, & Greene, 2010). The probability of risk-taking behaviors, combined with access to firearms and a combat setting, present service members with means and opportunity to act on suicidal urges. Furthermore, for a deployed service member, access to mental health care can be limited. This can pose a risk to safety in order to get a provider to the unit or to transport someone to a larger base and get the service member to a provider. Even if a provider can be accessed, being in a combat zone can create a number of therapy-interfering behaviors such as poor sleep hygiene, poor homework compliance, and various treatment communication difficulties (scheduling, wellness checks) (Bryan et al., 2010). While different branches have created different prevention plans, these interfering behaviors can occur in any branch.

Another interbranch trend which is widely recognized is the higher suicide rates among Army and Marine service members (Ayer, Ramchand, Geyer, Burgette, & Kofner, 2016; Department of Defense, 2016). It has been theorized that stress related to instances of improvised explosive devices (IEDs) could explain the higher occurrence of Army suicide rates. A significant relationship between attempted or completed suicide and IED explosions was found by Ursano et al. (2017). This could contribute to Army suicide rates, as it is possible that the Army encounters IEDs more frequently than other branches. An IED explosion could cause a military member to experience grief related to unit casualties, confronting their own mortality and fears about living in an unstable environment. Military leaders as well as the Department of Defense have put strategies in place to prevent and treat mental health symptoms related to stress which deployed service members may encounter.

The Army and Marine Corps both rely on noncommissioned officers called gatekeepers to identify and report behaviors which could indicate suicidal ideation (Ayer et al., 2016). While this could reinforce elements of social support, it also introduces human error (gatekeeper evaluation) with regard to the ability to address these behaviors, as well as the role of stigma as to the screening of suicidal behaviors. About two-thirds of Army and Marine gatekeepers (64% and 73%, respectively) reported receiving 1–10 hours of suicide prevention training, while 34% of Army gatekeepers and 24% of Marine gatekeepers reported receiving over 11 hours of suicide prevention training (Ayer et al., 2016). Of note is that these results only report on attendance of training, not comprehension, and the quality of these training sessions is often unstandardized. In Marine populations, it is noted that there may be more stigma related to mental health results with regard to untreated suicidal ideation. In a sample of over 1700 Marines, when asked why they do not seek mental health care, the top three responses cited were embarrassment and unit cohesion/leadership dynamics (VanSickle et al., 2016).

## *Cultural intersectionality*

As previously noted, access to healthcare when deployed can be scarce. However, after returning to civilian life, new veterans can still struggle to access healthcare. Mental health stigma which may develop during deployment could persist once veterans return. This combined with the separation from a veteran's unit can make a veteran less likely to seek healthcare (while making them more likely to experience psychopathology) (McDaniel, Thomas, Albright, Fletcher, & Shields, 2018). Further, mental health care can be lacking in rural areas for both veterans and civilians, reducing access to mental health care (Brooks, Dailey, Bair, & Shore, 2016). Another trend in civilian populations which can also be seen in veteran populations is that low-income veterans experience poorer mental health symptoms, though the direction of this relationship is unclear (LeHavot, O'Hara, Washington, Yano, & Simpson, 2015; McDaniel et al., 2018). Furthermore, an association has been found between being of minority or low-income status and under-utilizing Veterans Health Administration (VA) healthcare (LeHavot, O'Hara, Yano, & Simpson, 2016).

Perceived trust of a provider can influence a veteran's decision to seek mental healthcare. Communication, openness, and perception of provider competence are known factors which can influence trust (van den Berk-Clark & McGuire, 2014; Chanfreau-Coffinier et al., 2018). Pease, Billera and Gerard (2016) discuss preliminary recommendations for providers seeking to build rapport with veterans such as acknowledging veteran status as a salient cultural identity (gathering information about time served, branch, and duties). Further, providers should incorporate aspects of positive psychology to emphasize growth rather than focusing on the illness, as well as working to empower veterans throughout care to combat stigma associated with psychopathology (Pease et al., 2016).

Military members and veterans cite stigma related to mental health as a reason not to seek mental healthcare. Within the identity of military member or veteran, other culturally salient identities that are stigmatized can affect an individual's willingness to seek treatment. In civilian populations, LGBTQ (lesbian/gay/bisexual/transgender/questioning) individuals are at a higher risk of experiencing suicidal ideation than non-LGBTQ individuals (Blosnich, Mays, & Cochran, 2014; Matarazzo et al., 2014; Sexton et al., 2018; Wilder & Wilder, 2012). In a study conducted by Blosnich and colleagues (2014), 47% of sexual minority veterans reported experiencing suicidal ideation at some point in their lives, compared to 22% of heterosexual veterans, a significant difference. For veterans identifying as a gender and sexual minority, an association was found between an increased risk of developing psychopathology and reports of stigma or bullying during deployment (Sexton et al., 2018). Further, isolation, discrimination, minority stress, and abuse have been identified as risk factors for suicidal ideation in this population (Chen, Granato, Shipherd, Simpson, & Lehavot, 2017; Wilder & Wilder, 2012). These factors can also be experienced by heterosexual military and veteran populations, creating a doubly increased risk. A deployed service member may experience isolation from family, whereas a veteran may also experience isolation from their unit. On September 20, 2011, "Don't Ask, Don't Tell," (the concept of not asking or disclosing non-heterosexual orientation) was repealed (Wilder & Wilder, 2012). This *allowed* for open communication of a new identity; however, stigma was still present. Internalized heterosexism within a unit (or with family prior to or post-deployment) can contribute to distress associated with increased risk of suicidal ideation (Wilder & Wilder, 2012). Deployed service members may choose to conceal this identity to avoid heterosexism and other discrimination, but this concealment can also lead to distress. In military LGBTQ populations, decreased social support was associated with increased suicidal ideation (Matarazzo et al., 2014).

Identifying as a sexual or gender minority can lead to a new level of isolation in which the veteran/active duty military individual encounters discrimination from civilian populations for being a veteran/military and encounters discrimination among military circles for being a sexual or gender minority. In spite of this, transgender military service members may be drawn to the military due to a desire either to mimic traditional gender roles or to express the gender with which they identify in a socially acceptable way (Lehavot et al., 2016). While military and veteran populations may experience increased suicidal ideation, for transgender military/veterans, this rate is higher for both ideation and completed suicides (Lehavot et al., 2016). Past year suicidal ideation and lifetime experience of plan or attempt has been correlated with being a transgender male, and specific factors include military stigma and mental health symptoms (Lehavot et al., 2016).

In addition to cultural effects related to identifying as transgender, additional trends in suicide behaviors can also be seen. In civilian populations, a known gender trend is that women attempt suicide more than men, but men complete suicide more than women (Freeman et al., 2017). In veteran populations, female veteran completion rates are closing the gap between male and female veterans who die by suicide (Dorsey Holliman, Monteith, Spitzer, & Brenner, 2018). In discussing military suicide rates, consideration should be given to the prevalence of military sexual assault. Research has shown that veterans who have been sexually assaulted while serving are more likely to attempt suicide, as well as more likely to die from suicide (Monteith et al., 2016). Further, the prevalence of military sexual assault is higher in women than in men (endorsed 25% of the time and 1.3% of the time, respectively, on VA screeners) (Monteith et al., 2016). This provides some evidence that female veterans may experience suicidal ideation more than male veterans. In addition, male military service members and veterans also experience unique stressors related to their gender which can be associated with suicidal ideation, such as stress related to meeting the male gender role. This can manifest in interpersonal interactions, relationships, and acknowledgment/expression of emotion (Sterling et al., 2017).

## Suicide risk

In a recent research study of OIF and OEF veterans and active duty service members, as well as non-deployed service members and veterans, researchers assessed the impact of suicide among military personnel (LeardMann et al., 2013). In addressing the prevalence and impact of suicide among military personnel since 2005, LeardMann and colleagues (2013) considered covariates including physical functioning, adverse life events, mental health, and behavioral factors (e.g., posttraumatic stress disorder (PTSD), depression, manic-depressive disorder, panic/anxiety symptoms, and alcohol-related problems) in relation to suicide risk.. Overall, this research study proposed that factors related to deployment, (e.g., length, number, combat exposure) were not indicative of risk for a suicide attempt. Rather, suicide risk factors such as male sex and mental disorders, which are also risk factors among civilian populations, were consistent within this sample of military personnel. LeardMann and colleagues suggest that overall stress related to being in the military during years of war may be a specific risk factor for suicide (2013). In addition to military related stress, they also identified manic-depressive disorder, depression, and alcohol-use problems as being mental health problems significantly related to suicide risk. These findings indicate that individuals with these specific mental health problems are at increased risk for suicide and that conducting thorough assessments of military populations may result in better success in detecting the presence of suicidal ideation. Aside from mental health disorders, research has shown that stressful life events such as unsuccessful romantic relationships, financial issues, legal/disciplinary problems, stress related to work, and employment limitations due to a medical condition are all factors that increase the risk of suicide among Army and Marine Corps populations.

In addition to specific risk factors for suicide, there is also some controversy regarding the categorization of death by suicide among military personnel. Often, research studies utilize the Defense Medical Surveillance System (DMSS) to determine deaths by suicide among military personnel. The DMSS contains mortality information on each branch of the military, as well as demographic and deployment data on each individual (Eaton, Messer, Wilson, Hoge, 2006). One issue when utilizing this tool to identify deaths by suicide among military personnel is the presence of mortality among military service members being "accidents" or "undetermined," – potentially leading to the under-detection of deaths by suicide in this population. Furthermore, Eaton et al. (2006) found that among the deaths documented in the DMSS, the Navy had a greater proportion of deaths labeled as "undetermined" or "pending" than any other branch of military, suggesting that deaths by suicide may be higher in the Navy than what is documented in the DMSS.

## Post-deployment psychological assessment

Although there appear to be some similarities among risk for suicide between civilian and military populations, it is important to consider distinct differences between these groups and how suicide may impact military personnel uniquely. In 1997, the U.S. military made it mandatory for all military personnel returning to the United States from deployment to undergo mental health screening (Department of Defense, 1997). The Department of Defense proposed the change in policy to "expand the concept of joint deployment medical surveillance to a more comprehensive approach to monitoring and assessing health occurrences related to participation of Service members in deployments" (p. 1). Currently, a service member can opt to complete the post-deployment assessments either in theater or once they have returned home, and the post-deployment reassessment is then completed within 90–180 days (Warner, Appenzeller, Mullen, Warner, & Grieger, 2008). Despite these assessments being mandated by the Department of Defense, there continues to be debate regarding the implementation of mental health screenings among military personnel (Zamorski, 2011). Based on a survey completed by Warner et al. (2008) service members' attitude towards post-deployment mental health screening varies. Service members disclosed that they were more likely to respond truthfully on the mental health screening when the person administering the screening was a mental health provider in the unit,

unit medical providers and mental health providers from other units being the next groups of providers service members were more likely to be truthful with. A common theme within this data is that service members were reporting that they are overall more comfortable being truthful with personnel who were military and had been part of a unit. This information can be helpful to providers considering the pervasive stigma many service members experience regarding mental health (Vogt, 2011). Furthermore, service members outlined barriers to seeking mental health treatment as being fear of work-related impediments, being treated differently by those in leadership positions, and fear that other service members would not have confidence in them because they had sought out mental health treatment. Overall, post-deployment psychological screening can help providers address mental health needs of service members, so long as the provider administering the assessment is someone whom the service member feels they can trust.

## *Suicide prevention strategies*

Utilizing suicide prevention strategies among service members can vary depending on a person's unique stress factors, deployment status, and the branch of the military. Many people debate whether the data regarding civilians and mental health is sufficient when assessing military personnel with mental health issues (Zamorski, 2011). One major challenge when addressing suicide prevention strategies is that many suicide prevention programs lack empirical data to support the effectiveness of the program (Ramchand et al., 2011). Typically, there are two categories that suicide prevention interventions fall under: awareness and skill building, and those that provide suicide screening followed by mental health referrals (Ramchand et al., 2011). Ramchand et al. (2011) conducted an extensive literature review and identified six practices that suicide prevention strategies should include: (1) raise awareness and promote self-care, (2) identify those at high risk, (3) facilitate access to quality care, (4) provide quality care, (5) restrict access to lethal means, and (6) respond appropriately. Suicide prevention strategies can be specific to the deployment environment, branch of military, and veteran status. While considering the unique nature of these identities can be helpful when considering suicide prevention, there is also a breadth of research that focuses on non-specific suicide prevention strategies that are targeted to military personnel as a whole.

## *Deployment environment*

Typically, during deployment, there are various mental health providers including psychiatrists, psychologists, occupational therapists, social workers, chaplains, and military enlisted mental health technicians present (Hill, Johnson, & Barton, 2006). When service members experience suicidal ideation during deployment, this poses a significant challenge for mental health providers to be able to provide adequate care and safety to the service member. Providing suicide prevention strategies during deployment is often difficult given the service member's access to weapons, distance from mental health providers, and job requirements while deployed (Hill et al., 2006). If a provider chooses to evacuate a soldier due to a service member's risk of suicide, then they are also depleting the fighting force and potentially putting others in danger, although this is often the simplest solution and best way for a service member to access proper care (Hill et al., 2006). In a research study that was conducted with soldiers during Operation Enduring Freedom II (OEF II), soldiers who experienced suicidal ideation were provided with a treatment plan that was unique to the level of severity and risk of suicide (Hill et al., 2006). For those who were at lower risk of suicide, determined by the treatment plan, safety plan, and assessment, interventions consisted of speaking with the chaplain or regular visits with mental health providers (Hill et al.,

2006). Those who were determined to be at higher risk for suicides during deployment were put on unit watch as a suicide prevention strategy, which focuses on social support to ensure the soldier's safety (Hill et al., 2006).

The United States Army defined unit watch as, "a commander's program that is designed and implemented to protect at-risk Soldiers from self-harm and harm to others" (Gould, Kalafat, HarrisMunfakh, Kleinman, 2010, p. 13) and can be used across military branches during deployment. A unit watch consists of a soldier being supervised at all times, with emphasis on the soldier not being left alone at any point while also not inflicting undue harm or humiliation to the soldier on unit watch (Gould et al., 2010). Often, a unit watch involves the soldier's personal belongings being searched for harmful weapons, which are removed if found, prohibiting substance use, and limiting a soldier's contact with people who may have a negative influence on them (Payne, Hill, & Johnson, 2008). The United States military conceptualizes suicidal ideation during deployment as stress fatigue, and unit watch is a way for a soldier to remain safe and supported while also not experiencing the stigmatization within the unit that often occurs after a psychiatric hospitalization (Payne et al., 2008). While there are several benefits of implementing a unit watch, there are also various challenges when implementing this suicide prevention strategy.

One major factor that should be considered when utilizing unit watch is that it is not recommended to be the single form of suicide prevention (Payne et al., 2008). Psychological and psychopharmacological treatments should also be considered in conjunction with a unit watch (Payne et al., 2008). Second, there is a significant lack of data on unit watch and the safety and efficacy of this intervention strategy (Payne et al., 2008). Finally, while unit watch may result in less stigmatization by fellow soldiers and service members than a psychiatric hospitalization, many soldiers who undergo unit watch do continue to experience some level of stigma (Payne et al., 2008). Although a unit watch does pose some challenges, it is a useful method for mental health providers to implement procedures to reduce the risk of suicide while also maintaining the fighting force.

The suicide prevention strategies that are often implemented when a soldier is deployed often differ when compared with non-deployed service members (Bryan et al., 2010). A major contributing factor is exposure to war zone violence – which has been shown to increase the prevalence of suicide risk (Bryan et al., 2010). Service members who have been deployed are more likely to own a firearm than non-deployed service members – putting combat-related service members at greater risk for death by suicide, as they have access to a lethal means for suicide (Bryan et al., 2010). Veterans often report experiencing a sense of belonging due the closeness of their unit while deployed, but they are also often experiencing a significant level of disconnect from their loved ones and home (Bryan et al., 2010). Communication is often difficult and unreliable while deployed, which may further decrease a service member's feeling of belonging (Bryan et al., 2010). Finally, factors such as insomnia, lack of willingness to disclose suicidal ideation, and limited access to mental health resources are all contributing factors to suicide risk that are unique to service members who are deployed, as opposed to those who are non-deployed.

Overall, service members who are deployed likely experience stressors that are distinct from service members who are non-deployed. Mental health providers who are in a deployment environment experience unique challenges and barriers when working with service members who are experiencing suicidal ideation. Mental health providers are faced with options such as psychiatric hospitalization, which may be the easiest solution but potentially damaging to the unit as a whole and result in the service member experiencing stigma as a result. Unit watch is a suicide prevention strategy that is often employed when a service member's suicide risk is elevated, although it may still result in stigmatization from the unit.

## Branch-specific strategies

In response to the impact of suicide within the military as a whole, the branches within the military have established methods for increasing awareness, education, and training. Similarly, Veterans Affairs has also begun to establish individualized suicide prevention strategies in an effort to decrease stigma and augment access to mental health care and resources. In addition to these strategies, there is also a breadth of research on suicide prevention that is generalized to the military as a whole and non-specific to specific identities within the military. The Department of Defense sponsors suicide-prevention strategies across various military service branches (Ramchand et al., 2011). There is a committee that is part of the Department of Defense that focuses on suicide prevention in the effort to reduce the risk the risk of suicide among all military populations (Ramchand et al., 2011).

### *U.S. Army*

Suicide prevention methods within the army are often disseminated through public awareness campaigns and psychoeducation provided by leaders and soldiers within the military (Ramchand et al., 2011). The Army focuses primarily on utilizing soldiers to help support other soldiers in crises and promote resilience (Ramchand et al., 2011). They also implement suicide prevention activities throughout multiple levels within the Army (Ramchand et al., 2011), including the Ask, Care, Escort (ACE) program that works towards reducing stigma training, providing soldiers information on how to identify suicidal behaviors (Griffith & Bryan, 2018).

### *U.S. Marines*

The Marines utilize mental health providers and other personnel within leadership roles in the Marines to train them on how to identify Marines at risk for suicide and the necessary steps to get adequate support when experiencing suicidal ideation (Ramchand et al., 2011). They also provide training relating to being able to identify suicide risk within their annual training. Similar to the Army, the Marines also provide suicide awareness through public platforms (Ramchand et al., 2011). Last, the Marines also have behavioral health professionals embedded into Marine infantry regiments to increase support of Marines before, during, and after deployment (Ramchand et al., 2011).

### *U.S. Air Force*

Within the Air Force, suicide prevention strategies focus on the highest-ranking officials with regard to encouraging and inspiring cultural changes and the perception of suicide (Ramchand et al., 2011). The Air Force takes the initiative to train and educate all personnel about suicide as well as procedures for when a fellow service member is experiencing suicidal ideation (Ramchand et al., 2011). They have also established the Integrated Delivery System (IDS), which focuses on specific suicide prevention strategies (Ramchand et al., 2011).

### *U.S. Navy*

The Navy conceptualizes suicide on a continuum of risk, and they utilize this model to address early intervention strategies that increase stress regulation and management when experiencing stressful life events (Ramchand et al., 2011). Similar to the Marines and Army, the Navy employs public platforms to provide psychoeducation regarding suicide (Ramchand et al., 2011). Finally, the Navy has adopted strategies that incorporate behavioral health care providers into nontraditional roles to increase their presence within the naval community (Ramchand et al., 2011).

## *Veterans*

Given the impact of deaths by suicide among veterans, some researchers have called for an increase in suicide research and prevention strategies, including more robust training for all providers within VA settings (York, Lamis, Pope, & Egede, 2013). Several strategies to help provide resources to veterans who are experiencing suicidal ideation have been implemented, including the Veteran Crisis Line, Suicide Prevention Coordinators, and increased access to telehealth (York et al., 2013). Universal prevention strategies that are utilized when working with veterans focus on psychoeducation about suicide risk as well as screening for specific suicide risk factors (Bruce, 2010). In addition to these broader programs and suicide prevention strategies, when a veteran is identified as being at increased risk for suicide, the VA has developed formalized individualized suicide prevention strategies.

Within the VA, providers emphasize the use of evidence-based practice when treating a veteran with suicide ideation (York et al., 2013). Selective prevention strategies focus on providing care and support to veterans who are known to be at risk for suicide (Bruce, 2010). At this level, evidence-based psychiatric treatments are implemented with the goal to reduce symptoms of mental health disorders, especially those related to suicide and suicidal ideation (Bruce, 2010; York et al., 2013). Interventions focus specifically on suicide as well as the proximal risk factors present (Bruce, 2010). These interventions revolve around monitoring the safety of the veteran and utilizing an individualized safety plan, as well as potentially using pharmacotherapy and psychotherapies for suicide risk (Bruce, 2010). Various therapeutic intervention strategies are frequently used when working with veterans who have experienced trauma, including cognitive processing therapy (CPT), prolonged exposure therapy (PE), and eye-movement desensitization and reprocessing (EMDR) (Reisman, 2016). Although these treatment interventions are specific for PTSD, some research suggests that PTSD and suicide ideation and behavior are associated (Reisman, 2016) and therefore the reduction of PTSD symptoms will likely reduce suicidal ideation. The strategies applied at this level of intervention can be utilized when working with veterans from all war eras (Bruce, 2010). Providers can also utilize the veterans medical records to place a Patient Record Flag to make other providers aware that a specific veteran is experiencing suicidal ideation and may need additional support (York et al., 2013).

## *Non-specific methods*

While there are unique differences in how each branch of the military approaches suicide risks assessments, there are also some similarities among branches and military status. Use of the media has been implemented across branches to increase awareness of suicide risk, although these campaigns do not often emphasize self-care (Ramchand et al., 2011). The Army, Navy, and Marine Corps try to detect those who are at higher risk of suicide by having specific personnel who are tasked with this responsibility and then refer those individuals to the necessary level of care (Ramchand et al., 2011). The Air Force has a slightly different method by which they monitor suicide risk severity with individuals who are already thought to be at risk for suicide. Overall, the method most frequently used by the Department of Defense is primary prevention, which focuses on broad educational and awareness campaigns (Griffith & Bryan, 2018).

The Department of Defense and RAND Center for Military Health Policy Research conducted a study that addressed suicide risk assessment and best practices. The data collected as a result of that research study proposed 14 recommendations for all branches of the military with regard to suicide assessment and prevention: (1) track suicide and suicide attempts systematically and consistently; (2) evaluate existing programs and ensure that new programs contain an evaluation component when they are implemented; (3) include training in skills building, particularly help-seeking behavior, in programs and initiatives that raise awareness and promote self-care; (4) define the scope of what is relevant to preventing suicide and form partnerships with the agencies and organizations responsible for initiatives in other areas; (5) evaluate gatekeeper training; (6) develop prevention programs based on research and surveillance; selected and indicated programs should be based on clearly identified risk factors specific to military populations and to each service; (7) ensure that continuity of services and care are maintained when service members or their caregivers transition between installations in a process that respects service members' privacy and autonomy; (8) make service members aware of the benefits of accessing behavioral health care and specific policies and repercussions for accessing such care and conduct research to inform this communication; (9) make service members aware of the different types of behavioral health caregivers available to them, including information on caregivers' credentials, capabilities, and the confidentiality afforded by each; (10) improve coordination and communication between caregivers and service providers; (11) assess whether there is an adequate supply of behavioral health-care professionals and chaplains available to service members; (12) mandate training on evidence-based or state-of-the-art practices for behavioral health generally and in suicide risk assessment specifically for chaplains, health-care providers, and behavioral health-care professionals; (13) develop creative strategies to restrict access to lethal means among military service members or those indicated to be at risk of harming themselves; and (14) provide formal guidance to commanders about how to respond to suicides and suicide attempts (Ramchand et al., 2011). These 14 recommendations provide an informative foundation for those working with military populations on how to address suicide risk and reduce suicidal ideation.

The Department of Veterans Affairs and the Department of Defense combined efforts to address the negative impact of suicide by providing clinicians and mental health providers with guidelines and recommendations on suicide risk. Based on these guidelines, when working with a service member who is at risk for suicide, determining the level of care and appropriate setting is an important initial step (VA/DoD, 2013). Various treatment settings include inpatient programs, partial hospitalization, outpatient care, primary care, emergency departments, and deployment-related options, and the level of acute risk for suicide and individualized treatment plan should be taken into consideration when determining the treatment setting (VA/DoD, 2013). Once the appropriate treatment setting has been established for a service member, the implementation of a treatment modality specific to suicide is recommended. The VA and Department of Defense identified cognitive therapy and problem-solving therapy, with a specific emphasis on suicide prevention, as a specific strategy for many veterans in reducing the risk of suicide (2013). Cognitive therapy as an evidence-based therapy emphasizes techniques towards identifying thoughts and core beliefs around suicidal ideation and increasing a person's ability to cope with stressful life events (VA/DoD, 2013). The goal of problem-solving therapy is to augment a person's ability to successfully solve problems and identify coping skills, which typically leads to increase feelings of confidence and self-control (VA/DoD, 2013). Specific treatment modalities will depend on the presentation of the veteran's symptomology and case conceptualization. Another approach to reducing the risk of suicide among military personnel is the use of pharmacological treatments. The use of pharmacological treatments is largely dependent on the underlying mental health disorder and symptom presentation (VA/DoD, 2013) and may be utilized as a stand-alone treatment or in conjunction with psychotherapy.

Finally, in order to reduce the risk of suicide, particularly in service members who have a recent suicide attempt, follow-up and careful monitoring of suicidality is crucial in ensuring safety. The VA and Department of Defense consider the first 30 days after release from a psychiatric inpatient unit to be the highest risk with regard to vulnerability, followed by the first week, month, and year as a time period when a service member may be vulnerable and at risk for a suicide attempt (2013). Following such a discharge, service members should be followed by either a mental health or primary care provider who has a previously established relationship with the service member, and ongoing follow-up regarding suicide risk should be assessed regularly (VA/DoD, 2013). Following up with service members to determine their continued level of risk and ensuring safety is a key component of such a suicide prevention effort.

## *Postvention and prevention after death by suicide*

Research has shown that when a person dies by suicide, those close to that person are also at risk of dying by suicide (Ramchand et al. 2015). Although this topic represents an area in research where there are not a major number of studies, there are some recommendations for providing support to those who have been impacted by a loved one having died by suicide. Ramchand et al. (2015) propose that implementing general suicide screening and suicide prevention strategies can be helpful in determining who may be at increased risk for suicide after experiencing death by suicide by someone close to the veteran or military service member. Furthermore, these researchers found that it is often beneficial to have a strategy for discussing suicide, as well as a method to identify those who may be more impacted by a death by suicide. Considering the impact of suicide among service members, and the closeness that is often experienced between military personnel, suicide is likely negatively to impact many service members – even if they themselves are not experiencing suicidal ideation.

## Conclusion

As the impact of suicide within military populations and veterans continues to be a very serious issue, researchers have begun to address gaps in the literature with regard to suicide prevention. Griffith and Bryan propose the need for secondary suicide prevention strategies, which are strategies that focus on identifying those who are experiencing suicidal ideation and are at increased risk for a suicide attempt (2018). A large body of research on primary suicide prevention strategies, which includes suicide awareness and education on a broad level, currently exists. Additionally, tertiary suicide preventions strategies, such as the evidence-based treatment methodologies previously discussed, are also well represented in the literature and in practice (Griffith & Bryan, 2018). The ability more accurately and consistently to identify those who are at risk for suicide through secondary suicide prevention strategies may greatly impact the prevalence of suicide among military personnel and veterans.

Military populations, especially those who have experienced combat, are uniquely impacted by psychopathology and mental health stigma. Suicide rates within military populations have risen in recent years, emphasizing the need to better identify and address suicide within this population. Providers who work with various military populations should be aware of unique risk factors that various sub-populations and identities within the military experience in the context of suicide and suicidal ideation. Even within the various branches of the military, suicide prevention strategies vary and are dependent on the setting. While the clinical and research efforts to reduce the occurrence of deaths by suicide have grown, there is much to be done to better support military populations and veterans and to identify and care for those most at risk.

## References

Ayer, L., Ramchand, R., Geyer, L., Burgette, L., & Kofner, A. (2016). The influence of training, reluctance, efficacy, and stigma on suicide intervention behavior among NCOs in the Army and Marine Corps. *The Journal of Primary Prevention*, *37*(3), 287–302.

van den Berk-Clark, C., & McGuire, J. (2014). Trust in health care providers: Factors predicting trust among homeless veterans over time. *Journal of Health Care For The Poor and Underserved*, *25*(3), 1278–1290.

Blosnich, J. R., Mays, V. M., & Cochran, S. D. (2014). Suicidality among veterans: Implications of sexual minority status. *American Journal of Public Health*, *104*(S4), S535–S537.

Brooks, E., Dailey, N. K., Bair, B. D., & Shore, J. H. (2016). Listening to the patient: Women veterans' insights about health care needs, access, and quality in rural areas. *Military Medicine*, *181*(9), 976–981.

Bruce, M. L. (2010). Suicide risk and prevention in veteran populations. *Annals of the New York cademy of Sciences*, *1208*(1), 98–103.

Bryan, C. J., Kanzler, K. E., Durham, T. L., West, C. L., & Greene, E. (2010). Challenges and considerations for managing suicide risk in combat zones. *Military Medicine*, *175*(10), 713–718.

Bush, N. E., Reger, M. A., Luxton, D. D., Skopp, N. A., Kinn, J., Smolenski, D., & Gahm, G. A. (2013). Suicides and suicide attempts in the U.S. military, 2008-2010. *Suicide & Life-Threatening Behavior*, *43*(3), 262–273.

Chanfreau-Coffinier, C., Schweizer, C. A., Bean-Mayberry, B. A., Darling, J. E., Canelo, I., Yano, E. M., & Gordon, H. S. (2018). Mental health screening results associated with women veterans' ratings of provider communication, trust, and care quality. *Women's Health Issues*, *28*(5), 430–438.

Chen, J. A., Granato, H., Shipherd, J. C., Simpson, T., & Lehavot, K. (2017). A qualitative analysis of transgender veterans' lived experiences. *Psychology of Sexual Orientation and Gender Diversity*, *4*(1), 63–74.

Department of Defense. (2016). *Suicide Event Report* (Report: RefID: 0-A2345E0). Washington, DC: Department of Defense.

Department of Defense Instruction 6490.3. (1997). Implementation and Application of Joint Medical Surveillance for Deployments. Washington, DC: Department of Defense.

Dorsey Holliman, B. A., Monteith, L. L., Spitzer, E. G., & Brenner, L. A. (2018). Resilience, cultural beliefs, and practices that mitigate suicide risk among African American women veterans. *SAGE Open*, *8*(1), 1–10.

Eaton, K. M., Messer, S. C., Wilson, A. L. G., & Hoge, C. W. (2006). Strengthening the validity of population-based suicide rate comparisons: An illustration using US military and civilian data. *Suicide and Life-Threatening Behavior*, *36*(2), 182–191.

Freeman, A., Mergl, R., Kohls, E., Székely, A., Gusmao, R., Arensman, E., … Rummel-Kluge, C. (2017). A cross-national study on gender differences in suicide intent. *BMC Psychiatry*, *17*(1), 1–11.

Gould, M. S., Kalafat, J., HarrisMunfakh, J. L., & Kleinman, M. (2010). *Health Promotion, Risk Reduction, and Suicide Prevention* [Pamphlet]. (n.p.) Department of the Army, December 17, 2009.

Griffith, J., & Bryan, C. J. (2018). Preventing suicides in the US military. *Psychological Services*, *15*(3), 251–261.

Harmon, L. M., Cooper, R. L., Nugent, W. R., & Butcher, J. J. (2016). A review of the effectiveness of military suicide prevention programs in reducing rates of military suicides. *Journal of Human Behavior in the Social Environment*, *26*(1), 15–24.

Hill, J. V., Johnson, R. C., & Barton, R. A. (2006). Suicidal and homicidal soldiers in deployment environments. *Military Medicine*, *171*(3), 228–232.

Kennedy, C. H., & Zillmer, E. A. (2012). *Military Psychology: Clinical and Operational Applications*, 2nd ed. New York, NY: The Guilford Press.

LeardMann, C. A., Powell, T. M., Smith, T. C., Bell, M. R., Smith, B., Boyko, E. J., … Hoge, C. W. (2013). Risk factors associated with suicide in current and former US military personnel. *Jama*, *310*(5), 496–506.

Lehavot, K., O'Hara, R., Washington, D. L., Yano, E. M., & Simpson, T. L. (2015). Posttraumatic stress disorder symptom severity and socioeconomic factors associated with Veterans Health Administration use among women veterans. *Women's Health Issues*, *25*(5), 535–541.

Lehavot, K., Simpson, T. L., & Shipherd, J. C. (2016). Factors associated with suicidality among a national sample of transgender veterans. *Suicide & Life-Threatening Behavior*, *46*(5), 507–524.

Matarazzo, B. B., Barnes, S. M., Pease, J. L., Russell, L. M., Hanson, J. E., Soberay, K. A., & Gutierrez, P. M. (2014). Suicide risk among lesbian, gay, bisexual, and transgender military personnel and veterans: What does the literature tell us? *Suicide & Life-Threatening Behavior*, *44*(2), 200–217.

McDaniel, J. T., Thomas, K. H., Albright, D. L., Fletcher, K. L., & Shields, M. M. (2018). Mental health outcomes in military veterans: A latent growth curve model. *Traumatology*, *24*(3), 228–235.

Millegan, J., Delaney, E. M., & Klam, W. (2016). Responding to trauma at sea: A case study in psychological first aid, unique occupational stressors, and resiliency self-care. *Military Medicine, 181*(11), e1692–e1695.
Monteith, L. L., Bahraini, N. H., Matarazzo, B. B., Gerber, H. R., Soberay, K. A., & Forster, J. E. (2016). The influence of gender on suicidal ideation following military sexual trauma among veterans in the Veterans Health Administration. *Psychiatry Research, 244*, 257–265.
Payne, S. E., Hill, J. V., & Johnson, D. E. (2008). The use of unit watch or command interest profile in the management of suicide and homicide risk: Rationale and guidelines for the military mental health professional. *Military Medicine, 173*(1), 25–35.
Pease, J. L., Billera, M., & Gerard, G. (2016). Military culture and the transition to civilian Life: Suicide risk and other considerations. *Social Work, 61*(1), 83–86.
Ramchand, R., Acosta, J., Burns, R. M., Jaycox, L. H., & Pernin, C. G. (2011). The war within: Preventing suicide in the US military. *Rand Health Quarterly, 1*(1), 2.
Ramchand, R., Ayer, L., Fisher, G., Osilla, K. C., Barnes-Proby, D., & Wertheimer, S. (2015). Suicide postvention in the Department of Defense: Evidence, policies and procedures, and perspectives of loss survivors. *Rand Health Quarterly, 5*(2), 20.
Reisman, M. (2016). PTSD treatment for veterans: What's working, what's new, and what's next. *Pharmacy and Therapeutics, 41*(10), 623–634.
Sexton, M. B., Davis, M. T., Anderson, R. E., Bennett, D. C., Sparapani, E., & Porter, K. E. (2018). Relation between sexual and gender minority status and suicide attempts among veterans seeking treatment for military sexual trauma. *Psychological Services, 15*(3), 357–362.
Sterling, A. G. 4th, Bakalar, J. L., Perera, K. U., DeYoung, K. A., Harrington-LaMorie, J., Haigney, D., & Ghahramanlou-Holloway, M. (2017). Perspectives of suicide bereaved individuals on military suicide decedents' life stressors and male gender role stress. *Archives of Suicide Research: Official Journal of the International Academy for Suicide Research, 21*(1), 155–168.
Ursano, R. J., Kessler, R. C., Naifeh, J. A., Mash, H. H., Fullerton, C. S., Bliese, P. D., ... Stein, M. B. (2017). Frequency of improvised explosive devices and suicide attempts in the U.S. Army. *Military Medicine, 182*(3), e1697–e1703.
VA/DoD. (2013). Clinical practice guideline for assessment and management of patients at risk for suicide. *Department of Veterans Affairs, Department of Defense.* Retrieved from https://www.healthquality.va.gov/guidelines/MH/srb/VADODCP_suiciderisk_full.pdf
VanSickle, M., Werbel, A., Perera, K., Pak, K., DeYoung, K., & Ghahramanlou-Holloway, M. (2016). Perceived barriers to seeking mental health care among United States Marine Corps noncommissioned officers serving as gatekeepers for suicide prevention. *Psychological Assessment, 28*(8), 1020–1025.
Vogt, D. (2011). Mental health-related beliefs as a barrier to service use for military personnel and veterans: A review. *Psychiatric Services, 62*(2), 135–142.
Warner, C. H., Appenzeller, G. N., Mullen, K., Warner, C. M., & Grieger, T. (2008). Soldier attitudes toward mental health screening and seeking care upon return from combat. *Military Medicine, 173*(6), 563–569.
Wilder, H., & Wilder, J. (2012). In the wake of don't ask don't tell: Suicide prevention and outreach for LGB service members. *Military Psychology* (Taylor & Francis Ltd), *24*(6), 624–642.
York, J. A., Lamis, D. A., Pope, C. A., & Egede, L. E. (2013). Veteran-specific suicide prevention. *Psychiatric quarterly, 84*(2), 219–238.
Zamorski, M. A. (2011). Suicide prevention in military organizations. *International Review of Psychiatry, 23*(2), 173–180.

# 26

# MILITARY SEXUAL TRAUMA AND SUICIDAL SELF-DIRECTED VIOLENCE

## A narrative review and proposed agenda for future research

*Lindsey L. Monteith, Ryan Holliday, Tim Hoyt, and Nazanin H. Bahraini*

Military sexual trauma (MST) is a term used in the U.S. Department of Veterans Affairs (VA) that refers to psychological trauma resulting from sexual assault (SA) or sexual harassment (SH) occurring during one's military service (U.S. Government, 2014). National VA policy mandates universal MST screening of all veterans who present for healthcare in the Veterans Health Administration (VHA; U.S. Government, 2014). Although the precise wording has evolved over time, the current MST screening questions used in VHA are: "When you were in the military ... (1) did you ever receive unwanted, threatening, or repeated sexual attention (for example, touching, cornering, pressure for sexual favors, or inappropriate verbal remarks)? [SH]; and (2) did you have sexual contact against your will or when you were unable to say no (for example, after being forced or threatened or to avoid other consequences)? [SA]" Veterans who endorse a positive response to either question (i.e., screen positive) are offered the opportunity to discuss potential resources, including free VHA care for MST-related health conditions (Foynes et al., 2018). Implementation of universal MST screening within VHA has been successful, with 98.7% of 5,173,091 veterans screened to date; of those, 29.1% of females and 1.6% of males have screened positive (Department of Veterans Affairs Office of Mental Health & Suicide Prevention, 2018).

While the VA uses the aforementioned definition of MST, the U.S. Department of Defense (DoD) differentiates SH and SA. Specifically, the DoD defines SH as "conduct that involves unwelcome sexual advances, requests for sexual favors, and deliberate or repeated offensive comments or gestures of a sexual nature" when such conduct is made a condition of a person's career, interferes with work performance, or creates a hostile or offensive environment (DoD, 2018a, p. 9). DoD defines SA as "intentional sexual contact characterized by the use of force, threats, intimidation, or abuse of authority or when the victim does not or cannot consent" (DoD, 2017, p. 122). In 2005, the DoD established a Sexual Assault Prevention and Response Office (SAPRO) to curtail instances of SA and increase safety and recourse for service members who experience SA (DoD, 2005). Approximately 4.3% of women and 0.6% of men in the U.S. military report experiencing SA each year, one-third of whom make a formal report of the incident (DoD, 2018b). DoD reports further indicate that 21.4% of active duty women and 5.7% of active duty men indicate experiencing SH in the previous year (Davis, Grifka, Williams, & Coffey, 2017). The SAPRO has implemented several initiatives aimed at preventing SA, increasing reporting of incidents, and supporting service members in the aftermath thereof; nonetheless, SA continues to pose a significant and persistent problem (DoD, 2018b).

Results from a recent meta-analysis suggest that MST is likely more prevalent than what is suggested by formal reporting statistics (Wilson, 2018). Fear, stigma, avoidance, and concerns about sexuality and masculinity likely deter reporting and disclosure of MST (Andresen & Blais, 2019; Dardis, Reinhardt, Foynes, Medoff, & Street, 2018; O'Brien, Keith, & Shoemaker, 2015). Multi-faceted initiatives across institutions have led to burgeoning research on MST (Holliday, Maguen, Hoyt, Blais, & Monteith, 2018b), which portrays an array of negative outcomes associated with experiencing SH or SA during one's military service. MST is associated with posttraumatic stress disorder (PTSD), depression, eating disorders, and substance use disorders (Kimerling, Gima, Smith, Street, & Frayne, 2007; Klingensmith, Tsai, Mota, Southwick, & Pietrzak, 2014; Millegan, Wang, LeardMann, Miletich, & Street, 2016; Surís & Lind, 2008), as well as cardiovascular disease, chronic pulmonary and liver disease, genito-unrinary conditions, sexually transmitted infection, and sexual dysfunction (Kimerling et al., 2007; Surís, Holliday, Weitlauf, North, & the Veteran Safety Initiative Writing Collaborative, 2013; Turchik et al., 2012). Whereas these mental and physical health risks are extensive, the toll of experiencing MST extends beyond one's mental and physical health, also encompassing impaired social functioning and interpersonal relationships, decreased military readiness, lower socioeconomic status, heightened risk for unemployment and disability, and homelessness (Brignone et al., 2016; Brownstone, Holliman, Gerber, & Monteith, 2018; Millegan et al., 2016; Monteith, Gerber, Brownstone, Soberay, & Bahraini, 2019; Surís et al., 2013).

Whereas the aforementioned findings illuminate a variety of adverse experiences associated with MST, the extent to which MST is associated with suicide, suicide attempts, and suicidal ideation (SI) has been less studied until recently, despite the fact that veterans remain at heightened risk for suicide (Department of Veterans Affairs, 2018). Understanding whether MST is associated with these outcomes, for whom this relationship is the strongest, and potential mechanisms by which this occurs is essential. Such knowledge would also inform targeted suicide prevention efforts for service members and veterans who experience MST. In this chapter, we review research on MST and SI, suicide attempt, and suicide, followed by knowledge of factors associated with these outcomes among MST survivors. We discuss the strengths and limitations of the research conducted and delineate recommendations for future research. When possible, we differentiate between suicide-related constructs, utilizing the nomenclature developed by the Centers for Disease Control and Prevention (CDC) and mandated for use in the VA and DoD (Suicidal Self-Directed Violence Classification System [SDVCS]; Brenner et al., 2011; CDC, 2011).

## Is military sexual trauma associated with suicidal ideation, suicide attempts, and suicide?

### *Suicidal ideation*

The majority of research examining the association between MST and suicide-related outcomes has focused on SI, defined as "thoughts of engaging in suicide-related behavior" (CDC, 2011, p. 90). These studies, described in Table 26.1, have primarily been cross-sectional and reliant upon self-report measures of SI.

*Table 26.1* Studies examining whether military sexual trauma is associated with suicidal ideation

| *Authors* | *Sample* | *MST* | *SI* | *Findings* |
|---|---|---|---|---|
| Klingensmith et al. (2014) | **1,468 veterans** (1,312 males, 156 females) | **VA MST Screen** | **Modified PHQ-9 SI Item ≥1**[a] | MST was associated with SI, adjusting for age, sex, marital status, race/ethnicity, employment, enlistment, branch, and lifetime depression/PTSD screens (AOR = 2.19; 95% CI: 1.25, 3.84). |
| Monteith et al. (2016a) | **354 veterans in VA care** (310 men, 44 women) | **VA MST Screen** (EMR) | **BSS**[a] | MST was associated with SI (OR = 2.64, 95% CI: 1.03, 6.75, p = 0.043), adjusting for age, gender, combat, PTSD, depressive disorders, negative affect, and lifetime suicide attempt. Gender moderated this. |
| Lemaire and Graham (2011) | **1,740 OEF/OIF veterans** with initial VA MH screens | **VHA MST screen** (EMR); **DRRI SH Scale** | **Clinical evaluation:** current SI | Neither MST nor SH (DRRI) were associated with current SI in unadjusted analyses or during forward stepwise regression models (with Bonferroni correction). |
| Blais and Geiser (2019) | **1,189 female veterans and service members** | **VA MST Screening Questions** (modified) | **PHQ-9 SI Item** | Depression severity (B = 0.19, 95% CI = 0.12, 0.28, β = 0.11) and PTSD-related anhedonia (B = 0.10, 95% CI = 0.01, 0.20; β = 0.06) mediated association between SA (but not SH) and SI. Age, race, marital status, branch, deployment, and veteran status were covaried. |

(*Continued*)

*Table 26.1* (Continued)

| *Authors* | *Sample* | *MST* | *SI* | *Findings* |
|---|---|---|---|---|
| Khan et al. (2019) | **403 women veterans** with ≥1 appointment at urban VA | **Exposure to Military Service Stressors**[b] | **PHQ-9 SI Item ≥1**[a] | SH and SA correlated with SI. SH (OR = 2.7, 95% CI: 1.2, 5.8, $p \leq 0.01$), but not SA, significant when including perceived life threat, killing, witnessing killings/injury, seeing injured/dead bodies, loss, injury. |
| Gradus, Street et al. (2013) | **2,321 veterans** (1,188 women, 1,133 men) who deployed for OEF/OIF | **DRRI SH Scale** | **SBQ-R-SF** item: SI since last deployment[c] | SH associated with post-deployment SI for women (b = .089; β = .29, $p < 0.0001$) and men (b = .079, β = 0.07, $p < 0.0001$). Controlling for symptoms of depression, PTSD, and alcohol use, SH was significant for women (b = 0.026; β = .087, $p < 0.05$), but not men. |
| Gradus et al. (2017) | **2,161 OEF/OIF veterans** (1,062 men, 1,099 women) | **DRRI SH Scale**[a] | **SBQ-SF item:** SI since last deployment[c] | Re-analysis of data using machine learning. Excluded if suicidal plans/attempts. SH was a key factor in SI among women, but not men. SH interacted with probable PTSD and depression for women. |
| Monteith et al. (2015); Monteith, Menefee et al. (2016c) | **199 OEF/OIF/OND veterans** (171 men, 28 women) entering VA inpatient care | **DRRI SH Scale** | **BSS** | DST was associated with BSS, adjusting for age, gender, and combat experiences. More severe DST was associated with SI, whereas other types were not, adjusting for age and gender. |

(*Continued*)

*Table 26.1* (Continued)

| *Authors* | *Sample* | *MST* | *SI* | *Findings* |
|---|---|---|---|---|
| Monteith, Hoffmire et al. (2018) | **824 OEF/OIF/OND veterans** (480 men, 344 women) | **DRRI-2 SH Scale** | **C-SSRS** (SI severity, past 3 months) | DST associated with SI in bivariate analyses for men and women ($p's \leq 0.01$). Accounting for age, PTSD symptoms, possible depression, and lifetime suicide attempt, PDS moderated association between DST and SI in men, but mediated this association in women. |
| Schry et al. (2015) | **2,546 post-9/11 male** veterans, active duty, Reserve members | **TLEQ;** unwanted touching[d] | **BSS** | Males with MSA histories reported higher BSS scores, compared to full sample ($\chi^2 = 16.10, p < 0.001$) and controls ($\chi^2 = 5.74, p = 0.017$) without MSA and matched on other trauma-related exposures. |
| Bryan et al. (2015) | **464 military personnel and veterans enrolled in college** (70.7% male) | **LEC;** SA and unwanted/uncomfortable sexual experiences[d] | **SITBI (self-report version):** SI after joining military | MST associated with SI, but not when adjusting for age, gender, and pre-military sexual trauma. SA and unwanted sexual experiences associated with SI, but not when examining pre-military sexual trauma. Significant interaction of gender and unwanted sexual experiences. SA and unwanted sexual experiences associated with SI in unadjusted models for men, but not women; only unwanted sexual experiences associated with SI in adjusted models for men. |

*(Continued)*

*Table 26.1* (Continued)

| *Authors* | *Sample* | *MST* | *SI* | *Findings* |
|---|---|---|---|---|
| White et al. (2018) | **997 National Guard personnel** from Utah and Idaho (82.5% male) | **LEC;** SA or uncomfortable/unwanted sexual experiences[d] | **SITBI (self-report version):** lifetime SI | MST was associated with SI (OR = 3.3, 95% CI: 2.1, 5.1, $p < 0.001$), adjusting for gender, age, race (OR = 3.3, 95% CI: 2.1, 5.4 $p < 0.001$); probable MDD, PTSD, AUD (OR = 2.3, 95% CI: 1.3, 3.9, $p < 0.01$); and premilitary sexual trauma (OR = 2.0, 95% CI: 1.1, 3.5, $p < 0.05$). |
| DiMauro et al. (2018) | **255 women veterans** in monogamous relationships exposed to military trauma | **Open-ended response** "most stressful military trauma" | **PHQ-9 SI Item** | Participants reporting sexual (vs. non-sexual) trauma as their most stressful military trauma reported more severe SI ($F = 4.74, p < 0.05$). Indirect effect of trauma type on SI via sexual satisfaction. Trauma type moderated association of sexual functioning and SI. |
| Griffith (2019) | **12,567 Army National Guard** (gender not reported) | **URI item**: SH in unit in past 12 months | **URI item:** SI past 12 months | SH was associated with SI, $B = 1.64$, OR = 5.17, 95% CI: 3.51–7.61, $p < 0.001$. Units/companies with more SH had more soldiers with SI. SH within the unit moderated the effect of trust in unit leaders on SI. |

*Abbreviations*: MST = military sexual trauma; VA = Veterans Affairs; PHQ-9 = Patient Health Questionnaire-9; SI = suicidal ideation; AOR = adjusted odds ratio; CI = confidence interval; PTSD = posttraumatic stress disorder; BSS = Beck Scale for Suicide Ideation; OR = Odds Ratio; OEF = Operation Enduring Freedom; OIF = Operation Iraqi Freedom; VAMC = Veterans Affairs Medical Center; DRRI = Deployment Risk and Resilience Inventory; SH = sexual harassment; SBQ-R-SF = Suicidal Behaviors Questionnaire-Revised Short Form; OND = Operation New Dawn; C-SSRS = Columbia-Suicide Severity Rating Scale; PDS = post-deployment support; TLEQ = Traumatic Life Events Questionnaire; SA = sexual assault; LEC = Life Events Checklist; SITBI = Self-Injurious Thoughts and Behaviors Interview; NS = not significant; URI = Unit Risk Inventory.

a Was scored dichotomously.
b SH/SA questions similar to MST screen.
c Specific to OEF/OIF deployment.
d During military service.

### *Veterans Affairs Military Sexual Trauma Screen*

Some of these studies examined the association between MST with SI, relying upon VA MST screening results in veterans' electronic medical records or administering the MST screen specifically for research purposes. Two of these studies found that veterans with a positive MST screen were more than twice as likely to report recent SI (Klingensmith et al., 2014; Monteith, Bahraini, Matarazzo, Gerber, Soberay, & Forster, 2016a). In contrast, an earlier study, which relied upon clinical suicide risk evaluation data, did not find a significant association between MST screening results and current SI (Lemaire & Graham, 2011). One advantage to using the VA MST screen to assess for MST is its established validity (Mengeling et al., 2019) and that it allows for consistent measurement of MST through the universal definition used throughout VA. As such, studies using the VA MST screen offer important implications – suggesting that screening for SI may be indicated for veterans who screen positive for MST (Klingensmith et al., 2014; Monteith, Bahraini, Matarazzo, Gerber et al., 2016a). However, the VA MST screen does not differentiate MST type (SH vs. SA), unless examining items individually (Khan et al., 2019).

### *Deployment Risk and Resilience Inventory*

Studies of SI have also utilized the Deployment Risk and Resilience Inventory (DRRI) Sexual Harassment Scale (King, King, Vogt, Knight, & Samper, 2006), which was developed to assess SH experiences while deployed. An advantage of this measure is that it captures the frequency with which various experiences occurred and yields a continuous score – a potential advantage over the dichotomous VA MST screen. It also includes items that assess SA (e.g., being raped); however, these experiences are not differentially weighted (e.g., being raped is scored similarly to verbal harassment), potentially impacting inference from findings.

The studies utilizing the DRRI SH Scale reported herein surveyed previously deployed veterans who served following 9/11 in Operation Enduring Freedom (OEF), Operation Iraqi Freedom (OEF), and/or Operation New Dawn (OND). Such studies have typically found that scores on the DRRI SH Scale are associated with SI (Gradus, Street, Suvak, & Resick, 2013; Monteith, Menefee, Forster, Wanner, & Bahraini, 2015; Monteith et al., 2018), with only one exception (Lemaire & Graham, 2011). Moreover, this association has been reported for both men and women, using different measures of SI, and accounting for combat-related experiences, although some gender differences have been reported (Gradus, King, Galatzer-Levy, & Street, 2017). Interestingly, an exploratory study examined whether specific types of experiences on the DRRI SH Scale were associated with SI; items indicative of deployment-related SA (e.g., unwanted attempts to have sex, being forced to have sex), but not deployment-related SH (e.g., crude and offensive sexual remarks, being offered a reward or special treatment for sexual behavior), were associated with SI (Monteith, Menefee, Forster, & Bahraini, 2016c). These findings underscore the importance of suicide prevention efforts for survivors of deployment SA while highlighting the need for additional research to examine whether specific MST experiences are associated with SI.

### *Alternate measures of sexual harassment/sexual assault*

Finally, a select number of studies have utilized alternate means of assessing MST. These studies reported that SH (Griffith, 2019), SA (Schry et al., 2015), and unwanted sexual experiences or SA during military service (White et al., 2018) were associated with SI in various samples. Of note, a study with college student veterans and service members found differing results depending on the covariates included (Bryan, Bryan, & Clemans, 2015). However, many of these studies used measures initially developed to assess lifetime trauma history that were modified to assess MST or SH/SA specifically. While replication of results with alternative methods of assessment is an important component of science, the reliability and validity of the modified methods, which were not originally developed for these purposes, is unknown. Readers are encouraged to keep this caveat in mind when considering such findings.

### *Factors that moderate or mediate these associations*

Researchers have also begun to examine factors that moderate or mediate the association between MST and SI. For example, gender has been identified as an important moderating variable (Bryan et al., 2015; Monteith, Bahraini, Matarazzo, Soberay, & Smith, 2016b). Gender differences have also been identified with respect to factors that attenuate, exacerbate, or account for the association between MST and SI. Gradus, Street et al. (2013) found that, when accounting for mental health symptoms, the association between DRRI SH Scale scores with post-deployment SI was no longer significant for men, but remained significant for women, although SA and SH were not differentiated. Thus, results suggested a need to identify other factors accounting for this association in women. Building upon these results, Monteith, Hoffmire et al. (2018) examined if perceived unit support and post-deployment support moderated or mediated the association between DRRI SH Scale scores and recent SI, with different results obtained for men vs. women. Perceived post-deployment support moderated the association between DRRI SH Scale scores and SI for men, but mediated this association among women. Blais and Geiser (2019) found that depressive symptom severity and PTSD-related anhedonia mediated the association between SA and SI among female service members and veterans. Finally, DiMauro, Renshaw, and Blais (2018) reported that sexual satisfaction mediated the association between sexual trauma and SI. In sum, the association between MST and SI appears to be moderated by factors such as gender and post-deployment support, and mediated by mental health symptoms, post-deployment support, and sexual satisfaction, with some evidence of gender differences within these.

## *Suicide attempt*

Fewer studies have examined whether MST is associated with suicide attempts, defined as "non-fatal self-directed potentially injurious behavior with any intent to die as a result of the behavior" (CDC, 2011, p. 21). Results from extant studies reveal a fairly consistent pattern of results (Table 26.2).

Across multiple studies, associations between MST and suicide attempts were significant and robust. In two large studies of veterans in VHA care, veterans with positive MST screens were significantly more likely to have ICD-9 codes related to "suicide and intentional self-inflicted injury" diagnosed in their medical record, compared to those who screened negative (Kimerling et al., 2007; Pavao et al., 2013). Both studies stratified analyses by gender, adjusted for demographic variables, and included "suicide and intentional self-inflicted injury" as one of multiple health outcomes of interest. Yet the specific ICD-9 codes used to classify "intentional self-inflicted injury" were not reported. Thus, it is difficult to ascertain if all types of self-directed violence included were suicidal (i.e., suicide attempts) or the extent to which non-suicidal self-directed violence (e.g., intentionally injuring oneself without intent to die) was also included. Addressing these limitations, Klingensmith et al. (2014) conducted a nationally representative survey of veterans, including those not using VHA care, in which they asked participants if they had ever attempted suicide. MST was associated with being over two times as likely to report a

*Table 26.2* Studies examining if military sexual trauma is associated with suicide attempt

| *Authors* | *Sample* | *MST* | *Suicide Attempt* | *Findings* |
|---|---|---|---|---|
| Kimerling et al. (2007) | **4,325,768 veterans** (185,880 women, 4,319,888 men) with VHA outpatient visit | **MST screen** (VA EMR) | **ICD-9 codes** "suicide and intentional self-inflicted injury" | In unadjusted analyses and adjusting for age and race, MST was associated with "suicide and intentional self-inflicted injury" among women (AOR = 2.15; 99% CI: 1.45, 3.21) and men (AOR = 2.93; 99% CI: 2.22, 3.88). |
| Pavao et al. (2013) | **126,598 homeless veterans** (8,915 females, 117,683 males) who used VHA outpatient care | **MST screen** from VHA MST Support Team Data Archive | **ICD-9-CM codes** "suicide and intentional self inflicted injury" | Adjusting for age, race, ethnicity, and marital status, MST was associated with "suicide and intentional self inflicted injury" in females (AOR = 1.64; 95% CI: 1.37, 1.95) and males (AOR = 1.73; 95% CI: 1.56, 1.92). |
| Klingensmith et al. (2014) | **1,468 veterans** (1,312 males, 156 females) in NHRVS | **MST Screen** | **Lifetime attempt:** "Have you ever tried to kill yourself?" | MST associated with suicide attempt (AOR = 2.78; 95% CI: 1.53, 5.05), adjusting for age, sex, race/ethnicity, marital status, employment, enlistment, branch, and depression/PTSD. |
| Gradus, Shipherd et al. (2013) | **680 former U.S. Marine recruits** | **Sexual Experiences Questionnaire**[a] | **Two questions** for suicide attempt 10 years post-training | SH (but not SA) during training predicted post-training suicide attempt (OR = 2.8; 95% CI: 1.2, 6.6), in unadjusted analyses and controlling for gender. |
| Rosellini et al. (2017) | **4,238 female regular Army Soldiers** who served from 2004–2009 with SA, matched to 21,190 controls without | **Administratively-recorded SA** (criminal justice data unrestricted SA report) | **DoDSER & ICD-9-CM codes** (E950-959) in treatment records (attempts 12 months after SA) | Women with SA were more likely to attempt suicide (OR = 3.0; 95% CI, 2.5, 3.6) vs. controls and adjusting for follow-up months, etc. Significant in those without prior MH treatment (OR = 6.2; 95% CI = 1.4, 28.0), but among not women with prior outpatient/inpatient treatment. |

(*Continued*)

*Table 26.2* (Continued)

| *Authors* | *Sample* | *MST* | *Suicide Attempt* | *Findings* |
|---|---|---|---|---|
| Bryan et al. (2015) | **464 military personnel and veterans enrolled in college** (70.7% male) | **LEC;** SA and/or unwanted/ uncomfortable sexual experiences[b] | **SITBI** (self-report): suicide attempt after joining military | MST associated with suicide attempt, but not adjusting for age, gender, and premilitary sexual trauma. SA and unwanted sexual experiences NS in fully adjusted model (with pre-military sexual trauma). Gender interactions NS. Unwanted sexual experiences associated with attempts in males. SA NS. |
| White et al. (2018) | **997 National Guard personnel** from Utah or Idaho (82.5% male) | **LEC;** SA or uncomfortable/ unwanted sexual experiences[b] | **SITBI** (self-report): lifetime suicide attempt | MST associated with suicide attempt adjusting for gender, age, and race (AOR = 3.9, 95% CI: 1.8, 8.6, $p < 0.001$), but not with probable MDD, PTSD, and AUD. Similar results accounting for premilitary sexual victimization. |
| Griffith (2019) | **12,567 Army National Guard soldiers** in 180 company-sized units[c] | **URI item**: SH in unit in past 12 months | **URI item:** suicide attempt in past 12 months | SH associated with suicide attempt (OR = 15.40, 95% CI: 7.60, 31.22, $p < 0.001$). Units with more SH had more suicide attempts. SH in unit moderated effect of trust in unit leaders on suicide attempts. |

*Abbreviations*: MST = military sexual trauma; VHA = Veterans Health Administration; FY = fiscal year; VA = Veterans Affairs; EMR = electronic medical records; ICD = International Classification of Diseases; AOR = adjusted odds ratio; CI = confidence interval; NHRVS = National Health and Resilience in Veterans Study; PTSD = posttraumatic stress disorder; SH = sexual harassment; SA = sexual assault; DoDSER = Department of Defense Suicide Event Report; CM = clinical modification; MH = mental health; LEC = Life Events Checklist; SITBI = Self-Injurious Thoughts and Behaviors Interview; NS = not significant; MDD = major depressive disorder; AUD = alcohol use disorder; URI = unit risk inventory.

a Scored dichotomously.
b During military service
c Gender not reported.

lifetime suicide attempt, accounting for an array of factors; however, the focus on lifetime suicide attempts (which could have preceded MST) potentially limited these findings.

Other studies have analyzed suicide attempt variables within prescribed timeframes, such as "lifetime," "in the past year," "since military service," or following an index event. Such studies often used adapted measures of lifetime trauma exposure (e.g., Life Events Checklist) or military SH and SA with differing definitions from the VA MST screen (i.e., Sexual Experiences Questionnaire), or analyzed records of administratively recorded military SA reports. Mirroring the aforementioned studies on SI, replication and validation of measures specific to MST remain important considerations. Nonetheless, these studies indicate that SH is associated with suicide attempts, both in the past year (Griffith, 2019) and the decade following recruit training (Gradus, Shipherd, Suvak, Giasson, & Miller, 2013), and that female soldiers with administratively recorded SA are more likely to attempt suicide in the ensuing year (Rosellini et al., 2017).

Although most studies on MST and suicide attempt have reported significant associations, two studies obtained divergent results. White et al. (2018) reported that MST was associated with lifetime suicide attempt when accounting for gender, age, and race, but not when covarying for probable major depressive disorder, PTSD, and alcohol use disorder. Bryan et al. (2015) reported a significant bivariate association between MST and suicide attempts; however, adjusted associations were not significant, apart from the association between unwanted sexual experiences during military service with suicide attempts among male college students.

## Factors that moderate or mediate these associations

Unlike research on SI, only three studies have examined factors that moderate the association between MST and suicide attempt, and none have examined mediators of this association. Mirroring his findings on SI, Griffith (2019) found that the association between SH and suicide attempt was exacerbated in units/companies with more SH. Bryan et al. (2015) did not find gender to moderate the association between MST and suicide attempts, a departure from their SI findings. Lastly, Rosellini et al. (2017) examined if prior mental health treatment moderated the association between SA and suicide attempts. Indeed, SA was associated with suicide attempts for female soldiers who lacked a record of mental health treatment in the past year, but was not significant in those with prior treatment, underscoring mental health care as potentially protective after SA.

## Suicide

By comparison to research on SI and suicide attempt, research examining if MST is associated with suicide (i.e., "death caused by self-directed injurious behavior with any intent to die as a result of the behavior"; CDC, 2011) has been sparse (Table 26.3).

This is likely due to the inherent challenges of studying suicide, a low-base rate occurrence. Of the two published studies, both relied upon documentation of MST and suicide based on electronic medical records or databases. Significant associations were found between MST and suicide (Kimerling, Makin-Byrd, Louzon, Ignacio, & McCarthy, 2016). Importantly, several important factors were accounted for, further supporting the salience of MST as a potential risk factor for suicide, above and beyond psychiatric diagnoses and demographic characteristics. Gender-stratified analyses, which were significant for both men and women, revealed a more nuanced portrait of the role of MST as it relates to suicide. Street et al. (2015) did not directly aim to examine associations between SA and suicide; however, their findings suggested that SA

*Table 26.3* Studies examining whether military sexual trauma is associated with suicide

| *Authors* | *Sample* | *MST* | *Suicide* | *Findings* |
|---|---|---|---|---|
| Kimerling et al. (2016) | **4,325,768 veterans** (5,991,080 men, 360,774 women) with outpatient visit(s) to VHA facilities in FY 2007-2011 and MST screen | **VHA MST screen,** per VA National Patient Care Database | **ICD-10 suicide codes** (X60-X84,Y87.0), per National Death Index data (through CY 2011 or death) | MST was associated with increased risk for suicide in unadjusted and fully adjusted models. Adjusting for age, rurality, depression, bipolar disorder, schizophrenia, PTSD, SUD, anxiety disorders, and medical morbidity, the hazards ratio was 1.19 (95% CI: 1.01, 1.39) for men and 1.35 (95% CI: 1.01, 1.83) for women. |
| Street et al. (2015) | **975,057 active duty enlisted Regular Army soldiers** serving in 2004-2009 | **Administratively-recorded SA** in prior 12 months | **Army Forces Medical Examiner Tracking System** | The effect of the gender X deployment interaction in predicting suicide was reduced when accounting for SA and other variables. Results suggest that SA may partially explain females' elevated suicide risk while deployed. |

*Abbreviations*: MST = military sexual trauma; VHA = Veterans Health Administration; FY = fiscal year; VA = Veterans Affairs; ICD = International Classification of Diseases; CY = calendar year; PTSD = posttraumatic stress disorder; SUD = substance use disorder; CI = confidence interval; SA = sexual assault.

may partially explain female soldiers' elevated risk for suicide while deployed. Unsurprisingly, given that research on MST and suicide is in its infancy, no studies have examined factors that mediate this association.

## What factors are associated with suicidal ideation and attempts among military sexual trauma survivors?

As a growing number of studies report that MST is associated with SI, suicide attempts, and suicide, identifying factors that are associated with these outcomes among MST survivors is increasingly important. Studies conducted in this area to date (Table 26.4) are described next.

### *Suicidal ideation*

Several factors have been found to relate to SI among samples of MST survivors. These include psychological distress, depressive symptoms, and hazardous alcohol use (Monteith, Holliday,

*Table 26.4* Studies examining factors associated with suicidal ideation and suicide attempts in military sexual trauma survivors

| *Study* | *Sample* | *SI* | *Suicide Attempt* | *SI Findings* | *Suicide Attempt Findings* |
|---|---|---|---|---|---|
| Monteith et al. (2016b) | **49 veterans** reporting MST (31 women, 18 men) | **Post-MST** (SITBI) | **Post-MST** (SITBI) | Institutional betrayal not associated with post-MST SI. Similar results in subsample with MSA. | Institutional betrayal was associated with post-MST suicide attempt. Similar results in subsample with MSA. |
| Monteith et al. (2017) | **92 female veterans** with MST, entering VA inpatient care | **Past-week** (BSS) | | Accounting for lifetime suicide attempts and depressive/PTSD symptoms, PB, TB, and FAD associated with SI. TB NS in full model. | |
| Blais and Monteith (2019) | **311 female service members or veterans** with MST and Criterion A | **Past 2 weeks** (PHQ-9) | | MST- (vs. non–MST events; combat/deployment) as PTSD symptom source associated with SI. Similar results adjusting for PTSD/depression. Age, marital status, branch, race, deployment also covaried. | |
| Holliday et al. (2018a,b) | **107 veterans with MST** (65 females, 42 males) | **Past-week** (BSS) | | MST survivors with lifetime NSSI more likely to endorse recent SI. | |
| Monteith, Smith et al. (2018) | **115 veterans** (56 females, 59 males) with MST | **Past 2 weeks** (PHQ-9[a]) | **As adult** ("tried to kill" self at ≥18) | Psychological distress, hazardous alcohol use, and general disapproval from others associated with SI. Age, race/ethnicity, branch, lifetime trauma, social support, and loneliness in model. | Hazardous alcohol use and general disapproval associated with suicide attempt. Lifetime trauma, distress, social support, loneliness included in model. |

(*Continued*)

*Table 26.4* (Continued)

| *Study* | *Sample* | *SI* | *Suicide Attempt* | *SI Findings* | *Suicide Attempt Findings* |
|---|---|---|---|---|---|
| Sexton et al. (2018) | **277 veterans** seeking MST care (70.8% female) | | **Lifetime** (MINI-IV) | | Sexual/gender minority veterans had higher lifetime suicide attempt rates. |
| Monteith, Holliday et al. (2019) | **108 veterans** (66 women, 42 men) with MST | **Past-week** (BSS[a]); **post-MST** (SITBI) | **Post-MST** (SITBI) | Negative PTCI Self and pre-MST SI associated with past-week SI. Similar results covarying for depressive/PTSD symptoms. Negative PTCI Self, MSA, pre-MST SI, and childhood physical abuse associated with post-MST SI. Age and gender included. | Negative PTCI Self and pre-MST suicide attempt associated with post-MST suicide attempt, controlling for age and gender. |

*Abbreviations*: SI = suicidal ideation; MST = military sexual trauma; SITBI = Self-Injurious Thoughts and Behaviors Interview; OR = odds ratio; CI = confidence interval; MSA = military sexual assault; VA = Veterans Affairs; BSS = Beck Scale for Suicide Ideation; PB = perceived burdensomeness; TB = thwarted belonging; FAD = fearlessness about death; PTSD = posttraumatic stress disorder; NS = not significant; PHQ-9 = Patient Health Questionnaire-9; NSSI = non-suicidal self-injury; MINI = Mini-International Neuropsychiatric Interview; PTCI = Posttraumatic Cognitions Inventory.

a Scored dichotomously.

Schneider, Forster, & Bahraini, 2019; Monteith, Smith, Holliday, & Pietrzak, 2018). Female service members and veterans who endorse MST, as opposed to other military experiences (e.g., combat/deployment-related), as the source of their current PTSD symptoms are more likely to report experiencing recent SI (Blais & Monteith, 2019). Nonetheless, PTSD symptoms have not consistently been found to be associated with recent SI among MST survivors when accounting for other factors (Monteith, Bahraini, & Menefee, 2017; Monteith, Holliday et al., 2019). Consistent with findings by Monteith, Menefee et al. (2016c), military SA is also associated with post-MST SI (Monteith, Holliday et al., 2019). In addition, pre-MST SI, lifetime NSSI, and fearlessness about death have also been found to be associated with SI in this population (Holliday, Smith, & Monteith, 2018; Monteith et al., 2017; Monteith, Holliday et al., 2019). Finally, specific interpersonal beliefs and experiences, such as perceptions of burdensomeness, thwarted belonging, general disapproval from others following one's worst traumatic experience, and negative posttraumatic beliefs about oneself, are associated with SI among veterans who have experienced MST (Monteith et al., 2017; Monteith, Holliday et al., 2019; Monteith, Smith et al. 2018). Together, these studies suggest that MST survivors' SI may relate to specific MST-related characteristics and a broad range of psychiatric, interpersonal, and cognitive-affective factors.

### *Suicide attempt*

Efforts to identify factors associated with suicide attempts in MST survivors have been sparse. Monteith, Smith et al. (2018) examined factors associated with suicide attempt in adulthood using National Health and Resilience in Veterans Study (NHRVS) data. MST survivors reporting hazardous alcohol use and general disapproval from others were more likely to report attempting suicide as an adult. Another study found that perceptions of institutional betrayal were associated with post-MST suicide attempts (Monteith et al., 2016b); in a larger examination, pre-MST suicide attempts and negative posttraumatic cognitions about the self were most strongly associated with post-MST suicide attempt (Monteith, Holliday et al., 2019). Lifetime history of suicide attempt has also been found to be more prevalent among sexual/gender minority MST survivors compared to non-sexual/gender minority MST survivors (Sexton et al., 2018). These studies suggest a potentially complex etiology of suicide attempts among MST survivors, with additional research warranted.

## Limitations and future research

The increased number of publications on this topic is promising. At the same time, understanding the limitations of this body of research can help with interpreting findings and guiding subsequent research. Some of these limitations are specific to certain constructs or topics, while others generalize more broadly.

### *Study design and intent*

Many studies described earlier were conducted as secondary analyses, which at times resulted in methods not ideal for testing the associations of interest (e.g., excluding participants with a recent suicide attempt). In addition, MST, particularly SA, and suicide attempts are occurrences with relatively low base rates. As such, some studies may have been inadequately powered to draw generalizable findings regarding the association between MST with suicide attempts and may warrant replication in larger samples. Relatedly, for examinations of factors associated with SI or suicide attempts among MST survivors, samples tended to be small. This is an inherent challenge to conducting research on SI and SA with MST survivors, as low base rates of these phenomenon, coupled with underreporting due to stigma and distrust, can make it challenging to study the intersection of these experiences. Finally, most studies described were cross-sectional. While no study design is without limitations, longitudinal studies are needed to identify predictors of suicide-related outcomes following MST. A comprehensive understanding of risk among this population also requires tracking factors that increase and decrease risk across multiple timepoints. Longitudinal examination of these phenomena with larger samples, sufficiently powered and designed to address these critical questions, would be invaluable to advance understanding of distinct trajectories of risk among MST survivors over time, including identification of factors that increase or mitigate suicide risk.

## *Measurement of suicidal ideation and suicide attempt*

Research has also been limited with respect to assessment of SI and suicide attempt. The majority of studies have relied upon self-report measures of suicide attempt or SI, including single-item self-report measures originally designed for other purposes (e.g., depression screens). Assessing SI or suicide attempts via single-item self-reports can result in misclassification, reduced statistical power, and erroneous results (Hom, Joiner, & Bernert, 2016; Millner, Lee, & Nock, 2015). Although there is not presently a consensus "gold standard" measure for assessing SI and SA, several validated self-report measures do exist (c.f. Batterham et al., 2015) and are worth including in future studies.

Relatedly, few studies have utilized interview-based measures of SI or suicide attempt, even though such measures can help to ensure accurate classification of SI and suicide attempts given proper training in administration and interpretation of such measures. Thus, future research may benefit from utilizing such measures to avoid misclassification. Another issue concerns the temporality of SI and suicide attempt, particularly for studies that assessed lifetime SI or suicide attempts, which is potentially problematic when examining associations with MST, as MST could have followed, rather than preceded, SI or suicide attempts. Prospectively examining future suicide attempts or SI, or examining SI or suicide attempts that more reasonably could have occurred at some point following MST, is suggested as an alternative.

Finally, relying upon suicide attempts coded in medical or administrative records is potentially problematic (CDC, 2008; Hansen, 2018), as the reliability of such data is unknown; it is possible that only suicide attempts warranting medical attention (e.g., hospitalization) were coded. Relatedly, reporting the specific ICD codes included is essential for determining the extent to which suicide attempts are being assessed and for replication by other investigators.

## *Measurement of military sexual trauma*

A leading challenge to this research is the broad variations that have occurred in the measurement of MST, as well as varying definitions used to assess SH or SA. The VA MST screen (or variations thereof) has received preliminary psychometric support (Mengeling et al., 2019), but was developed to screen for MST rather than to delineate the impact of differing MST experiences. Relying solely upon MST screening results in VA medical records also runs the risk of false negatives (Divita et al., 2017; Gundlapalli et al., 2017), due to stigma associated with disclosure of MST.

The DRRI SH Scale can be useful for assessing the frequency of SH experiences during deployment; however, the fact that instructions are specific to deployment suggests that (unless modified) this measure is only applicable to those who have deployed – a subset of veteran and military populations. Most DoD reports on SH and SA rely solely on legal definitions with broad, encompassing categories that do not differentiate between qualitatively and contextually different incidents. These reports would be improved through a more detailed assessment of the specific events experienced by individuals, allowing for more tailored assessment and intervention in military units. In addition, validation of the DRRI SH scale has relied primarily on traditional psychometric methods; however, modern psychometric methods, such as Rasch analysis, may be particularly useful for enhancing measurement precision by weighting individual items based on their contributions to the underlying trait. Rasch analysis would also allow for examination of gender-based differential item functioning, which is critical to ensuring that items on the DRRI SH scale do not function differently for men and women. As it stands, difficulties in MST measurement remain a prominent challenge to advancing this body of research. Addressing this limitation would be monumental in overcoming these inherent methodological challenges.

## Gender

Studies took varying approaches to gender. Some adjusted for gender or sex as a covariate, likely due to insufficient numbers to conduct gender-stratified analyses. Many studies also stratified results by gender, which at times revealed important gender differences. These findings were further corroborated when examining gender as a potential moderator or conducting mediation analyses separately by gender. These results underscore the importance of continued research with sufficient numbers of men and women to stratify analyses by gender or examine potential moderation by gender. Finally, while several studies exclusively sampled females, only one quantitative study has occurred exclusively with men (Schry et al., 2015). This major limitation contributes to ongoing gaps in knowledge regarding the impact of MST among men and may inadvertently reinforce existing stereotypes and notions regarding MST (e.g., that MST only happens to women; Morris, Smith, Farooqui, & Surís, 2014). As men who experience MST may be at particularly elevated risk for SI and experience unique concerns that potentially place them at heightened risk (e.g., masculinity-related concerns; Monteith, Brownstone, Gerber, Soberay, & Bahraini, 2018), examination of such constructs in relation to SI and SA is needed.

## Covariates

Across published studies, a largely consistent pattern emerged: MST was generally associated with heightened risk for SI, suicide attempt, and suicide. Many studies that accounted for known correlates of SI, suicide attempt, and suicide (e.g., psychiatric symptoms or diagnoses; prior SI or suicide attempt) continued to report significant associations. Further, associations generally remained significant when accounting for other stressful or traumatic military-related experiences (e.g., combat exposure; Monteith et al., 2015; Monteith, Bahraini, Matarazzo, Gerber et al., 2016a) or when matching based on trauma exposure (Schry et al., 2015), with one exception (Khan et al., 2019). Research accounting for pre-military sexual victimization, however, has been mixed (Bryan et al., 2015; White et al., 2018), suggesting another potentially important avenue for future research.

## Samples

Many studies of veterans focused on those in VHA care. Yet suicide rates have been increasing more quickly among veterans not using VHA care (e.g., who dropped out or never used care; VA, 2018). Moreover, MST survivors' reasons for not using VHA care (e.g., institutional betrayal, distrust) may also heighten their risk for suicide (Holliday & Monteith, 2019). Thus, research specific to non-VHA veteran samples is needed. In addition, research with active and reserve service members has been fairly limited outside of congressionally mandated annual surveys by SAPRO (e.g., Department of Defense, 2018b). For example, five suicide deaths (1.7%) and 86 suicide attempts (6.9%) in the 2016 DoD Suicide Event Report recorded SA or SH in the 12 months prior to the suicidal self-directed violence (Pruitt et al., 2018). However, few other published studies on active or reserve military personnel have identified factors that influence this association. Low rates of formal reporting also limit the identification of at-risk samples during military service (Department of Defense, 2018b). Although restricted reporting options allow service members who experience SA and SH to receive services (Department of Defense, 2017), the protected nature of these reports precludes in-depth analysis of their relation to suicidal self-directed violence (Pruitt et al., 2018). As such, this knowledge is largely limited to veterans. This is an important limitation since active military personnel, who may have experienced SA or SH more recently, may differ in their risk for SI or suicide attempts, compared to veterans, who may have experienced MST years or even decades ago.

## *Broader comprehensive examinations*

MST is an unfortunately common potentially traumatic event among men and women who serve in the U.S. military, resulting in a number of mental and physical health consequences. MST has also been associated with SI, suicide attempt, and suicide across a number of studies. While several factors are proposed to partially account for this relationship, longitudinal research is needed to fully elucidate drivers of suicidal self-directed violence among this population (Holliday, Maguen et al., 2018). Utilizing prior research as a framework, considering psychiatric (e.g., specific diagnoses or symptoms) and psychosocial (e.g., distress, interpersonal functioning) factors following the experience of MST within the context of pre-military (e.g., childhood trauma) and sociodemographic (e.g., gender, sexual identity) factors will likely provide the best lens to interpret results and advance understanding of for whom, when, and under what circumstances MST increases risk for SI and suicide attempts.

Researchers can then conceptualize a comprehensive model for understanding drivers of suicide risk within this population, as well as to determine appropriate assessment and intervention. Much of the research described earlier focused on factors that have been established as correlates of suicide risk in other populations (e.g., depression, prior suicide attempts, and hazardous alcohol use). However, understanding the extent to which unique MST sequelae and characteristics (e.g., interpersonal and institutional responses to disclosure and reporting; MST-related care) are associated with SI and suicide attempt would be invaluable. In addition, the majority of interventional research specific to survivors of MST and suicide-related outcomes has been limited to secondary analyses of larger clinical trials.

For example, Wiblin, Holder, Holliday, and Surís (2018) examined correlates of suicide-related beliefs among male and female veterans with MST-related PTSD, noting the relation of these beliefs to depressive symptoms, psychosocial functioning, and negative posttraumatic-related beliefs – mirroring findings of some previously discussed studies. In the same sample, Holliday, Holder, Monteith, and Surís (2018a) reported that cognitive processing therapy, an evidence-based treatment for PTSD, was associated with decreases in suicide-related beliefs following treatment, with reductions sustained up to 6 months post-treatment. Yet this study occurred in the context of a single trial, was limited in sample size, and did not analyze other correlates of SI (e.g., hopelessness) or compare results to other evidence-based treatments (e.g., prolonged exposure). This trial also focused on veterans with PTSD and, similar to other PTSD trials, excluded those experiencing elevations in acute suicide risk, further precluding examination of interventions in the presence of elevated acute suicide risk (e.g., current SI with intent; Holliday, Holder, Olson-Madden, & Monteith, 2019).

Focusing exclusively on PTSD interventional research is a major limitation to understanding how to mitigate suicide risk in the MST population at large, as several psychiatric diagnoses are associated with MST, and SH may be more strongly associated with other psychiatric diagnoses (Williamson, Holliday, Holder, North, & Surís, 2017). As such, more comprehensive examination of assessment and intervention specific to MST survivors presenting with a spectrum of psychiatric diagnoses (e.g., transdiagnostic interventions, such as skills training in affective and interpersonal regulation) remains warranted. In addition, no studies have examined the acceptability, feasibility, or efficacy of suicide-specific interventions (e.g., safety planning, lethal means safety, cognitive therapy for suicide prevention) among MST survivors, which would provide valuable insight into whether such treatments are optimally suited for preventing suicide among survivors of MST. Finally, based on multiple factors that can impact MST-related research (e.g., initial non-disclosure of MST, seeking treatment outside of the military or VA settings), a multi-institutional response (e.g., DoD, VA, community-based settings) espousing collaboration is likely requisite (Holliday et al., 2018a,b).

## Conclusions

MST is a potentially traumatic event that is associated with heightened risk for experiencing SI, attempting suicide, and dying by suicide. Recent research has been integral in elucidating the nuances of these associations – for example, highlighting specific subgroups for whom this relationship is especially pronounced, as well as specific MST-related experiences (such as SA) that may particularly lend themselves to elevated risk for SI or suicide attempt. Moreover, a monumental step in this body of research has entailed beginning to elucidate factors that are associated with SI, as well as suicide attempts, among survivors of MST. Such studies have led to understanding of the broad array of cognitive-affective, interpersonal, and institutional factors that may be associated with SI and suicide attempts among MST survivors. Some of these factors have been identified as relevant in other populations as well, while others appear to be particularly salient to the experience of MST. A critical next step for future research in this area is to identify trajectories of SI and suicide attempt following MST and to discern intrapersonal, interpersonal, and institutional factors that predict such trajectories. Translating knowledge from such endeavors into actual intervention and prevention efforts (e.g., determining efficacious means of preventing suicidal self-directed violence among service members and veterans who have experienced MST) is equally important. While this line of research can present unique challenges and considerations, addressing these questions is essential for optimizing suicide prevention efforts for survivors of MST.

**Note**: This material is based upon work supported in part by the Department of Veterans Affairs (VA); the Rocky Mountain MIRECC for Suicide Prevention; and the VA Office of Academic Affiliations, Advanced Fellowship Program in Mental Illness Research and Treatment. The views expressed are those of the authors and do not necessarily represent the views or policy of the VA, the Department of Defense, or the United States Government.

## References

Andresen, F. J., & Blais, R. K. (2019). Higher self-stigma is related to lower likelihood of disclosing military sexual trauma during screening in female veterans. *Psychological Trauma: Theory, Research, Practice and Policy*, *11*(4), 372–378.

Batterham, P. J., Ftanou, M., Pirkis, J., Brewer, J. L., Mackinnon, A. J., Beautrais, A., … Christensen, H. (2015). A systematic review and evaluation of measures for suicidal ideation and behaviors in population-based research. *Psychological Assessment*, *27*(2), 501–512.

Blais, R. K., & Geiser, C. (2019). Depression and PTSD-related anhedonia mediate the association of military sexual trauma and suicidal ideation in female service members/veterans. *Psychiatry Research, 279*, 148–154.

Blais, R. K., & Monteith, L. L. (2019). Suicide ideation in female survivors of military sexual trauma: The trauma source matters. *Suicide and Life-Threatening Behavior, 49*(3), 643–652.

Brenner, L. A., Breshears, R. E., Betthauser, L. M., Bellon, K. K., Holman, E., Harwood, J. E., … & Nagamoto, H. T. (2011). Implementation of a suicide nomenclature within two VA healthcare settings. *Journal of Clinical Psychology in Medical Settings, 18*(2), 116–128.

Brignone, E., Gundlapalli, A. V., Blais, R. K., Carter, M. E., Suo, Y., Samore, M. H., … Fargo, J. D. (2016). Differential risk for homelessness among US male and female veterans with a positive screen for military sexual trauma. *JAMA Psychiatry, 73*(6), 582–589.

Brownstone, L. M., Holliman, B. D., Gerber, H. R., & Monteith, L. L. (2018). The phenomenology of military sexual trauma among women veterans. *Psychology of Women Quarterly, 42*(4), 399–413.

Bryan, C. J., Bryan, A. O., & Clemans, T. A. (2015). The association of military and premilitary sexual trauma with risk for suicide ideation, plans, and attempts. *Psychiatry Research, 227*(2-3), 246–252.

Centers for Disease Control and Prevention. (2008). Strategies to improve external cause-of-injury coding in state-based hospital discharge and emergency department data systems. *MMWR, 57*, 1–15.

Centers for Disease Control and Prevention. (2011). *Self-Directed Violence Surveillance: Uniform Definitions and Recommended Data Elements*. Retrieved from: https://www.cdc.gov/violenceprevention/pdf/Self-Directed-Violence-a.pdf

Dardis, C. M., Reinhardt, K. M., Foynes, M. M., Medoff, N. E., & Street, A. E. (2018). "Who are you going to tell? Who's going to believe you?" Women's experiences disclosing military sexual trauma. *Psychology of Women Quarterly, 42*(4), 414–429.

Davis, L., Grifka, A., Williams, K., & Coffey, M. (Eds.). (2017). *2016 Workplace and Gender Relations Survey of Active Duty Members*. Alexandria, VA: Office of People Analytics.

Department of Defense. (2005). *Sexual Assault Prevention and Response (SAPR) Program* (Department of Defense Instruction 6495.01). Washington, DC: Author.

Department of Defense. (2017). *Sexual Assault Prevention And Response (SAPR) Program Procedures* (Department of Defense Instruction 6495.02, incorporating Change 3). Washington, DC: Author.

Department of Defense. (2018a). *Harassment Prevention and Response in the Armed Forces* (Department of Defense Instruction 1020.03). Washington, DC: Author.

Department of Defense. (2018b). *Department of Defense Annual Report on Sexual Assault in the Military: Fiscal Year 2017*). Washington, DC: Author.

Department of Veterans Affairs. (2018). *VA National Suicide Data Report 2005-2016*. Retrieved from https://www.mentalhealth.va.gov/docs/data-sheets/OMHSP_National_Suicide_Data_Report_2005-2016_508.pdf

Department of Veterans Affairs Office of Mental Health & Suicide Prevention. (2018). *Military Sexual Trauma (MST) Screening Report, Fiscal Year 2018*). Retrieved from: https://vaww.vashare.vha.va.gov/sites/mst/reports-and-data/screening-and-treatment

DiMauro, J., Renshaw, K. D., & Blais, R. K. (2018). Sexual vs. non-sexual trauma, sexual satisfaction and function, and mental health in female veterans. *Journal of Trauma & Dissociation, 19*(4), 403–416.

Divita, G., Brignone, E., Carter, M. E., Suo, Y., Blais, R. K., Samore, M. H., … Gundlapalli, A. V. (2017). Extracting sexual trauma mentions from electronic medical notes using natural language processing. *Studies in Health Technology and Informatics, 245, 351*.

Foynes, M. M., Makin-Byrd, K., Skidmore, W. C., King, M. W., Bell, M. E., & Karpenko, J. (2018). Developing systems that promote veterans' recovery from military sexual trauma: Recommendations from the Veterans Health Administration national program implementation. *Military Psychology, 30*(3), 270–281.

Gradus, J. L., King, M. W., Galatzer-Levy, I., & Street, A. E. (2017). Gender differences in machine learning models of trauma and suicidal ideation in veterans of the Iraq and Afghanistan Wars. *Journal of Traumatic Stress, 30*(4), 362–371.

Gradus, J. L., Shipherd, J. C., Suvak, M. K., Giasson, H. L., & Miller, M. (2013). Suicide attempts and suicide among Marines: A decade of follow-up. *Suicide and Life-Threatening Behavior, 43*(1), 39–49.

Gradus, J. L., Street, A. E., Suvak, M. K., & Resick, P. A. (2013). Predictors of suicidal ideation in a gender-stratified sample of OEF/OIF veterans. *Suicide and Life-Threatening Behavior, 43*(5), 574–588.

Griffith, J. (2019). The sexul harassment-suicide connection in the U.S. military: Contextual effects of hostile work environment and trusted unit leaders. *Suicide and Life-Threatening Behavior, 49*(1), 41–53.

Gundlapalli, A.V., Brignone, E., Divita, G., Jones, A. L., Redd, A., Suo, Y., … Samore, M. H. (2017). Using structured and unstructured data to refine estimates of military sexual trauma status among US military veterans. *Studies in Health Technology and Informatics*, *238*, 128–131.

Hansen, A. (2018). Coding's impact on suicide prevention. *Journal of AHIMA*, *89*(8), 68–69.

Holliday, R., Holder, N., Monteith, L. L., & Surís, A. (2018a). Decreases in suicide cognitions after cognitive processing therapy among veterans with posttraumatic stress disorder due to military sexual trauma: A preliminary examination. *Journal of Nervous and Mental Disease*, *206*(7), 575–578.

Holliday, R., Holder, N., Olson-Madden, J. H., & Monteith, L. L. 2019. Treating posttraumatic stress disorder in the presence of acute suicide risk in veterans and active duty service members: A call for research. *Journal of Nervous and Mental Disease*. *207*(7), 611–614.

Holliday, R., Maguen, S., Hoyt, T., Blais, R. K., & Monteith, L. L. (2018b). Addressing military sexual assault and harassment through a collaborative, multi-institutional approach. *The Military Psychologist*, *33*(3), 6–9.

Holliday, R., & Monteith, L. L. (2019). Seeking help for the health sequelae of military sexual trauma: A theory-driven model of the role of institutional betrayal. *Journal of Trauma & Dissociation*, *20*(3), 340–356.

Holliday, R., Smith, N. B., & Monteith, L. L. (2018). An initial investigation of nonsuicidal self-injury among male and female survivors of military sexual trauma. *Psychiatry Research*, *268*, 335–339.

Hom, M. A., Joiner Jr, T. E., & Bernert, R. A. (2016). Limitations of a single-item assessment of suicide attempt history: Implications for standardized suicide risk assessment. *Psychological Assessment*, *28*(8), 1026–1030.

Khan, A. J., Li, Y., Dinh, J.V., Donalson, R., Hebenstreit, C. L., & Maguen, S. (2019). Examining the impact of different types of military trauma on suicidality in women veterans. *Psychiatry Research*, *274*, 7–11.

Kimerling, R., Gima, K., Smith, M.W., Street, A., & Frayne, S. (2007). The Veterans Health Administration and military sexual trauma. *American Journal of Public Health*, *97*(12), 2160–2166.

Kimerling, R., Makin-Byrd, K., Louzon, S., Ignacio, R.V., & McCarthy, J. F. (2016). Military sexual trauma and suicide mortality. *American Journal of Preventive Medicine*, *50*(6), 684–691.

King, L. A., King, D. W., Vogt, D. S., Knight, J., & Samper, R. E. (2006). Deployment Risk and Resilience Inventory: A collection of measures for studying deployment-related experiences of military personnel and veterans. *Military Psychology*, *18*(2), 89–120.

Klingensmith, K., Tsai, J., Mota, N., Southwick, S., Pietrzak, R. (2014). Military sexual trauma in US veterans: Results from the National Health and Resilience in Veterans Study. *Journal of Clinical Psychiatry*, *75*(10), 1133–1139.

Lemaire, C. M., & Graham, D. P. (2011). Factors associated with suicidal ideation in OEF/OIF veterans. *Journal of Affective Disorders*, *130*(1-2), 231–238.

Mengeling, M. A., Burkitt, K. H., True, G., Zickmund, S. L., Ono, S. S., Bayliss, N. K., … Sadler, A. G. (2019). Sexual trauma screening for men and women: Examining the construct validity of a two-item screen. *Violence and Victims*, *34*(1), 175–193.

Millegan, J., Wang, L., LeardMann, C. A., Miletich, D., & Street, A. E. (2016). Sexual trauma and adverse health and occupational outcomes among men serving in the US military. *Journal of Traumatic Stress*, *29*(2), 132–140.

Millner, A. J., Lee, M. D., & Nock, M. K. (2015). Single-item measurement of suicidal behaviors: Validity and consequences of misclassification. *PLOS ONE*, *10*(10), e0141606). doi:10.1371/journal.pone.0141606

Monteith, L. L., Bahraini, N. H., Matarazzo, B. B., Gerber, H. R., Soberay, K. A., & Forster, J. E. (2016a). The influence of gender on suicidal ideation following military sexual trauma among veterans in the Veterans Health Administration. *Psychiatry Research*, *244*, 257–265.

Monteith, L. L., Bahraini, N. H., Matarazzo, B. B., Soberay, K. A., & Smith, C. P. (2016b). Perceptions of institutional betrayal predict suicidal self-directed violence among veterans exposed to military sexual trauma. *Journal of Clinical Psychology*, *72*(7), 743–755.

Monteith, L. L., Bahraini, N. H., & Menefee, D. S. (2017). Perceived burdensomeness, thwarted belongingness, and fearlessness about death: Associations with suicidal ideation among female veterans exposed to military sexual trauma. *Journal of Clinical Psychology*, *73*(12), 1655–1669.

Monteith, L. L., Brownstone, L. M., Gerber, H. R., Soberay, K. A., & Bahraini, N. H. (2018). Understanding suicidal self-directed violence among men exposed to military sexual trauma: An ecological framework. *Psychology of Men and Masculinities*, *20*(1), 23–35.

Monteith, L. L., Gerber, H. R., Brownstone, L. M., Soberay, K. A., & Bahraini, N. H. (2019). The phenomenology of military sexual trauma among male veterans. *Psychology of Men & Masculinities*, *20*(1), 115–127.

Monteith, L.L., Hoffmire, C.A., Holliday, R., Park, C.L., Mazure, C.M., & Hoff, R.A. (2018). Do unit and post-deployment social support influence the association between deployment sexual trauma and suicidal ideation? *Psychiatry Research*, *270*, 673–681.

Monteith, L. L., Holliday, R., Schneider, A. L., Forster, J. E., & Bahraini, N. H. (2019). Identifying factors associated with suicidal ideation and suicide attempts following military sexual trauma. *Journal of Affective Disorders, 252*, 300–309.

Monteith, L. L., Menefee, D. S., Forster, J. E., & Bahraini, N. H. (2016c). A closer examination of sexual trauma during deployment: Not all sexual traumas are associated with suicidal ideation. *Suicide and Life-Threatening Behavior, 46*(1), 46–54.

Monteith, L.L., Menefee, D.S., Forster, J.E., Wanner, J., & Bahraini, N.H. (2015). Sexual trauma and combat during deployment: Associations with suicidal ideation among OEF/OIF/OND veterans. *Journal of Traumatic Stress, 28*(4), 283–288.

Monteith, L. L., Smith, N. B., Holliday, R., & Pietrzak, R. H. (2018). Psychiatric and interpersonal correlates of suicide ideation in military sexual trauma survivors: The National Health and Resilience in Veterans Study. *Chronic Stress*. Advance online publication. doi:10.1177/2470547018815901

Morris, E. E., Smith, J. C., Farooqui, S.Y., & Surís, A. M. (2014). Unseen battles: The recognition, assessment, and treatment issues of men with military sexual trauma. *Trauma, Violence, & Abuse, 15*(2), 94–101.

O'Brien, C., Keith, J., & Shoemaker, L. (2015). Don't tell: Military culture and male rape. *Psychological Services, 12*(4), 357–365.

Pavao, J., Turchik, J. A., Hyun, J. H., Karpenko, J., Saweikis, M., McCutcheon, S., … Kimerling, R. (2013). Military sexual trauma among homeless veterans. *Journal of General Internal Medicine, 28*(2), 536–541.

Pruitt, L. D., Smolenski, D. J., Bush, N. E., Skopp, N. A., Edwards-Stewart, A., & Hoyt, T.V. (2018). *Department of Defense Suicide Event Report (DoDSER): Calendar Year 2016 Annual Report* (Publication No. 0-A2345E0). Washington, DC: Defense Health Agency.

Rosellini, A. J., Street, A. E., Ursano, R. J., Chiu, W. T., Heeringa, S. G., Monahan, J., … Bliese, P. D. (2017). Sexual assault victimization and mental health treatment, suicide attempts, and career outcomes among women in the US Army. *American Journal of Public Health, 107*(5), 732–739.

Schry, A. R., Hibberd, R., Wagner, H. R., Turchik, J. A., Kimbrel, N. A., Wong, M., … Brancu, M. (2015). Functional correlates of military sexual assault in male veterans. *Psychological Services, 12*(4), 384–393.

Sexton, M. B., Davis, M. T., Anderson, R. E., Bennett, D. C., Sparapani, E., & Porter, K. E. (2018). Relation between sexual and gender minority status and suicide attempts among veterans seeking treatment for military sexual trauma. *Psychological Services, 15*(3), 357–362.

Street, A. E., Gilman, S. E., Rosellini, A. J., Stein, M. B., Bromet, E. J., Cox, K. L., … Lewandowski-Romps, L. (2015). Understanding the elevated suicide risk of female soldiers during deployments. *Psychological Medicine, 45*(4), 717–726.

Surís, A., Holliday, R., Weitlauf, J. C., North, C. S., & the Veteran Safety Initiative Writing Collaborative. (2013). Military sexual trauma in the context of veterans' life experiences. *Federal Practitioner, 30*(suppl. 3), 16s–20s.

Surís, A., & Lind, L. (2008). Military sexual trauma: A review of prevalence and associated health consequences in veterans. *Trauma, Violence, & Abuse, 9*(4), 250–269.

Turchik, J. A., Pavao, J., Nazarian, D., Iqbal, S., McLean, C., & Kimerling, R. (2012). Sexually transmitted infections and sexual dysfunctions among newly returned veterans with and without military sexual trauma. *International Journal of Sexual Health, 24*(1), 45–59.

U.S. Government. (2014). *Veterans' Benefits: Counseling and Treatment for Sexual Trauma, 38 USC §1720D*. Washington, DC. Retrieved from https://www.gpo.gov/fdsys/pkg/USCODE-2014-title38/pdf/USCODE-2014-title38-partII-chap17-subchapII-sec1720D.pdf

White, K. L., Harris, J. A., Bryan, A. O., Reynolds, M., Fuessel-Herrmann, D., & Bryan, C. J. (2018). Military sexual trauma and suicidal behavior among National Guard personnel. *Comprehensive Psychiatry, 87*, 1–6.

Wiblin, J., Holder, N., Holliday, R., & Surís, A. (2018). Predictors of unbearability, unlovability, and unsolvability in veterans with military-sexual-trauma-related posttraumatic stress disorder. *Journal of Interpersonal Violence*. Advance online publication. doi: 10.1177/0886260518777554

Williamson, M. L. C., Holliday, R., Holder, N., North, C. S., & Surís, A. (2017). A reexamination of military sexual trauma and posttraumatic stress disorder. *Psychiatric Annals, 47*(3), 134–138.

Wilson, L. C. (2018). The prevalence of military sexual trauma: A meta-analysis. *Trauma, Violence, & Abuse, 19*(5), 584–597.

# 27

# UNDERSTANDING SUICIDE AMONG FEMALE VETERANS

## A theory-driven approach

*Lindsey L. Monteith, Ryan Holliday, Diana P. Brostow, and Claire A. Hoffmire*

There has been concern regarding excess risk for suicide among U.S. veterans for many years (Gibbons, Brown, & Hur, 2012). Yet only recently have advances in veteran suicide surveillance efforts enabled accurate quantification of the magnitude of this risk, including for female veterans (Department of Veterans Affairs, 2017b), who constitute the fastest growing subpopulation of veterans, having increased from <1% in 2000 to 9.4% in 2015 (Frayne et al., 2018). Through these efforts, it has become clear that suicide is a substantial problem that disproportionally affects female veterans relative to female non-veterans. In 2016, the suicide rate for female veterans was 80% higher than that observed for non-veteran females (Department of Veterans Affairs, 2018a). Moreover, from 2005 to 2016, female veterans experienced a 34% increase in age-adjusted suicide rates (Figure 27.1; Department of Veterans Affairs, 2018b). Understanding subgroups of female veterans who are at particularly elevated risk for suicide, as well as factors associated with suicide risk in this population, is critical for addressing this growing and tragic phenomenon.

### Changing trends in relation to Veterans Health Administration use and age

#### *Recent use of Veterans Health Administration care*

Recent epidemiological work has revealed important subgroups of female veterans who have experienced notable changes in their suicide rates over time. For example, female veterans who used Veterans Health Administration (VHA) care in the year prior to their death experienced a 22% increase in age-adjusted suicide rates, compared to an increase of 31% among female veterans not using VHA care in the year preceding death (Department of Veterans Affairs, 2018b). Considering that approximately two-thirds of female veterans do not access VHA care (Department of Veterans Affairs, 2017a; Frayne et al., 2018), these differences are particularly salient. Moreover, these findings underscore the importance of elucidating the reasons for these differences while also determining effective means of preventing suicide in female veterans who do not use VHA care.

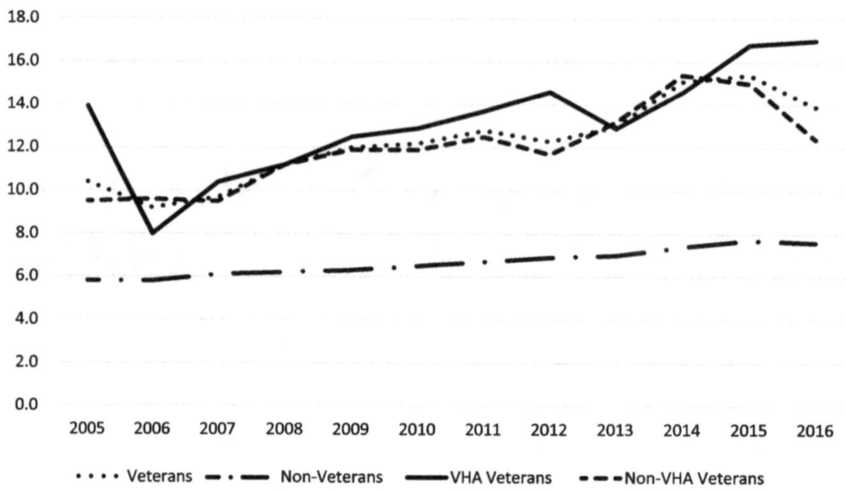


*Figure 27.1* Age-Adjusted Suicide Rates among Females (per 100,000) by veteran Status. This figure was created from publically available data in the Department of Veterans Affairs 2016 National Suicide Data Report Appendix (Department of Veterans Affairs, 2018b).

## *Age*

In addition to the aforementioned findings related to recent use of VHA care, several important findings have been reported with respect to age. From 2005 to 2016, suicide rates among females, including both veterans and non-veterans, were highest for those between 35 and 54 years of age (Figures 27.2a and 27.2b; Department of Veterans Affairs, 2018b).

Suicide rates also increased from 2005–2016 for all age groups of female veterans examined. However, from 2014 to 2016, a notable decline was observed for 35- to 54-year-old female veterans, which was not observed among non-veteran adult females. Further data are needed to determine

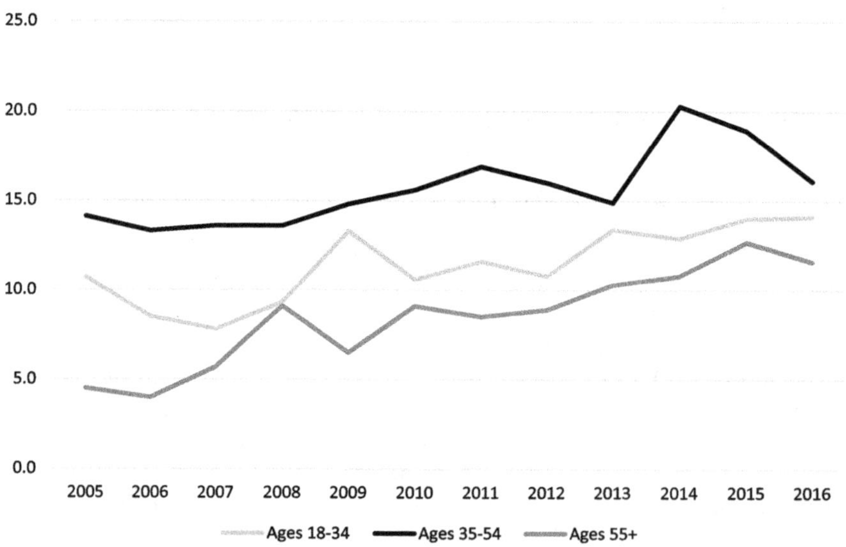


*Figure 27.2* (a) Age-Specific Female veteran Suicide Rates (per 100,000). This figure was created from publically available data in the Department of Veterans Affairs 2016 National Suicide Data Report Appendix (Department of Veterans Affairs, 2018b).

*(Continued)*

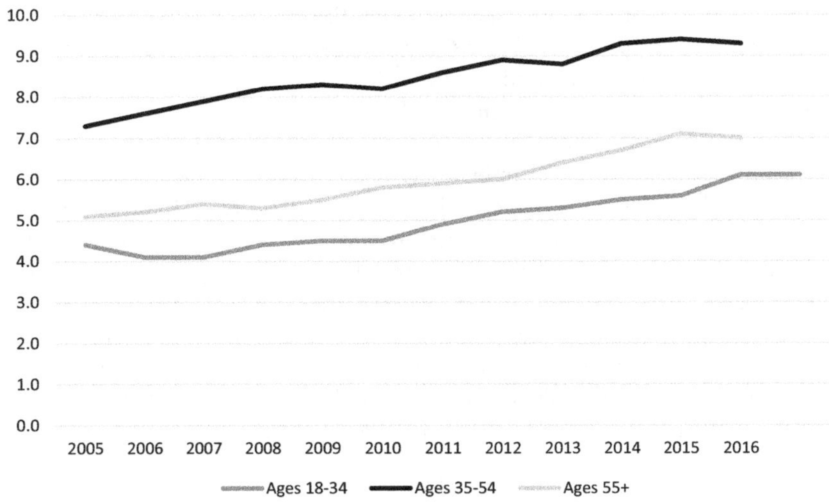


*Figure 27.2* (Continued) (b) Age-Specific Female Non-veteran Suicide Rates (per 100,000). This figure was created from publically available data in the Department of Veterans Affairs 2016 National Suicide Data Report Appendix (Department of Veterans Affairs, 2018b).

if this represents a true declining trend among female veterans 35–54 years of age or random fluctuation. In addition, among female veterans, those between 18–34 years of age were at higher risk for suicide than females who were 55 years of age and older; in contrast, this trend was reversed for non-veteran females. These findings highlight the need for suicide-focused research with female veterans, whose overall suicide risk and periods of heightened risk differ from female civilians.

The excess risk observed among young female veterans (relative to older female veterans) may be attributable to a number of factors, including experiences during and following military service. Younger women are also more likely to have served in Operation Enduring Freedom (OEF), Operation Iraqi Freedom (OIF), and Operation New Dawn (OND), a service era that has entailed increased combat exposure and involvement for female veterans (Street, Vogt, & Dutra, 2009). For post-9/11 female veterans, combat and deployment are often experienced concurrently with military sexual harassment and assault (LeardMann et al., 2013), as well as stress regarding family separation (Mattocks et al., 2012). Younger female veterans are also more likely to have recently transitioned from military service back to civilian life, which may pose unique challenges and psychosocial stress (Ahern et al., 2015; Burkhart & Hogan, 2015; Mattocks et al., 2012), potentially heightening risk for suicide (Pease, Billera, & Gerard, 2015). For example, although risk for suicide decreases during the first seven years following separation for OEF/OIF male veterans, female veterans experience a different pattern in their suicide risk following separation that remains "relatively constant" (Bullman, Hoffmire, Schneiderman, & Bossarte, 2015, p. 965). As such, there is a strong need for research aimed at understanding female veterans' military service, transition, and reintegration experiences in relation to suicide risk.

## Factors associated with suicidal self-directed violence and suicidal ideation among female veterans

### *Mental health conditions*

While the aforementioned findings suggest important subgroups of female veterans warranting increased attention, research has also focused on elucidating factors associated with suicidal self-directed violence (SDV) among female veterans. Prior research has largely focused on the relation of mental health diagnoses with risk for suicide and primarily among female veterans in VHA care. Taken in aggregate, findings from these studies generally mirror findings from the general population. For example, in a study of veterans who used VHA care, having a psychiatric diagnosis was associated with significantly increased risk of suicide (Ilgen et al., 2010). Moreover, all psychiatric diagnoses examined (i.e., bipolar disorder, schizophrenia, alcohol abuse/dependence, drug abuse/dependence, depression, posttraumatic stress disorder [PTSD], and other anxiety disorders) were associated with increased age-adjusted suicide risk among female veterans, which exceeded the risk associated with psychiatric diagnoses among male veterans (Ilgen et al., 2010). Other research has also confirmed that a substance use disorder diagnosis (i.e., alcohol, cocaine, amphetamine, cannabis, and sedatives) is associated with suicide risk among female veterans, although opioid use disorder appears to be a particularly robust suicide risk factor for female veterans (Bohnert, Ilgen, Louzon, McCarthy, & Katz, 2017). Thus, elucidating pathways by which substance use disorders relate to suicide risk among female veterans is essential (Chapman & Wu, 2014).

## *Stressful and traumatic experiences*

In addition to mental health conditions, stressful and potentially traumatic experiences also appear to be associated with suicide risk among female veterans. In particular, there is compelling evidence that military sexual trauma, which is experienced by approximately 38.4% of female veterans (Wilson, 2018), is associated with suicide, suicide attempts, and suicidal ideation among female veterans (Gradus, Street, Suvak, & Resick, 2013; Kimerling, Gima, Smith, Street, & Frayne, 2007; Kimerling, Makin-Byrd, Louzon, Ignacio, & McCarthy, 2016; Monteith et al., 2018). Other studies examining stressful or traumatic events among female veterans have primarily focused on suicidal ideation, rather than suicide or suicide attempts. For example, combat exposure, general harassment, and stressful post-battle experiences are also associated with post-deployment suicidal ideation among female veterans, although these associations are not significant when accounting for mental health symptoms (Gradus et al., 2013). Another recent study found that perceived life threat and sexual harassment during one's military service were both associated with suicidal ideation among female veterans (Khan et al., 2019).

Early life experiences, such as childhood sexual and physical abuse, also appear to be associated with suicidal ideation and suicide attempts among female veterans (Benda, 2005; Roy & Janal, 2006). In addition, intimate partner violence has been found to be associated with suicidal ideation (Brignone, Sorrentino, Roberts, & Dichter, 2018), which may be due to shared risk factors (Iovine-Wong et al., 2019). These findings are particularly important given that female veterans tend to experience a high prevalence of stressful and traumatic experiences across their lifespan (Zinzow, Grubaugh, Monnier, Suffoletta-Maierle, & Frueh, 2007). For example, not only are female veterans highly likely to experience traumas and stressors during their military service (e.g., sexual harassment and sexual assault), they are also likely to encounter pre- (e.g., childhood abuse) and post-military stressful and traumatic events (e.g., intimate partner violence) that may also impact their risk for suicide (Kelly, Skelton, Patel, & Bradley, 2011; Tiet, Finney, & Moos, 2006). Thus, continued examination of such factors and the methods by which they impact female veterans' risk for suicide remains pertinent. As comprehensive examination of female veterans' suicide risk in relation to the range of potentially stressful and traumatic events across the lifespan has been limited, a broader lifespan-focused approach is needed.

*Table 27.1* Overview of constructs relevant to each theory

| *Theory* | *Constructs* |
|---|---|
| Interpersonal-psychological theory | Perceived burdensomeness, thwarted belongingness, and acquired capability for suicide (i.e., fearlessness about death and pain tolerance) |
| Fluid vulnerability theory | Cognitive, affective, physiological, and behavioral vulnerabilities; precipitating stressors; and risk factors |
| Three-step theory | Pain; hopelessness; connectedness (protective); and dispositional, acquired, and practical capacity |
| Cultural model of suicide | Life stressors (minority stress, social discord, and social sanctions), idioms of distress, and cultural meanings of stressors and suicide (e.g., cultural sanctions and acceptability) |
| Minority stress model | External minority stress (e.g., discrimination, prejudice, rejection) and internal minority distress (e.g., internalization of such experiences) |

*Note:* This table summarizes the key constructs inherent to each theory discussed in this chapter.

### *Other factors*

A comprehensive review of all published research on factors associated with suicide, suicide attempts, and suicidal ideation among female veterans is beyond the scope of the present chapter. Nonetheless, it is worth noting that other factors have also been found to be associated with suicidal ideation among female veterans. Examples include sexual dysfunction (Blais, Monteith, & Kugler, 2018) and being divorced, separated, or widowed (Blosnich, Brenner, & Bossarte, 2016). As such, further elucidation of such factors, as well as a comprehensive, theory-driven model of female veteran suicide, remains crucial to appropriately conceptualize and prevent suicide among this subset of the veteran population.

## Applying current theories of suicide to female veterans

As previously noted, to date, research aimed at understanding factors associated with suicide among female veterans has been limited, primarily utilizing a diagnostic-based framework or examining specific types of stressful or traumatic events in isolation. However, the majority of women with mental health conditions do not die by suicide, suggesting that a transdiagnostic approach may be valuable for understanding factors underlying female veterans' risk for suicide. Theory-driven approaches to suicide can be useful for idiographic case conceptualization of suicide risk when working with female veterans. Nonetheless, research attempting to apply theories of suicide to female veterans has been limited. Below we describe several contemporary theories of suicide that could be applied to understanding suicide among female veterans. Key constructs relevant to each of these theories are also summarized in Table 27.1.

### *Interpersonal-psychological theory of suicide*

The interpersonal-psychological theory of suicide (IPTS) (Joiner, 2007) proposes that thwarted belongingness (an unmet psychological need to belong) and perceived burdensomeness (the belief that one's existence poses a burden to others) cause passive suicidal ideation, which can progress into suicidal desire when an individual experiences hopelessness regarding perceived burdensomeness and thwarted belongingness (Van Orden et al., 2010). Suicidal intent is posited to occur when an individual experiences both suicidal desire and lowered fear of death. Lastly, the IPTS proposes that suicidal self-directed violence (S-SDV) is most likely to occur in the context of elevated perceived burdensomeness, thwarted belongingness, lowered fear of suicide, and increased tolerance of pain (Van Orden et al., 2010). Thus, the acquired capability for suicide, which is composed of fearlessness about death and increased pain tolerance, is considered necessary for an individual to act upon suicidal desire.

The IPTS can be used to understand potential mechanisms by which specific experiences that are commonly experienced by female veterans relate to heightened suicide risk. For example, Van Orden et al. (2010) noted that childhood abuse, intimate partner violence, loneliness, and family conflict may increase thwarted belongingness. Indeed, research has found that childhood emotional abuse is associated with thwarted belongingness (Smith, Monteith, Rozek, & Meuret, 2018). Female veterans also are at risk for experiencing unemployment, homelessness, mental illness, and comorbid physical health diagnoses (Agha, Lofgren, VanRuiswyk, & Layde, 2000; Byrne, Montgomery, & Dichter, 2013; Hamilton, Williams, & Washington, 2015; Zinzow et al., 2007). These experiences may, in turn, exacerbate perceived burdensomeness (Silva, Ribeiro, & Joiner, 2015; Van Orden et al., 2010). In addition, repeated exposure to painful or fearful experiences, such as childhood abuse, suicide loss, combat exposure, and prior S-SDV are posited to lower fear of death and heighten physical pain tolerance (Van Orden et al., 2010). Such experiences are potentially relevant for many female veterans (Cerel et al. 2015; Street et al., 2009; Zinzow et al., 2007). Furthermore, childhood physical abuse and violent combat-related experiences have both been found to be associated with higher acquired capability for suicide (Bryan, & Cukrowicz, 2011; Smith et al., 2018), although these associations have not been examined in female veterans specifically.

Recent systematic reviews of the IPTS, which were based on an extensive number of studies with diverse samples (albeit few samples of female veterans), have concluded that there is strong support for certain components of the theory (e.g., the main effect of perceived burdensomeness and the interaction between perceived burdensomeness and thwarted belongingness in suicidal ideation), whereas other components of the IPTS have either been untested or generally not supported (Chu et al., 2017; Ma, Batterham, Calear, & Han, 2016). Research examining the IPTS in female veterans specifically remains limited to two studies. In a qualitative study of female veterans who had been deployed to combat zones in Iraq or Afghanistan, female veterans described feeling as if they did not belong – due to feeling like a minority in the male-dominated military setting, as well as feeling different from civilian women (Gutierrez et al., 2013). Female veterans also described feeling like a burden due to perceptions that they were not pulling their weight in relation to their male colleagues. Interpersonal difficulties, emotional numbing, substance use, and thoughts of self-directed violence were described in relation to acquired capability for suicide.

Monteith, Bahraini, and Menefee (2017) examined whether IPTS factors were associated with recent suicidal ideation among female veterans who had experienced military sexual trauma. Perceived burdensomeness, thwarted belongingness, and fearlessness about death were associated with suicidal ideation, accounting for past suicide attempts and recent PTSD and depressive symptoms. These studies provide initial support for the potential relevance of the IPTS to female veterans. However, results were based on relatively small and circumscribed samples, such as female veterans entering VHA inpatient treatment and post-9/11 female veterans. Thus, additional research is needed to examine if the IPTS is supported in other samples of female veterans, as well as to identify factors associated with perceived burdensomeness, thwarted belongingness, fearlessness about death, and elevated pain tolerance among female veterans.

## *Fluid vulnerability theory*

While additional theories of suicide have yet to be robustly examined among female veterans, alternative, complementary theories may be relevant for further understanding and conceptualizing female veterans' risk for suicide. Rudd (2006) proposed the fluid vulnerability theory (FVT), which emphasizes that "suicidal episodes" are dynamic, with heightened periods of suicide risk (i.e., the suicidal mode) that occur based on interactions between specific vulnerabilities, stressors, and risk factors, which can be cognitive, affective, physiological, or behavioral in nature (Wolfe-Clark & Bryan, 2017). A unique feature of the FVT is that it distinguishes between baseline and acute risk for suicide, positing that relatively minor stressors may trigger suicidal episodes for individuals at high baseline risk for suicide (i.e., those with more vulnerability). In addition, negative beliefs about oneself, others, and the future are theorized to comprise an individual's suicide-belief system (Rudd, 2006). At the present time, research on the FVT remains limited, and no research has examined this theory in female veterans specifically. However, given its focus on various systems (e.g., cognitive, affective, physiological, and behavioral), this theory could be examined to determine what the "suicidal mode" entails for female veterans, as well as to elucidate specific cognitive, affective, physiological, and behavioral factors which drive suicide risk.

## *Three-step theory*

Building upon prior theories of suicide, Klonsky and May (2015) proposed the three-step theory (3ST), which theorizes that pain and hopelessness lead to suicidal ideation, while connectedness (e.g., to a purpose or loved ones) potentially protects against escalation of suicidal ideation. The 3ST also emphasizes the importance of dispositional (e.g., genetics), acquired (c.f. Joiner, 2007), and practical components of suicide capacity, which facilitate the transition from suicidal ideation to S-SDV. The 3ST remains a relatively new theory that has received some initial support (c.f. Klonsky, Saffer, & Bryan, 2018), although it has not yet been tested in female veterans specifically.

Nevertheless, the specific emphasis on practical components of suicide capacity (i.e., concrete factors that facilitate S-SDV) may be highly relevant to understanding suicide among female veterans. Female veterans are more likely to own and have access to firearms, a highly lethal means of suicide that few individuals survive, compared to civilian women (Cleveland, Azrael, Simonetti, & Miller, 2017). Female veterans are also more likely to use firearms as their means of suicide, compared to civilian women (Department of Veterans Affairs, 2018a). Thus, future research could examine whether female veterans, in comparison to civilian women, are more likely to have elevated practical capacity surrounding firearms, which is essential for understanding potential ways in which female veterans' risk for suicide may be targeted. Future research on the 3ST could also examine other potential aspects of suicide capacity (e.g., dispositional, acquired), as well as sources of hopelessness, psychological pain, and connectedness in female veterans.

## *Cultural model of suicide*

Lastly, the cultural model of suicide (Chu, Goldblum, Floyd, & Bongar, 2010) emphasizes the role of culture in suicide-specific stressors and the expression of S-SDV. This model was developed based on a synthesis and analysis of prior suicide-focused research with different minority populations. The cultural model of suicide posits that culture influences idioms of distress (e.g., words used to communicate suicidal ideation, methods used to attempt suicide). Culture is also theorized to influence the specific stressors that result in S-SDV, with a particular emphasis on minority stress, cultural sanctions, and social discord. Lastly, cultural meanings of stressors and suicide are posited to influence both pain tolerance and S-SDV.

The cultural model of suicide emphasizes different facets of suicide risk (e.g., idioms of distress, culture) that are not explicitly included in other leading theories of suicide. Although this model has not been applied to female veterans, it would lend itself well to qualitative research (e.g., *How do female veterans experience and express SI? How do female veterans perceive suicide and the various stressors that they experience? What is the meaning of different methods of S-SDV among female veterans?*). Initial qualitative research has yielded valuable findings in relation to the potential context and precipitants of suicide among female veterans. For example, substance use disorders, lack of supportive relationships, non-military trauma, multiple prescriptions for sedatives, and insufficient patient-provider communication have been noted as common experiences among female veteran suicide decedents, based on qualitative review of VHA medical records (Denneson, Cromer, Jacobson, Teo, & Dobscha, 2016). Qualitative research also suggests that female veterans who call the Veterans Crisis Line often report concerns related to military sexual trauma, non-suicidal self-injury, mental health, relationships, and finances (Ramchand et al., 2016). In addition, female veterans less often directly asked for help, frequently presented with depressed affect, and generally reported familial responsibilities as their primary motivation to live. These studies may suggest potential life stressors (e.g., military sexual trauma, non-military trauma, lack of supportive relationships, financial problems) and idioms of distress (e.g., depressed affect, non-suicidal self-injury) to investigate in future research applying the cultural model of suicide to female veterans.

## The utility of alternative theories to understand female veteran suicide

Contemporary theories of suicide may be applied to understand suicidal SDV among female veterans. However, none of these theories were developed specifically for female veterans. Until researchers test these theories with female veterans specifically, it is not possible to determine the extent to which these theories sufficiently explain suicidal SDV and suicidal ideation among female veterans. Moreover, alternate theories have been developed in other contexts that may also be useful for understanding suicide among female veterans, despite not originating as theories of suicide. Below, we describe an example of how one such theory may be highly relevant to understanding suicide among female veterans.

### *Minority stress model*

The minority stress model (MSM) was initially developed to understand stress experienced by marginalized groups (e.g., sexual and racial-ethnic minorities; Brooks, 1981; Meyer, 1995). It provides an understanding of how individuals within a minority group may face chronically high levels of stress. Although the MSM has only recently been applied to understanding suicide and only in select populations, it may provide additional insight relevant to the context of female veterans' suicide risk (e.g., discrimination, intersectionality) that is not emphasized in most leading theories of suicide.

Female veterans are the overwhelming minority in military settings, in which cisgender men are the overwhelming majority. In addition, female veterans experience high rates of gender-based violence (e.g., sexual harassment and assault, intimate partner violence) during their military service (Wilson, 2018; Zinzow et al., 2007). Such experiences may be experienced as disempowering and silencing, and can lead to internalization and self-blame (Brownstone,

Holliman, Gerber, & Monteith, 2018). Some female veterans prematurely separate from military service after experiencing military sexual trauma (Dichter & True, 2015). However, experiences of gender-based violence and being a minority do not necessarily end following military separation. Research suggests that female veterans may continue to experience sexual harassment in VHA settings, as well as perceptions of VHA care as male-dominated, which may heighten distress and deter female veterans from using VHA care (Klap et al., 2019; Monteith et al., 2018; Street, Gradus, Stafford, & Kelly, 2007; Washington, Kleimann, Michelini, Kleimann, & Canning, 2007; Washington, Yano, Simon, & Sun, 2006; Wolff & Mills, 2016). This continued minority stress may pose a continued risk for suicide (e.g., by limiting access to evidence-based care). Thus, the MSM may help with understanding the ways in which various experiences of female veterans relate to suicide risk, especially among female veterans not using VHA care.

The MSM also emphasizes the intersectionality of various identities (e.g., race, ethnicity, sexual orientation), which may be useful for understanding why specific groups of veterans (e.g., those who identify as a sexual minority) are at further increased risk for suicide (Blosnich, Bossarte, & Silenzio, 2012; Blosnich, Gordon, & Fine, 2015). According to the MSM, female veterans believing themselves to be a marginalized minority within the context in which they work and live would be at increased risk for experiencing distress. The intersectionality of additional factors could potentially exacerbate such distress. For instance, female veterans who identify as a racial-ethnic, gender, and/or sexual minority may experience a greater sense of being stigmatized and may not identify as part of a larger group (e.g., a woman who identifies as a lesbian, Black female veteran may feel as if she is a minority among female veterans as a whole). Although suicide-focused research with racial or ethnic minority female veterans has been sparse (Dorsey Holliman, Monteith, Spitzer, & Brenner, 2018), recent applications of the MSM to veteran suicide (e.g., Tucker et al., 2019) have been informative and can be used to consider further applications of this model to other potentially marginalized populations. Additional research is needed to fully examine the extent to which the MSM can be applied to better understand suicide among female veterans.

## *Additional theories to consider*

While a comprehensive review of other theoretical models is beyond the scope of the present chapter, a few additional theories are worth briefly mentioning. Feminist theory has only sporadically been discussed in relation to suicide (e.g., Jaworski, 2010; Lester, 1990) and could be applied to better understand the ways in which female veterans' broader societal experiences relate to their risk for suicide. In addition, theory regarding institutional betrayal (Smith & Freyd, 2013, 2014) may provide insight into the potential role of institutional behaviors (or lack thereof) in risk for suicide, particularly in the context of military-related traumatic events (Monteith, Bahraini, Matarazzo, Soberay, & Smith, 2016).

## *Limitations, future research, and clinical implications*

Taken in aggregate, limited research and theory preclude comprehensive understanding and conceptualization of suicide risk among female veterans, all the more so when attempting to determine the most appropriate assessments and interventions. Research on veteran suicide has less commonly focused on female veterans, which may be due to small samples of female veterans in prior research and a tendency to focus on male veterans. Although multiple studies of veterans have included women, most suicide-focused research with veterans has been sex- or gender-neutral, rarely attempting to delineate gender- or sex-specific correlates of suicide or differences by sex or gender

(Hoffmire & Denneson, 2018). Some studies have conducted analyses specifically with men in their samples but have not been able to extend such analyses to women in their samples (e.g., Tiet et al., 2006). In addition, suicide-focused qualitative research with female veterans has been limited and has often utilized methods that did not entail directly interviewing female veterans (e.g., Denneson et al., 2016; Ramchand et al., 2016). This suggests an important and underutilized method for future research with female veterans. Finally, even though research suggests that several different mental health diagnoses and trauma-related factors are associated with suicide risk among female veterans, many of these studies solely examined discrete traumatic experiences (e.g., military sexual trauma) occurring at specific periods in female veterans' lives (e.g., during their military service) or focused on suicidal ideation as their outcome. As a result, overarching understanding of risk and protective factors for suicide among female veterans remains limited, and several important knowledge gaps remain. Given that the etiology of suicide is known to be complex and is likely influenced by interactions between different experiences across the lifespan, future research is needed to address these complexities. Examination of various stressful and traumatic experiences over female veterans' lifespans will require considerably larger sample sizes, as well as assessments at different time periods, in order to form a comprehensive understanding of suicide risk in this population.

As it stands, the lack of theory-driven research in female veterans also poses a critical knowledge gap, and future research will need to examine the hypotheses posed in the theories described above. Comparing the utility of various theories may be useful for determining the extent to which existing theories of suicide apply to female veterans or need to be modified. Future research into factors associated with suicide risk in female veterans must also expand beyond the existing literature and in doing so address the limitations of previous studies. Several important factors remain understudied, and, crucially, there is a need for a more robust theoretical understanding of the interrelationships between various socio-demographic and cultural characteristics as they pertain to female veterans' risk for suicide.

Though further research is needed to more clearly elucidate potential drivers of suicide risk among female veterans, the result of female veterans' excess suicide risk has major implications both in clinical settings and from a population-health perspective to reduce the suicide burden. As such, the number of veteran lives lost to suicide will continue to increase unless the rising rate in this population can be curbed. Importantly, as previously noted, the majority of female veterans do not access VHA care (Frayne et al., 2018). Though a portion of those who do access VHA care are seen for mental health or substance use disorder specialty care (Department of Veterans Affairs, 2018a), unfortunately, the majority are not. As such, developing effective, upstream suicide prevention approaches for female veterans that can be implemented in a variety of settings that female veterans more commonly access (e.g., primary care or women's health clinic settings) is critical. Simultaneously, ensuring access to evidence-based care for female veterans seen in outpatient and inpatient mental health settings remains essential. Lastly, the majority of studies on suicide risk among female veterans have focused on those using VHA care, despite the increasing need to better understand and reach female veterans who are not accessing VHA services (Department of Veterans Affairs, 2018a). To facilitate these efforts, research is needed that tests the acceptability, feasibility, and effectiveness of delivering suicide-specific interventions (e.g., Safety Planning, lethal means safety) to female veterans in both VHA and non-VHA settings. This remains a critical frontier for future research.

## Conclusion

Rising rates of suicide among female veterans underscore the need to better understand factors that explain both suicidal ideation and S-SDV among female veterans. Research to date has focused primarily on mental health diagnoses and specific trauma-related experiences. Few studies have focused on non-VHA using female veterans, despite their increasing suicide rates (Department of Veterans Affairs, 2018a). Several theories may be pertinent to identifying and understanding factors that relate to suicidal SDV among female veterans. These include the IPTS, FVT, 3ST, cultural model of suicide, and MSM. Qualitative research, as well as quantitative research with larger samples of female veterans, is strongly needed to facilitate understanding of suicide and how to prevent it among the growing population of female veterans.

## References


Agha, Z., Lofgren, R. P., VanRuiswyk, J. V., & Layde, P. M. (2000). Are patients at Veterans Affairs medical centers sicker? A comparative analysis of health status and medical resource use. *Archives of Internal Medicine, 160*(21), 3252–3257.

Ahern, J., Worthen, M., Masters, J., Lippman, S. A., Ozer, E. J., & Moos, R. (2015). The challenges of Afghanistan and Iraq veterans' transition from military to civilian life and approaches to reconnection. *PLOS ONE, 10*(7), e0128599.

Benda, B. B. (2005). Gender differences in predictors of suicidal thoughts and attempts among homeless veterans that abuse substances. *Suicide and Life-Threatening Behavior, 35*(1), 106–116.

Blais, R. K., Monteith, L. L., & Kugler, J. (2018). Sexual dysfunction is associated with suicidal ideation in female service members and veterans. *Journal of Affective Disorders, 226*, 52–57.

Blosnich, J. R., Bossarte, R. M., & Silenzio, V. M. (2012). Suicidal ideation among sexual minority veterans: Results from the 2005–2010 Massachusetts Behavioral Risk Factor Surveillance Survey. *American Journal of Public Health, 102*(S1), S44–S47.

Blosnich, J. R., Brenner, L. A., & Bossarte, R. M. (2016). Population mental health among U.S. military veterans: Results of the Veterans Health Module of the Behavioral Risk Factor Surveillance System, 2011–2012. *Annals of Epidemiology, 26*(8), 592–596.

Blosnich, J. R., Gordon, A. J., & Fine, M. J. (2015). Associations of sexual and gender minority status with health indicators, health risk factors, and social stressors in a national sample of young adults with military experience. *Annals of Epidemiology, 25*(9), 661–667.

Bohnert, K. M., Ilgen, M. A., Louzon, S., McCarthy, J. F., & Katz, I. R. (2017). Substance use disorders and the risk of suicide mortality among men and women in the US Veterans Health Administration. *Addiction, 112*(7), 1193–1201.

Brignone, E., Sorrentino, A. E., Roberts, C. B., & Dichter, M. E. (2018). Suicidal ideation and behaviors among women veterans with recent exposure to intimate partner violence. *General Hospital Psychiatry, 55*, 60–64.

Brooks, V. R. (1981). *Minority Stress and Lesbian Women*. Lexington, MA: Lexington Books.

Brownstone, L. M., Holliman, B. D., Gerber, H. R., & Monteith, L. L. (2018). The phenomenology of military sexual trauma among women veterans. *Psychology of Women Quarterly, 42*(4), 399–413.

Bryan, C. J., & Cukrowicz, K. C. (2011). Associations between types of combat violence and the acquired capability for suicide. *Suicide and Life-Threatening Behavior, 41*(2), 126–136.

Bullman, T., Hoffmire, C., Schneiderman, A., & Bossarte, R. (2015). Time dependent gender differences in suicide risk among Operation Enduring Freedom and Operation Iraqi Freedom veterans. *Annals of Epidemiology, 25*(12), 964–965.

Burkhart, L., & Hogan, N. (2015). Being a female veteran: A grounded theory of coping with transitions. *Social Work in Mental Health, 13*(2), 108–127.

Byrne, T., Montgomery, A. E., & Dichter, M. E. (2013). Homelessness among female veterans: A systematic review of the literature. *Women & Health, 53*(6), 572–596.

Cerel, J., Van De Venne, J. G., Moore, M. M., Maple, M. J., Flaherty, C., & Brown, M. M. (2015). Veteran exposure to suicide: Prevalence and correlates. *Journal of Affective Disorders, 179*, 82–87.

Chapman, S. L. C., & Wu, L. T. (2014). Suicide and substance use among female veterans: A need for research. *Drug and Alcohol Dependence, 136*, 1–10.

Chu, C., Buchman-Schmitt, J. M., Stanley, I. H., Hom, M. A., Tucker, R. P., Hagan, C. R., … Michaels, M. S. (2017). The interpersonal theory of suicide: A systematic review and meta-analysis of a decade of cross-national research. *Psychological Bulletin, 143*(12), 1313–1345.

Chu, J. P., Goldblum, P., Floyd, R., & Bongar, B. (2010). The cultural theory and model of suicide. *Applied and Preventive Psychology*, *14*(1–4), 25–40.

Cleveland, E. C., Azrael, D., Simonetti, J. A., & Miller, M. (2017). Firearm ownership among American veterans: Findings from the 2015 National Firearm Survey. *Injury Epidemiology*, *4*(1), 33.

Denneson, L. M., Cromer, R., Jacobson, L. E., Teo, A., & Dobscha, S. K. (2016). Female veterans who died by suicide: Qualitative analysis of medical records. *Military Behavioral Health*, *4*(3), 276–284.

Department of Veterans Affairs. (2017a). *The past, present and future of women veterans*. Retrieved from https://www.va.gov/vetdata/docs/SpecialReports/Women_Veterans_2015_Final.pdf

Department of Veterans Affairs. (2017b). *Suicide among veterans and other Americans, 2001–2014*. Retrieved from https://www.mentalhealth.va.gov/docs/2016suicidedatareport.pdf

Department of Veterans Affairs. (2018a). *VA National Suicide Data Report 2005–2016*. Retrieved from https://www.mentalhealth.va.gov/docs/data-sheets/OMHSP_National_Suicide_Data_Report_2005–2016_508.pdf

Department of Veterans Affairs. (2018b). *2016 National Suicide Data Report Appendix*. Retrieved from https://www.mentalhealth.va.gov/mentalhealth/suicide_prevention/data.asp

Dichter, M. E., & True, G. (2015). "This is the story of why my military career ended before it should have": Premature separation from military service among US women veterans. *Journal of Women & Social Work*, *30*(2), 187–199.

Dorsey Holliman, B. A., Monteith, L. L., Spitzer, E. G., & Brenner, L. A. (2018). Resilience, cultural beliefs, and practices that mitigate suicide risk among African American women veterans. *SAGE Open*, *8*(1), 1–10.

Frayne, S. M., Phibbs, C. S., Saechao, F., Friedman, S. A., Shaw, J. G., Romodan, Y., … Haskell, S. (2018). *Sourcebook: Women Veterans in the Veterans Health Administration. Volume 4: Longitudinal Trends in Sociodemographics, Utilization, Health Profile, and Geographic Distribution*. Retrieved from https://www.womenshealth.va.gov/WOMENSHEALTH/docs/WHS_Sourcebook_Vol-IV_508c.pdf

Gibbons, R. D., Brown, C. H., & Hur, K. (2012). Is the rate of suicide among veterans elevated? *American Journal of Public Health*, *102*(Suppl 1), S17–19.

Gradus, J. L., Street, A. E., Suvak, M. K., & Resick, P. A. (2013). Predictors of suicidal ideation in a gender-stratified sample of OEF/OIF veterans. *Suicide and Life-Threatening Behavior*, *43*(5), 574–588.

Gutierrez, P. M., Brenner, L. A., Rings, J. A., Devore, M. D., Kelly, P. J., Staves, P. J., … Kaplan, M. S. (2013). A qualitative description of female veterans' deployment-related experiences and potential suicide risk factors. *Journal of Clinical Psychology*, *69*(9), 923–935.

Hamilton, A. B., Williams, L., & Washington, D. L. (2015). Military and mental health correlates of unemployment in a national sample of women veterans. *Medical Care*, *53*(4 Suppl 1), S32–38.

Hoffmire, C. A., & Denneson, L. M. (2018). Concerning trends in suicide among women veterans point to need for more research on tailored interventions. *HSR&D Forum*, *6*(3), 9.

Ilgen, M. A., Bohnert, A. S., Ignacio, R. V., McCarthy, J. F., Valenstein, M. M., Kim, H. M., & Blow, F. C. (2010). Psychiatric diagnoses and risk of suicide in veterans. *Archives of General Psychiatry*, *67*(11), 1152–1158.

Iovine-Wong, P. E., Nichols-Hadeed, C., Thompson Stone, J., Gamble, S., Cross, W., Cerulli, C., & Levandowski, B. A. (2019). Intimate partner violence, suicide, and their overlapping risk in women veterans: A review of the literature. *Military Medicine*, *184*(5–6), e201–e210.

Jaworski, K. (2010). The gender-ing of suicide. *Australian Feminist Studies*, *25*(63), 47–61.

Joiner, T. E. (2007). *Why People Die by Suicide*. Cambridge, MA: Harvard University Press.

Kelly, U. A., Skelton, K., Patel, M., & Bradley, B. (2011). More than military sexual trauma: Interpersonal violence, PTSD, and mental health in women veterans. *Research in Nursing & Health*, *34*(6), 457–467.

Khan, A. J., Li, Y., Dinh, J. V., Donalson, R., Hebenstreit, C. L., & Maguen, S. (2019). Examining the impact of different types of military trauma on suicidality in women veterans. *Psychiatry Research*, *274*, 7–11.

Kimerling, R., Gima, K., Smith, M. W., Street, A., & Frayne, S. (2007). The Veterans Health Administration and military sexual trauma. *American Journal of Public Health*, *97*(12), 2160–2166.

Kimerling, R., Makin-Byrd, K., Louzon, S., Ignacio, R. V., & McCarthy, J. F. (2016). Military sexual trauma and suicide mortality. *American Journal of Preventive Medicine*, *50*(6), 684–691.

Klap, R., Darling, J. E., Hamilton, A. B., Rose, D. E., Dyer, K., Canelo, I., … Yano, E. M. (2019). Prevalence of stranger harassment of women veterans at Veterans Affairs Medical Centers and impacts on delayed and missed care. *Women's Health Issues*, *29*(2), 107–115.

Klonsky, E. D., & May, A. M. (2015). The three-step theory (3ST): A new theory of suicide rooted in the "ideation-to-action" framework. *International Journal of Cognitive Therapy*, *8*(2), 114–129.

Klonsky, E. D., Saffer, B. Y., & Bryan, C. J. (2018). Ideation-to-action theories of suicide: A conceptual and empirical update. *Current Opinion in Psychology*, *22*, 38–43.

LeardMann, C. A., Pietrucha, A., Magruder, K. M., Smith, B., Murdoch, M., Jacobson, I. G.,. . . Millennium Cohort Study Team. (2013). Combat deployment is associated with sexual harassment or sexual assault in a large, female military cohort. *Women's Health Issues*, *23*(4), e215–e223.

Lester, D. (1990). The study of suicide from a feminist perspective. *Crisis: The Journal of Crisis Intervention and Suicide Prevention*, *11*(1), 38–43.

Ma, J., Batterham, P. J., Calear, A. L., & Han, J. (2016). A systematic review of the predictions of the interpersonal–psychological theory of suicidal behavior. *Clinical Psychology Review*, *46*, 34–45.

Mattocks, K. M., Haskell, S. G., Krebs, E. E., Justice, A. C., Yano, E. M., & Brandt, C. (2012). Women at war: Understanding how women veterans cope with combat and military sexual trauma. *Social Science & Medicine*, *74*(4), 537–545.

Meyer, I. H. (1995). Minority stress and mental health in gay men. *Journal of Health and Social Behavior*, *36*(1), 38–56.

Monteith, L. L., Bahraini, N. H., Gerber, H. R., Dorsey Holliman, B., Schneider, A. L., Holliday, R., & Matarazzo, B. B. (2018). Military sexual trauma survivors' perceptions of Veterans Health Administration care: A qualitative examination. *Psychological Services*. Advance online publication. doi:10.1037/ser0000290

Monteith, L. L., Bahraini, N. H., Matarazzo, B. B., Soberay, K. A., & Smith, C. P. (2016). Perceptions of institutional betrayal predict suicidal self-directed violence among veterans exposed to military sexual trauma. *Journal of Clinical Psychology*, *72*(7), 743–755.

Monteith, L. L., Bahraini, N. H., & Menefee, D. S. (2017). Perceived burdensomeness, thwarted belongingness, and fearlessness about death: Associations with suicidal ideation among female veterans exposed to military sexual trauma. *Journal of Clinical Psychology*, *73*(12), 1655–1669.

Monteith, L. L., Hoffmire, C. A., Holliday, R., Park, C. L., Mazure, C. M., & Hoff, R. A. (2018). Do unit and post-deployment social support influence the association between deployment sexual trauma and suicidal ideation? *Psychiatry Research*, *270*, 673–681.

Pease, J. L., Billera, M., & Gerard, G. (2015). Military culture and the transition to civilian life: Suicide risk and other considerations. *Social Work*, *61*(1), 83–86.

Ramchand, R., Ayer, L., Kotzias, V., Engel, C., Predmore, Z., Ebener, P., … Haas, G. (2016). Suicide risk among women veterans in distress: Perspectives of responders on the Veterans Crisis Line. *Women's Health Issues*, *26*(6), 667–673.

Roy, A., & Janal, M. (2006). Gender in suicide attempt rates and childhood sexual abuse rates: Is there an interaction? *Suicide & Life-Threatening Behavior*, *36*(3), 329–335.

Rudd, M.D. (2006). Fluid vulnerability theory: A cognitive approach to understanding the process of acute and chronic suicide risk. In T. E. Ellis (Ed.), *Cognition and Suicide: Theory, Research, and Therapy* (pp. 355–368). Washington, DC: American Psychological Association.

Silva, C., Ribeiro, J. D., & Joiner, T. E. (2015). Mental disorders and thwarted belongingness, perceived burdensomeness, and acquired capability for suicide. *Psychiatry Research*, *226*(1), 316–327.

Smith, C. P., & Freyd, J. J. (2013). Dangerous safe havens: Institutional betrayal exacerbates sexual trauma. *Journal of Traumatic Stress*, *26*(1), 119–124.

Smith, C. P., & Freyd, J. J. (2014). Institutional betrayal. *American Psychologist*, *69*(6), 575–587.

Smith, N. B., Monteith, L. L., Rozek, D. C., & Meuret, A. E. (2018). Childhood abuse, the interpersonal–psychological theory of suicide, and the mediating role of depression. *Suicide and Life-Threatening Behavior*, *48*(5), 559–569.

Street, A. E., Gradus, J. L., Stafford, J., & Kelly, K. (2007). Gender differences in experiences of sexual harassment: Data from a male-dominated environment. *Journal of Consulting and Clinical Psychology*, *75*(3), 464–474.

Street, A. E., Vogt, D., & Dutra, L. (2009). A new generation of women veterans: Stressors faced by women deployed to Iraq and Afghanistan. *Clinical Psychology Review*, *29*(8), 685–694.

Tiet, Q. Q., Finney, J. W., & Moos, R. H. (2006). Recent sexual abuse, physical abuse, and suicide attempts among male veterans seeking psychiatric treatment. *Psychiatric Services*, *57*(1), 107–113.

Tucker, R. P., Testa, R. J., Reger, M. A., Simpson, T. L., Shipherd, J. C., & Lehavot, K. (2019). Current and military-specific gender minority stress factors and their relationship with suicide ideation in transgender veterans. *Suicide and Life-Threatening Behavior*, *49*(1), 155–166.

Van Orden, K. A., Witte, T. K., Cukrowicz, K. C., Braithwaite, S. R., Selby, E. A., & Joiner, T. E., Jr. (2010). The interpersonal theory of suicide. *Psychological Review*, *117*(2), 575–600.

Washington, D. L., Kleimann, S., Michelini, A. N., Kleimann, K. M., & Canning, M. (2007). Women veterans' perceptions and decision-making about Veterans Affairs health care. *Military Medicine*, *172*(8), 812–817.

Washington, D. L., Yano, E. M., Simon, B., & Sun, S. (2006). To use or not to use: What influences why women veterans choose VA health care. *Journal of General Internal Medicine, 21*(Suppl 3), S11–18.

Wilson, L. C. (2018). The prevalence of military sexual trauma: A meta-analysis. *Trauma, Violence, & Abuse, 19*(5), 584–597.

Wolfe-Clark, A. L., & Bryan, C. J. (2017). Integrating two theoretical models to understand and prevent military and veteran suicide. *Armed Forces & Society, 43*(3), 478–499.

Wolff, K. B., & Mills, P. D. (2016). Reporting military sexual trauma: A mixed-methods study of women veterans' experiences who served from World War II to the war in Afghanistan. *Military Medicine, 181*(8), 840–848.

Zinzow, H. M., Grubaugh, A. L., Monnier, J., Suffoletta-Maierle, S., & Frueh, B. C. (2007). Trauma among female veterans: A critical review. *Trauma, Violence, & Abuse, 8*(4), 384–400.

# 28

# RESILIENCE AND STRESS IN MILITARY COMBAT FLIGHT ENGINEERS

*Reoot Cohen-Koren, Dror Garbi, Shirley Gordon, Nirit Yavnai, Yifat Erlich Shoham, and Leah Shelef*

Israeli military combat evacuation and rescue helicopter crews include pilots, medical rescue staff and flight engineers. Just like the pilots and the medical rescue staff, the flight engineers, too, are considered aircrew fighters since they take part in military operations and evacuate wounded soldiers from fighting zones under conditions that are defined as life threatening.

While there are clear definitions for the job descriptions of pilots and medical rescue staff, flight engineers are responsible for a wide range of operations within the helicopter and out of it, both in the air and on the ground. While in the air, they are in charge of the technical operations and maintenance of the aircraft. On the ground, they take part in the evacuation of casualties. According to instructions, in dangerous circumstances, it is their duty to help the medical staff take care of the casualties, despite them not having any medical knowledge or prior preparation. The flight engineers are a critical and integral part of the helicopter rescue teams, who are exposed to the distressing sights, sounds and smells of wounded or dead people. War scenarios and exposure to traumatic events, combined with life-risking situations, increase the vulnerability of the individuals to emotional distress.

## Exposure to traumatic events

Experiencing a single traumatic event or repeated ones increases the risk of developing psychopathology (Mitchell & Everly, 2001; Thomas et al., 2010). Exposure to traumatic events may also contribute to posttraumatic stress disorder. Such exposure is a necessary criterion for a diagnosis of posttraumatic stress disorder (PTSD) (American Psychiatric Association, 2013). There are four sets of symptoms that characterize PTSD based on the DSM-V: intrusion (e.g., flashbacks, nightmares and recollections of the traumatic event), avoidance (dissociation and detachment, loss of interest in previously enjoyed activities), hyper-arousal (e.g., irritability, startle reaction, loss of concentration and sleep difficulties) and negative changes in thoughts and mood that occur or worsen following the experience of the traumatic event (e.g., 'The world is an evil place'; a negative emotional state such as shame, anger or fear) (American Psychiatric Association [APA], 2013).

It is also well known that exposure to others who have PTSD can contribute to the phenomenon known as secondary PTSD that occurs even without exposure to a traumatic event. Secondary PTSD is a condition that imitates posttraumatic symptoms such as exhaustion, increased alertness, avoidance and emotional numbness (Baird & Kracen, 2006; Boscarino, Figley, & Adams, 2004). A number of studies have found significant levels of secondary traumatization among care-givers like emergency unit nurses (Gates & Gillespie, 2008; Palgi, Ben-Ezra, Langer, & Essar, 2009).

Exposure to traumatic events, when untreated, may result in more severe symptoms and influence all areas of life (Solomon & Dekel, 2008). In addition, among both, civilians and soldiers, mental distress in the wake of exposure to traumatic events is often co-morbid with drug abuse (Breslau, Davis, & Schultz, 2003; Jacobsen, Southwick, & Kosten, 2001), alcohol consumption (Maes, Delmeire, Mylle, & Altamura, 2001) and suicide risk (Krysinska & Lester, 2010).

## Epidemiology of posttraumatic stress disorder

### *Civilians*

Posttraumatic stress disorder is the most frequent disorder that occurs after traumatic events (Galea, Nandi, & Vlahov, 2005; Neria, Nandi, & Galea, 2008). Although most people do not develop PTSD, many of those exposed to a traumatic event will show some of the symptoms, at least for a short period of time, with a variety of psychological consequences (American Psychiatric Association [APA], 2000). Based on the DSM-IV, it was found (n = 34,000), that lifetime PTSD prevalence among U.S. adults was between 6.4% and 6.8% (Kessler et al., 2005; Pietrzak, Goldstein, Southwick, & Grant, 2011). Similar percentages (n = 36,309) were found based on DSM-V criteria, namely 6.1% for lifetime PTSD prevalence and 4.7% current PTSD prevalence (Goldstein et al., 2016). Thus, the percentages of PTSD among U.S. adults did not change even when the DSM criteria changed.

### *Military staff and combat experience*

Combat exposure among military service staff has been repeatedly linked to a myriad of negative health consequences (Hoge et al., 2004). It can lead not only to PTSD (Hoge, Riviere, Wilk, Herrell, & Weathers, 2014; Vasterling et al., 2010) but to a variety of emotional distress disorders (Ferrier-Auerbachet, Erbes, Polusny, Rath, & Sponheim, 2010). Depression and anxiety are prominent amongst those diagnoses (Hoge et al., 2014; Lapierre, Schwegler, & LaBauve, 2007). Behavioral problems were also associated with emotional distress, particularly alcohol and substance abuse (Hoge et al., 2014; Petrakis et al., 2012) and suicide (Bryan, Goodman, Chappelle, Thompson, & Prince, 2018). Those emotional distress disorders were particularly associated with military personal who have been deployed to war zones (Chappelle et al., 2014; Ferrier-Auerbach et al., 2010). In studies conducted by the Mental Health Advisory Team in U.S. Army combat-deployed soldiers (Ferrier-Auerbach et al., 2010), 18% of participants (n = 2,677) reported some psychological problems, 16% met criteria for PTSD and 7% reported depression. In another study on National Guard soldiers deployed to Iraq, they found that 9% met the criteria for depression and 7% met the criteria for PTSD (Ferrier-Auerbach et al., 2010).

Combat experience can lead to occupational burnout as well (Bryan et al., 2018). A study among U.S. Air Force military personnel found that emotional exhaustion and lack of personal accomplishment, which constitute subscales of burnout, were correlated with depression, whereas cynicism was correlated with PTSD (Bryan et al., 2018).

As mentioned above, among civilians, the findings were similar even when referring to criteria of different versions of the DSM. Naturally, among soldiers who experienced combat conditions, the rates are much higher. In a study of 1,822 U.S. infantry soldiers, 946 of whom had been deployed to Iraq or Afghanistan, 13% (n = 224) demonstrated all the criteria of PTSD according to the DSM-IV and 12% (n = 216) according to the DSM-V. However, when considering only the results of the soldiers who had been exposed to combat, the percentages rise to 19% (n = 177) who fully meet DSM-IV PTSD-criteria and 18% (n = 165) for DSM-V criteria (Hoge et al., 2014).

It appears that after exposure to combat, the likelihood of emotional distress (Ramchand, Rudavsky, Grant, Tanielian, & Jaycox, 2015) and of PTSD at various levels (Creamer, Wade, Fletcher, & Forbes, 2011; Gates et al., 2012; Kok, Herrell, Thomas, & Hoge, 2012; Richardson, Frueh, & Acierno, 2010; Sofko, Currier, & Drescher, 2016) in combatants increases.

It is notable that a study in UK troops deployed to Iraq or Afghanistan found the prevalence of PTSD to be only 4% (Fear et al., 2010). Similarly, the prevalence of PTSD in Dutch infantry veterans who had served in Iraq was estimated at 4% (Engelhard et al., 2007). Hunt, Wessely, Jones, Rona and Greenberg (2014) indicate that the prevalence of reported PTSD is lower in the UK armed forces than in military forces of other countries such as the United States, Australia, and Canada.

The rates of PTSD found among Israeli veterans of the Yom Kippur War (Belenky, Noy, Solomon & Del Jones, 1985) and the First Lebanon War (Noy, Nardi & Solomon, 1986) were 10%–20%. In a recent study conducted on soldiers who participated in the second Lebanon war (July 12–August 14, 2006) and in operation 'Cast Lead' (December 27, 2008–January18, 2009), it was found that PTSD prevalence was similar among treatment-seeking veterans who had been deployed to high exposure combat-zones (59.5% vs. 61.4%) (Levi & Lubin, 2018).

## Distress among military flight engineers

While there are several medical studies on flight engineers, as far as we know, only one study deals with the psychological aspects of this population. This study was published in 1992 and presents the profile of flight engineers (n = 164) who were involved in a labor dispute. While the participants in that study were found to be committed to their careers and to identify with the goals of the company, the possibility of redundancy brought about emotions that were related to distrust in management, as well as depression, anxiety, psychosomatic stress symptoms, decrease in general well-being, strained family life and impaired relations with their spouses. The findings raised concerns about the possible effects on in-flight safety and organizational effectiveness (Visser & Van Staden, 1992).

One of the reasons for the scarcity of studies on flight engineers is probably the long-held belief that they are only technicians, unrelated to the medical teams' job and devoid of any interaction with the casualties.

The first survey on the effects of combat on flight engineers ever performed in the Israel Defense Forces (IDF) (albeit not published) took place 5 months after the end of the Second Lebanon War and was administered to 24 flight engineers (mean age 31.5; range 23–42). Most of them (73%) were married at the time the survey was done and served as career soldiers (82%). Average seniority in the unit was 9.2 years (range 2–23 years). Seventy-two percent of the flight engineers had participated in more than 10 medical evacuation events, and for 82%, that was their first war experience. Seventy-seven percent of the flight engineers reported that the harsh sights were moderately distressful. The most common coping mechanisms were ignoring and repressing the emotions evoked by the experience and concentrating on the task. A small minority among the participants reported that they found talking to friends about the experience could be helpful in coping with it (Moriya & Luria, 2016).

Another study among Israeli soldiers was done after the 'Protective Edge' operation (July 8, 2014–August 26, 2014). This study included 49 (40.2%) infantry soldiers, 24 (19.7%) pilots and flight engineers and 49 (40.2%) soldiers engaged in electronic warfare. The levels of emotional distress (e.g. PTSD symptoms, somatization and state anxiety) were higher in the infantry group compared to the other two groups (Shelef et al., 2018). However, since the flight engineers and pilots were bundled into one group, it is impossible to draw accurate conclusions regarding specifically the flight engineers.

This chapter's authors came to understand that there is a need for research that focuses on flight engineers. This resulted in a recent study (Cohen-Koren et al., n.d.), that included 112 flight engineers, the vast majority of whom were male (99.1%); 29.5% were between the ages of 18 and 30, 25% were between the ages of 31 and 40 and almost half of them (45.5%) were 41 years or older. Fifty percent (50.9%) had been in military service for over 20 years, another 25.0% had been in military service between 10 and 19 years and the rest between 0 and 9 years. The majority (66.1%) were career soldiers, with 34% at officer rank. Almost half of the participants (47.2%) had been exposed to 11 or more emotionally distressing sights. More than half (55.8%) had participated in more than four rescue missions that included fatalities. Only one participant was found to have PTSD (above the cutoff point in PCL-5), while 76.6% (n = 85) reported states of anxiety but did not surpass the PCL-% cutoff point. State anxiety was found in 23.6%. It is important to stress that this study was not done following a military war operation but rather during a time of regular activity in the IDF.

That study found that 13.4% suffered from mild depression (n = 15), with BDI scale scores ranging from 10 to 18. The point of intersection that indicates psychological distress (n = 7) was passed by 6.2% of the population, and 22.3% reported impairments in social functioning.

The highest percentage of flight engineers with anxiety (66.7%) was among those who reported participating in 1–5 missions with distressing sights. A similarly high percentage of anxiety sufferers (66.7%) was also found among those who were exposed to 1–3 events that included fatalities (p = 0.015). While the flight engineers did not demonstrate full-criteria PTSD (with one exception), they did report having multiple PTSD symptoms, as well as depression and anxiety. The authors would like to stress yet again the fact that this study did not follow and was not related directly to any military operation or war.

## Coping with stress

People try to cope in a variety of ways when faced with threats to their person or their life or to people who are close to them. Coping strategies are cognitive, emotional and behavioral efforts to cope with stressful situations. Literature identified two main coping categories: problem-focused coping and emotion-focused coping. In problem-focused coping, a person attempts to reduce or manage the problem by removing or changing the stressful factor. In emotion-focused coping, they try to reduce or eliminate the emotional stress and the emotions that arise following exposure to a traumatic event (Folkman, 2011; Helgeson, 2011; Sever, Somer, Ruvio & Soref, 2008; Shalev, 1996).

Currently, most accepted coping theory includes three coping strategies: problem-focused coping (active coping, planning and using instrumental support), emotion-focused coping (acceptance, positive reframing, humor, religion, using emotional support, self-distraction, venting) and non-adaptive coping (denial, substance use, behavioral disengagement and self-blame) (Carver, 1997; Carver, Scheier & Weintraub, 1989). In the above-mentioned study, Cohen-Koren and her colleagues (n.d.) found that flight engineers who were between the ages of 31–40 reported using significantly more emotion-focused coping strategies than other age groups. It was also found that the group that was exposed to 1–3 events with fatalities reported using significantly more emotion-focused coping strategies than participants who were exposed to 4 or more events with fatalities. Flight engineers who were exposed to 1–5 events with distressing sights (as well as flight engineers who were exposed to 1–3 events with fatalities) had higher scores in non-adaptive coping strategies as compared to flight engineers that were exposed to more such events. The explanation may be that this age group is under higher stress due to being career soldiers still in the process of building their military career while at the same time building a family life through marriage and childbirth.

According to Lazarus, & Folkman (1984), the more effective the coping strategies, the less the sense of stress and the more efficient the adaptation. Even though most stress factors lead to usage of both problem-focused and emotion-focused coping strategies, problem-focused coping strategies are used more often when people feel in control and can change the situation through their actions. Emotion-focused coping strategies are used primarily when people feel unable to effect a change in the situation (Lazarus, & Folkman, 1984). In the long run, emotion-focused coping, namely inactive coping, leads to more problems than problem-focused coping, namely active coping (Bal, Van Oost, De Bourdeauthuij & Crombez, 2003; Tolan, Gorman-Smith, Henry, Chung & Hunt, 2002).

## *Prevention of posttraumatic stress disorder in the army*

Yehuda, Vermetten, McFarlane & Lehrner (2014) raised an important point when it came to prevention of PTSD. They claimed that countries that do not take into account the cost of mental health outcomes as part of the cost of war may not make appropriate provisions for treatment.

The U.S. military has been active in fighting zones for many years. The attitude towards PTSD and its prevention has changed over those years and attention has turned from post-deployment to pre-deployment efforts in attempts to reduce the harmful effects of exposure to trauma (Hourani, Council, Hubal & Strange, 2011). In their literature review on primary prevention efforts to stem the outbreak of PTSD, Hourani and colleagues (2011) showed that the most promising preventive approaches are still those that utilize exposure strategies, especially when used in conjunction with psycho-education, stress reduction and increased skills training. The innovation of using preventive approaches was in their focusing on secondary prevention by reducing the impact of stressors, thus reducing the likelihood of developing PTSD among symptomatic personnel deployed to combat zones. Vermetten et al. (2014) compared PTSD prevention practices and organization in soldiers deployed to Afghanistan in the United Kingdom, Canada, Australia, the United States and the Netherlands. All the countries had to work to reduce stigma related to mental health problems. Training for 'social leadership' and 'paraprofessional peer supporters' were identified as important components that may help reduce the probability for developing PTSD. The advantage of these approaches is in their potential to speed up responses to mental health issues within the units, lowering dependency on mental health professionals (Vermetten et al., 2014). Castro (2014), elaborates in his article on the elements of the prevention programs mentioned in Vermetten et al.'s (2014) article. However, in his review, Castro (2014) claimed that the only mental health training program for military personnel to have undergone rigorous scientific validation is 'battle mind training'. The battle mind program is a cognitive and skill-based program that reframes transition difficulties as a failure to adapt skills learned in combat to the home environment. According to this program, mental health training is provided before, during and after deployment and could attenuate the adverse effects of combat. Since the first battle mind program publication, other resilience-based training programs have expanded upon its concepts while maintaining its core elements (Castro, 2014). Needless to say, these programs focused primarily on infantry combat soldiers and were mostly directed at combat veterans from Iraq and Afghanistan.

Thompson and Jetly (2014) suggest that scenario-based operational ethics training as part of high-intensity military field training may constitute an important addition to traditional military ethics education and training. Much has been written about the effectiveness of the 'debriefing' technique. Most studies have failed to find it clinically beneficial. Some studies even found that they caused worsening in the participants' condition (Bisson, 2003; Lewis, 2003). However, these studies were characterized by several methodological drawbacks which weaken validity and reliability. The studies that have found debriefing to be clinically efficient pointed out certain populations that would gain the most benefit from the intervention. In a large randomized trial of combat veterans returning from a year-long deployment to Iraq, a follow-up conducted 4 months after returning home found that psychological (battle mind) debriefing reduced the psychological symptoms associated with PTSD in those with high level combat exposure better than standard post-deployment stress education (Adler, Bliese, McGurk, Hoge & Castro, 2009). According to Castro (2014), the use of psychological debriefings, at least within a military context, appears to provide a useful early intervention tool for lessening the impact of combat and deployment stress on the mental health and well-being of service members.

### *Prevention of posttraumatic stress disorder in the Israeli army*

The IDF developed a cognitive behavior therapy-based training program for their soldiers called 'Magen' ('Shield' in Hebrew) that is mandatory and aims to prevent or reduce the prevalence of posttraumatic stress disorder. Soldiers are taught how to deliver mental first aid to fellow soldiers in the field. Magen is based on the SIX Cs model that aims to help shift the person from helplessness and passiveness to active effective functioning within minutes following a perceived traumatic event (Farchi et al., 2018). Preliminary evidence suggests that 'Magen' is efficacious in reducing PTSD symptoms (Ginat, Svetlizky, Barezin, Pharchi & Fruchter, 2015). 'Magen' is guided by three principles: 'Timing'—intended for immediate implementation, even when still under fire; 'Compatibility'—suiting the military culture, mission-focused and 'Simplicity'—easy to implement and use even by non-professionals.

The philosophy underlying the provision of mental health care in the IDF as well as 'Magen' is based on Salmon's (1919) principles: proximity—offering the care in close proximity to the combat unit even within the combat zone; immediacy—providing treatment right at the onset of emotional difficulties; expectancy—providing the treatment with the expectation for the patient to return to full duty at the front line within a short period of time; simplicity—utilization of short-duration, simple interventions (Ginat et al., 2015). These principles serve as a theoretical model for mental health care and were designed to preserve the readiness and numbers of military personnel on the battlefield by reducing the impact of traumatic and potentially debilitating experiences (Chappelle, & Lumley, 2006; Lamberg, 2004).

The ongoing Middle East conflict requires the IDF mental health department to provide training programs for prevention and treatment, thus ensuring minimal disruptions to the readiness of combat units. In a sample of 429 IDF soldiers who served in the 1982 Lebanon War, the front-line treatment for combat stress-reactions was found to yield significant reductions in PTSD and other psychiatric symptoms—reductions that persisted for up to 20 years post-combat (Solomon, Shklar, & Mikulincer, 2005).

The 'Magen' program is unique not only in using the soldiers in the field to provide immediate help on the battlefield to other soldiers, right when anxiety symptoms appear, but also in that it leads to usage of cognitive-focused strategies rather than emotion-focused strategies (Bal et al., 2003; Tolan et al., 2002). As is well known, focus on cognitive strategies may neutralize the emotional response to an overwhelming event and make it possible to increase one's level of functioning. As trauma is associated with increased amygdala activity and less with functions of the prefrontal cortex (PFC), using emotional communication may increase and worsen the effects of the perceived trauma (Admon et al., 2012, Hantman & Farchi, 2015). The more the amygdala is activated, the less the need for PFC activation in order to perform (Siddiqui, Chatterjee, Kumar, Siddiqui & Goyal, 2008). The 'Magen' program was designed to provide tools that activate the PFC, thus reducing amygdala activation.

In study on the entire maneuver component of an infantry brigade that consisted of 719 male soldiers aged 18–27 years, it was found that 4-month post-combat PTSD prevalence was higher (7.8%) in a control group that received no training compared to soldiers who received either eight or four sessions of attention bias modification training (6.7% and 2.6%, respectively). In general, this computer-based intervention was designed to mitigate stress-related symptoms by targeting disruptions in threat monitoring and thus mentally preparing the soldiers for traumatic combat events (Wald et al., 2016). These findings are supported by another study among IDF soldiers (Wald et al., 2017). One hundred thirty male IDF soldiers, from an infantry brigade which participated in a battle, received the attention bias modification training mentioned above. Findings revealed that solely combat exposure accounted for 4.6% of the variance from baseline to follow-up in stress-related symptoms. One conclusion was that this specific training has the potential to reduce risk for stress-related symptoms associated with combat exposure and may thus be used as a preventative intervention (Wald et al., 2017).

In recent years, clinical experience and data have begun to accumulate in the mental health department of the IDF. The 'Team Resource Intervention' technique is an immediate intervention technique that originated in the debriefing technique (Rose, Bisson, Churchill, Wessely, 2002) but instead of engaging traumatic injury, it emphasizes unit strength.

This intervention is taught by the commander of the unit and immediately after operational missions in order to help soldiers return to full functioning as quickly as possible. The team resource protocol is short in time and focused on five stages: (a) an explanation of the purpose of the team conversation; (b) thoughts and impressions, based on the debriefing stage (Rose et al., 2002); (c) assessment of mental resources (what helped to cope during the event and now); (d) summary, assessment of function, validation, normalization of responses and emphasis on values and norms and (e) finally, providing tasks to continue, directing attention to the future and following for a return to function. Moreover, the IDF Medical Corps have developed for this idea mental health guidelines requiring medical teams to undergo Resource Intervention after administering preventive intervention to casualties following a potentially stressful event (Knobler, Nachshoni, Jaffe, Peretz & Yehuda, 2007).

## *Prevention of posttraumatic stress disorder in the Israeli Air Force and mental preparation for flight engineers*

Readiness is the primary principle guiding the pre-combat PTSD prevention programs used in the IDF to prepare the ground combat force. Prevention programs for air crews, however, are few. The most prominent is administered to pilots and recently also to operators of remote piloted aircraft.

Professional debriefing is a predominant tool in the Israeli Air Force, performed prior to and following each flight. Its purpose is to help improve performance by learning from experience and drawing conclusions for the next flight.

The team resource, as practiced in the IDF, is also embedded in the air force among fighters and rescue teams. The main purpose of the team resource intervention among combat rescue teams is to investigate the circumstances of the event, analyze the medical team's functioning and draw the relevant conclusions and the manner of their implementation. The purpose of the guidelines is to encourage using the tool to help cope mentally and possibly prevent stress reactions and help in screening individuals who need further professional intervention for stress reactions. Currently, medical Team Resource Intervention is conducted as a routine after potentially traumatic events, especially after incidents involving fatalities. However, as mentioned earlier, even though flight engineers are part of the rescue team, they are not considered part of the medical team and therefore do not participate in the Team Resource Intervention following exposure to traumatic events. Although in a previous study by the authors, no PTSD was found among flight engineers, there were clear symptoms of anxiety that support the fact that this population should also be included in those receiving intervention after potentially traumatic events (Cohen-Koren et al., n.d.).

Recently, the Psychology branch of the Israeli Air Force has developed a Mental Training program for IDF flight engineers according to the findings of Cohen-Koren et al. (Ibid). The program comprises four components:

1. 'Personal Resilience'—A prevention component that is implemented in the flight engineers' training course, the purpose of which is to increase personal resilience and to help prepare them for the mental challenges of their work. This is a 2-day mental training program. The first day attempts to prepare the participants mentally for distressing sights by visiting a trauma room, watching the staff in action and attending a lecture by the head psychologist of the department about the mental implications of treating a severely wounded person. The second day, held by the clinical psychologist of the unit, teaches tools for coping with the experience of encountering fatalities, rescue under fire and with life-threatening situations while managing fear of failure prior to and during a rescue mission.
2. 'Resource Intervention'—The second phase of the program focuses on group interaction. This phase is based on the principles of a program developed for air combat teams, namely the Cognitive Inoculation program that was devised as preparation for air combat (Oz & Lurie, 2013). The principal of the Cognitive Inoculation program is: (1) strengthening the program participants' belief in their ability to cope in stressful situations; (2) creating a work-language, proper tools and legitimacy for dealing with the issue in all its aspects; (3) achieving group cohesion; (4) enriching the commander's 'toolkit' in a field that is not his core occupation; (5) defining the commander as a resource in general and particularly during combat, as well as upgrading his abilities. This part consists of a workshop run by the unit's clinical psychologist where all aircraft commanders are taught how to conduct a Resource Intervention after rescue missions that included fatalities or life-threatening situations. It should be noted that this program constitutes the first ever intervention program for flight engineers.
3. 'Squadron Strength'—This phase focuses on various groups and includes:
    a. A workshop for new flight engineers who were exposed to distressing sights or events that may have included corpse evacuation. In this half-day workshop, new flight engineers are given tools that help them handle the sights they have encountered during a rescue mission.
    b. A workshop for middle-aged flight engineers (31–40), those who were found to be at the highest risk of developing emotional distress (Cohen-Koren, n.d.). This part consists of half-day workshops during which participants are offered additional tools for managing stressful situations, fear of failure and encounters with fatalities alongside tools for balancing the stress that is created from the combination of work load with family life and rescue missions.
4. 'Monitoring'—This component of the program teaches how to preserve mental fitness as a routine. Included is administration of a short questionnaire (10 minutes) that tests distress. This questionnaire is routinely administered at the Air Medicine Unit and, when necessary, is followed by referral to targeted intervention by professional clinicians.

## Summary

The psychological branch of the Israeli Air Force started to internalize the fact that flight engineers are fully active participants in rescue operations. Hence, the psychological branch of the Israeli Air Force has developed a prevention program to be administered prior to potentially traumatic events and an intervention for thereafter in order to increase mental resilience and improve coping skills before an event occurs and reduce morbidities such as PTSD and other psychological distress symptoms after the event.

## References

Adler, A. B., Bliese, P. D., McGurk, D., Hoge, C. W., & Castro, C. A. (2009). Battlemind debriefing and battlemind training as early interventions with soldiers returning from Iraq: Randomization by platoon. *Journal of Consulting and Clinical Psychology*, 77(5), 928–940.

Admon, R., Lubin, G., Rosenblatt, J. D., Stern, O., Kahn, I., Assaf, M., & Hendler, T. (2012). Imbalanced neural responsivity to risk and reward indicates stress vulnerability in humans. *Cerebral Cortex*, *23*(1), 28–35.

American Psychiatry Association. (2000). *Diagnostic and Statistical Manual of Mental Disorders* (4th ed., Text revision). Washington, DC: Author.

American Psychiatric Association. (2013). *Diagnostic and Statistical Manual of Mental Disorders* (5th ed., DSM-5). Washington, DC: American Psychiatric Association.

Baird, K., & Kracen, A. C. (2006). Vicarious traumatization and secondary traumatic stress: A research synthesis. *Counselling Psychology Quarterly*, *19*(2), 181–188.

Bal, S., Van Oost, P., De Bourdeaudhuij, I., & Crombez, G. (2003). Avoidant coping as a mediator between self-reported sexual abuse and stress-related symptoms in adolescents. *Child Abuse & Neglect*, *27*(8), 883–897.

Belenky, G. L., Noy, S., Solomon, Z., & Del Jones, F. (1985). Psychiatric casualties (battle shock) in Israeli defense forces in the war in Lebanon June–September 1982. In P. Pichot, P. Berner, R. Wolf, & K. Thau (Eds.), *Psychiatry* (pp. 479–484). Boston, MA: Springer.

Bisson, J. I. (2003). Single-session early psychological interventions following traumatic events. *Clinical Psychology Review*, *23*(3), 481–499.

Boscarino, J. A., Figley, C. R., & Adams, R. E. (2004). Compassion fatigue following the September 11 terrorist attacks: A study of secondary trauma among New York City social workers. *International Journal of Emergency Mental Health*, *6*(2), 57–66.

Breslau, N., Davis, G. C., & Schultz, L. R. (2003). Posttraumatic stress disorder and the incidence of nicotine, alcohol, and other drug disorders in persons who have experienced trauma. *Archives of General Psychiatry*, *60*(3), 289–294.

Bryan, C., Goodman, T., Chappelle, W., Thompson, W., & Prince, L. (2018). Occupational stressors, burnout, and predictors of suicide ideation among US Air Force remote warriors. *Military Behavioral Health*, *6*(1), 3–12.

Carver, C. S. (1997). You want to measure coping but your protocol too long: Consider the brief cope. *International Journal of Behavioral Medicine*, *4*(1), 92–100.

Carver, C. S., Scheier, M. F., & Weintraub, J. K. (1989). Assessing coping strategies: A theoretically based approach. *Journal of Personality and Social Psychology*, *56*(2), 267–283.

Castro, C. A. (2014). The US framework for understanding, preventing, and caring for the mental health needs of service members who served in combat in Afghanistan and Iraq: A brief review of the issues and the research. *European Journal of Psychotraumatology*, *5*(1), 1–12.

Chappelle, W. L., McDonald, K. D., Prince, L., Goodman, T., Ray-Sannerud, B. N., & Thompson, W. (2014). Symptoms of psychological distress and post-traumatic stress disorder in United States Air Force "drone" operators. *Military Medicine*, *179*(Suppl. 8), 63–70.

Chappelle, W., & Lumley, V. (2006). Outpatient mental health care at a remote US air base in Southern Iraq. *Professional Psychology: Research and Practice*, *37*(5), 523–530.

Cohen-Koren, R., Garbi, D., Gordon, S., Yavnai, N., Erlich-Shoham, Y., & Shelef, L. (n.d.). *Predictors of emotional distress reported by combat military flight engineers*. Unpublished manuscript.

Creamer, M., Wade, D., Fletcher, S., & Forbes, D. (2011). PTSD among military personnel. *International Review of Psychiatry*, *23*(2), 160–165.

Engelhard, I. M., Van den Hout, M. A., Weerts, J., Arntz, A., Hox, J., & McNally, R. J. (2007). Deployment-related stress and trauma in Dutch soldiers returning from Iraq prospective study. *The British Journal of Psychiatry, 191*(2), 140–145.

Farchi, M., Levy, T. B., Gershon, B. B., Hirsch-Gornemann, M. B., Whiteson, A., & Gidron, Y. (2018). The SIX Cs model for immediate cognitive psychological first aid: From helplessness to active efficient coping. *International Journal of Emergency Mental Health and Human Resilience, 20*(2), 1–12.

Fear, N. T., Jones, M., Murphy, D., Hull, L., Iversen, A. C., Coker, B.,... Wessely, S. (2010). What are the consequences of deployment to Iraq and Afghanistan on the mental health of the UK armed forces? A cohort study. *The Lancet, 375*(9728), 1783–1797.

Ferrier-Auerbach, A. G., Erbes, C. R., Polusny, M. A., Rath, M., & Sponheim, S. R. (2010). Predictors of emotional distress reported by soldiers in the combat zone. *Journal of Psychiatric Research, 44*(7), 470–476.

Folkman, S. (2011). Stress, health, and coping: Synthesis, commentary, and future directions. In S. Folkman, (Ed.), *The Oxford Handbook of Stress, Health and Cooping* (pp. 453–462). New York, NY, US: Oxford University Press, Inc.

Galea, S., Nandi, A., & Vlahov, D. (2005). The epidemiology of post-traumatic stress disorder after disasters. *Epidemiologic Reviews, 27*(1), 78–91.

Gates, D. M., & Gillespie, G. L. (2008). Secondary traumatic stress in nurses who care for traumatized women. *Journal of Obstetric, Gynecologic & Neonatal Nursing, 37*(2), 243–249.

Gates, M. A., Holowka, D. W., Vasterling, J. J., Keane, T. M., Marx, B. P., & Rosen, R. C. (2012). Posttraumatic stress disorder in veterans and military personnel: Epidemiology, screening, and case recognition. *Psychological Services, 9*(4), 361–382.

Ginat, K., Svetlizky, V., Barezin, N., Pharchi, M., & Fruchter, E. (2015). MAGEN—Mental health enhancement in the front line. Paper presented at the *German Military Mental Health Conference*, Berlin.

Goldstein, R. B., Smith, S. M., Chou, S. P., Saha, T. D., Jung, J., Zhang, H.,... Grant, B. F. (2016). The epidemiology of DSM-5 posttraumatic stress disorder in the United States: Results from the National Epidemiologic Survey on Alcohol and Related Conditions-III. *Social Psychiatry and Psychiatric Epidemiology, 51*(8), 1137–1148.

Hantman, S., & Farchi, M. (2015). From helplessness to active coping in Israel: Psychological first aid. In E. M. Schott, & E. L. Weiss, (Eds.), *Transformative Social Work Practice* (pp. 467–484). Thousand Oaks: Sage Publications.

Helgeson, V. S. (2011). Gender, stress, and coping. In S. Folkman, (Ed.), *The Oxford Handbook of Stress, Health and Cooping* (pp. 63–85). New York, NY, US: Oxford University Press, Inc.

Hoge, C. W., Castro, C. A., Messer, S. C., McGurk, D., Cotting, D. I., & Koffman, R. L. (2004). Combat duty in Iraq and Afghanistan, mental health problems, and barriers to care. *New England Journal of Medicine, 351*(1), 13–22.

Hoge, C. W., Riviere, L. A., Wilk, J. E., Herrell, R. K., & Weathers, F. W. (2014). The prevalence of post-traumatic stress disorder (PTSD) in US combat soldiers: A head-to-head comparison of DSM-5 versus DSM-IV-TR symptom criteria with the PTSD checklist. *The Lancet Psychiatry, 1*(4), 269–277.

Hourani, L. L., Council, C. L., Hubal, R. C., & Strange, L. B. (2011). Approaches to the primary prevention of posttraumatic stress disorder in the military: A review of the stress control literature. *Military Medicine, 176*(7), 721–730.

Hunt, E. J., Wessely, S., Jones, N., Rona, R. J., & Greenberg, N. (2014). The mental health of the UK Armed Forces: Where facts meet fiction. *European Journal of Psychotraumatology, 5*(1). doi.org/10.3402/ejpt.v5.23617.

Jacobsen, L. K., Southwick, S. M., & Kosten, T. R. (2001). Substance use disorders in patients with posttraumatic stress disorder: A review of the literature. *American Journal of Psychiatry, 158*(8), 1184–1190.

Kessler, R. C., Berglund, P., Demler, O., Jin, R., Merikangas, K. R., & Walters, E. E. (2005). Lifetime prevalence and age-of-onset distributions of DSM-IV disorders in the National Comorbidity Survey Replication. *Archives of General Psychiatry, 62*(6), 593–602.

Knobler, H. Y., Nachshoni, T., Jaffe, E., Peretz, G., & Yehuda, Y. B. (2007). Psychological guidelines for a medical team debriefing after a stressful event. *Military Medicine, 172*(6), 581–585.

Kok, B. C., Herrell, R. K., Thomas, J. L., & Hoge, C. W. (2012). Posttraumatic stress disorder associated with combat service in Iraq or Afghanistan: Reconciling prevalence differences between studies. *The Journal of Nervous and Mental Disease, 200*(5), 444–450.

Krysinska, K., & Lester, D. (2010). Post-traumatic stress disorder and suicide risk: A systematic review. *Archives of Suicide Research, 14*(1), 1–23.

Lamberg, L. (2004). Military psychiatrists strive to quell soldiers' nightmares of war. *JAMA, 292*(13), 1539–1540.

Lapierre, C. B., Schwegler, A. F., & LaBauve, B. J. (2007). Posttraumatic stress and depression symptoms in soldiers returning from combat operations in Iraq and Afghanistan. *Journal of Traumatic Stress: Official Publication of the International Society for Traumatic Stress Studies, 20*(6), 933–943.

Lazarus, R. S., & Folkman, S. (1984). Coping and adaptation. In W. D. Gentry (Ed.), *The Handbook of Behavioral Medicine* (pp. 282–325). New York: Guilford.

Levi, O., & Lubin, G. (2018). Treatment seeking for posttraumatic stress in Israel Defense Forces veterans deployed in the Second Lebanon War (2006) and "Operation Cast Lead" in the Gaza Strip (2009): A comparative study. *Anxiety, Stress, & Coping, 31*(3), 328–337.

Lewis, S. J. (2003). Do one-shot preventive interventions for PTSD work? A systematic research synthesis of psychological debriefings. *Aggression and Violent Behavior, 8*(3), 329–343.

Maes, M., Delmeire, L., Mylle, J., & Altamura, C. (2001). Risk and preventive factors of post-traumatic stress disorder (PTSD): Alcohol consumption and intoxication prior to a traumatic event diminishes the relative risk to develop PTSD in response to that trauma. *Journal of Affective Disorders, 63*(1–3), 113–121.

Mitchell, J. T., & Everly, G. S. (2001). *The Basic Critical Incident Stress Management Course.* Ellicott City, MD: International Critical Incident Stress Foundation.

Moriya, G., & Luria, A. (2016). Group intervention for air-born mechanics following the Second Lebanon War and establishment of a prevention program—A case study. *Aviation Psychology* (pp. 149–179). Air Force Psychology Branch, Air Force/Ministry of Defence. [Hebrew]

Neria, Y., Nandi, A., & Galea, S. (2008). Post-traumatic stress disorder following disasters: A systematic review. *Psychological Medicine, 38*(4), 467–480.

Noy, S., Nardi, C., & Solomon, Z. (1986). Battle and military unit characteristics and the prevalence of psychiatric casualties. In N. A. Milgram, (Ed.), *Stress and Coping in Time of War: Generalizations From the Israeli Experience* (pp. 73–77). New York: Brunner/Mazel.

Oz, I. & Lurie, O. (2013). Aviation disaster crisis management—Multidimensional psychological intervention. In C. H. Kennedy, & G. G. Kay, (Eds.). *Aeromedical Psychology* (pp. 309–324). London: CRC Press.

Palgi, Y., Ben-Ezra, M., Langer, S., & Essar, N. (2009). The effect of prolonged exposure to war stress on the comorbidity of PTSD and depression among hospital personnel. *Psychiatry Research, 168*(3), 262–264.

Petrakis, I. L., Ralevski, E., Desai, N., Trevisan, L., Gueorguieva, R., Rounsaville, B., & Krystal, J. H. (2012). Noradrenergic vs serotonergic antidepressant with or without naltrexone for veterans with PTSD and comorbid alcohol dependence. *Neuropsychopharmacology, 37*(4), 996–1004.

Pietrzak, R. H., Goldstein, R. B., Southwick, S. M., & Grant, B. F. (2011). Prevalence and Axis I comorbidity of full and partial posttraumatic stress disorder in the United States: Results from Wave 2 of the National Epidemiologic Survey on Alcohol and Related Conditions. *Journal of Anxiety Disorders, 25*(3), 456–465.

Ramchand, R., Rudavsky, R., Grant, S., Tanielian, T., & Jaycox, L. (2015). Prevalence of, risk factors for, and consequences of posttraumatic stress disorder and other mental health problems in military populations deployed to Iraq and Afghanistan. *Current Psychiatry Reports, 17*(5), 37–48.

Richardson, L. K., Frueh, B. C., & Acierno, R. (2010). Prevalence estimates of combat-related post-traumatic stress disorder: Critical review. *Australian and New Zealand Journal of Psychiatry, 44*(1), 4–19.

Rose, S. C., Bisson, J., Churchill, R., & Wessely, S. (2002). Psychological debriefing for preventing post traumatic stress disorder (PTSD). *Cochrane Database of Systematic Reviews, 2*(2), 1–46.

Salmon, T. W. (1919). War neuroses and their lesson. *New York Medical Journal, 108*, 993–994.

Sever, I., Somer, E., Ruvio, A., & Soref, E. (2008). Gender, distress, and coping in response to terrorism. *Affilia, 23*(2), 156–166.

Shalev, A. Y. (1996). Stress versus traumatic stress: From acute homeostatic reaction to chronic psychopathology. In B. A. Van Der Kolk, A. C. McFarlane, & L. Weisaeth, (Eds.), *Traumatic Stress: The Effects of Overwhelming Experience on Mind, Body, and Society* (pp. 77–101). New York: Guilford Press.

Shelef, L., Fruchter, E., Gordon, S., Garbi, D., Yavnai, N., Haber, Y., … Shahar, G. (2018). Impairment among soldiers who were exposed to combat during military operation. *International Journal of Psychiatry, 3*(1), 1–5.

Siddiqui, S. V., Chatterjee, U., Kumar, D., Siddiqui, A., & Goyal, N. (2008). Neuropsychology of prefrontal cortex. *Indian Journal of Psychiatry, 50*(3), 202–208.

Sofko, C. A., Currier, J. M., & Drescher, K. D. (2016). Prospective associations between changes in mental health symptoms and health-related quality of life in veterans seeking posttraumatic stress disorder residential treatment. *Anxiety, Stress, & Coping, 29*(6), 630–643.

Solomon, Z., & Dekel, R. (2008). The contribution of loneliness and posttraumatic stress disorder to marital adjustment following war captivity: A longitudinal study. *Family Process*, *47*(2), 261–275.
Solomon, Z., Shklar, R., & Mikulincer, M. (2005). Frontline treatment of combat stress reaction: A 20-year longitudinal evaluation study. *American Journal of Psychiatry*, *162*(12), 2309–2314.
Thomas, J. L., Wilk, J. E., Riviere, L. A., McGurk, D., Castro, C. A., & Hoge, C. W. (2010). Prevalence of mental health problems and functional impairment among active component and National Guard soldiers 3 and 12 months following combat in Iraq. *Archives of General Psychiatry*, *67*(6), 614–623.
Thompson, M. M., & Jetly, R. (2014). Battlefield ethics training: integrating ethical scenarios in high-intensity military field exercises. *European Journal of Psychotraumatology*, *5*(1). Published online. doi: 10.3402/ejpt.v5.23668.
Tolan, P. H., Gorman-Smith, D., Henry, D., Chung, K. S., & Hunt, M. (2002). The relation of patterns of coping of inner-city youth to psychopathology symptoms. *Journal of Research on Adolescence*, *12*(4), 423–449.
Vasterling, J. J., Proctor, S. P., Friedman, M. J., Hoge, C. W., Heeren, T., King, L. A., & King, D. W. (2010). PTSD symptom increases in Iraq deployed soldiers: Comparison with nondeployed soldiers and associations with base line symptoms, deployment experiences, and postdeployment stress. *Journal of Traumatic Stress: Official Publication of The International Society for Traumatic Stress Studies*, *23*(1), 41–51.
Vermetten, E., Greenberg, N., Boeschoten, M. A., Delahaije, R., Jetly, R., Castro, C. A., & McFarlane, A. C. (2014). Deployment-related mental health support: Comparative analysis of NATO and allied ISAF partners. *European Journal of Psychotraumatology*, *5*(1), 1–19.
Visser, D., & Van Staden, F. (1992). The experience of work circumstances and stress; a profile of flight engineers in a labour dispute. *SA Journal of Industrial Psychology*, *18*(1), 1–6.
Wald, I., Bitton, S., Levi, O., Zusmanovich, S., Fruchter, E., Ginat, K., . . . Bar-Haim, Y. (2017). Acute delivery of attention bias modification training (ABMT) moderates the association between combat exposure and posttraumatic symptoms: A feasibility study. *Biological Psychology*, *122*, 93–97.
Wald, I., Fruchter, E., Ginat, K., Stolin, E., Dagan, D., Bliese, P. D., . . . Bar-Haim, Y. (2016). Selective prevention of combat-related post-traumatic stress disorder using attention bias modification training: A randomized controlled trial. *Psychological Medicine*, *46*(12), 2627–2636.
Yehuda, R., Vermetten, E., McFarlane, A. C., & Lehrner, A. (2014). PTSD in the military: Special considerations for understanding prevalence, pathophysiology and treatment following deployment. *European Journal of Psychotraumatology*, *5*(1). Published online. doi: 10.3402/ejpt.v5.25322.

# 29

# FAMILY-BASED PSYCHOLOGICAL INTERVENTIONS

## A heuristic approach

*Jyoti Mishra and Shobit Garg*

> Ayam nijah paroveti ganana laghuchetasam
> Udharacharitanam tu vasudhiva kutumbhakam
> (This is my own and that a stranger'—is the calculation of the narrow-minded. For the magnanimous hearts however, the entire earth is but a family)
>
> *(Hitopadesha, 1.3.71)*

As a social being, man has played the game of survival well and has evolved to the present level of civilization advancement ... and if there is one cohesive force that binds a group of people into unique set pieces of culture, cuisine and civility, then it is the institution of family, all the more replete with diversity, color and compassion in the Indian context—who took it a step further with the concept of "Vasudhaiv kutumb kam" taken right out of the Upanishads and ingrained it in our DNA. But in today's rapidly urbanizing world, one institution that still truly lives by this ancient shlok in its letter and spirits is the Armed Forces.

Military life, where deployment away from family is a fact of life, is tough; comes with passion, bravery and dedication and is entirely different from the civil life. For families, deployment fears are realistic and the effect of war can extend far beyond the deployment fear. A family that loses the active presence of a person serving in the military through separation faces significant challenges and stress in life. Along with the unspoken fear of military life, there are general day-to-day concerns like the lifestyle, frequent moves and lack of stability which create additional issues for the family to handle. Partners of service members who are deployed, for example, face the challenge of worrying about and supporting a loved one for an extended period without knowing if or when they will return. Not only the service member but also the family, spouses and children face a number of other challenges like emotional, social and, even more, day-to-day challenges before, during and after deployment. They prepare themselves for the absence of one of the important members in their family.

There are several stages of the emotional deployment cycle (Pincus, House, Christenson, & Adler, 2001), considering three broad periods of deployment and associated family stress, as the entire family is affected with a range of stressors and challenges throughout the deployment cycle. These stages are: (1) pre-deployment, (2) deployment and (3) post-deployment or reintegration

(Gewirtz, Erbes, Polusny, Forgatch & DeGarmo, 2011; Mohan & Arora, 2013). The *pre-deployment period* involves anticipatory worry and tension associated with imminent separation. The *deployment period* involves prolonged separation and single parenting when the deployed parent is absent. The *post-deployment period*, or *period of reintegration* after deployment, can bring additional challenges for two-parent families in re-establishing a parenting 'equilibrium'. There is increasing and compelling evidence describing the significant deleterious impact of the deployment cycle on family members, including children. The literature supports evidence-based engagement strategies, a successful process of engaging military families with young children in a home-based reintegration program designed to support parenting and strengthen parent-child relationships (Ross & Devoe, 2014). However, rates of engagement and service utilization in prevention and intervention services continue to lag far below apparent need among service members and their families because of both practical and psychological barriers. Existing evidence also suggests that military veterans with mental health disorders have poorer family functioning (McFarlane, 2009; Gates, Holowka, Vasterling, Keane, Marx & Rosen, 2012; Sayers, Farrow, Ross & Oslin, 2009; Creech & Misca, 2017). They report family problems such as feeling like a guest in their household, their children acting afraid or not being warm toward them or being unsure about their family role. Depression and posttraumatic stress disorder symptoms were both associated with higher rates of family reintegration problems (Sayers et al., 2009).

It is a fact that family members of deployed military personnel are often the 'silent ranks', supporting service members' military journey through all its stages. The potential challenges and mental health consequences of war-time deployments to service members and veterans are considerable, and increasing evidence suggests that families face diverse stressors at different periods in the deployment cycle (Pincus et al., 2001). Therefore, the psychological consequences of military life also apply to families of military personnel. Despite the known detrimental impact of military life on family members, very little emphasis has been given to family-based psychological interventions with their families. The mental health of military personnel and their families becomes essential for the growth and security of the nation. Therefore, the current chapter focuses on what these stressors are; how they affect the mental health of the serving person and his/her entire family and the psychotherapeutic approach to dealing with these demanding states for families of military personnel.

## Positive effect of military life on family

It is an acknowledged fact that the status of a 'warrior' has stood the test of time. History and society have both held the soldier in respect and offered him a unique status in society. Additionally, the military family is a close-knit family. It is a world in itself, full of its own curriculum, fun and responsibilities. The military not only looks after service members but also gives all the comfort to the family. They have access to major life securities like medical security, financial security and their own schools to promote children's education. The civilian life, on the other hand, lacks job satisfaction, and aimlessness in many leads to frustration and results in job-hopping. On the contrary, a sense of purpose, responsibility and pride negate any job dissatisfaction in the military.

## Negative effect of military life on family

Every good thing comes with some repercussions and so does the military life. It can be bravery for the nation but a lifetime of pain for the family members. Figures presented by the defense minister to the Lok Sabha in 2013 revealed that a total of 368 defense personnel committed suicide from the year 2010–2012 (Indian Military News, 2013). Along with the occupational stressor, job satisfaction (Sharma, 2015; Asnani et al., 2001), most suicides and cases of violence stemmed from marital discord, domestic problems, love affairs, property disputes or heavy debts rather than stress related to insurgency-affected regions (Suman, 2009). Approximately 2 million children have been affected by military deployment, where adjustment-related issues of children experiencing a parent's combat deployment have emerged only within the past 5–10 years (Gewirtz & Zamirt, 2014).

Military families are aware, but not every family is used to deployment; they may experience boundary ambiguity in terms of roles and task performances within the family. Findings from studies suggest high rates of depression in spouses of service members (Verdeli et al., 2011), as greater deployment exposure was related to impaired family functioning and marital instability (Lester et al., 2016). There has also been a significant impact of mothers' anxiety on the children of deployed parents (Palmer, 2008; McFarlane, 2009; Gates et al., 2012; Sayers et al., 2009, Creech & Misca, 2017). Additionally, reunification after deployment further poses challenges for families, like re-establishing roles and routines and the potentially necessary accommodation of combat-related injuries or psychological effects (Walsh et al., 2014). Reunification and deployment also pose significant levels of parenting stress and identified specific challenges, like difficulty reconnecting with children, adapting expectations from military to family life and co-parenting. Hence, fathers acknowledged regret about missing an important period in their child's development and indicated a strong desire to improve their parenting skills. They described a need for support in expressing emotions, nurturing and managing their tempers (Walsh et al., 2014).

Deployment stressors have a detrimental impact on the mental health outcomes of women in terms of adverse life events, war experiences, posttraumatic stress disorder (PTSD) and depression symptoms. They had difficulties in emotional regulation, parenting, couple adjustment and child functioning (Gewirtz, Pinna, Hanson, & Brockberg, 2014). These parental stress symptoms were associated with impairments in social-emotional adjustment in young children, increased anxiety in early childhood and adjustment problems in school-age children (Lester et al., 2016). On the other hand, female service members in the military also experience trauma, sexual assault, sexual harassment and other interpersonal stressors during deployment (Street, Vogt & Dutra, 2009; Dutra et al., 2011).

Despite the resiliency of many military families, young children can be especially vulnerable to stressors of parental deployment and family transition. This type of separation can constitute a developmental crisis for a young child (Paris, DeVoe, Ross & Acker, 2010), affects academic functioning, increases greater acting-out behavior and makes emotional adjustment worse for boys (Martinez & Forgatch, 2002; Prabhu, Prakash, Bhat, & Gambhir, 2011; Rossiter, D'Aoust & Shafer, 2016). Military parents of young children often bear repeated separations from their children, and these may disrupt the early parent-child relationship. Various qualitative studies have reported parental experiences of separation with their families during deployment-related adjustment challenges, disruptions of routines and responsibilities and heightened emotions on the part of each family member. Parents face difficulty in supporting the development of qualities like strength, confidence and self-sufficiency in their young children due to problems in dealing with the negative emotions and difficult behaviors that their children exhibit (Chandra, Martin, Hawkins & Richardson, 2010; Dayton, Walsh, Muzik, Erwin & Rosenblum, 2014; Paris et al., 2010; McFarlane, 2009; Willerton, Schwarz, Wadsworth & Oglesby, 2011).

## Psychological intervention for families

Stress is experienced and managed differently by each individual. For each person, stressful situations may vary but the psychological and physiological symptoms remain the same. Military life also adds day-to-day life stressors in addition to deployment and stress related to frequent shifts. In order to cope better in a stressful situation, it is important to address and understand the stressor and what it demands. Families need to understand how deployment can affect not only the serving member but also the family. Life in the military is a known phenomenon, however to many it differs when it comes to personal experiences. Therefore, "*Psycho-education*" (Schachman, 2010) about the process of pre, during and post deployment becomes essential. For a person who is a civilian, not being aware about the military life, the deployment process, and risk of life, it is difficult for them to live in this unpredictable life. Families may also need time to adapt to new roles resulting from the absence of the military personnel. Efforts should be made to prepare a person psychologically about the process as soon as they enter the military life. They should be educated about how these changes can affect them psychologically, emotionally and physically. Additionally, while dealing with the process, healthy coping becomes essential for the survival since stressful conditions may develop unhealthy copings like smoking, drinking and over or under eating etc. "*Building Resilience*" is an important part of healthy coping. It is an ability to cope successfully within stressful situation and return back to the previous healthy mental health condition. A person with good resilient ability maintains a more positive viewpoint and copes with stress more effectively. Some people are born with these resilience traits but these behaviors can also be learned. In Military life, the emphasis should be on strength based resilience building approach while working with the military families. This helps the family members to use their strength and values to cope better with deployment related stress. The intervention begins with identifying their past ways of coping with military related stressors as that will help in identifying and emphasizing existing effective practices within the families. "*Cognitive and behavioral therapy*" (CBT) approaches can be found effective in order to foster resilience. Developing a good social support system, building positive belief and self esteem among military personnel and family members, adjusting to the new changes, working on problem-solving and attribution style and emotional awareness facilitate building resilience. Moreover, cognitive behavioral therapy is effective as well as cost effective in comparison with alternative treatments (Durham et al., 2005). It helps in restructuring negative automated beliefs, as psychological health is a critical factor that needs to be addressed in military families. It aims to help a person restructure their cognition and replace maladaptive thoughts with thoughts that are more helpful in coping with stressful situations. Despite a high prevalence of mental health difficulties in the family members of armed forces veterans, little is known about their mental health treatment (Dimiceli, Steinhardt & Smith, 2010; Spelman, Hunt, Seal & Burgo-Black, 2012; Drummet, Coleman & Cable, 2003). Research also examined the heightened anxiety, social support and physical health of spouses of military service members and reported poorer overall health and more physical health comorbidities (Fields, Nichols, Martindale-Adams, Zuber, & Graney, 2012). Existing systematic reviews have found cognitive behavioral therapy and psychosocial interventions effective for the treatment of mental health disorders, mainly depression and anxiety in armed forces veterans as well as family members (Kitchiner, Roberts, Wilcox & Bisson, 2012).

Some *family therapy* techniques may help in alleviating stressful conditions that crop up because of unpredictable circumstances in military families. In military families, crisis does not regularly or predictably occur in the course of family development. Nevertheless when it comes, it puts lot of pressure onto the family. *Brief family crisis intervention* becomes imperative in these unpredictable circumstances. Adams (1991) articulated basic components to follow in brief family crisis intervention, where he talked about *identifying the family in crisis, the symptoms of each person in the family, what has precipitated the crisis, affective connection among the family members, relevant past coping resources, formulating the situation, providing family members with a cognitive grasp of what has happened, supporting affective integration and then mobilizing effective coping*. These steps can help the therapist teach family members to deal effectively with these unpredictable circumstances. Furthermore, *structural interventions* (Minuchin, 1974) may be useful in working with military families. Some structural family therapy techniques are designed to create new interactional sequences. For example, in *enactment*, the therapist guides family members to perform new patterns of behavior, allowing the therapist to observe how family members interact and see how the problem behavior is embedded in the interactional sequence of transactions. The therapist directly instructs family members to act in particular ways that are more useful. This disrupts existing patterns, testing the system's ability to adopt alternative and more functional rules. Another technique is *communication skill*, where family problems are conceptualized in terms of communication problems: 'we have trouble talking', 'my wife or family member does not understand', 'we simply don't seem to be on the same wavelength' or 'my child does not listen to anyone'. The problem can become worse when the discussion about managing the family situation occurs from distant modes like telephonically and not in person. As part of the assessment process, the therapist notes both the family's verbal and the nonverbal communications and discrepancies between the messages sent via each of these two channels and also from other modes (e.g., conversation via telephone). The therapist should also consider the clarity, directness and sufficiency of communications, both verbal and nonverbal, and the availability and openness of those to whom communications are addressed. The therapist provides direct instruction in communication skills, for example, listening attentively; reflecting or summarizing accurately; delivering clear non-blaming 'I' messages and being congruent in affect, body language and verbal content (McKay, Fanning & Paleg, 2006).

## Couples therapy

The instability associated with military life becomes a common ground for divorces. Various marital concerns like role expectations, domestic pressure, arguments, inadequate salary and sexual deprivation are the major causes of stress in military personnel's life (Mohan & Arora, 2013). Approximately 20% of couples are distressed, with marital satisfaction decreasing in the first decade of marriage (Bradbury, Fincham & Beach, 2000). Couples distress is strongly associated with emotional, substance abuse and health-related problems (Whisman & Uebelacker, 2006).

The relevant issues of couples can also be addressed through family interventions like couples therapy. In couples therapy, through a *problem-solving approach*, the first step is to make the couple agree on a clear definition of the problem. The couple then brainstorms possible solutions with the therapist, discussing the implications of each. The therapist leads the couple to consider all possibilities and prepares them to use their communication skills to prioritize each option. The therapist offers behaviorally specific feedback as the couple work through the decision-making process. When the couple has developed a solution, the therapist helps them out to develop a contract specifying the terms of the agreement and work accordingly. The therapist works on altering the couple's view of the presenting problem to be more objective and contextualized, decreasing emotion-driven communication, improving communication and promoting strengths. In couples therapy, *behavioral couples therapy* (Jacobson & Margolin, 1979; Stuart, 1980) is one of the techniques focusing on behavioral exchanges (Atkins, Dimidjian, & Christensen, 2003) and also incorporates communication skills training, problem solving and behavioral contracting (Johnson & Lebow, 2000), helping in overcoming the marital/couple conflicts among military personnel. *Role-playing* is another action technique that can be useful when verbal approaches prove ineffective. It can be especially valuable in families who intellectualize their problems. Having them act out scenes or events from their lives can facilitate change. For example, a family might be asked to act out what happens when father returns from deployment; here the therapist, like sculpting, through role-playing brings something of the reality of family life into the therapy session and brings the family into awareness about the faulty communication and problem-solving approaches. *Mindfulness practice* is another effective approach to dealing with stressful situations. It is the direction of attention towards one's ongoing experience, in a manner that is characterized by openness and acceptance (Bishop et al., 2004). Mindfulness interventions are derived from Eastern meditation practice. Mindfulness practices invite one to attend to one's thought processes, emotions and sensations without judgment. There is no effort to reform cognitions or reject particular emotions. Mindfulness practices can be used as a self-regulation practice. Attention to physiological sensations and breath encourages physiological relaxation.

Family therapy should be considered when there is a malfunctioning family group and the problems which therapy is to address are related to the functioning of the family. Family therapy aims to change the functioning of families. It is likely to be of value when the presenting problems concern children or adolescents, when families present complain that members have problems in relating to each other and when a family appears to be having difficulty making the changes required in the family. Family therapy is neither a 'cure-all' nor a treatment of last resort but an effective way of dealing with problems rooted in a dysfunctional family system. It may sometimes be usefully combined with the treatment of individual family members.

Spouses of military service members report increased depression and anxiety post-deployment as they work to reintegrate the family and service member. Reconnecting the family, renegotiating roles that have shifted, reestablishing communication patterns and dealing with mental health concerns are all tasks that spouses must undertake as part of reintegration. Study findings also suggest that a well-established, high-access intervention can help improve quality of life for military spouses who are struggling with reintegration of the service member and family (Nichols, Martindale-Adams, Graney, Zuber & Burns, 2013). So, adapting a mental health service to meet the needs of armed forces veterans and family members enhances acceptability and promotes mental health help-seeking (Farrand et al., 2018). Efforts to improve mental health service provision for family members are increasingly important, given that they represent an underserved group (Blaisure et al., 2016).

## *Psychological intervention for children*

Despite evidence that parenting has significant influence on children's functioning and that parenting may be impaired during stressful family transitions, there is a dearth of empirically supported psychological interventions tailored for military families reintegrating after deployment (Gewirtz et al., 2014; Walsh et al., 2014; DeVoe, Paris, Emmert-Aronson, Ross & Acker, 2017). In situations where families and military service personnel are confronted by more complex stressors, these parenting strategies may be important adjuncts to other treatment approaches.

1. Being a single non-deployed parent is exhausting. Help parents 'maintain routine practices' during the deployment cycle as rules and clear limits on children's behavior protect both children and parents through unpredictable transition phases. For a single parent, they become inconsistent, become more relaxed in setting limits and even maintain the same parenting during deployment or following reintegration. Non-deployed parents often sense an inability to stop children's unwanted behavior and often show anger and frustration in return. Suggesting parents maintain 'The Three Rs': the practice of rules, routines and rituals, may represent a means of maintaining family stability in the face of the stresses experienced before, during and after deployment (Sheppard, Malatras & Israel, 2010).
2. Parents need to practice *effective problem solving and coping mechanisms*. The therapist should work on family problem-solving as a tool to support children's coping, as it facilitates ways to reduce children's daily anxieties by increasing predictability and communication, particularly during times of transition.

3. Communication with the children becomes important for the parents in order to understand child psychology. The therapist should facilitate regular communication among family members. Depending upon the child's age and cognitive abilities, strategies should be planned accordingly as to how much information a child can take in. Role-playing situations can support parents to strengthen children's coping with anxiety at stressful times of deployment (Sherman & Sherman, 2009).
4. The therapist needs to make the parents understand that the impact of transitions is anxiety-provoking for children, too. *Minimizing child transitions* during the deployment cycle helps the child cope better as frequent moves to different new schools and making new peer circles on the one hand make the child more flexible in handling life situations but on the other hand might be very stressful for some children.
5. Inconsistent and double-bind communication is emotionally distressing and gives conflicting parenting messages to the children. It can be confusing, stressful and anxiety provoking for them. *Co-parenting*, united parenting, can facilitate a healthy approach towards children with defined goals before and during reintegration.
6. *Regulating emotions of parents*, as posttraumatic stress symptoms, depression and substance abuse have all been linked with impaired parenting practices (Gewirtz, Polusny, DeGarmo, Khaylis & Erbes, 2010). A stressed non-deployed spouse can easily transfer his or her stress onto the child. Providing parents with psychological strategies for encouragement of children using token systems, incentive charts, use of praise and tangible rewards becomes important as a counterbalance to the downward spiral of coercive cycles between parents and children.

*Parent Management Training program: Oregon Model* (PMTO, Patterson, 2005) is another of the strategies for supporting parenting in military families and may be delivered in the context of a variety of psychological services (i.e., individual or family therapy or prevention services). Generally, these strategies include: (1) building on the resilience of military families, (2) addressing family stress within the context of the deployment cycle and (3) offering strategies to enhance emotion regulation as a key to effective parenting. This program holds the core principles of providing information, teaching, practice and support regarding five effective parenting practices designed to reduce coercive tactics and promote positive parenting (Patterson, 2005; Reid, Patterson & Snyder, 2002). These practices are: (1) contingent skill encouragement, (2) limit-setting, (3) positive involvement, (4) monitoring children's activities and (5) effective family problem-solving. This model is theoretically based on the social interaction learning model, a well-validated model in improving parenting, child adjustment and family functioning like reduced maternal depression, reduced maternal substance use, reduced child substance use, increased income, reduced financial stress and lower rates of police arrests for youngsters and mothers (Forgatch et al., 2009; Patterson et al., 2004) and might be quite appropriate and applicable to military families. Prevention studies shortly after a critical transition showed that the strengthened parenting practices promoted long-term positive adjustment for children and their parents. These studies show the potential efficacy of PMTO in intervening with military families around the time of stressful military transitions such as preparing for or readjusting after a combat deployment (Forgatch, Patterson, DeGarmo & Beldavs, 2009; Reid et al., 2002; Forgatch & Kjøbli, 2016). The *After Deployment Adaptive Parenting Tools* (ADAPT) program is another model based on a PMTO group intervention (Parenting Through Change; Forgatch & DeGarmo, 1999; Forgatch et al., 2009) that has already been shown to be effective for exposure to stressful transitions and traumatic events (Gewirtz & Taylor, 2009). The ADAPT curriculum involves role-playing exercises and audio-visual material relevant to military families and culture. This program targets common post-deployment adjustment reactions like, irritability, hypervigilance and avoidance that can affect the family dynamics by doing psycho-education and a focus on emotional regulation within a parenting context. ADAPT parent training intervention was associated with linear reductions in PTSD symptoms and represents common pathways for both mothers and fathers (DeGarmo & Gewirtz, 2018; Chesmore, Piehler & Gewirtz, 2018).

Randomized controlled trials showed evidence-based parenting programs have been successful in promoting adaptive parenting practices among families exposed to stress and deployment (Gewirtz, DeGarmo & Zamir, 2018; Chesmore et al., 2018). Studies of children's resilience have revealed the critical role of effective parenting in healthy youth adaptation among families living in stressful circumstances (Masten, 2001). It provides a protective buffer for children and youth, particularly during times of adversity (Gewirtz, Forgatch & Wieling, 2008). Helping parents work together as a team to apply effective child-rearing practices reduces stress at home, provides a supportive context for the transition phase and improves children's resilience as well (Walsh et al., 2014). Hence, various studies supports and suggest the implications for intervention programs include the provision of parenting and self-care skills and inclusion of the father as parenting partner in the intervention. (Dayton et al., 2014; Forgatch & DeGarmo, 1999).

## Intervention for military personnel

Not only the family but also the service members themselves may face a range of challenges in reintegrating into family life long after a combat deployment, including adjustment issues of transitioning back to work and family life. Cases of suicide and fratricide in the armed forces are also being reported (Indian Military News, 2013). Family problems are a major source of stress, more than the fear of war and PTSD. Despite all these mental health concerns, minimal efforts have been made to provide psychological services to military personnel.

As growing evidence reports, some common mental health disorders like depression, generalized anxiety disorder, social anxiety, occupational stress, obsessive compulsive disorder, panic and post-traumatic stress disorder (PTSD) are highly prevalent in the armed forces and veterans (Hoge et al., 2004; NICE, 2011; Sharma, 2015), with depression, anxiety and alcohol misuse highly prevalent and more common than PTSD (Iversen & Greenberg, 2009; Iversen et al., 2011).Another common problem for armed forces veterans is anger that can inhibit psychosocial functioning and lead to loss of emotional support (Miller, 2006). Regardless of the prevalence of these difficulties, many armed forces veterans do not seek mental health treatment (Vogt, 2011). One of the causes of this may be the anticipation of stigma and discrimination (Sharp et al., 2015), negative perceptions regarding mental health services (Zinzow, Britt, McFadden, Burnette & Gillispie, 2012) and beliefs that mental health difficulties can only be handled by the individual him- or herself (Creamer et al., 2006); all these factors contribute to low rates of seeking mental health treatment (Iversen et al., 2005). Not seeking treatment may aggravate difficulties for some individuals, as those who are discharged from service with psychological problems are at high risk of experiencing ongoing ill health (Iversen and Greenberg, 2009). No existing review takes mental health into account.

The most effective treatments to promote serving members' mental health concerns are talk therapies, all of which are referred to as a form of cognitive-behavioral therapy. In the above sections, efficacies of evidence-based psychological therapies have been discussed in detail. The prolonged exposure (Steenkamp, Litz, Hoge & Marmar, 2015; Foa et al., 2018; Powers, Halpern, Ferenschak, Gillihan & Foa, 2010) and cognitive processing therapy (Chen et al., 2019; Holder, Holliday, Williams, Mullen, & Surís, 2018; Steenkamp et al., 2015) forms of cognitive behavior therapy also have the most scientific evidence for efficacy in treating PTSD in military personnel. Prolonged exposure exposes the person to gradually approach their trauma-related memories, feelings and situation with the aim that these memories become not dangerous and not to be avoided. Completing PE can not only reduce the symptoms of PTSD, it also can give the client a strong feeling of accomplishment that he/she was able to face many things that he or she thought he or she couldn't. In addition to PTSD, PE has been shown to help with depression as well. Cognitive processing therapy helps the person learn how to modify and challenge unhelpful beliefs related to the trauma and reduces ongoing negative effects on current life.

There are special sessions in CPT that target specific issues, including trust, power and control, safety, self-esteem, esteem for other people and intimacy. This therapy may also help in dealing with the family-related issues and stressors. In order to deal with stressors military personnel use nonprescription drugs or drink alcohol to try not to think about what happened or what will happen to them professionally and personally. Using drugs and alcohol in return can also increase anger, create problems at work and home and result in many other health hazards. Motivation enhancement therapy (MET) focuses on internally motivated change to stop drug use, while relapse prevention therapy (RPT) focuses on preventing relapse and continuing to motivate a person to remain abstinent. Evidence-based psychological and psychosocial interventions, such as motivation interviewing (Walker et al., 2017), cognitive-behavioral therapy and contingency management show promising results for decreasing drinking and alcohol dependence among this population (Hawkins, Grossbard, Benbow, Nacev & Kivlahan, 2012; Cooper, Chatters, Kaltenthaler & Wong, 2015).

## Management through alternative modes

Managing psychological and psychosocial issues face to face may not be feasible every time, especially in military life. However, an appreciation of newer technologies (like internet, web chat and telephones) and rapid modes of communication in today's time makes life less stressful in comparison to the past in the military. A deployed parent often remains absent from major family life events and responsibilities because of deployment in combat regions. A study highlighted the role of online communication with their partner via e-mail, instant messaging, Facebook, blogs and chat rooms with the purpose of restoring balance in their roles (Schachman, 2010). These efforts help in maintaining healthy and cordial relationships among the family members of military personnel and promote positive mental health as well. It was found that spouse depression and anxiety were decreased and perceived social support was increased during the course of the study where telephone support groups focused on helping spouses with basic reintegration process involving focus on education, skill building (communication skills, problem-solving training, cognitive behavioral techniques, stress management), support (Nichols et al., 2013) and web-based tele-health communication (Forgatch & Kjøbli, 2016). Various randomized control trials found significant effects of telephone-delivered cognitive behavioral therapy in reducing worry, generalized anxiety disorder symptoms and depressive symptoms (Brenes, Danhauer, Lyles, Hogan & Miller, 2015; Brenes, Danhauer, Lyles, Anderson & Miller, 2017) and in treatment of depression and those at risk of drinking (Hundt, Barrera, Robinson & Cully, 2014; Kitchiner et al., 2012; O'Shea, Watkins & Farrand, 2017).

## Recommendations

These recommendations are useful in addressing common deployment issues and related stress in military families. However, stressors that are severe in nature will require more intensive individual and family approaches and professional services. These recommendations are;

1. Emphasis should be given to educate the military personnel's family members, especially the spouse, about military life, as it is completely different from civil life.

2. There should be a counseling cell of mental health professionals in all units specially focusing on their family members' mental health.
3. Initiatives should be taken to provide mental health support to those family members who are staying home and not with the soldier.
4. Availability of mental health professionals to remote and border areas to help soldiers with healthy coping.
5. Promoting digital modes of communication for family members and military personnel.
6. Helping single parents work on parenting skills and deal with parenting challenges in the absence of the deployed parent.
7. Focus should also be given to the identification and management of marital conflict, as this may be a significant cause of distress for military personnel and their families as well.

It is important to note that these family and parenting strategies do not constitute treatment for severe individual distress faced by military personnel and their families and therefore are not a substitute for individual therapy for severe neurotic illnesses such as PTSD, depression or substance abuse. Similarly, families who face more significant challenges like injured service members, or those who have lost a parent or a spouse in combat, may require more intensive individual or family intervention like trauma-focused cognitive behavioral therapy and family therapies. However, these strategies may be important adjuncts to other treatment approaches.

## Conclusion

As the military has the responsibility for the safety of the nation, it becomes essential to look after their emotional, psychological and social well-being, and their families as an essential part of the approach. A considerable body of evidence highlights the need to pay increased attention to military families affected by military life and related stressors, as psychological intervention strategies show promising results for enhancing the well-being of military children and families. Additionally, helping parents, family members or caregivers to work together as a team to apply effective coping strategies reduces stress at home and provides a supportive context for the transition between the battlefront and the home front.

## References

Adams, J. (1991). Family crisis intervention and psychosocial care for children and adolescents. In C. S. Austad & W. H. Berman (Eds.), *Psychotherapy in Managed Health Care: The Optimal Use of Time & Resources* (pp. 111–125). Washington, DC: American Psychological Association.

Asnani, V., Pandey, U. D., Chaudhary, P. N., Singhal, S. N. P., Tripathi, R. K., & Boro, S. R. (2001). Stress and job satisfaction among soldiers operating in counter-insurgency areas. DIPR note no. 562, 3–33.

Atkins, D. C., Dimidjian, S., & Christensen, A. (2003). Behavioral couple therapy: Past, present, and future. In T. L. Sexton, G. R. Weeks, & M. S. Robbins (Eds.), *Handbook of Family Therapy: The Science and Practice of Working with Families and Couples* (pp. 281–302). New York: Brunner-Routledge.

Bishop, S. R., Lau, M., Shapiro, S., Carlson, L., Anderson, N. D., Carmody, J., ... Devins, G. (2004). Mindfulness: A proposed operational definition. *Clinical Psychology: Science and Practice, 11*(3), 230–241.

Blaisure, K. R., Saathoff-Wells, T., Pereira, A., MacDermid Wadsworth, S., & Dombro, A. L. (2016). *Serving Military Families: Theories, Research, and Application.* Hove, UK: Routledge.

Bradbury, T. N., Fincham, F. D., & Beach, S. R. H. (2000). Research on the nature and determinants of marital satisfaction: A decade in review. *Journal of Marriage and Family, 62*(4), 964–980.

Brenes, G. A., Danhauer, S. C., Lyles, M. F., Anderson, A., & Miller, M. E. (2017). Long-term effects of telephone-delivered psychotherapy for late-life GAD. *American Journal of Geriatric Psychiatry, 25*(11), 1249–1257.

Brenes, G. A., Danhauer, S. C., Lyles, M. F., Hogan, P. E., & Miller, M. E. (2015). Telephone-delivered cognitive behavioral therapy and telephone-delivered nondirective supportive therapy for rural older adults with generalized anxiety disorder: A randomized clinical trial. *JAMA Psychiatry*, *72*(10), 1012–1020.

Chandra, A., Martin, L. T., Hawkins, S. A., & Richardson, A. (2010). The impact of parental deployment on child social and emotional functioning: Perspectives of school staff. *Journal of Adolescent Health*, *46*(3), 218–223.

Chen, J. A., Fortney, J. C., Bergman, H. E., Browne, K. C., Grubbs, K. M., Hudson, T. J., & Raue, P. J. (2019). Therapeutic alliance across trauma-focused and non-trauma-focused psychotherapies among veterans with PTSD. *Psychological Services*. Epub ahead of print. doi: 10.1037/ser0000329. Retrieved from https://www.ncbi.nlm.nih.gov/pubmed/30742471.

Chesmore, A. A., Piehler, T. F., & Gewirtz, A. H. (2018). PTSD as a moderator of a parenting intervention for military families. *Journal of Family Psychology*, *32*(1), 123–133.

Cooper, K., Chatters, R., Kaltenthaler, E., & Wong, R. (2015). Psychological and psychosocial interventions for cannabis cessation in adults: A systematic review short report. *Health Technology Assessment*, *19*(56), 1–130.

Creamer, M., Carboon, I., Forbes, A. B., McKenzie, D. P., McFarlane, A. C., Kelsall, H. L., & Sim, M. R. (2006). Psychiatric disorder and separation from military service: A 10-year retrospective study. *American Journal of Psychiatry*, *163*(4), 733–734.

Creech, S. K., & Misca, G. (2017). Parenting with PTSD: a review of research on the influence of PTSD on parent-child functioning in military and veteran families. *Frontiers in Psychology*, *8*, Article No.1101. doi: 10.3389/fpsyg.2017.01101.

Dayton, C. J., Walsh, T. B., Muzik, M., Erwin, M., & Rosenblum, K. L. (2014). Strong, safe, and secure: Negotiating early fathering and military service across the deployment cycle. *Infant Mental Health Journal*, *35*(5), 509–520.

DeGarmo, D. S., & Gewirtz, A. H. (2018). A recovery capital and stress-buffering model for post-deployed military parents. *Frontiers in Psychology*, *9*, Article No:1832. doi: 10.3389/fpsyg.2018.01832.

DeVoe, E. R., Paris, R., Emmert-Aronson, B., Ross, A., & Acker, M. (2017). A randomized clinical trial of a postdeployment parenting intervention for service members and their families with very young children. *Psychological Trauma: Theory, Research, Practice, and Policy*, *9*(S1), 25–34.

Dimiceli, E. E., Steinhardt, M. A., & Smith, S. E. (2010). Stressful experiences, coping strategies, and predictors of health-related outcomes among wives of deployed military servicemen. *Armed Forces & Society*, *36*(2), 351–373.

Drummet, A. R., Coleman, M., & Cable, S. (2003). Military families under stress: Implications for family life education. *Family Relations*, *52*(3), 279–287.

Durham, R. C., Chambers, J. A., Power, K. G., Sharp, D. M., Macdonald, R. R., Major, K. A., Dow, M. G., & Gumley, A. I. (2005). Long-term outcome of cognitive behaviour therapy clinical trials in central Scotland. *Health Technology Assessment*, *9*(42), 1–174.

Dutra, L., Grubbs, K., Greene, C., Trego, L. L., McCartin, T. L., Kloezeman, K., & Morland, L. (2011). Women at war: Implications for mental health. *Journal of Trauma & Dissociation*, *12*(1), 25–37.

Farrand, P., Jeffs, A., Bloomfield, T., Greenberg, N., Watkins, E., & Mullan, P. (2018). Low intensity CBT: Mental health service acceptability for the armed forces veteran community. *Occupational Medicine*, *68*, 391–398.

Fields, J. A., Nichols, L. O., Martindale-Adams, J., Zuber, J., & Graney, M. (2012). Anxiety, social support, and physical health in a sample of spouses of OEF/OIF service members. *Military Medicine*, *177*(12), 1492–1497.

Foa, E. B., McLean, C. P., Zang, Y., Rosenfield, D., Yadin, E., Yarvis, J. S., .... Peterson, A. L. (2018). Effect of prolonged exposure therapy delivered over 2 weeks vs 8 weeks vs present-centered therapy on PTSD symptom severity in military personnel: A randomized clinical trial. *JAMA*, *319*(4), 354–364.

Forgatch, M. S., & DeGarmo, D. S. (1999). Parenting through change: an effective prevention program for single mothers. *Journal of Consulting & Clinical Psychology*, *67*(5), 711–724.

Forgatch, M. S., Patterson, G. R., DeGarmo, D. S., & Beldavs, Z. G. (2009). Testing the Oregon delinquency model with 9-year follow-up of the Oregon Divorce Study. *Development and Psychopathology*, *21*, 637–660.

Forgatch, M. S., & Kjøbli, J. (2016). Parent Management Training-Oregon model: Adapting intervention with rigorous research. *Family Process*, *55*(3), 500–513.

Gates, M. A., Holowka, D. W., Vasterling, J. J., Keane, T. M., Marx, B. P., & Rosen, R. C. (2012). Posttraumatic stress disorder in veterans and military personnel: Epidemiology, screening, and case recognition. *Psychological Services*, *9*(4), 361–382.

Gewirtz, A., Forgatch, M., & Wieling, E. (2008). Parenting practices as potential mechanisms for child adjustment following mass trauma. *Journal of Marital and Family Therapy*, *34*(2), 177–192.

Gewirtz, A., & Taylor, T. (2009). Participation of homeless and abused women in a parent training program: Science and practice converge in a battered women's shelter. In M. F. Hindsworth, & T. B. Lang (Eds.), *Community Participation and Empowerment* (pp. 97–114). Hauppage, NY: Nova Science Publishers, Inc.

Gewirtz, A. H., Polusny, M.A., DeGarmo, D. S., Khaylis, A., & Erbes, C. R. (2010). Posttraumatic stress symptoms among national guard soldiers deployed to Iraq: Associations with parenting behavior and couple adjustment. *Journal of Consulting & Clinical Psychology*, *78*(5), 599–610.

Gewirtz, A. H., Erbes, C. R. Polusny, M. A., Forgatch, M. S., & DeGarmo, D. S. (2011). Helping military families through the deployment process: Strategies to support parenting. *Professional Psychology: Research and Practice*, *42*(1), 56–62.

Gewirtz, A. H., & Zamir, O. (2014). The impact of parental deployment to war on children: The crucial role of parenting. *Advances in Child Development and Behavior*, *46*, 89–112.

Gewirtz, A. H., McMorris, B. J., Hanson, S., & Davis, L. (2014). Family adjustment of deployed and non-deployed mothers in families with a parent deployed to Iraq or Afghanistan. *Professional Psychology: Research and Practice*, *45*(6), 465–477.

Gewirtz, A. H., Pinna, K. L., Hanson, S. K., & Brockberg, D. (2014). Promoting parenting to support reintegrating military families: after deployment, adaptive parenting tools. *Psychological Services*, *11*(1), 31–40.

Gewirtz, A. H., DeGarmo, D. S., & Zamir, O. (2018). After deployment, adaptive parenting tools: 1-year outcomes of an evidence-based parenting program for military families following deployment. *Prevention Science*, *19*(4), 589–599.

Hawkins, E. J., Grossbard, J., Benbow, J., Nacev, V., & Kivlahan, D. R. (2012). Evidence-based screening, diagnosis, and treatment of substance use disorders among veterans and military service personnel. *Military Medicine*, *177*, 8–29.

Hoge, C.W., Castro, C.A., Messer, S. C., McGurk, D., Cotting, D. I., & Koffman, R. L. (2004). Combat duty in Iraq and Afghanistan, mental health problems, and barriers to care. *New England Journal of Medicine*, *351*(1), 13–22.

Holder, N., Holliday, R., Williams, R., Mullen, K., & Surís, A. (2018). A preliminary examination of the role of psychotherapist fidelity on outcomes of cognitive processing therapy during an RCT for military sexual trauma-related PTSD. *Cognitive Behaviour Therapy*, *47*(1), 76–89.

Hundt, N. E., Barrera, T. L., Robinson, A., & Cully, J. A. (2014). A systematic review of cognitive behavioural therapy for depression in veterans. *Military Mediine*, *179*(9), 942–949.

Indian Military News. (2013). *Indian Military News. 368 defense personnel committed suicide in 2010–2012.* Retrieved from, http://wp.me/px9JX-31T.

Iversen, A., Nikolaou, V., Greenberg, N., Unwin, C., Hull, L., Hotopf, M.,... Wessely, S. (2005). What happens to British veterans when they leave the armed forces? *The European Journal of Public Health*, *15*(2), 175–184.

Iversen, A., & Greenberg, N. (2009). Mental health of regular and reserve military veterans. *Advances in Psychiatric Treatment*, *15*(2), 100–106.

Iversen, A. C., van Staden, L., Hughes, J. H., Greenberg, N., Hotopf, M., Rona, R. J.,... & Fear, N.T. (2011). The stigma of mental health problems and other barriers to care in the UK Armed Forces. *BMC Health Services Research*, *11*(1), 31.

Jacobson, N. S., & Margolin, G. (1979). *Marital Therapy: Strategies Based on Social Learning and Behavior Exchange Principles*. New York: Brunner/Mazel.

Johnson, S., & Lebow, J. (2000). The "coming of age" of couple therapy. *Journal of Marital and Family Therapy*, *26*(1), 23–28.

Kitchiner, N., Roberts, N., Wilcox, D., & Bisson, J. (2012). Systematic review and meta-analyses of psychosocial interventions for veterans of the military. *European Journal of Psychotraumatology*, *3*(1), 19267.

Lester, P., Aralis, H., Sinclair, M., Kiff, C., Lee, K. H., Mustillo, S., & Wadsworth, S. M. (2016). The impact of deployment on parental, family and child adjustment in military families. *Child Psychiatry & Human Development*, *47*(6), 938–949.

Martinez, C. R. J., & Forgatch, M. S. (2002). Adjusting to change: Linking family structure transitions with parenting and boys' adjustment. *Journal of Family Psychology*, *16*(2), 107–117.

Masten, S. A. (2001). Ordinary magic: Resilience processes in development. *American Psychologist*, *56*(3), 227–238.

McFarlane, A. C. (2009). Military deployment: The impact on children and family adjustment and the need for care. *Current Opinion in Psychiatry*, *22*(4), 369–373.

McKay, M., Fanning, P., & Paleg, K. (2006). *Couple Skills: Making Your Relationship Work* (2nd ed.). Oakland, CA: New Harbinger Publications.

Miller, S. (2006). Anger and military veterans. *Columbia Social Work Review, 4*, 7–16.

Minuchin, S. (1974). *Families and Family Therapy*. Cambridge, MA: Harvard University Press.

Mohan, S., & Arora, N. (2013). Interrelationship between stress due to inadequate salary and rewards and marital discord: A study of Indian Army officers. *International Journal of Advanced Research in Management and Social Sciences (IJARMSS), 2*(12).

NICE. (2011). Common mental health problems: identification and pathways to care. NICE Guidelines. p. CG123.

Nichols, L. O., Martindale-Adams, J., Graney, M. J., Zuber, J., Burns, R. (2013). Easing reintegration: telephone support groups for spouses of returning Iraq and Afghanistan service members. *Health Communication, 28*(8), 767–777.

O'Shea, L., Watkins, E., & Farrand, P. (2017). Psychological interventions for the treatment of depression, anxiety, alcohol misuse or anger in armed forces veterans and their families: Systematic review and meta-analysis protocol. *Systematic Reviews, 6*(1), 112.

Palmer, C. (2008). A theory of risk and resilience factors in military families. *Military Psychology, 20*(3), 205–217.

Paris, R., DeVoe, E. R., Ross, A. M., & Acker, M. L. (2010). When a parent goes to war: Effects of parental deployment on very young children and implications for intervention. *American Journal of Orthopsychiatry, 80*(4), 610–618.

Patterson, G. R. (2005). The next generation of PMTO models. *The Behavior Therapist, 28*, 27–33.

Patterson, G. R., DeGarmo, D., & Forgatch, M. S. (2004). Systematic changes in families following prevention trials. *Journal of Abnormal Child Psychology, 32*, 621–633.

Pincus, S. H., House, R., Christenson, J., & Adler, L. E. (2001). The emotional cycle of deployment: A military family perspective. *US Army Medical Department Journal, 4*(5), 1–6.

Powers, M. B., Halpern, J. M., Ferenschak, M. P., Gillihan, S. J., & Foa, E. B. (2010). A meta-analytic review of prolonged exposure for posttraumatic stress disorder. *Clinical Psychology Review, 30*(6), 635–641.

Prabhu, H. R. A., Prakash, J., Bhat, P. S., & Gambhir, J. (2011). Study of events in serving personnel and its association with psychopathology in their children: A multicentric study. *MJAFI, 67*, 225–229.

Reid, J. B., Patterson, G. R., & Snyder, J. (2002). *Antisocial Behavior in Children and Adolescents: A Developmental Analysis and Model for Intervention*. Washington, DC: American Psychological Association.

Ross, A. M., & DeVoe, E. R. (2014). Engaging military parents in a home-based reintegration program: A consideration of strategies. *Health & Social Work, 39*(1), 47–54.

Rossiter, A. G., D'Aoust, R., & Shafer, M. R. (2016). Military serving at what cost? The effects of parental service on the well-being our youngest military members. *Annual Review of Nursing Research, 34*, 109–117.

Sayers, S. L., Farrow, V. A., Ross, J., & Oslin, D. W. (2009). Family problems among recently returned military veterans referred for a mental health evaluation. *Journal of Clinical Psychiatry, 70*(2), 163–170.

Schachman, K. A. (2010). Online fathering: The experience of first-time fatherhood in combat-deployed troops. *Nursing Research, 59*(1), 11–17.

Sharma, S. (2015). Occupational stress in the armed forces: An Indian army perspective. *IIMB Management Review, 27*(3), 185–195.

Sharp, M. L., Fear, N. T., Rona, R. J., Wessely, S., Greenberg, N., Jones, N., & Goodwin, L. (2015). Stigma as a barrier to seeking health care among military personnel with mental health problems. *Epidemiologic Reviews, 37*(1), 144–162.

Sheppard, S. C., Malatras, J. W., & Israel, A. C. (2010). The impact of deployment on US military families. *American Psychologist, 65*(6), 599–609.

Sherman, M. D., & Sherman, D. M. (2009). *My Story: Blogs by Military Teens*. Edina: Beaver's Pond Press.

Spelman, J. F., Hunt, S. C., Seal, K. H., & Burgo-Black, A. L. (2012). Post deployment care for returning combat veterans. *Journal of General Internal Medicine, 27*(9), 1200–1209.

Steenkamp, M. M., Litz, B. T., Hoge, C. W., & Marmar, C. R. (2015). Psychotherapy for military-related PTSD: A review of randomized clinical trials. *JAMA, 314*(5), 489–500.

Street, A. E., Vogt, D., & Dutra, L. (2009). A new generation of women veterans: Stressors faced by women deployed to Iraq and Afghanistan. *Clinical Psychology Review, 29*(8), 685–694.

Stuart, R. B. (1980). *Helping Couples Change: A Social Learning Approach to Marital Therapy*. New York: Guilford Press.

Suman, M. (2009). Indian army and management of stress. *Indian Defense Review, 24*(2).

Verdeli, H., Baily, C., Vousoura, E., Belser, A., Singla, D., & Manos, G. (2011). The case for treating depression in military spouses. *Journal of Family Psychology, 25*(4), 488–496.

Vogt, D. (2011). Mental health-related beliefs as a barrier to service use for military personnel an veterans: A review. *Psychiatric Services*, *62*(2), 135–142.

Walker, D. D., Walton, T. O., Neighbors, C., Kaysen, D., Mbilinyi, L., Darnell, J.,… Roffman, R. A. (2017). Randomized trial of motivational interviewing plus feedback for soldiers with untreated alcohol abuse. *Journal of Consulting & Clinical Psychology*, *85*(2), 99–110.

Walsh, T. B., Dayton, C. J., Erwin, M. S., Muzik, M., Busuito, A., & Rosenblum, K. L. (2014). Fathering after military deployment: Parenting challenges and goals of fathers of young children. *Health & Social Work*, *39*(1), 35–44.

Whisman, M. A., & Uebelacker, L. A. (2006). Impairment and distress associated with relationship discord in a national sample of married or cohabiting adults. *Journal of Family Psychology*, *20*(3), 369–377.

Willerton, E., Schwarz, R. L., Wadsworth, S. M., & Oglesby, M. S. (2011). Military fathers' perspectives on involvement. *Journal of Family Psychology*, *25*(4), 521–530.

Zinzow, H. M., Britt, T. W., McFadden, A. C., Burnette, C. M., & Gillispie, S. (2012). Connecting active duty and returning veterans to mental health treatment: Interventions and treatment adaptations that may reduce barriers to care. *Clinical Psychology Review*, *32*, 741–753.

# 30

# CHRONIC DISEASE RISKS AND SERVICE-RELATED POST-TRAUMATIC STRESS DISORDER IN MILITARY VETERANS

*Jeanne Mager Stellman and Steven D. Stellman*

Post-traumatic stress disorder (PTSD) and traumatic brain injury (TBI) have come to be recognized as the hallmark signatures of the Iraq conflict (2003–2011) and of the ongoing war in Afghanistan (Institute of Medicine, 2012). By contrast, the diagnosis of post-traumatic stress disorder did not even formally enter the Diagnostic and Statistical Manual of Mental Disorders until 1980 (American Psychiatric Association, 1980), seven years after the stand-down of U.S. troops in Vietnam, despite decades—one might even say millennia—of evidence demonstrating that exposure to traumatic conditions on the battlefield leads to serious, persistent mental health problems for many war veterans.

In many ways, the relationship between the traumatic combat exposures, development of post-traumatic stress disorder and co-morbid physical health conditions is following a similar trajectory: there are dozens of peer-reviewed studies showing that co-morbidity exists, but *recognition* of such co-morbidity is missing from general public health discourse on veteran health and veteran benefit programs. Of particular importance for both treatment of those with disorders, and for gaining further understanding of the relationships, is that fact that co-morbidity of physical illness is generally absent from medical practice guidelines, especially in relevant medical specialties. As noted by Koenen et al. (2017), the American Diabetes Association does not cite stress as a risk factor, nor does the American Heart Association consider there to be an established causal relationship between disease development and stress (as of 2013 in Koenen et al.).

In this chapter we review the literature on the relationship between traumatic military exposures and post-traumatic stress disorder and on the development and the incidence of co-morbid physical health conditions, diseases and mortality. We discuss various models that have been proposed for the relationship between trauma, post-traumatic stress disorder and physical health comorbidity. We also discuss some of the methodological issues that arise when designing and executing studies to test the strength and direction of these relationships. We provide supportive longitudinal data for co-morbidity derived from our cohort study of 3,403 Vietnam veterans who were randomly selected from American Legion membership and surveyed in 1984 and 1998. Finally, we provide suggestions for future research and for improvements in health services and benefits available to veterans.

## Reviews

A number of reviews of post-traumatic stress disorder as a precursor or predictor of a variety of physical illnesses have appeared within the past dozen years (Table 30.1). Most recently, Ryder, Azcarate and Cohen (2018) have provided an extensive review. Most reviews focus on cardiovascular diseases, varying mainly in the extent of including other co-morbidities. Some of the reviews also provide a perspective on biological mechanisms and biomarkers current at the time of publication. Bedi and Arora (2007), for example, summarize a large number of studies of stressors and heart disease, emphasizing the role of the sympathetic nervous system, such as disturbances in the sympathoadrenal and the hypothalamic-pituitary-adrenal axis, neuroendocrine responses to chronic stressors and allostatic load (see below for further discussion of biological responses). However, they provide little detail on individual studies. By contrast, Qureshi, Pyne, Magruder, Schulz and Kunik (2009) summarize, in-depth, both positive and negative findings of seven epidemiological studies that investigated associations between post-traumatic stress disorder and diseases of the heart, endocrine system (diabetes and thyroid disease), skin, digestive system (ulcers), genitourinary system and musculoskeletal system (arthritis), as well as neurological diseases (epilepsy, stroke), female reproductive system and respiratory system (asthma) (Qureshi et al., 2009). Pacella, Hruska and Delahanty (2013) carried out a formal meta-analysis of over 60 studies involving both veteran and civilian populations. Their results demonstrated significant relationships between post-traumatic stress disorder and a set of disease endpoints that include physical cardio-respiratory and gastrointestinal outcomes, as well as musculoskeletal pain, general health symptoms, medical conditions and health-related quality of life (Pacella et al., 2013). Studies of the impact of post-traumatic stress disorder on cardiovascular disease have also been presented by Levine, Levine and Levine (2014), while Cohen, Edmondson and Kronish (2015) summarize some of the same studies, but include depression along with post-traumatic stress disorder. Table 30.1 also includes narrower reviews: one on post-traumatic stress disorder and reproductive outcomes (Krulewitch, 2016) and one on post-traumatic stress disorder and diabetes (Vancampfort et al., 2016).

Given the emphasis on the potential role of external stressors in the etiology of cardiovascular diseases and the proliferation of potential explanatory mechanisms, it is not surprising that the majority of studies of post-traumatic stress disorder comorbid with physical illnesses focus on cardiovascular outcomes. Table 30.2 summarizes 11 studies published between 1994 and 2013 that present evidence for comorbid associations of cardiovascular diseases with diabetes, digestive disorders, musculoskeletal and other problems.

While epidemiological literature on physical ailments comorbid with post-traumatic stress disorder is largely dominated by cardiovascular diseases, a number of studies have addressed other endpoints as well. Post-traumatic stress disorder as a predictor of deaths due to suicide and accidental poisoning has been reported in Veteran Affairs (VA) studies (Bullman & Kang, 1994). In a nationally representative sample of veterans drawn in the mid-1980s and followed through 2000, veterans with post-traumatic stress disorder were at increased risk of death from multiple causes including total cardiovascular disease, cancer and external causes (e.g., motor vehicle accidents, accidental poisoning, suicide and homicide) (Boscarino, 2006).

Studies of reproductive outcomes among female veterans have also shown post-traumatic stress disorder effects. Female veterans of the Iraq and Afghanistan conflicts had significantly greater risks for a number of outcomes, including sexually transmitted infections, urinary

*Table 30.1* Review articles published since 2007 summarizing findings on post-traumatic stress disorder as a predictor or risk factor for cardiovascular disease and other physical health outcomes

| *Publication year* | *Reference* | *Disease or diseases covered* |
|---|---|---|
| 2007 | Bedi & Arora (2007) | Cardiovascular |
| 2009 | Qureshi et al. (2009) | Cardiovascular, diabetes, cancer, neurological, digestive, musculoskeletal |
| 2013 | Pacella et al. (2013) | Cardiovascular, digestive |
| 2014 | Levine et al. (2014) | Cardiovascular, diabetes, neurological |
| 2015 | Cohen, Edmondson & Kronish (2015) | Cardiovascular |
| 2016 | Krulewitch (2016) | Reproductive |
| 2016 | Vancampfort et al. (2016) | Diabetes |
| 2017 | Edmondson & von Kanel (2017) | Cardiovascular |
| 2017 | Koenen et al. (2017) | Cardiovascular |
| 2018 | Ryder, Azcarate, & Cohen (2018) | Cardiovascular, diabetes, musculoskeletal |

*Table 30.2* Studies of post-traumatic stress disorder and conditions comorbid with cardiovascular disease

| *Study* | *Design* | *Other endpoints* |
|---|---|---|
| Kawachi, Sparrow, Vokonas, & Weiss (1994) | Cohort | Cardiovascular and "sudden" death |
| Boscarino (1997) | Cross-sectional | Diabetes, digestive, musculoskeletal, neurological |
| Schnurr, Spiro, & Paris (2000) | Cohort | Diabetes, digestive, musculoskeletal, dermatological |
| Boscarino (2006) | Cohort | Cancer, all-cause mortality, external cause mortality |
| Boscarino (2008) | Cohort | Depression, substance use |
| O'Toole & Catts (2008) | Cohort | Digestive, dermatological, musculoskeletal, respiratory |
| Andersen, Wade, Possemato, & Ouimette (2010) | Cohort | Diabetes, neurological, digestive, musculoskeletal |
| Cohen, Marmar, Ren, Bertenthal, & Seal (2009) | Cross-sectional | Hypertension, diabetes, obesity |
| Cohen et al. (2010) | Cross-sectional | Health care utilization |
| Ahmadi et al. (2011) | Cross-sectional | Hypertension, diabetes, cardiovascular biomarkers, mortality |
| Wachen et al. (2013) | Cross-sectional | Diabetes, digestive, musculoskeletal, neurological, respiratory |

tract infections and inflammatory diseases of the pelvic organs, candida vaginitis, genital pain, endometriosis and infertility (Cohen et al., 2012). In a study of more than 16,000 deliveries covered by the Veterans Health Administration between 2000 and 2012, births to mothers with post-traumatic stress disorder were significantly more likely to experience a preterm birth (Shaw et al., 2014).

Besides concerns about diagnosed illnesses, there have been a number of studies comparing reports of multiple symptoms in veterans with post-traumatic stress disorder to those without post-traumatic stress disorder. These findings often emerge in studies that focus on a diagnosable disease outcome but that also include general questions on an array of physical symptoms. Taken individually, these associations may be difficult to categorize, but systematic study of such reports could lead to identification of new co-morbidities or complications in veterans with post-traumatic stress disorder. In the Normative Aging Study of veterans, Schnurr and Spiro (1999) reported higher post-traumatic stress disorder scores in veterans with poorer SF-36 physical and mental health. In a study of over 2,300 Gulf War veterans, post-traumatic stress disorder symptomatology was associated with symptomatic health problems in both women and men, independent of combat exposure. Similar results were observed in a sample of 1,030 Iraq and Afghanistan veterans from Mid-Atlantic states (Schry et al., 2015; Wagner, Wolfe, Rotnitsky, Proctor, & Erickson, 2000). In the cohort of American Legionnaires, surveyed in 1984 and 1998, significantly greater number of reports of five non-specific physical symptoms (fainting spells, fatigue, aches, cold, skin problems) and poorer family functioning (marital and sexual satisfaction) in those with symptoms of post-traumatic stress disorder was observed (Koenen, Stellman, Sommer, & Stellman, 2008).

In a study of 1,022 DVA and community clinic cardiac patients, those with post-traumatic stress disorder had a greater burden of physical symptoms, diminished quality of life and greater physical limitations, irrespective of comorbid depression (Cohen, Marmar, Ren, Bertenthal, & Seal, 2009). Among National Guard and Reserve members recently returned from Iraq and Afghanistan, post-traumatic stress disorder was a strong predictor of SF-36 measures of physical and mental health aggregate scores (Asnaani, Reddy, & Shea, 2014). A cross-sectional study of more than 5,500 Vietnam-era veterans showed poorer health functioning in all domains of the Veterans RAND 36-item health survey and the WHO Disability Assessment Schedule 2.0 among those with post-traumatic stress disorder (Goldberg et al., 2014). In the National Health and Resilience in Veterans Study, post-traumatic stress disorder was found to be a predictor of sleep disorder, obstructive lung diseases (asthma, COPD), osteoporosis or osteopenia and migraine (El-Gabalawy, Blaney, Tsai, Sumner, & Pietrzak, 2018).

## *Vietnam veteran mortality studies*

The evolution of epidemiological thinking regarding the impact of PTSD on long-term health may be illustrated in mortality studies. The earliest mortality studies of Vietnam veterans focused on physical health outcomes and environmental etiologies, giving little attention to mental health conditions such as PTSD as possible predictors or modifiers of mortality risks. Identification of increased suicide risks led to awareness of possible links to mental health conditions, including PTSD, leading to incorporation of measures of PTSD, with the most recent studies providing strong evidence for higher death rates among veterans with PTSD.

Shortly after the War, states with Vietnam veteran bonus payment programs carried out epidemiological analyses by linking bonus payment files to state death records. In Massachusetts, standardized proportional mortality ratios (PMRs) and mortality odds ratios (MORs) were both significantly elevated for soft-tissue sarcoma (STS), compared to Vietnam era veterans (PMR = 880, MOR = 5.16) as well as non-veterans. STS is a cancer linked to exposure to the herbicide Agent Orange and its dioxin contaminant (Kogan & Clapp, 1985). A comparison of West Virginia veterans who had served in Vietnam to non-veterans showed Vietnam veterans with significantly high PMRs for larynx cancer (429) and STS (429) and with elevated Hodgkin disease (208, not significant). Veterans who had served in Vietnam also had significantly higher risks for total lymphoma as well as Hodgkin disease and testicular cancer, and higher STS (not significant) (Holmes, 1986). The New York State Department of Health compared MORs among veterans with Vietnam service who died in New York State to veterans of the Vietnam era with no Vietnam service (total n = 4,558). The causes with the most elevated MORs and their

95% confidence intervals were non-motor vehicular injuries of transport (MOR = 2.18, 95% CI 1.19, 3.96), other accidents and burns (MOR = 1.37, (0.95, 1.98) and homicide (MOR = 1.59, 95% CI 0.86, 2.94) (Lawrence et al., 1985). In 1988, researchers in the Department of Veterans Affairs examined proportionate mortality rates among Army and Marine veterans and observed that Marines had significantly high PMRs for lung cancer (158) and non-Hodgkin lymphoma (210) (Breslin, Kang, Lee, Burt, & Shepard, 1988).

Beginning in 1987, cohort studies of Vietnam veterans began to show unexpectedly high risks of suicide and drug-related causes, leading to subsequent consideration of behavioral and mental health factors as potential risk factors for premature death. In 1987 the Centers for Disease Control began the Veteran Experience Study (VES) with a nationwide cohort of over 18,000 Vietnam-era veterans and reported excess mortality from suicide, motor vehicle accidents and accidental poisonings during the first five years post-discharge among those who served in Vietnam, with drug-related deaths more common among those with Vietnam service throughout the follow-up period ending in 1983 (CDC, 1987).

In one of the earliest cohort studies in which PTSD was assessed in Vietnam veterans, Bullman and Kang (1994) reported that veterans with PTSD were four times more likely to die from suicide (RR = 4.0, 95% CI 2.2–7.0) and from accidental poisoning (RR = 2.9, 95% CI 1.1–2.5). In a subsequent analysis of 34,534 U.S. Army veterans followed through 1991 for an average of 21 years, the same group reported a relative risk for suicide of 1.8 (95% CI 1.1–3.0) among veterans who were wounded more than once and hospitalized for a wound. No direct measurement of PTSD, however, was available (Bullman & Kang, 1996). In a study of 1,866 men admitted to a VA PTSD residential treatment program between 1990 and 1998 and followed through 1999, the risk of death was nearly twice that of the general population with a similar age and race distribution (SMR = 1.8, 95% CI = 1.5–2.2) and even higher for "behavioral causes" such as accidents, chronic substance abuse and suicide/homicide/police intervention (SMR for suicide = 4.0, 95% CI = 1.8–7.4) (Drescher, Rosen, Burling, & Foy, 2003).

The young age of a large proportion of those who served in the military during the Vietnam Era (median birth year 1947) implies that even a very large cohort must be followed for a long time in order to detect statistically meaningful increases in chronic diseases that primarily affect older persons. By 2000, the mean cohort age of the VES cohort was only 53 years. Nevertheless, follow-up through that year revealed all-cause mortality to be 7% higher in Vietnam compared to non-Vietnam veterans (95% CI = 0.97–1.18), with significantly increased risks for death from unintentional poisoning and drug-related causes (Boehmer, Flanders, McGeehin, Boyle, & Barrett, 2004). A subsequent analysis of the same cohort by Boscarino (2006), utilized data on PTSD for 7,924 Vietnam Theater and 7,364 Vietnam-era veterans that had been assessed among VES participants in telephone interviews in the mid-1980s. Significantly elevated hazard ratios were observed for total mortality (HR = 2.2, 95% CI = 1.7–2.7), cardiovascular mortality (HR = 1.7, 95% CI = 1.0–2.7), cancer mortality (HR = 1.9, 95% CI = 1.1–3.3) and external causes (HR = 2.3, 95% CI = 1.4–3.9). The latter category included motor vehicle accidents, accidental poisonings, suicides and homicides.

PTSD was determined in 1,946 participants in the VA Normative Aging Study cohort who were followed for up to 15 years through May 2001, using either the Mississippi Scale for Combat-Related PTSD (Keane, Caddell, & Taylor, 1988) or the related Keane scale (Litz et al., 1991). Even though few individuals met established criteria for PTSD using either scale, nevertheless age-adjusted relative risks for fatal CHD associated with a one-SD increase in symptom level were 1.30 (27 deaths, 95% CI 1.01–1.7) using the Mississippi scale and 1.13 (24 deaths, 95% CI 0.97–1.43) with the Keane scale (Kubzansky, Koenen, Spiro, Vokonas, & Sparrow, 2007).

All-cause (but not cause-specific) mortality was determined in the VA Large Health Survey of over 550,000 veterans surveyed by questionnaire in 1999 and followed for nine years. Prior diagnoses of PTSD and other mental health conditions were determined via linkage to diagnostic data in VA records. The adjusted hazard ratio for total mortality comparing those with PTSD to those without the condition was 1.02 (95% CI 1.01–1.04). However, it became non-significant after adjustment for psychiatric comorbidity such as schizophrenia, bipolar disorder and major depressive disorder (HR = 0.95, 95% CI 0.83–1.00) or for an array of medical comorbidities and adverse lifestyle behaviors such as obesity and smoking (HR = 0.91, 95% CI 0.86–0.98), which the authors interpreted as suggestive of mediation by the latter factors (Chwastiak, Rosenheck, Desai, & Kazis, 2010).

A mortality analysis was carried out in a cohort of 637 veterans not previously diagnosed with coronary artery disease who had undergone coronary artery calcium (CAC) screening for "clinical indications" in VA facilities. CAC measures plaque deposits and is associated with future risk of myocardial infarction and stroke; it is generally reported as a numerical score, with scores of 101 to 400 indicating moderate risk and over 400 high risk of a coronary event. The mean CAC score for veterans with PTSD was 448 (±472 SD) compared to 332 (±336) in those without PTSD ($p < 0.001$). The prevalence of PTSD increased in a dose-dependent manner with increasing CAC score. The mean follow-up period was 42 months, during which the proportion of veterans with PTSD who died was 17.1% compared to 10.4% of those without PTSD ($p < 0.003$). Kaplan-Meier curves showed substantially worsening survival with increasing CAC score (Ahmadi et al., 2011).

Most recently, Schlenger and colleagues (2015) analyzed the mortality experience of the National Vietnam Veteran Readjustment Study based on follow-up of 1,632 Theater veterans interviewed in 1987 and followed for up to 24 years. "High probability of PTSD" was assigned using the DSM III-R criteria in effect at the time of interview. The age-adjusted hazard ratio for association of PTSD with all-cause mortality was 2.27 (95% CI 1.34–3.84); additional adjustment for other demographic characteristics reduced it to 1.87 (95% CI 1.06–3.30). (Schlenger et al., 2015).

## *Biological basis for post-traumatic stress disorder—physical health risk relationships*

The term "fight-or-flight" to describe the "normal" fear response of human beings is now a part of the everyday vocabulary. In 1914, Walter Cannon first wrote of the phenomena associated with a fear or startle response based on observations carried out in his Harvard physiology laboratory. He described the elevated levels of adrenalin (corticosteroids) liberated by fear, rage, asphyxia and pain. He observed that the stress response also led to elevated levels of blood sugars, inhibition of gastric motions and increased secretion of gastric fluids, as well as to contraction of many blood vessels involved in gastrointestinal function. He wrote of more obvious physiological responses, like bristling of hair and, in general, put together various pathways in which the body marshaled its resources for responding to (fighting or fleeing from) a threat by increasing some functions and repressing others (e.g., increased muscle activation and decreased digestion), all in response to a purely emotional input (Cannon, 1914).

Much work on the physiological response to stress has, of course, been carried out in the past century since Cannon's pioneering paper. In broad strokes, the physiological responses described by Cannon are almost certainly intimately related to the elevated risks observed for many co-morbid physical health effects. Yehuda's laboratory has been very active in this area of study, and she has published extensively in the field. Her 2002 and 2009 reviews are useful here for a brief overview. Yehuda (2002; 2009) summarizes and updates the original Cannon formulation of the "normal fear response" biological reactions. At the heart of the stress response is the activation of the hypothalamic-pituitary-adrenal (HPA) axis. Activation of the HPA leads to elevation of both cortisol and corticotropin-releasing factor. These two components of the stress response make up a sensitive negative-feedback system where elevation of cortisol induces inhibitory responses in other parts of the endocrine system (the pituitary, the hypothalamus, the hippocampus and the amygdala). The inhibition of these systems, in turn, feeds into a negative feedback loop in which glucocorticoid receptors work to restore corticosteroid levels to the basal hormone levels at which they had been functioning. In a "normal" individual, the levels of both cortisol and corticotropin-releasing factor will return to normal basal levels after the stress has abated. People with post-traumatic stress disorders, however, often continue to exhibit low levels of cortisol and high levels of corticotropin-releasing factor, a combination that is also found in patients with major depression.

Other physiological and biochemical responses also occur. For example, the stress-response appears to also induce an arousal of the sympathetic nervous system, resulting in elevated levels of catecholamines. Blood sugar chemistry is altered, and here, too, complex feedback loops that have been related to increased risk for cardiovascular diseases are present. Immunological alterations have also been noted. A variety of chronic diseases are associated with alterations of normal functioning (homeostasis) in these basic functions.

## Conceptually framing the issue

The proposed pathways by which post-traumatic stress disorder can be associated with physical health are complex. Schnurr and Green (2004) have proposed a conceptual framework, shown in Figure 30.1, for "understanding relationships among trauma, posttraumatic stress disorder, and health outcomes" (Schnurr, Wachen, Green, & Kaltman, 2014). Among the many reasons underlying this complexity are:

- Individuals have personal, social and cultural factors that they bring to the traumatic situation. Such characteristics are likely to modify the nature and degree of post-traumatic stress disorder risk and of co-morbidity. Personal factors may include genetic differences that can affect both increased susceptibility to adverse effects of trauma and resilience in dealing with the trauma and its potential sequelae.
- Trauma exposure can, in and of itself, directly and separately lead to injury and other health effects, without the presence of post-traumatic stress disorder.
- Trauma can act indirectly to increase risk for co-morbidity by increasing risk for post-traumatic stress disorder which, in turn, leads to increased risk for co-morbid physical diseases. There can be a variety of pathways by this can occur, such as:
  - Biological alterations in immune function
  - Activation of various hormonal pathways associated with the stress response
  - Psychological disturbance, which, in turn, may lead to an increase in behavioral risk factors for developing disease, like smoking, drinking and poor diet
  - Continuing impediment to development of more healthful behaviors
  - Interference with normal family functioning and coping
  - Decreased likelihood of seeking and utilizing available health services, which, in turn, can lead to avoidably poorer health status and to failure to avail themselves of salutary treatment options

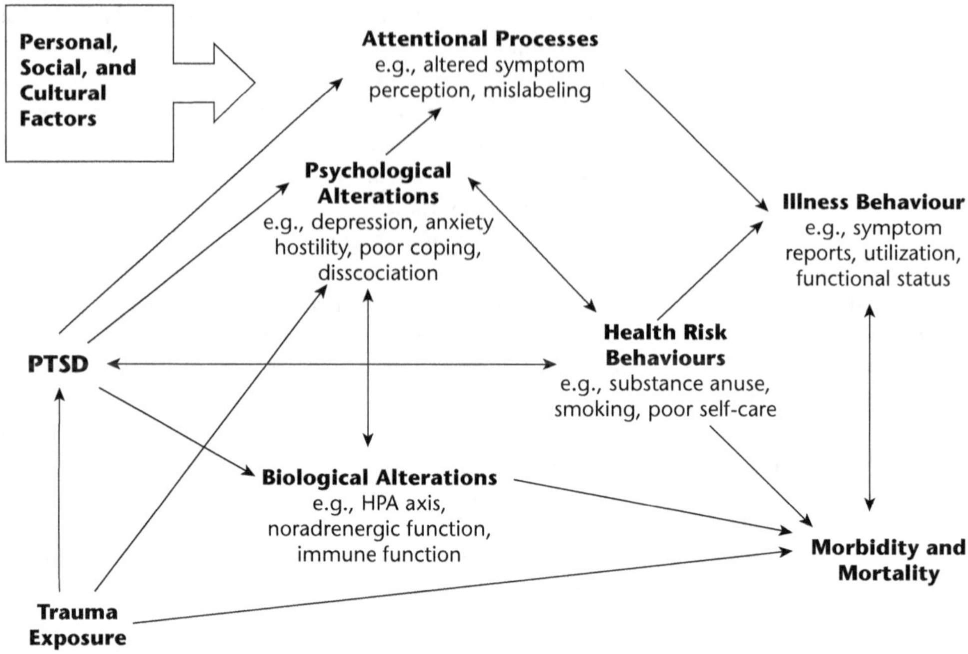


*Figure 30.1* A model relating traumatic exposure and post-traumatic stress disorder to physical health outcomes.

*Source:* From Schnurr and Green, 2004. (In the public domain.)

## *Methodological considerations in studies of post-traumatic stress disorder and physical health*

Koenen and co-workers (2017) reviewed data on the relationship between post-traumatic stress disorder and cardiometabolic disease (CMD) and presented a conceptual framework consistent with that of Schnurr and Green (2004). They also provided a comprehensive discussion of the difficulties of establishing causation, particularly since the directionality of many effects is not clear, noting, in fact, that bidirectionality can also exist. They posit their discussion on the relationship between cardiometabolic disease and post-traumatic stress disorder co-morbidity, but in fact these methodological considerations apply to virtually all chronic diseases. Adverse physical health outcomes can be modeled in at least five ways:

1. *Causation*—post-traumatic stress disorder directly causes a physical health condition through biological mechanisms associated with the stress response and its alteration of the HPA axis or stimulation of catecholamines, immune system changes and other factors.
2. *Confounding*—a traumatic risk factor explains the observed relationship between co-morbid chronic disease and post-traumatic stress disorder. They may both be independently associated with the traumatic exposure or some other factor, like socio-economic status, may be independently related to both post-traumatic health disorder and the physical health outcome in question.
3. *Reverse causation*—in which the physical health condition can increase risk for PTSD. For example, Koenen et al. (2017) note that cardiometabolic disease can increase levels of C-reactive protein, which can increase risk for post-traumatic stress disorder.

4. *Mediation*—in which post-traumatic stress disorder produces consequences that, in turn, lead to elevated risk for disease. For example, post-traumatic stress disorder is associated with risky health behaviors like smoking, which in turn lead to a variety of chronic diseases. When a condition acts as a mediator, special statistical techniques should be employed to determine whether PTSD itself makes a unique contribution to the elevated risk observed, beyond acting as a mediator.
5. *Bidirectional effects*—refers to a feedback loop model in which development of post-traumatic stress disorder increases risk for physical disease which, in turn, exacerbates reactions associated with trauma (e.g., trauma from experiencing a stroke) and hence symptoms of post-traumatic stress disorder are induced.

The sequence of events is often difficult to assess. For example, Kessler, Sonnega, Bromet, Hughes and Nelson, 1995, in their analyses of the National Comorbidity Survey, found that when co-morbid conditions arise, they are usually secondary to the onset of post-traumatic stress disorder. They note that the analytic approach to assessment of the impact of trauma-producing PTSD on subsequent physical conditions depends on how PTSD is to be treated in the model, that is, whether it is modeled as an independent causative factor, a mediator or as working through some other complex pathway. These more complex pathways will require statistical techniques beyond ordinary regression.

## Post-traumatic stress disorder as a multidimensional disorder

Another methodological issue to be considered is the multidimensionality of post-traumatic stress disorder. The symptoms of post-traumatic stress disorder generally fall into one of three clusters: re-experiencing the traumatic event, avoiding situations which are reminders of the event and hyperarousal. Re-experiencing of the event can manifest itself by nightmares and flashbacks. Avoidance of people, places or things reminiscent of the traumatic event can lead to social withdrawal in general and, of relevance to this discussion, to problems in interactions with the health care system. Hyperarousal is related to physiological responses that can include irritability, poor concentration and sleep disturbance. There may be an extreme startle response and hypervigilance. Recent exploration of the dimensionality of PTSD has led to proposals to define additional dimensions, in particular dysphoria (Frankfurt, Anders, James, Engdahl, & Winskowski, 2015), which would add further complexity.

The different dimensions of post-traumatic stress disorder are likely to be associated with the separate paths in the Schnurr and Green model (Schnurr & Green, 2004). Koenen et al. (2017) posit that it is likely to be useful to use dimensional components of post-traumatic stress disorder to analyze biological and behavioral effects separately.

Withdrawal from social interactions that are pathognomonic for post-traumatic stress disorder may affect utilization of health care services. For example, survivors of the World Trade Center disaster with post-traumatic stress disorder were significantly more likely to have reported unmet mental health care needs 10 to 11 years post-disaster compared to those without post-traumatic stress disorder, especially persons with severe conditions characterized by functional impairment and depression, and mental health service non-users displayed strong attitudinal barriers to seeking treatment (Ghuman, Brackbill, Stellman, Farfel, & Cone, 2014).

Much of the earlier literature on post-traumatic stress disorder in both military and civilian populations has been based on cross-sectional analysis of survey data or on baseline data from newly established cohort studies. As these cohorts have been followed over time, it has become possible to study the evolution or "trajectories" of post-traumatic stress disorder and other chronic health outcomes, as delineated by Koenen et al. (2017). For example, in their study of Vietnam-era veteran twins followed-up for up to 20 years, Magruder and colleagues classified 4,138 study participants in five mutually exclusive categories: never had post-traumatic stress disorder (86.0%), late onset (4.7%), late recovery (2.9%), early recovery (4.1%) and chronic (2.2%). Their study, however, had a 60% response rate at 20 years, and their estimates are accordingly weighted for non-response (Magruder et al., 2016).

Such so-called "distancing" raises possible concerns regarding bias due to selective non-response by participants in cohort studies of time course of post-traumatic stress disorder and its impact on chronic disease outcomes. Thus, in addition to differential utilization of healthcare services, there is likelihood that there will be a post-traumatic stress disorder-related response bias for participation in survey and medical examination-based research.

In the American Legion–Columbia University cohort ofVietnam veterans, a disproportionate number of the most chronically ill veterans were lost to follow-up in 1998. The prevalence of severe symptomatology in the non-responders at Time 2 was 11.1% at Time 1, compared to 9.5% at Time among responders to both waves. The mean PTSD scores in 1984 were significantly higher than those who responded at both times (Koenen et al., 2008). The extent and impact of this type of bias has also been estimated by Yu, Brackbill, Stellman, Ghuman and Farfel (2015), who compared PCL-derived post-traumatic stress disorder prevalence among World Trade Center survivors participating in up to three surveys over an 11-year period. Of the 67,670 Wave 1 (baseline) participants eligible for the two later follow-ups, 69% responded at Wave 2 and 63% responded at Wave 3, but the latter group included 6,682 "drop-ins" who were Wave 2 non-respondents. The baseline prevalence of post-traumatic stress disorder (defined as a PCL score of at least 44) was significantly greater among three-wave non-participants (17.9% vs. 15.2%). Both Wave 3 dropouts and drop-ins were significantly more likely to report poor/fair health as well as lower respiratory symptomatology at earlier waves. Nevertheless, there were no significant differences in odds ratios for association between specific traumatic exposures on 9/11 and post-traumatic stress disorder at Wave 3. The authors concluded that "despite a somewhat downward bias in prevalence estimates, attrition from the WTC Health Registry follow-up studies does not lead to serious bias in associations between 9/11 disaster exposures and key health outcomes."

The time course of post-traumatic stress disorder will also vary with the individual. Variations in time course are also likely to have an effect on risk of development of co-morbid physical health outcomes. Cross-sectional studies cannot satisfactorily test such effects. Longitudinal data from the Columbia University-American Legion Vietnam Veteran Health Study carried out on a cohort study of 3,403 Vietnam veterans can shed some light on the course of post-traumatic stress disorder and comorbidity. The cohort, first surveyed in 1984 (Snow, Stellman, Stellman, & Sommer, 1988) and then again in 1998 (Koenen et al., 2008), were compared as to risks for self-reported medically diagnosed heart disease, hypertension and other chronic illnesses among veterans who served in Vietnam with those who served elsewhere and by intensity of post-traumatic stress disorder among those serving in Vietnam, adjusting for potential confounders such as cigarette smoking and obesity. Post-traumatic stress disorder was significantly associated with diagnoses of heart disease and hypertension during the 15-year observation period, as well as a wide range of measures of health, well-being and physical function.

Data from this cohort also provide insight into the effect of time-course of post-traumatic stress disorder on comorbidity. Figure 30.2 illustrates the levels of self-reported symptoms of depression as a function of presence or absence of post-traumatic stress disorder in 1984 and 1998. Respondents who never reported post-traumatic stress disorder consistently report significantly lower levels of physical signs of depression than those with post-traumatic stress disorder at either time one or time two (intermittent/recurrent PTSD) versus those with post-traumatic stress

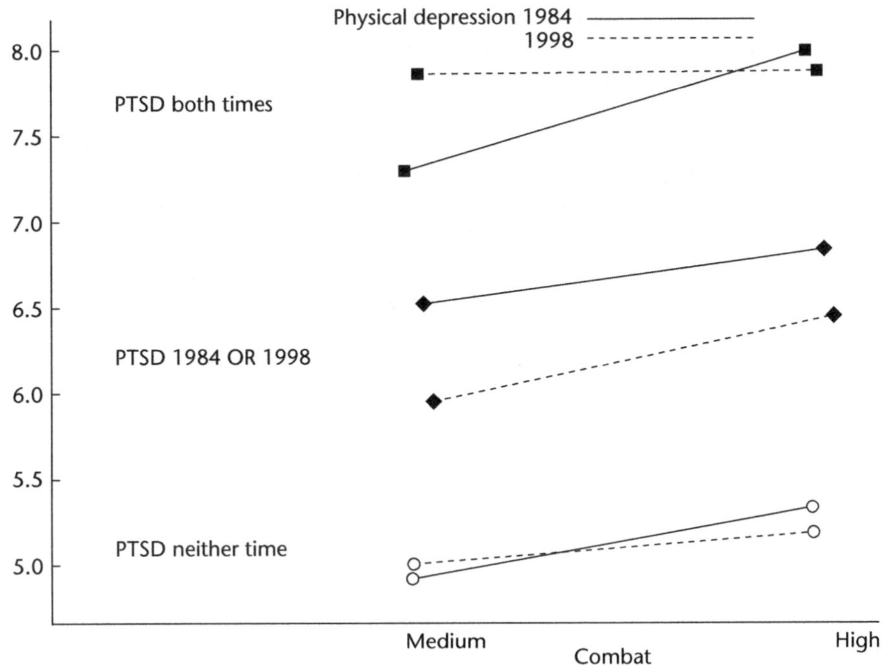


*Figure 30.2* The relationship between mean levels of physical signs of depression and level of military combat.

*Source*: Reported by respondents to both waves of the American Legion – Columbia University Vietnam Veterans Health Study (see Koenen et al., 2008 for a description of the cohort) who also reported medium and high levels of combat exposure. A similar pattern was observed for each of the health and well-being outcomes in the study.

disorder at both time periods, who scored significantly higher levels of distress than the other two groups. These data are consistent with the models described here and support the need for approaches to analyses that take both time and multidimensional nature of post-traumatic stress disorder into account. Similar patterns were observed for the other outcomes reported. In addition, substantial evidence has now been gathered demonstrating that people with levels of post-traumatic stress disorder symptoms who do not meet criteria for full-blown PTSD can still exhibit greater risk for co-morbid disease, in addition to showing elevated risk for functional impairment, suicidality and symptom severity.

Friedman, Keane and Resick (2014) have proposed that diagnoses of subsyndromal PTSD be added to the clinically accepted diagnostic criteria for the syndrome. Such a definition would assist in further studies since the revised definition of PTSD would lead to more consistent measurement in clinical and epidemiological settings than has been the case in the past. Friedman, Keane and Resick (2014) further posit that post-traumatic stress disorder may be a spectrum disorder in which "symptoms are distributed along a mild-to-severe continuum." They acknowledge that there is arbitrariness in the current definition of when PTSD is subsyndromal/partial and when it is full PTSD.

## Directions for future research and improvements in public health practice

Despite decades of studies demonstrating a relationship between stressful exposures, both traumatic and non-traumatic, on physical health, there has been remarkably little systematic inclusion of diagnosis and treatment of co-morbid conditions in either in the mental health arena or in the relevant physical health specialties. In the area of occupational health and stressful working conditions, the first government publication recognizing that physical sequelae could arise from stress conditions associated with the organization of work was published in 2002 by the NORA Organization of Work Team of the U.S. National Institute for Occupational Safety and Health (NORA, 2002).

The reluctance and professional inertia in addressing this issue and in pursuing a research agenda to further elucidate it is partially attributable to the methodological difficulties described earlier. There is also, undoubtedly, a political aspect. The costs and liabilities associated with recognizing that omnipresent stress conditions could be associated with physical disease could be astronomical. In the case of post-traumatic stress disorder, it has been suggested that advocacy by veteran groups and feminist activists played an important role in the formal recognition of the disorder in the DSM-III in 1980 (Friedman, Keane, & Resick, 2014). Recognition that PTSD could also be leading to chronic physical health diseases has not yet made its way into veteran benefit programs.

An ironic twist to the lack of recognition in many professional and governmental programs policies can be found in some governmental worker compensation programs. For example, in the State of Maryland, while PTSD is compensable under certain complicated conditions, it is *only* compensable if the claimant can establish coincident physical illness purported to be associated with the PTSD (personal communication, LR Stellman, JD).

Koenen et al. (2017) note that "despite mounting evidence post-traumatic stress disorder is not acknowledged as a risk factor for either cardiovascular disease (CVD) or type-2 diabetes (T2D)" and hence "neither systematic surveillance nor treatment is provided to individuals with post-traumatic stress disorder." The same general statements can be applied to virtually every other physical health problem described here.

While post-traumatic stress disorder is a major medical care and research focus in the Veterans Administration health care system, there are currently no established mechanisms by which a veteran with a co-morbid physically condition likely either to have been caused or exacerbated by post-traumatic stress disorder can automatically have his or her co-morbid conditions "recognized" by the Veterans Administration benefits system and hence be automatically eligible for medical and disability benefits.

The Agent Orange medical and benefit programs, available to all veterans who served in Vietnam and develop a disease recognized to be associated with exposure to Agent Orange and other military herbicides, is a model for a program that could be made available to veterans who have developed post-traumatic stress disorder associated with military service *and* a recognized comorbid physical health condition. (https://www.va.gov/disability/eligibility/hazardous-materials-exposure/agent-orange/, last accessed May 30, 2019). As with Agent Orange, a registry could be established for the systematic collection of traumatic military experiences, and, going beyond the Agent Orange Registry, there could be linkage to military records so that other at-risk individuals serving in similarly traumatized units could be identified, registered and followed. This ideal-world scenario is, of course, costly but is also long overdue.

In addition to developing information on the at-risk population and keeping track of those who enter the Veterans Administration system with symptoms of both post-traumatic stress exposure and chronic physical health problems, there could also be pro-active training of medical care providers to recognize co-morbid disease. Practice guidelines could be developed to assist in this process and relevant information entered in uniform fashion into the electronic medical record, thereby facilitating follow-up studies. Clearly there is an urgent need to establish surveillance systems, to carry out further research on risks and appropriate treatment that takes into account the complex causal pathways involved in these co-morbid processes. There is also a need to develop educational programs for medical practitioners and for veterans and their families so that treatment and medical adherence practices can be improved. Friedman, Keane and Resick (2014) outline innovative approaches to "top-down and bottom-up" mechanisms for both improving treatment and facilitating creative implementation strategies.

## References

Ahmadi, N., Hajsadeghi, F., Mirshkarlo, H. B., Budoff, M., Yehuda, R., & Ebrahimi, R. (2011). Post-traumatic stress disorder, coronary atherosclerosis, and mortality. *American Journal of Cardiolology, 108*(1), 29–33.

American Psychiatric Association. (1980). *Diagnostic and Statistical Manual of Mental Disorders* (3rd ed.). Washington, DC: American Psychiatric Association.

Andersen, J., Wade, M., Possemato, K., & Ouimette, P. (2010). Association between posttraumatic stress disorder and primary care provider-diagnosed disease among Iraq and Afghanistan veterans. *Psychosomatic Medicine, 72*(5), 498–504.

Asnaani, A., Reddy, M. K., & Shea, M. T. (2014). The impact of PTSD symptoms on physical and mental health functioning in returning veterans. *Journal of Anxiety Disorders, 28*(3), 310–317.

Bedi, U. S., & Arora, R. (2007). Cardiovascular manifestations of posttraumatic stress disorder. *Journal of the National Medical Association, 99*(6), 642–649.

Boehmer, T. K., Flanders, W. D., McGeehin, M. A., Boyle, C., & Barrett, D. H. (2004). Postservice mortality in Vietnam veterans: 30-year follow-up. *Archives of Internal Medicine, 164*(17), 1908–1916.

Boscarino, J. A. (1997). Diseases among men 20 years after exposure to severe stress: Implications for clinical research and medical care. *Psychosomatic Medicine, 59*(6), 605–614.

Boscarino, J. A. (2006). Posttraumatic stress disorder and mortality among U.S. Army veterans 30 years after military service. *Annals of Epidemiology, 16*(4), 248–256.

Boscarino, J. A. (2008). Psychobiologic predictors of disease mortality after psychological trauma: Implications for research and clinical surveillance. *Journal of Nervous & Mental Disease, 196*(2), 100–107.

Breslin, P., Kang, H. K., Lee, Y., Burt, V., & Shepard, B. M. (1988). Proportionate mortality study of US Army and US Marine Corps veterans of the Vietnam War. *Journal of Occupational Medicine, 30*(5), 412–419.

Bullman, T. A., & Kang, H. K. (1994). Posttraumatic stress disorder and the risk of traumatic deaths among Vietnam veterans. *Journal of Nervous & Mental Disease, 182*(11), 604–610.

Bullman, T. A., & Kang, H. K. (1996). The risk of suicide among wounded Vietnam veterans. *American Journal of Public Health, 86*(5), 662–667.

Cannon, W. B. (1914). The emergency function of the adrenal medulla in pain and the major emotions. *American Journal of Physiology-Legacy Content, 33*(2), 356–372.

CDC. (1987). Postservice mortality among Vietnam veterans. The Centers for Disease Control Vietnam Experience Study. *JAMA, 257*(6), 790–795.

Chwastiak, L. A., Rosenheck, R. A., Desai, R., & Kazis, L. E. (2010). Association of psychiatric illness and all-cause mortality in the National Department of Veterans Affairs Health Care System. *Psychosomatic Medicine, 72*(8), 817–822.

Cohen, B. E., Edmondson, D., & Kronish, I. M. (2015). State of the art review: Depression, stress, anxiety, and cardiovascular disease. *American Journal of Hypertension, 28*(11), 1295–1302.

Cohen, B. E., Gima, K., Bertenthal, D., Kim, S., Marmar, C. R., & Seal, K. H. (2010). Mental health diagnoses and utilization of VA non-mental health medical services among returning Iraq and Afghanistan veterans. *Journal of General Internal Medicine, 25*(1), 18–24.

Cohen, B. E., Maguen, S., Bertenthal, D., Shi, Y., Jacoby, V., & Seal, K. H. (2012). Reproductive and other health outcomes in Iraq and Afghanistan women veterans using VA health care: Association with mental health diagnoses. *Womens Health Issues, 22*(5), e461–e471.

Cohen, B. E., Marmar, C., Ren, L., Bertenthal, D., & Seal, K. H. (2009). Association of cardiovascular risk factors with mental health diagnoses in Iraq and Afghanistan war veterans using VA health care. *JAMA, 302*(5), 489–492.

Drescher, K. D., Rosen, C. S., Burling, T. A., & Foy, D. W. (2003). Causes of death among male veterans who received residential treatment for PTSD. *Journal of Traumatic Stress: Official Publication of the International Society for Traumatic Stress Studies, 16*(6), 535–543.

Edmondson, D., & von Kanel, R. (2017). Post-traumatic stress disorder and cardiovascular disease. *Lancet Psychiatry, 4*(4), 320–329.

El-Gabalawy, R., Blaney, C., Tsai, J., Sumner, J. A., & Pietrzak, R. H. (2018). Physical health conditions associated with full and subthreshold PTSD in U.S. military veterans: Results from the National Health and Resilience in Veterans Study. *Journal of Affective Disorders*, *227*, 849–853.

Frankfurt, S., Anders, S. L., James, L. M., Engdahl, B., & Winskowski, A. M. (2015). Evaluating the dimensionality of PTSD in a sample of OIF/OEF veterans. *Psychological Trauma: Theory, Research, Practice, and Policy*, 7(5), 430–436.

Friedman, Matthew, J., Keane, T. M., & Resick, P. A. (2014). *Handbook of PTSD, Second Edition: Science and Practice*. Retrieved from http://ebookcentral.proquest.com/lib/columbia/detail.action?docID=1691133.

Ghuman, S. J., Brackbill, R. M., Stellman, S. D., Farfel, M. R., & Cone, J. E. (2014). Unmet mental health care need 10–11 years after the 9/11 terrorist attacks: 2011–2012 results from the World Trade Center Health Registry. *BMC Public Health*, *14*(1), 491.

Goldberg, J., Magruder, K. M., Forsberg, C. W., Kazis, L. E., Ustun, T. B., Friedman, M. J., … Smith, N. L. (2014). The association of PTSD with physical and mental health functioning and disability (VA Cooperative Study #569: The course and consequences of posttraumatic stress disorder in Vietnam-era veteran twins). *Quality of Life Research*, *23*(5), 1579–1591.

Holmes, A. P. (1986). *West Virginia Vietnam-Era Veterans Mortality Study* (p. 30). Charleston: West Virginia Department of Health Vietnam-Era Veterans Mortality Study Committee.

Institute of Medicine. (2012). *Treatment for Posttraumatic Stress Disorder in Military and Veteran Populations: Initial Assessment*. Washington, DC: The National Academies Press.

Kawachi, I., Sparrow, D., Vokonas, P. S., & Weiss, S. T. (1994). Symptoms of anxiety and risk of coronary heart disease. The Normative Aging Study. *Circulation*, *90*(5), 2225–2229.

Keane, T. M., Caddell, J. M., & Taylor, K. L. (1988). Mississippi Scale for Combat-Related Posttraumatic Stress Disorder: Three studies in reliability and validity. *Journal of Consulting and Clinical Psychology*, *56*(1), 85–90.

Kessler, R. C., Sonnega, A., Bromet, E., Hughes, M., & Nelson, C. B. (1995). Posttraumatic stress disorder in the National Comorbidity Survey. *Archives of General Psychiatry*, *52*(12), 1048–1060.

Koenen, K. C., Stellman, S. D., Sommer, J. F., & Stellman, J. M. (2008). Persisting posttraumatic stress disorder symptoms and their relationship to functioning in Vietnam veterans: A 14-year follow-up. *Journal of Traumatic Stress: Official Publication of The International Society for Traumatic Stress Studies*, *21*(1), 49–57.

Koenen, K. C., Sumner, J. A., Gilsanz, P., Glymour, M. M., Ratanatharathorn, A., Rimm, E. B., … Kubzansky, L. D. (2017). Post-traumatic stress disorder and cardiometabolic disease: Improving causal inference to inform practice. *Psychological Medicine*, *47*(2), 209–225.

Kogan, M. D., & Clapp, R. W. (1985). Mortality among Vietnam veterans in Massachusetts, 1972–1983 (p. 29). *Massachusetts Office of Commissioner of Veterans Services Agent Orange Program Massachusetts Department of Public Health, Division of Health Statistics and Research*.

Krulewitch, C. J. (2016). Reproductive health of active duty women in medically austere environments. *Military Medicine*, *181*(1 Suppl), 63–69.

Kubzansky, L. D., Koenen, K. C., Spiro, A., Vokonas, P. S., & Sparrow, D. (2007). Prospective study of posttraumatic stress disorder symptoms and coronary heart disease in the Normative Aging Study. *Archieves of General Psychiatry*, *64*(1), 109–116.

Lawrence, C. E., Reilly, A. A., Quickenton, P., Greenwald, P., Page, W. F., & Kuntz, A. J. (1985). Mortality patterns of New York State Vietnam veterans. *American Journal of Public Health*, *75*(3), 277–279.

Levine, A. B., Levine, L. M., & Levine, T. B. (2014). Posttraumatic stress disorder and cardiometabolic disease. *Cardiology*, *127*(1), 1–19.

Litz, B. T., Penk, W. E., Walsh, S., Hyer, L., Blake, D. D., Marx, B., … Bitman, D. (1991). Similarities and differences between MMPI and MMPI-2 applications to the assessment of posttraumatic stress disorder. *Journal of Personally Assessment*, *57*(2), 238–253.

Magruder, K. M., Goldberg, J., Forsberg, C. W., Friedman, M. J., Litz, B. T., Vaccarino, V., … Smith, N. L. (2016). Long-term trajectories of PTSD in Vietnam-era veterans: The course and consequences of PTSD in twins. *Journal of Traumatic Stress*, *29*(1), 5–16.

NORA Organization of Work Team Members. (2002). *The Changing Organization of Work and the Safety and Health of Working People* (No. DHHS (NIOSH) Publication Number 2002-116). Cincinnati: National Institute for Occupational Safety and Health.

O'Toole, B. I., & Catts, S. V. (2008). Trauma, PTSD, and physical health: An epidemiological study of Australian Vietnam veterans. *Journal of Psychosomatic Research*, *64*(1), 33–40.

Pacella, M. L., Hruska, B., & Delahanty, D. L. (2013). The physical health consequences of PTSD and PTSD symptoms: A meta-analytic review. *Journal of Anxiety Disorders*, *27*(1), 33–46.

Qureshi, S. U., Pyne, J. M., Magruder, K. M., Schulz, P. E., & Kunik, M. E. (2009). The link between posttraumatic stress disorder and physical comorbidities: A systematic review. *Psychiatric Quarterly*, *80*(2), 87–97.

Ryder, A. L., Azcarate, P. M., & Cohen, B. E. (2018). PTSD and physical health. *Current Psychiatry Reports*, *20*(12), 116. doi: 10.1007/s11920-018-0977-9.

Schlenger, W. E., Corry, N. H., Williams, C. S., Kulka, R. A., Mulvaney-Day, N., DeBakey, S., … Marmar, C. R. (2015). A prospective study of mortality and trauma-related risk factors among a nationally representative sample of Vietnam veterans. *American Journal of Epidemiology*, *182*(12), 980–990.

Schnurr, P. P., & Green, B. L. (2004). Understanding relationships among trauma, posttraumatic stress disorder, and health outcomes. In P. P. Schnurr & B. L. Green (Eds.), *Trauma and Health: Physical Health Consequences of Exposure to Extreme Stress* (pp. 247–275). Washington, DC: American Psychological Association.

Schnurr, P. P., & Spiro, A. (1999). Combat exposure, posttraumatic stress disorder symptoms, and health behaviors as predictors of self-reported physical health in older veterans. *The Journal of Nervous and Mental Disease*, *187*(6), 353–359.

Schnurr, P. P., Spiro, A., & Paris, A. H. (2000). Physician-diagnosed medical disorders in relation to PTSD symptoms in older male military veterans. *Health Psychol*, *19*(1), 91–97.

Schnurr, P. P., Wachen, J. S., Green, B. L., & Kaltman, S. (2014). Chapter 28: Trauma exposure, PTSD, and physical health. In M. J. Friedman, T. M. Keane, & P. A. Resick (Eds.), *Handbook of PTSD—Second Edition: Science and Practice*. New York, NY: Guilford Press.

Schry, A. R., Rissling, M. B., Gentes, E. L., Beckham, J. C., Kudler, H. S., Straits-Troster, K., & Calhoun, P. S. (2015). The relationship between posttraumatic stress symptoms and physical health in a survey of U.S. veterans of the Iraq and Afghanistan era. *Psychosomatics*, *56*(6), 674–684.

Shaw, J. G., Asch, S. M., Kimerling, R., Frayne, S. M., Shaw, K. A., & Phibbs, C. S. (2014). Posttraumatic stress disorder and risk of spontaneous preterm birth. *Obstetrics & Gynecology*, *124*(6), 1111–1119.

Snow, B. R., Stellman, J. M., Stellman, S. D., & Sommer, J. F. (1988). Post-traumatic stress disorder among American Legionnaires in relation to combat experience in Vietnam: Associated and contributing factors. *Environmental Research*, *47*(2), 175–192.

Vancampfort, D., Rosenbaum, S., Ward, P. B., Steel, Z., Lederman, O., Lamwaka, A.V., … Stubbs, B. (2016). Type 2 diabetes among people with posttraumatic stress disorder: Systematic review and meta-analysis. *Psychosomatic Medicine*, *78*(4), 465–473.

Wachen, J. S., Shipherd, J. C., Suvak, M., Vogt, D., King, L. A., & King, D. W. (2013). Posttraumatic stress symptomatology as a mediator of the relationship between warzone exposure and physical health symptoms in men and women. *Journal of Traumatic Stress*, *26*(3), 319–328.

Wagner, A. W., Wolfe, J., Rotnitsky, A., Proctor, S. P., & Erickson, D. J. (2000). An investigation of the impact of posttraumatic stress disorder on physical health. *Journal of Traumatic Stress*, *13*(1), 41–55.

Yehuda, R. (2002). Post-traumatic stress disorder. *New England Journal of Medicine*, *346*(2), 108–114.

Yehuda, R. (2009). Status of glucocorticoid alterations in post-traumatic stress disorder. *Annals of the New York Academy of Sciences*, *1179*, 56–69.

Yu, S., Brackbill, R. M., Stellman, S. D., Ghuman, S., & Farfel, M. R. (2015). Evaluation of non-response bias in a cohort study of World Trade Center terrorist attack survivors. *BMC Rresearch Notes*, *8*, 42. doi: 10.1186/s13104-015-0994-2.

# 31

# NEUROTRANSMITTER AND NEUROTROPHIC BIOMARKERS IN COMBAT-RELATED POSTTRAUMATIC STRESS DISORDER

*Gordana Nedic Erjavec, Matea Nikolac Perkovic, Dubravka Svob Strac, Lucija Tudor, and Nela Pivac*

## Posttraumatic stress disorder

Posttraumatic stress disorder (PTSD) is a severe mental disorder that requires direct or indirect exposure to a single traumatic event or prolonged exposure to stressful events (American Psychiatric Association, 2013). Traumatic experience might also be induced by witnessing a stressful event that happened to family members or a close friend or exposure to consequences of traumatic events in professional workers (APA, 2013; Pai, Suris & North, 2017). PTSD develops only in vulnerable individuals after traumatic exposure. Vulnerability to PTSD is still not completely understood, but is assumed to be associated with altered connectivity and disturbed functions of various CNS regions and altered immune function as well as disturbed neurotransmitter and neuroendocrine responses. The interaction between these biological/genetic factors and psychological, cognitive, emotional, environmental and psychosocial factors contributes to vulnerability to PTSD.

The biggest challenge in PTSD research is to find biomarkers of vulnerability or resilience, especially in soldiers. It is important to recognize biological characteristics associated with higher vulnerability to combat trauma in veterans (Frank, Zamorski, Lee & Colman, 2018), as vulnerability factors might impact individual responses and disturb unity in military platoons.

Characteristic symptom clusters in PTSD, according to the DSM-5 (APA, 2013; Rosebrock, Arditte Hall, Rando, Pineles & Liverant, 2018; Ito, Takebayashi, Suzuki & Horikosh, 2018) and described therein are: re-experiencing, avoidance, numbing and hyper-arousal. They should last minimally one month after traumatic exposure. However, in addition to these, individuals with PTSD frequently develop other symptoms that affect clinical picture, course of illness and treatment response. These symptoms include agitated and psychotic symptoms (Compean & Hamner, 2019) or suicidal behaviour.

Although people are exposed frequently to various traumatic events (Bisson, Cosgrove & Roberts, 2015), only a subset of individuals will develop PTSD (Kessler et al., 2017). As opposed to the prevalence of trauma, the prevalence of PTSD is less than 9% in civilians (Atwoli, Stein,

Koenen & McLaughlind, 2015), while the prevalence in combat exposed veterans ranges between 11%–30% in the United States (Dursa, Reinhard, Barth & Schneiderman, 2014; Kang, Natelson, Mahan, Lee & Murphy, 2003; Kintzle, Barr, Corletto & Castro, 2018) and 18%–40% in Croatian (Komar & Vukusic, 1999) war veterans.

## Biomarkers in posttraumatic stress disorder

There are different risk factors that predispose individuals to PTSD. Finding these risk factors is the major goal in biomarker research, which is focused on finding reliable and easily available, validated and specific biomarkers related to the vulnerability to develop PTSD (Lehrner & Yehuda, 2014). Biomarkers of PTSD are related to different symptoms and behaviours which occur in PTSD and develop due to disturbances in diverse biological systems. PTSD in military veterans presents with a more severe time course, with pronounced hyper-arousal and intrusive symptoms (Guina, Nahhas, Sutton & Farnsworth, 2018).

Biomarkers are measurable and quantifiable biological parameters that should be used as key tools to provide crucial information on the complex molecular mechanisms of the disorder. They can be classified as diagnostic biomarkers, disease prognosis biomarkers and biomarkers for monitoring the clinical or therapeutic response (Nikolac Perkovic et al., 2017). In PTSD there are numerous types of physiological markers (such as neuroimaging and psychophysiological measures and behavioural and neurocognitive responses), but the majority of biomarkers are different parameters that can be determined in peripheral bio-fluids. Since a variety of markers were proposed for PTSD, an attempt has been made to organize these metadata in one PTSD Biomarker Database (Domingo-Fernandez et al., 2019). However, specific and validated biomarkers for PTSD are still missing (Lehrner & Yehuda, 2014). Most biomarkers are related to the major neurotransmitters dopamine, noradrenaline (Pan, Kaminga, Wen & Liu, 2018) and serotonin, but also to the major stress response system, i.e. the hypothalamic-pituitary-adrenal (HPA) axis, immune system and metabolic and glycomic markers (Kao, Stalla, Stalla, Wotjak & Anderzhanova, 2015; Olff & van Zuiden, 2017; Nedic Erjavec et al., 2018; Konjevod et al., 2019, VanDyke, Burton, Hanidovic & Burke, 2017). Biomarkers have military applications (i.e. they might be used to confirm PTSD diagnosis and severity, to predict risk, for validation of trauma exposure, to predict recovery, to verify the recovery and for subtyping PTSD) but there must also be consideration of the potential ethical, social, clinical and legal implications (Lehrner & Yehuda, 2014).

## Dopaminergic markers in posttraumatic stress disorder

One of the most elaborated hypotheses of the neurobiological underpinning of PTSD is a dysfunction of the dopaminergic system (Lee, Wang & Tsien, 2016). The dopamine signalling system regulates several neurological processes, including fear-conditioning in different brain areas (Aubry, Serrano & Burghardt, 2016). Neuroimaging studies reveal dopamine dysregulations in individuals with PTSD as a response to stressful stimuli (Charney, 2004). Also, a higher response of the dopaminergic system after the stress exposure often occurs, and it is associated with PTSD symptoms such as restlessness, nightmares, fear memories and impulsivity (Pezze & Feldon, 2004), as well as alterations in memory and other cognitive functions, anhedonia and hypervigilance (Sher et al., 2005). Systematic levels of dopamine were measured in order to investigate how they relate to PTSD. In a study dealing with PTSD in mothers of childhood cancer survivors (Glover et al., 2003), an increased concentration of urinary dopamine in women with PTSD was found compared to values in women without PTSD and in healthy control women. However, another study (Osuch et al., 2009), and a recent meta-analysis of seven studies accounting for a total of 656 participants (Pan et al., 2018), reported no significant changes in dopamine levels between individuals with PTSD and healthy control subjects.

Genetic variations in the dopaminergic system involved in signalling transduction may affect the ability to cope with stress stimuli in people who have been exposed to traumatic events (Cornelis, Nugent, Amstadter & Koenen, 2010). A SNP in dopamine receptor D2 (DRD2) gene (rs6277) has been reported as significantly associated with PTSD where war veterans with PTSD were more likely C allele carriers than healthy control subjects (Voisey et al., 2009). Another polymorphism (rs1800497) of the DRD2 gene was significantly associated with PTSD according to meta-analysis, which examined six studies with a total of 597 PTSD patients and 1155 controls (Li et al., 2016). Further on, a long (seven or more repeats) allele (considering VNTR in the third exon of dopamine receptor D4 [DRD4] gene) was found to be associated with significantly intense avoidance/numbing symptoms and PTSD symptoms in general, but not with the PTSD diagnosis itself, among flood survivors (Dragan & Oniszczenko, 2009). There are contradictory reports about the association of PTSD and a variable number of tandem repeats (VNTR) in a dopamine transporter gene (DAT1). According to some studies, there is an increased risk of PTSD in grown-up individuals (Segman et al., 2002) and preschool children (Drury, Theall, Keats & Scheeringa, 2009) with 9 repeats (9R) of 40-bp sequence when compared with individuals/children carrying 10 repeats-allele (10R). On the other hand, a study carried out on a group of people suffering from PTSD after exposure to an earthquake found no association between DAT1 VNTR and PTSD risk (Bailey et al., 2010). Due to the scarce presence of DAT in prefrontal cortex (PFC), dopamine degradation in this area mostly depends on catechol-O-methyltransferase (COMT) activity, which is altered by its functional polymorphism rs4680 (Val108/158Met substitution), also affecting the overall dopaminergic signalling (Goldberg & Weinberger, 2004). Namely, the Met allele is associated with reduced COMT activity and dopamine degradation, causing higher dopamine availability in the PFC (Chen et al., 2004). Therefore, it is not surprising that there is a certain association of COMT rs4680 and PTSD, manifested either in reduced hippocampal volume in Val homozygotes (Hayes et al., 2017) or in better cognitive performance in veterans with PTSD (Mestrovic et al., 2018), increased endocrine stress reactivity (Alexander et al., 2011) and deficits in fear conditioning (Deslauriers et al., 2018) in Met carriers or homozygotes. On the other hand, a meta-analysis (Li et al., 2016) including five studies dealing with COMT rs4680 found no association between this polymorphism and PTSD risk. In complex diseases such as PTSD, some genetic effects are often more pronounced in certain combinations with other genetic or environmental factors. For example, it was reported that men carrying a combination of Met/Met COMT rs4680 genotype and 10R/10R DAT1 VNTR have higher cortisol responses and impaired stress recovery during public speaking (Alexander et al., 2011).

Another important dopamine degrading enzyme is monoamine oxidase (MAO) that exists in two isoforms, MAO-A and MAO-B, and participates in the regulation of important processes such as mood, emotions and behaviour (Edmondson, Binda, Wang, Upadhyay, & Mattevi, 2009). Its role in the pathophysiology of different mental and neurodegenerative disorders was also reported (Adolfsson, Gottfries, Oreland, Wiberg & Winblad, 1980; Brunner, Nelen, Breakefield, Ropers & van Oost, 1993; Sandler, Glover, Clow & Jarman, 1993). Considering the role of MAO in PTSD, literature data are inconsistent. In contrast to one earlier study (Davidson, Lipper, Kilts, Mahorney & Hammett, 1985) that found platelet MAO-B activity to be reduced in PTSD, another study reported no difference in platelet MAO-B activity between war veterans with or without PTSD (Pivac, Muck-Seler, Sagud & Jakovljevic, 2002) or increased platelet MAO-B activity in veterans with psychotic PTSD compared to veterans with non-psychotic features (Pivac et al., 2007; Svob Strac et al., 2016). Additionally, altered MAO-B activity was found in platelets of veterans with more severe PTSD symptoms according to the Clinician Administered PTSD Scale (CAPS) and in veterans with agitation assessed by the Positive and Negative Syndrome Scale (PANSS) (Svob Strac et al., 2016). Polymorphisms of MAO-A and MAO-B genes were found to be associated with different pathologies (Bortolato & Shih, 2011), but a lack of association of the most studied MAO-B (rs1799836) (Pivac et al., 2007) and MAO-A (MAOA-uVNTR) (Svob Strac et al., 2016) polymorphisms with combat related PTSD was reported. A methylation of MAO-A gene exonI/intronI region was recently suggested as a disease status and severity (according to CAPS) marker of PTSD in men (Ziegler et al., 2018). Namely, a hypermethylation of 3 CpGs was found in male subjects with current PTSD when compared with remitted PTSD patients and healthy control subjects, while symptom severity significantly correlated with MAO-A methylation (Ziegler et al., 2018).

It seems that there are several potential dopaminergic biomarkers of PTSD, but still more extensive studies are needed to overcome the effects of severity of PTSD symptoms, race and ethnicity and type and duration of traumatic events on overall mechanisms of development and course of PTSD.

## Serotonergic biomarkers of posttraumatic stress disorder

Various studies have investigated the role of serotonergic system in PTSD. A variety of data suggested that serotonergic dysregulation might be related to PTSD symptoms such as hypervigilance, exaggerated startle, irritability, impulsivity, aggression, intrusive memories, depressed mood and suicidality (Southwick et al., 1999). Findings supporting the role of serotonin (5-HT) in PTSD include clinical efficacy of selective serotonin reuptake inhibitors (SSRIs), (Marshall, Beebe, Oldham & Zaninelli, 2001; Martenyi, Brown, Zhang, Prakash & Koke, 2002; Davidson, Rothbaum, van der Kolk, Sikes & Farfel, 2001) and blunted prolactin response to 5-HT–releasing drug fenfluramine (Davis, Clark, Kramer, Moeller & Petty, 1999), as well as exaggerated reactivity to m-chloro-phenyl-piperazine, the 5-HT agonist (Southwick et al., 1997).

A number of studies have addressed serotonergic function in PTSD by investigating the peripheral markers. Since blood platelets store 5-HT and possess many other serotonergic components, including serotonin transporter (5-HTT), enzyme MAO-B, as well as various 5-HT receptors, they are considered an easily available, limited peripheral model of central 5-HT function (Muck-Seler & Pivac, 2011). Studies measuring 5-HT uptake into platelets (Cicin-Sain et al., 2000; Mellman & Kumar, 1994) and platelet 5-HT concentrations (Muck-Seler et al., 2003; Pivac et al., 2002), as well as [3H] paroxetine (Maguire, Norman, Burrows, Hopwood & Morris, 1998) or [3H] imipramine (Weizman et al., 1996) binding to platelet membranes demonstrated no alteration of peripheral 5-HT function in PTSD. On the other hand, diminished 5-HT activity was suggested by studies demonstrating lower platelet paroxetine binding (Arora, Fichtner, O'Connor & Crayton, 1993; Maes et al., 1999), decreased 5-HT concentration in platelets (Guo et al., 2016, Li et al., 2016) and decreased platelet-poor plasma 5-HT levels (Spivak et al., 1999) in PTSD patients.

Lower platelet 5-HT concentration was also found in suicidal PTSD patients compared with non-suicidal patients or healthy controls (Kovacic, Henigsberg, Pivac, Nedic & Borovecki, 2008). These results might explain different findings (Guo et al., 2016, Li et al., 2016) from the literature, since suicidality is frequently associated with PTSD and the presence of suicidal behaviour might have affected platelet 5-HT data. On the other hand, higher platelet 5-HT concentration was observed in PTSD veterans with psychotic features compared to veterans with PTSD who did not develop psychotic symptoms (Pivac et al., 2006) and in non-depressed veterans with PTSD and early insomnia compared to veterans with PTSD who did not develop insomnia (Kovacic Petrovic et al., 2019). Therefore, the presence of psychotic symptoms or insomnia affected platelet 5-HT concentration in veterans with PTSD.

Regarding genetic variants of the components of the serotonergic system, most of the PTSD studies focused on 5-HTT encoded by the *SLC64A* gene. Reduced 5-HTT binding has been observed in the amygdala of PTSD patients (Murrough et al., 2011). In a PET study, Frick et al. (2016) found both higher and lower 5-HTT and neurokinin-1 (NK1) receptor availability in several brain regions of PTSD patients; however, lower expression overlap between 5-HTT and NK1 receptors has been associated with higher PTSD symptom severity.

Frequently investigated *SLC64A* promoter polymorphic region (5-HTTLPR) has two common functional variants. The S (short) allele, in comparison to L (long) allele, is associated with lower gene transcription, decreased 5-HTT activity and lower 5-HT reuptake from the synaptic cleft (Morey et al., 2011; Kolassa, Illek, Wilker, Karabatsiakis & Elbert, 2015), resulting in increased amygdala reactivity to emotional stimuli and consequently higher fear conditioning (Hariri et al., 2002). However, since a third functional allele $L_G$ has been identified (Nakamura, Ueno, Sano & Tanabe, 2000), the 5-HTTLPR polymorphism is now considered triallelic. This $A > G$ substitution at nucleotide 6 of the first of two extra 22-bp repeats in the L allele results in a transcriptional level comparable with that of the S allele. Many studies found an association of 5-HTTLPR polymorphism with PTSD (Goenjian et al., 2012; Kilpatrick et al., 2007; Koenen et al., 2009a; Lee et al., 2005), which was consistently observed in combat related studies (Kimbrel et al., 2015; Liu et al., 2015; Wang et al., 2011). However, the reported findings have been contradictory (Gressier et al., 2013; Kolassa et al., 2010; Kovacic Petrovic, Nedic Erjavec, Nikolac Perkovic, Peraica & Pivac, 2016; Navarro-Mateu, Escámez, Koenen, Alonso & Sánchez-Meca, 2013). For example, in some studies, increased risk of PTSD was associated with the 5-HTTLPR low expression S allele (Lee et al., 2005; Kolassa et al., 2010; Kilpatrick et al., 2007; Koenen et al., 2009a, Xie et al., 2009; Mellman et al., 2009), whereas other studies report association with the high expression L allele (Thakur, Joober & Brunet, 2009; Grabe et al., 2009). In addition, the LL genotype of 5-HTTLPR was associated with more severe early insomnia in veterans with PTSD compared to veterans with PTSD who did not develop early insomnia (Kovacic Petrovic et al., 2019). Nevertheless, recently, Zhao et al. (2017) performed a meta-analysis of 14 studies with 15,883 subjects and found strong evidence that 5-HTTLPR significantly influenced the relationship between stress and PTSD. Another meta-analysis demonstrated an association between the SS genotype and PTSD only in high trauma-exposed participants (Gressier et al., 2013). The review of meta-analyses on PTSD candidate gene research confirmed the association of 5-HTTLPR with PTSD, but only in an interaction with environment (Sheerin, Lind, Bountress, Nugent & Amstadter, 2017). Namely, for 5-HTTLPR, a significant $G \times E$ effect was found to be associated with PTSD. The 5-HTTLPR genotype interacted with adult traumatic events and childhood adversity to increase the risk for PTSD, especially in individuals exposed to high rates of both types of trauma (Xie et al., 2009). Consistently with the plasticity gene hypothesis (Belsky et al., 2009), the S allele was associated with increased risk of PTSD when combined with high rates of crime and unemployment, whereas it was associated with decreased PTSD risk in low-risk environments (Koenen et al., 2009a). Various animal studies also demonstrated that 5-HTT gene variations are associated with changes in serotonergic function following early stressful experiences (Murphy et al., 2001; Bennett et al., 2002).

In addition, *SLC6A4* methylation was shown to modify the effect of the number of traumatic events on PTSD risk. Koenen et al. (2011) reported that individuals with more traumatic events were at increased risk for PTSD, but only at lower *SLC6A4* methylation levels, whereas at higher methylation levels, subjects with more traumatic events were protected from PTSD. Methylation patterns of 5-HTTLPR have also been associated with an increased risk of unresolved loss or trauma, but only for individuals with the L allele (Van, Caspers, Bakermans-Kranenburg, Beach & Philibert, 2010), suggesting an interaction between methylation density and the 5-HTT genotype. Such 5-HTTLPR genetic–epigenetic interactions have been shown to influence circulating cortisol levels in response to stress (Alexander et al., 2014).

Moreover, Bharadwaj et al. (2016) suggested rs363276 polymorphism, located in intron 14 of the *SLC18A2* gene, coding for a transporter of monoamines serotonin, dopamine and noradrenaline, as a new candidate polymorphism associated with PTSD risk.

Some studies suggested a significant role of 5-HT receptors in PTSD. The G allele of the -1438A/G substitution polymorphism (rs6311), located in the promoter region of 5-HT2A receptor gene, was significantly associated with increased risk of PTSD (Mellman et al., 2009; Lee, Kwak, Paik, Kang & Lee, 2007). Additionally, another 5-HTR2A genetic variant (rs7997012) was shown to moderate the association between PTSD severity and reduced default mode network (DMN) connectivity (Miller et al., 2016). In a PET study, Sullivan et al. (2013) observed higher brainstem and forebrain 5-HT1A binding in PTSD. However, no change in 5-HT1A receptor binding was found in a study by Bonne et al. (2005).

Regarding PTSD research assessing the 5-HT synthesis and metabolism, observed association of PTSD symptoms with the T allele of rs2108977 and rs11178997 polymorphisms, located in the *TPH1* and *TPH2* genes, respectively, suggested that variants in the genes coding for tryptophan hydroxylase (TPH), the enzyme involved in 5-HT synthesis, represent risk factors for PTSD symptoms (Goenjian et al., 2012). In subjects with war-related chronic PTSD, the CSF concentrations of 5-hydroxyindoleacetic acid (5-HIAA), a main 5-HT metabolite, tend to decline during presentation of a trauma-related video, suggesting its association with laboratory-induced PTSD symptoms (Geracioti et al. 2013).

In conclusion, these data summarize an important role of serotonergic system, its components and genetic variations of the 5-HT system's components in the development of PTSD and its symptoms.

## Neurotrophic factors in posttraumatic stress disorder

During development and in adulthood, neurotrophins have an important role in regulating neurogenesis, differentiation, maintenance and growth of neuronal cells. Neurotrophins, like nerve growth factor (NGF) and brain-derived neurotrophic factor (BDNF), are also the key modulators of brain plasticity and their expression is under strong influence of psychophysical stress (Bothwell, 2014). BDNF is widely expressed in various types of tissues and cells, including the limbic system where it is involved in the regulation of stress and fear responses (Peters, Dieppa-Perea, Melendez & Quirk, 2010; Rosas-Vidal, Do-Monte, Sotres-Bayon & Quirk, 2014). Studies involving animal models (Faure, Uys, Marais, Stein & Daniels, 2007) support the theory of BDNF over-expression, in plasma and in hippocampus, as a compensatory mechanism which is induced by the traumatic stress (Zhang et al., 2014). Brain-derived neurotropic factor (BDNF), although not a conventional CVD risk factor, has also drawn attention. Namely, low peripheral levels of BDNF were associated with an increased risk of stroke (Pikula et al., 2013) and was proposed as a biomarker of heart failure (Takashio et al., 2015). In line with this, higher serum BDNF levels were protective, suggesting decreased risk of CVD (Kaess et al., 2015). Studies have reported either decreased BDNF levels in patients with different types of non-combat related trauma (Dell'Osso et al., 2009; Stratta et al., 2016), no difference in BDNF levels in subjects after traffic accidents (Su et al., 2015; van den Heuvel, Suliman, Malan- Müller, Hemmings & Sedat, 2016), and elevated BDNF levels in subjects severely injured in traffic accidents (Matsuoka, Nishi,

Noguchi, Kim & Hashimoto, 2013), in those with trauma not specified (Hauck et al., 2010) and in young male veterans with PTSD (Blessing et al., 2017). It seems that peripheral BDNF levels are higher after the trauma, and decrease thereafter (Hauck et al., 2010). Higher methylation within the BDNF promoter I region, in peripheral leukocytes, was reported in veterans with PTSD, even 45 years after combat exposure, relative to non-psychiatric controls (Kim et al., 2017).

Recent findings propose that up-regulated BDNF release might help to prevent the development of PTSD since it induces fear extinction and mediates more successful trauma processing (Dincheva, Lynch & Lee, 2016). Data from animal studies suggest that stress, along with BDNF up-regulation, leads towards startle response abnormality (Zhang et al., 2014), one of the main symptoms of PTSD, i.e. hyperarousal (Jovanovic, Norrholm, Sakoman, Esterajher & Kozarić-Kovacić, 2009).

Different genome-wide association studies (GWAS) and candidate gene studies (Koenen, Amstadter & Nugent, 2009b; Miller &Wiener, 2014) have pointed out the importance of BDNF in the pathology of PTSD. A common SNP in the BDNF gene, which results in a valine to methionine substitution at position 66 (Val66Met), has been associated with poorer memory performance and reduced hippocampal volumes (Brooks et al., 2014; Zhang et al., 2006). Compared to Val/Val homozygotes, the Met allele carriers show less activity-dependent release of BDNF in the hippocampus (Hashimoto et al., 2008; Miller & Wiener, 2014; Sanchez et al., 2011). Study by Zhang and colleagues (2014) detected a higher frequency of Met/Met genotype carriers and a twofold higher frequency of Met allele carriers in subjects diagnosed with PTSD, compared to individuals without PTSD symptoms, which fits with the findings of smaller hippocampal volume in Met allele carriers (Bueller et al., 2006; Harrisberger et al., 2014; Szeszko et al., 2005). Results also indicate that Met allele carriers have poorer memory performance compared to Val/Val genotype carriers (Hariri et al., 2003). Our study further supports the interaction between BDNF Val66Met polymorphism and PTSD, showing a higher frequency of Met allele carriers in veterans with psychotic PTSD, compared non-psychotic veterans with PTSD or veterans without PTSD (Pivac et al., 2012). BDNF Met allele was also associated with exaggerated startle, linking Met/Met genotype to hyper-arousal vulnerability (Zhang et al., 2014). However, there are also studies that suggest no association between PTSD and BDNF Val66Met polymorphism (Zhang et al., 2006), which emphasizes the need for further research evaluating the genetic association between BDNF and PTSD, also taking into account BDNF central/peripheral levels and epigenetic patterns that influence BDNF gene expression. In addition, as demonstrated by Zhang and colleagues (2014), it is very important to have in mind different environmental conditions, trauma exposure type and trauma severity, time passed since exposure, treatment approach, and age.

In the last decade BDNF blood levels (plasma and serum levels) were proposed as potential biomarkers in PTSD diagnosis (Angelucci et al., 2014; Grassi-Oliveira, Stein, Lopes, Teixeira & Bauer, 2008; Hauck et al., 2010; Matsuoka et al., 2013). Data suggested that individuals diagnosed with PTSD have significantly higher BDNF serum levels, compared to controls (Hauck et al., 2009; Hauck et al., 2010), with BDNF levels being higher in Met allele carriers then in Val/Val genotype carriers (Harris et al., 2006). However, some studies showed opposite results (Dell'Osso et al., 2009). These contradictory results might be a consequence of differences in the sample population and different time points during the course of the disease in which the analyses were done. Nevertheless, peripheral BDNF levels could contribute to development of less invasive diagnostics in PTSD (Zhang et al., 2014).

NGF is another key member of neurotrophin family (Bothwell, 2014), but there is still not enough effort to investigate the diagnostic potential of NGF in PTSD in details. Study by Tural and colleagues (2018) showed lower serum NGF levels in individuals with chronic PTSD, compared to trauma exposed non-PTSD individuals. In addition, the authors reported that after 12 weeks of treatment with escitalopram, the NGF production was restored, hippocampal activity was decreased and there was a significant improvement in Clinician-Administered PTSD Scale (CAPS) scores in individuals with PTSD (Tural, Tamer Aker, Turan Sodan, Ünver & Akansel, 2018).

In summary, neurotrophic system, with BDNF and NGF and genetic variants of the BDNF are associated with PTSD. However, the clear relationship still needs to be elucidated.

## Conclusion

The most frequent data on the biomarkers in PTSD are biomarkers related to components of the dopaminergic, serotonergic and neurotrophic systems, especially BDNF. It is difficult to conclude on the specificity of these biomarkers since in the literature there are mixed results regarding the changes of biomarker findings in PTSD. Namely, the same biomarkers were reported to be increased, decreased, or unchanged in PTSD. The comparison with corresponding control samples might also contribute to these divergent findings, since for veterans with PTSD studies should include combat exposed veterans who did not develop PTSD. More data regarding PTSD biomarkers will lead to improved knowledge on the biological underpinning of PTSD. Notwithstanding the alterations in dopaminergic, serotonergic and neurotrophic systems, development of PTSD depends on the complex interactions between these and other biological factors, environmental and trauma-related factors, psychological factors, presence or absence of early stress, and/or social and family support. This review focused on the most frequently studied markers of the dopaminergic, serotonergic and BDNF systems. However, due to the complexity of PTSD, efforts of the future studies should be to offer a panel of biomarkers that will more appropriately reveal alterations in PTSD.

**Note:** The work was supported by the Croatian Science Foundation, project number IP-2014-09-4289.

## References

Adolfsson, R., Gottfries, C. G., Oreland, L., Wiberg, A., & Winblad, B. (1980). Increased activity of brain and platelet monoamine oxidase in dementia of Alzheimer type. *Life Science, 27*(12), 1029–1034.

Alexander, N., Osinsky, R., Mueller, E., Schmitz, A., Guenthert, S., Kuepper, Y., & Hennig, J. (2011). Genetic variants within the dopaminergic system interact to modulate endocrine stress reactivity and recovery. *Behavioural Brain Research, 216*(1), 53–58.

Alexander, N., Wankerl, M., Hennig, J., Miller, R., Zänkert, S., Steudte-Schmiedgen, S., ... Kirschbaum, C. (2014). DNA methylation profiles within the serotonin transporter gene moderate the association of 5-HTTLPR and cortisol stress reactivity. *Translational Psychiatry, 4*(9), e443–e443.

American Psychiatric Association. (2013). *Diagnostic and Statistical Manual of Mental Disorders* (5th ed.). Washington, DC: American Psychiatric Press Inc.

Angelucci, F., Ricci, V., Gelfo, F., Martinotti, G., Brunetti, M., Sepede, G.., ... Caltagirone, C. (2014). BDNF serum levels in subjects developing or not post-traumatic stress disorder after trauma exposure. *Brain Cognition, 84*, 118–122.

Arora, R. C., Fichtner, C. G., O'Connor, F., & Crayton, J. W. (1993). Paroxetine binding in the blood platelets of post-traumatic stress disorder patients. *Life Sciences, 53*(11), 919–928.

Atwoli, L., Stein, D. J., Koenen, K. C., & McLaughlind, K. A. (2015). Epidemiology of posttraumatic stress disorder: prevalence, correlates and consequences. *Current Opinion in Psychiatry, 28*(4), 307–311.

Aubry, A. V., Serrano, P. A., & Burghardt, N. S. (2016). Molecular mechanisms of stress-induced increases in fear memory consolidation within the amygdala. *Frontiers in Behavioral Neuroscience, 10*(191), 1–10.

Bailey, J. N., Goenjian, A. K., Noble, E. P., Walling, D. P., Ritchie, T., & Goenjian, H. A. (2010). PTSD and dopaminergic genes, DRD2 and DAT, in multigenerational families exposed to the Spitak earthquake. *Psychiatry Research, 178*(3), 507–510.

Belsky, J., Jonassaint, C., Pluess, M., Stanton, M., Brummett, B., & Williams, R. (2009). Vulnerability genes or plasticity genes? *Molecular Psychiatry, 14*(8), 746–754.

Bennett, A. J., Lesch, K. P., Heils, A., Long, J. C., Lorenz, J. G., Shoaf, S. E., … Higley, J. D. (2002). Early experience and serotonin transporter gene variation interact to influence primate CNS function. *Molecular Psychiatry*, 7(1), 118–122.

Bharadwaj, R. A., Jaffe, A. E., Chen, Q., Deep-Soboslay, A., Goldman, A. L., Mighdoll, M. I., … Kleinman, J. E. (2016). Genetic risk mechanisms of posttraumatic stress disorder in the human brain. *Journal of Neuroscience Research, 96*(1), 21–30.

Bisson, J. I., Cosgrove, S., & Roberts, N. P. (2015). Post-traumatic stress disorder. *British Medical Journal, 351*, h6161–h6161.

Blessing, E. M., Reus, V., Mellon, S. H., Wolkowitz, O. M., Flory, J. D., Bierer, L., … Marmar, C. R. (2017). Biological predictors of insulin resistance associated with posttraumatic stress disorder in young military veterans. *Psychoneuroendocrinology, 82*, 91–97.

Bonne, O., Bain, E., Neumeister, A., Nugent, A. C., Vythilingam, M., Carson, R. E., … Charney, D. S. (2005). No change in serotonin type 1A receptor binding in patients with posttraumatic stress disorder. *American Journal of Psychiatry, 162*(2), 383–385.

Bortolato, M., & Shih, J. C. (2011). Behavioral outcomes of monoamine oxidase deficiency: Preclinical and clinical evidence. *International Review of Neurobiology, 100*, 13–42.

Bothwell, M. (2014). NGF, BDNF, NT3, and NT4. *Handbook of Experimental Pharmacology, 220*, 3–15.

Brooks, S. J., Nilsson, E. K., Jacobsson, J. A., Stein, D. J., Fredriksson, R., Lind, L., & Schiöth, H. B. (2014). BDNF polymorphisms are linked to poorer working memory performance, reduced cerebellar and hippocampal volumes and differences in prefrontal cortex in a Swedish elderly population. *PLOS ONE, 9*, e82707.

Brunner, H. G., Nelen, M., Breakefield, X. O., Ropers, H. H., & van Oost, B. A. (1993). Abnormal behavior associated with a point mutation in the structural gene for monoamine oxidase A. *Science, 262*(5133), 578–580.

Bueller, J. A., Aftab, M., Sen, S., Gomez-Hassan, D., Burmeister, M., & Zubieta, J. K. (2006). BDNFVal66Met allele is associated with reduced hippocampal volume in healthy subjects. *Biological Psychiatry, 59*, 812–815.

Charney, D. S. (2004). Psychobiological mechanisms of resilience and vulnerability to extreme stress: Relevance to prevention and treatment of anxiety disorders. *Biological Psychiatry, 55*, 1S–1S.

Chen, J., Lipska, B. K., Halim, N., Ma, Q. D., Matsumoto, M., Melhem, S., … Weinberger, D. R. (2004). Functional analysis of genetic variation in catechol-O-methyltransferase (COMT): Effects on mRNA, protein, and enzyme activity in postmortem human brain. *American Journal of Human Genetics, 75*(5), 807–821.

Cicin-Sain, L., Mimica, N., Hranilovic, D., Balija, M., Ljubin, T., Makarić, G., … Jernej, B. (2000). Posttraumatic stress disorder and platelet serotonin measures. *Journal of Psychiatric Research, 34*(2), 155–161.

Compean, E., & Hamner, M. (2019). Posttraumatic stress disorder with secondary psychotic features (PTSD-SP): Diagnostic and treatment challenges. *Progress in Neuro-Psychopharmacology and Biological Psychiatry, 88*, 265–275.

Cornelis, M. C., Nugent, N. R., Amstadter, A. B., & Koenen, K. C. (2010). Genetics of post-traumatic stress disorder: Review and recommendations for genome-wide association studies. *Current Psychiatry Reports, 12*(4), 313–326.

Davidson, J. R., Rothbaum, B. O., van der Kolk, B. A., Sikes, C. R., & Farfel, G. M. (2001). Multicenter, double-blind comparison of sertraline and placebo in the treatment of posttraumatic stress disorder. *Archives of General Psychiatry, 58*(5), 485–492.

Davidson, J., Lipper, S., Kilts, C. D., Mahorney, S., & Hammett, E. (1985). Platelet MAO activity in posttraumatic stress disorder. *American Journal of Psychiatry, 142*(11), 1341–1343.

Davis, L. L., Clark, M., Kramer, G. L., Moeller, F. G., & Petty, F. (1999). D-fenfluramine challenge in posttraumatic stress disorder. *Biological Psychiatry, 45*(7), 928–930.

Dell'Osso, L., Carmassi, C., Del Debbio, A., Catena Dell'Osso, M., Bianchi, C., da Pozzo, E., … Piccinni, A. (2009). Brain-derived neurotrophic factor plasma levels in patients suffering from post-traumatic stress disorder. *Progress in Neuro-Psychopharmacology and Biological Psychiatry, 33*, 899–902.

Deslauriers, J., Acheson, D. T., Maihofer, A. X., Nievergelt, C. M., Baker, D. G., Geyer, M. A., … Marine Resiliency Study Team. (2018). COMT val158met polymorphism links to altered fear conditioning and extinction are modulated by PTSD and childhood trauma. *Depression and Anxiety, 35*(1), 32–42.

Dincheva, I., Lynch, N. B., & Lee, F. S. (2016). The role of BDNF in the development of fear learning. *Depression and Anxiety, 33*, 907–916.

Domingo-Fernandez, D., Provost, A., Marin-Llao, J., Lasseter, H., Diaz, K., Daskalakis, N., … Haas, M. (2019). PTSD biomarker database: Deep dive meta-database for PTSD biomarkers, visualizations, and analysis tools. *BioRxiv*. Advance online publication. doi: https://doi.org/10.1101/547901.

Dragan, W. L., & Oniszczenko, W. (2009). The association between dopamine D4 receptor exon III polymorphism and intensity of PTSD symptoms among flood survivors. *Anxiety, Stress & Coping, 22*(5), 483–495.

Drury, S. S., Theall, K. P., Keats, B. J. B., & Scheeringa, M. (2009). The role of the dopamine transporter (DAT) in the development of PTSD in preschool children. *Journal of Traumatic Stress, 22*(6), 534–539.

Dursa, E. K., Reinhard, M. J., Barth, S. K. & Schneiderman, A. I. (2014). Prevalence of a positive screen for PTSD among OEF/OIF and OEF/OIF-era veterans in a large population-based cohort. *Journal of Traumatic Stress, 27*, 542–549.

Edmondson, D. E., Binda, C., Wang, J., Upadhyay, A. K., & Mattevi, A. (2009). Molecular and mechanistic properties of the membrane-bound mitochondrial monoamine oxidases. *Biochemistry, 48*(20), 4220–4230.

Faure, J., Uys, J. D., Marais, L., Stein, D. J., & Daniels, W. M. (2007). Early maternal separation alters the response to traumatization: Resulting in increased levels of hippocampal neurotrophic factors. *Metabolic Brain Disease, 22*, 183–195.

Frank, C., Zamorski, M. A., Lee, J. E., & Colman, I. (2018). Deployment-related trauma and post-traumatic stress disorder: Does gender matter? *European Journal of Psychotraumatology, 9*(1), 1486123.

Frick, A., Ahs, F., Palmquist, A. M., Pissiota, A., Wallenquist, U., Fernandez, M., … Fredrikson, M. (2016). Overlapping expression of serotonin transporters and neurokinin-1 receptors in posttraumatic stress disorder: A multi-tracer PET study. *Molecular Psychiatry, 21*, 1400–1407.

Goenjian, A. K., Bailey, J. N., Walling, D. P., Steinberg, A. M., Schmidt, D., Dandekar, U., & Noble, E. P. (2012). Association of TPH1, TPH2, and 5HTTLPR with PTSD and depressive symptoms. *Journal of Affective Disorders, 140*(3), 244–252.

Geracioti, T. D. Jr., Jefferson-Wilson, L., Strawn, J. R., Baker, D. G., Dashevsky, B. A., Horn, P. S., & Ekhator, N. N. (2013). Effect of traumatic imagery on cerebrospinal fluid dopamine and serotonin metabolites in posttraumatic stress disorder. *Journal of Psychiatric Research, 47*(7), 995–998.

Glover, D. A., Powers, M. B., Bergman, L., Smits, J. A., Telch, M. J., & Stuber, M. (2003). Urinary dopamine and turn bias in traumatized women with and without PTSD symptoms. *Behavioural Brain Research, 144*(1–2), 137–141.

Goldberg, T. E., & Weinberger, D. R. (2004). Genes and the parsing of cognitive processes. *Trends in Cognitive Sciences, 8*(7), 325–335.

Grabe, H. J., Spitzer, C., Schwahn, C., Marcinek, A., Frahnow, A., Barnow, S., … Rosskopf, D. (2009). Serotonin transporter gene (SLC6A4) promoter polymorphisms and the susceptibility to posttraumatic stress disorder in the general population. *American Journal of Psychiatry, 166*(8), 926–933.

Grassi-Oliveira, R., Stein, L. M., Lopes, R. P., Teixeira, A. L., & Bauer, M. E. (2008). Low plasma brain-derived neurotrophic factor and childhood physical neglect are associated with verbal memory impairment in major depression—A preliminary report. *Biological Psychiatry, 64*, 281–285.

Gressier, F., Calati, R., Balestri, M., Marsano, A., Alberti, S., Antypa, N., & Serretti, A. (2013). The 5-HTTLPR polymorphism and posttraumatic stress disorder: A meta-analysis. *Journal of Traumatic Stress, 26*(6), 645–653.

Guina, J., Nahhas, R. W., Sutton, P., & Farnsworth, S. (2018). The influence of trauma type and timing on PTSD symptoms. *The Journal of Nervous and Mental Disease, 206*, 72–76.

Guo, J. C., Tian, Z. L., Wang, X. D., Guo, M., Li, M. Y., Gao, Y. S., … You, L. (2016). Post-traumatic stress disorder after typhoon disaster and its correlation with platelet 5-HT concentrations. *Asian Pacific Journal of Tropical Medicine, 9*(9), 913–915.

Hariri, A. R., Mattay, V. S., Tessitore, A., Kolachana, B., Fera, F., Goldman, D., … Weinberger, D. R. (2002). Serotonin transporter genetic variation and the response of the human amygdala. *Science, 297*(5580), 400–403.

Hariri, A. R., Goldberg, T. E., Mattay, V. S., Kolachana, B. S., Callicott, J. H., Egan, M. F., & Weinberger, D. R. (2003). Brain-derived neurotrophic factor val66met polymorphism affects human memory-related hippocampal activity and predicts memory performance. *Journal of Neuroscience, 23*, 6690–6694.

Harris, S. E., Fox, H., Wright, A. F., Hayward, C., Starr, J. M., Whalley, L. J., & Deary, I. J. (2006). The brain-derived neurotrophic factor Val66Met polymorphism is associated with age-related change in reasoning skills. *Molecular Psychiatry, 11*, 505–513.

Harrisberger, F., Spalek, K., Smieskova, R., Schmidt, A., Coynel, D., Milnik, A., … Borgwardt, S. (2014). The association of the BDNF Val66Met polymorphism and the hippocampal volumes in healthy humans: A joint meta-analysis of published and new data. *Neuroscience & Biobehavioral Reviews, 42*, 267–278.

Hashimoto, R., Moriguchi, Y., Yamashita, F., Mori, T., Nemoto, K., Okada, T., ... Ohnishi, T. (2008). Dose-dependent effect of the Val66Met polymorphism of the brain-derived neurotrophic factor gene on memory-related hippocampal activity. *Neuroscience Research, 61*, 360–367.

Hauck, S., Gomes, F., Silveira Júnior Ede, M., Almeida, E., Possa, M., & Ceitlin, L. H. (2009). Serum levels of brain-derived neurotrophic factor in acute and posttraumatic stress disorder: A case report study. *Brazilian Journal of Psychiatry, 31*, 48–51.

Hauck, S., Kapczinski, F., Roesler, R., de Moura, S. E. J., Magalhães, P.V., Kruel, L. R., ... Ceitlin, L. H. (2010). Serum brain-derived neurotrophic factor in patients with trauma psychopathology. *Progress in Neuro-Psychopharmacology & Biological Psychiatry, 3*, 459–462.

Hayes, J. P., Logue, M. W., Reagan, A., Salat, D., Wolf, E. J., Sadeh, N., & Miller, M. W. (2017). COMT Val158Met polymorphism moderates the association between PTSD symptom severity and hippocampal volume. *Journal of Psychiatry & Neuroscience, 42*(2), 95–102.

Ito, M., Takebayashi, Y., Suzuki, Y. & Horikoshi, M. (2018). Posttraumatic stress disorder checklist for DSM-5: Psychometric properties in a Japanese population. *Journal of Affective Disorders, 247*, 11–19.

Jovanovic, T., Norrholm, S. D., Sakoman, A. J., Esterajher, S., & Kozarić-Kovačić, D. (2009). Altered resting psychophysiology and startle response in Croatian combat veterans with PTSD. *International Journal of Psychophysiology, 71*, 264–268.

Kaess, B. M., Preis, S. R., Lieb, W., Beiser, A. S., Yang, Q., Chen, T. C., ... Stewart, A. F. (2015). Circulating brain-derived neurotrophic factor concentrations and the risk of cardiovascular disease in the community. *Journal of the American Heart Association, 4*, e001544.

Kang, H. K., Natelson, B. H., Mahan, C. M., Lee, K.Y., & Murphy, F. M. (2003). Post-traumatic stress disorder and chronic fatigue syndrome-like illness among Gulf War veterans: A population-based survey of 30,000 veterans. *American Journal of Epidemiology, 157*, 141–148.

Kao, C.Y., Stalla, G., Stalla, J., Wotjak, C. T., & Anderzhanova, E. (2015). Norepinephrine and corticosterone in the medial prefrontal cortex and hippocampus predict PTSD-like symptoms in mice. *European Journal of Neuroscience, 419*, 1139–1148.

Kessler, R. C., Aguilar-Gaxiola, S., Alonso, J., Benjet, C., Bromet, E. J., Cardoso, G., ... Koenen K. C. (2017). Trauma and PTSD in the WHO World Mental Health Surveys. *European Journal of Psychotraumatology, 8*(5), 1353383.

Kilpatrick, D. G., Koenen, K. C., Ruggiero, K. J., Acierno, R., Galea, S., Resnick, H. S., ... Gelernter, J. (2007). The serotonin transporter genotype and social support and moderation of posttraumatic stress disorder and depression in hurricane-exposed adults. *American Journal of Psychiatry, 164*(11), 1693–1699.

Kim, T.Y., Kim, S. J., Chung, H. G., Choi, J. H., Kim, S. H., & Kang, J. I. (2017). Epigenetic alterations of the BDNF gene in combat-related post-traumatic stress disorder. *Acta Psychiatrica Scandinavica, 135*(2), 170–179.

Kimbrel, N. A., Morissette, S. B., Meyer, E. C., Chrestman, R., Jamroz, R., Silvia, P. J., ... Young, K. A. (2015). Effect of the 5-HTTLPR polymorphism on posttraumatic stress disorder, depression, anxiety, and quality of life among Iraq and Afghanistan veterans. *Anxiety, Stress, & Coping, 28*(4), 456–466.

Kintzle, S., Barr, N., Corletto, G., & Castro, C. A. (2018). PTSD in U.S. veterans: The role of social connectedness, combat experience and discharge. *Healthcare (Basel), 6*(3). pii: E102. Advance online publication. doi: 10.3390/healthcare6030102.

Koenen, K. C., Aiello, A. E., Bakshis, E., Amstadter, A. B., Ruggiero, K. J., Acierno, R., ... Galea, S. (2009a). Modification of the association between serotonin transporter genotype and risk of posttraumatic stress disorder in adults by country-level social environment. *American Journal of Epidemiology, 169*(6), 704–711.

Koenen, K. C., Uddin, M., Chang, S. C., Aiello, A. E., Wildman, D. E., Goldmann, E., & Galea, S. (2011). SLC6A4 methylation modifies the effect of the number of traumatic events on risk for post-traumatic stress disorder. *Depression & Anxiety, 28*(8), 639–647.

Koenen, K. C., Amstadter, A. B., & Nugent N. R. (2009b). Gene-environment interaction in post-traumatic stress disorder: An update. *Journal of Traumatic Stress, 22*, 416–426.

Kolassa, I. T., Ertl, V., Eckart, C., Glockner, F., Kolassa, S., Papassotiropoulos, A., ... Elbert, T. (2010). Association study of trauma load and SLC6A4 promoter polymorphism in posttraumatic stress disorder: Evidence from survivors of the Rwandan genocide. *The Journal of Clinical Psychiatry, 71*(5), 543–547.

Kolassa, I. T., Illek, S., Wilker, S., Karabatsiakis, A., & Elbert, T. (2015). Neurobiological findings in post-traumatic stress disorder. In U. Schnyder, & M. Cloitre, (Eds.), *Evidence Based Treatments for Trauma-Related Psychological Disorders* (pp. 63–86). Cham: Springer.

Komar, Z., & Vukusic, H. (1999). Post-traumatic stress disorder in Croatian war veterans; incidence and psychosocial features. In D. Dekaris, & A. Sabioncello, (Eds.), *New Insights in Post-traumatic Stress Disorder (PTSD)* (pp. 42–44). Proceedings (Abstracts, references and illustrations). Zagreb: Croatian Academy of Science and Arts.

Konjevod, M., Tudor, L., Svob Strac, D., Nedic Erjavec, G., Barbas, C., Nikolac Perkovic, … Pivac, N. (2019). Metabolomic and glycomic findings in posttraumatic stress disorder. *Progress in Neuro-Psychopharmacology and Biological Psychiatry*, *88*, 181–193.

Kovacic Petrovic, Z., Nedic Erjavec, G., Nikolac Perkovic, M., Peraica, T., & Pivac, N. (2016). No association between the serotonin transporter linked polymorphic region polymorphism and severity of posttraumatic stress disorder symptoms in combat veterans with or without comorbid depression. *Psychiatry Research*, *244*, 376–381.

Kovacic Petrovic, Z., Nedic Erjavec, G., Nikolac Perkovic, M., Svob Strac, D., Peraica, T., Tudor, L., & Pivac, N. (2019). The association between serotonin transporter polymorphism, platelet serotonin concentration and insomnia in non-depressed veterans with posttraumatic stress disorder. *Psychiatria Danubina*, *31*(1), 78–87.

Kovacic, Z., Henigsberg, N., Pivac, N., Nedic, G., & Borovecki, A. (2008). Platelet serotonin concentration and suicidal behavior in combat related posttraumatic stress disorder. *Progress in Neuro-Psychopharmacology & Biological Psychiatry*, *32*(2), 544–551.

Lee, H. J., Kwak, S. K., Paik, J. W., Kang, R. H., & Lee, M. S. (2007). Association between serotonin 2A receptor gene polymorphism and posttraumatic stress disorder. *Psychiatry Investigation*, *4*(2), 104–108.

Lee, H. J., Lee, M. S., Kang, R. H., Kim, H., Kim, S. D., Kee, B. S., … Paik, I. H. (2005). Influence of the serotonin transporter promoter gene polymorphism on susceptibility to posttraumatic stress disorder. *Depression & Anxiety*, *21*(3), 135–139.

Lee, J. C., Wang, L. P., & Tsien, J. Z. (2016). Dopamine rebound-excitation theory: Putting brakes on PTSD. *Frontiers in Psychiatry*, 7, 163.

Lehrner, A., & Yehuda, R. (2014). Biomarkers of PTSD: Military applications and considerations. *European Journal of Psychotraumatology*, *5*. doi: 10.3402/ejpt.v5.23797.

Li, L. Z., Bao, Y. J., He, S. B., Wang, G., Guan, Y. L., Ma, D. X., & Yang, J. Y. (2016). The association between genetic variants in the dopaminergic system and posttraumatic stress disorder: A meta-analysis. *Medicine*, *95*(11), e3074. doi: 10.1097/MD.0000000000003074.

Li, L., Li, M. X., Pan, L. H., Wang, G. M., Guo, M., Fu, L. Q., … Xie, M. X. (2016). Comparative analysis of platelet 5-HT concentrations in Han and Li patients with post-traumatic stress disorder. *Genetics and Molecular Research*, *15*(3). DOI: 10.4238/gmr.15038265.

Liu, Y., Garett, M. E., Dennis, M. F., Green, K. T., VA Mid-Atlantic MIRECC Registry Workgroup, Ashley-Koch, A. E., … Kimbrel, N. A. (2015). An examination of the association between 5-HTTLPR, combat exposure, and PTSD diagnosis among U.S. veterans. *PLOS ONE*, *10*(3), e0119998.

Maes, M., Lin, A. H., Verkerk, R., Delmeire, L., Van Gastel, A., Van der Planken, M., & Scharpé, S. 1999. Serotonergic and noradrenergic markers of post-traumatic stress disorder with and without major depression. *Neuropsychopharmacology*, *20*(2), 188–197.

Maguire, K., Norman, T., Burrows, G., Hopwood, M., & Morris, P. (1998). Platelet paroxetine binding in post-traumatic stress disorder. *Psychiatry Research*, 77(1), 1–7.

Marshall, R. D., Beebe, K. L., Oldham, M., & Zaninelli, R. (2001). Efficacy and safety of paroxetine treatment for chronic PTSD: A fixed-dose, placebo-controlled study. *American Journal of Psychiatry*, *158*(12), 1982–1988.

Martenyi, F., Brown, E. B., Zhang, H., Prakash, A., & Koke, S. C. (2002). Fluoxetine versus placebo in posttraumatic stress disorder. *Journal of Clinical Psychiatry*, *63*(3), 199–206.

Matsuoka, Y., Nishi, D., Noguchi, H., Kim, Y., & Hashimoto, K. (2013). Longitudinal changes in serum brain-derived neurotrophic factor in accident survivors with posttraumatic stress disorder. *Neuropsychobiology*, *68*, 44–50.

Mellman, T. A., & Kumar, A. M. (1994). Platelet serotonin measures in posttraumatic stress disorder. *Psychiatry Research*, *53*(1), 99–101.

Mellman, T. A., Alim, T., Brown, D. D., Gorodetsky, E., Buzas, B., Lawson, W. B., … Charney, D. S. (2009). Serotonin polymorphisms and post-traumatic stress disorder in a trauma exposed African–American population. *Depression & Anxiety*, *26*(11), 993–997.

Havelka Mestrovic, A., Tudor, L., Nikolac Perkovic, M., Nedic Erjavec, G., Kozaric Petrovic, Z., Svob Strac, D., … Pivac, N. (2018). Significant association between catechol-O-methyltransferase (COMT) Val(158/108) Met polymorphism and cognitive function in veterans with PTSD. *Neuroscience Letters*, *666*, 38–43.

Miller, J. K., & Wiener, J. M. (2014). PTSD recovery, spatial processing, and the val66met polymorphism. *Frontiers in Human Neuroscience*, *26*(8), 100. doi: 10.3389/fnhum.2014.00100.

Miller, M. W., Sperbeck, E., Robinson, M. E., Sadeh, N., Wolf, E. J., Hayes, J. P., … McGlinchey, R. (2016). 5-HT2A gene variants moderate the association between PTSD and reduced default mode network connectivity. *Frontiers in Neuroscience*, *10*, 299.

Morey, R., Hariri, A., Gold, A., Hauser, M., Munger, H., Dolcos, F., & McCarthy, G. (2011). Serotonin transporter gene polymorphisms and brain function during emotional distraction from cognitive processing in posttraumatic stress disorder. *BioMed Central Psychiatry*, *11*, 76–89.

Muck-Seler, D., & Pivac, N. (2011). Serotonin. *Periodicum Biologorum*, *113*(1), 29–41.

Muck-Seler, D., Pivac, N., Jakovljevic, M., Sagud, M., & Mihaljevic-Peles, A. (2003). Platelet 5-HT concentration and comorbid depression in war veterans with and without posttraumatic stress disorder. *Journal of Affective Disorders*, *75*(2), 171–179.

Murphy, D. L., Li, Q., Engel, S., Wichems, C., Andrews, A., Lesch, K. P., & Uhl, G. (2001). Genetic perspectives on the serotonin transporter. *Brain Research Bulletin*, *56*(5), 487–494.

Murrough, J. W., Huang, Y., Hu, J., Henry, S., Williams, W., Gallezot, J. D., … Neumeister, A. (2011). Reduced amygdala serotonin transporter binding in posttraumatic stress disorder. *Biological Psychiatry*, *70*(11), 1033–1038.

Nakamura, M., Ueno, S., Sano, A., & Tanabe, H. (2000). The human serotonin transporter gene linked polymorphism (5-HTTLPR) shows ten novel allelic variants. *Molecular Psychiatry*, *5*(1), 32–38.

Navarro-Mateu, F., Escámez, T., Koenen, K. C., Alonso, J., & Sánchez-Meca, J. (2013). Meta-analyses of the 5-HTTLPR polymorphisms and post-traumatic stress disorder. *PLOS ONE*, *8*(6), e66227.

Nedic Erjavec, G., Konjevod, M., Nikolac Perkovic, M., Svob Strac, D., Tudor, L., Barbas, C., Grune, T., … Pivac, N. (2018). Short overview on metabolomic approach and redox changes in psychiatric disorders. *Redox Biology*, *14*, 178–186.

Nikolac Perkovic, M., Nedic Erjavec, G., Svob Strac, G., Uzun, S., Kozumplik, O., & Pivac, N. (2017). Theranostic biomarkers for schizophrenia. *International Journal of Molecular Sciences*, *18*(4), 733. Published online. doi: 10.3390/ijms18040733.

Olff, M., & van Zuiden, M. (2017). Neuroendocrine and neuroimmune markers in PTSD: Pre-, peri- and post-trauma glucocorticoid and inflammatory dysregulation. *Current Opinion in Psychology*, *14*, 132–137.

Osuch, E. A., Benson, B. E., Luckenbaugh, D. A., Geraci, M., Post, R. M., & McCann, U. (2009). Repetitive TMS combined with exposure therapy for PTSD: A preliminary study. *Journal of Anxiety Disorder*, *23*(1), 54–59.

Pai, A., Suris, A. M., & North, C. S. (2017). Posttraumatic stress disorder in the DSM-5: controversy, change, and conceptual considerations. *Behavioral Sciences*, 7(1), 7. Doi: https://doi.org/10.3390/bs7010007.

Pan, X., Kaminga, A. C., Wen, S. W., & Liu, A. (2018). Catecholamines in post-traumatic stress disorder: A systematic review and meta-analysis. *Frontiers in Molecular Neuroscience*, *11*, 450. doi: 10.3389/fnmol.2018.00450.

Pezze, M. A., & Feldon, J. (2004). Mesolimbic dopaminergic pathways in fear conditioning. *Progress in Neurobiology*, *74*(5), 301–320.

Peters, J., Dieppa-Perea, L. M., Melendez, L. M., & Quirk G. J. (2010). Induction of fear extinction with hippocampal-infralimbic BDNF. *Science*, *328*, 1288–1290.

Pikula, A., Beiser, A. S., Chen, T. C., Preis, S. R., Vorgias, D., DeCarli, C., … Seshadri, S. (2013). Serum brain-derived neurotrophic factor and vascular endothelial growth factor levels are associated with risk of stroke and vascular brain injury: Framingham Study. *Stroke*, *44*, 2768–2775.

Pivac, N., Knezevic, J., Kozaric-Kovacic, D., Dezeljin, M., Mustapic, M., Rak, D., & Muck-Seler, D. (2007). Monoamine oxidase (MAO) intron 13 polymorphism and platelet MAO-B activity in combat-related posttraumatic stress disorder. *Journal of Affective Disorders*, *103*(1–3), 131–138.

Pivac, N., Kozaric-Kovacic, D., Mustapic, M., Dezeljin, M., Borovecki, A., Grubisic-Ilic, M., & Muck-Seler, D. (2006). Platelet serotonin in combat related posttraumatic stress disorder with psychotic symptoms. *Journal of Affective Disorders*, *93*(1–3), 223–227.

Pivac, N., Muck-Seler, D., Sagud, M., & Jakovljevic, M. (2002). Platelet serotonergic markers in posttraumatic stress disorder. *Progress in Neuro-Psychopharmacology & Biological Psychiatry*, *26*(6), 1193–1198.

Pivac, N., Kozaric-Kovacic, D., Grubisic-Ilic, M., Nedic, G., Rakos, I., Nikolac, M., … Muck-Seler, D. (2012). The association between brain-derived neurotrophic factor Val66Met variants and psychotic symptoms in posttraumatic stress disorder. *World Journal of Biological Psychiatry*, *13*, 306–311.

Rosas-Vidal, L. E., Do-Monte, F. H., Sotres-Bayon, F., & Quirk G. J. (2014). Hippocampal-prefrontal BDNF and memory for fear extinction. *Neuropsychopharmacology*, *39*, 2161–2169.

Rosebrock, L. A., Arditte Hall, K. A., Rando, A., Pineles, S. L. & Liverant, G. I. (2018). Rumination and its relationship with thought suppression in unipolar depression and comorbid PTSD. *Cognitive Therapy and Research*, *43*(1), 226–235.

Sanchez, M. M., Das, D., Taylor, J. L., Noda, A., Yesavage, J. A., & Salehi, A. (2011). BDNF polymorphism predicts the rate of decline in skilled task performance and hippocampal volume in healthy individuals. *Translational Psychiatry*, *1*, 10.

Sandler, M., Glover, V., Clow, A., & Jarman, J. (1993). Monoamine oxidase-B, monoamine oxidase-B inhibitors, and Parkinson's disease. A role for superoxide dismutase? *Advances in Neurology*, *60*, 238–241.

Segman, R. H., Cooper-Kazaz, R., Macciardi, F., Goltser, T., Halfon, Y., Dobroborski, T., & Shalev, A. Y. (2002). Association between the dopamine transporter gene and posttraumatic stress disorder. *Molecular Psychiatry*, 7(8), 903–907.

Sheerin, C. M., Lind, M. J., Bountress, K., Nugent, N. R., & Amstadter, A. B. (2017). The genetics and epigenetics of PTSD: Overview, recent advances, and future directions. *Current Opinion in Psychology*, *14*, 5–11.

Sher, L., Oquendo, M. A., Li, S., Burke, A. K., Grunebaum, M. F., Zalsman, G., & Mann, J. J. (2005). Higher cerebrospinal fluid homovanillic acid levels in depressed patients with comorbid posttraumatic stress disorder. *European Neuropsychopharmacology*, *15*(2), 203–209.

Southwick, S. M., Krystal, J. H., Bremner, J. D., Morgan, C. A. III, Nicolaou, A. L., Nagy, L. M., ... Charney, D. S. (1997). Noradrenergic and serotonergic function in posttraumatic stress disorder. *Archives of General Psychiatry*, *54*(8), 749–758.

Southwick, S. M., Paige, S., Morgan, C. A. III, Bremner, J. D., Krystal, J. H., & Charney, D. S. (1999). Neurotransmitter alterations in PTSD: Catecholamines and serotonin. *Seminars in Clinical Neuropsychiatry*, *4*(4), 242–248.

Spivak, B., Vered, Y., Graff, E., Blum, I., Mester, R., & Weizman, A. (1999). Low platelet-poor plasma concentrations of serotonin in patients with combat-related posttraumatic stress disorder. *Biological Psychiatry*, *45*(7), 840–845.

Stratta, P., Sanità, P., Bonanni, R. L., de Cataldo, S., Angelucci, A., Rossi, R., ... Rossi, A. (2016). Clinical correlates of plasma brain-derived neurotrophic factor in post-traumatic stress disorder spectrum after a natural disaster. *Psychiatry Research*, *244*, 165–167.

Su, S., Xiao, Z., Lin, Z., Qiu, Y., Jin, Y., & Wang, Z. (2015). Plasma brain-derived neurotrophic factor levels in patients suffering from post-traumatic stress disorder. *Psychiatry Research*, *229*, 365–369.

Sullivan, G. M., Ogden, R. T., Huang, Y. Y., Oquendo, M. A., Mann, J. J., & Parsey, R. V. (2013). Higher *in vivo* serotonin-1a binding in posttraumatic stress disorder: A PET study with [11C]WAY-100635. *Depression & Anxiety*, *30*(3), 197–206.

Svob Strac, D., Kovacic Petrovic, Z., Nikolac Perkovic, M., Umolac, D., Nedic Erjavec, G., & Pivac, N. (2016). Platelet monoamine oxidase type B, MAOB intron 13 and MAOA-uVNTR polymorphism and symptoms of post-traumatic stress disorder. *Stress*, *19*(4), 362–373.

Szeszko, P. R., Lipsky, R., Mentschel, C., Robinson, D., Gunduz-Bruce, H., Sevy, S., .... Malhotra, A. K. (2005). Brain-derived neurotrophic factor val66met polymorphism and volume of the hippocampal formation. *Molecular Psychiatry*, *10*, 631–636.

Takashio, S., Sugiyama, S., Yamamuro, M., Takahama, H., Hayashi, T., Sugano, Y., ... Anzai, T. (2015). Significance of low plasma levels of brain-derived neurotrophic factor in patients with heart failure. *The American Journal of Cardiology*, *116*(2), 243–249.

Thakur, G. A., Joober, R., & Brunet, A. (2009). Development and persistence of posttraumatic stress disorder and the 5-HTTLPR polymorphism. *Journal of Traumatic Stress*, *22*(3), 240–243.

Tural, Ü., Tamer Aker, A., Turan Sodan, H., Ünver, H., & Akansel, G. (2018). Neurotrophic factors and hippocampal activity in PTSD. *PLOS ONE*, *13*, e0197889.

van den Heuvel, L., Suliman, S., Malan-Müller, S., Hemmings, S., & Seedat, S. (2016). Brain-derived neurotrophic factor Val66Met polymorphism and plasma levels in road traffic accident survivors. *Anxiety Stress Coping*, *29*, 616–629.

Van, I. M. H., Caspers, K., Bakermans-Kranenburg, M. J., Beach, S. R., & Philibert, R. (2010). Methylation matters: Interaction between methylation density and serotonin transporter genotype predicts unresolved loss or trauma. *Biological Psychiatry*, *68*(5), 405–407.

VanDyke, L. M., Burton, E., Hamidovic, A. & Burge, M.R. (2017). Hormonal aspects of post-traumatic stress disorder. *Endocrinology Diabetes and Metabolis Journal*, *1*(2), 1–10.

Voisey, J., Swagell, C. D., Hughes, I. P., Alorris, C. P., van Daal, A., Noble, E. P., & Lawford, B. R. (2009). The DRD2 gene 957C>T polymorphism is associated with posttraumatic stress disorder in war veterans. *Depression and Anxiety*, *26*(1), 28–33.

Wang, Z., Baker, D. G., Harrer, J., Hamner, M., Price, M., & Amstadter, A. (2011). The relationship between combat-related posttraumatic stress disorder and the 5-HTTLPR/rs25531 polymorphism. *Depression & Anxiety*, *28*(12), 1067–1073.

Weizman, R, Laor, N, Schujovitsky, A, Wolmer, L, Abramovitz-Schnaider, P, Freudstein-Dan, A, & Rehavi, M. (1996). Platelet imipramine binding in patients with posttraumatic stress disorder before and after phenelzine treatment. *Psychiatry Research*, *63*(2–3), 143–150.

Xie, P., Kranzler, H. R., Poling, J., Stein, M. B., Anton, R. F., Brady, K., … Gelernter, J. (2009). Interactive effect of stressful life events and the serotonin transporter 5-HTTLPR genotype on post-traumatic stress disorder diagnosis in 2 independent populations. *Archives of General Psychiatry*, *66*(11), 1201–1209.

Zhang, H., Ozbay, F., Lappalainen, J., Kranzler, H. R., van Dyck, C. H., Charney, D. S., … Gelernter, J. (2006). Brain derived neurotrophic factor (BDNF) gene variants and Alzheimer's disease, affective disorders, posttraumatic stress disorder, schizophrenia, and substance dependence. *American Journal of Medical Genetics Part B: Neuropsychiatric Genetics*, *141B*, 387–393.

Zhang, L., Benedek, D. M., Fullerton, C. S., Forsten, R. D., Naifeh, J. A., Li, X. X., ….. Ursano, R. J. (2014). PTSD risk is associated with BDNF Val66Met and BDNF overexpression. *Molecular Psychiatry*, *19*, 8–10.

Zhao, M., Yang, J., Wang, W., Ma, J., Zhang, J., Zhao, X., … Yang, Y. (2017). Meta-analysis of the interaction between serotonin transporter promoter variant, stress, and posttraumatic stress disorder. *Scientific Reports*, 7(1), 16532.

Ziegler, C., Wolf, C., Schiele, M. A., Feric Bojic, E., Kucukalic, S., Sabic Dzananovic, E., & Domschke, K. (2018). Monoamine oxidase A gene methylation and its role in posttraumatic stress disorder: first evidence from the south eastern Europe (SEE)-PTSD study. *International Journal of Neuropsychopharmacology*, *21*(5), 423–432.

# 32

# NEUROENDOCRINE AND IMMUNE BIOMARKERS OF POSTTRAUMATIC STRESS DISORDER IN COMBAT VETERANS

*Nela Pivac, Marcela Konjevod, Marina Sagud, Suzana Uzun, and Oliver Kozumplik*

## Posttraumatic stress disorder

Posttraumatic stress disorder (PTSD) is a trauma- and stress-related disorder, frequent in modern society in civilians as well as in soldiers. It develops as a consequence of the exposure to a traumatic event(s), usually combat in military personnel, but only in a portion of vulnerable soldiers. Risk factors for PTSD include different neuroanatomical, neuroendocrine, immune, psychological, molecular, genetic, cognitive, emotional, environmental and psychosocial factors and the interaction between them. Therefore, biomarkers of vulnerability to develop PTSD in soldiers and different symptoms and comorbidities that occur in combat related PTSD are important research targets in military medicine. The quest to identify biomarkers is a major question in PTSD research, i.e. to detect and identify factors of vulnerability (risk factors) or factors of resilience (protective factors) in military veterans. This issue is especially important in the military, since there is still an unfulfilled goal to find validated and confirmed biomarkers that would predict PTSD development in soldiers exposed to combat trauma, which might also foresee the behaviour and responses in soldiers under combat stress (Frank, Zamorski, Lee & Colman, 2018).

Diagnostic criteria for PTSD according to the DSM-5 (American Psychiatric Association, 2013; Ito, Takebayashi, Suzuki & Horikosh, 2018; Rosebrock, Arditte Hall, Rando, Pineles & Liverant, 2018) are: Criterion A: stressor; Criterion B: intrusion symptoms; Criterion C: avoidance; Criterion D: negative alterations in cognition and mood; Criterion E: alterations in arousal and reactivity; Criterion F: duration of more than 1 month; Criterion G: functional significance (symptoms create distress or functional impairment) and Criterion H: exclusion (symptoms are not due to medication, substance use or other illness). There are two subtypes of PTSD: (1) dissociative subtype: in addition to meeting criteria for diagnosis, an individual experiences high levels of depersonalization and derealization. Depersonalization is an experience of being an outside observer of or detached from oneself (e.g., feeling as if "this is not happening to me" or one were in a dream). Derealization is an experience of unreality, distance or distortion (e.g., "things are not real") and (2) delayed subtype is presented when full diagnostic criteria are not met until at least six months after the trauma(s), although onset of symptoms may occur immediately (Weathers, 2017; Yehuda et al., 2015). The diagnosis of PTSD has been criticized on numerous grounds, but principally for three reasons: (1) the alleged pathologizing of normal events, (2) the inadequacy of Criterion A and (3) symptom overlap with other disorders. The diagnosis of PTSD also represents a challenge since in the military, due to malingering (Ali, Jabeen & Alam, 2015); veterans might exaggerate or describe false symptoms because of the possible benefits, i.e. financial or personal gains, or forensic purposes. On the other hand, PTSD might be under-diagnosed due to denial or minimalizing of symptoms since a diagnosis of PTSD might affect veterans' employment status and/or promotion in military service and later in civilian life (Lehrner & Yehuda, 2014).

Although exposure to trauma is common throughout the world (Bisson, Cosgrove & Roberts, 2015), the majority of people will not develop PTSD; the prevalence of PTSD development differs among different populations, while the type of traumatic exposure significantly affects its development (Kessler et al., 2017). The lifetime prevalence of PTSD in the general population is 1.3%–8.8% (Atwoli, Stein, Koenen & McLaughlind, 2015). The prevalence of PTSD is much higher in the military. A prevalence of 11%–30% of war-related traumatic experiences has been recorded among U.S. combat veterans (Dursa, Reinhard, Barth & Schneiderman, 2014; Kang, Natelson, Mahan, Lee & Murphy, 2003). This prevalence differs according to the area of service, and the estimated prevalence of PTSD is 30% in Vietnam veterans, 10% in Gulf War veterans, 15% in veterans serving in Iraq and 11% in veterans returning from Afghanistan (Kintzle, Barr, Corletto & Castro, 2018). The estimated prevalence of PTSD in Croatian veterans is 18%–40% (Komar & Vukusic, 1999; Priebe et al., 2010).

## Biomarkers in posttraumatic stress disorder

PTSD is associated with diverse symptoms, and these different symptoms are all linked with disturbances in different biological, psychological, social and neurological functions. Combat-related PTSD has some special characteristics since it is associated with higher total PTSD severity and arousal symptom intrusions (Guina, Nahhas, Suttom & Farnsworth, 2018), often with traumatic brain injury and other blast injuries, resulting in the "poly-trauma" concept and treatment resistance (Corvalan & Klein, 2011). Dysregulation of many biological systems is associated with PTSD, but to date, there are still no specific, validated or sensitive biomarkers for PTSD (Lehrner & Yehuda, 2014).

Biomarkers are non-invasive and objective measures of patient diagnosis, prognosis and treatment (Nikolac Perkovic et al., 2017). They should be determined in easily available fluids such as in saliva, blood, urine and cerebrospinal fluid. However, there are no valid biomarkers for PTSD, and some proposed biomarkers (Lehrner & Yehuda, 2014) still lack specificity and sensitivity. Therefore, there is a high priority for clearly defined and replicated valid biomarkers to discriminate between persons susceptible or resilient to the development of PTSD. Such biomarkers might provide insights into the disease pathogenesis, and can be used for prevention (if related to vulnerability to develop PTSD). In the military, they might even help in selection of resilient individuals who are less likely to develop PTSD when exposed to combat stressors. Biomarkers might also be used for development of targeted treatments and interventions. Biomarkers of PTSD related to the neuroendocrine system (the hypothalamic-pituitary-adrenal [HPA] axis), are CRH, ACTH and cortisol; biomarkers of the immune system are C-reactive protein /CRP/ and other cytokines: interleukin 6 /IL-6/, interferon-$\gamma$) (Kao, Stalla, Stalla,

Wotjak & Anderzhanova, 2015; Konjevod et al., 2019; Nedic Erjavec et al., 2018; Olff & van Zuiden, 2017; VanDyke, Burton, Hanidovic & Burke, 2017).

## Biomarkers of posttraumatic stress disorder comorbid with cardiovascular diseases

Reliable diagnosis of PTSD is sometimes difficult to achieve (Lehrner & Yehuda, 2014) since PTSD is rarely an isolated disorder and is frequently comorbid with other somatic (Britvic et al., 2015) and psychiatric (Flory, 2015) disorders. In the military, PTSD is very often associated with mild traumatic brain injury (TBI) (Lehrner & Yehuda, 2014). Among the somatic comorbidities in veterans, the most frequent are: cardiovascular, dermatological, musculoskeletal, pulmonary and metabolic diseases (Britvic et al., 2015). Therefore, combat-related PTSD increases the likelihood of developing somatic diseases (Britvic et al., 2015). In Croatia, PTSD veterans with cardiovascular diseases (CVD), myocardial infarction, arrhythmia and other CVD were more common than in general population (Britvic et al., 2015). It has been known for a long time that severe trauma, such as combat stress, affects the heart and cardiovascular system (Tennant, 1982). Prospective studies have consistently found that individuals with PTSD, without known CVD, have increased risk of future CVD events. These findings are evident for both war-related (Ahmadi et al., 2015; Beristianos, Yaffe, Cohen & Byres, 2016; Britvic et al., 2015; Scherrer et al., 2010) and civilian PTSD (Gradus et al., 2015; Jordan et al., 2013). Middle-aged American war veterans with PTSD had higher risk for congestive heart failure, myocardial infarction and peripheral vascular disease than veterans without PTSD (Beristianos et al., 2016). Patients from the Veteran's Administration diagnosed with PTSD were at increased risk of incident myocardial infarction (Scherrer et al., 2010). Similarly, PTSD was associated with myocardial infarction, stroke, ischaemic stroke and venous thromboembolism in a Danish population, with the strongest association in the youngest age group of up to 39 years of age (Gradus et al., 2015). This association existed in the absence of depression, alcohol abuse or other substance dependences (Gradus et al., 2015). Veterans without known coronary heart disease (CHD) had the highest risk of cardiovascular mortality, especially if they had PTSD comorbid with mild TBI (Ahmadi et al., 2015). The presence of PTSD was associated with myocardial ischemia in a large cohort of otherwise asymptomatic veterans (Turner, Neylan, Schiller, Li & Cohen, 2013). PTSD was most strongly associated with incident CVD among veterans of 50 to 59 years of age (Scherrer et al., 2019). Despite large advances in prevention and early recognition and treatment of CVD, these conditions are still the leading cause of death worldwide, while PTSD is considered a fast-track to a premature CVD (Wentworth et al., 2013; Wolf & Schnurr, 2016). These results demonstrate the importance for physicians and healthcare providers to regularly assess cardiovascular (CV) health in combat veterans and other populations with high levels of trauma exposure, regardless of the age or other demographic variables. The connection between PTSD with the increased CVD is still not clear. Risk factors for CVD are numerous and interrelated (Sagud et al., 2017), and CVD biomarkers are elevated in patients with PTSD. Common pathways for both PTSD and CVD include sympathetic over-activation and inflammation (Brudey et al., 2015). Hypertension is related to chronic autonomic activation and HPA axis dysregulation (Kibler, Tursich, Ma, Malcolm & Greenbarg, 2014). Hypertension was found to be the most common component of the metabolic syndrome in middle-aged veterans with PTSD (Rosenbaum et al., 2015). Subjects with PTSD, including both war-related and civilian PTSD, had almost twice the risk for metabolic syndrome than individuals from the general population (Rosenbaum et al., 2015), and this was evident even in young veterans (Blessing et al., 2017). The presence of metabolic syndrome was also associated with a severity of PTSD symptoms (Heppner et al.,

2012). Individuals with PTSD also manifested other indicators of autonomic over-activity, such as increased muscle sympathetic nerve activity in response to virtual reality combat exposure (Park et al., 2017). The association between PTSD and incident CVD might be explained by the presence of numerous comorbid conditions, such as obesity, dyslipidaemia, hypertension, smoking, depression and sleep disturbances (Scherrer et al., 2019). In Croatian war veterans with PTSD, increased median BMI, cholesterol and triglyceride levels were found in individuals with the median age of 56 years who did not have any CVD and who were not taking cardiovascular drugs or statins (Tudor et al., 2018).

Chronic peripheral, low-grade inflammation is manifested in elevated circulatory levels of CRP and pro-inflammatory cytokines, suggesting immune dysfunction in PTSD (Hori & Kim, 2019; Kibler et al., 2014; Lindqvist et al., 2017). Several cross-sectional studies reported an association between elevated CRP levels and PTSD (Canetti, Russ, Luborsky, Gerhart & Hobfoll, 2014; Farr et al., 2015), particularly with re-experiencing and avoiding symptoms (Canetti et al., 2014; Rosen et al., 2017), overall PTSD severity (Farr et al., 2015; Michopolus et al., 2015; Rosen et al., 2017) and chronicity (Solomon et al., 2017). Meta-analysis confirmed higher IL-6, interleukin 1β (IL-1β), tumour necrosis factor-α (TNF-α) and interferon γ levels in a PTSD group compared to healthy controls (Passos et al., 2015). The duration of PTSD was positively correlated with IL-1β and PTSD severity with IL-6 levels (Passos et al., 2015). These findings show bidirectional association, since PTSD is associated with immune activation, and chronicity and severity of PTSD contribute to more pronounced inflammation. The link between increased levels of pro-inflammatory cytokines and CVD is, among others, increased platelet and endothelial activation (Michaud et al., 2013). Moreover, endothelial activation and/or endothelial damage contribute to the increased risk of arterial thrombosis in PTSD patients (Robicsek, Makhoul, Klein, Brenner & Sarig, 2011). The particularly important substances in the development and acceleration of atherosclerosis are IL-1β and IL-6 (Loppnow, Buerke, Werdan & Rose-John, 2011). This association might be in part explained by risk habits such as smoking and alcohol abuse (Dennis et al., 2014).

Brain-derived neurotropic factor (BDNF), although not a conventional CVD risk factor, has also drawn attention. Namely, low peripheral levels of BDNF were associated with an increased risk of stroke (Pikula et al., 2013) and were proposed as a biomarker of heart failure (Takashio et al., 2015). In line with this, higher serum BDNF levels were protective, suggesting decreased risk of CVD (Kaess et al., 2015). The most frequently studied polymorphisms on the BDNF gene, Val66Met and C270T, were not significantly associated with BMI or plasma lipid levels in Croatian war veterans with combat-related PTSD (Tudor et al., 2018). Namely, there was no association between BDNF Val66Met and BMI values, as well as plasma total cholesterol, triglycerides, HDL and LDL cholesterol levels in veterans with PTSD (Tudor et al., 2018).

All these studies delineated the need to advance our understanding of the underlying mechanisms by which PTSD contributes to the development of CVD. Unhealthy lifestyle, such as smoking, poor diet and physical inactivity, substantially contributes to increased CVD risks. The well-established association between PTSD and CVD may be at least partly due to poor diet, sedentary lifestyle, obesity and smoking (van den Berk-Clark et al., 2018). Patients with PTSD had higher smoking rates (Kelly, Jensen & Sofuoglu, 2015; Sagud et al., 2018) and nicotine dependence levels (Sagud et al., 2018), more difficulties in smoking cessation and stronger withdrawal symptoms than individuals not diagnosed with PTSD (Kelly et al., 2015). Moderate/heavy nicotine dependence was associated with more severe negativity, hyper-arousal and depression in veterans with PTSD (Sagud et al., 2018). Veterans with PTSD were also more frequently smokers (Dennis et al., 2014) and had higher prevalence of nicotine dependence when compared to veterans without PTSD (Beristianos et al., 2016). This was confirmed in twin studies, where twins with PTSD were more frequently smokers than twins without PTSD, despite no differences in BMI, HDL cholesterol and systolic blood pressure (Goetz et al., 2014). In addition, patients with PTSD frequently abuse alcohol, mainly to self-medicate anxiety and hyper-arousal (Edmondson, Kronish, Shaffer, Falzon & Burg, 2013). Other life-style risk factors include sedentary life-style and consumption of unhealthy food. Sedentary behaviour is related to other CVD risk factors, such as increased blood pressure, insulin resistance and high cholesterol levels (Brock, King, Wofford & Harell, 2005). Decreased physical activity and unhealthy eating habits in patients with PTSD were related to obesity (Hall, Hoerster & Yancy, 2015). In Croatian war veterans, BMI did not differ between veterans with or without PTSD and age-matched healthy control subjects (Kozaric Kovacic et al., 2009), but this study detected a high prevalence of obesity in a healthy group, and therefore veterans with PTSD and healthy groups had similar BMI values in the overweight range (Kozaric Kovacic et al., 2009). Recent meta-analysis confirmed the association between PTSD and increased BMI (Suliman et al., 2016). Although the aforementioned studies reveal increased CVD risk factors in individuals with PTSD, this relationship appears to be complex and not fully elucidated. The severity, symptom clusters and duration of PTSD symptoms, but not a PTSD diagnosis, contribute to increased CVD risks (Sagud et al., 2017).

To the best of our knowledge, there are no data on the effects of common treatment methods to decrease CVD risk markers in patients with PTSD. Unlike in the PTSD population, current literature provides evidence of effectiveness of different interventions, such as CBT and transcendental meditation, on reducing recurrent CVD events in cohorts with CHD, while data on primary prevention are lacking (Cohen, Edmondson & Kronish, 2015). Specifically, despite the well-known relationship between the presence of PTSD and increased CVD risk, there are no data on the influence of any treatment on CVD risk factors in patients with PTSD, which is recognized in the current literature as an impetus for the future research (Koenen et al., 2017). Progressive muscular relaxation had beneficial effects on sleep and depression in a small study on patients with PTSD (Blanaru et al., 2012). While establishing efficacious treatment for PTSD is an important goal, showing that such treatment would also improve cardio-metabolic health is a key outstanding question in the literature (Koenen et al., 2017).

Early detection and effective management of hypertension, type 2 diabetes mellitus, depression, anxiety, sleep disorders and other CVD risk factors, particularly smoking, might help to mitigate the risk of CVD in patients with PTSD. Thus, the recognition of CVD risk markers which might cause silent ischemia is important to prevent its further progression to potentially fatal CVD events. All aforementioned risk markers might be modifiable risk factors for CHD (Edmondson et al., 2013). Since at present it is unknown that any specific treatment which targets PTSD symptoms might alleviate CVD risks, future treatment trials addressing both PTSD symptoms and markers of CVD are urgently needed.

## Biomarkers of the hypothalamic-pituitary-adrenal axis in posttraumatic stress disorder

PTSD is associated with disturbed regulation and altered function of the major stress response system, the hypothalamic–pituitary–adrenal (HPA) axis (VanDyke et al., 2017; Yehuda et al., 2015). Release of all hormones of the HPA axis activation cascade, i.e. CRH, ACTH and glucocorticoid hormones (cortisol), is altered in PTSD. Although acute stress elicits an increase in these hormones, in PTSD there are heterogeneous and divergent findings regarding basal measurements of CRH, ACTH and cortisol (VanDyke et al., 2017). In normal conditions (stress-free conditions), the sympathetic nervous system (SNS) interacts with the HPA axis to produce response to external stimuli: the HPA axis is activated with the subsequent release of CRH, arginine vasopressin, ACTH and cortisol, and these hormones control their release via negative feedback. In situations of extreme stress and trauma, the SNS response activates fast release of noradrenaline, adrenaline and dopamine, with activation of the HPA axis and increased cortisol release. Cortisol binds in normal stress-free conditions to mineralocorticoid receptors and exerts negative feedback of CRH, arginine vasopressin and ACTH, but in stressful situations, it also binds to glucocorticoid receptors (Daskalakis, McGill, Lehrner & Yehuda, 2015; de Kloet et al., 2007). In extreme stress situations and in PTSD, this negative feedback control is disturbed and the whole system is out of balance, which leads to exaggerated responses to normal stimuli and disturbed glucocorticoid and CRH signalling (Osório, Probert, Jones, Young & Robbins, 2017; VanDyke et al., 2017; Yehuda et al., 2015). This insufficient signalling is responsible for diverse effects such as the inability of HPA downregulation and increased negative feedback sensitivity in PTSD but also consolidation of the traumatic memories and deficits in fear conditioning and inability of fear extinction (Osório et al., 2017; Yehuda et al., 2015).

Conclusions of literature review regarding HPA axis alterations in PTSD are mixed, but most of the studies reveal attenuated cortisol in subjects with PTSD; however, these findings depend on the time of the day when cortisol was measured, on the biological fluid in which cortisol was measured and confounding variables such as age, sex, and medication use, as well as the time past after traumatic experience (VanDyke et al., 2017). In contrast, in Croatian war veterans, plasma cortisol levels were increased both in veterans with PTSD as well as in veterans who did not develop PTSD but were exposed to the same traumatic experiences when compared to non-exposed healthy control subjects (Pivac, Muck-Seler, Jakovljevic, Ljubicic & Crncevic-Orlic, 1998). Contrarily to these data, no significant differences in basal blood concentrations of cortisol and ACTH were found in subjects with current PTSD, lifetime PTSD, trauma controls and healthy subjects (Savic, Knezevic, Damjanovic, Spiric & Matic, 2012). ACTH levels did not differ between subjects with or without PTSD (Kellner, Yassouridis, Hubner, Baker & Wiedemann, 2003; Muhtz, Wester, Yassouridis, Wiedemann & Kellner, 2008). The data using salivary cortisol in PTSD are more uniform, since most of the data detected decreased salivary cortisol in PTSD compared to control subjects (Pan, Wang, Wu, Wen & Liu, 2018; VanDyke et al., 2017). CRH has numerous roles in the CNS, and it modulates glutamatergic, dopaminergic, serotonergic and noradrenergic neurotransmission (Risbrough & Stein, 2006). CRH was found to be elevated in cerebrospinal fluid (CSF) of Vietnam veterans with PTSD as compared to healthy controls after single sampling (Bremner et al., 1997). This result was confirmed later after serial CSF sampling (Baker et al., 1999), but there is a lack of novel studies that would corroborate these findings. Increased CRH concentrations might be associated with hyper-arousal in PTSD patients (Bremner et al., 1997). Higher CRH in PTSD contributes to various disturbances and symptoms due to the widespread localization of CRH neurons and CRHR1 and CRHR2 receptors in different brain regions; there are complex and bidirectional interactions between glucocorticoids and CRH that are responsible for the disturbed fear and stress responses and divergent symptoms in PTSD (Raglan, Schmidt & Schulkin, 2017).

Components of the HPA axis are regulated by a number of genes, and their polymorphisms might affect HPA axis activity and the cascade of events that happens if this regulation is disturbed in PTSD (Mehta & Binder, 2012). The most common risk genetic variants associated with PTSD are the gene variants encoding glucocorticoid receptors (NR3C1). These polymorphisms (TthIII I, ER22/23EK, N363S, BclI and GR-9 beta) are important since they affect cortisol release and glucocorticoid receptor sensitivity and resistance. Genetic variants of the mineralocorticoid receptors (NR3C2), I180V and -2G/C, affect stress response and plasma cortisol level. A SNP within the CRH binding protein (CRHBP) locus, rs10473984, was significantly associated with higher plasma ACTH levels and increased dexamethasone suppression of ACTH (Binder et al., 2010). Other frequently studied genetic variants are those coding for CRHR1R and a co-chaperone of the glucocorticoid receptors (FKBP5), encoding FK506-binding protein 5 (Yehuda et al., 2015). An association between a SNP within the CRHR1 gene, rs110402, and cortisol levels was detected in subjects traumatized in childhood (Osório et al., 2017). FKBP5 is important since it regulates glucocorticoid receptor sensitivity and elicits less efficient nuclear translocation: after it binds to the receptor complex, cortisol binds with lower affinity to this receptor (Yehuda et al., 2015). PTSD is associated with downregulation of FKBP5 gene expression (Daskalakis et al., 2015). Genetic variants of FKBP5 are linked with increased FKBP5 protein expression, resulting in impaired binding of glucocorticoids to glucocorticoid receptors and consequently in disrupted negative feedback of the HPA axis. A recent meta-analysis confirmed the significant gene-by environment (GxE) effects of the FKBP5 variants (rs1360780, rs3800373, rs9296158, rs9470080) with traumatic event exposure on PTSD (Hawn et al., 2019).

Therefore, these data suggest that PTSD is associated with imbalance and dysregulation of the HPA axis and consequently with altered homeostasis: blunted cortisol secretion; low glucocorticoid signalling; impaired negative feedback, leading to the allostatic load and overreactions to moderate stressful stimuli (Daskalakis et al., 2015). All these alterations result in altered neurotransmission regulated by dopamine, noradrenaline, serotonin, brain derived neurotrophic factor (BDNF), and others, and in altered immune response.

## Biomarkers of the immune system in PTSD

Besides dysregulation of the HPA axis, PTSD is also associated with low-level chronic inflammation, and altered cytokine levels (Gill, Saligan, Woods & Page, 2009; Gola et al., 2013; Altemus, Dhabhar & Yang, 2006). Such dysregulation of immune system represents a risk for several metabolic disorders, including diabetes type 1 and 2, arthritis, cardiovascular and autoimmune diseases (Gill et al., 2009; Britvic et al., 2015). The interaction between stress and alterations in inflammatory response system (Maes et al., 1999a,b) leads to altered levels of humoral and cell-mediated cytokines (Altemus et al., 2006). Cytokines have multiple roles as the signalling molecules that mediate cross-talk between the peripheral immune system and central nervous system (Furtado & Katzman, 2015). Therefore, normal cytokine function is important for maintaining homeostasis. The immune system is associated with the HPA axis, as it maintains homeostasis and protects against inflammatory processes. Thus, dysregulation of the HPA axis and macrophage activation both stimulate pro-inflammatory cytokines that promote inflammation. Several studies showed that pro-inflammatory cytokines, including TNF-$\alpha$, interleukin-1$\alpha$ (IL-1$\alpha$), IL-1$\beta$ and IL-6, are associated with PTSD (Altemus, Dhabhar & Yang, 2006; Neigh & Ali, 2016; Wang et al., 2019). TNF-$\alpha$ and IL-6, which are responsible for symptoms associated with infection, can pass through the blood brain barrier and affect the HPA axis (Gill et al., 2009; Rohleder, Joksimovic, Wolf & Kirscbaum, 2004). There are mixed results on the role of cytokines in PTSD. Individuals with PTSD of both genders showed increases of IL-6 and TNF-$\alpha$ levels, while PTSD patients with comorbidities, such as depression, had even greater HPA axis dysregulation compared to controls (Gill et al., 2009). In addition, interleukin-2 (IL-2), interleukin-4 (IL-4), interleukin-8 (IL-8) and interleukin-10 (IL-10) were also significantly increased in individuals with PTSD compared to healthy controls (Furtado & Katzman, 2015). However, opposite results are also reported. IL-4 was decreased in PTSD subjects, while TNF-$\alpha$ and interferon-$\gamma$ were increased in female but not in male PTSD patients

(Furtado & Katzman, 2015; Gill, Vythilingam & Page, 2008). Cytokines IL-1β, IL-6 and TNF-α are released after stimulation of microglia or macrophages and have a major role in acute phase reaction. IL-1β is a cytokine that plays an important role in the inflammatory and immune system. It is associated with several behavioural impairments, including sleep alterations, eating disorders and lack of energy (Maes et al., 1999a,b). In addition, it has a role in regulation of the HPA axis, expression of serotonin transporter and catecholaminergic system (Maes et al., 1999a,b). IL-1β is positively correlated with several PTSD symptoms, including avoidance and re-experiencing, and with certain comorbidities (Spivak et al., 1997; Waheed, Dalton, Wesemann, Ibrahim & Himmerich, 2018). IL-1β is a crucial cytokine since it is released and promotes release of other cytokines, including IL-6 (Jones & Thomsen, 2012). Increased levels of IL-6 are associated with PTSD development and vulnerability (Gill, Luckenbaugh, Charney & Vythilingam, 2010). IL-6 has a major role in dopamine regulation in the hippocampus and maintains the survival of dopaminergic neurons (Gill et al., 2009). Increased levels of IL-6 affect vascular endothelium and may be associated with stroke, myocardial infarction, arrhythmias or cardiac death (Gil et al., 2010). As PTSD is frequently comorbid with different psychiatric and somatic disorders, comorbidities might affect IL-6 levels. Namely, patients with PTSD without depression had a different IL-6 profile compared to patients with PTSD comorbid with depression (Gil et al., 2010). Pro-inflammatory cytokines, IL-1β and IL-6, affect the HPA axis, stimulate its activity and alter cortisol levels (Gill et al., 2008). T-helper cell 1 (Th1) cytokines IL-2 and IFN-γ are increased among PTSD patients, while inhibitory mediators TGF-β, IL-4 and IL-10 levels were decreased among PTSD patients (Wang, Mandel, Levingston & Young, 2016; Wang & Young, 2016). However, several studies found that IL-4 and IL-10 levels were increased in PTSD patients (Guo et al., 2012; Wang & Young, 2016). Out of the inflammatory mediators, TNF-α, IL-1, IL-17 and IL-6 were increased (Hoge et al., 2009; Wang et al., 2016; Wang & Young, 2016). It is proposed that Th1 and inflammatory cytokines are increased, while inhibitory mediators are decreased in patients with PTSD (Wang et al., 2016). Levels of IL-6, IL-1β and TNF-α were elevated in PTSD patients even months after traumatic events (Jones & Thomsen, 2012). However, a similar level of inflammatory cytokines in women that recovered from PTSD and non-traumatized women was also observed (Wang & Young, 2016). Cytokine levels can be altered due to certain health conditions, including cancer or other neuropsychiatric disorders, such as depression (Wang et al., 2016). Alterations in the immune system are reflected not only in the impaired cytokine levels, but also in the composition of immune cells. Th1 and Th17 pro-inflammatory cells are increased, while anti-inflammatory cells, CD4+PD-L1+ cells, were decreased in patients with PTSD (O'Donovan et al., 2015). PTSD is associated with enhanced risk for development of autoimmune disorders due to biological impairments caused by chronic low-level inflammation (O'Donovan et al., 2015). Patients with PTSD have reduced gene methylation and impaired gene expression in immune cells (O'Donovan et al., 2015). Moreover, genes that encode for TNF- IL-16 and Il-18 had enhanced expression in PTSD patients compared to healthy controls, while genes that encode for IL-8 were demethylated in PTSD patients (Zass, Hart, Seedat, Hemmings & Malan-Müller, 2017). Chronic inflammation characteristic of PTSD stimulates the immune system to release CRP. CRP is released as a result of the acute stress, infection or impairments in the immune system. However, chronic stress and depression might also cause release of CRP due to chronic inflammation (Heath et al., 2013). Typical PTSD symptoms, such as re-experiencing/flashbacks, might affect CRP levels, which are increased in PTSD subjects (Heath et al., 2013). An altered immune system, together with an altered HPA axis, increases the risk for autoimmune diseases in PTSD. Individuals with PTSD have a higher risk to develop autoimmune disorders compared to subjects without psychiatric disorder, while females are more prone to developing autoimmune disorder compared to males (O'Donovan et al., 2015). Chronic inflammation characteristic of PTSD stimulates reactive oxygen species that cause accelerated aging through telomere impairment (Passos et al., 2015).

In brief, the immune system, together with the HPA axis, is dysregulated in PTSD, and although the results from the literature are inconsistent, immune alterations in PTSD reveal immunological imbalance and a low grade pro-inflammatory state in PTSD (Wang, Caughron & Young, 2017).

Most of the reported data reveal increased plasma levels of pro-inflammatory cytokines such as IFN-$\gamma$, IL-6, TNF-$\alpha$, and IL-17 (Wang et al., 2017).

## Conclusion

Pronounced dysregulation of the HPA axis and immune system is present in PTSD. PTSD is often comorbid with other psychiatric as well as somatic disorders. Among the somatic comorbidities, cardiovascular and metabolic are the most frequent, and the most serious, since severe trauma, such as combat stress, affects the cardiovascular system. This review focused on the neuroendocrine and immune markers, which are common for PTSD as well as in PTSD comorbid with CVD. Biomarkers should be easily accessible and determined in easily available body fluids such as blood, saliva or urine. They might help in improving clinical care and treatment. Since PTSD is a complex and polygenic disorder, associated with different comorbid disorders, impaired stress response and altered immune systems might only describe part of the potential patho-mechanisms underlying PTSD.

**Note:** The work was supported by the Croatian Science Foundation, project number IP-2014-09-4289.

## References

Ahmadi, N., Hajsadeghi, F., Yehuda, R., Anderson, N., Garfield, D., Ludmer, C., & Vaidya, N. (2015). Traumatic brain injury, coronary atherosclerosis and cardiovascular mortality. *Brain Injury*, *29*(13–14), 1635–1641.

Ali, S., Jabeen, S., & Alam, F. (2015). Multimodal approach to identifying malingered posttraumatic stress disorder: A review. *Innovations in Clinical Neurosciences*, *12*(1–2), 12–20.

Altemus, M., Dhabhar, F.S., & Yang, R. (2006). Immune function in PTSD. *Annals New York Academy of Sciences*, *1071*, 167–183.

American Psychiatric Association. (2013) *Diagnostic and Statistical Manual of Mental Disorders, Fifth Edition*. Washington, DC: American Psychiatric Press Inc.

Atwoli, L., Stein, D.J., Koenen, K.C., & McLaughlind, K.A. (2015). Epidemiology of posttraumatic stress disorder: Prevalence, correlates and consequences. *Current Opinion in Psychiatry*, *28*(4), 307–311.

Baker, D.G., West, S.A., Nicholson, W.E., Ekhator, N.N., Kasckow, J.W., Hill, K.K., ... Geracioti, T.D. (1999). Serial corticotropin-releasing hormone levels and adrenocortical activity in combat veterans with posttraumatic stress disorder. *American Journal of Psychiatry*, *156*, 585–588.

Beristianos, M.H., Yaffe, K., Cohen, B., & Byers, A.L. (2016). PTSD and risk of incident cardiovascular disease in aging veterans. *American Journal of Geriatric Psychiatry*, *24*(3), 192–200.

Binder, E.B., Owens, M.J., Liu, W., Deveau, T.C., Rush, A.J., Trivedi, M.H., ... Nemeroff, C.B. (2010). Association of polymorphisms in genes regulating the corticotropin-releasing factor system with antidepressant treatment response. *Achieves in General Psychiatry*, *67*(4), 369–379.

Bisson, J.I., Cosgrove, S., & Roberts, N.P. (2015). Post-traumatic stress disorder. *British Medical Journal*, *351*, h6161.

Blanaru, M., Bloch, B., Vadas, L., Arnon, Z., Ziv, N., Kremer, I., & Haimov, I. (2012). The effects of music relaxation and muscle relaxation techniques on sleep quality and emotional measures among individuals with posttraumatic stress disorder. *Mental Illness*, *4*(2), e13.

Blessing, E. M., Reus, V., Mellon, S. H., Wolkowitz, O.M., Flory, J. D., Bierer, L., ... Marmar, C.R. (2017). Biological predictors of insulin resistance associated with posttraumatic stress disorder in young military veterans. *Psychoneuroendocrinology*, *82*, 91–97.

Bremner, J. D., Licinio, J., Darnell, A., Krystal, J. H., Owens, M. J., Southwick, S. M., ... Charney, D. S. (1997). Elevated CSF corticotropin-releasing factor concentrations in posttraumatic stress disorder. *American Journal of Psychiatry, 154*, 624–629.

Britvic, D., Anticevic, V., Kaliterna, M., Lusic, L., Beg, A., Brajevic-Gizdic, I., ... Pivac, N. (2015). Comorbidities with PTSD among combat veterans; 15 years postwar analysis. *International Journal of Clinical and Healthy Psychology, 15*, 81–92.

Brock, C.M., King, D.S., Wofford, M.R., & Harrell, T.K. (2005). Exercise, insulin resistance, and hypertension: A complex relationship. *Metabolic Syndrome and Related Disorders, 3*, 60–65.

Brudey, C., Park, J., Wiaderkiewicz, J., Kobayashi, I., Mellman, T. A., & Marvar, P. J. (2015). Autonomic and inflammatory consequences of posttraumatic stress disorder and the link to cardiovascular disease. *American Journal of Physiology – Regulatory, Integrative and Comparative Physiology, 309*(4), 315–321.

Canetti, D., Russ, E., Luborsky, J., Gerhart, J. I., & Hobfoll, S. E. (2014). Inflamed by the flames? The impact of terrorism and war on immunity. *Journal of Traumatic Stress, 27*, 345–352.

Cohen, B.E., Edmondson, D., & Kronish, I.M. (2015). State of the art review: Depression, stress, anxiety, and cardiovascular disease. *Americal Journal of Hypertension, 28*(11), 1295–1302.

Corvalan, J.C., & Klein, D. (2011). PTSD: Diagnosis, evolution, and treatment of combat-related psychological/psychiatric injury. *Missouri Medicine, 108*(4), 296–303.

Daskalakis, N.P., McGill, M.A., Lehrner, A., & Yehuda, R. (2015). Endocrine aspects of PTSD: Hypothalamic-pituitary-adrenal (HPA) axis and beyond. In C.R. Martin, V.R. Preedy, & V.B. Patel (Eds.), *Comprehensive Guide to Post-Traumatic Stress Disorder* (pp. 245–260). Switzerland: Springer International Publishing.

De Kloet, C.S., Vermetten, E., Heijnen, C.J., Geuze, E., Lentjes, E.G., & Westenberg, H.G. (2007). Enhanced cortisol suppression in response to dexamethasone administration in traumatized veterans with and without posttraumatic stress disorder. *Psychoneuroendocrinology, 32*, 215–226.

Dennis, P.A., Ulmer, C.S., Calhoun, P.S., Sherwood, A., Watkins, L.L., Dennis, M.F., & Beckham, J.C. (2014). Behavioural health mediators of the link between posttraumatic stress disorder and dyslipidemia. *Journal of Psychosomatic Research*, 77(1), 45–50.

Dursa, E.K., Reinhard, M.J., Barth, S.K., & Schneiderman, A.I. (2014). Prevalence of a positive screen for PTSD among OEF/OIF and OEF/OIF-era veterans in a large population-based cohort. *Journal of Traumatic Stress, 27*, 542–549.

Edmondson, D., Kronish, I.M., Shaffer, J.A., Falzon, L., & Burg, M.M. (2013). Posttraumatic stress disorder and risk for coronary heart disease: A meta-analytic review. *American Heart Journal, 166*, 806–814.

Farr, O. M., Ko, B.J., Joung, K. E., Zaichenko, L., Usher, N., Tsoukas, M., ... Mantzoros, C. S. (2015). Posttraumatic stress disorder, alone or additively with early life adversity, is associated with obesity and cardiometabolic risk. *Nutrition, Metabolism & Cardiovascular Diseases, 25*, 479–488.

Flory, J. D. (2015). Comorbidity between post-traumatic stress disorder and major depressive disorder: Alternative explanations and treatment considerations. *Dialogues in Clinical Neuroscience, 17*, 141–150.

Frank, C., Zamorski, M.A., Lee, J. E., & Colman, I. (2018). Deployment-related trauma and post-traumatic stress disorder: Does gender matter?. *European Journal of Psychotraumatology, 9*(1), Advance online publication. doi: 10.1080/20008198.2018.1486123

Furtado, M., & Katzman, M. A. (2015). Neuroinflammatory pathways in anxiety, posttraumatic stress, and obsessive compulsive disorders. *Psychiatry Research, 229*, 37–48.

Gill, J., Luckenbaugh, D., Charney, D., & Vythilingam, M. (2010). Sustained elevation of serum interleukin-6 and relative insensitivity to hydrocortisone differentiates posttraumatic stress disorder with and without depression. *Biological Psychiatry, 68*, 999–1006.

Gill, J. M., Saligan, L., Woods, S., & Page, G. (2009). PTSD is associated with an excess of inflammatory immune activities. *Perspectives in Psychiatric Care, 45*(4), 262–277.

Gill, J., Vynthilingam, M., & Page, G.G. (2008). Low cortisol, high DHEA, and high levels of stimulated TNF-α, and IL-6 in women with PTSD. *Journal of Traumatic Stress, 21*(6), 530–539.

Goetz, M., Shah, A., Goldberg, J., Cheema, F., Shallenberger, L., Murrah, N.V., ... Vaccarino, V. (2014). Posttraumatic stress disorder, combat exposure, and carotid intima-media thickness in male twins. *American Journal of Epidemiology, 180*(10), 989–996.

Gola, H., Engler, H., Sommershof, A., Adenauer, H., Kolassa, S., Schedlowski, M., ... Kolassa, I-T. (2013). Posttraumatic stress disorder is associated with an enhanced spontaneous production of pro-inflammatory cytokines by peripheral blood mononuclear cells. *BMC Psychiatry, 13*(1), 40. doi: 10.1186/1471-244X-13-40.

Gradus, J.L., Farkas, D.K., Svensson, E., Ehrenstein, V., Lash, T.L., Milstein, A., ... Sørensen, H.T. (2015). Associations between stress disorders and cardiovascular disease events in the Danish population. *British Medical Journal, 5*(12), e009334.

Guina, J., Nahhas, R.W., Sutton, P., & Farnsworth, S. (2018). The influence of trauma type and timing on PTSD symptoms. *The Journal of Nervous and Mental Disease, 206*, 72–76.

Guo, M., Liu, T., Guo, J. C., Jiang, X. L., Chen, F., & Gao, Y. S. (2012). Study on serum cytokine levels in posttraumatic stress disorder patients. *Asian Pacific Journal of Tropical Medicine, 5*(4), 323–325.

Hall, K.S., Hoerster, K. D., & Yancy, W. S. (2015). Post-traumatic stress disorder, physical activity, and eating behaviors. *Epidemiologic Reviews, 37*, 103–115.

Hawn, S. E., Sheerin, C. M., Lind, M. J., Hicks, T. A., Marraccini, M. E., Bountress, K., … Amstadter, A. B. (2019). GxE effects of FKBP5 and traumatic life events on PTSD: A meta-analysis. *Journal of Affective Disorders, 243*, 455–462.

Heath, N.M., Chesney, S.A., Gerhart, J.I., Goldsmith, R.E., Luborsky, J.L., Stevens, N.R., & Hobfoll, S.E. (2013). Interpersonal violence, PTSD and inflammation: Potential psychogenic pathways to higher C-reactive protein levels. *Cytokine, 63*, 172–178.

Heppner, P.S., Lohr, J.B., Kash, T.P., Jin, H., Wang, H., & Baker D.G. (2012). Metabolic syndrome: Relative risk associated with post-traumatic stress disorder (PTSD) severity and antipsychotic medication use. *Psychosomatics, 53*(6), 550–558.

Hoge, E.A., Brandstetter, K., Moshier, S., Pollack, M.H., Wong, K.K., & Simon, N.M. (2009). Broad spectrum of cytokine abnormalities in panic disorder and posttraumatic stress disorder. *Depression and Anxiety, 26*, 447–455.

Hori, H., & Kim, Y. (2019). Inflammation and post-traumatic stress disorder. *Psychiatry and Clinical Neurosciences, 73*(4), 143–153.

Ito, M., Takebayashi, Y., Suzuki, Y., & Horikoshi, M. (2018). Posttraumatic stress disorder checklist for DSM-5: Psychometric properties in a Japanese population. *Journal of Affective Disorders, 247*, 11–19.

Jones, K.A., & Thomsen, C. (2012). The role of innate immune system in psychiatric disorders. *Molecular and Cellular Neuroscience, 53*, 52–62.

Jordan, H.T., Stellman, S.D., Morabia, A., Miller-Archie, S.A., Alper, H., Laskaris, Z., … Cone, J.E. (2013). Cardiovascular disease hospitalizations in relation to exposure to the September 11, 2001 World Trade Center disaster and posttraumatic stress disorder. *Journal of the American Heart Association, 2*(5), e000431.

Kaess, B. M., Preis, S. R., Lieb, W., Beiser, A. S., Yang, Q., Chen, T. C., … Stewart, A. F. (2015). Circulating brain-derived neurotrophic factor concentrations and the risk of cardiovascular disease in the community. *Journal of the American Heart Association, 4*, e001544.

Kang, H.K., Natelson, B. H., Mahan, C.M., Lee, K.Y., & Murphy, F. M. (2003). Post-traumatic stress disorder and chronic fatigue syndrome-like illness among Gulf War veterans: A population-based survey of 30,000 veterans. *American Journal of Epidemiology, 157*, 141–148.

Kao, C.Y., Stalla, G., Stalla, J., Wotjak, C. T., & Anderzhanova, E. (2015). Norepinephrine and corticosterone in the medial prefrontal cortex and hippocampus predict PTSD-like symptoms in mice. *European Journal of Neuroscience, 419*, 1139–1148.

Kellner, M., Yassouridis, A., Hübner, R., Baker, D. G., & Wiedemann, K. (2003). Endocrine and cardiovascular responses to corticotropin-releasing hormone in patients with posttraumatic stress disorder: A role for atrial natriuretic peptide? *Neuropsychobiology, 47*, 102–108.

Kelly, M.M., Jensen, K.P., & Sofuoglu, M. (2015). Co-occurring tobacco use and posttraumatic stress disorder: Smoking cessation treatment implications. *The American Journal on Addictions, 24*, 695–704.

Kessler, R.C., Aguilar-Gaxiola, S., Alonso, J., Benjet, C., Bromet, E.J., Cardoso, G., … On behalf of the WHO World Mental Health Survey Collaborators. (2017). Trauma and PTSD in the WHO World Mental Health Surveys. *European Journal of Psychotraumatology, 8*(5). doi: 10.1080/20008198

Kibler, J. L., Tursich, M., Ma, M., Malcolm, L., & Greenbarg, R. (2014). Metabolic, autonomic and immune markers for cardiovascular disease in posttraumatic stress disorder. *World Journal of Cardiology, 6*, 455–461.

Kintzle, S., Barr, N., Corletto, G., & Castro, C.A. (2018). PTSD in U.S. veterans: The role of social connectedness, combat experience and discharge. *Healthcare (Basel), 6*(3), 102. doi: 10.3390/healthcare6030102.

Koenen, K.C., Sumner, J.A., Gilsanz, P., Glymour, M.M., Ratanatharathorn, A., Rimm, E. B., … Kubzansky, L.D. (2017). Post-traumatic stress disorder and cardiometabolic disease: Improving causal inference to inform practice. *Psychological Medicine, 47*(2), 209–225.

Komar, Z., & Vukusic, H. (1999). Post-traumatic stress disorder in Croatian war veterans; incidence and psychosocial features. In D. Dekaris, & A. Sabioncello (Eds.), *New Insights in Post-Traumatic Stress Disorder (PTSD)* (pp. 42–44). *Proceedings (Abstracts, references and illustrations)*. Croatian Academy of Science and Arts, Zagreb.

Konjevod, M., Tudor, L., Svob Strac, D., Nedic Erjavec, G., Barbas, C., Zarkovic, N., … Pivac, N. (2019). Metabolomic and glycomic findings in posttraumatic stress disorder. *Progress in Neuro-Psychopharmacology and Biological Psychiatry*, *88*, 181–193.

Kozaric-Kovacic, D., Ilic, M.G., Romic, Z., Vidovic, A., Jendricko, T., & Pivac N. (2009). Body mass index in male Caucasian veterans with or without posttraumatic stress disorder. *Progress in Neuro-Psychopharmacology and Biological Psychiatry*, *33*, 1447–1450.

Lehrner, A., & Yehuda, R. (2014). Biomarkers of PTSD: Military applications and considerations. *European Journal of Psychotraumatology*, *5*. Online publication. doi: 10.3402/ejpt.v5.23797

Lindqvist, D., Mellon, S.H., Dhabhar, F.S., Yehuda, R., Grenon, S.M., Flory, J.D., … Wolkowitz, O.M. (2017). Increased circulating blood cell counts in combat-related PTSD: Associations with inflammation and PTSD severity. *Psychiatry Research*, *258*, 330–336.

Loppnow, H., Buerke, M., Werdan, K., & Rose-John, S. (2011). Contribution of vascular cell-derived cytokines to innate and inflammatory pathways in atherogenesis. *Journal of Cellular and Molecular Medicine*, *15*, 484–500.

Maes, M., Lin, A., Delmeire, L., Van Gastel, A., Kenis, G., De Jongh, R., & Bosman, E. (1999a). Elevated serum interleukin-6 (IL-6) and IL-6 receptor concentrations in posttraumatic stress disorder following accidental man-made traumatic events. *Biological Psychiatry*, *45*, 833–839.

Maes, M., Lin, A. H., Verkerk, R., Delmeire, L., Van Gastel, A., Van der Planken, M., & Scharpé, S. (1999b). Serotonergic and noradrenergic markers of post-traumatic stress disorder with and without major depression. *Neuropsychopharmacology*, *20*(2), 188–197.

Mehta, D., & Binder, E.B. (2012). Gene x environment vulnerability factors for PTSD: The HPA-axis. *Neuropharmacology*, *62*, 654–662.

Michaud, M., Balardy, L., Moulis, G., Gaudin, C., Peyrot, C., Vellas, B.,… Nourhashemi, F. (2013). Proinflammatory cytokines, aging, and age-related diseases. *Journal of the American Medical Directors Association*, *14*, 877–882.

Michopoulos, V., Rothbaum, A.O., Jovanovic, T., Almli, L.M., Bradley, B., Rothbaum, B.O., … Ressler, K.J. (2015). Association of CRP genetic variation and CRP level with elevated PTSD symptoms and physiological responses in a civilian population with high levels of trauma. *American Journal of Psychiatry*, *172*, 353–362.

Muhtz, C., Wester, M., Yassouridis, A., Wiedemann, K., & Kellner, M. (2008). A combined dexamethasone/corticotropin-releasing hormone test in patients with chronic PTSD–first preliminary results. *Journal of Psychiatric Research*, *42*, 689–693.

Nedic Erjavec, G., Konjevod, M., Nikolac Perkovic, M., Svob Strac, D., Tudor, L., Barbas, C., … Pivac, N. (2018). Short overview on metabolomic approach and redox changes in psychiatric disorders. *Redox Biology*, *14*, 178–186.

Neigh, G.N., & Ali, F.F. (2016). Co-morbidity of PTSD and immune system dysfunction: Opportunities for treatment. *Current Opinion in Pharmacology*, *29*, 104–110.

Nikolac Perkovic, M., Nedic Erjavec, G., Svob Strac, G., Uzun, S., Kozumplik, O., & Pivac, N. (2017). Theranostic biomarkers for schizophrenia. *International Journal of Molecular Sciences*, *18*(4), 733. doi: 10.3390/ijms18040733

O'Donovan, A., Cohen, B.E., Seal, K.H., Bertenthal, D., Margaretten, M., Nishimi, K., & Neylan, T.C. (2015). Elevated risk for autoimmune disorders in Iraq and Afghanistan veterans with posttraumatic stress disorder. *Biological Psychiatry*, *77*(4), 365–374.

Olff, M., & van Zuiden, M. (2017). Neuroendocrine and neuroimmune markers in PTSD: Pre-, peri- and post-trauma glucocorticoid and inflammatory dysregulation. *Current Opinion in Psychology*, *14*, 132–137.

Osório, C., Probert, Z., Jones, E., Young, A.H., & Ian Robbins, I. (2017). Adapting to stress: Understanding the neurobiology of resilience. *Behavioral Medicine*, *43*(4), 307–322.

Pan, X., Wang, Z., Wu, X., Wen, S.W., & Liu, A. (2018). Salivary cortisol in post-traumatic stress disorder: A systematic review and meta-analysis. *BioMed Central Psychiatry*, *18*(1):324.

Park, J., Marvar, P.J., Liao, P., Kankam, M.L., Norrholm, S.D., Downey, R.M., … Rothbaum, B.O. (2017). Baroreflex dysfunction and augmented sympathetic nerve responses during mental stress in veterans with post-traumatic stress disorder. *The Journal of Physiology*, *595*(14), 4893–4490.

Passos, I.C., Vasconcelos-Moreno, M.P., Costa, L.G., Kunz, M., Brietzke, E., Quevedo, J., … Kauer-Sant'Anna, M. (2015). Inflammatory markers in post-traumatic stress disorder: A systematic review, meta-analysis, and meta-regression. *Lancet Psychiatry*, *2*, 1002–1012.

Pikula, A., Beiser, A.S., Chen, T.C., Preis, S.R., Vorgias, D., DeCarli, C., … Seshadri, S. (2013). Serum brain-derived neurotrophic factor and vascular endothelial growth factor levels are associated with risk of stroke and vascular brain injury: Framingham Study. *Stroke*, *44*, 2768–2775.

Pivac, N., Muck-Seler, D., Jakovljevic, M., Ljubicic, D., & Crncevic-Orlic Z. (1998). Hypothalamic-pituitary-adrenal axis function in posttraumatic stress disorder in Croatian war veterans. *Psychiatria Danubina, 10*, 103–104.

Priebe, S., Bogic, M., Ajdukovic, D., Franciskovic, T., Galeazzi, G.M., Kucukalic, A., ... Schutzwohl, M. (2010). Mental disorders following war in the Balkans: A study in 5 countries. *Archives of General Psychiatry, 67*, 518–528.

Raglan, G.B., Schmidt, L.A., & Schulkin J. (2017). The role of glucocorticoids and corticotropin-releasing hormone regulation on anxiety symptoms and response to treatment. *Endocrine Connections, 6*, 1–7.

Risbrough, V.B., & Stein, M.B. (2006). Role of corticotropin releasing factor in anxiety disorders: A translational research perspective. *Hormones and Behavior, 50*(4), 550–561.

Robicsek, O., Makhoul, B., Klein, E., Brenner, B., & Sarig, G. (2011). Hypercoagulation in chronic post-traumatic stress disorder. *The Israel Medical Association Journal, 13*(9), 548–552.

Rohleder, N., Joskimovic, Lj., Wolf, J.M., & Kirschbaum, C. (2004). Hypocortisolism and increased glucocorticoid sensitivity of pro-inflammatory cytokine production in Bosnian war refuges with posttraumatic stress disorder. *Biological Psychiatry, 55*, 745–751.

Rosebrock, L.A., Arditte Hall, K.A., Rando, A., Pineles, S.L., & Liverant, G.I. (2018). Rumination and its relationship with thought suppression in unipolar depression and comorbid PTSD. *Cognitive Therapy and Research, 43*(1), 226–235.

Rosen, R.L., Levy-Carrick, N., Reibman, J., Xu, N., Shao, Y., Liu, M., ... Galatzer-Levy, I.R. (2017). Elevated C-reactive protein and posttraumatic stress pathology among survivors of the 9/11 World Trade Center attacks. *Journal of Psychiatric Research, 89*, 14–21.

Rosenbaum, S., Vancampfort, D., Steel, Z., Newby, J., Ward, P. B., & Stubbs, B. (2015). Physical activity in the treatment of Post-traumatic stress disorder: A systematic review and meta-analysis. *Psychiatry Research, 230*, 130–136.

Sagud, M., Jaksic, N., Vuksan-Cusa, B., Loncar, M., Loncar, I., Mihaljevic Peles, A.M., ... Jakovljevic, M. (2017). Cardiovascular disease risk factors in patients with posttraumatic stress disorder (PTSD): A narrative review. *Psychiatria Danubina, 29*(4), 421–430.

Sagud, M., Petrovic, B., Vilibic, M., Mihaljevic Peles, A., Vuksan-Cusa, B., Rados, I., ... Trkulja, V. (2018). The relationship among psychopathology, religiosity, and nicotine dependence in Croatian war veterans with posttraumatic stress disorder. *Croatian Medical Journal, 59*(4), 165–177.

Savic, D., Knezevic, G., Damjanovic, S., Spiric, Z., & Matic, G. (2012). The role of personality and traumatic events in cortisol levels–where does PTSD fit in?. *Psychoneuroendocrinology, 37*, 937–947.

Scherrer, J.F., Chrusciel, T., Zeringue, A., Garfield, L.D., Hauptman, P.J., Lustman, P.J., ... True, W.R. (2010). Anxiety disorders increase risk for incident myocardial infarction in depressed and nondepressed Veterans administration patients. *American Heart Journal, 159*(5), 772–779.

Scherrer, J.F., Salas, J., Cohen, B.E., Schnurr, P.P., Schneider, F.D., Chard, K.M., ... Lustman, P.J. (2019). Comorbid conditions explain the association between posttraumatic stress disorder and incident cardiovascular disease. *Journal of the American Heart Association, 8*(4): e011133.

Solomon, Z., Levin, Y., Assayag E.B., Furman, O., Shenhar-Tsarfaty, S., Berliner, S., & Ohry, A. (2017). The implication of combat stress and PTSD trajectories in metabolic syndrome and elevated C-reactive protein levels: A longitudinal study. *The Journal of Clinical Psychiatry, 78*(9), 1181–1186.

Spivak, B., Shohat, B., Mester, R., Avraham, S., Gil-Ad, I., Bleich, A., ... Weizman, A. (1997). Elevated levels of serum interleukin-1β in combat-related posttraumatic stress disorder. *Biological Psychiatry, 42*, 345–348.

Suliman, S., Anthonissen, L., Carr, J., du Plessis, S., Emsley, R., Hemmings, S.M., ... Seedat, S. (2016). Posttraumatic stress disorder, overweight, and obesity: A systematic review and meta-analysis. *Harvard Review of Psychiatry, 24*(4), 271–293.

Takashio, S., Sugiyama, S., Yamamuro, M., Takahama, H., Hayashi, T., Sugano, Y., ... Ogawa, H. (2015). Significance of low plasma levels of brain-derived neurotrophic factor in patients with heart failure. *American Journal of Cardiology, 116*, 243–249.

Tennant, C. (1982). Psychosocial stress and ischaemic heart disease: An evaluation in the light of the diseases' attribution to war service. *Australian New Zeland Journal of Psychiatry, 16*(2), 31–36.

Tudor, L., Konjevod, M., Nikolac Perkovic, M., Svob Strac, D., Nedic Erjavec, G., Uzun, S., ... Pivac, N. (2018). Genetic variants of the brain-derived neurotrophic factor and metabolic indices in veterans with posttraumatic stress disorder. *Frontiers in Psychiatry, 9*, 637.

Turner, J.H., Neylan, T.C., Schiller, N.B., Li, Y., & Cohen, B.E. (2013). Objective evidence of myocardial ischemia in patients with posttraumatic stress disorder. *Biological Psychiatry, 74*, 861–866.

van den Berk-Clark, C., Secrest, S., Walls, J., Hallberg, E., Lustman, P.J., Schneider, F.D., & Scherrer, J.F. (2018). Association between posttraumatic stress disorder and lack of exercise, poor diet, obesity, and co-occuring smoking: A systematic review and meta-analysis. *Health Psychology, 37*(5), 407–416.

VanDyke, L.M., Burton, E., Hamidovic, A., & Burge, M.R. (2017). Hormonal aspects of post-traumatic stress disorder. *Endocrinology Diabetes and Metabolism Journal, 1*(2), 1–10.

Waheed, A., Dalton, B., Wessemna, U., Ibrahim, M.A.A., & Himmerich, H. (2018). A systematic review of interleukin-1β in post-traumatic stress disorder: Evidence from human and animal studies. *Journal of Interferon & Cytokine Research, 38*(1), 1–11.

Wang, Z., Caughron, B., & Young, M.R.I. (2017). Posttraumatic stress disorder: An immunological disorder? *Frontiers in Psychiatry, 8*, 222.

Wang, Z., Mandel, H., Levingston, C.A., & Young M.R.I. (2016). An exploratory approach demonstrating immune skewing and a loss of coordination among cytokines in plasma and saliva of Veterans with combat PTSD. *Human Immunology*, 77, 652–657.

Wang, W., Wang, L., Xu, H., Cao, C., Liu, P., Luo, S., … Zhang, X. (2019). Characteristics of pro- and anti-inflammatory cytokines alteration in PTSD patients exposed to a deadly earthquake. *Journal of Affective Disorders, 248*, 52–58.

Wang, Z., & Young, M.R.I. (2016). PTSD, a disorder with and immunological component. *Frontiers in Immunology*, 7(219), 1–6.

Weathers, F.W. (2017). Redefining posttraumatic stress disorder for DSM-5. *Current Opinion in Psychology, 14*(122), 122–126.

Wentworth, B.A., Stein, M.B., Redwine, L.S., Xue, Y., Taub, P.R., Clopton, P., … Maisel, A.S. (2013). Post-traumatic stress disorder: A fast track to premature cardiovascular disease? *Cardiology in Review, 21*, 16–22.

Wolf, E.J., & Schnurr, P.P. (2016). PTSD-related cardiovascular disease and accelerated cellular aging. *Psychiatric Annals, 46*, 527–532.

Yehuda, R., Hoge, C.W., McFarlane, A.C., Vermetten, E., Lanius, R.A., Nievergelt, C.M., … Hyman, S.E. (2015). Post-traumatic stress disorder. *Nature Review, 15057.*

Zass, L.J., Hart, S.A., Seedat, S., Hemmings, S.M.J., & Malan-Müller, S. (2017). Neuroinflammatory genes associated with post-traumatic stress disorder: Implications for comorbidity. *Psychiatric Genetics*, 27(1), 1–16.

# 33

# MORAL INJURY IN SERVICE MEMBERS AND VETERANS

*Sheila Frankfurt, Alanna Coady, Breanna Grunthal, Stephanie Ellickson-Larew, and Brett T. Litz*

Litz et al. (2009) defined moral injury as the lasting psychological, emotional, spiritual, behavioral and social impact of exposure to acts that transgress deeply held moral beliefs and expectations. The idea that service members (SMs) and veterans can suffer in a lasting way psychically by events that transgress deeply held moral beliefs has captured the attention of veterans and SMs and the larger public (e.g., Press, 2018), as well as clinicians and researchers who work with these populations. However, there are major unanswered questions about the construct validity of moral injury. For example, there is no paradigmatic way of defining moral injury as an outcome and the parameters of what should be considered potentially morally injurious events are not clear. It is also not clear whether moral injury has incremental explanatory or clinical validity beyond posttraumatic stress disorder (PTSD; APA, 2013). This chapter will review empirical evidence about potentially morally injurious events, the moral injury syndrome, and causal models of moral injury; examine the potential similarities and differences between moral injury and PTSD; and conclude by considering issues in the field and avenues for research and construct development.

## Potentially morally injurious events

There are putatively two broad ways of categorizing potentially morally injurious events (PMIEs). The first entails personal transgressions such as perpetrating or failing to prevent morally compromising acts (PMIE-Self). The second involves being the victim of or subject to another's behavior or bearing witness to or learning about events that transgress deeply held beliefs (PMIE-Other). In the military context, Shay (1994) emphasized moral injury events that are a "betrayal of what's right, by someone who holds legitimate authority, in a high-stakes situation" (PMIE-Betrayal).

There are also two broad methods of operationalizing PMIEs for research and clinical purposes. The first method entails asking SMs and veterans to report exposure to high-magnitude military and warzone events that might be construed as PMIEs and assessing their relative contribution to various mental and behavioral health outcomes, after accounting for potentially traumatic events (PTEs). These events typically involve engaging in violence, such as killing, engaging in, or failing to prevent excessive violence or atrocities (i.e., PMIEs-Self), or witnessing harm to civilians or prisoners (i.e., PMIEs-Other). This method is analogous to the testing the conditional probability of developing PTSD following different types of PTEs (e.g., Anders, Frazier, & Frankfurt, 2011).

The second method entails using an existing moral injury events measure to assess reports of experiences that putatively violated deeply held moral beliefs or expectations. The two measures of moral injury events are the *Morally Injurious Experiences Scale* (MIES; Nash et al., 2013) and the *Moral Injury Questionnaire-Military Version* (Currier, Holland, Drescher, & Foy, 2015a). The MIES is a 9-item measure that assesses PMIEs-Self (e.g., "I acted in ways that violate my own moral code or values."), PMIEs-Other (e.g., "I saw things that were morally wrong."), and PMIEs-Betrayal (e.g., "I feel betrayed by leaders I once trusted."). The MIQM is a 20-item measure that includes both appraisals (e.g., "I did things that betrayed my personal values.") and events (e.g., "I saw/was involved in the death(s) of an innocent in the war.") and is summed to create a total exposure score. These measures unfortunately conflate reports of exposure to morally injurious events and the outcomes of these events (e.g., MIES item 4; *I am troubled by having … violated my own moral code*; Frankfurt & Frazier, 2016).

The prevalence of PMIEs likely varies by branch, theater, era, and military status (e.g., active duty vs. veteran), as well as by method of assessing PMIEs. In general, it appears that approximately 40%–60% of U.S. Army and Marine veterans of post-9/11 wars reported killing combatants and approximately 10%–30% reported killing noncombatants as part of their military service; a minority of veterans report engaging in or witnessing excessive violence or atrocities (see Frankfurt & Frazier, 2016, for review). In a nationally representative sample of post-9/11 U.S. veterans surveyed using the MIES, about 25% reported PMIE-Others and PMIE-Betrayals and about 10% reported PMIE-Selfs; further, approximately 15% reported exposure to two or more types of PMIEs (Wisco et al., 2017). Although combat exposure appears to be associated with a higher likelihood of exposure to PMIEs (Frankfurt & Frazier, 2016), it should not be presumed that all combat veterans have experienced PMIEs. Similarly, PMIEs do not necessarily entail moral injury or lasting harms, but there is evidence that PMIEs may have uniquely deleterious effects over and above that of life-threatening PTEs and general combat exposure.

## Moral injury

Litz and colleagues (2009) proposed that moral injury is a complex trauma syndrome, characterized by chronic reexperiencing, avoidance, and emotional numbing symptoms, accompanied with self-condemnation and dysphoria, as well as self-harming and self-handicapping behaviors. Although there are no firm boundaries around the putative moral injury syndrome, a few broad domains of impact have been consistently examined. We next review research on the association between exposure to PMIEs and PTSD and other psychiatric disorders, suicidal ideation and behaviors, substance abuse, and social and functional impairment.

### *Posttraumatic stress disorder and other psychiatric disorders*

#### *Event-defined potentially morally injurious events*

A number of studies have examined whether reporting killing or participating in excessive violence is associated with PTSD and depression. Killing had a small albeit significant positive association with PTSD, but not depression, in samples of U.S. First Gulf War veterans and Iraq War soldiers and Canadian service members (Maguen et al., 2010; Maguen et al., 2011; Richardson, King, Shnaider, & Elhai, 2017). Exposure to (including participation in) excessive violence or atrocities was positively correlated with PTSD or depression, even after accounting for the relative contribution of combat exposure, in samples of U.S. Vietnam War veterans and Canadian Afghanistan War veterans (Beckham, Feldman, & Kirby, 1998; Dennis et al., 2017; Nazarov, Fikretoglu, Liu, Thompson, & Zamorski, 2018).

### *Studies using the national Vietnam Veterans Readjustment Survey*

The National Vietnam Veterans Readjustment Survey [NVVRS], a high-quality, population-based epidemiological study of Vietnam veterans (Kulka et al., 1990), has been reanalyzed to examine the psychological consequences of reports of killing or participating in excessive violence relative to general combat exposure. Veterans who endorsed killing or participating in atrocities reported significantly higher PTSD symptoms, with a large effect, than those who reported neither; these large effects remained after accounting for general combat exposure and potential PTSD symptom over reporting (MacNair, 2002). Killing non-enemies (e.g., the elderly, women, and children) and the enemy predicted PTSD after accounting for combat (although the pattern of significance was mixed across two PTSD scales [i.e., the MMPI and Mississippi PTSD scale]); however, killing did not predict depression (Maguen et al., 2009). Harming civilians or prisoners was associated with a higher likelihood of developing PTSD (63% of harmers vs. 15% of non-harmers), and veterans who had harmed others in the context of a high combat exposure deployment and had preexisting psychological vulnerabilities (e.g., childhood physical abuse) had a nearly 100% rate of developing PTSD (Dohrenwend, Yager, Wall, & Adams, 2013). The likelihood of harming others was more strongly correlated with combat exposure than psychological vulnerabilities (e.g., childhood abuse). A complicated model that tested killing and committing excessive violence alongside other warzone stressors found mixed results: killing predicted PTSD, but neither committing atrocities, life-threat-based stressors, nor environmental stressors did after accounting for killing (Fontana & Rosenheck, 1999).

### *Measure-defined potentially morally injurious events*

Total MIES scores predicted a small increased incidence of a current mental disorder in a nationally representative sample of U.S. post-9/11 veterans (Wisco et al., 2017). Total MIES scores also were correlated with PTSD, depression, anxiety, and negative affectivity among war-zone deployed Marines, with small to moderate effects (Nash et al., 2013).

PMIE-Self scores had weak to moderate positive correlations with PTSD in samples of combat-exposed post-9/11 veterans (Frankfurt et al., 2018) and war-zone deployed Marines (Jordan, Eisen, Bolton, Nash, & Litz, 2017). Contrary to theory, PMIE-Self scores did not correlate with either PTSD or depression in samples of Air Force psychiatric outpatients or Army National Guard members (Bryan et al., 2015). Yet PMIE-Self scores, but not PMIE-Other or PMIE-Betrayal scores, predicted a small increased risk of current mental disorder in U.S. post-9/11 veterans (Wisco et al., 2017).

PMIE-Other scores were positively correlated with PTSD in Air Force psychiatric outpatients (Bryan et al., 2015) and Army National Guard members (Bryan et al., 2015). Contrary to theory, PMIE-Other scores were moderately *negatively* associated with depression in Air Force psychiatric outpatients and were not associated with depression in Army National Guard members (Bryan et al., 2015).

PMIE-Betrayal scores, a special potentially stressful class of PMIE-Other, were very weakly to moderately correlated with PTSD in combat-exposed post-9/11 veterans (Frankfurt et al., 2018), war-zone deployed Marines (Jordan et al., 2017), Air Force psychiatric outpatients, and

Army National Guard members (Bryan et al., 2015). PMIE-Betrayal was moderately correlated with depression in combat-exposed post-9/11 veterans (Frankfurt et al., 2018).

Total MIQM scores predicted PTSD and depressive symptoms with moderate effect, after accounting for combat exposure or demographic variables, in a sample of U.S. Iraq and Afghanistan war veteran college students (Currier et al., 2015a). Although subsequent studies have used the MIQM to study the relation between PMIEs and PTSD and depression, correlation tables or point estimates were not published (e.g., Currier, et al., 2017).

## Suicidal ideation and behavior

### Event-defined potentially morally injurious events

Reporting killing appeared to increase the liability of psychiatric problems to suicidal ideation among U.S. Vietnam and post-9/11 Army and National Guard veterans (Maguen, Luxton et al., 2011; Maguen et al., 2012; Kline, Weiner, Interian, Scherbakov, & St. Hill, 2016). For example, endorsing killing increased the likelihood of reporting morbid thoughts and suicidal ideation by 50%–200% for veterans with comorbid depression, PTSD, and alcohol problems compared to veterans with these psychiatric complaints and no history of killing (Kline, et al., 2016). In a sample of heavily combat-exposed U.S. Vietnam War veterans, participation in wartime atrocities contributed to suicidal ideation via associations with PTSD, depression, and guilt, but these psychological factors made a larger contribution to suicidal ideation than reports of participation in atrocities (Dennis et al., 2017). Reports of feeling responsible for someone's death did not predict suicidal ideation after accounting for combat, PTSD, and depression in a sample of Canadian veterans (Richardson et al., 2017). In a qualitative study, U.S. combat veterans described how their suicidal ideation was a sign of their self-condemnation and a recompense for wartime killing (Purcell, Koenig, Bosch, & Maguen, 2016).

### Measure-defined potentially morally injurious events

Total MIES scores predicted a small increased risk of current suicidal ideation but not suicide attempts in post-9/11 U.S. veterans (Wisco et al., 2017). Among U.S. Air Force psychiatric outpatients, suicide attempters reported higher PMIE-Self and PMIE-Other scores than either the suicide ideation-only group or no-suicidality group, but the suicide ideation-only group and the no-suicidality group reported statistically equivalent PMIE scores (Bryan, Bryan, Morrow, Etienne, & Ray-Sannerud, 2014). However, the full picture of Bryan et al.'s (2014) results was mixed and contrary to established knowledge about suicide risk. PMIE-Self scores *positively* predicted suicidal ideation, but PMIE-Betrayal scores *negatively* predicted suicidal ideation; most strikingly, depression was not associated with suicidal ideation.

MIQM scores predicted suicide risk after accounting for demographic variables in community samples of Iraq and Afghanistan War veterans; however, the bivariate correlation was nonsignificant (Currier, Holland & Malott, 2015b).

## Substance abuse

### Event-defined potentially morally injurious events

Reporting killing was associated with increased alcohol consumption or problematic alcohol use, after accounting for combat exposure, in samples of U.S. Gulf War and Iraq War veterans (Killgore et al., 2008; Maguen et al., 2010; Maguen et al., 2011). U.S. Vietnam War veterans who reported atrocities reported more heroin and marijuana use than veterans who did not report participating in atrocities (Yager, Laufer, & Gallops, 1984). In a qualitative study on the impact of killing, combat veterans said that they use alcohol to suppress and numb their feelings related to memories of killing in war (Purcell et al., 2016).

### *Measure-defined potentially morally injurious events*

A modified MIQM was used to assess the relationship between PMIEs and substance use among veterans enrolled in college (Battles, Kelley, Jinkerson, Hamrick, & Hollis, 2019; Davies, Prince, Bravo, Kelley, & Crain, 2019). In one analysis, MIQM scores were weakly correlated with alcohol use and misuse among male, but not female, combat veterans (Battles et al., 2019). In another analysis, MIQM scores were very strongly correlated with alcohol misuse and moderately correlated with drug use in both males and females (Davies et al., 2019). The interpretation of these studies is constrained by the use of relatively idiosyncratic and psychometrically unvalidated revisions to the MIQM. It is also not clear what accounts for the major difference in correlations in different subsets of apparently the same parent study. To date, the MIES has not been used to assess the relationship between PMIEs and substance use.

## ***Social and functional impairment***

### *Event-defined potentially morally injurious events*

Reports of killing non-combatants, but not combatants, were associated with functional impairment, after accounting for reports of combat exposure, in the NVVRS (Maguen et al., 2009). Two qualitative studies examined the impact of PMIEs on social functioning (McCormack & Ell, 2017; Purcell et al., 2016). U.S. combat veterans reported that they numb and suppress their emotions to cope with their memories of killing, and recognize that this numbing comes at the expense of being close to their loved ones and being part of important familial events, such as funerals (Purcell et al., 2016). Australian combat veterans reported being reluctant to share intimate details of their combat experiences, in part because they feel dehumanized by committing PMIEs and in part because they feel misunderstood or judged for their actions after returning to civilian life and consequently feel isolated and disconnected from others (McCormack & Ell, 2017).

### *Measure-defined potentially morally injurious events*

MIQM scores were moderately positively correlated with work and social impairment in Iraq and Afghanistan War veteran college students (Currier et al., 2015a,b). The MIES has not been used to predict social or functional impairment.

## ***Other collateral effects***

Other plausible consequences of PMIEs have been proposed, but not directly examined in a moral injury framework. These additional collateral effects include severe recklessness, demoralization, and self-handicapping (proposed by Litz et al., 2009); loss of trust in oneself, others, or transcendental/spiritual beings (proposed by Jinkerson, 2016); embitterment; and enduring personality changes after catastrophic experiences (Shay, 1994). Research is needed to explore to what extent or how these proposed outcomes are related to PMIE exposure.

## Moral injury causal framework

Litz et al. (2009) proposed a working causal framework to describe how moral injury develops. When veterans cannot square their wartime actions with their deeply held beliefs about right and wrong conduct, they may experience severe psychological conflict. Some veterans may develop global, stable, negative attributions about themselves and the world to account for their PMIEs. If these attributions involve significant and dispositional self-blame, incapacitating shame, guilt, or anxiety may result. These painful emotions and a fear of being judged by others may then cause veterans to withdraw from interpersonal relationships and social support, depriving them of positive, caring experiences that might disconfirm their negative expectations of themselves and others. Shay (1994) highlighted the role of anger, particularly the ways in which anger in response to leadership betrayals and the loss of close comrades drives social withdrawal. In this way, social withdrawal contributes to a global sense of self-condemnation and lack of forgiveness, which further intensifies and entrenches shame, guilt, and anxiety and leads to moral injury syndrome (e.g., traumatic symptoms, self-harm).

A small body of literature has tested one of the main assertions of this framework: that PMIEs are associated with guilt, shame, and anger and these emotions drive the complicated clinical picture that is called moral injury. Other assertions, such as the role of intrapsychic psychological conflict or negative attributions, have been subject to relatively little empirical testing. We will review literature on the association between event- and measure-defined PMIE and guilt, shame, and anger and on the proposed causal pathway from PMIE through guilt/shame/anger to mental and behavioral health problems.

## *Guilt, shame, and anger*

### *Event-defined potentially morally injurious events*

Event-defined PMIEs (e.g., killing) moderately correlated with guilt among U.S. veterans (Frankfurt, Frazier, & Engdahl, 2017). Reporting exposure to (including participation in) atrocities was moderately correlated with trauma-related guilt and was weakly correlated with hostility, among U.S. Vietnam War veterans (Dennis et al., 2017). U.S. Vietnam veterans who reported participating in or observing abusive violence reported significantly higher combat-related guilt scores, with a large effect (Marx et al., 2010). Reporting killing, either enemy combatants or noncombatants, predicted anger, relationship problems, and interpersonal violence in U.S. Vietnam and Iraq War veterans, even after accounting for combat exposure (Maguen et al., 2009; Maguen et al., 2010). In a qualitative study on post-deployment anger, three (13%) veterans described PMIEs, such as being betrayed by their chain of command, as the source of their anger (Worthen & Ahern, 2014). However, it is difficult to parse the relative contribution of PMIEs to anger and aggressiveness from that of general combat exposure. For example, in Worthen and Ahern (2014)'s small qualitative study on anger, 46% of participants ($n = 6$) named the loss of structure following separation from the military as their major source of post-deployment anger. In a clinical sample of heavily combat-exposed U.S. Vietnam War male veterans seeking PTSD specialty care ($N = 603$), reports of combat exposure and reports of exposure to atrocities had nearly equivalent correlations with hostility and aggression (Dennis et al., 2017); in this study, combat exposure and reports of exposure to (including participation in) atrocities were very highly correlated, and 91% of the sample reported exposure to atrocities.

### *Measure-defined potentially morally injurious events*

PMIEs, measured using the MIES, demonstrated a mixed pattern of correlations with guilt, shame, and anger. PMIE-Self and PMIE-Betrayal scores were weakly correlated with guilt/shame (measured using one item) and anger, but also dissociation, among combat-exposed Marines (Jordan et al., 2017). Among Army National Guard members, PMIE-Self and PMIE-Betrayal but not PMIE-Other scores were associated with anger (Bryan et al., 2015). Contrary to the moral injury model, PMIEs (-Self, -Other, -Betrayal) were not correlated with guilt or shame among Air Force psychiatric patients (Bryan et al., 2015).

### *Causal framework tests*

Guilt partially accounted for the relationship between event-defined PMIEs (e.g., killing, exposure to or participation in atrocities) and PTSD, depression, and general psychological disturbance among U.S. combat veterans (Dennis et al., 2017; Fontana, Rosenheck, & Brett, 1992; Frankfurt et al., 2017; Marx et al., 2010). PMIE-Self MIES scores were associated with PTSD and depression via guilt/shame (Frankfurt et al., 2018; Jordan et al., 2017; Lancaster, 2018) and PMIE-Betrayal MIES scores were associated with PTSD and depression via anger (Jordan et al., 2017; Lancaster, 2018). However, the cross-relationships (i.e., PMIE-self with PTSD/ depression via anger, PMIE-Betrayal with PTSD/depression via guilt/shame) have not been consistently tested. The perceived wrongness of one's actions predicted anger, guilt, and shame over and above general combat exposure in an online sample of combat veterans (Lancaster & Erbes, 2017). A measure of military-related guilt, shame, and anger (i.e., the Expressions of Moral Injury Scale [EMIS]) was moderately correlated with the MIQM and was highly correlated with PTSD and depression, but these relationships were not tested simultaneously (Currier, et al., 2017).

### ***Other causal processes***

Following the publication of Litz et al., (2009), researchers and clinicians proposed novel working causal models of moral injury. These include a meaning-making model (Kopacz et al., 2019), a functional model (Farnsworth, Drescher, Evans, & Walser, 2017), and spiritual and religious difficulties models. The meaning-making model proposes that moral injury results from an inability to reconcile one's cognitive and emotional causal appraisals and attributions about PMIEs with one's global meaning system (Kopacz et al., 2019). The functional model proposes that moral injury is the suffering that results from dysfunctional strategies for coping with moral pain, but that moral pain, such as intense PMIE-related guilt and shame, is itself normal and nonpathological (Farnsworth et al., 2017). The role of spiritual and religious difficulties in the moral injury constellation is ambiguous. Spiritual and religious difficulties have been situated as a mediator between PMIEs and PTSD and depression (Evans et al., 2018), as a correlate of moral injury (Currier, Foster, & Isaak, 2019), or even as constituting moral injury (Harris et al., 2018). Various definitions of spirituality and religiousness, which alternately emphasize how spiritual life bears on meaning-making (Currier et al., 2015a,b), religious affiliation (Koenig et al., 2018), or one's relationship with God (Harris et al., 2018), complicate the picture further. Thus far, these alternative causal models have not been synthesized with either the original Litz et al., (2009) model or with each other.

In sum, the development of clear working causal processes of moral injury is constrained by definitional ambiguity and lack of integration across different perspectives, as well as the complex interrelationships between multiple psychological domains.

## Posttraumatic stress disorder and moral injury

Among U.S. veterans, 11%–20% of post-9/11 veterans have PTSD in any given year, and around 30% of U.S. Vietnam War veterans will have had PTSD in their lifetimes (Veterans Affairs, 2018). Although moral injury is often contrasted with PTSD (e.g., Jinkerson, 2016), it is unclear whether moral injury has incremental validity over PTSD or is even a separable construct from PTSD. We will briefly review the PTSD diagnosis and then consider the purported differences between PTSD and moral injury.

### *Posttraumatic stress disorder*

To meet *DSM-5* criteria for PTSD, someone must first report a potentially traumatic event (i.e., meet Criterion A; APA, 2013). Criterion A is defined in terms of relatively objective event characteristics: exposure to actual or threatened death, serious injury, or sexual violence via either direct experience, witnessing the event in person, or learning about such an event occurring to a close family member or close friend, or experiencing repeated or extreme exposure to aversive details of the traumatic event as part of one's job. PTSD syndrome is defined in terms of reexperiencing symptoms (e.g., unwanted memories or flashbacks; Criterion B), avoidance symptoms (e.g., of trauma-related thoughts or feelings; Criterion C), negative changes in cognitions and mood (e.g., negative affect; Criterion D), and alterations in arousal and reactivity symptoms (e.g., irritability or aggression; Criterion E). To receive a PTSD diagnosis, at least six symptoms (one from Criteria B and C, two from Criteria D and E) must be experienced for at least one month, cause social, occupational, and functional impairment, and not be better accounted for by substance use or other medical diagnoses.

A few issues with the PTSD diagnosis need to be kept in mind when comparing moral injury with PTSD. First, there is no field-wide agreement on one working causal model for PTSD development. PTSD was previously believed to be a fear-conditioning disorder, as reflected by its inclusion in *DSM-IV-TR's* Fear and Anxiety Disorders chapter (APA, 2000). Currently, the field is agnostic as to how PTSD develops, as reflected by its inclusion in *DSM-5*'s Trauma and Other Stressor-Related Disorders chapter (APA, 2013). Relatedly, there is no field-wide agreement on why potentially traumatic events are traumatizing, or even which events should be considered potentially traumatic. Second, the *DSM-5* PTSD diagnosis includes many non-trauma-specific symptoms that reflect dysphoria or depression such as Criterion D.4, *Persistent negative emotional state* (Rasmussen, Verkuilen, Jayawickreme, Wu, & McCluskey, 2019). In fact, Criterion D includes six symptoms of major depressive disorder (e.g., negative mood, loss of interest in activities). Thus, which mental and behavioral health problems should be considered post-traumatic is both diagnostically and conceptually unclear. Thirdly, PTSD symptoms are fungible and there are 636,120 possible symptom combinations that can result in a PTSD diagnosis (Galatzer-Levy & Bryant, 2013). Using *DSM-5* criteria, people can actually meet PTSD diagnostic criteria without classic traumatic symptoms like flashbacks, nightmares, or intrusive memories but instead with primarily dysphoric or negative affect symptoms (e.g., guilt, shame, or anger). Thus, the uncertainty, complexity, and heterogeneity of the PTSD field must be considered as part of the ground against which moral injury is contrasted.

## Posttraumatic stress disorder and moral injury

The relationship between PTSD and moral injury is challenging to understand for reasons ancillary to those already considered. First, moral injury's etiology closely resembles that of PTSD. Second, moral injury and PTSD share defining symptomatic features. Third, efforts to distinguish PTSD from moral injury often rely on weak accounts of both conditions. Fourth, PTSD and moral injury are different kinds of things, such that comparisons sit on uneven ground.

### Etiology

Both PTSD and moral injury are predicated on the occurrence of a sufficiently disturbing event that catalyzes mental and behavioral health problems. The PTSD diagnosis requires that an event surpasses a threshold of dangerousness (i.e., Criterion A), which demarcates between painful but merely stressful events and traumatic events. Moral injury requires an event be sufficiently transgressive or morally troubling to be a PMIE. Litz & Kerig (2019) proposed a continuum that broadly distinguishes relatively common and expected *moral challenges* from higher-magnitude *moral stressors* from the exceedingly rare and debilitating *morally injurious events.* However, neither established threshold nor criteria demarcates PMIEs from events which, albeit troubling, are better thought of as mundane cruelty or unkindness.

The event characteristics that define PTEs and PMIEs (e.g., dangerousness vs. transgressiveness) are neither orthogonal nor mutually exclusive. Schemes to classify distinct military trauma types found that many veterans' worst traumas involved both danger and transgression, for example, "Service member gave an order to fire on attacking insurgents that resulted in a civilian being shot" (p. 802, Litz et al., 2018; Stein et al., 2012). Additionally, some life-threatening traumas, such as interpersonal traumas that involve sexual assault committed by close loved ones, likely involve transgression of deeply held norms of right and wrong, and so could be considered PMIE-Others. Trying to partial out or disentangle the differential contribution of these event characteristics may miss the forest for the trees, particularly in a combat context where some level of life threat can be assumed. Thus, at this point, it may be premature to rely on differences in types of causal events to distinguish between PTSD and moral injury.

### Symptoms

Complicating matters further, PTSD and moral injury resemble each other at the symptom level. For example, Litz et al.'s (2009) conceptual working model of moral injury includes trauma-related reexperiencing, avoidance, and emotional numbing, that is, PTSD's Criterion B and C. Guilt, shame, and anger are proposed to be catalysts of moral injury development, but these are also symptoms of PTSD's Criterion D (and depression). Thus, moral injury and PTSD are likely to have similar clinical presentations.

### Strawman accounts of posttraumatic stress disorder and moral injury

Some efforts to distinguish PTSD from moral injury have compared outdated fear-based conditioning models of PTSD to facile but simplistic guilt- and shame-based models of moral injury (e.g., Frankfurt, et al., 2017). One frequently cited study purported to empirically distinguish PTSD (which was characterized by reexperiencing, memory loss, and flashbacks) from moral injury (which was characterized by shame, regret, sorrow, and alienation; Bryan, Bryan, Roberge, Leifker, & Rozek, 2017). However, it is equally plausible that the foregoing study identified a PTSD Criterion B factor (which was labeled "PTSD"), a Criterion D factor (which was labeled "moral injury"), and a third factor of items from assorted measures (anger, depression, insomnia, substance abuse, and nightmares, which was labeled "PTSD & MI"). Shared measurement variance may have accounted for some portion of these findings because "PTSD" and "moral injury" were measured using items from different scales. Importantly, other research has found that PMIEs were more strongly associated with reexperiencing symptoms than life-threat traumas (e.g., Stein et al., 2012), which, speculatively, may reflect the profound inner conflict these types of events produce. Thus, a simple delineation of PTSD from moral injury vis-à-vis reexperiencing symptoms would seem to be lacking construct validity. Because of the significant conceptual and symptomatic overlap between moral injury and PTSD, rational consideration, not solely empirical research, is needed to settle this matter.

### *Further distinguishing posttraumatic stress disorder and moral injury*

Complicating all this further is that PTSD and moral injury are, in ontological terms, different types of things. PTSD is a psychiatric diagnosis that provides a rich description of purportedly unique event-related phenomena. The *DSM-5* is designed to describe specific syndromes but not, beyond Criterion A, to provide reasons for syndromes. Moral injury was not originally proposed as a novel diagnosis and instead resembles something like an inferential clinical heuristic. "Moral injury" provides a compelling explanation for why some veterans and service members seem to be haunted by particularly violent or gruesome combat experiences, to have more complex, intractable mental and behavioral health problems, and to be a poorer fit for single-incident, life-threat-based trauma treatments. Efforts to clarify moral injury in terms of PTSD will flounder unless this ontological difference is properly conceptualized and handled.

## Discussion

Whether moral injury should be an independent mental health diagnoses, or whether moral injury should be a clinical heuristic for treatment planning and intervention, remains to be decided. Our literature review identified a few key findings, limitations of the extant body of research and methods of studying moral injury, and ideas for future research.

### *Key findings*

The reviewed literature presented a few broad findings. Reporting killing and exposure to or, participation in, atrocities, were most consistently associated with increased PTSD and depression. The MIES and MIQM had, in general, an inconsistent pattern of associations with PTSD and depression. PMIEs do not by themselves increase the risk of suicidality, but these experiences increase the liability of known risk factors such as PTSD and depression. PMIEs appear to be associated with guilt and are speculatively associated with shame and anger; however, these latter emotions are examined less frequently. In general, most moral injury propositions are untested.

There were also a few surprising patterns of results. First, in studies when killing combatants and noncombatants were disaggregated and tested simultaneously, some indicated killing the enemy but *not* noncombatants was associated with psychiatric distress (Killgore et al., 2008), whereas in other studies it was vice versa (Maguen et al., 2009). Second, in a number of studies, PMIEs were associated with PTSD but either were *not* associated with depression or were *negatively* associated with depression, even though PTSD and depression were highly correlated (Bryan et al., 2014; Maguen et al., 2009, 2010; Richardson et al., 2017). It is unknown whether these results reflect genuine multifinality or if these are sample-specific findings due to measurement or research design characteristics (e.g., small sample sizes or multicollinearity between PTSD and depression). Traditional statistical analytic methods may have difficulty modeling the dynamic or mutually informative relationships between co-occurring PMIEs or highly related psychological phenomena (e.g., PTSD and depression).

## Issues in the moral injury field and future directions

Overall, it is difficult to draw generalizable conclusions from the existing body of literature due to issues related to defining PMIEs, to identifying the moral injury syndrome, and to developing causal models of moral injury.

### Potentially morally injurious events

We organized our literature review around the two complementary, but parallel, methods of operationalizing PMIEs: either by a specific event or by a measure assessing perceived transgression. These bodies of moral injury research reflect two fundamentally different notions of PMIEs: whether moral injury develops because veterans themselves or others *violated* deeply held morals, or because they *feel* that they themselves or others violated deeply held morals. Neither extant PMIE measurement approach is wholly adequate, in large part because the field has not agreed on how PMIEs should be conceptualized.

The first approach examined the effect of violent combat events that seem to be more disturbing than other combat experiences, because these events are speculatively PMIEs (e.g., killing). This literature makes up the empirical and conceptual background of the moral injury field. However, most of these studies were not framed in terms of the moral injury model and so may have limited applicability. On the one hand, it might be overreaching or overly pacifist to assume that killing and committing atrocities' wrongness is the reason why these events are so psychologically destructive. On the other hand, it might be disingenuous malpractice to ignore veterans' accounts of how they are haunted or tormented by these actions (e.g., Thomas, 2004).

The second approach examined the relationship between appraisals of moral transgression and mental and behavioral health outcomes. On the one hand, these studies are explicitly testing hypotheses derived from the moral injury model using measures of PMIEs (i.e., MIES, MIQM). However, these measures do not have established clinical norms, and the average reported scores of both measures are quite low, which suggests a relatively rare incidence of PMIEs. Thus, findings using the MIES and MIQM might not be generalizable to heavily PMIE-exposed, or severely morally injured samples.

Different approaches to modifying or refining the PMIE category could be considered. One way could be defining PMIEs in terms of potentially verifiable events, similarly to how PTEs are defined in terms of life-threat. It may be worth considering whether PMIEs could be defined in terms of interpersonal harm (real or imagined) to another person, along the lines of the interpersonal PTEs, but from the opposite angle examining effects of the trauma on the actor(s). Alternatively, the PMIE-Other category could be excluded, because this ground may be covered by the PTSD diagnosis. As it is, it will be difficult to test whether PMIEs and PTEs have differential impacts on mental health when these categories are not mutually exclusive or exhaustive. There is room to test other approaches to refining the PMIE category that we have not considered here. In any event, how the field decides to define PMIEs will direct moral injury construct development, as well as shape the psychology field's (and greater society's) idea of transgression, moral behavior, and moral health.

## *Moral injury syndrome*

A major barrier to drawing strong conclusions about moral injury is the lack of firm boundaries around exactly what moral injury is or agreement in the field about where to draw these boundaries. Litz and colleagues (2009) proposed the first conceptual model of moral injury. Although this remains highly cited, many variations of this definition have been circulated (see Hodgson & Carey, 2017, for review), and there is little guidance around how to synthesize these varied definitions. Currently, a large, international effort to comprehensively map the moral injury domain from the ground up using qualitative data from interviews with veterans, active duty service members, and mental health providers is underway (Yeterian et al., 2019); hopefully, this project will produce a phenomenologically and ecologically valid model of moral injury.

Relevant individual differences, such as developmental history and early life experiences, personality characteristics, and values and beliefs, need to be included in moral injury theorizing and research. Little clinical guidance on how to conceptualize moral injury or formulate models of individual variability in responses to PMIEs is currently available. The moral injury field will benefit from empirical and conceptual research that robustly examines how the impact of PMIEs differs according to an individual's background and personality.

The life course and changes over time and social context of the complex, difficult emotions that follow PMIEs need to be better understood. For example, in a qualitative study on the psychological impact of killing, a majority of veterans acknowledged feeling some satisfaction, exhilaration, or gratifying omnipotence in the immediate aftermath of killing; the sense of intense conflict and being haunted intensified after deployment or separating from the military (Purcell et al., 2016). Accordingly, the time point of assessment must be considered when interpreting results of studies because the affective valence and behavioral manifestations of moral injury likely changes. As Purcell suggests, "The consequences of killing [and likely all PMIEs] are neither static nor singular, they resonate differently in the lives of individual veterans and often evolve over time" (Purcell et al., 2016, p. 1088).

## ***Moral injury causal frameworks***

What it means to be a "mechanism" of moral injury is unclear, as is how to distinguish between a moral injury causal process vs. a moral injury symptom. For example, the role of guilt and shame in moral injury poses a conundrum: guilt and shame are elements of PTSD and depression (APA, 2013), which are themselves symptoms of the moral injury outcome. Yet, guilt and shame are also putative mechanisms of moral injury, confounding the relationships between PTSD, depression, moral injury, and these particular moral emotions. Thus, definitional instability complicates whether aspects of moral injury should be regarded as mechanisms or outcomes.

We caution researchers against precipitous confirmatory model testing, because the moral injury domain remains roughly mapped. For example, dissociation was not tested as a potential mediator of PMIEs and PTSD/depression because it was assumed that dissociation only mediates PTEs and PTSD/depression (Jordan et al., 2017; Lancaster, 2018). In previous research, however, reporting killing was independently predictive of dissociation even after accounting for combat (i.e., a presumed PTE) (Maguen et al., 2009). Along these lines, it seems unlikely that anger is only a response to PMIE-Betrayal and not PMIE-Self, yet exclusivity of causal pathways is often presumed. These overly stringent models are likely due to the need for non-overlapping, mechanistic causal pathways in statistical modeling, but these statistical tests may reify simplistic or unrepresentative accounts of moral injury.

Instead, tests of the moral injury should remain exploratory, draw on clinical literature as well as empirical research, and be based on SMs and veterans' first-hand accounts of the impact of PMIEs. The field would benefit enormously from more qualitative investigations of the phenomenology and lived experience of moral injury from the perspective of veterans and SMs (c.f., Purcell et al., 2016). Research methodology that captures the dynamic, multidirectional, and mutually constitutive relationships between PMIE-related emotions, thoughts, and behaviors may create more representative and clinically useful models of psychological suffering. Such research may help elucidate these complex psychological relationships in ways that deepen not only understanding of trauma but also complexity of interacting intrapsychic, environmental, and social factors in psychological disorder and health.

## Summary

In sum, moral injury presents the notion that war, and military service in general, is rife with moral complexity, and that some military traumas are disturbing because the involve moral violations, rather than, or in addition to, life-threat. The moral injury construct is expansive and covers conceptual ground shared by other existing mental health constructs, such as PTSD and depression. It will take major collaborative efforts from clinicians and researchers across diverse fields to move the moral injury field forward and deepen our understanding of war-related psychological suffering.

## Author's note

This research was supported by the U.S. Department of Veterans Affairs (VA), the Central Texas Veterans Healthcare System, and a Career Development Award – 1 (IK1RX002427-01A2) to Dr. Frankfurt from the Rehabilitation Research and Development Service of the VA Office of Research and Development. The views expressed in this article are those of the authors and do not necessarily reflect the position or policy of VA or our academic affiliates.

## References

American Psychiatric Association. (2000). *Diagnostic and Statistical Manual of Mental Disorders*, (4th ed.). American Psychiatric Association: Chicago.

American Psychiatric Association. (2013). *Diagnostic and Statistical Manual of Mental Disorders*, (5th ed.). American Psychiatric Association: Chicago.

Anders, S., Frazier, P., & Frankfurt, S. (2011). Variations in Criterion A and PTSD rates in a community sample of women. *Journal of Anxiety Disorders*, *25*(2), 176–184.

Battles, A., Kelley, M., Jinkerson, J., Hamrick, H., & Hollis, B. (2019). Associations among exposure to potentially morally injurious experiences, spiritual injury, and alcohol use among combat veterans. *Journal of Traumatic Stress*, *32*(3), 405–413.

Beckham, J., Feldman, M., & Kirby, A. (1998). Atrocities exposure in Vietnam combat veterans with chronic posttraumatic stress disorder: Relationship to combat exposure, symptom severity, guilt, and interpersonal violence. *Journal of Traumatic Stress*, *11*(4), 777–785.

Bryan, C., Bryan, A. O., Anestis, M., Anestis, J. C., Green, B., Etienne, N., … Ray-Sannerud, B. (2015). Measuring moral injury: Psychometric properties of the Moral Injury Events Scale in two military samples. *Assessment*, *23*(5), 557–570.

Bryan, A., Bryan, C., Morrow, C., Etienne, N., & Ray-Sannerud, B. (2014). Moral injury, suicidal ideation, and suicide attempts in a military sample. *Traumatology*, *20*(3), 154–160.

Bryan, C., Bryan A., Roberge, E., Leifker, F., & Rozek, D. (2017). Moral injury, posttraumatic stress disorder, and suicidal behavior among National Guard personnel. *Psychological Trauma: Theory, Research, Practice, & Policy*, *10*(1), 36–45.

Currier, J., Farnsworth, J., Drescher, K., McDermott, R., Sims, B., & Albright, D. (2017). Development and evaluation of the Expressions of Moral Injury Scale–Military Version. *Clinical Psychology & Psychotherapy, 25*(3), 474–488.

Currier, J., Foster, J., & Isaak, S. (2019). Moral injury and spiritual struggles in military veterans: A latent profile analysis. *Journal of Traumatic Stress, 32*(3), 393–404.

Currier, J., Holland, J. M., Drescher, K., & Foy, D. (2015a). Initial psychometric evaluation of the Moral Injury Questionnaire–Military version. *Clinical psychology & psychotherapy, 22*(1), 54–63.

Currier, J., Holland, J., & Malott, J. (2015b). Moral injury, meaning making, and mental health in returning veterans. *Journal of Clinical Psychology, 71*(3), 229–240.

Davies, R., Prince, M., Bravo, A., Kelley, M., & Crain, T. (2019). Moral injury, substance use, and posttraumatic stress disorder symptoms among military personnel: An examination of trait mindfulness as a moderator. *Journal of Traumatic Stress, 32*(3), 414–423.

Dennis, P., Dennis, N., Van Voorhees, E., Calhoun, P., Dennis, M., & Beckham, J. (2017). Moral transgression during the Vietnam War: A path analysis of the psychological impact of veterans' involvement in wartime atrocities. *Anxiety, Stress, and Coping, 30*(2), 188–201.

Dohrenwend, B., Yager, T., Wall, M., & Adams, B. (2013). The roles of combat exposure, personal vulnerability, and involvement in harm to civilians or prisoners in Vietnam-war-related posttraumatic stress disorder. *Clinical Psychological Science, 1*(3), 223–238.

Evans, W., Stanley, M., Barrera, T., Exline, J., Pargament, K., & Teng, E. (2018). Morally injurious events and psychological distress among veterans: Examining the mediating role of religious and spiritual struggles. *Psychological Trauma: Theory, Research, Practice, and Policy, 10*(3), 360–367.

Farnsworth, J., Drescher, K., Evans, W., & Walser, R. (2017). A functional approach to understanding and treating military-related moral injury. *Journal of Contextual Behavioral Science, 6*(4), 391–397.

Fontana, A., & Rosenheck, R. (1999). A model of war zone stressors and posttraumatic stress disorder. *Journal of Traumatic Stress, 12*(1), 111–126.

Fontana, A., Rosenheck, R., & Brett, E. (1992). War zone traumas and posttraumatic stress disorder symptomatology. *The Journal of Nervous and Mental Disease, 180*(12), 748–755.

Frankfurt, S., DeBeer, B., Morissette, S., Kimbrel, N., La Bash, H., & Meyer, E. (2018). Mechanisms of moral injury following military sexual trauma and combat in post-9/11 U.S. war veterans. *Frontiers in Psychiatry.* doi:10.3389/fpsyt.00520.

Frankfurt, S., & Frazier, P. (2016). A review of research on moral injury in combat veterans. *Military Psychology, 28*(5), 318–330.

Frankfurt, S., Frazier, P., & Engdahl, B. (2017). Indirect relations between transgressive acts and general combat exposure and moral injury. *Military Medicine, 182*(11–12), e1950–1956.

Galatzer-Levy, I., & Bryant, R. (2013). 636,120 ways to have posttraumatic stress disorder. *Perspectives on Psychological Science, 8*(6), 651–662.

Harris, J., Usset, T., Voecks, C., Thuras, P., Currier, J., & Erbes, C. (2018). Spiritually integrated care for PTSD: A randomized controlled trial of "Building Spiritual Strength". *Psychiatry Research, 267*, 420–428.

Hodgson, T., & Carey, L. (2017). Moral injury and definitional clarity: Betrayal, spirituality and the role of chaplains. *Journal of Religion and Health, 56*(4), 1212–1228.

Jinkerson, J. (2016). Defining and assessing moral injury: A syndrome perspective. *Traumatology, 22*(2), 122–130.

Jordan, A., Eisen, E., Bolton, E., Nash, W., & Litz, B. (2017). Distinguishing war-related PTSD resulting from perpetration- and betrayal based morally injurious events. *Psychological Trauma: Theory, Research, Practice, and Policy, 9*(6), 627–634.

Killgore, W., Cotting, D., Thomas, J., Cox, A., McGurk, D., Vo, A., … Hoge, C. (2008). Post-combat invincibility: Violent combat experiences are associated with increased risk-taking propensity following deployment. *Journal of Psychiatric Research, 42*(13), 1112–1121.

Kline, A., Weiner, M., Interian, A., Scherbakov, A., & St. Hill, L. (2016). Morbid thoughts and suicidal ideation in Iraq War veterans: The role of direct and indirect killing in combat. *Depression and Anxiety, 33*(6), 473–482.

Koenig, H. G., Youssef, N. A., Ames, D., Oliver, J. P., Teng, E. J., Haynes, K., … Pearce, M. (2018). Moral injury and religiosity in US veterans with posttraumatic stress disorder symptoms. *The Journal of Nervous and Mental Disease, 206*(5), 325–331.

Kopacz, M., Lockman, J., Lusk, J., Bryan, C., Park, C., Sheu, S., & Gibson, W. (2019). How meaningful is meaning making? *New Ideas in Psychology, 54*, 76–81.

Kulka, R., Schlenger, W., Fairbank, J., Hough, R., Jordan, B., Marmar, C., & Weiss, D. (1990). *Brunner/Mazel Psychosocial Stress Series, No. 18. Trauma and the Vietnam War Generation: Report of Findings from the National Vietnam Veterans Readjustment Study*. Brunner/Mazel: Philadelphia, PA, US.

Lancaster, S. (2018) Negative outcomes after morally injurious experiences: A replication and extension. *Psychological Trauma: Theory, Research, Practice, and Policy*, *10*(4), 456–462.

Lancaster, S., & Erbes, C. (2017). Importance of moral appraisals in military veterans. *Traumatology*, *23*(4), 317–322.

Litz, B., Contractor, A., Rhodes, C., Dondanville, K., Jordan, A., Resick, P., … Peterson, A. (2018). Distinct trauma types in military service members seeking treatment for posttraumatic stress disorder. *Journal of Traumatic Stress*, *31*(2), 286–295.

Litz, B., & Kerig, P. (2019). Introduction to the special issue on moral injury: Conceptual challenges, methodological issues, and clinical applications. *Journal of Traumatic Stress*, *32*(3), 341–349.

Litz, B., Stein, N., Delaney, E., Lebowitz, L., Nash, W., Silva, C., & Maguen, S. (2009). Moral injury and moral repair in war veterans: A preliminary model and intervention strategy. *Clinical Psychology Review*, *29*(8), 695–706.

MacNair, R. (2002). Perpetration-induced traumatic stress in combat veterans. *Peace and Conflict: Journal of Peace Psychology*, *8*(1), 63–72.

Maguen, S., Lucenko, B., Reger, M., Gahm, G., Litz, B., Seal, K., … Marmar, C. (2010) The impact of reported direct and indirect killing on mental health symptoms in Iraq war veterans. *Journal of Traumatic Stress*, *23*(1), 86–90.

Maguen, S., Luxton, D., Skopp, N., Gahm, G., Reger, M., Metzler, T., & Marmar, C. (2011a). Killing in combat, mental health symptoms, and suicidal ideation in Iraq war veterans. *Journal of Anxiety Disorders*, *25*(4), 563–567.

Maguen, S., Metzler, T., Bosch, J., Marmar, C., Knight, S., & Neylan, T. (2012). Killing in combat may be independently associated with suicidal ideation. *Depression and Anxiety*, *29*(11), 918–923.

Maguen, S., Metzler, T., Litz, B., Seal, K., Knight, S., & Marmar, C. (2009). The impact of killing in war on mental health symptoms and related functioning. *Journal of Traumatic Stress*, *22*(5), 435–443.

Maguen, S., Vogt, D., King, L., King, D., Litz, B., Knight, S., & Marmar, C. (2011b). The impact of killing on mental health symptoms in Gulf War veterans. *Psychological Trauma: Theory, Research, Practice, and Policy*, *3*(1), 21–26.

Marx, B., Foley, K., Feinstein, B., Wolf, E., Kaloupek, D., & Keane, T. (2010). Combat-related guilt mediates the relations between exposure to combat-related abusive violence and psychiatric diagnoses. *Depression and Anxiety*, *27*(3), 287–293.

McCormack, L., & Ell, L. (2017). Complex psychosocial distress postdeployment in veterans: Reintegration identity disruption and challenged moral integrity. *Traumatology*, *23*(3), 240–249.

Nash, W., Marino Carper, T., Mills, M., Au, T., Goldsmith, A., & Litz, B. (2013). Psychometric evaluation of the Moral Injury Events Scale. *Military medicine*, *178*(6), 646–652.

Nazarov, A., Fikretoglu, D., Liu, A., Thompson, M., & Zamorski, M. (2018). Greater prevalence of post-traumatic stress disorder and depression in deployed Canadian armed forces personnel at risk for moral injury. *Acta Psychiatrica Scandinavica*, *137*(4), 342–354.

Press, E. (2018) The wounds of the drone warrior. *New York Times Magazine*.

Purcell, N., Koenig, C., Bosch, J., & Maguen, S. (2016). Veterans' perspectives on the psychosocial impact of killing in war. *The Counseling Psychologist*, *44*(7), 1062–1099.

Rasmussen, A., Verkuilen, J., Jayawickreme, N., Wu, Z., & McCluskey, S. (2019). When did posttraumatic stress disorder get so many factors? Confirmatory factor models since DSM-5. *Clinical Psychological Science*, 7(2), 234–248.

Richardson, D., King, L., Shnaider, P., & Elhai, J. (2017). Adverse combat experiences, feeling responsible for death, and suicidal ideation in treatment-seeking veterans and actively serving Canadian armed forces members. *Journal of Military, Veteran, and Family Health*, *3*(1), 34–40.

Shay, J. (1994). *Achilles in Vietnam: Combat and the Undoing of Character*. Scribner: New York.

Stein, N., Mills, M., Hall, K., Mendoza, C., Borah, A., Resick, P., & Litz, B. (2012). A scheme for categorizing traumatic military events. *Behavior modification*, *36*(6), 787–807.

Thomas, C. (2004). *At Hell's Gate: A Soldier's Journey*. Shambhala Publications: Boulder, CO, USA.

U.S. Department of Veterans Affairs. (2018). *How common is PTSD in veterans?* Retrieved from https://www.ptsd.va.gov/understand/common/common_veterans.asp

Wisco, B., Marx, B., May, C., Martini, B., Krystal, J., Southwick, S., & Pietrzak, R. (2017). Moral injury in U.S. combat veterans: Results from the national health and resilience in veterans study. *Depression and Anxiety*, *34*(4), 340–347.

Worthen, M., & Ahern, J. (2014). The causes, course, and consequences of anger problems in veterans returning to civilian life. *Journal of Loss and Trauma, 19*(4), 355–363.

Yager, T., Laufer, R., & Gallops, M. (1984). Some problems associated with war experience in men of the Vietnam generation. *Archives of General Psychiatry, 41*(4), 327–333.

Yeterian, J., Berke, D., Carney, J., McIntyre-Smith, A., St. Cyr, K., King, L.,… Litz, B., & Members of the Moral Injury Outcomes Project Consortium. (2019). Defining and measuring moral injury: Rationale, design, and preliminary findings from the moral injury outcome scale consortium. *Journal of Traumatic Stress, 32*(3), 363–372.

# 34

# STUDENT SERVICE MEMBERS/ VETERANS' MENTAL HEALTH ON CAMPUS

## Risk and resources

*Dan Nyaronga*

### On campus mental health risk and challenges for student service members/veterans (SSM/V)

As of 2012 at least one-half million student service members/veterans (SSM/Vs) have used their post-9/11 GI bill benefits (Sander, 2012), and these numbers are expected to increase as the United States continues to draw down its forces in Afghanistan. However, a fair number of students who matriculated into higher education post-military service are suffering from unhealed wounds of war, both physical and mental (Church, 2009), as well as dropout risk and failure to successfully complete degrees (Nyaronga & Toma, 2015). Zivin, Eisenberg, Gollust, and Golberstein (2009) report that one-third of American SSM/Vs have at least one mental health issue that persists over two years and fewer than half have received treatment. For example, research on SSM/Vs by Rudd Goudling, and Bryan (2011) found a high level of overall psychological distress among SSM/Vs, with 35% experiencing severe anxiety, 24% experiencing severe depression, and 46% reporting significant symptoms of PTS (Rudd et al., 2011). In addition, there is considerable evidence indicating that PTS commonly co-occurs with other psychiatric problems, most often with major depression, anxiety, and substance abuse (e.g., Keane & Wolfe, 1990; Rudd et al., 2011; Stewart, 1996). A study by Campbell & Riggs (2015) involving interviews with Operation Enduring Freedom/ Operation Iraqi Freedom (OEF/OIF) veterans also found PTS and depression as common themes throughout their exploratory interviews. Furthermore, another study by Lipka (2011) indicated that half of the SSM/V population in the United States has contemplated suicide, as compared to 17% of general student population (Rudd et al., 2011). PTS, in particular, has previously been found to be associated with increased risk for suicidal thinking and suicide attempts (Jakupcak, Cook, Imel, Fontana, Rosenheck, & McFall, 2009; Sareen, Cox, Stein, Afifi, Fleet, & Asmundson, 2007). Given the mental health challenges among trauma-exposed SSM/Vs on college campus, knowledge about interference in academic performance and school dropout is critical to inform outreach and intervention strategies for university and college campuses.

## Trauma related distress and school dropout among college/university-enrolled student service members/veterans

Trauma exposure and trauma-related distress have been found to be associated with higher likelihood of dropping out of college, academic difficulties, lower GPA, low educational self-efficacy, academic motivation, and campus alienation (Bachrach & Read, 2012; Barry, Whiteman, & MacDermid Wadsworth, 2012; Boyraz, Granda, Baker, Tidwell, & Waits, 2016; Boyraz, Horne, Owens, & Armstrong, 2013; Duncan, 2000; Elliott, Gonzalez, & Larsen, 2011; Jordan, Combs, & Smith, 2014). More specifically, general anxiety was found to have a direct negative effect and a greater impact on SSM/Vs' academic adjustment and achievements. General anxiety symptoms consist of worry, sleep and concentration difficulties, irritability, feeling restless or keyed up, and fatigue (Gerdes & Mallinckrodt, 1994). The symptoms may impede study efforts, completion of coursework, and class attendance (Campbell & Riggs, 2015; Milanak, Gros, Magruder, Brawman-Mintzer, & Frueh, 2013). In addition, PTS has been significantly associated with being in a physical altercation and high-risk drinking for college-enrolled OEF/OIF veterans (Elliot et al., 2011; Ness, Rocke, Harrist, & Vroman, 2014). Furthermore, alcohol misuse is a serious issue among SSM/Vs on and off campus, especially men and those with greater combat exposure (Santiago, Wilk, Milliken, Castro, Engel, & Hoge, 2010), who may use alcohol to self-medicate their mental health issues (Bray, Marsden, & Peterson, 1991; Elliot, 2015).

Studies indicate that 30%–40% of SSM/Vs do not complete their degree programs (Cate, 2013), and they tend to have lower grade point averages (GPAs) than their non-SSM/V student counterparts (Durdella & Kim, 2012; Elliot, 2015). Also, in a comparison sample, Boyraz et al. (2016) found a significant difference in college drop-out rates between SSM/Vs who had PTS (35%) and those who did not have PTS (20.8%). In another sample of SSM/Vs, Schonfeld, Braue, Stire, Gum, Cross, and Brown (2015) revealed that 28% of enrolled SSM/Vs reported having difficulty adjusting to the college life. The adjustment problems were significantly associated with mental health challenges (Schonfeld et al., 2015). Common themes in SSM/V challenges include interpersonal difficulties in social relationships (e.g., feeling isolated or lack of connection with friends, family members, and college peers; Campbell & Riggs, 2015), difficulty in transitioning to the role of student, feeling overwhelmed, and problems concentrating on academic studies. Women who have been sexually assaulted report that, as a result of the assault, they dropped a class, changed majors, changed universities, and moved residences (Lindquist, Barrick, Krebs, Crosby, Lockard, & Sanders-Phillips, 2013). Moreover, dropping classes, changing majors, or transferring to a new institution is also associated with greater educational costs, longer completion times, and other negative consequences that may impact SSM/Vs' financial, professional, and social functioning (Artime, Buchholz, & Jakupcak, 2018).

A longitudinal study by Johnson, Graceffo, Hayes, and Locke (2014) on service utilization by SSM/Vs at college/university counseling centers from 2010 to 2012 reveals that only 2% of SSM/Vs sought treatment from college counseling services and noted no change over time in service utilization. Given the mental health challenges, the steady increase in SSM/Vs enrolled in college, the interference in academic performance among trauma-exposed SSM/Vs, and the absence of increase in treatment seeking on campus, it is critical to understand the barriers to seeking mental health services in order to develop intervention strategies for college campuses.

## Barriers to seeking mental health services by student service members/veterans on college campuses

In a study by Bonar, Bohnert, Walters, Ganoczy, and Valenstein (2015) to assess barriers to seeking treatment among SSM/Vs, the most commonly reported barrier was the desire to avoid record of mental health treatment on their military record (43%), followed by treatment being embarrassing (31%), lack of confidentiality (29%), harmful to their career (29%), difficult to get an appointment (27%), costs too much (21%), and being seen as weak (26%) (Artime, Buchholz, & Jakupcak, 2018; Danish & Antonides, 2009).

Traditionally, institutions require students to self-identify if they have a disability and are in need of accommodations. However, unlike their non-SSM/V peers with documented disabilities who learned to seek accommodations in colleges, SSM/Vs with disabilities have less time to learn how to navigate the system (Vance & Miller, 2009). Many of their wounds are also not readily noticeable to faculty and staff. This makes self-advocacy even more important for SSM/Vs, but that is complicated by the culture they come from, a highly structured military environment, where admitting to such difficulties carries stigma and is seen as a weakness (Danish & Antonides, 2009; Pietrzak, Johnson, Goldstein, Malley, & Southwick, 2009; Warner, Appenzeller, Grieger, Belenkiy, Breitbach, Parker, & Hoge, 2011). The military environment promotes strong beliefs and values that may benefit SSM/Vs in their college coursework. However, such an environment may also set them apart from the rest of the college and contribute to a sense of alienation and frustration (DiRamio, Ackerman, & Mitchell, 2008; Rumann & Hamrick, 2010). In particular, military culture values strength and independence, and stigma regarding mental illness is high (Kim, Britt, Klocko, Riviere, & Adler, 2011), both of which may foster an avoidant coping style in response to psychological pain.

Studies suggest that avoidant coping contributes to substantial risk for dysfunction, yet SSM/Vs with this coping style may be reluctant to seek help with their difficulties from professors, mentors and academic advisors, administrators, and mental health providers because they do not want to be viewed as weak, a burden, or dependent on anyone for help (Romero, Riggs, & Ruggero, 2015). Avoidant coping is an attempt to turn away from adverse experiences and emotions and may take the form of denying or minimizing the result or effects of a specific problem or choosing not to deal with the problem (Boden, Bonn-Miller, Vujanovic, & Drescher, 2012). According to Feeny and Foa (2005), individuals with PTS use an avoidant coping style in an attempt to control their trauma-related intrusive thoughts, experiences, and emotions related to PTS. Others, like Whiteman and Barry (2011) conceptualize avoidant coping, especially among SSM/Vs, as a behavior characterized by the likelihood of consuming more alcohol than non-SSM/Vs and reluctance to seek counseling (Bonar & Domenici, 2011). Studies indicate that an avoidant coping style can aid individuals in managing their daily activities post-crisis; however, engaging in avoidant coping over an extended period of time could lead to mental health problems (Barlow, 2002; Holahan & Moos, 1987). Given the mental health challenges and interference in academic performance among trauma-exposed SSM/Vs, knowledge about available mental health services and resources is critical to inform intervention strategies on campuses.

## Mental health resources and services for student service members/veterans on college campuses

### *Problem-focused coping*

It allows an individual to deal with the stressor causing distress (Folkman & Lazarus, 1985) and adapt or remove the source of stress via one's behavior (Lazarus & Folkman, 1984) and may be particularly useful in a military context that encourages taking swift action to resolve obstacles and minimize threats. Problem-focused coping is also adaptive in the college environment, thus allowing SSM/Vs to take responsibility to ensure their own academic success (Romero, Riggs, & Ruggero, 2015). That means SSM/Vs are encouraged to talk to campus doctors or health officials about their health status in order to initiate early intervention.

### Screening for early intervention

Screening includes being asked about military and SSM/V status, combat exposure, depression, anxiety, and posttraumatic stress symptoms. Identifying and reaching those in need is a critical first step, particularly for suicide prevention. If SSM/V-specific orientations are conducted, it would provide greater potential for mass screening efforts. Based on interviews with SSM/Vs who had served in Iraq or Afghanistan about the adjustment to college life, DiRamio et al. (2008) recommend a holistic approach for assisting SSM/Vs. The approach includes implementing a mandatory orientation, which makes use of mentors or transition coaches, especially senior SSM/V colleagues from veteran organizations on campus.

### Outreach

Studies suggest the need for university disability service offices and counseling centers to ensure that SSM/Vs have a clear understanding of the help and services available at their disposal and the process by which they can receive the services. On the clinical service side, studies suggest the need for counseling centers to think in a strategic fashion about the potential for significant demand or needs from SSM/Vs, design interventions and outreach programs specifically aimed at SSM/Vs, provide access to mental health services on campus (Campbell & Riggs, 2015), offer long-lasting treatment benefits with a structured protocol, or offer appropriate referrals to a different mental health service provider. In addition, having spent so much time in a structured military environment and ascribing to military values, SSM/Vs may feel frustrated by the different values on campus and often conflicting values at institutions of higher education (Durdella & Kim, 2012), as well as by the assumptions and values of peers and faculty regarding SSM/V military service. They might also feel isolated from their non-SSM/V peers, given differences in age and experiences. As such, counseling centers and expert clinicians are encouraged to help SSM/Vs explore their pent-up emotions upon return from service to the civilian life and in navigating college life. Furthermore, studies encourage SSM/Vs to make use of the counseling center services on campus.

### Cultural competency

Studies suggest the need for university officials, administrators, and primary health care and mental health service providers to be culturally competent to work with military undergraduates. In addition, studies encourage a deeper understanding of potential difficulties related to the transition from the military environment to college life, the influence of military culture and deployment experiences on the individual (pre-deployment or post-deployment), the culture and subcultures (based on the different branches and different ranks) of the military (Wurster, Rinaldi, Woods, & Liu, 2013), and possible demographic factors such as age, gender, and marital status. As such, college counseling centers are encouraged to hire service providers with first-hand experience with SSM/V issues, who can provide training on military culture and the unique challenges that the SSM/V faces. New hires with no military background are also encouraged to complete cultural competency training.

### Cultural competency training

In order to implement relevant, effective supports for SSM/Vs, studies encourage mental health practitioners who are in contact with SSM/Vs to understand the nature of combat-related neurobehavioral symptoms and warning signs (including PTS, depression, substance abuse, and suicidal ideation). In addition, they are encouraged to understand the nature of clinical problems and how the symptoms may affect learning. Furthermore, they are encouraged to respond in a caring and effective manner, hence facilitating the transition to clinical care, if needed (Rudd et al., 2011). As such, specialized training in the behavioral sequelae of TBI, PTS, and other mental health challenges is seen as relevant for campus personnel who work directly with SSM/Vs (Kennedy, Krause, & Turkstra, 2008). The training may also include education regarding military culture, the deployment cycle, and specific military related stressors, as well as screening and clinical care for SSM/Vs. Health practitioners are also encouraged to know the various resources available for SSM/Vs, including: workshops offered and educational materials provided by various organizations (e.g., the National Center for PTS, Defense and Veterans Brain Injury Center). Such trainings and resources can help prepare counselors or health practitioners to evaluate military undergraduates for clinical problems, including signs by using validated clinical assessment tools, and offer evidence-based on-campus treatment (Bonar & Domenici, 2011). Campus health and counseling officials are encouraged to utilize workshops and readings on topics such as military culture, challenges of the deployment cycle, reaction to combat trauma, and effective counseling interventions for SSM/Vs (Rudd et al., 2011). Officials should not only be equipped with knowledge regarding symptoms and treatment strategies for PTS and other SSM/V mental health challenges (Borsari et al., 2017), they should also be able to provide an extensive referral list to other health providers, including Veterans Administration hospitals. Colleges and universities are also encouraged to provide cultural sensitivity training workshops to both the student body and faculty (Nyaronga & Toma, 2015). Studies indicate that without adequate training, many students or faculty may be less equipped to identify and connect with SSM/V students. For example, identifying most veterans on campus as "wounded" or "disabled" may send a message that they are "different" and/or incapable of getting better (Bonar & Domenici, 2011).

## Social support

The institutions of higher education are asked to consider implementing mechanisms to foster peer social support among SSM/Vs, such as one-on-one peer mentoring or discussion groups led by SSM/Vs themselves (Nyaronga & Toma, 2015). Emotional support from SSM/V peers also helps with academic adjustment and provides a buffer against mental health challenges (Nyaronga & Toma, 2015; Whiteman, Barry, Mroczek, & Macdermid Wadsworth, 2013). As such, institutions of higher education must be ready to enact holistic approaches to minimize the stress, isolation, and difficulties experienced by SSM/Vs (DiRamio et al., 2008). Broader strategies on reducing SSM/V mental health concerns on campuses may include having a SSM/V center within Veteran Affairs offices on campus, thus creating opportunities for SSM/Vs to network with each other for social support, participate in group therapy, or share similar experiences with friends and colleagues who served in the military (Nyaronga & Toma, 2015), that is, allowing SSM/Vs to feel more identified with their academic institutions and ensure that they are connected with SSM/V members who have similar military experiences and potentially similar social class or worldviews. Peer support has been recommended to help individuals with mental illness develop a sense of community in the process of recovery (Mead, Hilton, & Curtis, 2001). In addition, social support from peers is an established predictor of academic success among first-generation, especially ethnic minority, SSM/Vs (Dennis, Phinney, & Chuateco, 2005; Nyaronga & Toma, 2015).

Studies suggest that all efforts to reach out to and assist SSM/Vs should be sensitive to subgroup differences within the military or veteran population, especially in referral cases. It is critical that health officials or counselors make appropriate contact with the culture and subcultures (e.g., based on the different branches and different ranks) of the military and possible demographic factors such as age, gender, and marital status (Elliot, 2015). For example, Veterans Affairs (VA) or National Guard referrals may be appropriate to help SSM/V members, as the VA has a "vet-to-vet" peer-support program for veterans with serious mental illness facilitated by a peer leader, just like the National Guard with a "buddy-to-buddy" program to help members address their mental health issues (Greden et al., 2010).

### *Hire more health providers with clinical background, skills, and cultural competency training*

Combat-related trauma requires specific treatment approaches. However, training and acquiring skills take years, especially for many clinicians at university or college campuses who may be outside the department of Veterans Affairs system, have limited exposure to the VA clinical models, and may not be up to date in terms of developing and implementing new clinical skills for supporting SSM/V members on campus. As such, there is a need to hire more health providers with clinical background, skills, and cultural competency training. These recommendations are consistent with efforts by the Department of Defense and Department of Veterans Affairs to increase the pool of mental health providers who can effectively treat service members, veterans, and their families.

## Conclusion

This review gives institutions of higher learning insights into identifying and supporting SSM/V needs, especially those with mental health challenges. This may be done by accurately targeting existing resources to the more at-risk SSM/V populations on campus. These studies are not without limitations. First, a number of studies (Elliot, 2015; Nyaronga & Toma, 2015; Romero, Riggs, & Ruggero, 2015; Rudd et al., 2011) used cross sectional design, which may have introduced the potential for common-method variance and prevent causal conclusions. Second, a number of studies (Artime et al., 2018; Ness et al., 2014; Nyaronga & Toma, 2015) had small sample sizes, which in a way could have affected power and limited further analysis to detect any effects that are particularly small. Third, a number of studies (Campbell & Riggs, 2015; Elliot, 2015; Ness et al., 2014; Nyaronga & Toma, 2015; Romero et al., 2015) limited their sample to enrolled SSM/Vs in single institutions with little information from other campuses on how other colleges perceive and behave toward SSM/Vs on campus. The studies also excluded spouses of SSM/Vs, school administrators, and faculty in their samples. The relative ethnic homogeneity of a sample limits the ability to generalize to more diverse groups, thus compromising the representativeness of the overall sample. Furthermore, the studies have not assessed the length of SSM/V enrollments at their current institutions. Clearly, the amount of time in a given environment is a significant factor in the outcome or could potentially confound the results. Fourth, a number of studies (Ness et al., 2014; Nyaronga & Toma, 2015; Romero et al., 2015; Whiteman et al., 2013) have used self-report data that addresses only individual SSM/V perceptions, and not the actual risk and resource in relationship to the outcomes; as such, the findings may be inflated by common method variance (Campbell & Fiske, 1959) and the results may be a mixed bag, especially in samples where majority of SSM/V participants were nontraditional students, who tend to be older, be married, have children, and be employed. Fifth, some studies had self-selection bias (Whiteman et al., 2013; Rudd et al., 2011) and others used non-randomized samples (Eliot, 2015). In addition, some studies (e.g., Bonar & Domenici, 2011; Wurster et al., 2013) were qualitative in nature with small cases, which would not by any means cover as much as a representative sample with respect to challenges that SSM/Vs face in college campuses.

Despite research limitations, the mentioned research/studies demonstrated the unique mental health challenges of SSM/Vs on college campuses and provided a comprehensive list of mental health services which would support retention and improved academic performance for SSM/Vs on campuses. The highlighted list of accommodations may depend on individual circumstances and needs to be determined on a case-by-case basis. As such, university or college counselors may benefit from greater coordination and collaboration with a wider network of mental health experts familiar with military-related care (Ness et al., 2014), as well as with military-undergraduate student leadership on campus, Student Affairs office, Disability Services office, and Financial Aid office to enhance the overall experience for military undergraduates on campus and play a pivotal role in the university community to reduce stressors and help SSM/Vs with readjustment challenges.

## References

Artime, T. M., Buchholz, K. R., & Jakupcak, M. (2018). Mental health symptoms and treatment utilization among trauma-exposed college students. *Psychological Trauma: Theory, Research, Practice, and Policy, 11*(3), 274–282.

Bachrach, R. L., & Read, J. P. (2012). The role of posttraumatic stress and problem alcohol involvement in university academic performance. *Journal of Clinical Psychology, 68*, 843–859.

Barlow, D. H. (2002). *Anxiety and Its Disorders: The Nature and Treatment of Anxiety and Panic* (2nd Ed.). New York: Guilford Press.

Barry, A. E., Whiteman, S. D., & MacDermid Wadsworth, S. (2012). Implications of posttraumatic stress among military-affiliated and civilian students. *Journal of American College Health, 60*, 562–573.

Boden, M. T., Bonn-Miller, M. O., Vujanovic, A. A., & Drescher, K. (2012). A prospective investigation of changes in avoidant and active coping and posttraumatic stress disorder symptoms among military veteran. *Journal of Psychopathology and Behavioral Assessment, 34*, 433–439.

Bonar, E. E., Bohnert, K. M., Walters, H. M., Ganoczy, D., & Valenstein, M. (2015). Student and nonstudent National Guard service members/veterans and their use of services for mental health symptoms. *Journal of American College Health, 63*, 437–446.

Bonar, T. C., & Domenici, P. L. (2011). Counseling and connecting with the military undergraduate: The intersection of military service and university life. *Journal of College Student Psychotherapy, 25*, 204–219.

Borsari, B., Yurasek, A., Miller, M. B., Murphy, J. G., McDevitt-Murphy, M. E., Martens, M. P., & Carey, M. G. D. B. (2017). Student service members/veterans on campus: Challenges for reintegration. *American Journal of Orthopsychiatry, 87*(2), 166–175.

Boyraz, G., Granda, R., Baker, C. N., Tidwell, L. L., & Waits, J. B. (2016). Posttraumatic stress, effort regulation, and academic outcomes among college students: A longitudinal study. *Journal of Counseling Psychology, 63*, 475–486.

Boyraz, G., Horne, S. G., Owens, A. C., & Armstrong, A. P. (2013). Academic achievement and college persistence of African American students with trauma exposure. *Journal of Counseling Psychology, 60*, 582–592.

Bray, R. M., Marsden, M. E., & Peterson, M. R. (1991). Standardized comparisons of the use of alcohol, drugs, and cigarettes among military personnel and civilians. *American Journal of Public Health, 81*(7), 865–869.

Campbell, D.T., & Fiske, D.W. (1959). Convergent and discriminant validation by the multitrait-multimethod matrix. *Psychological Bulletin, 56*, 81–105.

Campbell, R., & Riggs, S. A. (2015). The role of psychological symptomatology and social support in the academic adjustment of previously deployed student veterans. *Journal of American College Health, 6*(7), 473–481

Cate, C. A. (2013, March 04). Exploring student veterans' post-secondary completion rates with two national surveys. *Student Veterans of America Research Brief, 1*(1). Retrieved from http://www.studentveterans.org/images/Documents/Research_Brief_2013_1.pdf

Church, T. E. (2009). Returning veterans on campus with war related injuries and the long road back home. *Journal of Postsecondary Education and Disability, 22*, 43–52.

Danish, S. J., & Antonides, B. J. (2009). What counseling psychologists can do to help returning veterans. *The Counseling Psychologist, 37*, 1076–1089.

Dennis, J. M., Phinney, J. S., & Chuateco, L. I. (2005). The role of motivation, parental support, and peer support in the academic success of ethnic minority first-generation college students. *Journal of College Student Development, 46*, 223–236.

DiRamio, D., Ackerman, R., & Mitchell, R. L. (2008). From combat to campus: Voices of student-veterans. *Journal of Student Affairs Research and Practice, 45*, 73–102.

Duncan, R. D. (2000). Childhood maltreatment and college drop-out rates. *Journal of Interpersonal Violence, 15*, 987–995.

Durdella, N., & Kim, Y. K. (2012). Understanding patterns of college outcomes among student veterans. *Journal of Studies in Education, 2*, 109–129.

Elliot, M. (2015). Predicting problems on campus: An analysis of college student veterans. *Analyses of Social Issues and Public Policy, 15*(1), 105–126.

Elliott, M., Gonzalez, C., & Larsen, B. (2011). U.S. military veterans transition to college: Combat, PTSD, and alienation on campus. *Journal of Student Affairs Research and Practice, 48*, 279–296.

Feeny, N. C., & Foa, E. B. (2005). Cognitive vulnerability to PTSD. In L. B. Alloy & J. H. Riskind (Eds.), *Cognitive Vulnerability to Emotional Disorders* (pp. 285–301). Mahweh, NJ, US: Erlbaum Publishers.

Folkman, S., & Lazarus, R. S. (1985). If it changes it must be a process: Study of emotion and coping during three stages of a college examination. *Journal of Personality and Social Psychology, 48*, 150–170.

Gerdes, H., Mallinckrodt, B. (1994). Emotional, social, and academic adjustment of college students: A longitudinal study of retention. *Journal of Counseling and Development, 72*, 281–288.

Greden, J. F., Valenstein, M., Spinner, J., Blow, A., Gorman, L. A., Dalack, G. W., Markus, S., & Kees, M. (2010). Buddy-to-buddy, a citizen soldier peer support program to counteract stigma, PTSD, depression, and suicide. *Annals of the New York Academy of Sciences, 1208*(1), 90–97.

Holahan, C. J., & Moos, R. H. (1987). Risk, resistance, and psychological distress: A longitudinal analysis with adults and children. *Journal of Abnormal Psychology, 96*, 3–13.

Jakupcak, M., Cook, J., Imel, Z., Fontana, A., Rosenheck, R., & McFall, M. (2009). Posttraumatic stress disorder as a risk factor for suicidal ideation in Iraq and Afghanistan war veterans. *Journal of Traumatic Stress, 22*, 303–306.

Johnson, M. C., Graceffo, J. M., Hayes, J. A., & Locke, B. D. (2014). Examining treatment-seeking college students with and without military experience and trauma histories. *Journal of College Counseling, 17*, 260–270.

Jordan, C. E., Combs, J. L., & Smith, G. T. (2014). An exploration of sexual victimization and academic performance among college women. *Trauma, Violence & Abuse, 15*, 191–200.

Keane, T. M., & Wolfe, J. (1990). Comorbidity in post-traumatic stress disorder: An analysis of community and clinical studies. *Journal of Applied Social Psychology, 20*, 1776–1788.

Kennedy, M., Krause, M., & Turkstra, L. (2008). An electronic survey about college experiences after traumatic brain injury. *Neuro Rehabilitation, 23*, 511–520.

Kim, P. Y., Britt, T. W., Klocko, R. P., Riviere, L. A., & Adler, A. B. (2011). Stigma, negative attitudes about treatment, and utilization of mental health care among soldiers. *Military Psychology, 23*, 65–81.

Lazarus, R. S., & Folkman, S. (1984). *Stress, Appraisal, and Coping*. New York: Springer.

Lindquist, C. H., Barrick, K., Krebs, C., Crosby, C. M., Lockard, A. J., & Sanders-Phillips, K. (2013). The context and consequences of sexual assault among undergraduate women at Historically Black Colleges and Universities (HBCUs). *Journal of Interpersonal Violence, 28*, 2437–2461.

Lipka, S. (2011, August 04). Half of student veterans have contemplated suicide, study shows. *The Chronicle of Higher Education*. Retrieved from http://chronicle.com/article/Half-of-Student-Veterans-Have/128524/

Mead, S., Hilton, D., & Curtis, L. (2001). Peer support: A theoretical perspective. *Psychiatric Rehabilitation Journal, 25*(2), 134–141.

Milanak, M. E., Gros, D. F., Magruder, K. M., Brawman-Mintzer, O., & Frueh, B. C. (2013). Prevalence and features of generalized anxiety disorder in Department ofVeteran Affairs primary care settings. *Psychiatry Research, 209*, 173–179.

Ness, B. M., Rocke, M. R., Harrist, C. J., & Vroman, K. G. (2014). College and combat trauma: An insider's perspective of the post-secondary education experience shared by service members managing neurobehavioral symptoms. *Neuro Rehabilitation, 35*, 147–158.

Nyaronga, D., & Toma, R. (2015). Student service members/veterans' mental health: Who is at a greater risk for posttraumatic stress symptoms on college campuses? *Military Psychology, 27*(4), 223–231.

Pietrzak, R. H., Johnson, D. C., Goldstein, M. B., Malley, J. C., & Southwick, S. M. (2009). Perceived stigma and barriers to mental health care utilization among OEF-OIF veterans. *Psychiatric Services, 60*(8), 1118–1122.

Romero, D. H., Riggs, S. A., & Ruggero, C. (2015). Coping, family social support, and psychological symptoms among student veterans. *Journal of Counseling Psychology, 62*(2), 242–252.

Rudd, M. D., Goudling, J., & Bryan, C. J. (2011). Student veterans: A national survey exploring psychological symptoms and suicide risk. *Professional Psychology, Research and Practice*, *42*, 354–360.

Rumann, C. B., & Hamrick, F. A. (2010). Student veterans in transition: Re-enrolling after war zone deployments. *The Journal of Higher Education*, *81*, 431–458.

Sander L. (2012, March 16). At half a million and counting, veterans cash in on Post-9/11 GI Bill. *Chronicle of Higher Education*, pp. A10–A11.

Santiago, P., Wilk, J., Milliken, C., Castro, C., Engel, C., & Hoge, C. (2010). Screening for alcohol misuse and alcohol-related behaviors among combat veterans. *Psychiatric Services*, *61*(6), 575–581.

Sareen, J., Cox, B. J., Stein, M. B., Afifi, T. O., Fleet, C., & Asmundson, G. J. (2007). Physical and mental comorbidity, disability, and suicidal behavior associated with posttraumatic stress disorder in a large community sample. *Psychosomatic Medicine*, *69*, 242–248.

Schonfeld, L., Braue, L. A., Stire, S., Gum, A. M., Cross, B. L., & Brown, L. M. (2015). Behavioral health and adjustment to college life for student service members/veterans. *Journal of American College Health*, *63*, 428–436.

Stewart, S. H. (1996). Alcohol abuse in individuals exposed to trauma: A critical review. *Psychological Bulletin*, *120*, 83–112.

Vance, M. L., & Miller, W. K. (2009). Serving wounded warriors: Current practices in postsecondary education. *Journal of Postsecondary Education and Disability*, *22*, 18–35.

Warner, C., Appenzeller, G., Grieger, T., Belenkiy, S., Breitbach, J., Parker, J., Hoge, C. (2011). Importance of anonymity to encourage honest reporting in mental health screening after combat deployment. *Archives of General Psychiatry*, *68*, 1065–1071.

Whiteman, S. D., & Barry, A. E. (2011). A comparative analysis of student service member/veteran and civilian student drinking motives. *Journal of Student Affairs Research and Practice*, *48*, 297–313.

Whiteman, S. D., Barry, A. E., Mroczek, D. K., & Macdermid Wadsworth, S. (2013). The development and implications of peer emotional support for student service members/veterans and civilian college students. *Journal of Counseling Psychology*, *60*, 265–278.

Wurster, K. G., Rinaldi, A. P., Woods, T. S., & Liu, W. M. (2013). First-generation student veterans: Implications of poverty for psychotherapy, *Journal of Clinical Psychology*, *69*(2), 127–137.

Zivin, K., Eisenberg, D., Gollust, S. E., & Golberstein, E. (2009). Persistence of mental health problems and needs in a college student population. *Journal of Affective Disorders*, *117*(3), 180–185.

# 35

# RESILIENCE AND CLINICAL ISSUES IN SURVIVAL BEHAVIOR UNDER ISOLATION AND CAPTIVITY

*Vasile Marineanu*

Survival, evasion, resistance and escape (SERE) psychology, as a specialized field of both clinical psychology and operational military psychology, aims to provide training programs within SERE courses and to work through specialized intervention with recovered personnel who have gone through isolation incidents, missing persons, hostages, prisoners of war (POWs), escapees etc., who are undergoing a process of reintegration into society, family and professional life.

Although, at present, the number of cases requiring the intervention of a SERE psychologist is rather low, the consequences of the lack of adequately trained professionals in this specialty are unacceptable given the characteristics of the current security environment when we face threats at any time: terrorist attacks, abductions or hostage-taking. Moreover, to fill a vacuum of knowledge and of practical skills needed for survival in hostile environments around the world, such as major cities and other urban areas where dangers can easily be overlooked, a new area has emerged under the name Urban SERE, a civilian version of the traditional SERE domain, rooted in military experience.

SERE psychologists base their specialized knowledge, skills and techniques on evidence-based theories about human behavior. These include research from over the past 30 years of experience in risk management and specialized intervention for traumatic stress management, used both in SERE programs and in working with recovered personnel who have survived isolation incidents, missing persons, detainees, captured and/or escaped personnel.

At present, in most NATO member states, SERE activities are being carried out in line with NATO Standard 7196 (STANAG 7196, 2007). They aim at familiarizing military personnel with the core elements of SERE psychology, principles, processes and procedures of this particular field of clinical and operational applications of psychology. Thus, participants are expected to understand both captivity and the isolation environment, how military training programs can contribute to build resilience, which are the risks associated with SERE training programs and how to mitigate them. At the same time, they are familiarized with the principles of the reintegration processes, the most frequent problems that may arise during this process and the support needed by families of those who go through a situation of isolation or captivity.

## Individual reactions during a survival event

The most common psychological reactions following a survival event, according to Leach (1994), include panic, paralysis anxiety, perceptual distortions, denial, depressive reactions, hyperreactivity, behavioral stereotypes, anger, guilt and psychic fall. Usually, most people who survive a disaster or a potential traumatic event develop similar psychological and emotional reactions. Most individuals will probably understand these reactions as normal responses to abnormal events and will quickly return to their previous operation. Some will remain more affected, with emotional sequelae, which will require more time and possibly professional treatment to return to the previous functional state. Eventually, a small percentage may never recover at all.

That is why it is quite difficult to predict who and what will experience and who will have long-term problems. In addition, this problem becomes even more complicated because of the well-intentioned trend of many mental health professionals to pathologize these normal reactions to traumatic stress. Many of these reactions are, in essence, the expression of defensive mechanisms, but they become problematic when they begin to substantially interfere with survival efforts or significantly inhibit adaptive behavior. Not everyone will experience these reactions at the same level of intensity and/or the same duration, but the growth in the intensity and the duration might be indicators that the person needs clinical attention.

Unfortunately, the more peritraumatic and posttraumatic risk factors, the higher probability to develop long-term clinical implications, no matter how well the person is prepared for the survival event through relevant stress inoculation training.Therefore, the extent to which these reactions will have a negative impact on recovery depends on a number of individual personality factors, the nature of survival training, the individual or group reactions during the event and the reintegration/recovery efforts following the event.The most common clinical issues related to SERE events are described below.

## Clinical issues related to SERE events

It is known that POW have, over time, experienced brutality, torture, coercion, loneliness and isolation, among many other forms of deprivation and exploitation. Each of these experiences was aimed at stressing the dependence of the captive on those in whose custody they were and, through these deprivations, to obtain maximum exploitation of the prisoners.The immediate and lifetime effect of these experiences is so big that it can rarely be said to be overstated.

The military personnel captured and detained as POWs manifest significantly high rates of physical and emotional trauma, posttraumatic stress disorder (PTSD) symptoms and other mental health problems.Thus, during the Second World War, about half of the captured soldiers in Germany and Japan developed PTSD, which remained symptomatic throughout their lives (Doran, Hoyt, Hiller Lauby & Morgan, 2012). Sutker and Allain (1996) suggest that between 88% and 96% of prisoners during the Korean War have experienced psychological and behavioral problems related to captivity.

Also, among POW during World War II, extremely high mortality rates, cognitive difficulties such as memory deficits, spatial orientation problems, less efficient planning skills and problems related to impulse control were reported. Some of these problems are supposed to be related to severe malnutrition often experienced by POW; those who lost 35% or more of the body weight during captivity had the highest degree of memory, verbal and visual learning deficiency (Sutker, Allain, & Johnson, 1993). Compared to veterans who have not experienced captivity, POWs have had multiple adaptation disorders, alcohol abuse problems, depressive disorders, anxiety disorders, compulsive eating, difficulties in couple relationships, musculoskeletal and gastrointestinal disorders, as well as premature aging (Doran et al., 2012).

## Effects of captivity on health

The effects of trauma on the health of POWs and hostages, in general, have been widely discussed in the psychological literature in various contexts. Thus, a relatively recent meta-analysis (Steel et al., 2009) shows that former POWs face serious psychological consequences, including PTSD. The same meta-analysis emphasizes the importance of exploring the concept of war torture and PTSD for non-Western populations and areas. In addition, it is worth mentioning that most of the studies in this field were based on samples composed exclusively or predominantly of male populations (Saab, Chaaya, Doumit & Farhood, 2003). These studies have attempted to explain PTSD in relation to exposure to war trauma and their severity (Ghaddar, Elsouri, & Abboud, 2016), and similar studies have explored general mental health and quality of life amongst refugees exposed to torture and traumatic events (Carlsson, Mortensen & Kastrup, 2006).

At the same time, a systematic literature review on the development and maintenance of PTSD after exposure to war trauma and torture has highlighted the importance of focusing attention on female prisoners, as female sex and advanced age are considered risk factors in the development of PTSD symptoms (Johnson & Thompson, 2008). Moreover, the differences between men and women victims of war torture, depending on the types of torture and their psychological consequences, have been highlighted in various studies on the population of former Yugoslavia. Thus, women with a higher risk for certain types of torture, such as sexual torture, have recorded higher scores than men in depression, anxiety and PTSD (Spiric et al., 2010). Also, women who have been subject to war torture in Nepal have reported, to a greater extent than men, symptoms of PSTD, persistent somatoform disorder of pain and dissociation disorders of amnesia and conversion (Van Ommeren et al., 2001).

In a study of 67 Lebanese prisoners exposed to torture, held in the Khiam detention center during Israeli occupation in Lebanon (1981–1989) and released between 1982 and 1997, the prevalence of PTSD and cardiovascular disease was high, even almost one or two decades after the release. Thus, the prevalence of PTSD was 28.4%, comparable to that of political prisoners tortured in Lebanon (30.6%), in Occupied Palestinian Territory and in Germany (33%). However, the prevalence of cardiovascular disease, adjusted by age, has been relatively high compared to the population of the same age of women who have been previously exposed to torture (16.42%–6%). The risk of developing PTSD increased with the length of detention, exposure to beatings and the threat of rape, and declined proportionally to the strength of religious beliefs, confirming once again the role of social support as a protective factor against PTSD. The risk of developing cardiovascular problems increased in proportion to the duration of detention, age and exposure to water boarding (Ghaddar, Elsouri, & Abboud, 2016).

## Long-term implications of captivity for mortality and health

Captivity during armed conflicts involves a variety of stressors and represents a traumatic experience perceived as particularly intense and prolonged. This experience often includes isolation, torture, humiliation, starvation, poor hygiene and lack of medical treatment. Former POWs have a high risk of developing a wide range of mental disorders, particularly PTSD and depression (Solomon et al., 2014).

Several studies on World War II (WWII) and Vietnam war survivors reported higher rates of morbidity among former prisoners of war compared to veterans who were not in captivity (Nice,

Garland, Hilton, Baggett, & Mitchell, 1996). Similarly, a study on Australian prisoners from WWII who were captured by the Japanese found that they reported more psychosomatic symptoms, had a higher number of medical conditions, used more drugs and were more prone to deficient functionality compared to a control group made up of veterans who were not in captivity (Creasey et al., 1999). In this respect, the findings of the study showed that the relationship between captivity and self-assessed health is mediated by the symptoms of PTSD and depression and are consistent with other studies which found that PTSD and depressive symptoms influence the relationship between exposure to trauma and physical health (Byers, Covinsky, Barnes, & Yaffe, 2012; Campbell, Greeson, Bybee, & Raja, 2008). Other studies on former military personnel with psychological and behavioral problems have shown that people diagnosed with PTSD and depression are at higher risk of cardiovascular disease, including stroke, and have weaker health condition than those without PTSD or depression (Kang, Bullman, & Taylor, 2006).

According to a study conducted by Solomon et al. (2014), former prisoners of war have an increased risk of premature mortality, with a near four times higher mortality rate compared to non-captive veterans. Some studies (Engdahl, Harkness, Eberly, Page, & Bielinski, 1993) suggest that the long-term health of people experiencing captivity is a direct result of torture and harsh physical conditions (e.g. poor hygiene, lack of food), which can make ex-prisoners more vulnerable to diseases (Creasey et al., 1999). However, a shorter duration of captivity (between 6 weeks and 8 months) and mild conditions in which captives are held may limit this increased risk of premature mortality (Solomon et al., 2014).

The relationship between captivity and the state of health of former POWs as well as their mortality rate could be explained by the impact of prolonged psychological stress on physiological functioning. McEwen (1998) has suggested that prolonged stress exposure can cause the body to hyperactivate various physiological systems, a condition encountered in chronic stress known as the allostatic load. This can lead to permanent physiological changes, including mild inflammation, elevated levels of cortisol, insulin response and low immunity, which predisposes the body to asthma and various infections, the most common being in the urinary tract. In addition, stress exposure was associated with telomere length wear, a region of the DNA molecule located at the end of each chromosome, which is involved in aging and longevity.

## *Posttraumatic stress disorder and the change in perception of pain*

Torture in captivity may take both physical form (severe beatings, burns, penetrating injuries, suspensions, electrical shocks, etc.), affecting vulnerable parts of the body as well as psychological torture (isolation, humiliation, false executions, etc.). Therefore, survivors of torture have, over time, an increased risk of physical and emotional distress, as well as of developing psychiatric disorders and dysfunctional behaviors, including PTSD. In addition to the association of PTSD with chronic pain, posttraumatic stress among survivors of severe trauma was also associated with altered pain perception, including changes in sensitivity and reactivity to painful stimuli. Thus, the presence or absence of PTSD can mediate the perception of pain (Defrin, Lahav & Solomon, 2017).

PTSD symptoms show an interesting evolution over time. Prospective studies evaluating the longitudinal course of PTSD among major trauma survivors describe four main trajectories (Solomon, Horesh, Ein-Dor & Ohry, 2012), (1) increased resilience: lack of intense reactions to traumatic stress and lack of clinical or subclinical PTSD; (2) recovery: baseline clinical or subclinical levels, followed by PTSD remission; (3) delayed onset PTSD: an asymptomatic initial period, followed by PTSD and (4) chronic PTSD: clinical or subclinical levels of PTSD, continuously present, with severe traumatic stress reaction since the onset of trauma.

Although many indicators of PTSD have been evaluated longitudinally, it is not yet certain that dysfunctional modulation of pain in patients with chronic or delayed onset PTSD precedes or stems from chronic pain or both develop simultaneously. In addition, the sensory profile may be influenced by drugs, which could condition dysfunctional pain modulation or pain threshold (Bar et al., 2011), although this possibility is unlikely. Overall, it appears that the evolution, intensity and duration of PTSD symptoms and the resulting distress are, rather than mere trauma exposure, involved in the association between trauma and altered pain perception/modulation. Thus, the higher the intensity and duration of PTSD, the greater the risk of chronic and irreversible changes in the pain perception system (Defrin et al., 2017).

## *Personification of torture*

Starting from the idea of personification of pain, several studies (Tsura, Shaharb, Defrinc, Lahava & Ginzburga, 2017; Van der Kolk, & McFarlane, 2012; Van der Kolk, 1994) suggest that the traumatic ramifications and later posttraumatic stress are not limited only to emotional distress and dis-adaptive behavior, but it seems that trauma interferes with the fundamental way of interpreting body signals and with the ways in which body experiences interact with oneself. It is known that people with chronic conditions that exhibit some symptoms, such as pain, over time tend to attribute these symptoms to human characteristics in a way that is relevant to their self, a process named personification of the disease (Epley, Waytz, Akalis & Cacioppo, 2008).

Pain is thus transformed from a sensory attribute and an external obstacle into an internal phenomenon that encompasses personal meanings and attributes. In other words, individuals develop a kind of relationship with their chronic medical condition that expresses how the self and pain interact. This transformation, in which chronic pain changes from an external experience an internal representation, suggests that the personification of pain can become both a risk factor for the psychological and behavioral problems as well as a factor of resilience to these problems. Therefore, traumatic experiences may be stored in the somatic memory and, hence, the healing process is rooted in the processing of body experiences. In this regard, studies show that catastrophic interpretations of body signals are considered pain-maintaining factors (Sullivan, 2012), including chronic pain, associated with modulation of dysfunctional pain and PTSD.

In this context, there should be an effort to reduce the symptoms of PTSD as a mechanism to protect against the development of torture personification and efforts to focus on identifying subjective somatic experience and how it contributes to building the personification of chronic symptoms. Starting from this idea, clinicians can help individuals adapt their narrative/personification to their problems so as to turn it into a more adaptive construction that redefines the sense of control of the person, independence and self-worth. However, additional research is needed to clarify how and which of the therapeutic methods should be used in the personification of pain to transform the self so that it is no more victimized by pain, but becomes stronger, more accepted and freed from traumatic experiences.

## *Effects of captivity on family life*

Most of the literature on POWs has generally focused on studying their biographies and the political and strategic aspects associated with their release. Only in the last 30 years, the attention of the scientists has turned to the families of POWs (Boss, 2006). The family of a captive person has often been described as the "waiting family," which coexists between hope and despair (Hunter, 1983). The captive father is present-absent in family life, because although it seems to be part of it, in the day-to-day life it is, in fact, absent. Boss (1999) called this *ambiguous loss situation*, who described it in 2004, as an experience felt as a clear and sharp pain in the form of physical loss (e.g., due to captivity or divorce) or emotional/cognitive loss in which the absent person continues to be physically present but emotionally absent (e.g., in Alzheimer). The world, which seemed safe, familiar and stable before the loss, is left in the past, while the present and the future are wrapped in a fog of uncertainty.

In many cases, society expects that the family of a captive, hostage or POW enter into a state of mourning. However, people cannot express themselves when dealing with an ambiguous loss, rather being stuck in time, in an uncertain situation linked to the thought that the loved ones cannot return, hoping at the same time and anticipating their return. Often, an indicator of ambiguous loss is the creation of a limit ambiguity, which family members describe as a situation where it is not known who is part of the family (Boss, 2006). A clear death is undoubtedly painful, but funeral rituals can happen when there is a body, and family and friends come together to commemorate the person who died (Boss, & Yeats, 2014).

Because the current explanatory models of loss and pain can be applied only partially to the psychological processes associated with the ambiguous loss faced by the families of captives, Ben-Asher and Shalev (2015) suggest two models that can provide general frameworks for understanding these processes.

The first is related to the stages of *pain/grief and mourning* (Bowlby, 1980). Although, in recent years, researchers have raised questions about Bowlby's model, pointing out that the pain process can not necessarily be delimited (Rubin, Malkinson & Witztum, 2012), in particular, *the first stage of the model, shock and numbness*, seems to be applicable to the families of captives, who respond to the news of captivity by *shock* and *non-acceptance*. This stage is characterized by outbursts of crying, intense tension, feelings of anger and panic. At this stage, the family may stick to routine behavior or, on the contrary, may experience extreme changes in behavior and functioning.

*The second stage, yearning and searching*, also corresponds to the experiences of the captive's families. While individuals who suffer from the death of a loved one are finally able to accept the loss and seek that person less and less in time, captives' families are often engaged in an intensive prolonged search process, which in many cases is accompanied by a desire to engage in intensive public actions. Although, in the end, there may be a decrease in the intensity of the search process, this reduction occurs very slowly and is accompanied by doubt and ambiguity (Dempsey & Baago, 1998).

Although the trauma that captives' families face does not have the legitimacy of the final loss present in those who face death, it is still possible for the family to experience *the third stage of pain, disorganization*. This stage is characterized by despair and disintegration of the individual's internal order. At this stage, the grieving individual begins to realize that the loss really happened. The stage is characterized by a sense of emptiness, apathy and depression, resulting from the need to move away from previous patterns of thinking and behaving. The individual who had been separated from the loved one suffers from pain and grief, but these feelings differ from those experienced in the earlier stages; they become part of the new routine of the mourning person (Rubin, 2000).

*The last stage of the model, the reorganization*, occurs if the pain for the person's death is likely to deviate from the ambiguous loss. At this stage, the person learns to take on new roles and to form a new vision of the world in order to adapt to the loss. For captives' families, the ambiguity of loss can prevent this kind of transformation. In some cases, the uncertainty that the loss occurred may even completely prevent the family from experiencing mourning. Moreover, families faced with the captivity of a member hope that the situation will be overturned and everything will return to normal one day. However, as suggested by the participants in Ben-Asher and Shalev's (2015) study, the hope that the return of captive parent will erase his absence is never fulfilled, but remains a simple fantasy. First of all, the father that the children remember is not the same father who returns. In this sense, some children may have a renewed sense of parent loss, which, as they realize, will never return. Secondly, when the father returns, he often stays physically or emotionally absent, for example, because of the need to undergo intense physical rehabilitation or because of a struggle to reintegrate into a family of people who now seem to be essentially foreigners.

The second model that can also help to understand the experience of ambiguous loss of a captive parent is the context of *complicated grief*. Complicated grief (DSM-V diagnosis: Other Specified Trauma and Stressor-Related Disorder, 309.89, example 5, American Psychiatric Association, 2013) is generally described as a situation where the usual mourning process is interrupted, such that an individual who has lost a loved one cannot move on with his life.

Particularly, the bereaved cannot accept the fact of the loved one's death, is unable to trust others following the loss, avoids closeness with people with whom he was close in the past and feels that his life and his future are meaningless. There are several risk factors that can influence the likelihood of a child who has lost a parent to develop complicated grief: the child's relationship with the deceased parent, the circumstances of his death, the multiple loss experience and the individual characteristics of the child.

Similarly, to children who suffer from complicated grief, those who face ambiguous loss experience prolonged sadness, embarrassment, anxiety and uncertainty; their loss is permanently felt and remains unresolved, being trapped in a deadlock. In the case of families experiencing the captivity of a member, however, the above-mentioned experiences occur in the absence of a clear, definitive information about the beloved's death—the ambiguous loss situation. In fact, this ambiguity and the expectations regarding the return of the person are those that hinder the natural course of mourning. This not only prevents children from continuing their lives, but the whole family, blocking communication between its members, eventually depriving the child of an adequate support system.

Boss and Carnes (2012) found that those who suffer such losses have strange, repetitive dreams about missing loved ones. Therefore, Ben-Asher and Shalev (2015) suggests that ambiguous loss is a special case of complicated grief involving traumatic elements, and Boss and Yeats (2014), in a study of complicated pain, identified six principles that could help families who experience ambiguous loss to overcome this blockage: finding meaning, tempering mastery, rebuilding identity, normalizing ambivalence, revising attachment and discovering hope.

## *Factors influencing the family's ability to rehabilitate*

During captivity, both the captive person and the family imagine a perfect reunion following release. In reality, however, due to the difficult process of family reintegration, routine and changes suffered by all family members, cultural shock and disappointment in general, the experience of reunion is much different and more complex than it would have been imagined by all involved in this process. Families need time to reconnect, to get to know each other again after the transformations, to renegotiate the roles and manage the mutual expectations and independence of each member. Also, the psychological and physical state of the repatriated captive affects the family's ability to rehabilitate (Hunter-King, 1993).

However, even if some parents released from captivity were forced, for example, to spend a lot of time in hospitals and thus continued to be absent from the lives of their children, and sometimes the father's return led to a certain degree of destabilization of the balance developed during the absence of his family; the return of the parent was an essential variable that eliminated the ambiguity of the child's life. Thus, it was noticed that after the return of the captive parents, the children had better school results, adapted better to the community, and showed a higher level of independence than children whose fathers did not return from captivity (Ben-Asher & Shalev, 2015).

On the other hand, some people experience the captivity of the parent as a catalyst for personal growth, reinforcing self-confidence and trust in the people around them. In this respect, Hunter (1983) noted that some families describe the period of captivity as an opportunity to grow, to strengthen mental power and to work more independently and maturely. This is in line with the concept of *post-traumatic growth* that occurs when a person who has suffered a severe crisis has the experience of having high personal and spiritual abilities after accepting prolonged pain and suffering and express the individual's ability to recover, to give new meaning to their own existence, followed by a sense of personal growth, strength and value.

The first reactions that occur after a traumatic event are often an indicator of what will happen in the future. Seligman (2002) describes the ability of people for hedonistic adaptation, wherein they resume their routine while adapting to changes (good or bad). Basically, all accounts of people who have learned the news of a parent's captivity refer to their first reactions or to the first actions taken by adults around them. These initial responses set the tone for the families' subsequent modes of behavior during the period of the father's absence. Some families enter a frozen state where family and personal activities stop waiting for the father to return, while others choose to continue their lives. Such changes in family routine are crucial to its functioning. Other people who have experienced such experiences describe the period of captivity as a sensory experience, to be in a fog (e.g., a black cloud surrounding everything). Such sensations are common in people who have experienced loss and bereavement (Braun & Berg, 1994). The family silence on captivity theme is somewhat in line with the *double-wall concept*, a term introduced by Bar-On (1994) to describe the silence of Holocaust survivors: parents do not say, wanting to forget and protect their children from painful memories, and children do not ask, as an expression of sensitivity to the need for silence of their parents.

Jensen, Martin and Watanabe (1996) studied the effects of the captivity of one of the parents, who were in Iraq during the Desert Storm operation, on American children and learned that more than half of the parents who remained at home (usually mothers) reported that many of their children faced depression, behavioral problems or difficulties at school, requiring treatment for these problems. According to the study, the most influential variable relating to the emotional state of children was the functioning level of the parent who remained at home, who, in the partner's absence, had to take over the roles of the other. Thus, Jensen et al. (1996) noted that many of the children who did not have behavioral problems had depressive symptoms, and with the return of the absent parent, when the situation was apparently remedied, behavioral problems, frustration and anger occurred.

A study conducted by Ben-Asher and Shalev (2015) presents the situation of Israeli adults, children of former Egyptian prisoners of war (1969–1973), showing that the long-term effects of captivity and their inner experience associated with the absence of the father create long-term effects, which also manifest in the period after his return. The findings of the study can be explained by *ambiguous loss theories* and by the theories of *loss and grief*. The author's recommendations aim at recognizing ambiguous loss as a significant stressful situation and providing social and professional assistance programs in order to build a life that prevents a frozen stage and continues both during the absence of the father and after his return.

In general, there are several practical recommendations resulting from ambiguous loss studies, which should be recognized as a stressful situation requiring professional and social assistance for family members. It is extremely important that family relationships (including intergenerational relationships) be strengthened within the family and, at the same time, create new relationships outside the family. Family members can be encouraged to find meaning in terms of ambiguous loss (a possibility would be to engage in charity work for the benefit of others who had similar experiences*), to create ceremonies and rituals* that constantly build and maintain family and social ties, such as family events and celebrations, and young people, in particular, to look for benefits in *dyadic interactions with a parent or colleague* or even in *individual therapy sessions* (Boss & Yeats, 2014). On the other hand, Boss (1999) emphasizes the practical nature of learning to live in uncertainty, to avoid self-blaming and blaming others and the power of realistic positive thinking. While a parent is absent, it is also important to strengthen the position of the other parent as a mediator and establish a routine that is as normal as possible.

In conclusion, Ben-Asher and Shalev (2015) recommend three directions of emotional support to the families of captive individuals: (1) familiarization with the stress-associated reactions, (2) providing help in setting new routines tailored to the new situation and (3) encouraging families to seek support from non-formal systems. Support interventions can encourage the present parent to talk with their children about their concerns and their difficulty in coping with the changes associated with the absence period. Also, the parent who remains at home must be aware of the many ways in which children can express their feelings and consider them, if appropriate, as particular ways of healthy behavior. The parent who is left with the family can also benefit from therapy in order to strengthen his/her own stress management skills and become a role model so that children do not feel responsible for their extreme emotional reactions. Finally, parents should be provided with guidance on the importance of children's school requirements and on the maintenance of professional and family responsibilities while creating a sense of control over their living space.

### *Secondary traumatization*

There are several authors who suggest that the experience of captivity affects not only the person who directly experienced that situation, but also those in his/her near environment (Hunter, 1988). This seems to be particularly the case of former prisoner's partners, as they often serve as a central element of support for their partners when they return home. There are several studies in this respect (Dekel, 2007) that showed how PTSD symptoms of former POWs may affect the emotional states of their wives, a process called secondary traumatization (ST). The term ST has been used to indicate the situation of people who have come into contact with traumatized persons and experience emotional distress and PTSD-like symptoms similar to those expressed by the survivor (Figley, 1995).

There are quite a few studies showing that ST in life partners is correlated with the severity of PTSD symptoms to war veterans and former POWs (Dekel, 2007). A possible mechanism that could contribute to this association is the partner's belief system regarding the PTSD-affected husband and the manner in which he/she interprets the situations, which affects the actual functioning of individuals (Beck, 2010). In recent years, however, some studies have shown that wives' perception of the severity of veterans PTSD symptoms is positively associated with the psychological and marital distress of spouses, and the perceptions of wives both on this aspect and on the burden to live with someone with PTSD are mediating factors of the posttraumatic effects of veterans with PTSD on the emotional distress of their wives (Renshaw, Rodebaughc & Rodrigues, 2010).

Literature dealing with the issue of veterans and former prisoners of war suggests that operational stress and/or situation of captivity is not limited to causing emotional distress and PTSD-related symptoms but inducing a wider range of manifestations. Thus, the findings of the study conducted by Zerach, Greene & Solomon (2015) showed that wives of former war prisoners reported a worse state of general health, higher somatization levels, higher ST levels, higher levels of perceived PTSD symptoms in their spouses and lower levels of marital adjustment compared to the wives of a control veterans' group.

ST resulting from the experience of living with a former POW suffering from PTSD could increase the subjective experience of a fragile general health state, higher levels of psychosomatic problems experienced and the number of physical illnesses faced by close relatives of former POWs. Indeed, this experience of taking care of people who have been traumatized is one of the challenges faced by the partners of veterans with PTSD (Beckham, Lytle & Feldman, 1996).

Also, Zerach et al. (2015) found that the level of marital adjustment moderated the relationship between ST and the well-being state of wives so that if a wife suffers from high levels of ST and also has a low marital adjustment, she reports a low level of general health. In families with a high level of marital adjustment, the husband plays an important role in supporting the wife to find adaptive coping strategies and provides tangible support, such as meal preparation and sharing the burden of childcare. In this case, the stress model as buffer (Cohen & McKay, 1984) could explain the moderating role of marital adjustment. In a strong marriage, a life partner can be a source of emotional support, reducing the effects of stress, even when he is a source of stress. This conclusion suggests a possible area of intervention addressing both psychological and social components within couple relationships to reduce the psychological and physiological burden of wives and to increase the relationship with their traumatized partners.

Given the fact that traumatic events can have long-term consequences not only for survivors of trauma but also for other significant persons in their lives, this has raised the problem of intergenerational inheritance of war traumas for veterans' descendants, as ST (Zerach et al., 2015). This has attracted the interest of several authors, but most of the studies have focused on young children and adolescents and based their conclusions on parents' reports (Rosenheck & Fontana, 1998). A transversal study of the children of former prisoners during the war between Iran and Iraq revealed a higher prevalence of depression and general anxiety compared to an average adult group (Razavi, Razavi-Ratki, Nojomi, & Namiranian, 2012). Although these studies provide insight into the long-term effects of captivity trauma on children of former POWs, their transversal study design limits the possibility of understanding the mechanisms of intergenerational transmission of captivity trauma.

A study conducted by Zerach, Kanat-Maymon, Aloni & Solomon (2016) shows that 40 years after the war between Israel and Egypt, known as the Yom Kippur War (1969–1973), the descendants of former POWs are at increased risk for ST symptoms as compared to the adult descendants of control group of veterans. Additionally, adult descendants whose parents suffer from PTSD are at higher risk for ST symptoms. These results are consistent with case studies (Rosenheck & Nathan, 1985), with empirical studies (Ahmadzadeh & Malekian, 2004), with systematic literature review (Dekel & Goldblatt, 2008) and meta-analyses (Lambert, Holzer, & Hasbun, 2014) that highlight significant associations between parents' PTSD symptoms and psychological and behavioral problems of children. Over the years, the post-traumatic stress of fathers, former POWs, both plays a role as a linear mediator between captivity and ST of their adult descendants, as well as a nonlinear risk factor of chronic and/or delayed PTSD. The authors suggest that severe interpersonal prolonged trauma of captivity triggers two parallel and interdependent processes. The first is *intrapsychic* in its nature and is reflected in the post-traumatic changes that ex-prisoners experience, and the second is *interpersonal* in nature and is reflected in damaged family relationships and in parenting practices. Both processes can put children at risk to develop ST symptoms. The results also show that depressive symptoms, both PTSD independent and comorbid PTSD, predict ST symptoms in adult descendants of captivity survivors (Zerach et al., 2015).

In other cases, for example, individuals who were exposed to the 9/11 terrorist attack, *posttraumatic stress* manifested by parents, but not depression, was associated with the severity of ST symptoms in their descendants (Fairbrother, Stuber, Galea, Fleischman, & Pfefferbaum, 2003). However, in 97 of the survivors, *parental depression* (but not PTSD) predicted ST symptomatology in their children (Zatzick et al., 2006). At the same time, people with PTSD symptoms may share a common variance with "negative" depressive symptoms. For example, the symptoms of *active avoidance* and *emotional paralysis*, which are known to be related to *negative parenting practices* and non-satisfaction (Ruscio, Weathers, King, & King, 2002), could explain the mediating role of depression between parental captivity and ST symptoms of their descendants (Zerach et al., 2015).

Moreover, considering that PTSD and depressive symptoms may be part of a common construction of general traumatic stress (Ginzburg, Ein-Dor & Solomon, 2010), the shared links between PTSD and depressive symptoms may be, in fact, a reflection of a factor that is responsible for intergenerational transmission of captivity trauma. This results in the need to examine and monitor descendants of former captive individuals who may be affected by the negative impact of their parents' experiences in captivity and post-traumatic stress, experiences that may also have consequences on family-related relationships and maladaptive practices of parents (Zerach et al., 2016).

## Factors of vulnerability and resilience

Studies (e.g. Lazarus, 1966, 1990; Lazarus & Folkman, 1984) have consistently supported the idea that realistic positive evaluations can provide a certain level of protection against stress and resilience as a positive psychological adaptation process to significant stressors that affect the development, the evolution or even the survival. On the contrary, the negative cognitive assessment of the threat and/or lack of perceived control are risk factors and are influenced by the subjective experience of stress. Lazarus (1966) was among the first to notice that when people perceive a situation as negative or threatening, they experience psychological distress as a direct result of their own negative assessment.

### *Preparing for extreme stress situations*

Recent studies suggest that individuals may be inoculated against the negative influence of extreme stress. Operating in hostile environments induce, in general, a high amount of stress due to sleep deprivation, adverse climatic conditions, fatigue, starvation, dehydration, discomfort and physical fatigue. Such stressors affect both psychological and physical functioning (Lieberman, Georgelis, Maher, & Yeghiayan, 2005), with potentially serious consequences, especially for those performing high-risk missions. According to the stress inoculation theory (Meichenbaum, 1985), resilience to stress can be developed by specific training.

Inoculation involves both a process of stress conceptualization under certain conditions that approximate significant stressors in the real world and a process of developing specific abilities to properly manage the situations in which such stressors emerge, especially when they are reexperienced. Thus, the proper management of a wider spectrum of stressors involves a reconceptualization of stressors by reevaluating newly acquired coping options. Therefore, adaptive coping mechanisms can be developed using stress-controlled exposure, relevant to operational reality (Stetz et al., 2007). According to this model, the training should be designed in a way that, by exposure to different levels of stress, the psychological and biological mechanisms of adaptive coping are activated in order to prevent the occurrence of panic and/or cognitive "suffocation," without causing emotional overwhelming/exhaustion. In this regard, there are considerable efforts to develop training strategies and scenarios that can improve training in ways that reduce the risk of panic. To this end, it may be useful to train individuals to recognize early signs of panic and to offer strategies to prevent panic attack.

## *Coping mechanisms*

Zakowski, Hall, Cousino-Klein and Baum (2001) found that coping strategies tend to be congruent with the assessment of the situation. Thus, a person's way of dealing with a stressful situation will tend to reflect his own assessment of the situation. Positive evaluations are more often associated with positive results and negative evaluations with less successful results. Positive evaluation seems to mediate both subjective results (e.g., self-assessment of well-being) and objective results (for example, task performance). When individuals perceive an event in positive (but realistic) manner, they tend to cope more effectively, enjoy positive emotions and experience greater confidence (Janis, 1983; Skinner & Brewer, 2002).

There is a robust part of literature that examines the extent to which resilient individuals respond to stress. Thus, *hardiness*, an individual characteristic contributing to the explanation of resilience along with other factors, describes a set of *dispositional features* including a strong sense of *self-confidence* and *self-efficacy*, *a locus of internal control* (Rotter, 1954) and the belief that *life has meaning and purpose* (Kobasa, 1979). Generally speaking, resilient individuals are better able to work well under stress (Westman, 1990) and are less susceptible to stress-related illnesses (Kobasa, 1979; Kobasa & Puccetti, 1983; Pengilly & Dowd, 2000). Research also shows that resilient individuals are more likely to engage in problem-solving strategies, while less resilient individuals tend to adopt strategies focused on emotions and avoidance (Pollock, 1989; Williams, Wiebe, & Smith, 1992).

## *Gender differences in coping and training strategies*

Differences between men and women in using coping strategies after extremely stressful or traumatic situations can explain possible differences between PTSD rates for men and women. Thus, there is evidence that some coping strategies based on emotions and avoidance are used more frequently by women than by men. These coping strategies include self-blaming, denial and rumination (Matud, 2004, Nolen-Hoeksma, 2001, 2012; Nolen-Hoeksma & Aldao, 2011).

In the only comparative study between men and women on coping styles in relation to emotional distress, conducted during a SERE course (Schmied et al., 2015) at the Centre for Security Forces SERE Learning Site West, located in San Diego, California, significant differences in the use of specific coping strategies have been identified. Thus, women tend to rely on self-blaming, denial and positive refocus more than men. Although statistically insignificant, women's scores in employing self-distraction and planning strategies were also higher than those of men. The results are consistent with the idea that women are more likely to use all types of coping strategies, both those classified as non-adaptive (e.g. self-blaming, denial) and adaptive (e.g., positive refocusing; Nolen-Hoeksma, 2012).

A possible explanation is that women can perceive stressful events as more threatening than men, which increases their need to deal with situations, or perhaps men, because of stigmatization, avoid reporting the use of such coping strategies, especially those associated with emotional expression, in an effort to meet gender expectations (Schmied et al., 2015).

It should also be emphasized that only two coping strategies, *self-blaming* and *denial*, have significantly mediated the relationship between gender and emotional distress during the SERE training. Two other strategies, *behavioral disengagement* and *self-distraction*, were directly associated with emotional distress at a later assessment after 24 hours, but not *planning*. In line with previous research, these results indicate that the use of maladaptive coping strategies may increase the risk of PTSD, but the use of adaptive coping strategies can prevent the occurrence of dysfunctional psychological outcomes (Nolen-Hoeksma, 2012).

Therefore, the results of this study indicate that the personnel involved in high-risk missions could benefit from training programs aimed at developing specific skills in order to ensure coping strategies before entering simulation exercises with realistic stressors and/or before high-risk missions. Moreover, these data suggest that coping abilities must be gender-adapted to meet the unique needs of men and women (Schmied et al., 2015).

## Conclusion

Most of the good practice guidelines for traumatic stress management (e.g. Department of Veterans Affairs/Department of Defense *Clinical Practice Guideline for Management of Post-traumatic Stress*, 2010) recommend a series of primary prevention activities by providing realistic training programs. These should include a *simulated* or *real exposure* to traumatic stimuli that could be encountered in the situations where the training is being made for (e.g. shooting, training for survival or training in simulated captivity). This principle of primary prevention can be applied to a multitude of professional roles (for example, those who are supposed to be involved in the handling of corpses can be trained in morgue-type environments) and is consistent with classical conditioning theories which support the idea that this exposure can help reduce the physiological arousal or anxiety associated with certain traumatic stimuli.

At the same time, a realistic training program also contributes to enhancing the perceived capacity to deal with potential traumatic events and the situations after that. Instruction and practice for using a variety of coping abilities (e.g., stress inoculation training, problem solving, assertiveness and cognitive restructuring) can be useful in training the personnel to tolerate stressful work environments. In addition, people can be prepared to cope with acute stress reactions, which are common after exposure to a potentially traumatic situation. Such training experiences help maximize expectations to control the traumatic situations and the physical and emotional sequels that they produce.

The use of positive role models specific to leaders and peers is also an effective tool for increasing the feeling that people can cope with the situation. The training program should include specific practical actions by calling for adaptive coping mechanisms to change the perspective of the threatening situation in a positive learning experience. Without such a positive learning action, simulating stressful situations or stimuli can induce feelings of helplessness, transforming the training program itself into something traumatic.

Therefore, one of the inherent roles of the SERE psychologist in this training is to increase the ability of the instructors to recognize and prevent the emergence of feelings of helplessness as a consequence of the training program. The SERE psychologist brings expertise into the training environment that can be useful in developing confidence and self-efficacy, the opposite of the feeling of helplessness. Moreover, when stress control principles are applied correctly, when students get stressed during training, they can be remodeled to complete their training and increase their self-efficacy.

At the same time, adequate management of traumatic stress can be complemented by secondary and tertiary prevention programs, especially for people who recover after certain isolation/captive survival incidents. In this respect, there are promising studies suggesting new approaches to early interventions (e.g., emergency unit protocols), mainly used to calm and stabilize patients with acute stress reactions and to process sensory and intrusive motor fragments. Used before memory consolidation, these interventions can reduce associative connections to past trauma, preventing accumulation of traumatic memories, promoting adaptive integration, expressed in self-assertion, adaptive coping, resilience and other functionality indicators. These positive changes, commonly known as post-traumatic growth, are often associated with adaptive emotional and problem-resolving ways of coping, with the personal satisfaction of social support, with accepting the situation and positive reinterpretation and generally experiencing functional emotional states. At the same time, rumination, intrusion and avoidance, although elements that could be interpreted as maintaining factors to posttraumatic stress disorder, are associated at this stage with posttraumatic growth (Linley & Joseph, 2004) as indicators of cognitive processing needed to build a more elaborate and coherent image of trauma that can be integrated into the autobiographical memory of the person.

## References

Ahmadzadeh, G., & Malekian, A. (2004). Aggression, anxiety, and social development in adolescent children of war veterans with PTSD versus those of non-veterans. *Journal of Research in Medical Sciences, 5*, 231–234.

American Psychiatric Association. (2013). *Diagnostic and Statistical Manual of Mental Disorders (5th ed.–DSM-V)*. Washington, DC.

Bar, K. J., Terhaar, J., Boettger, M. K., Boettger, S., Berger, S., & Weiss, T. (2011). Pseudohypoalgesia on the skin: A novel view on the paradox of pain perception in depression. *Journal of Clinical Psychopharmacology, 31*, 103–107.

Bar-On, D. (1994). *Fear and Hope*. Tel Aviv: Hakibbutz Hameuchad.

Beck, J. (2010). *Psihoterapie cognitivă: Fundamente şi perspective*, trad. Elena Crăciunescu, Teodora Rohian, Editura RTS, Cluj-Napoca.

Beckham, J. C., Lytle, B. L., & Feldman, M. E. (1996). Caregiver burden in partners ofVietnam War veterans with posttraumatic stress disorder. *Journal of Consulting and Clinical Psychology, 64*(5), 1068–1072.

Ben-Asher, S., & Shalev, R. (2015). Perceptions of the absence and reintegration of a prisoner of war father from the perspective of his children as adults. *Journal of Child & Adolescent Trauma, 8*, 265–276.

Boss, P. (1999). *Ambiguous Loss: Learning to Live with Unresolved Grief.* Cambridge: Harvard University Press.

Boss, P. (2006). *Loss, Trauma, and Resilience: Therapeutic Work with Ambiguous Loss*. New York: W. W. Norton.

Boss, P., & Carnes, D. (2012). The myth of closure. *Family Process, 51*(4), 456–469.

Boss, P., & Yeats, J. R. (2014). Ambiguous loss: A complicated type of grief when loved ones disappear. *Cruse Bereavement Care, 33*, 63–69.

Bowlby, J. (1980). *Loss: Sadness & Depression. Attachment and Loss (vol. 3)*; (International psycho-analytical library no.109). London: Hogarth Press.

Braun, M. J., & Berg, D. H. (1994). Meaning reconstruction in the experience of parental bereavement. *Death Studies, 18*, 105–129.

Byers, A. L., Covinsky, K. E., Barnes, D. E., & Yaffe, K. (2012). Dysthymia and depression increase risk of dementia and mortality among older veterans. *American Journal of Geriatric Psychiatry, 20*(8), 664–672.

Campbell, R., Greeson, M. R., Bybee, D., & Raja, S. (2008). The cooccurrence of childhood sexual abuse, adult sexual assault, intimate partner violence, and sexual harassment: A mediational model of posttraumatic stress disorder and physical health outcomes. *Journal of Consulting and Clinical Psychology, 76*(2), 194–207.

Carlsson, J. M., Mortensen, E. L., & Kastrup, M. (2006). Predictors of mental health and quality of life in male tortured refugees. *Nordic Journal of Psychiatry, 60*(1), 51–57.

Cohen, S., & McKay, G. (1984). Social support, stress, and the buffering hypothesis: A theoretical analysis. *Handbook of Psychology and Health, 4*, 253–267.

Creasey, H., Sulway, M. R., Dent, O., Broe, G. A., Jorm, A., & Tennant, C. (1999). Is experience as a prisoner of war a risk factor for accelerated age-related illness and disability? *Journal of the American Geriatrics Society, 47*, 60–64.

Defrin, R., Lahav, Y., & Solomon, Z. (2017). Dysfunctional pain modulation in torture survivors: The mediating effect of PTSD. *The Journal of Pain, 18*(1), 1–10.

Dekel, R. (2007). Posttraumatic distress and growth among wives of prisoners of war: The contribution of husbands' posttraumatic stress disorder and wives' own attachment. *American Journal of Orthopsychiatry* 77(3), 419–426.

Dekel, R., & Goldblatt, H. (2008). Is there intergenerational transmission of trauma? The case of combat veterans' children. *American Journal of Orthopsychiatry*, *78*, 281–289.

Dempsey, M., & Baago, S. (1998). Latent grief: The unique and hidden grief of cares of loved ones with dementia. *American Journal of Alzheimer's Disease*, *13*, 84–91.

Department of Veterans Affairs/Department of Defense. (2010). *Clinical Practice Guideline for Management of Post-traumatic Stress*. Retrieved from https://www.healthquality.va.gov/guidelines/MH/ptsd/cpg_PTSD-full-201011612.PDF.

Doran, A. P., Hoyt, G., Hiller Lauby, M. D., & Morgan, C. A. III. (2012). Survival, evasion, resistance, and escape (SERE) training: Preparing military members for the demands of captivity. In C. H. Kennedy & E. A. Zillmer (Eds.), *Military Psychology, Second Edition: Clinical and Operational Applications* (pp. 241–261). New York, NY, US: Guilford Press.

Engdahl, B. E., Harkness, A. R., Eberly, R. E., Page, W. F., & Bielinski, J. (1993). Structural models of captivity trauma, resilience, and trauma response among former prisoners of war 20 to 40 years after release. *Social Psychiatry and Psychiatric Epidemiology*, *28*, 109–115.

Epley, N., Waytz, A., Akalis, S., & Cacioppo, J. T. (2008). When we need a human: Motivational determinants of anthropomorphism. *Social Cognition*, *26*(2), 143–155.

Fairbrother, G., Stuber, J., Galea, S., Fleischman, A. R., & Pfefferbaum, B. (2003). Post-traumatic stress reactions in New York City children after the September 11, 2001, terrorist attacks. *Ambulatory Pediatrics*, *3*(6), 304–311.

Figley, C. R. (1995). *Compassion Fatigue: Coping with Secondary Traumatic Stress Disorder in Those Who Treat the Traumatized*. London: Routledge.

Ghaddar, A., Elsouri, G., & Abboud, Z. (2016). Torture and long-term health effects among Lebanese female political prisoners. *Journal of Interpersonal Violence*, *31*(3), 500–514.

Ginzburg, K., Ein-Dor, T., & Solomon, Z. (2010). Comorbidity of posttraumatic stress disorder, anxiety and depression: A2 0-year longitudinal study of war veterans. *Journal of Affective Disorders*, *123*(1), 249–257.

Hunter, E. J. (1983). Captivity: The family in waiting. In C. R. Figley & H. I. McCubbin (Eds.), *Stress and the Family, II: Coping with Catastrophe* (pp. 166–184) New York: Brunner/Mazel.

Hunter, E. J. (1988). The psychological effects of being a prisoner of war. In *Human Adaptation to Extreme Stress: From the Holocaust to Vietnam* (pp. 157–170) New York, NY, US: Plenum Press.

Hunter-King, E. J. (1993). Long-term effects on children of a parent missing in wartime. In F. W. Kaslow (Ed.). *The Military Family in Peace and War* (pp. 48–65) New York: Springer.

Janis, I. (1983). The patient as decision maker. In D. Gentry (Ed.). *Handbook of Behavioral Medicine*. New York: Guilford.

Jensen, P. S., Martin, D., & Watanabe, H. (1996). Children's response to separation during operation desert storm. *Journal of the American Academy of Child and Adolescent Psychiatry*, *35*, 433–441.

Johnson, H., & Thompson, A. (2008). The development and maintenance of posttraumatic stress disorder (PTSD) in civilian adult survivors of war trauma and torture: A review. *Clinical Psychology Review*, *28*, 36–47.

Kang, H. K., Bullman, T. A., & Taylor, J. W. (2006). Risk of selected cardiovascular diseases and posttraumatic stress disorder among former World War II prisoners of war. *Annals of Epidemiology*, *16*, 381–386.

Kobasa, S. C. (1979). Stressful life events, personality and health: An enquiry into hardiness. *Journal of Personality and Social Psychology*, *37*(1), 1–11.

Kobasa, S. C., & Puccetti, M. C. (1983). Personality and social resources in stress resistance. *Journal of Personality and Social Psychology*, *45*(4), 839–850.

Lambert, J. E., Holzer, J., & Hasbun, A. (2014). Association between parents' PTSD severity and children's psychological distress: A meta-analysis. *Journal of Traumatic Stress*, *27*(1), 9–17.

Lazarus, R. S. (1966). *Psychological Stress and the Coping Process*. New York: McGraw-Hill.

Lazarus, R. S. (1990). Theory based stress measurement. *Psychological Inquiry*, *1*, 3–13.

Lazarus, R. S., & Folkman, S. (1984). *Stress, Appraisal and Coping*. New York: Springer.

Leach, J. (1994). *Survival Psychology*. New York: New York University Press.

Lieberman, H. R., Georgelis J. H., Maher T. J., & Yeghiayan S. K. (2005). Tyrosine prevents effects of hyperthermia on behavior and increases norepinephrine. *Physiology & Behavior*, *84*, 33–38.

Linley, P. A., & Joseph, S. (2004). Positive change following trauma and adversity: A review. *Journal of Traumatic Stress*, *17*, 11–21.

Matud, P. (2004). Gender differences in stress and coping styles. *Personality and Individual Differences*, *27*, 1401–1415.

McEwen, B. S. (1998). Stress, adaptation, and disease: Allostasis and allostatic load. *Annals of the New York Academy of Sciences*, *840*, 33–44.

Meichenbaum, D. (1985). *Stress Inoculation Training*. New York: Pergamon Press.

Nice, D. S., Garland, C. F., Hilton, S. M., Baggett, J. C., & Mitchell, R. E. (1996). Long-term health outcomes and medical effects of torture among US Navy prisoners of war in Vietnam. *Comment. The journal of the American Medical Association, 276*, 416–417.

Nolen-Hoeksma, S. (2001). Gender differences in depression. *Current Directions in Psychological Science, 10*, 173–176.

Nolen-Hoeksma, S. (2012). Emotion regulation and psychopathology: The role of gender. *Annual Review of Clinical Psychology, 8*, 161–187.

Nolen-Hoeksma, S., & Aldao, A. (2011). Gender and age differences in emotion regulation strategies and their relationship to depressive symptoms. *Personality and Individual Differences, 51*, 704–708.

Pengilly, J. W., & Dowd, E. T. (2000). Hardiness and social support as moderators of stress. *Journal of Clinical Psychology, 56*(6), 813–820.

Pollock, S. E. (1989). The hardiness characteristic: A motivating factor in adaptation. *Advanced Nursing Science, 11*, 53–62.

Razavi, S. H., Razavi-Ratki, S. K., Nojomi, M. M., & Namiranian, N. (2012). Depression and general anxiety in the prisoner of war's children: A cross sectional study. *Medical Journal of the Islamic Republic of Iran, 26*, 179–184.

Renshaw, K. D., Rodebaughc, T. L., & Rodrigues, C. S. (2010). Psychological and marital distress in spouses of Vietnam veterans: Importance of spouses' perceptions. *Journal of Anxiety Disorders, 24*, 743–750.

Rosenheck, R., Fontana, A. (1998). Transgenerational effects of abusive violence on the children of Vietnam combat veterans. *Journal of Traumatic Stress, 11*, 731–742.

Rosenheck, R., Nathan, P. (1985). Secondary traumatization in the children of Vietnam veterans with post-traumatic stress disorder. *Hospital & Community Psychiatry, 36*, 538–539.

Rotter, J. B. (1954). *Social Learning and Clinical Psychology*. Englewood Cliffs, NJ: Prentice Hall.

Rubin, S. S. (2000). Psychodynamic perspectives on treatments with the bereaved: Modifications of the therapeutic-transference paradigm. In R. Malkinson, S. S. Rubin, & E. Witztum (Eds.), *Traumatic and Nontraumatic Loss and Bereavement: Clinical Theory and Practice* (pp. 117–141). Madison: Psychosocial Press.

Rubin, S. S., Malkinson, R., & Witztum, E. (2012). *Working with the Bereaved: Multiple Lenses on Loss and Mourning*, New York: Routledge.

Ruscio, A. M., Weathers, F. W., King, L. A., & King, D. W. (2002). Male war-zone veterans' perceived relationships with their children: The importance of emotional numbing. *Journal of Traumatic Stress, 15*(5), 351–357.

Saab, B. R., Chaaya, M., Doumit, M., & Farhood, L. (2003). Predictors of psychological distress in Lebanese hostages of war. *Social Science & Medicine, 57*, 1249–1257.

Schmied, E. A., Padilla, G. A., Thomsena, C. J., Hiller Lauby, M. D., Harrisa, E., & Taylora, M. K. (2015). Sex differences in coping strategies in military survival school. *Journal of Anxiety Disorders, 29*, 7–13.

Seligman, M. E. P. (2002). *Authentic Happiness: Using the New Positive Psychology to Realize Your Potential for Lasting Fulfilment*. New York: Free Press.

Skinner, N., & Brewer, N. (2002). The dynamics of threat and challenge appraisals prior to stressful achievement events. *Journal of Social and Personality Psychology, 83*, 678–692.

Solomon, Z., Greene, T., Ein-Dor, T., Zerach, G., Benyamini, Y., & Ohry, A. (2014). The long-term implications of war captivity for mortality and health. *Journal of Behavioural Medicine, 37*, 849–859.

Solomon Z., Horesh D., Ein-Dor T., & Ohry, A. (2012). Predictors of PTSD trajectories following captivity: A 35-year longitudinal study. *Psychiatry Research, 199*, 188–194.

Spiric, Z., Opacic, G., Jovic, V., Samardzic, R., Knezevic, G., Mandic-Gajic, G., & Todorovic, M. (2010). Gender differences in victims of war torture: Types of torture and psychological consequences. *Vojnosanitetski Pregled, 67*, 411–418.

STANAG 7196 (SD5). The NATO Survival, Evasion, Resistance, Escape (SERE) Training Standard, March, 2007.

Steel, Z., Chey, T., Silove, D., Marnane, C., Bryant, R. A., & Van Ommeren, M. A. (2009). Association of torture and other potentially traumatic events with mental health outcomes among populations exposed to mass conflict and displacement: A systematic review and meta-analysis. *Journal of the American Medical Association, 302*, 537–549.

Stetz, M., Thomas, M., Russo, M., Stetz, T., Widzunas, R., McDonald, J., … Romano, J. (2007). Stress, mental health and cognition: A brief review of relationships and countermeasures. *Aviation, Space, and Environmental Medicine, 78*(5), 252–260.

Sullivan, M. J. (2012). The communal coping model of pain catastrophizing: Clinical and research implications. *Canadian Psychology/ Psychologie Canadienne, 53*(1), 32–41.

Sutker, P. B., & Allain, A. N., Jr. (1996). Assessment of PTSD and other mental disorders in World War II and Korean Conflict POW survivors and combat veterans. *Psychological Assessment, 8*(1), 18–25.
Sutker, P. B., Allain, A. N., Jr., & Johnson, J. L. (1993). Clinical assessment of long-term cognitive and emotional sequelae to World War II prisoner-of-war confinement: Comparison of pilot twins. *Psychological Assessment, 5*(1), 3–10.
Tsura, N., Shaharb, G., Defrinc, R., Lahava, Y., & Ginzburga, K. (2017). Torturing personification of chronic pain among torture survivors. *Journal of Psychosomatic Research, 99*, 155–161.
Van der Kolk, B. A. (1994). The body keeps the score: Memory and the evolving psychobiology of posttraumatic stress. *Harvard Review of Psychiatry, 1*, 253–265.
Van der Kolk, B. A., & McFarlane, A. C. (Eds.) (2012). *Traumatic Stress: The Effects of Overwhelming Experience on Mind, Body, and Society*. New York, NY: Guilford Press.
Van Ommeren, M. I., de Jong, J. T., Sharma, B., Komproe, I., Thapa, S. B., & Cardeña, E. (2001). Psychiatric disorders among tortured Bhutanese refugees in Nepal. *Archives of General Psychiatry, 58*, 475–482.
Westman, M. (1990). The relationship between stress and performance: The moderating effect of hardiness. *Human Performance, 3*(3), 141–155.
Williams, P. G., Wiebe, D. J., & Smith, T. W. (1992). Coping processes as mediators of the relationship between hardiness and health. *Journal of Behavioral Medicine, 15*, 237–255.
Zakowski, S. G., Hall, M. H., Cousino-Klein, H., & Baum, A. (2001). Appraised control, coping, and stress in a community sample: A test of the goodness-of-fit hypothesis. *Annals of Behavioral Medicine, 23*, 158–165.
Zatzick, D., Russo, J., Grossman, D. C., Jurkovich, G., Sabin, J., Berliner, L., & Rivara, F. (2006). Posttraumatic stress and depressive symptoms, alcohol use, and recurrent traumatic life events in a representative sample of hospitalized injured adolescents and their parents. *Journal of Pediatric Psychology, 31*, 377–387.
Zerach, G., Greene, T., & Solomon, Z. (2015). Secondary traumatization and self-rated health among ex-POW's wives: The moderating role of marital adjustment. *Journal of Health Psychology, 20*(2), 222–235.
Zerach, G., Kanat-Maymon, Y., Aloni, R., & Solomon, Z. (2016). The role of fathers' psychopathology in the intergenerational transmission of captivity trauma: A twenty three-year longitudinal study. *Journal of Affective Disorders, 190*, 84–92.

# INDEX

*Note*: Page numbers followed by "*n*" indicate footnotes, **Bold** page number denotes Table. *Italic* page number denotes Figure.

For Product Safety Concerns and Information please contact our EU representative GPSR@taylorandfrancis.com
Taylor & Francis Verlag GmbH, Kaufingerstraße 24, 80331 München, Germany

www.ingramcontent.com/pod-product-compliance
Lightning Source LLC
LaVergne TN
LVHW080112011026
849160LV00013B/1078
*9780367237073*